T0327329

FORMULAS FOR DYNAMICS, ACOUSTICS AND VIBRATION

FORMULAS FOR DYNAMICS, ACOUSTICS AND VIBRATION

Robert D. Blevins

WILEY

This edition first published 2016

© 2016 John Wiley & Sons, Ltd

Registered office
John Wiley & Sons Ltd, The Atrium, Southern Gate, Chichester, West Sussex, PO19 8SQ, United Kingdom

For details of our global editorial offices, for customer services and for information about how to apply for permission to reuse the copyright material in this book please see our website at www.wiley.com.

The right of the author to be identified as the author of this work has been asserted in accordance with the Copyright, Designs and Patents Act 1988.

All rights reserved. No part of this publication may be reproduced, stored in a retrieval system, or transmitted, in any form or by any means, electronic, mechanical, photocopying, recording or otherwise, except as permitted by the UK Copyright, Designs and Patents Act 1988, without the prior permission of the publisher.

Wiley also publishes its books in a variety of electronic formats. Some content that appears in print may not be available in electronic books.

Designations used by companies to distinguish their products are often claimed as trademarks. All brand names and product names used in this book are trade names, service marks, trademarks or registered trademarks of their respective owners. The publisher is not associated with any product or vendor mentioned in this book.

Limit of Liability/Disclaimer of Warranty: While the publisher and author have used their best efforts in preparing this book, they make no representations or warranties with respect to the accuracy or completeness of the contents of this book and specifically disclaim any implied warranties of merchantability or fitness for a particular purpose. It is sold on the understanding that the publisher is not engaged in rendering professional services and neither the publisher nor the author shall be liable for damages arising herefrom. If professional advice or other expert assistance is required, the services of a competent professional should be sought

Library of Congress Cataloging-in-Publication Data

Blevins, Robert D.
Formulas for dynamics, acoustics and vibration / Robert D. Blevins.
pages cm
Includes bibliographical references and index.
ISBN 978-1-119-03811-5 (cloth)
1. Engineering mathematics–Formulae. 2. Dynamics–Mathematics. 3. Vibration–Mathematical models. I. Title.
TA332.B59 2015
620.001′51–dc23

2015015479

A catalogue record for this book is available from the British Library.

Set in 10/12pt TimesLTStd by SPi Global, Chennai, India

1 2016

Contents

Preface

This book is an illustrated compilation of formulas for solving dynamics, acoustics, and vibration problems in engineering. Over 1000 formulas for solution of practical problems are presented in 60 tables for quick reference by engineers, designers, and students.

The majority of the formulas are exact solutions for dynamics, acoustics, and vibrations. Their origin lies in the development of calculus and its application to physics by Sir Isaac Newton and Gottfried Wilhelm Leibnitz in the 17th century. Sir Isaac Newton (1642–1727) formulated the inverse square law of gravitation and proved all three of Kepler's laws of planetary motion. Leonhard Euler (1707–1783) derived the equations of rigid body dynamics and wave propagation. In 1732, Daniel Bernoulli described the normal modes of hanging chains in terms of Bessel functions. Gustav Kirchhoff developed the theory of plate vibrations in 1850. The modern age of exact classical dynamic solutions began with publication of Lord Rayleigh's *The Theory of Sound* (first edition 1877), followed by A. E. H. Love's *A Treatise on the Mathematical Theory of Elasticity* (first edition 1892–1893), Horace Lamb's *The Dynamical Theory of Sound* (first edition 1910), and Stephen Timoshenko's *Vibration Problems in Engineering* (first edition 1928). These books are foundations of vibration analysis and they are still in print.

Formulas in this book span the technical literature from the second edition of *The Theory of Sound* (1894) to technical journals of the 21st century. Most were generated by modern contributors including Arthur W. Leissa, H. Max Irvine, Phil McIver, Werner Soedel, Daniel J. Gorman, and W. Kitpornchai. Systems too large and complex for exact solution by formulas are analyzed approximately with numerical methods (Appendices A and B) that are formulated, interpreted, and checked with exact classical formulas and reasoning.

As structures become lighter and more flexible, they become more prone to vibration and dynamic failure. Vibration failures include wind induced vibration of the Tacoma Narrows suspension bridge in 1940 and the steam flow-induced radiation leaks from vibrating tubes that shut down the San Onofre Nuclear Generating Station in 2012. Dynamic and vibration analysis differs from static analysis by including time and inertia. Dynamic analysis starts with identification of a flexible system in a dynamic environment. The next step is calculation of the system mass, stiffness, natural frequency, mode shape, and transmissibility. The third step is calculation of dynamic response to the time-dependent loads. Then design changes can be made to increase reliability and reduce noise and vibration.

Methods to reduce unwanted noise and vibration fatigue failures include the following: (1) increase stiffness, (2) increase damping to reduce amplitude, (3) reduce excitation with load paths to ground, (4) detune natural frequencies from excitation frequencies, (5) reduce stress concentrations and installation preload, (6) use fatigue- and wear-resistant materials, and (7) inspect and repair to limit propagating damage. Increasing stiffness solves many vibration problems, but the size and weight of the stiffened structure may not meet design goals. Design optimization of light, flexible systems requires dynamic analysis and test.

Formulas and data for dynamic analysis are presented in the tables. The text discusses examples, explanation, and some derivations. Chapter 1 provides definitions, symbols, units, and geometric properties. Chapter 2 provides dynamics of point masses and rigid bodies. Chapters 3 through 5 provide natural frequencies and mode shapes for elastic beams, plates, shells, and spring–mass systems. Chapter 6 provides fluid and acoustic solutions to the wave equation and added mass. Chapter 7 presents formulas for the response of elastic structures to sinusoidal, transient, and random loads. Chapter 8 presents properties of structural solids, liquids, and gases that support the formulas provided in Chapters 1 through 7. The property data is given in both SI (metric) and US customary (ft-lb) units. Appendices A and B present approximate and numerical methods for natural frequency and time history analysis.

The formulas are ready to apply with pencils, spreadsheets, and digital devices. There are 35 worked-out examples. Supporting content and software is available at www.aviansoft.com. The table formats were developed by Raymond J. Roark. Many results from my previous formulas book are included and updated with knowledge gleaned over the intervening 30 years. Many contributed to this book. The author would like to thank György Szász, Dr. Hwa-Wan Kwan, Mark Holcomb, Ryan Bolin, Larry Julyk, Alex Rasche, and Keith Rowley for reviewing chapters. Beocky Yalof skillfully edited the original manuscript. This book is dedicated to the individuals who developed the solutions herein.

1

Definitions, Units, and Geometric Properties

1.1 Definitions

Acceleration The rate of change of velocity. The second derivative of displacement with respect to time.

Added mass The mass of fluid entrained by a vibrating structure immersed in fluid. The natural frequency of vibration of a structure surrounded by fluid is lower than that of the structure vibrating in a vacuum owing to the added mass of fluid (Tables 6.9 and 6.10).

Amplitude The maximum excursion from the equilibrium position during a vibration cycle.

Antinode Point of maximum vibration amplitude during free vibration in a single mode. See node.

Attenuation, acoustic Difference in sound or vibration between two points along the path of energy propagation. Also see damping, insertion loss.

Bandwidth The range of frequencies through which vibration energy is transferred.

Beam Slender structure whose cross section and deflection vary along a single axis. Beams support tension, compression, and bending loads. Shear deformations are negligible compared to bending deformations in slender beams.

Boundary condition Time-independent constraints that represent idealized structural interfaces, such as zero force, displacement, velocity, rotation, or pressure.

Broad band A process consisting of a large number of component frequencies, none of which is dominant, distributed over a broad frequency band, usually more than one octave. Also see narrow band and tone.

Formulas for Dynamics, Acoustics and Vibration, First Edition. Robert D. Blevins.
© 2016 John Wiley & Sons, Ltd. Published 2016 by John Wiley & Sons, Ltd.

Bulk modulus of elasticity The ratio of the hydrostatic stress, equal in all directions, to the relative change in volume it produces. $B = E/[3(1-\nu)]$ for elastic isotropic materials where E is the modulus of elasticity and ν is Poisson's ratio. $B = \rho c^2$ for fluids where ρ is the mean density and c is the speed of sound.

Cable A uniform, one-dimensional structure that can bear only tensile loads parallel to its own axis. A cable is a massive string. It has zero bending rigidity and it stretches in response to tensile loads. Also see chain.

Cable modulus The change in longitudinal stress (axial force divided by cross-sectional area) divided by the change in longitudinal strain produced by the stress. A solid rod has a cable modulus equal to the modulus of elasticity of the rod material. The cable modulus of a woven cable is typically about 50% of that of the modulus of its fibers.

Center of gravity The point on which a body balances. The center of mass. The sum of the gravitational moments created by the elements of mass is zero about the center of mass.

Center of percussion The point on a rigid body that does not accelerate when the body is impulsively loaded.

Centrifugal force Outward reaction of a mass on a rotating body away from the axis of rotation. Centrifugal (adjective) means "outward from center."

Centripetal acceleration Acceleration of a point on a rotating body toward the center of rotation. Centripetal (adjective) means "toward the axis of rotation."

Centroid Volumetric center of a volume or area. The center of mass and centroid are the same for a homogeneous body.

Chain A uniform, massive one-dimensional structure that bears only tensile loads parallel to its own axis. No bending or shear loads are borne. In contrast to cables, ideal chains do not extend under tensile loads.

Concentrated mass (point mass) A point in space with finite mass and zero rotational inertia.

Consistent units A unit system in which Newton's second law, force equals mass times acceleration, is identically satisfied without additional dimensional factors. See Table 1.2.

Coriolis acceleration Accelerations induced on a moving particle in a rotating system, after the French engineer-mathematician Gustave-Gaspard Coriolis.

Crest factor See peak-to-rms ratio.

Damping The ability of a system to absorb vibration energy. Damping limits resonant vibration amplitude and causes free vibrations to decay with time.

Damping factor A nondimensional measure of the damping of a system equal to $1/(2\pi)$ times the natural logarithm of the ratio of the amplitude of one cycle to the amplitude of the following cycle during free vibration.

Decibel (dB) Sound pressure level in decibels is 10 times the logarithm, to the base 10, of the ratio of the mean square sound pressure to a reference mean square sound pressure.

Deformation The displacement of a structure from its reference or equilibrium position.

Density Mass per unit volume.

Divergence (1) Unstable torsion motion caused by aerodynamic forces that overcome structural stiffness, and, (2) spreading of sound waves propagating from a source.

Dynamic amplification factor Also called *dynamic load factor* (*DLF*), and *magnification factor*. The maximum dynamic response amplitude of a single degree of freedom elastic system to a dynamic force divided by the static response to a steady force with the same magnitude.

Eigen German for "own characteristic." Eigenvalue is a scalar solution to a homogeneous linear equation of motion. Natural frequency is eigenvalue. Eigen vector is the associated spatial mode shape.

Elastic A material or structure whose deformations increase linearly with increasing load. Most practical structures are elastic, or approximately elastic, for loads below the onset of yielding or buckling.

Flutter Unstable divergent oscillation caused by aerodynamic forces.

Force As defined by Newton, force is proportional to mass times acceleration, Equation 1.1.

Forced vibration Vibration of a system in response to an external periodic force.

Free vibration Vibrations in the absence of external loads. Free vibrations take place after an elastic system is released from a displacement.

Frequency The number of times a periodic motion repeats itself per unit time. Vibration frequency is the number of sinusoidal periods per unit times in either units of cycles per second, called Hertz, or in cycles per minute, which is rpm, or in radians per second.

Fundamental mode The lowest natural frequency and mode shape of an elastic system.

Harmonic motion Simple harmonic motion is sinusoidal in time about an equilibrium point.

Harmonics Motion at integer multiples of a frequency.

Hertz Hertz as the unit of cycles per second was adopted by the General Conference on Weights and Measures in 1960. Its name honors Heinrich Hertz, a pioneer investigator of electromagnetic waves.

Impedance (1) Fluid mechanic impedance is ratio of pressure to fluid velocity, (2) mechanical impedance is ratio of force to velocity, (3) step impedance is the ratio of pressure differential across to velocity through a component.

Impulse The force multiplied by the time increment and integrated over the time interval during which the force acts. It has units of force-seconds. Impulse produces change in momentum. Rotational impulse is torque integrated over time and has units of force-length-seconds.

Inertial frame A set of coordinates that does not accelerate; a frame in which Newton's second law holds.

Insertion loss The change in sound pressure level between two points in an acoustic circuit when a component is inserted between the two. Also see transmission loss. Generally expressed in decibels.

Jerk The rate of change of acceleration.

Kinematics Motion within geometric constraints.

Kinetic energy The energy of mass in motion. $\frac{1}{2}MV^2$ is the kinetic energy of a mass M with velocity V.

Mass ratio The weight of a structure divided by the weight of a circumscribed cylinder of the surrounding fluid.

Membrane A thin, massive, elastic uniform sheet that can support only tensile loads in its own plane. A membrane can be flat like a drumhead or curved like a soap bubble. A one-dimensional membrane is a cable. A massless, one-dimensional, membrane is a string. The term membrane is also used for elastic systems without bending.

Modal density The number of modes of vibration with natural frequencies in a specified frequency band. See Table 6.6.

Mode shape (eigenvector) A dimensionless shape function defined over the space of a structure that describes the relative displacement of any point as the structure vibrates in a single mode. A mode shape is independent of time. There is a unique mode shape for each natural frequency of the structure. Any deformation of the structure, consistent with the boundary conditions, can be expressed as a linear sum of mode shapes.

Modulus of elasticity (Young's modulus) The ratio of normal stress to the normal strain it produces in a material. The modulus of elasticity has units of pressure. A material is *isotropic* if the modulus of elasticity is independent of direction. Also see bulk modulus.

Moment See torque.

Momentum Mass times its velocity vector. Rotational momentum is the polar mass moment of inertia of a body times its rotational velocity about an axis.

Moment of inertia of a body	The sum of the products obtained by multiplying each element of mass within a body by the square of its distance from a given axis.
Moment of inertia of an area	The sum of the products obtained by multiplying each element of area by the square of its distance from a given axis.
Narrow band	Vibration or sound process whose frequency components fall within a narrow band, generally less than one-third octave, so that a single peak follows each zero crossing with positive slope. Also see broad band and tone.
Natural frequency (eigenvalue)	The frequency at which a linear elastic structure will freely vibrate in free vibration. Continuous or multimass structures have multiple natural frequencies. The lowest of these is the fundamental natural frequency. See Section 3.3.
Neutral axis	The axis of zero bending stress through the cross section of a beam. The neutral axis of homogeneous beams passes through the centroid of the cross section. See Section 4.1.
Node	Point on a structure that does not deflect during vibration in a mode. Antinode is a point on a structure with maximum deflection during vibration in a single mode. Also, a point in space.
Noise	Multifrequency acoustic pressure.
Nonstructural mass	Mass without a corresponding stiffness. See particle.
Octave, one-third octave	A logarithmic frequency scale originating in musical notation. The octave band is a frequency range where the upper frequency is twice the lower frequency. The octave bands are subdivided into three one-third-octave bands, with the ratio between the upper and lower limits of each one-third-octave band being $2^{1/3}$.
Orthotropic	A material whose properties have two mutually perpendicular planes of symmetry. The material properties are direction dependent. A lamina of parallel fibers has orthotropic material properties.
Particle, point mass, concentrated mass	A point in space with finite mass and zero rotational inertia.
Peak-to-rms ratio	The ratio of the maximum value above the mean to the root-mean square value, about the mean, of a data time history. The peak-to-rms ratio of a sine wave is $2^{1/2}$ and the peak-to-rms ratio of a Gaussian time history approaches infinity. Also called crest factor.
Period of vibration	The reciprocal of frequency. Period is the time in seconds to complete one cycle of oscillation.
Phase angle	The angle, relative to 360°, at a point in time between two harmonic waves with the same frequency.

Plate A thin, flat, two-dimensional elastic structure that conforms to a two-dimensional surface. A plate has mass and supports bending loads. A plate without bending rigidity is a membrane.

Poisson's ratio The ratio of the lateral shrinkage to the longitudinal expansion of a bar of a given material that has been placed under a uniform axial load. Poisson's ratio is often near 0.3 for metals and 0.4 for rubber-like materials. It is dimensionless. A material with a Poisson's ratio of 0.5 has constant volume during loading.

Potential energy Stored energy. Potential energy is the negative of work. Mgh is the potential energy created by raising mass M by height h where g is the acceleration due to gravity.

Power spectral density The mean square value of a process, within a specified frequency band, divided by the width of that band. Also see octave.

Product of inertia of a body The sum of the products obtained by multiplying each element of mass of a body by the product of its distances from two mutually perpendicular axes. Table 1.6.

Product of inertia of an area The sum of the products obtained by multiplying each element of area of a section by the product of its distances from two mutually perpendicular axes. Table 1.5.

Radius of gyration of a body The square root of the quantity formed by dividing the mass moment of inertia of a body by the mass of the body.

Radius of gyration of an area The square root of the quantity formed by dividing the area moment of inertia of a section by the area of the section.

Random vibration A multifrequency process, described by its statistical properties.

Resonance Response to an external periodic force having the same frequency as the natural frequency of the system. The amplitude of vibration will become larger than the static response to the same force for dampine factors less than $1/2^{1/2}$.

Response The response of a system is the motion, or other output, resulting from dynamic excitation of a system.

Response spectrum Maximum response to a given transient load, often plotted as a function of the natural frequency and damping.

Restitution coefficient The ratio of the velocity of two objects after a collision to their velocity ratio before the collision, relative to the center of mass. The restitution coefficient is zero for a perfectly plastic collision. It is maximum of two for elastic collision.

Rigid body A body whose deformations are negligible.

Root-mean-square The square root of the average, over many cycles of vibration, of the square of a time history of vibration.

Rotary inertia The inertia associated with rotation of a structure about an axis. The sum of the products of elements of mass of a body times their velocity times the distance from the axis of rotation.

Rotor A body that spins about a fixed axis.

Seiching The system of waves in a harbor produced as the harbor responds sympathetically to waves in the open sea. Also see sloshing.

Sidereal day 86,164 s (23.93 h) Earth's period of rotation with respect to distant stars, which is 4 min shorter than the sun's 24-h solar day because of the earth's daily advance in its orbit about the sun.

Shear beam A beam whose deformation in shear substantially exceeds the flexural (bending) deformation.

Shear coefficient A dimensionless quantity, dependent on the shape of the cross section of a beam that is introduced into beam theory to account for the nonuniform distribution of shear stress and shear strain over the cross section. See Section 4.1, Table 4.11.

Shear modulus The rate of change in shear stress of a material that produces a unit shear strain. For isotropic elastic material the shear modulus is $G = E/[2(1 + \nu)]$.

Shell A thin elastic structure defined by a curved surface. A curved plate is a shell. A shell without bending rigidity is a curved membrane.

Shock Vibration imposed suddenly and over a period of time comparable to or shorter than the natural period of vibration.

Sloshing Surface waves in a liquid-filled basin, Table 6.7.

Sonic fatigue The vibration of plate and shell structures induced by fluctuating pressure on their surfaces, Tables 7.3, 7.4, 7.5.

Sound pressure level (SPL) Twenty times the logarithm to base 10 of the rms acoustic pressure relative to a reference pressure, decibels. See Appendix C.

Specific impulse The thrust produced by a rocket motor, divided by the initial weight of fuel, times the time in seconds the fuel burns. It is a function of the fuel composition and combustion temperature. See Section 2.2.

Spectrum The distribution of vibration amplitude or energy versus frequency.

Speed of sound The speed at which small pressure fluctuations propagate through an infinite fluid or solid. See Table 6.1.

Spring constant The change in load on a linear elastic structure divided by the change in deformation that results. The torsional spring constant is the change in torque divided by the change in angular position in radians. See Table 3.2.

Spring-mass system A body or mass on a massless elastic suspension.

Stress Force on a unit area. Stress has units of pressure.

String A massless one-dimensional structure defined by a straight line that bears a uniform tension. A string cannot bear bending or compression. A massive string is a cable.

Tone (pure) A sound or vibration at a single frequency. Also see broad band and narrow band.

Torque Torque, or moment, is the vector cross product of force with a vector from a reference axis of rotation. It has units of force-length. Torque produces angular acceleration. See section 2.3.

Torsion coefficient Change in torque on a linear elastic shaft divided by the change in angular deformation, in radians, that results, per unit length of shaft. See Tables 3.2, 4.15.

Transient vibration Vibration that develops over a limited time.

Transmission loss The change in sound pressure level of a forward propagating sound wave between two points along its path. Also see insertion loss.

Vector Quantity with magnitude and direction such as velocity.

Vibration Oscillation in time. See free vibration, pure tone, and random vibration.

Viscosity The ability of a fluid to resist shearing deformation. The viscosity of a Newtonian (linear) fluid is defined as the ratio between the shear stress applied to a fluid and the shearing strain that results. Kinematic viscosity is defined as viscosity divided by fluid density; it has units of length squared over time.

Wave length The distance in space that a propagating wave travels in one period of oscillation. Wave length is the speed of propagation divided by the frequency.

White noise A noise or vibration whose spectral density is constant over all frequencies. Pink noise is distributed over a finite frequency range.

Work The integral of the scalar dot product of force vector and incremental displacement vector during the force application. Work has units of force times length (energy). Rotational work is the integral of torque times the incremental angle of rotation. See Section 2.2.

1.2 Symbols

General nomenclature for the book is in Table 1.1. In addition the heading of each table contains the nomenclature that applies to that table. These symbols and abbreviations are consistent with engineering usage and technical literature. Vectors are written in bold face type (**B**). Mode shapes have an over tilde (~) to denote that they are independent of time.

The Greek letter λ is used for wave length and longitude; it is also used for the dimensionless natural frequency parameters of beams, plates, and shells, where it generally appears with a subscript. The symbol I is used for area moments of inertia and J is used for mass moments of inertia. The overworked symbols t and T are used for time, oscillation period, transpose of a matrix, tension, and thickness. Definitions in the tables clarify usage.

Table 1.1 Nomenclature

Symbol	Units	Definition
A	length^2	area
B	force / length^2	bulk modulus
C	(location)	center of gravity
C	length^4	torsion constant
c	length/time	speed of sound
e	dim'lss	2.71828 = Exp(1)
E	force / length^2	modulus of elasticity
f	1/time (Hertz)	frequency, f_n = natural frequency
g	length / time^2	acceleration due to gravity
I	length^4	area moment of inertia
i	dim'lss	imaginary constant $(-1)^{½}$
J	mass × length^4	mass moment of inertia
J	dimensionless	Bessel function of first kind
K	dimensionless	shear coefficient
k	force / length	deflection spring constant
k	moment / radian	torsional spring constant
L	length	Length
M	mass	Mass or Mach number
m	mass / length	mass per unit length
M	force × length	Moment
P	-	point in space
P	force	Load
p	force / length^2	Pressure
S	force / length	tension per unit length of edge
$S_p(f)$	rms^2 / Hz	power spectral density of quantity
T	force	Tension
T	time	period of vibration
t	time	Time
x, y, z	length	mutually orthogonal displacement
X,Y,Z	length	Displacements
Y	dimensionless	Bessel function of second kind
ω	radians	Angle
ε	length / length	Strain
γ	dimensionless	ratio of specific heats of a gas
γ	mass / length^2	mass per unit area
ν	dimensionless	Poisson's ratio
π	dimensionless	3.141592653
θ	radians	angle of rotation
ρ	mass / length^3	Density
σ	dimensionless	beam mode shape parameter
σ	force / length^2	Stress
ω	radians / second	circular frequency, $2\pi f$
ζ	dimensionless	fraction of critical damping

Table 1.1 Nomenclature, continued

Subscripts, superscripts, and bars

Symbol	Units	Definition
i,j,k,l,m,n	dimensionless	counting integers, 1,2,3...
rms	-	root-mean-square
Vector	(bold face)	Vector
$\overline{\text{Avg}}$	(over bar)	average over time
$\tilde{y}$	(over tilde)	mode shape, dimensionless
$[X]^T$	-	Transpose of matrix [X]
A×B	(×)	vector cross product, Eq. 1.24
A•B	(•)	vector dot product, Eq. 1.25
|X|		determinant or magnitude of X
X′(t)	-	ordinary derivative, dX/dt
$\dot{x}(t)$	over dot (.)	derivative with respect to time, t

Abbreviations

Quantity	Abbreviation	Quantity	Abbreviation
Centigrade degrees	°C	logarithm, natural base e	ln
centimeter	cm	mega-Pascal	MPa
cubic centimeters	cc	Meter	m
decibel	dB	millimeter	mm
decaNewton	daN	Newton	N
degree	deg	Kelvin degrees	°K
Fahrenheit degrees	°F	Pascal	Pa
feet	ft	Pound	lb
gram	g	power spectral density	$S_p(f)$
gravity's acceleration	g_c	Radian	rad
Hertz = cycles /second	Hz	Rankine degrees	°R
inch	in.	root-mean-square	rms
Bessel function of 1st kind	J(x)	second	s
complete elliptic integral 2nd	E(x)	Sound Pressure Level	SPL
kilogram	kg	times 10 raised to power x	E+x
logarithm base 10	$\log_{10}$	ton, metric	Te

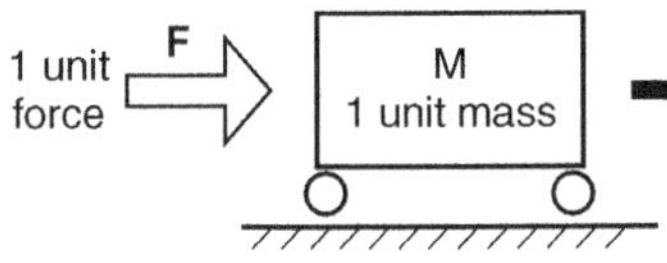

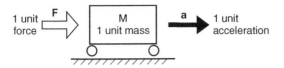

Figure 1.1 Newton's second law in consistent units, $F = Ma$

1.3 Units

The change in motion [of mass] is proportional to the motive force. – Newton [1].

The formulas presented in this book give the correct results with consistent sets of units. In *Consistent Units*, Newton's second law is identically satisfied without factors: One unit of force equals one unit of mass times one unit of acceleration, Figure 1.1.

$$F = Ma \tag{1.1}$$

Table 1.2 presents sets of consistent units that identically satisfy $F = Ma$. These units are used internationally in professional engineering. They are recommended for use with this book.

Table 1.2 Consistent sets of engineering units

System	Force	Mass	Length	Time	Pressure	Density	g_c (a)
1 m-k-s (SI)	Newton	kilogram	meter	second	Pascal (b)	$\frac{\text{kilogram}}{\text{meter}^3}$	$9.807\,\frac{\text{meter}}{\text{second}^2}$
2 c-g-s	dyne (c)	gram	centimeter	second	$\frac{\text{dyne}}{\text{centimeter}^2}$	$\frac{\text{gram}}{\text{centimeter}^3}$	$980.7\,\frac{\text{centimeter}}{\text{second}^2}$
3 cm-kg-s	kilogram	$\frac{\text{kilogram}}{g_c}$	centimeter	second	$\frac{\text{kilogram}}{\text{centimeter}^2}$	$\frac{\text{kilogram}}{g_c\text{-centimeter}^3}$	$980.7\,\frac{\text{centimeter}}{\text{second}^2}$
4 mm-N-s	Newton	metric ton (f)	millimeter	second	megaPascal (d)	$\frac{\text{metric ton}}{\text{millimeter}^3}$	$9807\,\frac{\text{millimeter}}{\text{second}^2}$
5 mm-dN-s	deca-Newton (e)	10^4 kilogram (f)	millimeter	second	$\frac{\text{decaNewton}}{\text{millimeter}^2}$	$\frac{10^4\text{ kilogram}}{\text{millimeter}^3}$	$9807\,\frac{\text{millimeter}}{\text{second}^2}$
6 ft-lb-s	pound	slug (g)	foot	second	$\frac{\text{pound}}{\text{foot}^2}$	$\frac{\text{slug}}{\text{foot}^3}$	$32.17\,\frac{\text{foot}}{\text{second}^2}$
7 in.-lb-s	pound	$\frac{\text{pound}}{g_c}$	inch	second	$\frac{\text{pound}}{\text{inch}^2}$	$\frac{\text{pound}}{g_c\text{-inch}^3}$	$386.1\,\frac{\text{inch}}{\text{second}^2}$

(a) The General Conference of Weights and Measures defined the standard acceleration of gravity at sea level and latitude 45 degrees as 9.80665 m/s^2 (32.1740 ft/s^2) in 1901. Acceleration of gravity increases with latitude from 9.780 m/s^2 at 0 degree latitude to 9.832 m/s^2 at 90 degree latitude (Refs [2, 3, 4, 5] Section 1.4).
(b) Pascal = Newton / meter^2.
(c) dyne = 0.00001 Newton. dyne / centimenter^2 = 0.1 Pa.
(d) Mega Pascal = 10^6 Newton/m^2 = 1 Newton / mm^2 = 1000000 Pascals = 10 bar = 0.1 hectobar.
(e) 1 daNewton =1 dN = 10 Newtons. daNewton / mm^2 = 10^7 Pascals = 100 bar = 1 hectobar. Tables 1.3, 1.4.
(f) Metric ton (Te) =1000 kg =10^6 gm. 10000 kg = 10 metric tons (Te).
(g) slug = pound / g_c. It has units of lb-s^2/ft. 1 slug weighs 32.17 lb on the surface of the earth.

Mass and force have different consistent units. In SI units (Systeme International of the International Organization for Standardization), meter is the unit of length, the unit of mass is kilogram, and second is the unit of time [2, 3, 4]. The SI *consistent unit of force* called *Newton* is kilogram-meter/second squared. Substituting these units in Equation 1.1 shows that 1 N force accelerates 1 kg mass at 1 m/s^2. The General Conference of Weights and Measures defined the Newton unit of force in 1948 and the Pascal unit of pressure, 1 N/m^2, in 1971.

It is the author's experience that inconsistent units are the most common cause of errors in dynamics calculation. See Refs [5–8]. While lack of an intuitive feel for dyne, Newton, or slug may be the reason to convert the final result of a calculation to a convenient customary unit in which mass and force have same customary units, it is important to remember that formulas derived from Newton's laws discussed in this book and most engineering software require the consistent units shown in Table 1.2 to produce correct results.

One Newton force is about the weight of a small apple. If this apple is made into apple butter and spread over a table 1 m^2 then resultant pressure is 1 Pa, which is a small pressure. There is a plethora of pressure units in engineering. Zero decibels (dB) pressure at 1000 Hz is the threshold of human hearing (Section 6.1); it is 20 μPa (20×10^{-6} Pa, 2.9×10^{-9} psi), which is a very small pressure. Stress in structural materials is measured in units of ksi (1000 psi, 6.894×10^6 Pa), MPa(10^6 Pa), decaNewton/mm^2, and hectobar (both 10^7 Pa), which are all large pressures. One hectobar stress is 500 billion times greater than 0 dB pressure.

Standard prefixes for decimal unit multipliers and their abbreviations are in Table 1.3. Table 1.4 has conversion factors; ASTM Standard SI 10-2002 [3], Taylor [4], and Cardarelli [9] provide many more.

Example 1.1 Force on mass

A 1 gram mass accelerates at 1 ft/s^2. What force is on the mass?

Solution: Newton's second law (Eq. 1.1) is applied. Consistent units are required. Gram-foot-seconds is not a consistent set of units. To make the calculation in SI units, case 1 of Table 1.2, grams are converted to kilograms and feet are converted to meters. The conversion factors in Table 1.4 are 1 gram = 0.001 kg, and 1 ft = 0.3048 m so in SI 1 ft/s^2 is 0.3048 m/s^2. Equation 1.1 gives the force that accelerates 1 g at 1 ft/s^2.

$$F = Ma = 0.001\ \text{kg}\ (0.3048\ \text{m}/\text{s}^2) = 0.0003048\ \text{kg–m}/\text{s}^2 = 0.0003048\ \text{N}$$

Table 1.3 Decimal unit multipliers

femto (f)	pico (p)	nano (n)	micro (μ)	milli (m)	centi (c)	deci (d)	deca (da)	hecto (h)	kilo (k)	mega (M)	giga (G)	tera (T)
10^{-15}	10^{-12}	10^{-9}	10^{-6}	10^{-3}	10^{-2}	10^{-1}	10^{1}	10^{2}	10^{3}	10^{6}	10^{9}	10^{12}

Example: 1 hectobar = 1 hbar = 100 bar = 100000 mbar = 10 megaPascal = 10 MPa. Ref. [3–5]

Table 1.4 Conversion factors

cc = cubic centimeter; DecaNewton = 10 Newtons; mile = US statute mile, unless otherwise noted; Pascal = 1 N/m^2; pound mass = pound, avoirdupois. Refs [3, 4, 5, 9]. See Table 1.2 for consistent sets of units.

Mass Units		
To convert from	to	multiply by
gram	kilogram	1. E-3
pound	kilogram	0.45359237
slug	kilogram	14.59390
slinch (=lbf-sec^2/in)	kilogram	175.1319
ton, long	kilogram	1016.047
ton, short	kilogram	907.1847
ton, metric (=tonne)	kilogram	1000
pound	gram	453.59237
gram	pound	2.204624 E-3
kilogram	pound	2.204624
slug (=lbf-sec^2/ft)	pound	32.17
slinch (=lbf-sec^2/in)	pound	386.1
ton, long	pound	2240
ton, metric (=tonne)	pound	2204.622
ton, short	pound	2000
10^4 kg (=10 metric tons)	pound	22046.22
Force Units		
To convert from	to	multiply by
decaNewton	Newton	10
kilogram	Newton	9.806650
pound	Newton	1/0.22481
pound	decaNewton	1/2.2481
dyne	Newton	1. E-5
decaNewton	pound	2.2481
dyne	pound	2.22481 E-6
gram	dyne	980.6
kilogram	pound	2.204624
Newton	pound	0.22481
Length Units		
To convert from	to	multiply by
inch	centimeter	2.54
inch	millimeter	25.4
centimeter	inch	0.3937
millimeter	inch	0.03937
centimeter	foot	0.03208
inch	foot	1/12

Table 1.4 Conversion factors, continued

Length Units (continued)

To convert from	to	multiply by
meter	foot	3.280833
mile, nautical	foot	6076.12
mile, U.S. statute	foot	5280
millimeter	foot	0.0032808
yard	foot	3
foot	inch	12
meter	inch	39.37
millimeter	inch	1/25.4
centimeter	meter	0.01
foot	meter	0.3048
inch	meter	0.0254
kilometer	meter	1000
mile, nautical	meter	1852
mile, U.S. statute	meter	1609.347
mile, U.S. statute	kilometer	1.609347
millimeter	meter	0.001
yard	meter	0.9144

Density Units

To convert from	to	multiply by
gram / cc	kilogram / meter^3	1000
gram / millimeter^3	kilogram / meter^3	10^6
gram /millimeter^3	gram/ cubic centimeter	1000
Te(=10^3 kg))/ millimeter^3	kilogram / meter^3	10^{12}
pound / foot^3	kilogram / meter^3	16.018346
pound / inch^3	kilogram / meter^3	2.76799 E+4
pound / inch^3	gm / cc	1/0.03612729
slug / foot^3	kilogram / meter^3	5.153788
metric ton/ mm^3	kilogram/mm^3	10E-9
gram / cc	pound / foot^3	62.42879
kilogram / meter^3	pound / foot^3	0.06242879
Te (=10^3 kg)/ millimeter^3	pound / foot^3	6.242879 E10
pound / inch^3	pound / foot^3	1728
slug / foot^3	pound / foot^3	32.17
gram / cc	pound / inch^3	0.03612729

Table 1.4 Conversion factors, continued

Density Units, continued		
To convert from	to	multiply by
kilogram / meter^3	pound / inch^3	3.612729 E-5
Te(10^3 kg) / millimeter^3	pound / inch^3	3.612729 E7
pound / foot^3	pound / inch^3	1/1728
slug / foot^3	pound / inch^3	1.186169 E-2
slinch / inch^3	pound / inch^3	386.1
Pressure Units		
To convert from	to	multiply by
atmosphere	kilogram / centimeter^2	1.0333
bar	kilogram / centimeter^2	10.1971
Pascal	kilogram / centimeter^2	1.019716 E-5
pound / foot^2	kilogram / centimeter^2	0.00048825
pound / inch^2	kilogram / centimeter^2	0.070309
kilogram / millimeter^2	kilogram / centimeter^2	100
pound / inch^2	dyne / centimeter^2	68947.57
atmosphere, standard	Pascal = 1 N/meter^2	1.013250 E+5
Bar	Pascal	1. E +5
dyne / centimeter ^{2}	Pascal	0.1
hectobar	Pascal	1. E +7
kilogram / centimeter^2	Pascal	9.806650 E+4
kilogram / meter^2	Pascal	9.80665
kilogram / millimeter^2	Pascal	9.806650 E+6
megaPascal (MPa)	Pascal	1 E+6
megaPascal (MPa)	Pound/ inch^2	145.0377
millibar	Pascal	100
Newton / meter^2	Pascal	1
Newton / millimeter^2	Pascal	1E+6
column H2O one mm high	Pascal	9.80665
column H2O one mm high	kilogram / meter^2	1
column Hg one mm high	Pascal	133.322
column one inch H2O high	Pascal	249.0889
decaNewton / millimeter^2	Pascal	1. E +7
pound / foot^2	Pascal	47.88026
pound / inch^2	Pascal	6894.757
Torr	Pascal	133.322
decaNewton / millimeter^2	kilogram / millimeter^2	1/0.9806650
atmosphere, standard	pound / foot^2	2116.215
Bar	MPa	0.1
1000 pound / inch^2 (ksi)	MPa	6.894757

Table 1.4 Conversion factors, continued

Pressure Units, continued		
To convert from	to	multiply by
Bar	pound / foot^2	2088.542
kilogram / meter^2	pound / foot^2	1/4.882428
Pascal	pound / foot^2	0.02088544
atmosphere, standard	pound / inch^2	14.69594
Bar	pound / inch^2	14.50377
column H2O one inch high	pound / inch^2	0.036127
column Hg one inch Hg	pound / inch^2	0.49115
column Hg one mm high	pound / inch^2	0.01933661
decaNewton / centimeter^2	pound / inch^2	14.50355
kilogram / mm^2	pound / inch^2	1422.3
Millibar	pound / inch^2	0.01450377
megaPascal (MPa)	pound / inch^2	145.0377
megaPascal (MPa)	1000 lb/ inch^2 (ksi)	0.1450377
Newton / millimeter^2	pound / inch^2	145.0377
Pascal	pound / inch^2	1.450377 E-4
pound / foot^2	pound / inch^2	1/144
hectobar	pound / inch^2	1450.377
Pressure Units (decibels)	Also see Table 6-1	
To convert from	to	Formula
decibels rel. 20 micro Pa	Pascal, rms	20E-6 Pa$\times 10^{dB/20}$
decibels rel. 20 micro Pa	psi, rms	2.9E-9 psi$\times 10^{dB/20}$
Pascals, root-mean-square	decibels, dB	$20 \log_{10}(p_{rms} / 20\times10^{-6}$ Pa)
Velocity Units		
To convert from	to	multiply by
foot / second	meter / second	1/3.280833
kilometer / hour	meter / second	1/3.6
meter / second	foot / second	3.280833
kilometer / hour	foot / second	0.9113425
knots	meter / second	0.5144444
knots	feet / second	1.687810
miles / hour	kilometer / hour	1.609347
miles / hour	meter / second	0.44704
miles / hour	feet / second	1.466667
miles / hour	knots	0.86897

Table 1.4 Conversion factors, continued

Temperature Units

To convert from	to	Formula
degrees Fahrenheit, °F	degrees Centigrade, °C	°C = (5/9)(°F – 32)
degrees Centigrade, °C	degrees Fahrenheit, °F	°F = (9/5) °C + 32
degrees Centigrade, °C	degrees Kelvin, °K	°K = °C + 273.15
degrees Fahrenheit, °F	degrees Rankine, °R	°R = °F + 459.69
degrees Kelvin, °K	degrees Centigrade, °C	°C = °K – 273.15
degrees Rankine, °R	degrees Fahrenheit, °F	°F = °R – 459.69
degrees Kelvin, °K	degrees Fahrenheit, °F	°F = (9/5) (°K–273.16) +32

Energy Units

To convert from	to	multiply by
inch-pound	Newton-meter (Joule)	0.1129848
foot-pound	Newton-meter	1.355818
Newton-meter	foot-pound	0.73756
BTU (international)	Newton-meter	1055.056
calorie	Joule = 1 Newton-meter	4.184
Newton-meter	Joule	1
horse power (US)	ft-lb/s	550
horse power (US)	watt = 1 N-m/s	745.6999
watt	Joule/s	1

Volume Units

To convert from	to	multiply by
cubic centimeter (cc)	liter	0.001
cubic centimeter (cc)	cubic inch	0.061024
liter	cubic centimeter (cc)	1000
cubic meter	liter	1000
barrel (US petroleum)	cubic meter	0.158987
barrel (US petroleum)	cubic inch	9702
gallon US	cubic inch	231
gallon US	cubic meter	0.003785412
gallon US	liter	3.785412
cubic foot	cubic inch	1728
cubic foot	liter	28.316

Standard Earth Acceleration, g = 9.80665 m^2/s

To convert from	to	multiply by
acceleration, m^2/s	g	1/9.80665
acceleration, ft^2/s	g	1/32.174
acceleration, cm^2/s	g	980.0665
acceleration, in^2/s	g	1/386.089

For calculation in US customary lb-ft-s units, case 6 in Table 1.2, grams are converted to slugs by converting grams to pounds then pounds to slugs.

$$1\text{ g} = 0.0022045\text{ lb }(1\text{ slug } / \ 32.17\text{ lb}) = 0.00006852\text{ slug}$$

$$F = Ma = 0.00006852\text{ slug }(1\text{ ft}/\text{s}^2) = 0.00006852\text{ lb}$$

The results imply the relationship between Newtons and pounds: 1 N/1 lb = 0.00006852/ 0.0003049 = 0.2248. See Table 1.4.

1.4 Motion on the Surface of the Earth

The earth can be modeled as a spinning globe with a 6380 km (3960 miles) equatorial radius (Figure 1.2) that revolves daily about its polar axis. (Geophysical models of earth are discussed in Refs [5] and [10–14].) Owing to its rotation, the earth's surface is not an inertial frame of reference. As one walks in a line on the surface of the earth, one is actually walking along a circular arc because the surface of the earth is curved. Further, the earth is rotating under one's feet. Accelerations induced by the earth's curvature and rotations are important for predicting weather, weighing gold, and launching projectiles.

Point P is on the surface of the earth at radius $r = 6{,}380{,}000$ m, longitude λ and polar angle θ as shown in Fig. 1.2. Its circumferential angular velocity is the sum of the rotation of the earth about polar axis ($\Omega = 7.272 \times 10^{-5}$ rad/s) and $d\lambda/dt$. When P is stationary with respect to the earth's surface, $d\lambda/dt = d\theta/dt = 0$. The velocity and acceleration of P with respect to the center of the earth are given in spherical coordinates in case 4 of Table 2.1 with these values.

$$\begin{aligned}
\mathbf{v} &= r(d\theta/dt)\mathbf{n}_\theta + r(\Omega + d\lambda/dt)\sin\theta\mathbf{n}_\lambda \\
\mathbf{v}_p &= 464.0\sin\theta\mathbf{n}_\lambda,\ \text{m/s} \\
\mathbf{a} &= -[r(d\theta/dt)^2 + r(\Omega + d\lambda/dt)^2\sin^2\theta]\mathbf{n}_r \\
&\quad + [rd^2\theta/dt^2 - r(\Omega + d\lambda/dt)^2\sin\theta\cos\theta]\mathbf{n}_\theta \\
&\quad + [rd^2\lambda/dt^2\sin\theta + 2r(\Omega + d\lambda/dt)\cos\theta d\theta/dt]\mathbf{n}_\lambda \\
\mathbf{a}_p &= -0.03374\sin^2\theta\mathbf{n}_r - 0.01687\sin 2\theta\mathbf{n}_\theta,\ \text{m/s}^2
\end{aligned} \tag{1.2}$$

On the equator, $\theta = 90$ degrees, the earth's surface velocity is 464.0 m/s (1670 km/hr, 1520 ft/s, 1038 mph). The inward radial acceleration of 0.03374 m/s^2 towards the center of the earth results in a 0.34% reduction in gravity. This explains the popularity of the equator for launching satellites. Objects weigh less on the equation than near the poles. For example, gold weighs 0.12% more at the mine in Nome Alaska (65.4 degrees N latitude, $\theta = 24.6$ deg) than at the bank in San Francisco (35.7 degrees N latitude, $\theta = 54.3$ deg).

Now consider that particle P moves freely at constant radius with an initial west-to-east velocity $\mathbf{v}_\lambda = r\ d\lambda/dt\ \mathbf{n}_\lambda$ and north-to-south velocity $\mathbf{v}_\theta = r\ d\theta/dt\ \mathbf{n}_\theta$, with respect to the surface of the earth. The polar and latitudinal components of accelerations are set to zero,

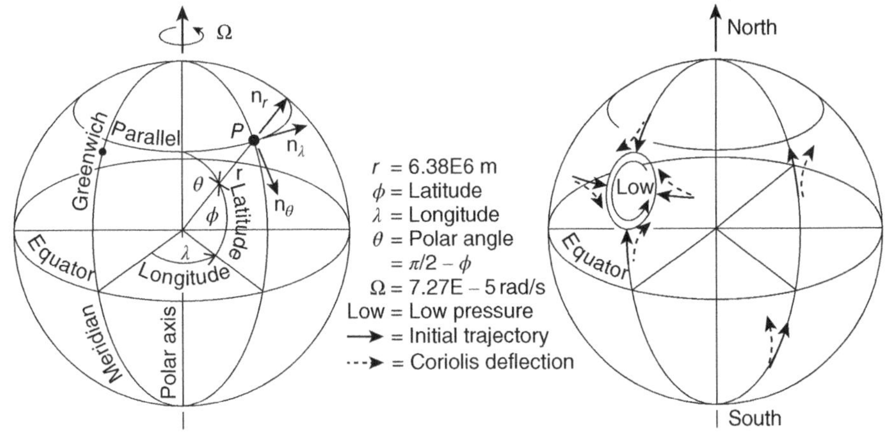

Figure 1.2 Motion of the surface of rotating earth and Coriolis deflection of moving particles relative to the earth. Latitude is zero at the equator.

$\mathbf{a}_\theta = \mathbf{a}_\lambda = 0$ for the freely moving particle. The previous kinematic equations are solved for the latitudinal and longitudinal angular accelerations relative to the earth.

$$d^2\theta/dt^2 = (\Omega + d\lambda/dt)^2 \sin\theta\cos\theta$$
$$d^2\lambda/dt^2 = -2(\Omega + d\lambda/dt)(d\theta/dt)\cos\theta/\sin\theta \tag{1.3}$$

These equations show that an initial west-to-east angular velocity ($d\lambda/dt$) induces a north-to-south acceleration ($d^2\theta/dt^2$) and a north-to-south velocity ($d\theta/dt$) produces an east-to-west acceleration ($-d^2\lambda/dt^2$) in the northern hemisphere. Thus, freely moving particles veer to their right in the Northern Hemisphere ($\theta < 90°$) and to the left in the Southern Hemisphere ($\theta > 0°$), as seen by an earth-based observer. These induced motions are named *Coriolis* accelerations, after the French engineer-mathematician Gustave-Gaspard Coriolis.

Figure 1.2 shows Coriolis accelerations spin air flowing inwards toward a region of low pressure. As a result hurricanes spin counterclockwise in the northern hemisphere and typhoons spin clockwise in the southern hemisphere. Coriolis forces affect ocean currents. Coriolis deflected tides carry migrating shad fish counterclockwise around Canada's Bay of Fundy [15].

1.5 Geometric Properties of Plane Areas

Table 1.5 has formulas for the geometric properties of plane areas [16–19]. Figure 1.3a shows a bounded plane area in the x–y plane with *area* A and *centroid* C at point x_c, y_c.

$$A = \int_A dx\,dy, \qquad x_c = \frac{1}{A}\int_A x\,dA, \qquad y_c = \frac{1}{A}\int_A y\,dA \tag{1.4}$$

Table 1.5 Properties of plane sections

Notation: A = cross-sectional area; C = centroid of area; K(x), E(x) = complete elliptical integrals of first and second kind; I_x = area moment of inertia about x-axis; I_y = area moment of inertia about y-axis; I_{xy} = area product of inertia about x and y axes; $I_p = I_x + I_y$ = polar area moment of inertia about z-axis; I_{xc}, I_{yc}, I_{xcyc} = area moment of inertia about axes through centroid; t = thickness; P = perimeter; x_C = distance from x-axis to centroid; y_C = distance from y-axis to centroid; θ = angle, radian. Also see Table 1.6 and Eqs 1.4 through 1.11 Refs [16–19].

Section	Area A and Centroid C	Area Moments of Inertia Ix, Iy	Area Products of Inertia Ixy
1. Right Triangle 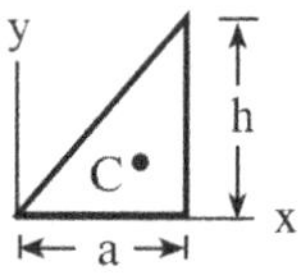	$x_C = \frac{2}{3}a$ $y_C = \frac{1}{3}h$ $A = \frac{1}{2}ah$	$I_{x_C} = \frac{ah^3}{36}$ $I_{y_C} = \frac{a^3h}{36}$ $I_x = \frac{ah^3}{12}$ $I_y = \frac{a^3h}{4}$ $I_{P_C} = I_{x_C} + I_{y_C} = \frac{ah^3 + a^3h}{36}$	$I_{x_C y_C} = \frac{a^2h^2}{72}$ $I_{xy} = \frac{a^2h^2}{8}$ Also see Table 1.7.
2. General Triangle 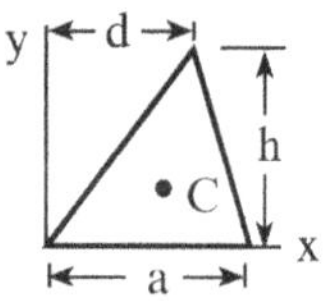	$x_C = \frac{1}{3}(a+d)$ $y_C = \frac{1}{3}h$ $A = \frac{1}{2}ah$ For both d>a and a>d.	$I_{x_C} = \frac{ah^3}{36}$ $I_{y_C} = \frac{ah}{36}\left(h^2 - ad + d^2\right)$ $I_x = \frac{ah^3}{12}$ $I_y = \frac{ah}{12}\left(a^2 + ad + d^2\right)$	$I_{x_C y_C} =$ $\frac{ah^2}{72}(2d-a)$ $I_{xy} =$ $\frac{ah^2}{24}(a+2d)$ see Table 1.7.
3. Square 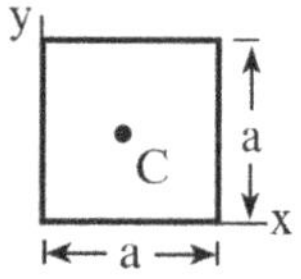	$x_C = \frac{1}{2}a$ $y_C = \frac{1}{2}a$ $A = a^2$	$I_{x_C} = I_{y_C} = \frac{a^4}{12}$, $I_{pc} = \frac{a^4}{6}$ $I_x = I_y = \frac{a^4}{3}$, $I_p = \frac{2a^4}{3}$	$I_{x_C y_C} = 0$ $I_{xy} = \frac{a^4}{4}$
4. Rectangle 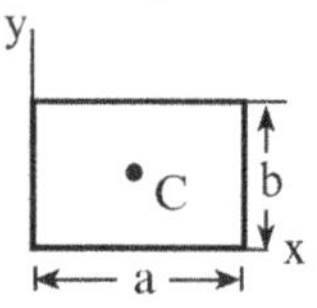	$x_C = \frac{1}{2}a$ $y_C = \frac{1}{2}b$ $A = ah$	$I_{x_C} = \frac{ab^3}{12}$ $I_{y_C} = \frac{a^3b}{12}$ $I_x = \frac{ab^3}{3}$ $I_y = \frac{a^3b}{3}$ $I_{P_C} = I_{x_C} + I_{x_C} = \frac{ab^3 + a^3b}{12}$	$I_{x_C y_C} = 0$ $I_{xy} = \frac{a^2b^2}{4}$
5. Parallelogram 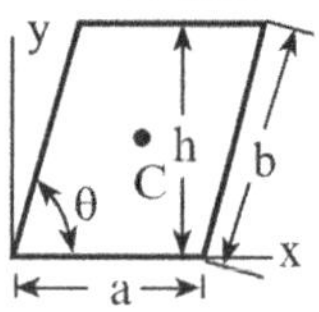	$x_C = \frac{1}{2}(a + b\cos\theta)$ $y_C = \frac{b}{2}\sin\theta = \frac{h}{2}$ $A = ab\cos\theta = ah$ $\theta = \arcsin\frac{h}{b}$	$I_{x_C} = ab^3\sin^3\theta/12 = ah^3/12$ $I_{y_C} = \frac{ab}{12}\sin\theta(a^2 + b^2\cos^2\theta)$ $I_x = (ab^3\sin^3\theta)/3 = ah^3/3$ $I_y = ab\sin\theta(a + b\cos\theta)^2/3$ $-(a^2b^2\sin\theta\cos\theta)/6$	$I_{x_C y_C} =$ $\frac{ab^3}{12}\sin^2\theta\cos\theta$ $h = b\sin\theta$

Table 1.5 Properties of plane sections, continued

Section	Area A and Centroid C	Area Moments of Inertia Ix, Iy	Area Products of Inertia Ixy
6. Inclined Rectangle 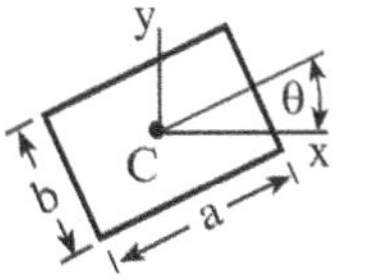	$x_C = 0$ $y_C = 0$ $A = ab$ Also see case 47.	$I_{x_C} = \frac{ab}{12}(a^2 \sin^2\theta + b^2 \cos^2\theta)$ $I_{y_C} = \frac{ab}{12}(a^2 \cos^2\theta + b^2 \sin^2\theta)$ $I_{p_C} = I_{x_C} + I_{y_C} = \frac{a^3 b + ab^3}{12}$	$I_{x_C y_C} =$ $\frac{ab}{24}(b^2 - a^2)\sin 2\theta$
7. Diamond 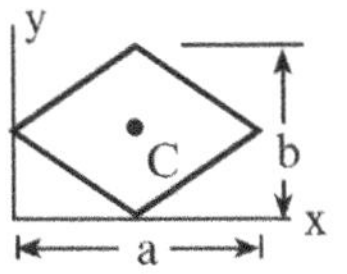	$x_C = \frac{a}{2}$ $y_C = \frac{b}{2}$ $A = \frac{ab}{2}$	$I_{x_C} = \frac{ab^3}{48}$ $I_{y_C} = \frac{a^3 b}{48}$	$I_{x_C y_C} = 0$
8. Hollow Rectangle 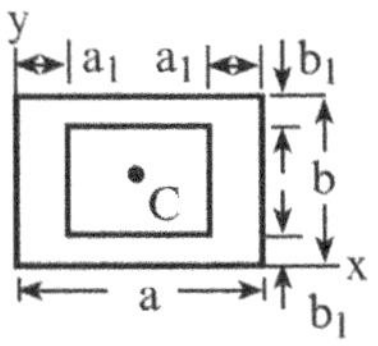	$x_C = \frac{1}{2}a$ $y_C = \frac{1}{2}b$ $A = 2ab_1 + 2a_1 b - 4a_1 b_1$	$I_{x_C} = \frac{ab^3 - (a - 2a_1)(b - 2b_1)^3}{12}$ $I_{y_C} = \frac{a^3 b - (a - 2a_1)^3 (b - 2b_1)}{12}$	$I_{x_C y_C} = 0$
9. Thin Hollow Rectangle 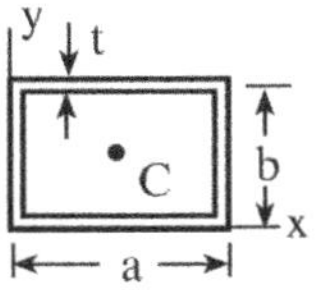	$x_C = \frac{1}{2}a$ $y_C = \frac{1}{2}b$ $A = 2(a + b)t$ $a, b >> t$	$I_{x_C} = \frac{tb^3}{6} + \frac{tab^2}{2}$ $I_{y_C} = \frac{ta^3}{6} + \frac{ta^2 b}{2}$, $I_{pC} = t(a+b)^3$ $I_x = \frac{2tb^3}{3} + tab^2$	$I_{x_C y_C} = 0$ $I_{pc} = \frac{t(a^3 + b^3)}{6} + \frac{tab(a^2 + b^2)}{2}$
10. Channel 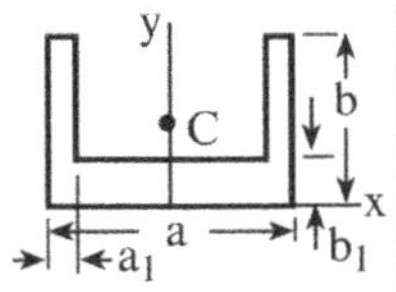	$x_C = 0$ $y_C = \frac{2a_1 b^2 + ab_1^2 - 2a_1 b_1^2}{2A}$ $A = 2a_1 b + ab_1 - 2a_1 b_1$	$I_x = \frac{2}{3}a_1 b^3 + \frac{1}{3}(a - 2a_1)b_1^3$ $I_y = I_{y_C} = \frac{a^3 b - (b - b_1)(a - 2a_1)^3}{12}$	$I_{x_C y_C} = I_{xy} = 0$
11. Thin Channel 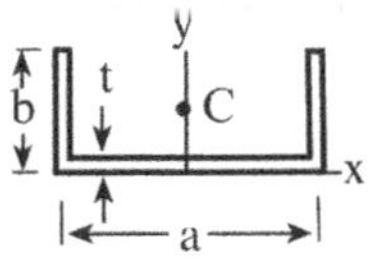	$x_C = 0$ $y_C = \frac{b^2}{a + 2b}$ $A = (a + 2b)t$ $a, b >> t$	$I_x = \frac{2tb^3}{3}$ $I_y = I_{y_C} = \frac{ta^3}{12} + \frac{ta^2 b}{2}$ $I_p = I_x + I_y$	$I_{x_C y_C} = I_{xy} = 0$

Table 1.5 Properties of plane sections, continued

Section	Area A and Centroid C	Area Moments of Inertia Ix, Iyy	Area Products of Inertia Ixy
12. Tapered Channel 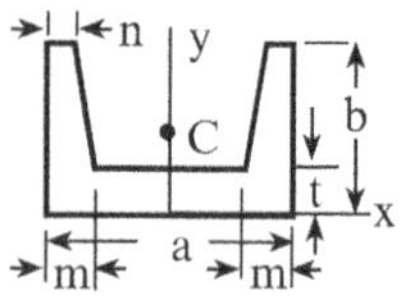	$x_C = 0$ $y_C = \frac{1}{6A}(6b^2n +$ $2(b-t)(m-n)(b+2t)$ $+3(a-2m)t^2)$ $A = at + (b-t)(m+n)$	$I_x = \frac{1}{3}(2nb^3 + (a-2m)t^3 + \frac{m-n}{2}(b-t)\bullet$ $(b^2+2bt+3t^2))$ $I_y = \frac{a^3t}{12} + \frac{b-t}{6}\Big(n^3 + 3n(a-n)^2 + \frac{m-n}{6}\bullet$ $(2(m-n)^2 + (3a-2m-4n)^2)\Big)$ $I_p = I_x + I_y, \quad I_{xy} = 0$	
13. I Section 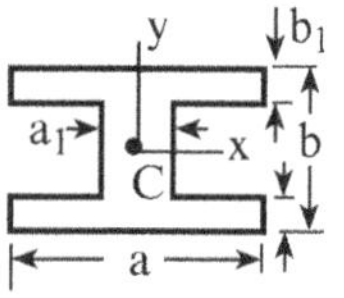	$x_C = 0$ $y_C = 0$ $A = 2ab_1 + a_1b - 2a_1b_1$	$I_{x_C} = \frac{1}{12}\left(ab^3 - (a-a_1)(b-2b_1)^3\right)$ $I_{y_C} = \frac{1}{12}\left(2a^3b_1 + (b-2b_1)a_1^3\right)$ $I_{x_Cy_C} = 0,\ I_p = I_x + I_y,$	
14. Tapered I 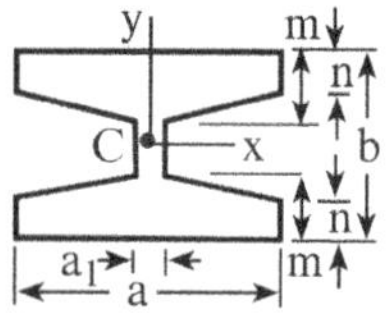	$x_C = 0$ $y_C = 0$ $A = a_1b + (a-a_1)(m+n)$	$I_{x_C} = \frac{1}{12}\left(ab^3 - \frac{(a-a_1)}{8(m-n)}(c^4 - e^4)\right)$ $I_{y_C} = \frac{1}{12}\left(2a^3n + ea_1{}^3 + \frac{(m-n)}{2(a-a_1)}(a^4 - a_1^4)\right)$ where $c = b - 2n,\ e = b - 2m.\ I_{x_Cy_C} = 0$	
15. Thin I Section 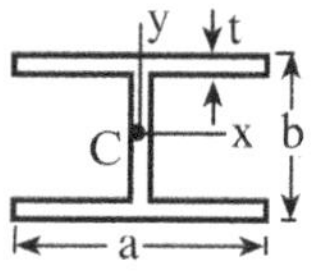	$x_C = 0$ $y_C = 0$ $A = (2a+b)t$ $a, b >> t$	$I_{x_C} = \frac{1}{12}tb^3 + \frac{1}{2}tab^2$ $I_{y_C} = ta^3/6$ $I_{p_C} = \frac{1}{12}tb^3 + \frac{1}{2}tab^2 + \frac{1}{6}ta^3$	$I_{x_Cy_C} = 0$
16. T Section 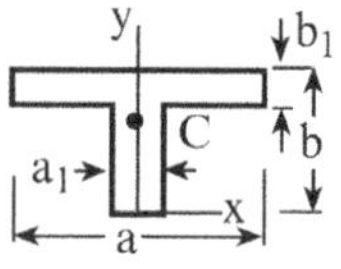	$x_C = 0$ $y_C = b - \frac{a_1b^2 + ab_1^2 - a_1b_1^2}{2A}$ $A = a_1b + ab_1 - a_1b_1$	$I_x = \frac{1}{3}(ab^3 - (a-a_1)(b-b_1)^3)$ $I_y = I_{y_C} = \frac{a^3b_1 + (b-b_1)a_1{}^3}{12}$ $I_p = I_x + I_y$	$I_{x_Cy_C} = 0$
17. Angle Section 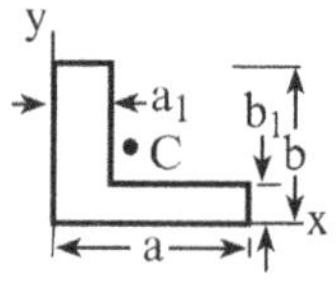	$x_C = \frac{a^2b_1 + a_1{}^2(b-b_1)}{2A}$ $y_C = \frac{a_1b^2 + b_1^2(a-a_1)}{2A}$ $A = a_1b + ab_1 - a_1b_1$	$I_x = \frac{1}{3}((a-a_1)b_1^3 + a_1b^3)$ $I_y = \frac{1}{3}(a^3b_1 + (b-b_1)a_1^3)$ $I_{x_C} = \frac{1}{3}(ay_C^3 - (a-a_1)\bullet$ $(y_C - b_1)^3 + (b-y_C)^3a_1)$	$I_{xy} = \frac{1}{4}(a^2b_1^2 +$ $a_1^2b^2 - a_1^2b_1^2)$ $I_{x_Cy_C} = I_{xy} - \frac{a^2b^2A}{4(b+b}$

Table 1.5 Properties of plane sections, continued

Section	Area A and Centroid C	Area Moments of Inertia Ix, Iy	Area Products of Inertia Ixy
18. Thin Angle Section 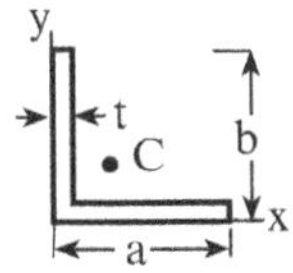	$x_C = \frac{a^2}{2(a+b)}$ $y_C = \frac{b^2}{2(a+b)}$ $A = (a+b)t$ $a, b >> t$	$I_{x_C} = b^3t/3 - tb^4/(4(a+b))$ $I_{y_C} = a^3t/3 - ta^4/(4(a+b))$ $I_x = \frac{1}{3}b^3t$ $I_y = \frac{1}{3}a^3t,\ I_p = \frac{t}{3}(a^3+b^3)$	$I_{xy} = \frac{1}{4}(a^2+b^2)t^2$ $I_{x_Cy_C} = -\frac{a^2b^2t}{4(a+b)}$
19. Z Section 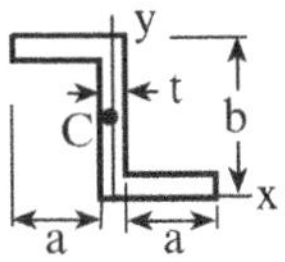	$x_C = 0$ $y_C = b/2$ $A = bt + 2at$ for thin section, $a,b >> t$ neglect t^2, t^3 in formulas.	$I_{x_C} = t(b^3 + 6ab^2 - 12abt + 8at^2)/12$ $I_{y_C} = I_y = t(8a^3 + 12a^2t + 6at^2 + bt^2)/12$ $I_x = t(b^3 + 3ab^2 - 3abt + 2at^2)/3$ $I_{x_Cy_C} = -a^2bt - (b-a)at^2 + at^3$	
20. Trapezoid 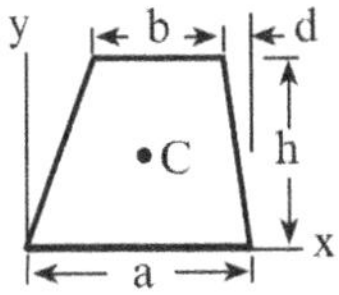	$x_C = \frac{1}{3(a+b)}(2a^2 + 2ab - ad - 2bd - b^2)$ $y_C = \frac{h}{3}\left(\frac{a+2b}{a+b}\right)$ $A = \frac{h}{2}(a+b)$	$I_{x_C} = h^3(a^2 + 4ab + b^2)/[36(a+b)]$ $I_{y_C} = h[a^4 + b^4 + 2ab(a^2+b^2)/[36(a+b)]$ $-d(a^3 + 3a^2b - 3ab^2 - b^3) + d^2(a^2 + 4ab + b^2)]$ $I_{x_Cy_C} = h^2[b(3a^2 - 3ab - b^2) + a^3$ $- d(2a^2 + 8ab + 2b^2)]/[72(a+b)]$ $I_x = h^3(a+3b)/12$	
21. Regular Polygon with n Sides 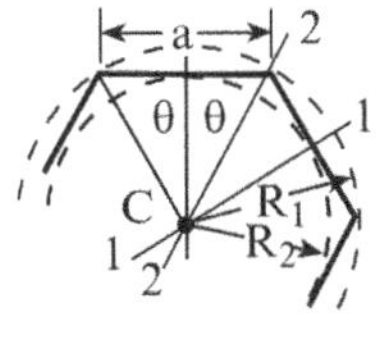	$R_1 = \frac{a}{2\sin\theta}$ $R_2 = \frac{a}{2\tan\theta}$ $\theta = 180/n$, degree $A = \frac{1}{4}a^2 n \cot\theta$	$I_1 = \frac{A(6R_1^2 - a^2)}{24}$ $I_2 = \frac{A(12R_2^2 + a^2)}{48}$	n = number sides = 3,4,5.. $\theta = 180/n$, degrees For n=3, $\theta = 60°$ $R_1 = a/3^{1/2}$, $R_2 = a/12^{1/2}$, $A = a^2 3^{1/2}/4$, $I_1 = I_2 = a^4 3^{1/2}/96$
22. Hollow Regular Polygon with n Sides 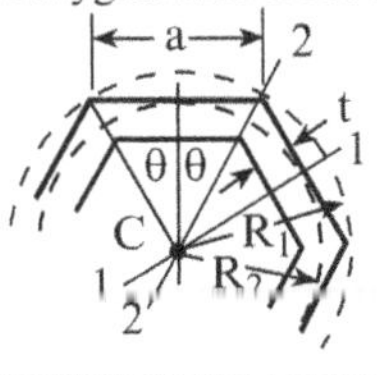	$R_1 = \frac{a}{2\sin\theta}$ $R_2 = \frac{a}{2\tan\theta}$ $\theta = 180/n$, (deg) $A = nat\left(1 - \frac{t}{a}\tan\theta\right)$	$I_1 = I_2 = \frac{na^3t}{8}\left(\frac{1}{3} + \frac{1}{\tan^2\theta}\right) \cdot \left[1 - \frac{3t}{a}\tan\theta + 4\left(\frac{t}{a}\tan\theta\right)^2 - 2\left(\frac{t}{a}\tan\theta\right)^3\right]$	n = number of sides = 3,4,5..
23. Circle 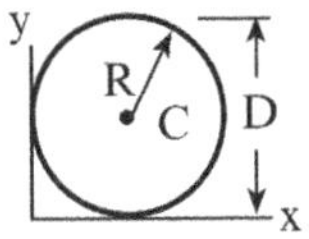	$x_C = R = D/2$ $y_C = R = D/2$ $A = \pi R^2 = \frac{\pi}{4}D^2$ $P = 2\pi R = \pi D$	$I_{x_C} = I_{y_C} = \frac{1}{4}\pi R^4 = \frac{1}{64}\pi D^4$ $I_{PC} = \frac{1}{2}\pi R^4 = \frac{1}{32}\pi D^4$ $I_x = I_y = \frac{5}{4}\pi R^4 = \frac{5}{64}\pi D^4$	$I_{x_Cy_C} = 0$ $I_{xy} = \pi R^4 = \frac{\pi}{16}D^4$

Table 1.5 Properties of plane sections, continued

Section	Area A and Centroid C	Area Moments of Inertia Ix, Iy	Area Products of Inertia Ixy
24. Semicircle 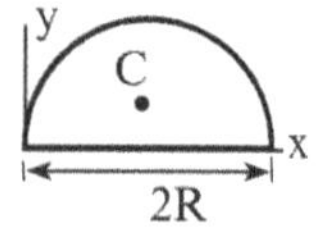	$x_C = R$ $y_C = \frac{4R}{3\pi}$ $A = \frac{1}{2}\pi R^2$	$I_{x_C} = \frac{R^4(9\pi^2 - 64)}{72\pi}$ $I_{y_C} = \frac{\pi R^4}{8}$ $I_x = \frac{\pi R^4}{8}$ $I_y = \frac{5\pi R^4}{8}$	$I_{x_C y_C} = 0$ $I_{xy} = \frac{2}{3} R^4$
25. Circular Sector 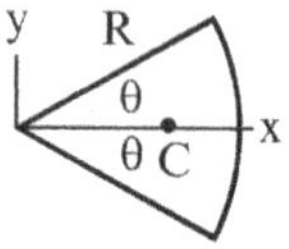	$x_C = \frac{2R}{3}\frac{\sin\theta}{\theta}$ $y_C = 0$ $A = R^2\theta$	$I_x = \frac{R^4}{4}(\theta - \sin\theta\cos\theta)$ $I_y = \frac{R^4}{4}(\theta + \sin\theta\cos\theta)$	$I_{x_C y_C} = 0$ $I_{xy} = 0$
26. Crescent 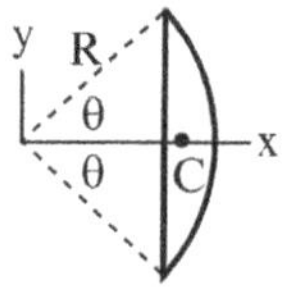	$x_C = \frac{2R}{3}\left(\frac{\sin^3\theta}{\theta - \sin\theta\cos\theta}\right)$ $y_C = 0$ $A = R^2(\theta - \sin\theta\cos\theta)$	$I_x = \frac{R^4}{4}(\theta - \sin\theta\cos\theta - \frac{2}{3}\sin^3\theta\cos\theta)$ $I_y = \frac{R^4}{4}(\theta - \sin\theta\cos\theta + 2\sin^3\theta\cos\theta)$ $I_{x_C y_C} = 0$, $I_{xy} = 0$, $I_p = I_x + I_y$	
27. Annulus 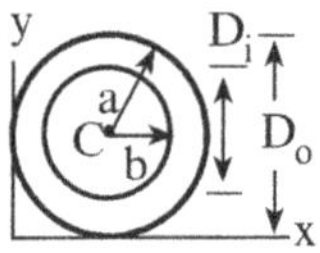	$x_C = a$ $y_C = a$ $A = \pi(a^2 - b^2)$ $= \frac{\pi}{4}(D_o^2 - D_i^2)$	$I_{x_C} = I_{y_C} = \frac{\pi}{4}(a^4 - b^4)$ $= \frac{\pi}{64}(D_o^4 - D_i^4)$ $I_x = I_y = \frac{5}{4}\pi a^4 - \pi a^2 b^2 - \frac{\pi}{4} b^4$	$I_{x_C y_C} = 0$ $I_{xy} = \pi a^2(a^2 - b^2)$ $I_{pC} = \frac{\pi}{2}(a^4 - b^4)$
28. Semi-Annulus 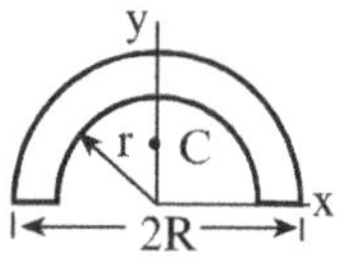	$x_C = 0$ $y_C = \frac{4R}{3\pi}\left(\frac{r}{R} + \frac{R}{R+r}\right)$ $A = \frac{\pi}{2}(R^2 - r^2)$	$I_x = \frac{\pi}{8}(R^4 - r^4)$ $I_y = I_x$ $I_p = \frac{\pi}{4}(R^4 - r^4)$	$I_{x_C y_C} = 0$ $I_{xy} = 0$
29. Sector Annulus 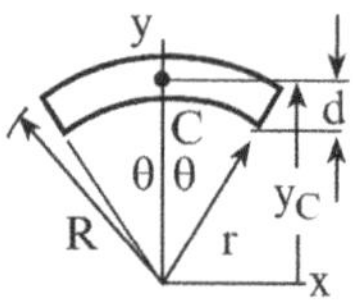	$x_C = 0$ $y_C = \frac{2\sin\theta}{3\theta}\left(\frac{R^3 - r^3}{R^2 - r^2}\right)$ $d = y_C - r\sin\theta$ $A = (R^2 - r^2)\theta$	$I_{x_C} = \frac{1}{4}(R^4 - r^4)(\theta + \sin\theta\cos\theta) - \frac{4}{9}\frac{(R^3 - r^3)^2}{(R^2 - r^2)}\frac{\sin^2\theta}{\theta}$ $I_x = \frac{1}{4}(R^4 - r^4)(\theta + \sin\theta\cos\theta)$ $I_{y_C} = I_y = \frac{1}{4}(R^4 - r^4)(\theta - \sin\theta\cos\theta)$	
30. Thin Annulus 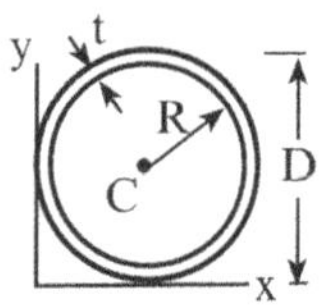	$x_C = R = D/2$ $y_C = R = D/2$ $A = 2\pi Rt = \pi Dt$ $D = 2R$ $R >> t$	$I_{x_C} = I_{y_C} = \pi R^3 t = \frac{\pi}{8} D^3 t$ $I_x = I_y = 3\pi R^3 t = \frac{3\pi}{8} D^3 t$ $I_{p_C} = 2\pi R^3 t = \frac{\pi}{4} D^3 t$	$I_{x_C y_C} = 0$ $I_{xy} = 2\pi R^3 t = \frac{\pi}{4} D^3 t$

Table 1.5 Properties of plane sections, continued

Section	Area A and Centroid C	Area Moments of Inertia Ix, Iy	Area Products of Inertia Ixy
31. Sector Thin Annulus 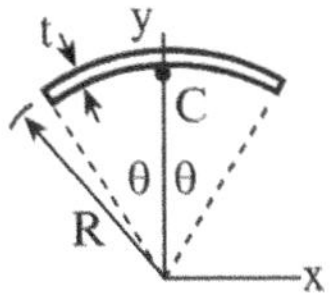	$x_C = 0$ $y_C = R\frac{\sin\theta}{\theta}$ $A = 2\theta Rt$ $R >> t$	$I_{x_C} = R^3t(\theta + \sin\theta\cos\theta$ $-(2\sin^2\theta)/\theta)$ $I_x = R^3t(\theta + \sin\theta\cos\theta)$ $I_{y_C} = I_y = R^3t\ (\theta - \sin\theta\cos\theta)$	$I_{x_C y_C} = 0$ $I_{xy} = 0$ $I_p = R^3t\theta$
32. Corner Complement 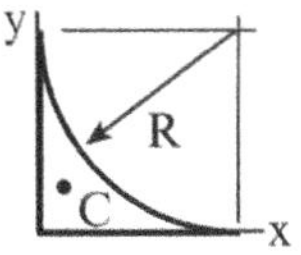	$x_C = \frac{10-3\pi}{3(4-\pi)}R$ $y_C = \frac{10-3\pi}{3(4-\pi)}R$ $A = (1-\pi/4)R^2$	$I_{x_C} = I_{y_C} = \frac{9\pi^2 - 84\pi + 176}{144(4-\pi)}R^4$ $I_x = I_y = \left(1 - \frac{5\pi}{16}\right)R^4$	$I_{xy} = \left(\frac{19}{24} - \frac{\pi}{4}\right)R^4$
33. Ellipse 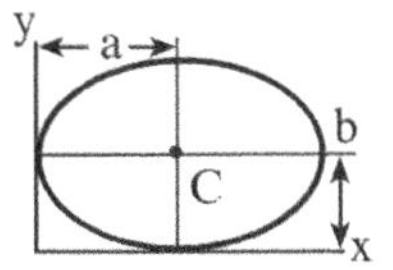	$x_C = a \quad A = \pi ab \quad I_{x_C} = \frac{\pi}{4}ab^3 \quad I_{y_C} = \frac{\pi}{4}a^3b \quad I_{x_C y_C} = 0 \quad I_{pc} = \frac{\pi}{4}ab(a^2+b^2)$ $y_C = b \quad I_x = \frac{5\pi}{4}ab^3 \quad I_y = \frac{5\pi}{4}a^3b \quad I_{xy} = \pi a^2b^2$ Ref. [16] Perimeter $P = 4aE\left[\sqrt{1-b^2/a^2}\right] \approx \pi\left(3(a+b) - \sqrt{(3a+b)(a+3b)}\right)$, $a > b$		
34. Semi Ellipse 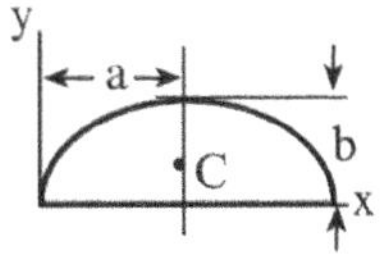	$x_C = a$ $y_C = \frac{4b}{3\pi}$ $A = \frac{\pi}{2}ab$	$I_{x_C} = \frac{ab^3}{72\pi}\left(9\pi^2 - 64\right)$ $I_{y_C} = \frac{\pi}{8}a^3b$ $I_x = \frac{\pi}{8}ab^3 \quad I_y = \frac{5\pi}{8}a^3b$	$I_{x_C y_C} = 0$ $I_{xy} = \frac{2}{3}a^2b^2$ $I_p = I_x + I_y$
35. Quarter Ellipse 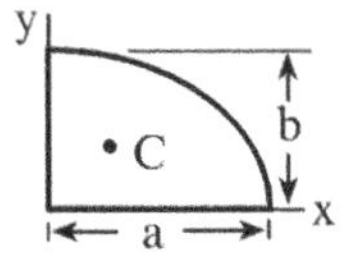	$x_C = \frac{4}{3\pi}a$ $y_C = \frac{4}{3\pi}b$ $A = \pi ab/4$	$I_{x_C} = ab^3\left(\frac{\pi}{16} - \frac{4}{9\pi}\right)$ $I_{y_C} = a^3b\left(\frac{\pi}{16} - \frac{4}{9\pi}\right)$ $I_x = \pi ab^3/16,\ I_y = \pi a^3b/16$	$I_{xy} = \frac{a^2b^2}{4}$ $I_p = I_x + I_y$
36. Hollow Ellipse 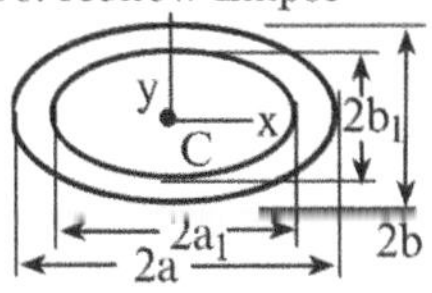	$x_C = 0$ $y_C = 0$ $A = \pi(ab - a_1b_1)$	$I_{x_C} = \frac{\pi}{4}\left(ab^3 - a_1b_1^3\right)$ $I_{y_C} = \frac{\pi}{4}\left(a^3b - a_1^3b_1\right)$	$I_{x_C y_C} = 0$ $I_{P_C} = \frac{\pi}{4}(ab(a^2+b^2)$ $a_1b_1(a_1^2 + b_1^2))$
37. Thin Hollow Ellipse 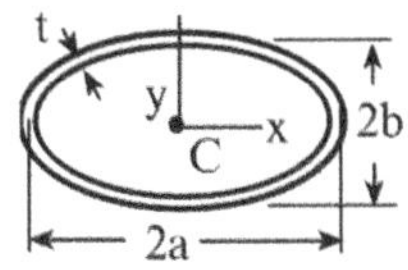	$x_C = 0$ $y_C = 0$ $A = \pi t(a+b)$ $a, b >> t$	$I_{x_C} = I_x = \frac{\pi b^2 t}{4}(a+3b)$ $I_{y_C} = I_y = \frac{\pi a^2 t}{4}(b+3a)$	$I_{x_C y_C} = 0$ $I_{P_C} = I_{x_C} + I_{y_C}$

Table 1.5 Properties of plane sections, continued

Section	Area A and Centroid C	Area Moments of Inertia Ix, Iy	Area Products of Inertia Ixy
38. Parabola, $x = a(y/b)^2$ 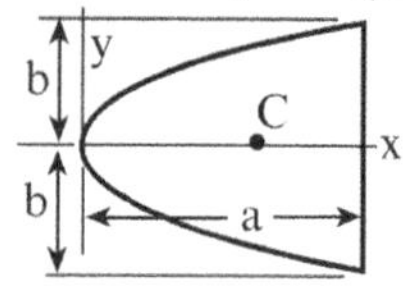	$x_C = \frac{3}{5}a$ $y_C = 0$ $A = \frac{4}{3}ab$	$I_{x_C} = I_x = \frac{4}{15}ab^3$ $I_{y_C} = \frac{16}{175}a^3b$ $I_y = \frac{4a^3b}{7}$	$I_{x_C y_C} = 0$ $I_p = I_x + I_y$
39. Semi-Parabola $x = a(y/b)^2$ 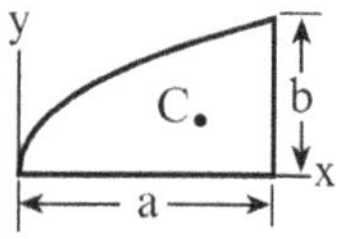	$x_C = \frac{3}{5}a$ $y_C = \frac{3}{8}b$ $A = \frac{2}{3}ab$	$I_{xC} = \frac{19}{480}ab^3$ $I_x = \frac{2}{15}ab^3$ $I_y = \frac{2}{7}a^3b$	$I_{xy} = \frac{1}{6}a^2b^2$
40. n^{th} Semi-Parabola $x=a(y/b)^n$, $n=1/2$ shown 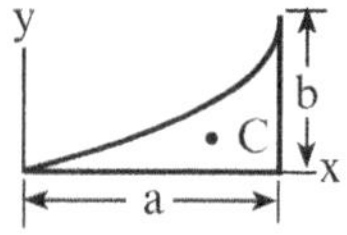	$x_C = \frac{n+1}{n+2}a$ $y_C = \frac{n+1}{2(2n+1)}b$ $A = \frac{1}{n+1}ab$	$I_x = \frac{1}{3(3n+1)}ab^3$ $I_y = \frac{1}{n+3}a^3b$	$I_{xy} = \frac{1}{4n+4}a^2b^2$ $n = 1/2$ for parabola
41. NACA Symmetric 4-Digit Airfoil 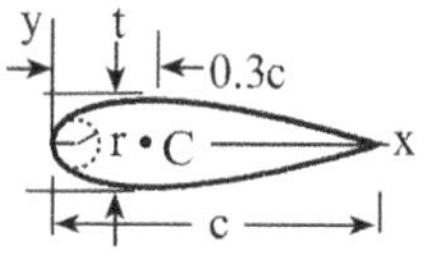	$x_C = 0.4204c$ $y_C = 0$ $A = 0.685tc$ thickness t at $x = 0.30c$ radius $r = 1.1019t^2/c$	$I_{x_C} = I_x = 0.03941ct^3$ $I_{y_C} = 0.03782c^3t$ $I_y = 0.1589c^3t$ $I_{xy} = 0$	airfoil profile Ref. [18], $y = \pm 5t(0.2969X^{1/2}$ $-0.1260X - 0.3516X^2 +$ $0.2843X^3 - 0.1015X^4)$ where $X = x/c$.
42. NACA Symmetric 4-Digit Airfoil 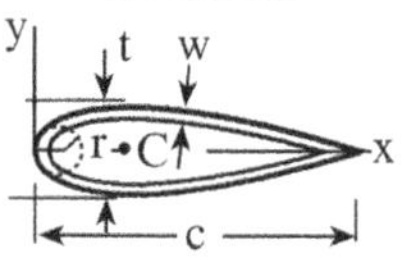	For thin skin airfoil, $c >> t >> w$ $x_C = 0.5c$ $y_C = 0$ $A = Pw,\ P \approx 2c + t^2/c$	$I_{x_C} = I_x = 0.2754wct^2$ $I_{y_C} = (1/6)c^3w$ $I_y = (2/3)c^3w$	Profile is given above. $I_{xy} = 0$ Approximate formulas
43. Sine Wave Stiffener 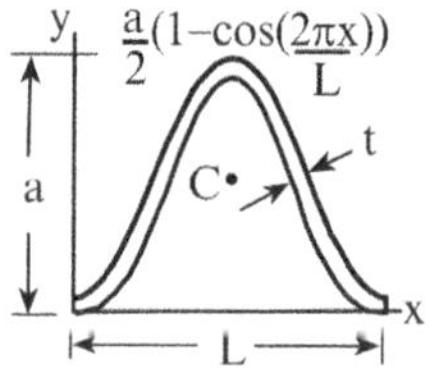	For thin skin, $L >> t$ $x_C = L/2,\ y_C = a/2$ $A = \frac{2tL}{\pi}E\left(-\frac{\pi^2a^2}{L^2}\right)$ $\approx tL[1 + 4.4(a/L)^2]^{1/2}$	$I_x = \frac{tL^3}{6\pi^3}\left[(1+\frac{\pi^2a^2}{L^2})K\left(-\frac{\pi^2a^2}{L^2}\right) - (1-\frac{\pi^2a^2}{L^2})E\left(-\frac{4\pi^2a^2}{L^2}\right)\right] = \alpha_1 tL^3$ $\alpha_1 \approx 0.124(a/L)^{1.5} + 0.664(a/L)^3,\ I_{x_C y_C} = 0$ K(x), E(x) = complete elliptic integrals	

Table 1.5 Properties of plane sections, continued

Section	Area A and Centroid C	Area Moments of Inertia Ix, Iy	Area Products of Inertia Ixy
44. Symmetric Sandwich 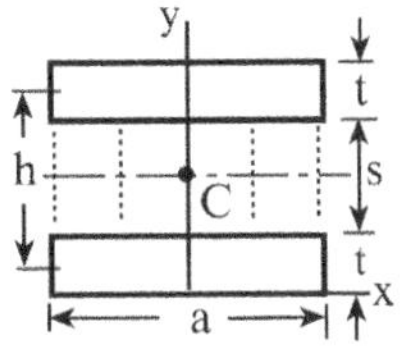	$x_C = 0$ $y_C = t + s/2$ $A = 2at$ $h = s + t$ The two identical rectangles are separated by a spacer.	$I_{x_C} = at\left(\frac{1}{2}s^2 + st + \frac{2}{3}t^2\right) = at\left(\frac{1}{2}h^2 + \frac{1}{6}t^2\right)$ $I_{y_C} = ta^3/6$ $I_x = at\left(s^2 + 3st + \frac{8}{3}t^2\right) = at\left(h^2 + th + \frac{2}{3}t^2\right)$ $I_{xy} = 0$	
45. Unequal Parallel Rectangles 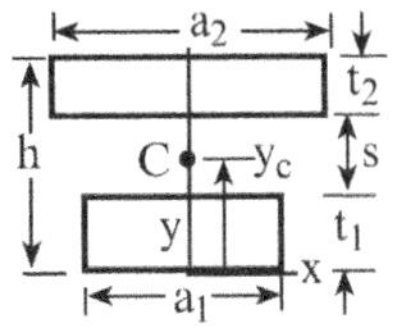	$x_C = 0,\ y_C = \dfrac{a_1t_1^2/2 + a_2t_2(h - t_2/2)}{A}$ $A = a_1t_1 + a_2t_2$ $s = h - t_1 - t_2$ Section is symmetric about y axis.	$I_{x_C} = \frac{a_1t_1^3}{12} + a_1t_1(y_C - t_1/2)^2 + \frac{a_2t_2^3}{12}$ $+ a_2t_2(h - y_C - t_2/2)^2$ $I_{y_C} = \frac{1}{12}(t_1a_1^3 + t_2a_2^3)$ $I_x = \frac{1}{3}a_1t_1^3 + \frac{1}{12}a_2t_2^3 + a_2t_2(h - t_2/2)^2,\ I_{xy} = 0$	
46. Translated Section 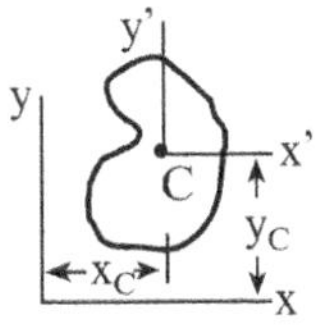	Area = A x' and y' axes are through centroid C and they are parallel to x and y axes, respectively.	$I_x = I_{x_C} + y_C^2 A$ $I_y = I_{y_C} + x_C^2 A$ I_x and I_y about x-y axes, I_{xc} and I_{yc} about x'-y' axes.	$I_{xy} = I_{x_C y_C} + x_C y_C A$
47. Rotated Section 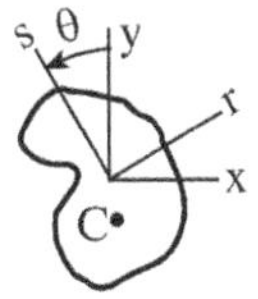	The r and s axes are rotated counterclockwise with respect to the x and y axes. Also see Eq. 1.8.	$I_x = I_r\cos^2\theta +$ $I_s\sin^2\theta + I_{rs}\sin 2\theta$ $I_y = I_r\sin^2\theta +$ $I_s\cos^2\theta - I_{rs}\sin 2\theta$ $I_p = I_x + I_y = I_r + I_s$	$I_{xy} = I_{rs}\cos 2\theta$ $+ \frac{1}{2}(I_s - I_r)\sin 2\theta$ Note: $I_{xy} = 0$ and I_x is minimum for $2\theta = \arctan 2I_{rs}/(I_r - I_s)$
48. General Plane Area 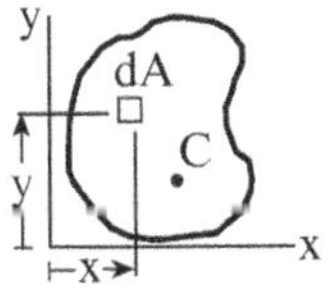	$x_C = \frac{1}{A}\int_A x\,dA$ $y_C = \frac{1}{A}\int_A y\,dA$ $A = \int_A dx\,dy$	$I_x = \int_A y^2 dA$ $I_y = \int_A x^2 dA$ $I_p = I_x + I_y$ $= \int_A (x^2 + y^2)dA$	$I_{xy} = \int_A xy\,dA$ Note mass per unit length $m = \rho A$ where ρ = material density

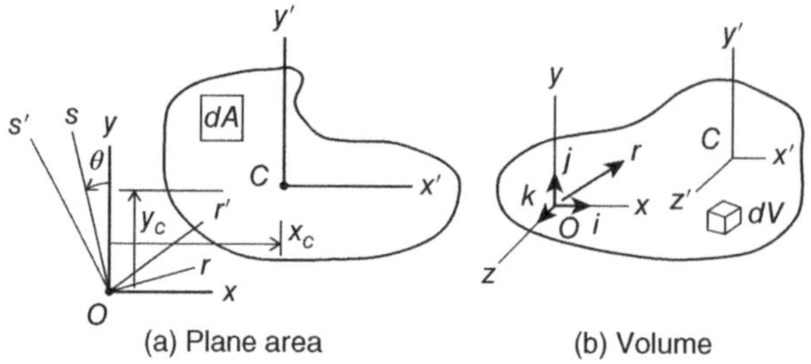

Figure 1.3 A plane section with centroid (C) and rotated and translated coordinate systems and a solid body with a translated coordinate system and a rotated vector **r**

The z-axis is perpendicular to the x–y plane. The element of area is $dA = dxdy$. *Area moments of inertia* about the x-axis (I_x) and y-axis (I_y), the *area product of inertia* (I_{xy}) about the x–y-axes, and the *polar area moment of inertia* about the z-axis ($I_p = I_{zz}$) are integrals over the area.

$$\begin{aligned} I_x &= \int_A y^2 dA, \quad I_y = \int_A x^2 dA, \quad I_{xy} = \int_A xy\, dA, \\ I_p &= I_{zz} = I_x + I_y = \int_A (x^2 + y^2) dA \end{aligned} \tag{1.5}$$

The symbol I is used for area moments of inertia and J is used for mass moments of inertia. I_x is the integral of the square of distance (y^2) along the y-axis from the x-axis times the elemental area. The *radius of gyration* for each axis is defined.

$$r_x = (I_x/A)^{1/2}, \; r_y = (I_y/A)^{1/2} \tag{1.6}$$

I_{xc}, I_{yc}, and I_{pc} are the area moments of inertia about axes with origin at the centroid C, Figure 1.3a. $I_{xy} = 0$ if the body is symmetric about either axis.

Parallel axis theorem transforms moments of inertia about the centroid, I_{xc}, I_{yc}, and I_{xcyc}, to moments of inertia about the offset parallel axes x, y, Figure 1.3a.

$$\begin{aligned} I_x &= I_{xc} + y_c^2 A, \quad I_y = I_{yc} + x_c^2 A, \quad I_{xy} = I_{xcyc} + x_c y_c A \\ I_p &= I_x + I_y = I_{pc} + (x_c^2 + y_c^2) A \end{aligned} \tag{1.7}$$

Translation away from the centroid increases the moment I_x and I_y [20].

The r-axis and the orthogonal s-axis shown in Figure 1.3a are rotated counterclockwise by the angle θ with respect to the x- and y-axes. The area moments of inertia about *rotated axes* are,

$$\begin{aligned} I_r &= I_x \cos^2\theta + I_y \sin^2\theta - I_{xy} \sin 2\theta \\ I_s &= I_y \cos^2\theta + I_x \sin^2\theta + I_{xy} \sin 2\theta \\ I_{rs} &= I_{xy} \cos 2\theta - (1/2)(I_{yy} - I_{xx}) \sin 2\theta \end{aligned} \tag{1.8}$$

The sum of the area moments of inertia about two perpendicular axes is independent of the rotation of the axes.

$$I_z = I_p = I_x + I_y = I_r + I_s \tag{1.9}$$

More coordinate transformations are in Table 2.2.

Principal axes are two mutually perpendicular axes about which the product of inertia is zero. The angle of the principal axes is found from Equation 1.8 with $I_{rs} = 0$.

$$\theta = \left(\frac{1}{2}\right) \arctan\left[\frac{2I_{xy}}{(I_y - I_x)}\right] \tag{1.10}$$

Substituting this θ into Equation 1.8 gives the *principal area moments of inertia,* which are the maximum and minimum moments of inertia about rotated axes.

$$\begin{aligned} I_{\max} &= \left(\frac{1}{2}\right)(I_x + I_y) + \left(\frac{1}{2}\right)[(I_x - I_y)^2 + 4I_{xy}^2]^{1/2} \\ I_{\max} &= \left(\frac{1}{2}\right)(I_x + I_y) - \left(\frac{1}{2}\right)[(I_x - I_y)^2 + 4I_{xy}^2]^{1/2} \end{aligned} \tag{1.11}$$

If the two principal area moments of inertia are equal, then the moments of inertia are independent of axis rotation.

Geometric properties of complex areas are obtained by subdividing (meshing) the areas into elementary sections, usually triangles or rectangles, and summing their properties in space about the global coordinate system. See Equations 1.7, 1.8, and Example 1.3. Areas and moments of inertia of standardized aluminum, steel, and timber sections are presented in Refs [21–23]. Section 4.1 discusses application to beam bending theory.

Example 1.2 Area and moment of inertia of pipe section

Compute the area and area moment of inertia about the axis *A-A* through the centroid of the pipe section on the left-hand side of Figure 1.4.

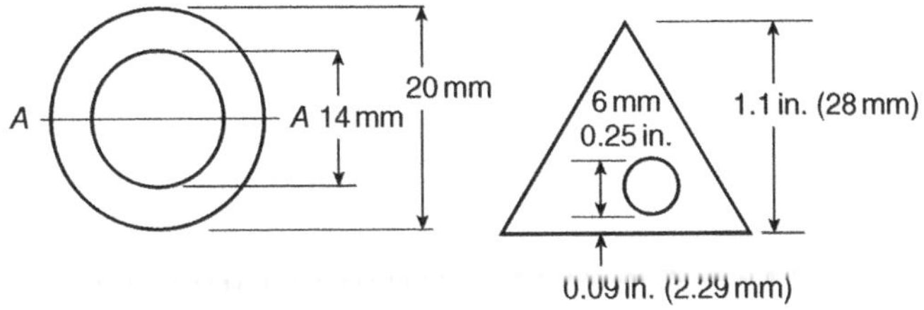

Figure 1.4 Geometric sections for Examples 1.3 and 1.4

Solution: The area and moment of inertia of the pipe section with an axis through its centroid are in case 27 of Table 1.5.

$$A = \pi(a^2 - b^2), \quad I_{xc} = \left(\frac{\pi}{4}\right)(a^4 - b^4)$$

Figure 1.4 shows the outer diameter $a = 20$ mm (0.7874 in.) and the inner diameter $b = 14$ mm (0.5512 in.). The radii are half these values, $a = 10$ mm (0.3937 in.), $b = 7$ mm (0.275 in.). Substituting these values into the above formulas gives,

$$A = 160.2\ \text{mm}^2\ (0.2483\ \text{in.}^2), I_{xc} = 5968\ \text{mm}^4 (0.01434\ \text{in.}^4)$$

The pipe section can also be computed with the thin annulus approximation given in case 30 of Table 1.5.

$$A = 2\pi R t, \qquad I_{x_C} = \pi R^3 t$$

The thickness $t = a - b = 3$ mm is half the difference in the diameters. The average radius $R = (a + b)/2 = 8.5$ mm (0.335 in.). Substituting into the previous formulas gives approximate values: $A = 150.2$ mm^2 (0.2328 in.^{2}), $I_{xc} = 5788$ mm^4 (0.0139 in.^{4}).

Example 1.3 Area and moment of inertia of a triangle

A triangle with a hole in it is shown on the right-hand side of Figure 1.4. Compute the cross-sectional area and area moment of inertia about the axis along the base of the triangle.

Solution: The area and area moment of inertia of the triangle with a hole are equal to the area and area moment of inertia of the triangle less the area and area moment of inertia of the hole. The area and moment of inertia of the triangle for an axis along its base are in case 2 of Table 1.5.

$$A = (1/2)bh^2, \qquad I_x = (1/12)bh^3$$

For our case, $b = 0.9$ in. (22.80 mm) and $h = 1.1$ in. (27.94 mm),

$$A = 0.4950\ \text{in.}^2\ (287.7\ \text{mm}^2), \qquad I_{xc} = 0.09982\ \text{in.}^4\ (41548\ \text{mm}^4)$$

The area of the 0.25 in. (6.35 mm) diameter hole is computed from case 23 of Table 1.5 using a radius $R = 0.25$ in.$/2 = 0.125$ in.: $A_{\text{hole}} = \pi R^2 = 0.04909$ in.^{2} (31.67 mm^2). The net area is the difference between the triangle and the hole: $A_{\text{net}} = 0.4950 - 0.04909 = 0.4459$ in.^{2} (287.7 mm^2).

The area moment of inertia of the triangle with the hole is the area moment of inertia of the triangle less the area moment of inertia of the hole, which is offset by $y_C = 0.09 + 0.125 = 0.215$ in. (5.46 mm), cases 23 and 45 of Table 1.5.

$$I_{\text{xhole}} = I_{x_C} + y_c^2 A = \left(\frac{1}{4}\right)\pi R^4 + y_c^2 \pi R^2 = 0.0001917 + 0.002269 = 0.002461\ \text{in.}^4$$

$$I_{\text{net}} = 0.09982 - 0.002461\ \text{in.}^4 = 0.09736\ \text{in.}^4\ (40520\ \text{mm}^4)$$

1.6 Geometric Properties of Rigid Bodies

Table 1.6 [16–18] has formulas for geometric properties of homogeneous rigid bodies. These are based upon classical solutions, such as Ref. 16; also see Section 1.7.

Table 1.6 Properties of homogeneous solids

Notation: A = cross sectional area; C = centroid (center of mass); J_x = mass moment of inertia about axis parallel to x axis; J_y = mass moment of inertia about axis parallel to y axis; J_z = mass moment of inertia about axis parallel to z axis through center of mass; J_{xy} = mass product moment of inertia; J_{xc}, J_{yc}, J_{zc}, J_{xcyc} = mass moments of inertia about axes through centroid; M = mass = ρV; P = perimeter of section; t = thickness; S = lateral surface area; V = volume; x_C = distance from x axis to center of mass; y_C = distance from y axis to center of mass; z_C = distance from z axis to center of mass; t = mass density. Also see Table 1.7 Refs [16, 17, 19].

Solid Geometry	Volume V and Center of Mass C	Mass Moments Inertia Jx, Jy, Jz	Mass Products of Inertia Jxy
1. Thin Rod 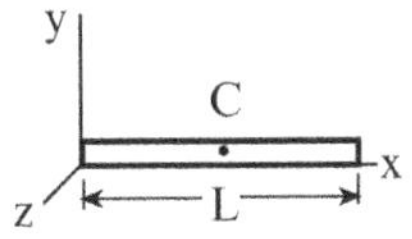	$x_C = \frac{L}{2}$ $y_C = 0$ $V = M/\rho = AL$ A = area cross section	$J_{x_C} = J_x = 0$ $J_{y_C} = J_{z_C} = \frac{1}{12}ML^2$ $J_y = J_z = \frac{1}{3}ML^2$	$J_{x_C y_C,\,etc} = 0$ $J_{xy,\,etc} = 0$ Also see Table 1-7.
2. Circular Arc Rod 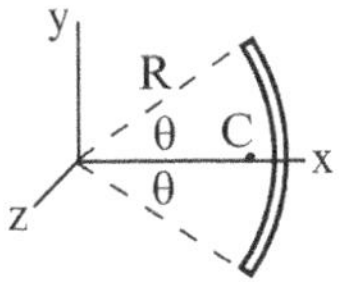	$x_C = \frac{\sin\theta}{\theta}R$ $y_C = 0$ $z_C = 0$ $V = M/\rho = 2\theta AR$ A = area cross section	$J_{x_C} = J_x = \frac{\theta - \sin\theta\cos\theta}{2\theta}MR^2$ $J_y = \frac{\theta + \sin\theta\cos\theta}{2\theta}MR^2$ $J_z = MR^2$	$J_{x_C y_C,\,etc} = 0$ $J_{xy,\,etc} = 0$
3. Thin Hoop 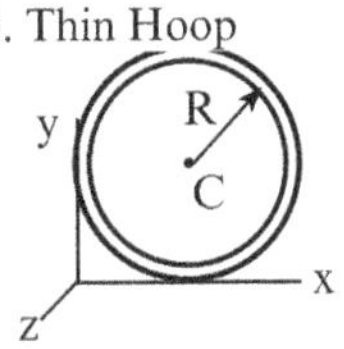	$x_C = R$ $y_C = R$ $z_C = 0$ $V = M/\rho = 2\pi AR$ A = area of section	$J_{x_C} = J_{y_C} = MR^2/2$ $J_{z_C} = MR^2$ $J_x = J_y = 3MR^2/2$ $J_z = 3\,MR^2$	$J_{x_C y_C,\,etc} = 0$ $J_{xy} = MR^2$ $J_{xz} = J_{yz} = 0$
4. Right Tetrahedron 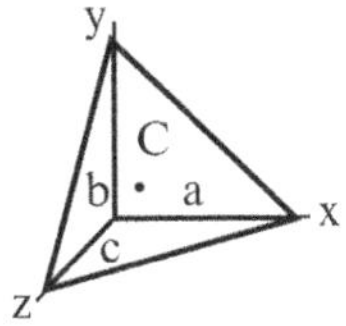 See case 4 Table 1-7.	$x_C = \frac{a}{4}$ $y_C = \frac{b}{c}$ $z_C = \frac{c}{4}$ $V = M/\rho = \frac{1}{6}abc$	$J_{x_C} = \frac{3}{80}M(b^2+c^2)$ $J_{y_C} = \frac{3}{80}M(a^2+c^2)$ $J_{z_C} = \frac{3}{80}M(a^2+b^2)$ $J_{x_C z_C} = -\frac{1}{80}Mac$ $J_x = \frac{1}{10}M(b^2+c^2)$ $J_y = \frac{1}{10}M(a^2+c^2)$ $J_z = \frac{1}{10}M(a^2+b^2)$ $J_{xz} = \frac{1}{20}Mac$	
5. Right Rectangular Pyramid 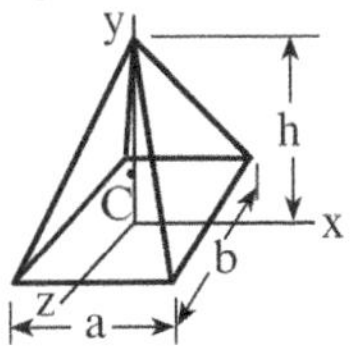	$x_C = 0$ $y_C = \frac{h}{4}$ $z_C = 0$ $S = ab + a\sqrt{h^2 + (b/2)^2}$ $+ b\sqrt{h^2 + (a/2)^2}$ $V = M/\rho = abh/3$	$J_{x_C} = \frac{M}{80}(4b^2 + 3h^2)$ $J_{y_C} = J_y = \frac{M}{20}(a^2 + b^2)$ $J_x = \frac{M}{20}(b^2 + 2h^2)$ $J_z = \frac{M}{20}(a^2 + 2h^2)$	$J_{x_C y_C} = J_{x_C z_C} = 0$ $J_{y_C z_C} = 0,$ $J_{xy} = J_{xz} = J_{yz} = 0$

Table 1.6 Properties of homogeneous solids, continued

Solid Geometry	Volume V, Mass M Center of Mass C	Mass Moments Inertia Jx, Jy, Jz	Mass Products Inertia Jxy
6. Right Angle Wedge 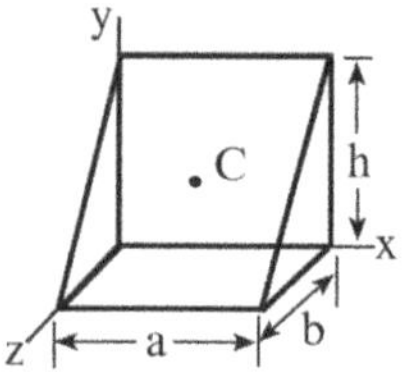	$x_C = a/2$ $y_C = h/3$ $z_C = b/3$ $V = \frac{M}{\rho} = \frac{1}{2}abh$	$J_x = \frac{M}{6}(b^2 + h^2)$ $J_y = \frac{M}{6}(a^2 + 2b^2)$ $J_z = \frac{M}{6}(2a^2 + h^2)$	$J_{xz} = \frac{M}{12}bh$ $J_{xy} = \frac{M}{4}ah$ $J_{yz} = \frac{M}{4}ab$ Also see Table 1.7
7. Cube 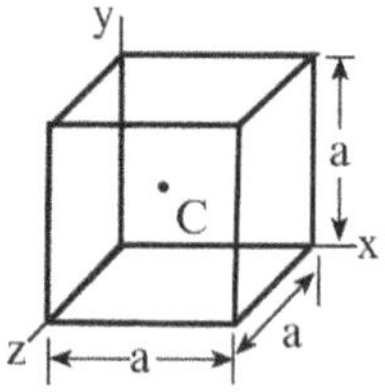	$x_C = \frac{a}{2}$ $y_C = \frac{a}{2}$ $z_C = \frac{a}{2}$ $V = M/\rho = a^3$	$J_{x_C} = J_{y_C} = J_{z_C} = \frac{1}{6}Ma^2$ $J_x = J_y = J_z = \frac{2}{3}Ma^2$	$J_{x_Cy_C} = J_{x_Cz_C} = 0$ $J_{y_Cz_C} = 0,$ $J_{xz} = \frac{1}{4}Ma^2 = J_{xy} = J_{yz}$
8. Rectangular Prism 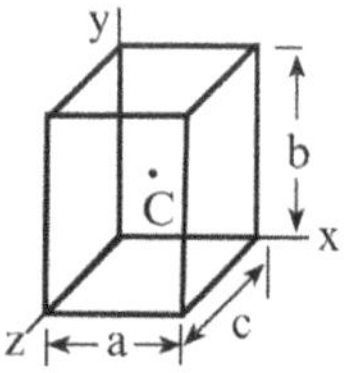	$x_C = \frac{a}{2}$ $y_C = \frac{b}{2}$ $z_C = \frac{c}{2}$ $V = abc,\ M = \rho V$	$J_{x_C} = \frac{M}{12}(b^2 + c^2)$ $J_{y_C} = \frac{M}{12}(a^2 + c^2)$ $J_{z_C} = \frac{M}{12}(a^2 + b^2)$ $J_x = \frac{M}{3}(b^2 + c^2)$	$J_{x_Cy_C} = J_{x_Cz_C} = 0$ $J_{xy} = \frac{1}{4}Mab$ $J_{xz} = \frac{1}{4}Mac$ $J_{yz} = \frac{1}{4}Mbc$
9. Torus 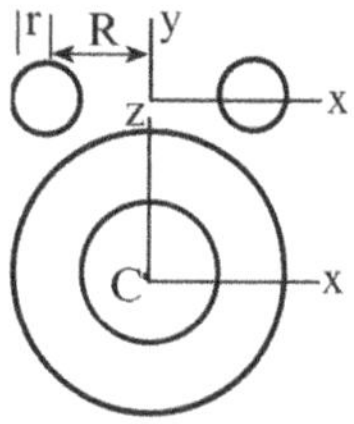	$x_C = 0$ $y_C = 0$ $z_C = 0$ $S = 4\pi^2 Rr$ $V = M/\rho = 2\pi^2 Rr^2$	$J_{x_C} = J_{z_C} = \frac{M}{2}(R^2 + \frac{5}{4}r^2)$ $J_{y_C} = J_y = \frac{M}{4}(4R^2 + 3r^2)$	$J_{x_Cy_C} = J_{x_Cz_C} = 0$ $J_{y_Cz_C} = 0,$ $J_{xy} = J_{xz} = J_{yz} = 0$
10. Sector of Hollow Torus 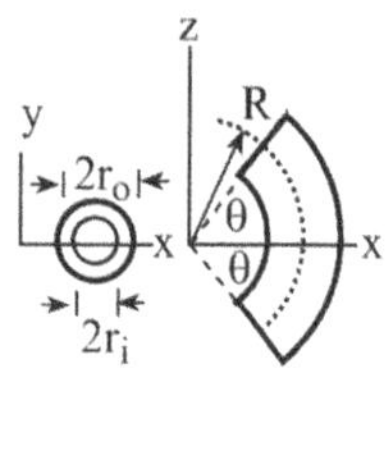	$x_C = \frac{\sin\theta}{\theta}\left(R + \frac{r_o^2 + r_i^2}{4R}\right)$ $y_C = 0$ $z_C = 0$ $S = 2\pi\theta R r_o$ $V = 2\pi R\theta(r_o^2 - r_i^2)$ $M = \rho V$, Ref. [16]	$J_{x_C} = \frac{M}{16\theta}\left(4R^2(2\theta - \sin 2\theta) + (r_o^2 + r_i^2)(10\theta - 3\sin 2\theta)\right)$ $J_{y_C} = \frac{M}{4\theta}\left(4R^2\theta + 3(r_o^2 + r_i^2)\theta - K\right)$, $J_{z_C} =$ $\frac{M}{16\theta}\left[4R^2(2\theta + \sin 2\theta) + (r_o^2 + r_i^2)(10\theta + 3\sin 2\theta) - 4K\right]$ where $K = \frac{2\sin^2\theta}{\theta}\left(2R^2 + r_o^2 + r_i^2 + \frac{(r_o^2 + r_i^2)^2}{8R^2}\right)$	

Table 1.6 Properties of homogeneous solids, continued

Solid Geometry	Volume V, Mass M, Center of Mass C	Mass Moments of Inertia Jx, Jy, Jz	Mass Products of Inertia Jxy
11Right Circular Cone 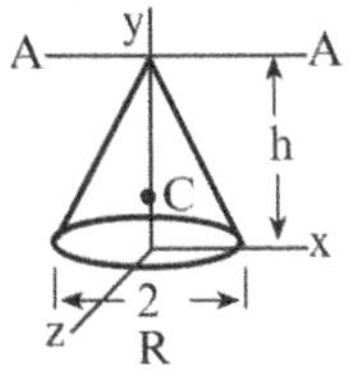	$x_C = 0$ $y_C = \frac{h}{4}$ $z_C = 0$ $S = \pi R(R + \sqrt{R^2 + h^2})$ $V = \frac{M}{\rho} = \frac{\pi}{3} R^2 h$	$J_{x_C} = J_{z_C} = \frac{3}{80} M(4R^2 + h^2)$ $J_{y_C} = J_y = \frac{3}{10} MR^2$ $J_x = J_z = \frac{1}{20} M(3R^2 + 2h^2)$ $J_{AA} = \frac{3}{20} M(R^2 + 4h^2)$	$J_{x_C y_C} = 0$ $J_{x_C z_C} = 0$ $J_{y_C z_C} = 0$ $J_{xy} = 0$ $J_{xz} = 0$ $J_{yz} = 0$
12. Frustrum of Cone 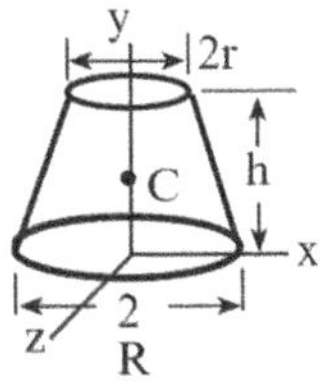	$x_C = 0$ $y_C = \frac{h}{4} \frac{R^2 + 2Rr + 3r^2}{R^2 + Rr + r^2}$ $z_C = 0$ $S =$ $\pi(R + r)\sqrt{(R - r)^2 + h^2}$ $V = \frac{M}{\rho} = \frac{\pi h}{3}(R^2 + Rr + r^2)$	$J_{y_C} = J_y = \frac{3M}{10}\left(\frac{R^5 - r^5}{R^3 - r^3}\right)$ $J_z = J_x =$ $\frac{M}{20}\left(\frac{2h^2(R^2 + 3Rr + 6r^2)}{R^2 + Rr + r^2} + \frac{3(R^4 + R^3 r + R^2 r^2 + Rr^3 + r^4)}{R^2 + Rr + r^2}\right)$	$J_{x_C y_C} = 0$ $J_{x_C z_C} = 0$ $J_{y_C z_C} = 0$ $J_{xy} = J_{xz} = 0$ $J_{yz} = 0$ $M = \rho V$
13. Right Circular Cylinder 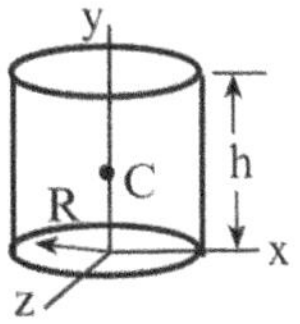	$x_C = 0$ $y_C = \frac{h}{2}$ $z_C = 0$ $S = 2\pi Rh$ $V = M/\rho = \pi R^2 h$	$J_{x_C} = J_{z_C} = \frac{M}{12}(3R^2 + h^2)$ $J_{y_C} = J_y = \frac{M}{2}R^2$ $J_x = J_z = \frac{M}{12}(3R^2 + 4h^2)$	$J_{x_C y_C} = 0$ $J_{x_C z_C} = 0$ $J_{y_C z_C} = 0$ $J_{xy} = 0$ $J_{xz} = J_{yz} = 0$
14. Hollow Right Circular Cylinder 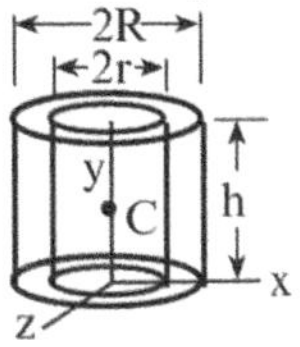	$x_C = 0$ $y_C = \frac{h}{2}$ $z_C = 0$ $V = M/\rho = \pi(R^2 - r^2)h$	$J_{x_C} = J_{z_C} = \frac{M}{12}(3R^2 + 3r^2 + h^2)$ $J_{y_C} = J_y = \frac{M}{2}(R^2 + r^2)$ $J_x = J_z = \frac{M}{12}(3R^2 + 3r^2 + 4h^2)$	$J_{x_C y_C} = J_{x_C z_C} = 0$ $J_{y_C z_C} = 0$ $J_{xy} = J_{xz} = J_{yz} = 0$
15. Thin Hollow Right Circular Cylinder 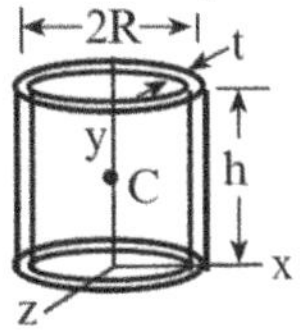	$x_C = 0$ $y_C = \frac{h}{2}$ $z_C = 0$ $V = M/\rho = 2\pi Rht$ $R >> t$	$J_{x_C} = J_{z_C} = \frac{M}{12}(6R^2 + h^2)$ $J_{y_C} = J_y = MR^2$ $J_x = J_z = \frac{M}{6}(3R^2 + 2h^2)$	$J_{x_C y_C} = J_{x_C z_C} = 0$ $J_{x_C y_C} = 0$ $J_{xy} = J_{xz} = J_{yz} = 0$

Table 1.6 Properties of homogeneous solids, continued

Solid Geometry	Volume, Mass M Center of Mass C	Mass Moments Inertia Jx, Jy, Jz	Mass Products of Inertia Jxy
16. Extruded Section From z=0 to z=h 	$x_C = \text{Section}\, x_C$ $y_C = \text{Section}\, y_C$ $z_C = h/2$ $V = M/\rho = Ah$ $A = \text{Section Area}$	$J_x = MI_x/A + Mh^2/3$ $J_y = MI_y/A + Mh^3/3$ $J_z = M(I_x + I_y)/A$ I_x, I_y = Area moments of inertia of cross section in x-y plane, Table 2.1	$J_{xy} = MI_{xy}/A$ $J_{xz} = Mhx_C$ $J_{yz} = Mhy_C$
17. Inclined Extrusion 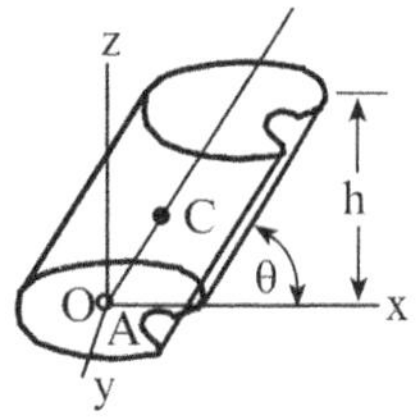	$x_C = (h/2)\sin\theta$ $y_C = 0$ $z_C = h/2$ $S = Ph$ $V = M/\rho = Ah$ A= x-y section area O is centroid of area P is perimeter of area A	$J_x = \rho h I_{xC} + \rho A h^3/3$ $J_y = \rho h I_{yC} +$ $\rho A h^3 (1 + \tan^2\theta)/3$ $J_z = \rho h (I_{x_C} + I_{y_C}) +$ $\rho A h^3 \tan^2\theta/3$ I_{xC} and I_{yC} are area moments inertia in x-y plane, Table 1.5, about area centroid O.	$J_{xz} =$ $\frac{\rho}{3} h^3 A \tan\theta$ $J_{xz} = J_{yz} = 0$ Note: Ends are cut parallel to x-y plane. θ is angle of extrusion.
18. Cylinder Section 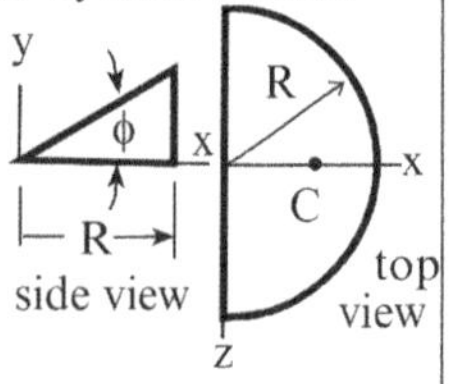	$x_C = \frac{3}{16}\pi R$ $y_C = \frac{3}{16}\pi R \tan\phi$ $z_C = 0$ $V = (2/3)R^3 \tan\phi$	$J_x = \frac{M}{5} R^2 \left(1 + 2\tan^2\phi\right)$ $J_y = \frac{3M}{5} R^2$ $J_z = \frac{2M}{5} R^2 \left(1 + \tan^2\phi\right)$	$J_{xy} = \frac{3M}{20} R^2 \tan\phi$ $J_{xz} = J_{yz} = 0$
19. Sphere 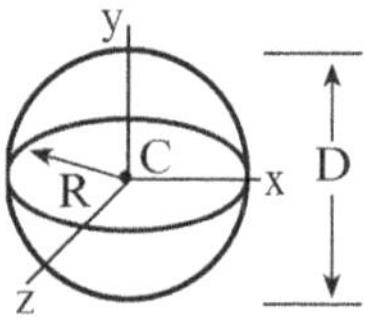	$x_C = 0$ $y_C = 0$ $z_C = 0$ $S = 4\pi R^2 = \pi D^2$ $V = \frac{M}{\rho} = \frac{4\pi}{3} R^3 = \frac{\pi}{6} D^3$	$J_{x_C} = J_{y_C} = J_{z_C} = \frac{2}{5} MR^2$ $= \frac{1}{10} MD^2$	$J_{x_C y_C} = 0$ $J_{x_C z_C} = 0$ $J_{y_C z_C} = 0$
20. Hollow Sphere 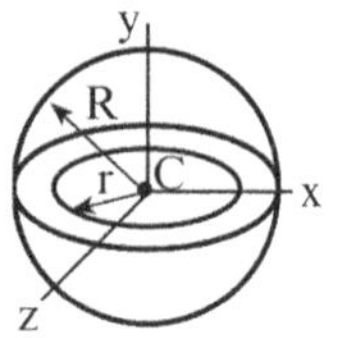	$x_C = 0$ $y_C = 0$ $z_C = 0$ $V = \frac{M}{\rho} = \frac{4\pi}{3}(R^3 - r^3)$ $= 4\pi R^2 t,\ R - r << R$	$J_{x_C} = J_{z_C} = J_{y_C} =$ $= \frac{2}{5} M \left(\frac{R^5 - r^5}{R^3 - r^3}\right)$ $= \frac{2}{3} MRt,\ t = R - r$ for $R >> R - r$	$J_{x_C z_C} = 0$ $J_{x_C y_C} = 0$ $J_{y_C z_C} = 0$

Table 1.6 Properties of homogeneous solids, continued

Solid Geometry	Volume V, Mass M Center of Mass C	Mass Moments Inertia Jx, Jy, Jz	Mass Products of Inertia Jxy
21. Ellipsoid 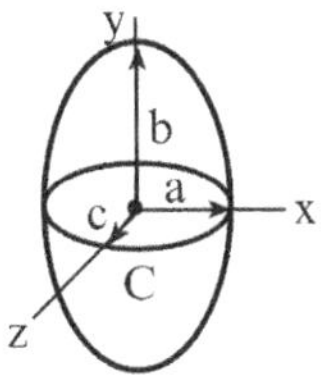	$x_C = 0$ $y_C = 0$ $z_C = 0$ $V = \frac{M}{\rho} = \frac{4}{3}\pi abc$ S = See Ref. [5], 9.148. sphere if a = b = c	$J_{x_C} = \frac{M}{5}(b^2 + c^2)$ $J_{y_C} = \frac{M}{5}(a^2 + b^2)$ $J_{z_C} = \frac{M}{5}(a^2 + b^2)$	$J_{xy} = J_{xz} = 0$ $J_{yz} = 0$ Equation of surface: $\frac{x^2}{a^2} + \frac{y^2}{b^2} + \frac{z^2}{c^2} = 1$
22. Elliptic Paraboloid 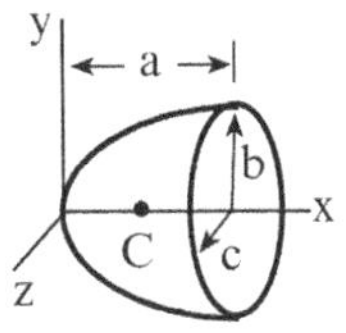	$x_C = \frac{2}{3}a$ $y_C = 0$ $z_C = 0$ $V = \frac{M}{\rho} = \frac{\pi}{2}abc$	$J_{x_C} = J_x = \frac{M}{6}(b^2 + c^2)$ $J_{y_C} = \frac{M}{18}(a^2 + 3c^2)$ $J_{z_C} = \frac{M}{18}(a^2 + 3b^2)$ $J_y = \frac{M}{6}(3a^2 + c^2)$ $J_z = (M/6)(3a^2 + b^2)$	$J_{xy} = J_{xz} = 0$ $J_{yz} = 0$ Equation of surface: $\frac{x}{a} = \frac{y^2}{b^2} + \frac{z^2}{c^2}$
23. Translated Body 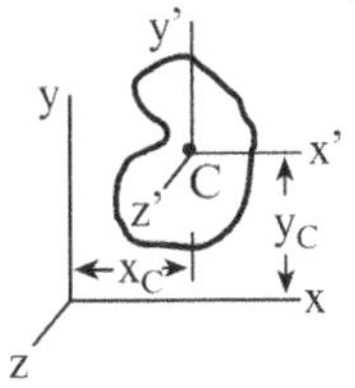	Mass = M x', y', and z' axes are through centroid C. They are parallel to x, y, and z axes, respectively.	$J_x = J_{x_C} + M(y_C^2 + z_C^2)$ $J_y = J_{y_C} + M(x_C^2 + z_C^2)$ $J_z = J_{z_C} + M(x_C^2 + y_C^2)$	$J_{xy} = J_{x_C y_C} + M x_C y_C$ $J_{yz} = J_{y_C z_C} + M y_C z_C$ $J_{xz} = J_{x_C z_C} + M x_C z_C$
24. Rotated Body 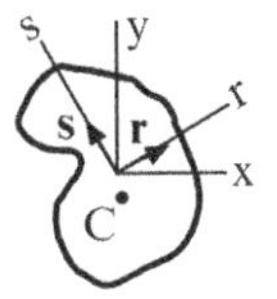	r and s axes are perpendicular and they pass through the origin of the x, y, and z axes. Their directions are specified by unit vectors **r** and **s**. $\mathbf{r} = l\mathbf{i} + m\mathbf{j} + n\mathbf{k}$ $\mathbf{s} = l'\mathbf{i} + m'\mathbf{j} + n'\mathbf{k}$	$J_r = l^2 J_x + m^2 J_y + n^2 J_z - 2lm J_{xy} - 2ln J_{xz} - 2mn J_{yz}$ $J_{rs} = -ll' J_x - mm' J_y - nn' J_z + (lm' + ml') J_{xy}$ $+ (ln' + l'n) J_{xz} + (mn' + nm') J_{yz}$ l, m, n are scalars that specify the direction of the unit vector **r** in the direction of the r axis. l', m', n' are scalars that specify the direction of the s axis Ref. [19]	
25. General Body 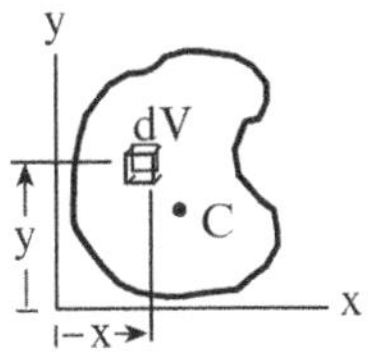	$x_C = \frac{1}{V}\int_V x\,dV$ $y_C = \frac{1}{V}\int_V y\,dV$ $z_C = \frac{1}{V}\int_V z\,dV$ $V = \frac{M}{\rho} = \int_V dx\,dy\,dz$	$J_x = \rho\int_V (y^2 + z^2)dV$ $J_y = \rho\int_V (x^2 + z^2)dV$ $J_z = \rho\int_V (x^2 + y^2)dV$	$J_{xy} = \rho\int_V xy\,dV$ $J_{xz} = \rho\int_V xz\,dV$ $J_{yz} = \rho\int_V yz\,dV$

The *mass M*, *volume V*, and location of the *center of mass C*, also the *center of gravity*, or centroid, are found by integrating element of mass $dM = \rho dV = \rho dxdydz$ over the body.

$$M = \int_V \rho dV, \qquad V = \int_V dV = \int_V dxdydx \tag{1.12a}$$

$$x_c = \frac{1}{M}\int_V \rho x dV, \quad y_c = \frac{1}{M}\int_V \rho y dV, \quad z_c = \frac{1}{M}\int_V \rho z dV \tag{1.12b}$$

The *mass moments of inertia* J_{xx}, J_{yy}, J_{zz} and the *mass products of inertia* about the x-, y-, and z-axes are

$$J_x = \int \rho(y^2 + z^2)dV \quad J_y = \int_V \rho(x^2 + z^2)dV \quad J_z = \int_V \rho\,(x^2 + y^2)dV \tag{1.13a}$$

$$J_{xy} = \int_V \rho xy dV \quad J_{xz} = \int_V \rho xz dV \quad J_{yz} = \int_V \rho yz dV \tag{1.13b}$$

The reader is cautioned that some authors (Refs [12, 13] and see Eq. 2.36) define the mass products of inertia (J_{xy}, J_{xz}, J_{yz}) as the negative of these expressions. The density ρ has units of mass per unit volume. If ρ is constant, as is the case in Tables 1.5, 1.6, and 1.7, the center of mass coincides with the centroid of the volume, density can be taken outside the integrals, the centroid is center of mass and mass is density times volume, $M = \rho V$.

Radius of gyration is the square root of the moment of inertia divided by the mass.

$$r_x = \left(\frac{J_x}{M}\right)^{1/2}, \quad r_y = \left(\frac{J_y}{M}\right)^{1/2}, \quad r_z = \left(\frac{J_z}{M}\right)^{1/2} \tag{1.14}$$

Parallel axis theorem transforms mass moments of inertia about the center of mass, J_{xc}, J_{yc}, J_{zc}, Figure 1.3b, to mass moments of inertia about the parallel axes offset by x_c, y_c, z_c.

$$J_x = J_{x_c} + M(y_c^2 + z_c^2), \quad J_y = J_{y_c} + M(x_c^2 + z_c^2), \quad J_z = J_{z_c} + M(x_c^2 + y_c^2) \tag{1.15a}$$

$$J_{xy} = J_{x_c y_c} + M x_c y_c, \quad J_{xz} = J_{x_c z_c} + M x_c z_c, \quad J_{yz} = J_{y_c z_c} + M y_c z_c \tag{1.15b}$$

Translation of axes away from the center of mass increases the moments of inertia J_x, J_y, J_z, (Eq. 1.15a); products of inertia (Eq. 1.15b) may increase or decrease.

Rotated unit vector **r** with origin O (Fig. 1.3b) is defined relative to unit-magnitude vectors **i**, **j**, and **k** in the x-, y-, and z-directions, respectively.

$$\mathbf{r} = a_x\mathbf{i} + a_y\mathbf{j} + a_z\mathbf{k}, \quad a_x^2 + a_y^2 + a_z^2 = 1 \tag{1.16}$$

The coefficients a_x, a_y, a_z are cosines of the angles between the rotated vector and the base coordinates. Look ahead to Equation 1.22. Mass moments of inertia about the rotated axis are

$$J_a = a_x^2 J_x + a_y^2 J_y + a_z^2 J_z - 2a_x a_y J_{xy} - 2a_x a_z J_{xz} - 2a_y a_z J_{yz} \tag{1.17}$$

The unit magnitude vector **s**,

$$\mathbf{s} = b_x\mathbf{i} + b_y\mathbf{j} + b_z\mathbf{k}, \quad b_x^2 + b_y^2 + b_z^2 = 1 \tag{1.18}$$

Table 1.7 Properties of elements described by points and vectors

Notation: A = cross sectional area; C = centroid; **OC** = vector from origin to center of mass (centroid of volume); $\mathbf{J_n}$ = mass moment of inertia about axis through origin in the direction of the unit vector **n**; M = mass, ρV; **n** = unit length vector through point O in direction of reference axis; O = origin of axes; **OP** = vector from point O to point P, etc.; P, Q, R, T, etc. = nodes (points) in space with Cartesian coordinates (x_P, y_P, z_P) etc.; S = surface area; t = thickness of plate; V = volume of body; |X| = magnitude or determinant of X; ρ = mass density; • = vector dot product, Eq. 1.25; × = vector cross product, Eq. 1.24 Refs [16, 24–27]

Geometry	Properties with Point Coordinates	Properties with Vectors
1. Point Mass at P 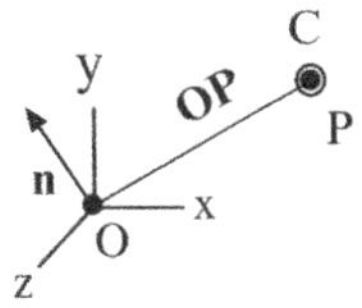	$x_C = x_P,\ y_C = y_P,\ z_C = z_P$ Mass = M, $\mathbf{OP} = x_p\mathbf{i} + y_p\mathbf{j} + z_p\mathbf{k}$ Mass moments of inertia about O, $J_x = M(y_P^2 + z_P^2)$ $J_{xy} = Mx_Py_P$ $J_y = M(x_P^2 + z_P^2)$ $J_{xz} = Mx_Px_P$ $J_z = M(x_P^2 + y_P^2)$ $J_{yz} = My_Pz_P$	Center of mass, **OC** = **OP** Mass moment of inertia about an axis through origin in the direction of the unit vector **n**, $J_n = M\lvert\mathbf{OP}\times\mathbf{i}\rvert^2$
2. Straight Rod 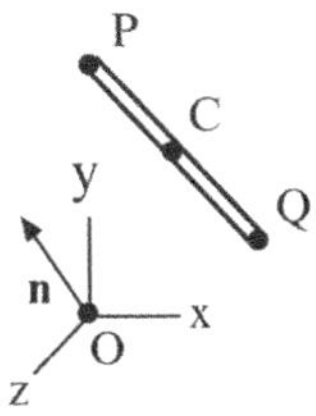	M = ρAL, where length L = $\left((x_P - x_Q)^2 + (y_P - y_Q)^2 + (z_P - z_Q)^2\right)^{1/2}$ A = cross sectional area, $A << L^2$ $x_C = (x_P + x_Q)/2,\ \ y_C = (y_P + y_Q)/2$ $z_C = (z_P + z_Q)/2$ $J_x = \frac{M}{3}(y_P^2 + y_Py_Q + y_Q^2 + z_P^2 + z_Pz_Q + z_Q^2)$ $J_{xy} = \frac{M}{6}(2x_Py_P + 2x_Qy_Q + x_Py_Q + x_Qy_P)$ J_y, J_z, J_{yz}, J_{xz} can be obtained from above by substituting x, y, or z for x and/or y.	$\mathbf{OC} = \frac{1}{2}(\mathbf{OP} + \mathbf{OQ}),\quad M = \rho A\lvert\mathbf{PQ}\rvert$ $J_n = \frac{M}{24}\bullet$ $\left[\left\lvert\left((3^{1/2}-1)\mathbf{OP} + (3^{1/2}+1)\mathbf{OQ}\right)\times\mathbf{n}\right\rvert^2\right.$ $\left.+\left\lvert\left((3^{1/2}+1)\mathbf{OP} + (3^{1/2}-1)\mathbf{OQ}\right)\times\mathbf{n}\right\rvert^2\right]$ Note: mass and inertia of a rod can be represented by two masses, mass M/2, located $\pm\ L/\sqrt{12}$ along the rod from the center of the rod where L is the length of the rod.
3. Thin Triangular Plate 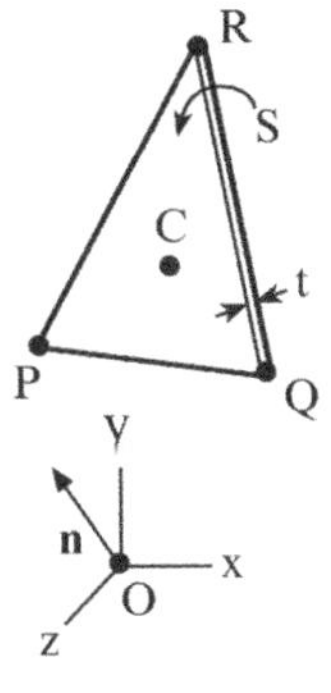 S = area of one side M = mass of plate	Mass = ρSt, where t = thickness $x_C = (x_P + x_Q + x_R)/3$ $y_C = (y_P + y_Q + y_R)/3$ $z_C = (z_P + z_Q + z_R)/3$ $S = \frac{1}{2}\begin{vmatrix} 1 & 1 & 1 \\ x_Q - x_P & y_Q - y_P & z_Q - z_P \\ x_R - x_P & y_R - y_P & z_R - z_P \end{vmatrix}$ $J_x = \frac{M}{12}((y_P + y_Q)^2 + (y_P + y_R)^2 +$ $(y_R + y_Q)^2 + (z_P + z_Q)^2 +$ $(z_P + z_R)^2 + (z_R + z_Q)^2)$ $J_{xy} = \frac{M}{12}((x_P + x_Q)(y_P + y_Q) +$ $(x_P + x_R)(y_P + y_R) +$ $(x_Q + x_R)(y_Q + y_R))$	$\mathbf{OC} = \frac{1}{3}(\mathbf{OP} + \mathbf{OQ} + \mathbf{OR}),\ M = \rho tS$ $S = \frac{1}{2}\lvert\mathbf{PQ}\times\mathbf{PR}\rvert$, each side $J_\mathbf{n} = \frac{M}{12}\left(\lvert(\mathbf{OP} + \mathbf{OR})\times\mathbf{n}\rvert^2\right.$ $+\lvert(\mathbf{OP} + \mathbf{OQ})\times\mathbf{n}\rvert^2$ $\left.+\lvert(\mathbf{OQ} + \mathbf{OR})\times\mathbf{n}\rvert^2\right)$ Mass and moment of inertia of thin triangle can be represented by three masses of size M/3 located midway between vertices Ref. [24]. J_y, J_z, J_{xz} can be obtained from the formulas for J_x and J_{xy} by substituting x, y, or z for x and/or y.

Table 1.7 Properties of elements described by points and vectors, continued

<table>
<tr><th>Geometry</th><th>Properties with Point Coordinates</th><th>Properties with Vectors</th></tr>
<tr><td>4. Tetrahedron (4-nodes)
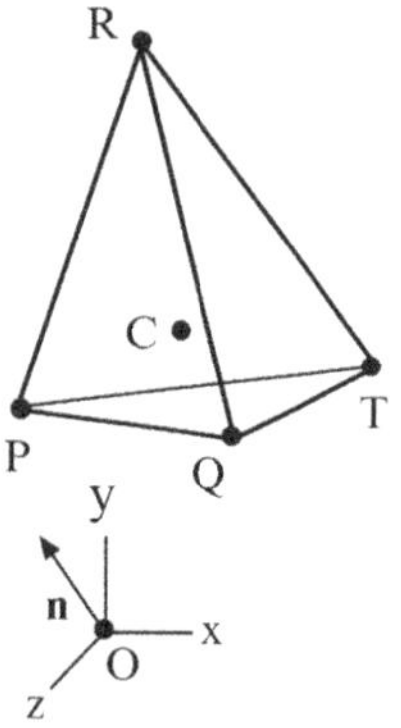
x_P, y_P, z_P are Cartesian coordinates of point P. OP is vector from point O to point P.</td><td>$M = \rho V$
$x_C = (x_P + x_Q + x_R + x_T)/4$
$y_C = (y_P + y_Q + y_R + y_T)/4$
$z_C = (z_P + z_Q + z_R + z_T)/4$
$S = \frac{1}{2}\begin{vmatrix} 1 & 1 & 1 \\ x_Q - x_P & y_Q - y_P & z_Q - z_P \\ x_R - x_P & y_R - y_P & z_R - z_P \end{vmatrix}$
$V = \frac{1}{6}\begin{vmatrix} x_T - x_P & y_T - y_P & z_T - z_P \\ x_Q - x_P & y_Q - y_P & z_Q - z_P \\ x_R - x_P & y_R - y_P & z_R - z_P \end{vmatrix}$
$J_x = \frac{M}{10}(y_P^2 + y_Q^2 + y_R^2 + y_T^2 + y_P y_Q + y_P y_R + y_Q y_R + y_P y_T + y_Q y_T + y_R y_T + z_P^2 + z_Q^2 + z_R^2 + z_T^2 + z_P z_Q + z_P z_R + z_Q z_R + z_P z_T + z_Q z_T + z_R z_T)$
$J_{xy} = \frac{M}{20}(2x_P y_P + 2x_Q y_Q + 2x_R y_R + 2x_T y_T + x_Q y_P + x_R y_P + x_T y_P + x_P y_Q + x_R y_Q + x_T y_Q + x_P y_R + x_Q y_R + x_T y_R + x_P y_T + x_Q y_T + x_R y_T)$</td><td>$\mathbf{OC} = \frac{1}{4}(\mathbf{OP} + \mathbf{OQ} + \mathbf{OR} + \mathbf{OT})$
$S = \frac{1}{2}\lvert \mathbf{PR} \times \mathbf{PT} \rvert + \frac{1}{2}\lvert \mathbf{PR} \times \mathbf{PQ} \rvert + \frac{1}{2}\lvert \mathbf{PT} \times \mathbf{PQ} \rvert + \frac{1}{2}\lvert \mathbf{QR} \times \mathbf{QT} \rvert$
$V = \frac{1}{6}\lvert (\mathbf{PQP} \times \mathbf{R}) \bullet \mathbf{PT} \rvert$
$J_n = \frac{M}{10}(\lvert \mathbf{OP} \times \mathbf{n} \rvert^2 + \lvert \mathbf{OQ} \times \mathbf{n} \rvert^2 + \lvert \mathbf{OR} \times \mathbf{n} \rvert^2 + \lvert \mathbf{OT} \times \mathbf{n} \rvert^2 + (\mathbf{OP} \times \mathbf{n}) \bullet (\mathbf{OT} \times \mathbf{n}) + (\mathbf{OQ} \times \mathbf{n}) \bullet (\mathbf{OR} \times \mathbf{n}) + (\mathbf{OQ} \times \mathbf{n}) \bullet (\mathbf{OT} \times \mathbf{n}) + (\mathbf{OP} \times \mathbf{n}) \bullet (\mathbf{OQ} \times \mathbf{n}) + (\mathbf{OP} \times \mathbf{n}) \bullet (\mathbf{OR} \times \mathbf{n}) + (\mathbf{OQ} \times \mathbf{n}) \bullet (\mathbf{OR} \times \mathbf{n}) + (\mathbf{OP} \times \mathbf{n}) \bullet (\mathbf{OT} \times \mathbf{n}))$
J_n is about n axis.
J_y, J_z, J_{yz}, J_{xz} can be obtained from formulas for J_x and J_{xy} by substituting y and/or z for x and y
Refs [25-27]</td></tr>
<tr><td>5. 6-Node Wedge
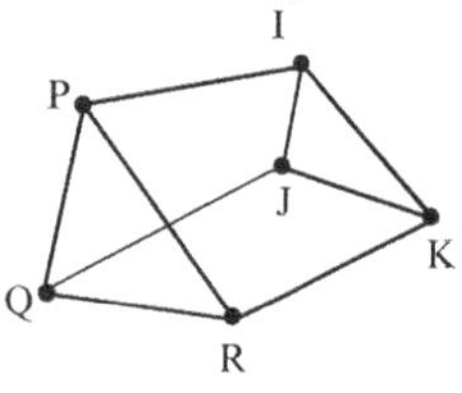</td><td>The 6-node wedge is subdivided into 3 tetrahedrons, PQRJ, PRJI, and RKIJ.
The properties of each tetrahedron are computed by the formulas in case 4. Their moments of inertia and volumes are summed in space to give the moments of inertia and volume of the wedge, case 6 of Table 1.6.</td><td>Wedge divided into 3 tetrahedrons.
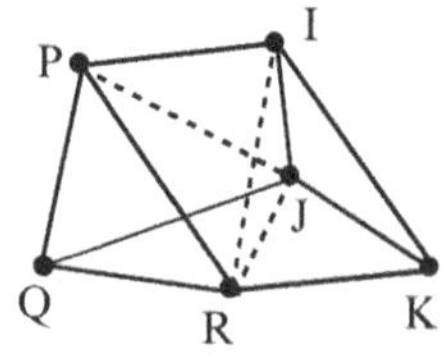</td></tr>
<tr><td>6. 8-Node Brick
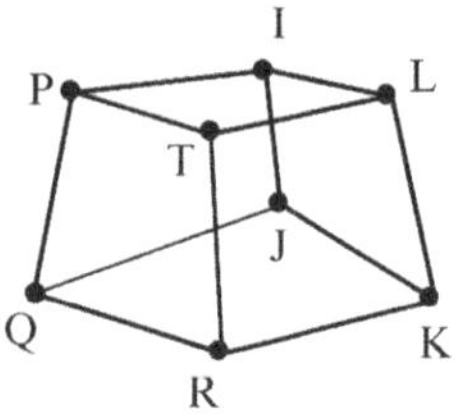</td><td>The 8-node brick is subdivided into two 6-node wedges, PQRIJK, PTRILK. The properties of each wedge are computed from 3 tetrahedrons, cases 4 and 5.

The results are summed in space to give the moments of inertia and volume of the brick Ref. [27].</td><td>Brick divided into two wedges.
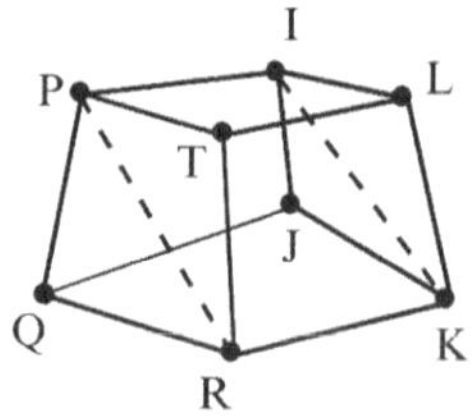</td></tr>
</table>

is perpendicular to **a** so $\mathbf{r} \bullet \mathbf{s} = a_x b_z + a_y b_y + a_z b_z = 0$ (see Eq. 1.25). The mass product of inertia with respect to the rotated **r–s**-axes is [19].

$$J_{rs} = -a_x b_x J_x - a_y b_y J_y - a_z b_z J_z + (a_x b_y + a_y b_x)\, J_{xy} + (a_x b_z + a_z b_x)\, J_{xz} + (a_y b_z + a_z b_y) J_{yz} \tag{1.19}$$

The general expression for the six mass moments of inertia in a rotated coordinate system (Eqs 1.19 and 1.22) is a tensor summation [19].

Principal axes of a solid body ([19], p. 558) are found by solution of the cubic equation that results from setting the determinant (|..|) of the matrix on the left-hand side of the following eigenvalue problem to zero.

$$\begin{bmatrix} (J_x - J) & J_{xy} & J_{xz} \\ J_{yx} & (J_y - J) & J_{yz} \\ J_{zx} & J_{yz} & (J_z - J) \end{bmatrix} \begin{Bmatrix} a \\ b \\ c \end{Bmatrix} = 0 \tag{1.20}$$

The three principal moments of inertia are J_1, J_2, and J_3. Products of inertia are zero about principal axes. Substituting these into the equation and solving gives the associated axis vector direction cosines (Eq. 1.16) of the principal axis.

The mass moments of inertia of thin planar slices from a homogeneous body are equal to their two-dimensional area moments of inertia (Table 1.5) times density. Two-dimensional areas times density extrude into the third dimension to create mass moments of inertia, cases 16 and 17 of Table 1.6 and the following example. Also see Section 1.7.

Example 1.4 Geometric properties of a steel pipe

A steel pipe made with density $\rho = 8$ g/cc (0.289 lb/in.^{3}) extends along the x-axis from $x = 0$ to $x = 250$ mm (9.843 in.). The center of the cross section, shown in Figure 1.4, coincides with the x-axis. Compute the pipe mass and its mass moments of inertia about its center of gravity.

Solution: Case 14 of Table 1.6 has the mass and mass moments of inertia of a tube that extends out the y-axis in terms of its outer radius $R = 10$ mm (0.3934 in.), the inner diameter, $r = 7$ mm (0.276 in.), and length $h = 250$ mm (9.84 in.). Substituting the x-axis for the y-axis, these formulas give

$$M = \rho\pi(R^2 - r^2)h = 320.4\ \text{g} = 0.3204\ \text{kg} = 0.706\ \text{lb}$$

$$J_x = \frac{M}{2}(R^2 + r^2) = 23870\ \text{gm} - \text{mm}^2 = 2.387 \times 10^{-5}\ \text{kg} - \text{m}^2 = 0.08152\ \text{lb} - \text{in}^2$$

$$J_y = \frac{M}{12}(3R^2 + 3r^2 + h^2) = 1.681 \times 10^6\ \text{g} - \text{mm}^2 = 1.681 \times 10^{-3}\ \text{kg} - \text{m}^2 = 5.744\ \text{lb} - \text{in}^2$$

The pipe extends out the x-axis from $x = y = 0$; y is perpendicular to x. This solution is in case 16 of Table 1.6.

1.7 Geometric Properties Defined by Vectors

Table 1.7 has the geometric properties of straight-sided polygonal planar areas and homogeneous volumes in terms of vectors to *nodes* at their vertices rather than relative dimensions. These formulas are based on Refs [16, 24–27].

A one-node object is a point mass. A two-node object is a rod. A three-node object is a triangle. A four-node solid is a tetrahedron, which is a pyramid-like solid with four vertices (nodes) and four triangular sides. A wedge is a six-node solid. A brick is an eight-node solid. The brick can be subdivided into two wedges each of which is further subdivided into three tetrahedrons (cases 4–6 of Table 1.7 [27]).

Vectors have magnitude and direction. Boldface (**r**) denotes vector. **i**, **j**, and **k** are the unit vectors in the *x*-, *y*-, and *z*-directions, respectively. The vector **R** can be defined by its starting coordinates (x_1, y_1, z_1) and ending coordinates (x_2, y_3, z_2), by its Cartesian components (r_x, r_y, r_z), or by its magnitude $(d^2 = r_x^{\ 2} + r_y^{\ 2} + r_z^{\ 2})$ times a unit vector **r**.

$$\mathbf{R} = r_x\mathbf{i} + r_y\mathbf{j} + r_z\mathbf{k} \tag{1.21}$$

$$\begin{aligned}\mathbf{r} = \frac{\mathbf{R}}{d} &= \left(\frac{r_x}{d}\right)\mathbf{i} + \left(\frac{r_y}{d}\right)j + \left(\frac{r_z}{d}\right)\mathbf{k} \\ &= \frac{(x_2 - x_1)}{d}\mathbf{i} + \frac{(y_2 - y_1)}{d}\mathbf{j} + \frac{(z_2 - z_1)}{d}\mathbf{k} \\ &= \cos\alpha\ \mathbf{i} + \cos\beta\ \mathbf{j} + \cos\gamma\ \mathbf{k}\end{aligned} \tag{1.22}$$

R and **r** make angles α, β, and γ with respect to the positive *x*-, *y*-, and *z*-axes, respectively. *Direction cosines* are the cosines of these angles and $\cos^2\alpha + \cos^2\beta + \cos^2\gamma = 1$.

The vectors **R** and **S** are defined by their Cartesian components.

$$\mathbf{R} = R_x\mathbf{i} + R_y\mathbf{j} + R_z\mathbf{k}, \quad \mathbf{S} = S_x\mathbf{i} + S_y\mathbf{j} + S_z\mathbf{k} \tag{1.23}$$

Cross product of two vectors is a vector [25].

$$\mathbf{R} \times \mathbf{S} = (R_yS_z - R_zS_y)\mathbf{i} + (R_zS_x - R_xS_z)\mathbf{j} + (R_xS_y - R_yS_x)\mathbf{k} = \begin{vmatrix}\mathbf{i} & \mathbf{j} & \mathbf{k} \\ R_x & R_y & R_z \\ S_x & S_y & S_z\end{vmatrix} \tag{1.24}$$

Here "| · |" means determinant. The magnitude of the vector cross product is $|\mathbf{r}||\mathbf{s}|\sin\theta$, where θ is the acute angle between vectors; it is the area of the parallelogram defined by bringing the bases of the vectors together. The cross product direction is perpendicular to the plane of the two vectors using the right-hand rule: vectors **r**, **s**, and $\mathbf{r} \times \mathbf{s}$ are a right-handed orthogonal triad $\mathbf{r} \times \mathbf{s} = -\mathbf{s} \times \mathbf{r}$. If $\mathbf{r} \times \mathbf{s} = 0$ then **r** and **s** are parallel.

Scalar product, or *dot product*, of two vectors is the scalar sum of the product of their components [12].

$$\mathbf{R} \bullet \mathbf{S} = R_xS_x + R_yS_y + R_zS_z \tag{1.25}$$

The magnitude of the dot product is $|\mathbf{r}| \cdot |\mathbf{s}| \cos\theta$, where θ is the acute angle between the two vectors; if $\mathbf{R} \bullet \mathbf{S} = 0$ then the vectors are perpendicular. The dot product is communicative, $\mathbf{R} \bullet \mathbf{S} = \mathbf{S} \bullet \mathbf{R}$. The magnitude of **R** is $(\mathbf{R} \bullet \mathbf{R})^{1/2}$.

The angle between two vectors is $\theta = \cos^{-1}(\mathbf{R} \bullet \mathbf{S})/(|\mathbf{R}||\mathbf{S}|)$.

Straight line is defined by the Cartesian coordinates (x,y,z) of a point x_0, y_0, z_0 on the line and the vector $\mathbf{a} = a_x\mathbf{i} + a_y\mathbf{j} + a_z\mathbf{k}$ in the direction of the line. Equations of the line are ([16], p. 243).

$$\begin{array}{lll} \text{canonical} & (x-x_0)/a_x = (y-y_0)/a_y = (z-z_0)/a_x & \\ \text{parametric} & x-x_0 = a_x t,\ y-y_0 = a_y t,\ z-z_0 = a_z t, \quad -\infty < t < \infty & (1.26) \\ \text{vector} & \mathbf{r}-\mathbf{r}_0 = \mathbf{a}t,\ \textit{with}\ \mathbf{r}_0 = x_0\mathbf{i} + y_0\mathbf{j} + z_0\mathbf{k} \quad -\infty < t < \infty & \end{array}$$

The minimum distance from a straight line to a point off the line is $|\mathbf{u} \times \mathbf{a}|/|\mathbf{a}|$ ([16], p. 245), where $\mathbf{a}$ is the line vector with direction cosines a_x, a_y, a_y, and $\mathbf{u}$ is a vector from a point on the line to the point off the line.

Equation of a Plane. If nonparallel vectors $\mathbf{a}$, $\mathbf{b}$, and $\mathbf{c}$ lie in a plane then the vector equation of the plane is $\mathbf{a} \bullet (\mathbf{b} \times \mathbf{c}) = 0$. If vectors $\mathbf{x}_1$, $\mathbf{x}_2$, and $\mathbf{x}_3$, are to points on the plane then $\mathbf{n} = (\mathbf{x}_1 - \mathbf{x}_2) \times (\mathbf{x}_2 - \mathbf{x}_3)$ is a vector perpendicular to the plane. If $\mathbf{r}$ is a vector on the plane then equation of the plane is $\mathbf{r} \bullet \mathbf{n} = D$, which is equivalent to the scalar equation $ax + by + cz = d$. The minimum distance from a plane to an off-plane point is $D = \mathbf{n} \bullet (\mathbf{x}_o - \mathbf{x})/|\mathbf{n}|$ where $\mathbf{x}$ is a vector to a point on the plane and $\mathbf{x}_o$ is the vector to the off-plane point [16].

References

[1] Newton, I., 1729 Sir Isaac Newton's Mathematical Principles of Natural Philosophy and his System of the World, translated into English by Andrew Motte in 1729 and revised by Florian Cajori, University of California Press, 1960. See pages [13] and 645 (note 2).

[2] McCutchen, C. W., "SI" Equals Imbecilic, American Physical Society News, vol. 10(9), 2001, pp. 4–5.

[3] American National Standard for Use of the International System of Units (SI): The Modern Metric System, 2002, IEEE/ASTM SI 10-2002, ASTM, West Conshohocken, PA, USA.

[4] Taylor, B.N., 1995, Guide for the Use of the International System of Units (SI), NIST Special Publication 811, U.S. Department of Commerce, National Institute of Standards and Technology.

[5] Linde, D. R. (ed.), Handbook of Chemistry and Physics, 78th ed., CRC Press, NY, 1998, pp. 1–21, 14-9.

[6] Adiutori, E. F., Fourier, Mechanical Engineering, vol. 127, 2005 pp. 30–31.

[7] Berman, B., Mixed-up Measurements, Discover 2000, p. 44.

[8] International Organization for Standardization, ISO 1000:1992, SI Units and Recommendations for the Use of their Multiples and of Certain Other Units.

[9] Cardarelli, F., and M. J. Shields, Encyclopaedia of Scientific Units, Weights and Measures: The SI Equivalents and Origins, Springer Verlag, N.Y., 2004.

[10] Kaye, G. W. C., and T. H. Laby, 1995, Tables of Physical and Chemical Constants, Longman Scientific & Technical, 6th ed., Section 2.7.4, John Wiley, N.Y.

[11] Cox, C.M., and B.F. Chao, Detection of a Large-Scale Mass Redistribution in the Terrestrial System Since 1998, Science, vol. 297, pp. 831–833, 2002.

[12] Greenwood, D.T., Principles of Dynamics, 2nd edition, Prentice Hall, Englewod Cliffs, N.J., 1988.

[13] Boiffier, J-T, The Dynamics of Flight, the Equations, John Wiley, N.Y., 1998.

[14] Thomson, W.T., Introduction to Space Dynamics, Dover, N.Y., 1986.

[15] McPhee, J., 2002, Founding Fish, Farrar, Straus and Giroux, N.Y., p. 109.

[16] Rektorys, K., Survey of Applicable Mathematics, The M.I.T Press, Cambridge, MA, p. 325, 1969.

[17] Myers, J.A., Handbook of Equtions for Mass and Area Properties of Various Geometrical Shapes, U.S. Naval Test Station, China Lake, California, AD274936, April 1962.

[18] Abbott, I. H., and A. E. von Doenhoff, Theory of Wing Sections, Dover, N.Y., p. 113, 1959.

[19] Shames, I. H., Engineering Mechanics, 4th ed., Prentice-Hall, Englewood Cliffs, N.J., 1997.

[20] AIAA Aerospace Design Engineers Guide, American Institute of Aeronautics and Astronautics, 1987.
[21] Aluminum Construction Manual, Aluminum Association, Washington D.C., 1975.
[22] Manual of Steel Construction, American Institute of Steel Construction, Chicago, Ill., 1992.
[23] American Institute of Timber Construction, Timber Construction Manual, 4th ed., John Wiley, N.Y., 1994.
[24] Faxen, H., Mekanik, Aktiebolaget Seelig & Co., Stockholm, 1952.
[25] McDougle, P., Vector Algebra, Wadsworth Publishing Co., Belmont, CA, 1973.
[26] Bowyer, A., and Woodwark, J., A Programmer's Geometry, Butterworths, London, 1983, pp. 118–120.
[27] Zienkiewicz, O. C., 1977, The Finite Element Method, 3rd ed., McGraw-Hill, NY, p. 142, Appendices 4, 5.

2

Dynamics of Particles and Bodies

This chapter presents formulas for the dynamics of particles, planets, and rotating bodies. These solutions trace back to Galileo's experiments with falling bodies and Kepler's measurements of planetary motion. Newton developed his laws of motion and gravity upon their shoulders. Leonard Euler used Newton's laws to formulate the equations of rigid body dynamics. Their analyses are called *classical dynamics*.

2.1 Kinematics and Coordinate Transformations

Table 2.1 [1, 2] has kinematic solutions. *Kinematics* is motion within geometric constraints. Kinematics includes the clever application of trigonometry to mechanisms and surfaces. Table 2.2 presents coordinate transformations, which resolve points and vectors into translated and rotated coordinates [1, 3–6]. They have general application.

For example, consider a mass M on a string length R that orbits a fixed anchor with tangential velocity $V = R\ d\theta/dt$ (Figure 2.1a). The acceleration (case 2 of Table 2.1) is radially inward, opposite to the unit outward radial vector $\mathbf{n}_r$. θ is the angular position:

$$\mathbf{a} = -R\left(\frac{d\theta}{dt}\right)^2 \mathbf{n}_r = -\frac{V^2}{R}\mathbf{n}_r \tag{2.1}$$

Once acceleration is known from kinematic analysis, the force on ball is computed by Newton's second law (Eqs 1.1 and 2.4). Vectors are in boldface type (**a**):

$$\mathbf{F} = M\mathbf{a} = MR\left(\frac{d\theta}{dt}\right)^2 \mathbf{n}_r \tag{2.2}$$

Using terms coined by Huygens and Newton, respectively, the mass inward acceleration is *centripetal acceleration*. The radial outward reaction of the ball against the string is a *centrifugal force*.

Example 2.1 Ball rolls on the earth

A student releases a frictionless ball on the surface of the earth (Figure 2.1b) at radius r and longitude λ. The earth rotates at $\Omega = 7.292 \times 10^{-5}$ rad/s, once per sidereal day (23.93 h) about the polar axis. What is the acceleration of the ball? Which way does the ball roll?

Formulas for Dynamics, Acoustics and Vibration, First Edition. Robert D. Blevins.
© 2016 John Wiley & Sons, Ltd. Published 2016 by John Wiley & Sons, Ltd.

Table 2.1 Kinematic motion

Notation: a = acceleration; b = parameter of spiral; $\mathbf{n}_r$ = unit normal vector r direction, etc.; P = point; **P** = vector to point P; R, r = radius from center; **r** = radius vector; s(t) = distance along fixed path; t = time; v = velocity of point; V = velocity magnitude; v_x = lateral velocity; v_y = vertical velocity; x = lateral position; X = lateral position on path; y = vertical position; Y = vertical position on path; ρ, θ = angular position, radians; ω = rate of rotation, rad/sec Refs [1, 2]

Motion of Point P	Position P, Velocity v, Accel. a	Comments and Special Cases
1. Cartesian Motions 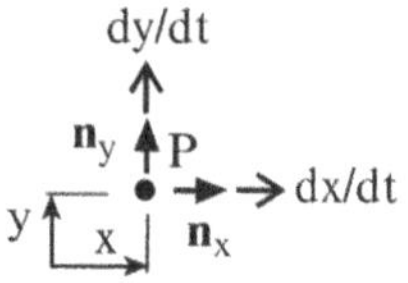	$\mathbf{P} = x\mathbf{n}_x + y\mathbf{n}_y$ $\mathbf{v} = \frac{d\mathbf{P}}{dt} = \frac{dx}{dt}\mathbf{n}_x + \frac{dy}{dt}\mathbf{n}_y$ $\mathbf{a} = \frac{d^2\mathbf{P}}{dt^2} = \frac{d^2x}{dt^2}\mathbf{n}_x + \frac{d^2y}{dt^2}\mathbf{n}_y$	$\mathbf{n}_x$ = unit vector in x direction $\mathbf{n}_y$ = unit vector in y direction
2. Circular Motion 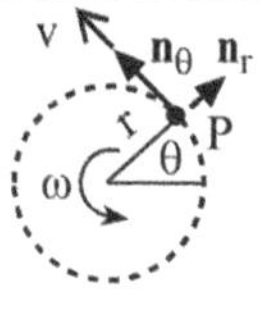	$\mathbf{P} = R\mathbf{n}_r$ radius R = constant $\mathbf{v} = d\mathbf{P}/dt = v\mathbf{n}_\theta = R\omega\mathbf{n}_\theta = \boldsymbol{\omega}\times\mathbf{R}$ $\mathbf{a} = \frac{d^2\mathbf{P}}{dt^2} = -\omega^2R\mathbf{n}_r + R\frac{d\omega}{dt}\mathbf{n}_\theta$ Rotational velocity $\omega = \frac{d\theta}{dt}$, rad./sec	Vector force on particle mass M, $\mathbf{F} = M\mathbf{a} = -MR\omega^2\mathbf{n}_r + MR\frac{d\omega}{dt}\mathbf{n}_\theta$ Unit radial vector = $\mathbf{n}_r$ Unit circumferential vector = $\mathbf{n}_\theta$ See case 3. Ref. [1], p. 39.
3. Motion in Cylindrical Coordinates 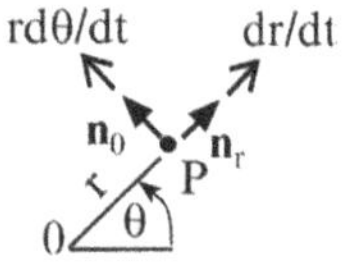	$\mathbf{r} = r\mathbf{n}_r$ = vector to point P Unit radial vector = $\mathbf{n}_r$ Ref. [1], p. 35 $\mathbf{v} = \frac{d\mathbf{r}}{dt} = \frac{dr}{dt}\mathbf{n}_r + r\frac{d\theta}{dt}\mathbf{n}_\theta$ Unit circumferential vector = $\mathbf{n}_\theta$ $\mathbf{a} = \frac{d^2\mathbf{r}}{dt^2} = \left(\frac{d^2r}{dt^2} - r\left(\frac{d\theta}{dt}\right)^2\right)\mathbf{n}_r + \left(r\frac{d^2\theta}{dt^2} + 2\frac{dr}{dt}\frac{d\theta}{dt}\right)\mathbf{n}_\theta$	
4. Motion in Spherical Coordinates 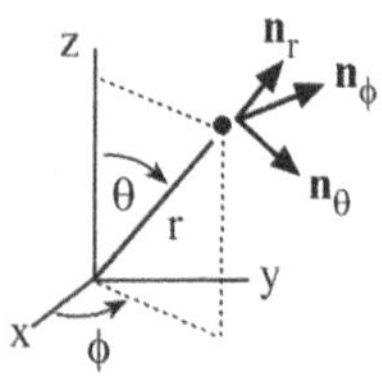	$\mathbf{r} = r\mathbf{n}_r$, vector to point P Ref. [1], p. 36 $\mathbf{v} = \frac{d\mathbf{r}}{dt} = \frac{dr}{dt}\mathbf{n}_r + r\frac{d\theta}{dt}\mathbf{n}_\theta + r\frac{d\phi}{dt}\sin\theta\,\mathbf{n}_\phi$ $\mathbf{a} = \frac{d^2\mathbf{r}}{dt^2} = \left(\frac{d^2r}{dt^2} - r\left(\frac{d\theta}{dt}\right)^2 - r\left(\frac{d\phi}{dt}\sin\theta\right)^2\right)\mathbf{n}_r + \left(r\frac{d^2\theta}{dt^2} + 2\frac{dr}{dt}\frac{d\theta}{dt}\right.$ $\left. - r\left(\frac{d\phi}{dt}\right)^2\sin\theta\cos\theta\right)\mathbf{n}_\theta + \left(r\frac{d^2\phi}{dt^2}\sin\theta + 2\frac{dr}{dt}\frac{d\phi}{dt}\sin\theta + 2r\frac{d\phi}{dt}\frac{d\theta}{dt}\cos\theta\right)\mathbf{n}_\phi$	
5. Motion on a Path 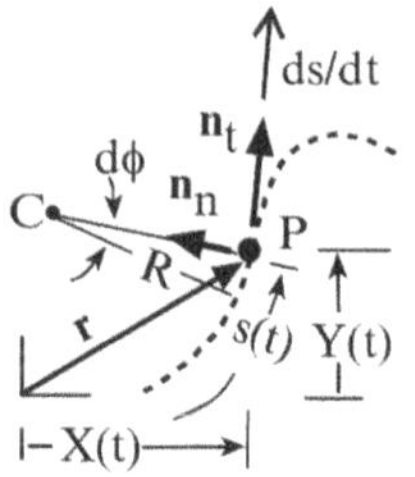C = instantaneous ctr s = distance along path **r** = vector to point P	$\mathbf{v} = \frac{d\mathbf{r}}{dt} = \frac{ds}{dt}\mathbf{n}_t = R\frac{d\phi}{dt}\mathbf{n}_t = \frac{d\boldsymbol{\phi}}{dt}\times\mathbf{R}$ $\mathbf{a} = \frac{d^2\mathbf{r}}{dt^2} = \frac{d^2s}{dt^2}\mathbf{n}_t + \frac{d\phi}{dt}\frac{ds}{dt}\mathbf{n}_n$, Radius curve $R = \left\|\frac{d\mathbf{n}_t}{ds}\right\|^{-1} = \left\|\frac{d^2\mathbf{r}}{ds^2}\right\|^{-1}$ vector normal path $\mathbf{n}_n = Rd^2\mathbf{r}/ds^2$ vector tangent to path $\mathbf{n}_t = d\mathbf{r}/ds$ $d\phi/dt$ = rate rotation about center C $d\mathbf{n}_t/ds = \mathbf{n}_n/R,\ d\phi = ds/R$	Two-dimensional motion X(t),Y(t) $R = \frac{(X'^2+Y'^2)^{3/2}}{\|X'Y''-X''Y'\|} = \frac{\left[1+(\frac{dY}{dX})^2\right]^{3/2}}{\|d^2Y/dX^2\|}$ coordinates of instantaneous center $x_C = X - \frac{dY}{dX}\frac{1+(dY/dX)^2}{d^2Y/dX^2}$ $y_C = Y + \frac{1+(dY/dX)^2}{d^2Y/dX^2}$ X′=dX/dt Ref. [1], p 37; Ref. [2], p. 235.

Table 2.1 Kinematic motion, continued

Motion of Point P	Position P, Velocity v, Accel. a	Comments and Special Cases
6. Spiral Motion 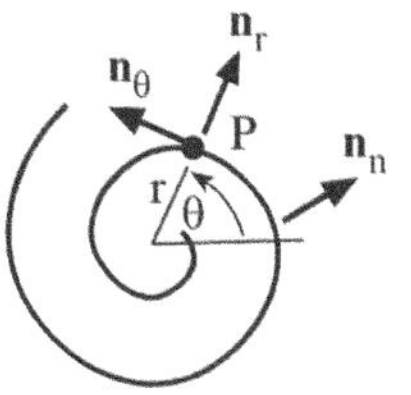	$\mathbf{P} = r\mathbf{n}_r = r_o e^{-b(t)}\mathbf{n}$ $\mathbf{v} = \frac{d\mathbf{P}}{dt} = -rb\frac{d\theta}{dt}\mathbf{n}_r + r\frac{d\theta}{dt}\mathbf{n}$ $\mathbf{a} = \frac{d^2\mathbf{P}}{dt^2} = \left(-ra\frac{d^2\theta}{dt^2} - r(1-a^2)\left(\frac{d\theta}{dt}\right)^2\right)\mathbf{n}_r + \left(r\frac{d^2\theta}{dt^2} - 2rb\left(\frac{d\theta}{dt}\right)^2\right)\mathbf{n}_\theta$	Radius of logarithmic spiral, $r = r_o e^{-b\theta(t)}$ Ref. [2], p. 40 r_o = constant, b = constant Vector normal to spiral $\mathbf{n}_n = \frac{1}{\sqrt{1+b^2}}\mathbf{n}_r + \frac{b}{\sqrt{1+b^2}}\mathbf{n}_\theta$ For constant velocity v_o $\frac{d\theta}{dt} = \frac{v_o}{r\sqrt{1+b^2}}, \ \frac{d^2\theta}{dt^2} = \frac{v_o^2}{r^2(1+b^2)}$
7. Motion in Helix 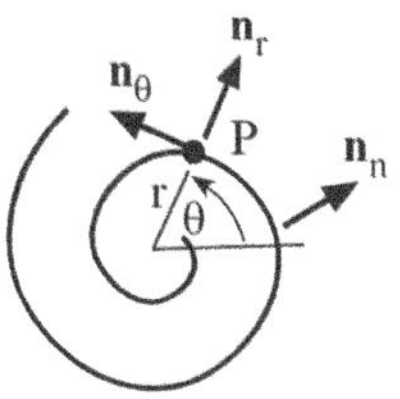	$\mathbf{P} = R\mathbf{n}_\phi + \beta R\phi\mathbf{n}_z$ $= R\cos\phi\mathbf{n}_x + R\sin\phi\mathbf{n}_y + \beta R\phi\mathbf{n}_z$ $\mathbf{v} = \frac{d\mathbf{P}}{dt} = R\frac{d\phi}{dt}\mathbf{n}_\phi + \phi R\frac{d\phi}{dt}\mathbf{n}_z$ $= (-R\sin\phi\mathbf{n}_x + R\cos\phi\mathbf{n}_y + \beta R\mathbf{n}_z)\frac{d\phi}{dt}$ $\mathbf{a} = \frac{d^2\mathbf{P}}{dt^2}$ $= -R\left(\frac{d\phi}{dt}\right)^2\mathbf{n}_r + R\frac{d^2\phi}{dt^2}\mathbf{n}_\phi + \beta R\frac{d^2\phi}{dt^2}\mathbf{n}_z$ $= (-R\cos\phi\mathbf{n}_x - R\sin\phi\mathbf{n}_y + \beta R\mathbf{n}_z)\frac{d^2\phi}{dt^2}$	R = constant cylinder radius β = constant helix angle Radius of curvature $= R(1+\beta^2)$ Length in one turn along helix $= 2\pi R\sqrt{1+\beta^2}$ Ref. [1], p. 40; Ref. [2], p. 213
8. Point on Rolling Circle Radius R 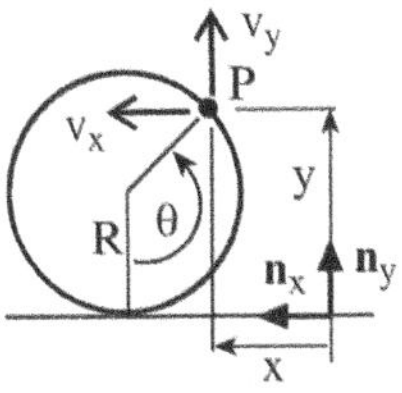	$\mathbf{P} = x\mathbf{n}_x + y\mathbf{n}_y = R(\theta - \sin\theta)\mathbf{n}_x + R(1-\cos\theta)\mathbf{n}_y$ $\mathbf{v} = \frac{d\mathbf{P}}{dt} = v_x\mathbf{n}_x + v_y\mathbf{n}_y = R(1-\cos\theta)\frac{d\theta}{dt}\mathbf{n}_x + R\sin\theta\frac{d\theta}{dt}\mathbf{n}_y$ $\mathbf{a} = \frac{d^2\mathbf{P}}{dt^2} = R\left[1-\cos\theta\frac{d^2\theta}{dt^2} + \sin\theta\left(\frac{d\theta}{dt}\right)^2\right]\mathbf{n}_x + R\left[\sin\theta\frac{d^2\theta}{dt^2} + \cos\theta\left(\frac{d\theta}{dt}\right)^2\right]\mathbf{n}_y$	
alternately 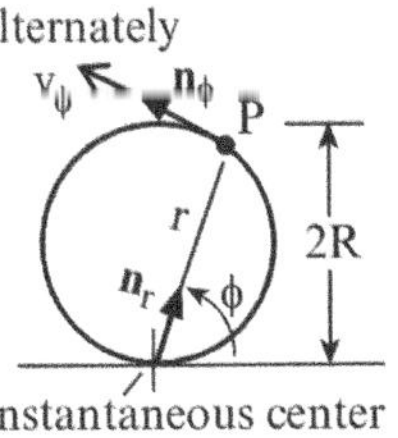 instantaneous center	In terms of instantaneous center using cylindrical coordinates, case 3. Radius $r = 2R\sin(\phi/2)$ from the instantaneous center. $\mathbf{P} = r\mathbf{n}_r$ $\mathbf{v} = \frac{d\mathbf{P}}{dt} = r\frac{d\phi}{dt}\mathbf{n}_\phi$ Circle does not slip relative to ground plane. $d\theta/dt = d\phi/dt$. Path of point on circumference describes cycloid Ref. [2], p.165–166	

Table 2.1 Kinematic motion, continued

Motion of Point P	Position P, Velocity v, Accel. a	Comments and Special Cases
9. Crank Shaft with Connecting Rod 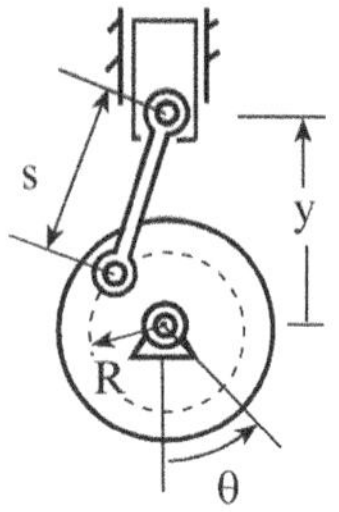	$y = R\cos\theta + \left(s^2 - (R\sin\theta)^2\right)^{1/2}$, θ=0 is top, s = length connect rod $$\frac{dy}{dt} = -\left(R\sin\theta + \frac{R^2\cos\theta\sin\theta}{\left(s^2-(R\sin\theta)^2\right)^{1/2}}\right)\frac{d\theta}{dt}$$ $$\frac{d^2y}{dt^2} = -R\cos\theta\left(\frac{d\theta}{dt}\right)^2 - R\sin\theta\frac{d^2\theta}{dt^2}$$ $$+\left[\frac{R^2(\sin^2\theta-\cos^2\theta)}{\left(s^2-(R\sin\theta)^2\right)^{1/2}} - \frac{R^4\cos^2\theta\sin^2\theta}{\left(s^2-(R\sin\theta)^2\right)^{3/2}}\right]\left(\frac{d\theta}{dt}\right)^2 - \frac{R^2\cos\theta\sin\theta}{\left(s^2-(R\sin\theta)^2\right)^{1/2}}\frac{d^2\theta}{dt^2}$$	
10. Point P Fixed in a Rigid Body 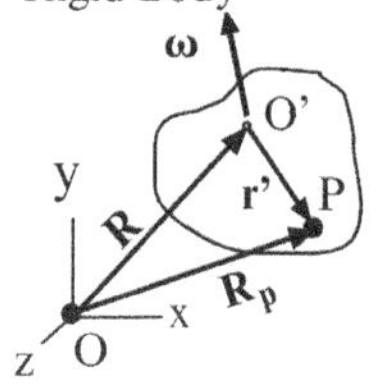	$\mathbf{P} = \mathbf{R}_p = \mathbf{R} + \mathbf{r}'$ $(\lvert\mathbf{r}'\rvert = \text{constant})$ $$\frac{d\mathbf{P}}{dt} = \frac{d\mathbf{R}_p}{dt} = \frac{d\mathbf{R}}{dt} + \mathbf{r}'\times\boldsymbol{\omega}$$ $$\frac{d^2\mathbf{R}_p}{dt^2} = \frac{d^2\mathbf{R}}{dt^2} + \frac{d\boldsymbol{\omega}}{dt}\times\mathbf{r}' + \boldsymbol{\omega}\times(\boldsymbol{\omega}\times\mathbf{r}')$$ $\mathbf{r}'$, $\mathbf{R}$, $\mathbf{R}_p$, and $\boldsymbol{\omega}$ are measured at each point in time in the inertial xyz frame.	The constant length vector **r'** is from point O' to point P. The position vector **R** goes from the inertial origin O to the point O' that is fixed in the rigid body. **r'** is fixed in the body. The rigid body rotates with rotation velocity ω about an axis defined by the vector **ω** that passes through the point O'. Ref. [1], pp. 44–50.
11. Point P in a Rotating and Translating Coordinate System 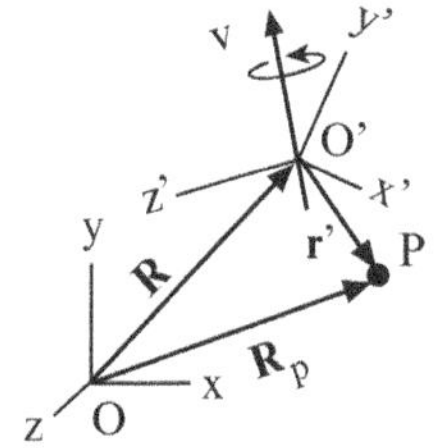 The frame x'y'z' rotates with angular velocity vector **ω** with respect to the inertial xyz frame.	$\mathbf{R}_p = \mathbf{R} + \mathbf{r}'$ $$\frac{d\mathbf{R}_p}{dt} = \frac{d\mathbf{R}}{dt} + \mathbf{r}'\times\boldsymbol{\omega} + \left[\frac{d\mathbf{r}'}{dt}\right]_r$$ See case 10 for $\lvert r'\rvert$= constant. $$\frac{d^2\mathbf{R}_p}{dt^2} = \frac{d^2\mathbf{R}}{dt^2} + \frac{d\boldsymbol{\omega}}{dt}\times\mathbf{r}' + \boldsymbol{\omega}\times(\boldsymbol{\omega}\times\mathbf{r}') + \left[\frac{d^2\mathbf{r}'}{dt^2}\right]_r + 2\boldsymbol{\omega}\times\left[\frac{d\mathbf{r}'}{dt}\right]_r$$ $$\left[\frac{d\mathbf{r}}{dt}\right]_r = \frac{dr_x}{dt}\mathbf{i}' + \frac{dr_y}{dt}\mathbf{j}' + \frac{dr_z}{dt}\mathbf{k}',\quad \left[\frac{d^2\mathbf{r}}{dt^2}\right]_r = \frac{d^2r_x}{dt^2}\mathbf{i}' + \frac{d^2r_y}{dt^2}\mathbf{j}' + \frac{d^2r_z}{dt^2}\mathbf{k}'$$ $\mathbf{R}_p$, r, **ω** and $\mathbf{r}' = r_x\mathbf{i}' + r_y\mathbf{j}' + r_z\mathbf{k}'$ are in their instantaneous position measured in the inertial xyz system. **i'**, **j'**, and **k'** are the unit vectors along x' a, y' and z' axes. i', j', and k' are resolved into the inertial xyz system by transform in Table 2.2. [dr/dt] and [d^2r/dt^2] are time derivatives computed in the rotating x'y'z' system. Also see Table 2.5.	Ref. [2], p. 50
12. Two Sliders in Track With Rigid Rod 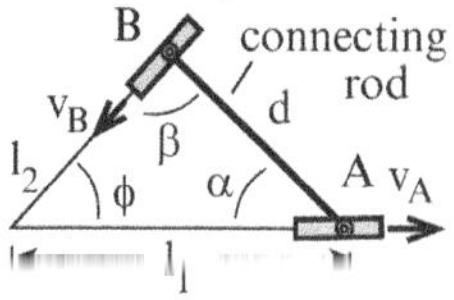	$v_B = v_A \cos\alpha / \cos\beta$ d = length of connecting rod. Sliders ride in fixed tracks, φ = constant v_B = 0 when α = 90 degrees	

Table 2.2 Coordinate transformations

Notation: i, j, k = unit vectors in direction of X, Y, and Z axes; $\mathbf{i}'$, $\mathbf{j}'$, $\mathbf{k}'$ = unit vectors in directions of x, y, and z axes; [T] = 2×2 or 3×3 transformation matrix, see properties in case 8; X, Y, Z = right-handed orthogonal position coordinates, usually earth or inertial coordinates; X_o, Y_o, Z_o = coordinates of origin of x, y, z systems with respect to X, Y, Z system; X_p, Y_p, Z_p = coordinates of point P in X, Y, Z; x, y, z = right-handed local body orthogonal coordinate system that is translated and rotated with respect to X, Y, Z system; x_p, y_p, z_p = coordinates of point P in x, y, z; α_x, α_y, α_z = counter clockwise rotation about the x, y, z axes, respectively; θ = angular position, radians; superscript T denotes transpose; overdot (.) denotes derivative with respect to Refs [2–6].

Rotation x,y,z System Relative to X,Y,Z	Forward and Inverse Transformation Between Coordinate Systems	Transformation Matrix [T] {X} = [T] {x}
1. Rotation +translation in Two Dimensions 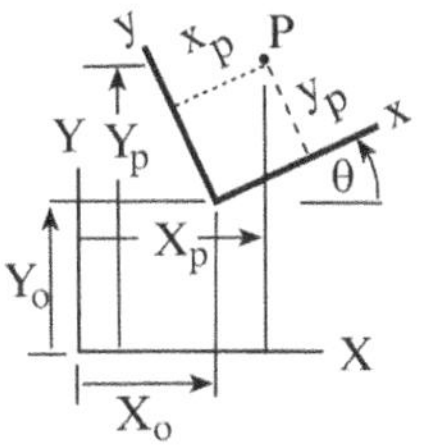	$X_p = X_o + x_p\cos\theta - y_p\sin\theta$ $Y_p = Y_o + x_p\sin\theta + y_p\cos\theta$ Equivalently, in matrix form, $\begin{pmatrix} X_P - X_o \\ Y_P - Y_o \end{pmatrix} = \begin{bmatrix} \cos\theta & -\sin\theta \\ \sin\theta & \cos\theta \end{bmatrix} \begin{pmatrix} x_P \\ y_P \end{pmatrix}$ Inverse transformation $\begin{pmatrix} x_P \\ y_P \end{pmatrix} = \begin{bmatrix} \cos\theta & \sin\theta \\ -\sin\theta & \cos\theta \end{bmatrix} \begin{pmatrix} X_P - X_o \\ Y_p - Y_o \end{pmatrix}$	$\begin{bmatrix} \cos\theta & -\sin\theta \\ \sin\theta & \cos\theta \end{bmatrix}$ (x,y,z) coordinates are local, body fixed coordinates of a point in the body, such as the fore, aft, port and starb'rd of a ship or airplane and (X,Y,Z) are earth inertial coordinates. $x_p = (X_p - X_o)\cos\theta$ $+ (Y_p - Y_o)\sin\theta$
2. Three Dimensional Rotation about x 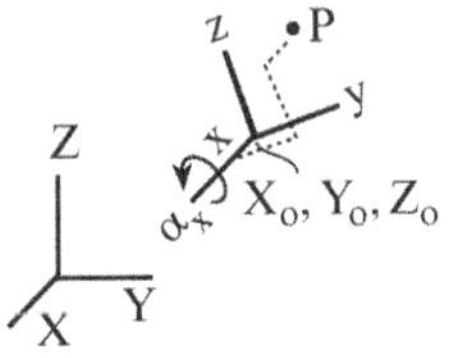	$\begin{pmatrix} X_P - X_o \\ Y_P - Y_o \\ Z_P - Z_o \end{pmatrix} = \begin{bmatrix} 1 & 0 & 0 \\ 0 & \cos\alpha_x & -\sin\alpha_x \\ 0 & \sin\alpha_x & \cos\alpha_x \end{bmatrix} \begin{pmatrix} x_P \\ y_P \\ z_P \end{pmatrix}$ Inverse Transformation $\begin{pmatrix} x_P \\ y_P \\ z_P \end{pmatrix} = \begin{bmatrix} 1 & 0 & 0 \\ 0 & \cos\alpha_x & \sin\alpha_x \\ 0 & -\sin\alpha_x & \cos\alpha_x \end{bmatrix} \begin{pmatrix} X_P - X_o \\ Y_P - Y_o \\ Z_P - Z_o \end{pmatrix}$	$\begin{bmatrix} 1 & 0 & 0 \\ 0 & \cos\alpha_x & -\sin\alpha_x \\ 0 & \sin\alpha_x & \cos\alpha_x \end{bmatrix}$ Note X_o, Y_o, and Z_o are coordinates of the origin of the x, y, z system measured in the X, Y, Z Ref. [4], p. 19.
3. Three Dimensional Rotation about y 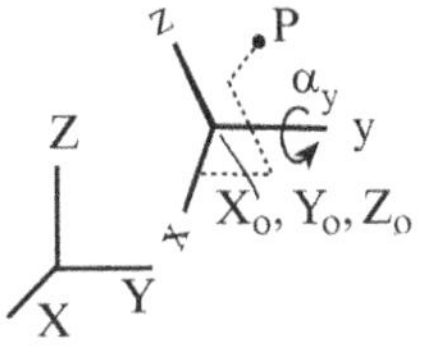	$\begin{pmatrix} X_P - X_o \\ Y_P - Y_o \\ Z_P - Z_o \end{pmatrix} = \begin{bmatrix} \cos\alpha_y & 0 & \sin\alpha_y \\ 0 & 1 & 0 \\ -\sin\alpha_y & 0 & \cos\alpha_y \end{bmatrix} \begin{pmatrix} x_P \\ y_P \\ z_P \end{pmatrix}$ Inverse Transformation $\begin{pmatrix} x_P \\ y_P \\ z_P \end{pmatrix} = \begin{bmatrix} \cos\alpha_y & 0 & -\sin\alpha_y \\ 0 & 1 & 0 \\ \sin\alpha_y & 0 & \cos\alpha_y \end{bmatrix} \begin{pmatrix} X_P - X_o \\ Y_P - Y_o \\ Z_P - Z_o \end{pmatrix}$	$\begin{bmatrix} \cos\alpha_y & 0 & \sin\alpha_y \\ 0 & 1 & 0 \\ -\sin\alpha_y & 0 & \cos\alpha_y \end{bmatrix}$ Note X_o, Y_o, and Z_o are coordinates of the origin of the x, y, z system measured in the X, Y, Z system. Ref. [4], p. 19.
4. Three Dimensional Rotation about z 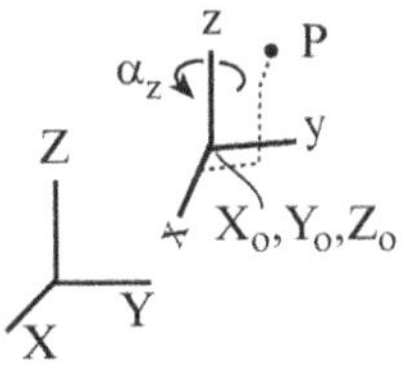	$\begin{pmatrix} X_P - X_o \\ Y_P - Y_o \\ Z_P - Z_o \end{pmatrix} = \begin{bmatrix} \cos\alpha_z & -\sin\alpha_z & 0 \\ \sin\alpha_z & \cos\alpha_z & 0 \\ 0 & 0 & 1 \end{bmatrix} \begin{pmatrix} x_P \\ y_P \\ z_P \end{pmatrix}$ Inverse Transformation $\begin{pmatrix} x_P \\ y_P \\ z_P \end{pmatrix} = \begin{bmatrix} \cos\alpha_z & \sin\alpha_z & 0 \\ -\sin\alpha_z & \cos\alpha_z & 0 \\ 0 & 0 & 1 \end{bmatrix} \begin{pmatrix} X_P - X_o \\ Y_P - Y_o \\ Z_P - Z_o \end{pmatrix}$	$\begin{bmatrix} \cos\alpha_z & -\sin\alpha_z & 0 \\ \sin\alpha_z & \cos\alpha_z & 0 \\ 0 & 0 & 1 \end{bmatrix}$ Note X_o, Y_o, and Z_o are coordinates of the origin of the x, y, z system measured in the X, Y, Z system. Ref. [4], p. 20.

Table 2.2 Coordinate transformations, continued

Rotation x,y,z System Relative to X,Y,Z	Forward and Inverse Transform Between Coordinate Systems	Comments
5. Multiple Rotations plus Translation 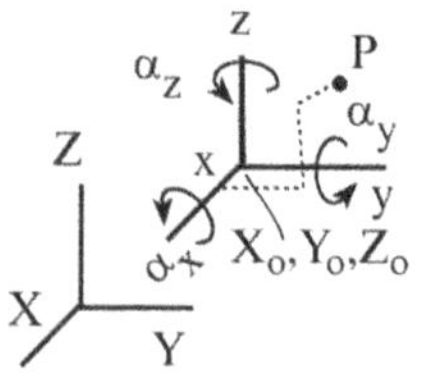	$\begin{pmatrix} X_P - X_o \\ Y_P - Y_o \\ Z_P - Z_o \end{pmatrix} = [T_3][T_2][T_1] \begin{pmatrix} x_P \\ y_P \\ z_P \end{pmatrix}$ Inverse Transformation $\begin{pmatrix} x_P \\ y_P \\ z_P \end{pmatrix} = \left[[T_3][T_2][T_1]\right]^T \begin{pmatrix} X_P - X_o \\ Y_P - Y_o \\ Z_P - Z_o \end{pmatrix}$	The X,Y,Z coordinates of a point are obtained from its local x,y,z coordinates by consecutive rotations about local x, y, or z axes and off setting its origin by X_o, Y_o, Z_o. $[T_i]$, i = x, y, or z in order of transforms, cases 2, 3, and 4.
6 Euler Angle Transform Aerospace Convention The inertial earth X,Y,Z coordinates are obtained from the local body x,y,z coordinates as follows. 1. Rotate ψ about z then 2. Rotate θ aboutrotated y axis, then 3. Rotate ϕ about rotated x axis, then 4. Translate origin by X_o, Y_o, Z_o	Coordinate transformation $\begin{pmatrix} X_P - X_o \\ Y_P - Y_o \\ Z_P - Z_o \end{pmatrix} = \begin{bmatrix} \cos\theta\cos\psi & (\cos\psi\sin\theta\sin\psi - \cos\phi\sin\psi) & (\cos\phi\cos\psi\sin\theta + \sin\phi\sin\psi) \\ \cos\theta\sin\psi & (\cos\phi\cos\psi + \sin\phi\sin\theta\sin\psi) & (\cos\phi\sin\theta\sin\psi - \cos\psi\sin\phi) \\ -\sin\theta & \cos\theta\sin\phi & \cos\theta\cos\phi \end{bmatrix} \begin{pmatrix} x_P \\ y_P \\ z_P \end{pmatrix}$ Angular Velocity transformation $\begin{pmatrix} \omega_x \\ \omega_y \\ \omega_z \end{pmatrix} = \begin{bmatrix} \cos\theta\cos\psi & -\sin\psi & 0 \\ \cos\theta\sin\psi & \cos\psi & 0 \\ -\sin\theta & 0 & 1 \end{bmatrix} \begin{pmatrix} \dot\phi \\ \dot\theta \\ \dot\psi \end{pmatrix}$	Ref. [4], p. 29, 234, Ref. [2], p. 357 Twelve Euler transforms exist. $\psi = \theta = \phi = 0$ gives idenity matrices.
7. Transform Vector **R** 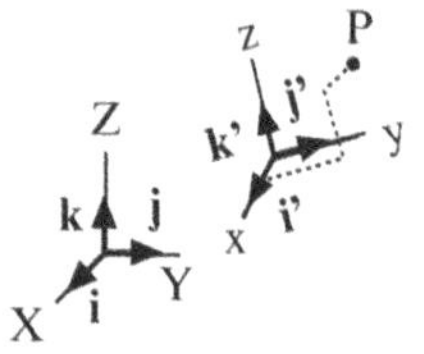 **R** and **R'** are vectors to the same point P	$\mathbf{R'} = x_p \mathbf{i'} + y_p \mathbf{j'} + z_p \mathbf{k'}$ rotated system $\mathbf{R} = X_p \mathbf{i} + Y_p \mathbf{j} + Z_p \mathbf{k}$ in fixed system $\begin{pmatrix} X_P - X_o \\ Y_P - Y_o \\ Z_P - Y_o \end{pmatrix} = \begin{bmatrix} \mathbf{i \cdot i'} & \mathbf{i \cdot j'} & \mathbf{i \cdot k'} \\ \mathbf{j \cdot i'} & \mathbf{j \cdot j'} & \mathbf{j \cdot k'} \\ \mathbf{k \cdot i'} & \mathbf{k \cdot j'} & \mathbf{k \cdot k'} \end{bmatrix} \begin{pmatrix} x_P \\ y_P \\ z_P \end{pmatrix}$ Inverse Transformation $\begin{pmatrix} x_P \\ y_P \\ z_P \end{pmatrix} = \begin{bmatrix} \mathbf{i \cdot i'} & \mathbf{j \cdot i'} & \mathbf{k \cdot i'} \\ \mathbf{i \cdot j'} & \mathbf{j \cdot j'} & \mathbf{k \cdot j'} \\ \mathbf{i \cdot k'} & \mathbf{j \cdot k'} & \mathbf{k \cdot k'} \end{bmatrix} \begin{pmatrix} X_P - X_o \\ Y_P - Y_o \\ Z_P - Y_o \end{pmatrix}$	**i, j, k** are unit vectors in X,Y, Z coordinates. **i', j', k'** are unit vectors in x, y, z coordinates. **i • j'**, etc. = direction cosines (•) is vector dot product, Eq. 1.25. Ref. [5], p. 29.
8. Properties of Cartesian Coordinate Transforms [T] From (x,y,z) to (X,Y,Z) $\begin{pmatrix} X_p \\ Y_p \\ Z_p \end{pmatrix} = [T]^{-1} \begin{pmatrix} x_p \\ y_p \\ z_p \end{pmatrix}$	From (X,Y,Z) to (x,y,z) $\begin{pmatrix} x_p \\ y_p \\ z_p \end{pmatrix} = [T] \begin{pmatrix} X_p \\ Y_p \\ Z_p \end{pmatrix}$	$[T]^{-1} = [T]^T/\|T\|$. In table $\|T\| = 1$ $[T]^T$ matrix transform of [T] $[T_{ij}]^T = [T_{ji}]$, $[T][T]^{-1} = [1]$ $[[T_1][T_2]]^T = [T_2]^T[T_1]^T$

Table 2.2 Coordinate transformations, continued

Coordinate Systems	Coordinate Transformation

9. Cylindrical to Cartesian Coordinates

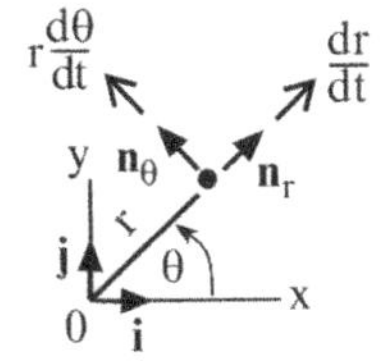

Position

$$x = r\cos\theta \qquad y = r\sin\theta \qquad z = z$$
$$r = \sqrt{x^2 + y^2} \qquad \theta = \tan^{-1} y/x$$
$$\mathbf{n}_r = \mathbf{i}\cos\theta + \mathbf{j}\sin\theta \qquad \mathbf{n}_\theta = -\mathbf{i}\sin\theta + \mathbf{j}\cos\theta$$

Velocity

$$\dot{x} = \dot{r}\cos\theta - r\dot{\theta}\sin\theta \qquad \dot{y} = \dot{r}\sin\theta + r\dot{\theta}\cos\theta \qquad \dot{z} = \dot{z}$$
$$\mathbf{v} = \dot{r}\,\mathbf{n}_r + r\dot{\theta}\,\mathbf{n}_\theta \qquad \dot{\theta} = \tan^{-1}\frac{x\dot{y} - y\dot{x}}{x^2 + y^2}$$
$$\mathbf{i} = \mathbf{n}_r\cos\theta - \mathbf{n}_\theta\sin\theta$$

Acceleration

$$\ddot{x} = (\ddot{r} - r\dot{\theta}^2)\cos\theta - (r\ddot{\theta} + \dot{r}\dot{\theta})\sin\theta$$
$$\ddot{y} = (\ddot{r} - r\dot{\theta}^2)\sin\theta + (r\ddot{\theta} + \dot{r}\dot{\theta})\cos\theta$$
$$\ddot{z} = \ddot{z} \quad \text{Ref. [6].}$$
$$\mathbf{j} = \mathbf{n}_r\sin\theta + \mathbf{n}_\theta\cos\theta$$

---

10. Cartesian to Cylindrical Coordinates x_o, y_o = location of origin for cylindrical system. Ref. [6].

Position

$$r = \sqrt{(x - x_o)^2 + (y - y_o)^2}, \qquad \theta = \tan^{-1}[(y - y_o)/(x - x_o)]$$

Velocity

$$\frac{dr}{dt} = \frac{(x - x_0)dx/dt + (y - y_o)dy/dt}{[(x - x_o)^2 + (y - y_o)^2]^{1/2}}, \qquad \frac{d\theta}{dt} = \frac{(y - y_0)dx/dt - (x - x_o)dy/dt}{(x - x_o)^2 + (y - y_o)^2}$$

Acceleration

$$\frac{d^2r}{dt^2} = \frac{(x - x_0)\dfrac{d^2x}{dt^2} + (y - y_o)\dfrac{d^2y}{dt^2} + \left(\dfrac{dx}{dt}\right)^2 + \left(\dfrac{dy}{dt}\right)^2}{[(x - x_o)^2 + (y - y_o)^2]^{1/2}} - \frac{\left((x - x_0)\dfrac{dx}{dt} + (y - y_o)\dfrac{dy}{dt}\right)^2}{[(x - x_o)^2 + (y - y_o)^2]^{3/2}}$$

$$\frac{d^2\theta}{dt^2} = \frac{(y - y_0)\dfrac{d^2x}{dt^2} - (x - x_o)\dfrac{d^2y}{dt^2}}{(x - x_o)^2 + (y - y_o)^2} - \frac{4\left(x - x_0)(y - y_o)\left(\left(\dfrac{dx}{dt}\right)^2 - \left(\dfrac{dy}{dt}\right)^2\right) + \dfrac{dx}{dt}\dfrac{dy}{dt}\left((y - y_0)^2 - (x - x_o)^2\right)\right)}{[(x - x_o)^2 + (y - y_o)^2]^2}$$

---

11. Spherical to Cartesian Coordinates

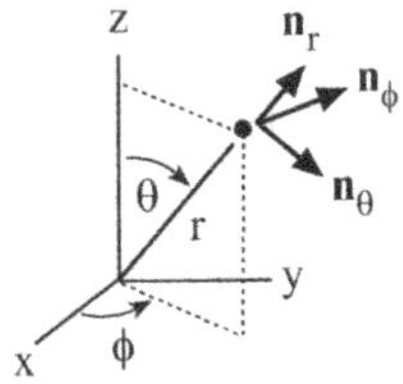

Position

$$x = r\sin\theta\cos\phi \qquad y = r\sin\theta\sin\phi \qquad z = r\cos\theta$$
$$r = \sqrt{x^2 + y^2 + z^2}$$
$$\theta = \cot^{-1}\left(z/\sqrt{x^2 + y^2}\right)$$
$$\phi = \tan^{-1} y/x$$

Velocity

$$\dot{x} = \dot{r}\sin\theta\cos\phi + r\dot{\theta}\cos\theta\cos\phi - r\dot{\phi}\sin\theta\sin\phi$$
$$\dot{y} = \dot{r}\sin\theta\sin\phi + r\dot{\theta}\cos\theta\sin\phi + r\dot{\phi}\sin\theta\cos\phi$$
$$\dot{z} = \dot{r}\cos\theta - r\dot{\theta}\sin\theta$$
$$dr/dt = (x\dot{x} + y\dot{y} + z\dot{z})/\sqrt{x^2 + y^2 + z^2}$$
$$\frac{d\theta}{dt} = \frac{(x\dot{x} + y\dot{y})z - \dot{z}(x^2 + y^2)}{\sqrt{x^2 + y^2}(x^2 + y^2 + z^2)}$$
$$\frac{d}{dt}\phi = (x\dot{y} - y\dot{x})/(x^2 + y^2)$$

Acceleration

$$\ddot{x} = (\ddot{r} - r\dot{\theta}^2 - r\dot{\phi}^2)\sin\theta\cos\phi + (2\dot{r}\dot{\theta} + r\ddot{\theta})\cos\theta\cos\phi - (r\ddot{\phi} + 2\dot{r}\dot{\phi})\sin\theta\sin\phi - 2r\dot{\theta}\dot{\phi}\cos\theta\sin\phi$$
$$\ddot{y} = (\ddot{r} - r\dot{\theta}^2 - r\dot{\phi}^2)\sin\theta\sin\phi + (2\dot{r}\dot{\theta} + r\ddot{\theta})\cos\theta\sin\phi + (r\ddot{\phi} + 2\dot{r}\dot{\phi})\sin\cos\phi + 2r\dot{\theta}\dot{\phi}\cos\theta\cos\phi$$
$$\ddot{z} = (\ddot{r} - r\dot{\theta}^2)\cos\theta - (r\ddot{\theta} + 2\dot{r}\dot{\theta})\sin\theta \qquad \text{Ref. [6].}$$

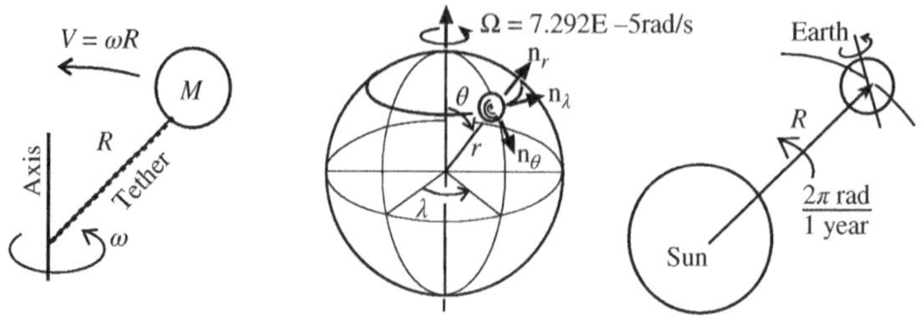

(a) Ball on a string (b) Ball on rotating earth (c) Earth orbits the sun

Figure 2.1 Three cases of rotation

Solution: Initially, the ball is stationary with respect to the surface of the earth: $d^2\theta/dt^2 = d^2\lambda/dt^2 = d\theta/dt = 0$ are zero, where θ is the polar angle. However, the earth's surface has rotational velocity $d\lambda/dt = \Omega$, which generates two spherical components of centripetal acceleration (case 4 of Table 2.1):

$$\begin{aligned} &\text{Radial acceleration:} \quad a_r = -r\Omega^2 \sin^2\theta \\ &\text{Tangential polar acceleration:} \, a_\theta = -r\Omega^2 \sin\theta \cos\theta \end{aligned} \tag{2.3}$$

The radial acceleration is sustained by gravity. The tangential acceleration of the ball toward the pole is not supported. So for zero net tangential force, the released ball accelerates in the opposite direction. The ball accelerates toward the equator.

2.2 Newton's Law of Particle Dynamics

2.2.1 *Constant Mass Systems*

Table 2.3 [1, 8–15] shows the dynamic solutions for compact masses, called particles, without rotational inertia. Cases 1–8 of Table 2.3 indicate particle trajectories, cases 9 to 20 are particles under gravity, and cases 20 through 28 are particle collisions. Solutions for particles and springs, viscous damping, velocity squared drag, and friction are also shown in Table 2.3.

Equations in Table 2.3 are based upon Newton's second law: force F on a particle equals the particle mass M times its *acceleration* vector $a = dv/dt$ (Figure 1.1, Section 1.2).

$$F = Ma \tag{2.4}$$

Newton's second law applies when (1) the accelerations are measured in an inertial frame of reference, a nonaccelerating, nonrotating, coordinate system fixed to distant stars, (2) velocities are small compared to the speed of light, and (3) consistent sets of units (Table 1.2) are used. Table 2.3 has the applications of Newton's second law to particles and to translation of the center of mass of rigid bodies (Section 2.3).

Momentum is mass M times its velocity vector $\mathbf{v}$:

$$\text{Momentum} = M\mathbf{v} \tag{2.5}$$

Table 2.3 Particle dynamics

Notation: a = acceleration; e = coefficient of restitution, or e = nat. log, or as subscript exit; F = force; F_D = drag force; g = acceleration due to gravity (Table 1.2); k = spring constant, Table 3.1; M = mass; **n** = unit normal vector; r = radius from center; S = surface; t = time; T = time period; u, v = velocity; V = velocity magnitude; x, y = position; ρ = density; μ = dimensionless friction coefficient. Consistent sets of units are given in Table 1.2 Refs [1, 8–15].

Particle Motion	Force F, Acceleration a, Velocity v Position x	Comments and Definitions
1. Force Accelerates Mass from Rest 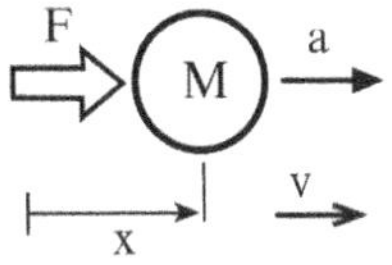 Newton's 2nd law.	$F = M(d^2x/dt^2)$ $a = \frac{d^2x}{dt^2} = \frac{F}{M} = constant$ $v = dx/dt = at$ $x = \frac{1}{2}at^2$ Initial conditions at t = 0: x = 0, v = 0	At time t, $t = v/a = (2x/a)^{1/2}$ $a = \frac{v}{t} = \frac{v^2}{2x} = \frac{F}{M}$ $x = \frac{v^2}{2a}$
2. Force Accelerates Mass with Initial Velocity V_o 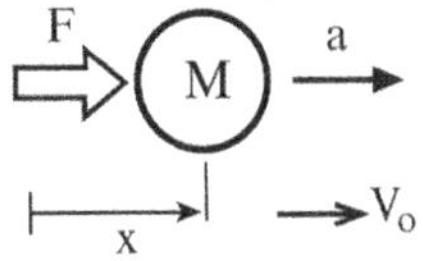 	$F = M(d^2x/dt^2)$ $a = \frac{d^2x}{dt^2} = \frac{F}{M} = constant$ $v = dx/dt = v = at + V_o$ $x = \frac{1}{2}at^2 + V_o t$ Intial conditions at t = 0: $x = 0, v = V_o$	$t = \frac{1}{a}\left(-V_o + \sqrt{V_o^2 + 2ax}\right)$ $= (v - V_o)/a$ $a = \frac{v - V_o}{t} = \frac{v^2 - V_o^2}{2x} = \frac{F}{M}$ $v = \sqrt{V_o^2 + 2ax}$ $x = v(v - V_o)/(2a)$
3. Force Increases Linearly in Time Accelerates Mass 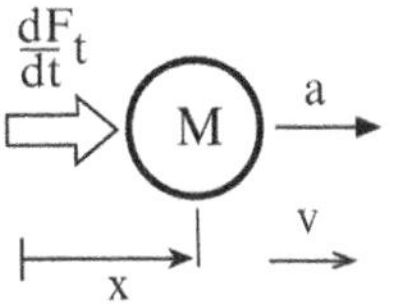 	$F = (dF/dt)t = Ma$, $dF/dt = \text{constant}$ $a = \frac{d^2x}{dt^2} = \frac{(dF/dt)}{M}t$ $v = \frac{dx}{dt} = \frac{1}{2}\frac{(dF/dt)}{M}t^2$ $x = \frac{1}{6}\frac{(dF/dt)}{M}t^3$ at t = 0: x = v = F = 0	$t = \left(\frac{6Mx}{dF/dt}\right)^{1/3}$ $v = \left(\frac{9x^2}{2M}\frac{dF}{dt}\right)^{1/3}$ $x = \frac{2^{1/2}}{3}\left(\frac{Mv^3}{dF/dt}\right)^{1/2}$
4. Force Vector Does Work on a Mass 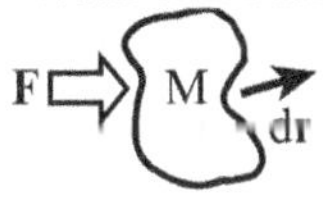 	$\text{Work} = \int_A^B \mathbf{F} \bullet d\mathbf{r} = \frac{1}{2}MV_B^2 - \frac{1}{2}MV_A^2$ V_A = velocity at point A, $V_A>0$ V_B = velocity at point B, $V_B>0$	**F** = force vector **dr** = vector element of path length Power = **F•v** • = vector dot product, Eq. 1.25.

Table 2.3 Particle dynamics, continued

Particle Motion	Force F, Acceleration a, Velocity v, Position x	Comments and Definitions
5. Varying Force Accelerates Mass General Case 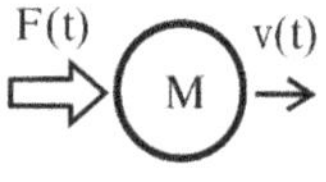	$F(t) = Ma$ $a = \frac{d^2x}{dt^2} = \frac{F(t)}{M}$ $v = \frac{dx}{dt} = \frac{1}{M}\int_0^t F(t)dt + v_0$ $x = \frac{1}{M}\int_0^t\left(\int_0^t F(t)dt\right)dt + v_0 t + x_0$ at t=0: $x = x_0, v = v_0$	Reduces to case 1 for constant force. Generally, F and a are vectors while mass is a constant scalar. $\text{Impulse} = \int_{t_1}^{t_2} F(t)dt = M(v_2 - v_1)$ For small times steps Δt $F\Delta t = M\Delta v, F = M\Delta v / \Delta t$, See Appendix B for integration.
6. Force Accelerates Variable Mass 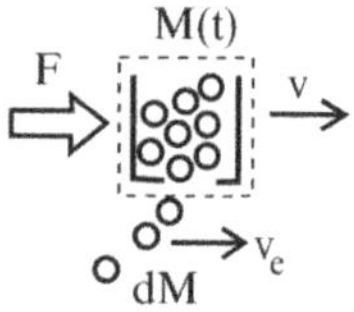	$F = \frac{dM}{dt}(v - v_e) + M\frac{dv}{dt}$ $a = \frac{F}{M} + \frac{v_e - v}{M}\frac{dM}{dt}$ $v_e - v$ =relative velocity at which mass leaves system. Note system gains mass for dM/dt>0.	For constant dM/dt: $M_2 = M_1 + (dM/dt)t_2$ and for F=0 and v_e=0 the velocity change between initial and final time, $v_2/v_1 = M_1/M_2$. Ref. [1], p. 168; Ref. [11] See Eq. 2.23.
7. Force on Deformable, Control Volume 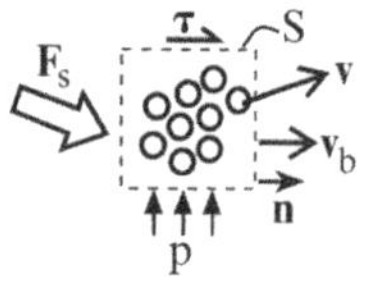 **n** = unit outward normal to control surface S.	$F_s - \int_S p\mathbf{n}dS + \int_S \boldsymbol{\tau} dS + \int_V \rho\mathbf{g}dV$ $= \frac{d}{dt}\int_V \rho\mathbf{v}dV + \int_S \rho\mathbf{v}(\mathbf{v} - \mathbf{v}_b)\bullet\mathbf{n}dS$ $\mathbf{F}_s$ = mechanical force vector on control volume V. p = pressure on S ρ = fluid density, **τ** =shear stress on S **v** = fluid velocity vector; **g** = gravity acceleration vector	Fluid has velocity **v**. Fluid passes through control surface S that has velocity $\mathbf{v}_b$ and bounds the control volume V. Left hand side is sum of external forces on control volume and right hand side is rate of change of momentum within control volume Ref. [1]
8. Mass M with Initial Velocity V_o and Velocity Resistance $F_R = \rho vsu$ 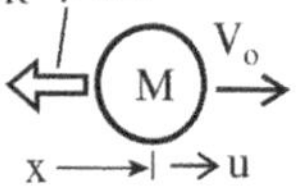	$Ma = -F_R,\ F_R = \mathit{resistance} = \rho vsu$ $a = \frac{d^2x}{dt^2} = -\frac{V_o}{L_c}u$ $u = \frac{dx}{dt} = V_o e^{-V_o t/L_c}$ $x = L_c\left[1 - e^{-V_o t/L_c}\right]$	ρ= fluid density Ref. [8] ν = fluid kinematic viscosity s = drag length, 3πD sphere dia D $L_c = \frac{V_o}{\rho vs}$ = distance for $u = \frac{V_o}{e}$ e = natural logarithm = 2.718 at t=0: $x(0) = 0, dx/dt = V_o$
9. Mass M with Initial Velocity V_o and Velocity Squared Drag $F_D = \frac{1}{2}\rho u^2 AC_D$ 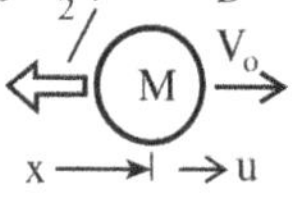	$Ma = -F_D,\ F_D = \mathit{drag} = \frac{1}{2}\rho Au^2 C_D$ $a = \frac{d^2x}{dt^2} = -\frac{u^2}{L_c}$ $u = \frac{dx}{dt} = \frac{V_o}{1 + V_o t / L_c}$ $x = L_c \log_e(1 + V_o t / L_c)$ At t=0: $x(0) = 0, dx/dt = V_o$	C_D = dim'less drag coefficient ≈ 1 for sphere Ref. [8] ρ = fluid density A = reference cross section area $= \pi D^2/4$ for sphere diameter D $L_c = \frac{M}{\rho AC_D}$, at $x = L_c$, $u = \frac{V_o}{2}$

Table 2.3 Particle dynamics, continued, Gravity

Additional Notation: a = acceleration; e = restitution coefficient, dimensionless; F = force; FD = drag force; M = mass; g = acceleration due to gravity, Table 1.2; t = time; v = lateral velocity; u, v = velocity; V = velocity magnitude; x = position; y = vertical position; μ = dimensionless friction coefficient. Consistent sets of units are given in Table 1.2.

Particle Motion	Force F, Acceleration a, Velocity v, Position x	Comments and Definitions
10. Mass Falls from Rest 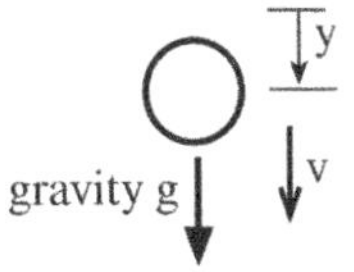 	$M\frac{d^2y}{dt^2} = F = Mg,\ \frac{d^2y}{dt^2} = g$ $v = \frac{dy}{dt} = -gt$ $y = \frac{1}{2}gt^2$ Result is independent of mass. Initial conditions at t=0: y = 0, v = 0.	Velocity after falling distance h $v = \sqrt{2gh}$ Time to fall distance h $t = \sqrt{\frac{2h}{g}}$ Resistance of *air is neglected*. See cases 11 and 12 for air resistance.
11. Mass Falls with Initial Velocity V_o 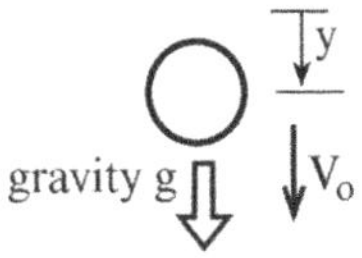 	$M\frac{d^2y}{dt^2} = F = Mg,\ a = \frac{d^2y}{dt^2} = g$ $v = dy/dt = gt + V_o t$ $y = \frac{1}{2}gt^2 + V_o t + y_o$ At t=0: y = y_o, v = V_o Result is independent of mass.	V_o is positive downward. For $V_o < 0$, maximum height is, $y_{max} = y_o + V_o^2/(2g)$, at $t = -V_o/g$ Result is independent of mass. Air resistance *is neglected*; see cases 12 and 13.
12. Mass Falls from Rest Resistance proportional To Velocity 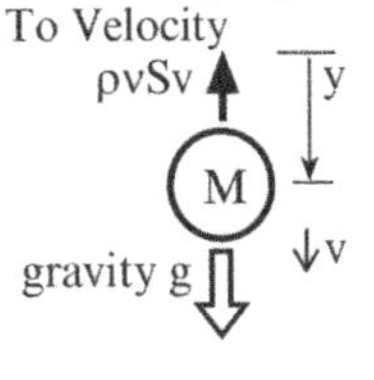 	$Ma = Mg - F_R$ $F_R = \rho v s v = $ *resistance* $a = \frac{d^2y}{dt^2} = ge^{-(g/V_T)t}$ $v = \frac{dy}{dt} = V_T\left(1 - e^{-(g/V_T)t}\right)$ $y = -\frac{V_T^2}{g}\left[1 - e^{-(g/V_T)t}\right] + V_T t$	ρ = fluid density Ref. [8], in part ν = kinematic viscosity s= drag length = 6πR for sphere with radius R Terminal Velocity $V_T = \frac{Mg}{\rho v s}$ Initial Conditions at t=0, $y(0) = 0, dy/dt = 0.$
13. Mass Falls from Rest Velocity Squared Drag 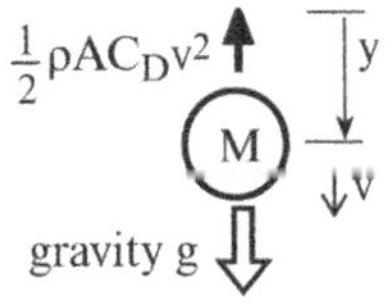 	$Ma = Mg - F_D$ $F_D = (1/2)\rho v^2 AC_D = $ *drag* $a = \frac{d^2y}{dt^2} = \frac{g}{M}(1 - \frac{v^2}{V_T^2})$ $v = \frac{dy}{dt} = V_T \tanh(gt/V_T)$ $y = \frac{V_T^2}{2g}\log_e(\cosh\frac{gt}{V_T})$	C_D = drag coef. ≈ 1 for sphere ρ = fluid density Ref. [8], in part A = reference cross section area $= \pi D^2/4$ for sphere diameter D Terminal velocity $V_T = \sqrt{\frac{2Mg}{\rho AC_D}}$ Initial Conditions at t = 0, $y(0) = 0, dy/dt = 0.$

Table 2.3 Particle dynamics, continued, Gravity

Particle Motion	Force F, Acceleration a, Velocity v, Position x	Comments and Definitions
14. Particle Slides on Frictionless Plane 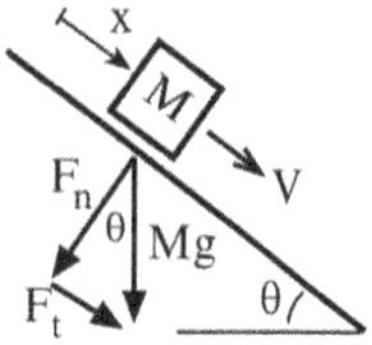	$M(d^2y/dt^2) = F_t = Mg\sin\theta$ $a = \frac{d^2x}{dt^2} = g\sin\theta$ $v = \frac{dx}{dt} = gt\sin\theta + V_o$ $x = \frac{1}{2}gt^2\sin\theta + V_o t$ At $t = 0$, $x = x_o$, $V = V_o$	$F_n = Mg\cos\theta$ $F_t = Mg\sin\theta$ Maximum height if $V_o < 0$, $x_{max} = x_o - \frac{V_o^2}{2g\sin\theta}$, at $t = \frac{V_o}{g\sin\theta}$ Frictionless incline. See next case for incline with friction. See case 17 of Table 2.6 for rolling.
15. Particle Slides on Plane with Friction 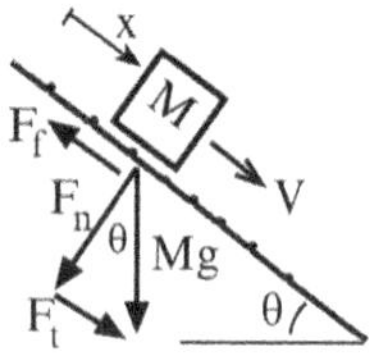	$Ma = F_t - F_f = Mg\sin\theta - \mu Mg\cos\beta$ $a = \frac{d^2x}{dt^2} = g\sin\theta \mp \mu g\cos\theta$ (a) $v = \frac{dx}{dt} = (\sin\theta \mp \mu\cos\theta)gt + V_o$ $x = \frac{1}{2}gt^2(\sin\theta - \mu\cos\theta) + V_o t + x_o$ – for V>0 (downhill), + for V<0	Friction force, $Fr = -\mu Mg\cos\theta$, opposes velocity v. Dimensionless friction coefficient $0 < \mu <= 0.3$ often used. Tangent gravity force, $Ft = Mg\sin\theta$. Gravity normal force, $Fn = Mg\cos\theta$. Critical angle $\tan\theta = \mu$.
16. Particle Trajectory with Initial Velocity V_o and Launch Angle β 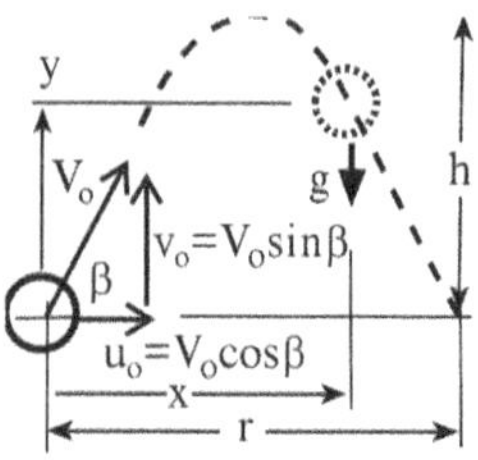	Vertical acceleration $a_y = \frac{d^2y}{dt^2} = -g$ Vertical velocity $v = \frac{dy}{dt} = -gt + v_o$ Horizontal vel. $u = u_o = \text{constant}$ Vert. position $y = -\frac{1}{2}gt^2 + v_o t + y_o$ Horizontal position $x = u_o t + x_o$ At $t = 0$, $x = x_o$, $y = y_o$ $u = u_o = V_o\cos\rho, v = v_o = V_o\sin\theta$ $V_o = \sqrt{u_o^2 + v_o^2}$, $\theta = \tan^{-1} v_o/u_o$	Maximum height above launch, $h = y_o + \frac{V_o^2\sin^2\beta}{2g}$, at $t = \frac{V_o\sin\beta}{g}$ Range at return to x=0. Range, $r = \frac{\sin 2\beta}{g}V_o^2$ Max range = V_o^2/g for $\beta = 45$ deg. Time of flight $T = (2V_o\sin\beta)/g$ Resistance of *air is neglected*-see Case 18 for air resistance
17. Trajectory through Fixed Point 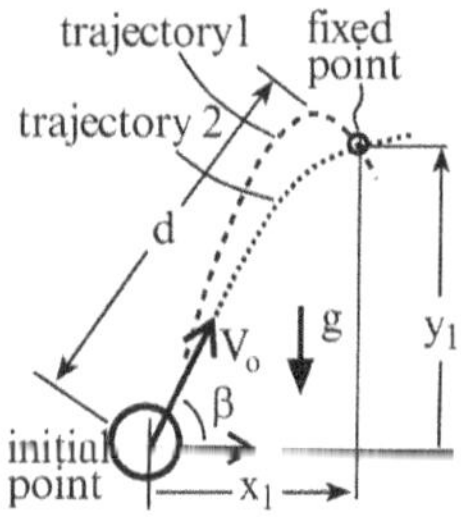	Initial vel. V_o must be greater than a minimum value to reach x_1,y_1 if $y_1>0$. $V_o^2\big\|_{min} \geq gy_1 + gd$ where $d = \sqrt{x_1^2 + y_1^2}$ Two trajectories pass through x_1,y_1. Time from launch to reach point x_1,y_1. $t_i^2 = \frac{2V_o^2 - 2gy_1}{g^2} \pm \frac{2\sqrt{V_o^4 - 2gy_1V_o^2 - g^2x_1^2}}{g^2}$, $i = 1,2$ Launch angle, $\beta = \tan^{-1}[(y_1 + (1/2)gt_i^2)/x_1]$, $i = 1.2$ Trajectory details are computed in previous case. For maximum x_1 on a slope $\tan\alpha = y_1/x_1$ given V_o the launch angle is $\beta = \alpha/2 + \pi/4$ where + applies if $\alpha > 0$. Resistance of air is neglected.	

Table 2.3 Particle dynamics, continued, Gravity

Particle Motion With Resistance	Force F, Acceleration a, Velocity v, Position x	Definitions and Initial Conditions
18. Particle Trajectory with Initial Velocity V_o Launch Angle βo and Velocity Squared Drag 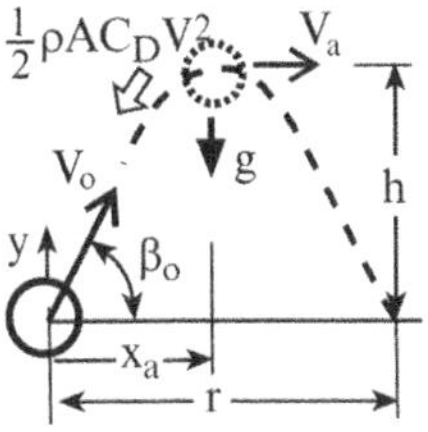	$Ma_x = -\frac{1}{2}\rho V^2 AC_D \cos^2\beta$ = horizonal acceleration Ref. [9] $Ma_y = -Mg + \frac{1}{2}\rho V^2 AC_D \lvert\sin\beta\rvert \sin\beta$ = vertical acceleration Maximum height above launch $(x = x_a)$, $h = \dfrac{V_o^2 \sin^2\beta_o}{g(2+(V_o/V_T)^2 \sin^2\beta_o}$ horizontal velocity at apex $V_a = \dfrac{V_o \cos\beta_o}{\sqrt{1+(V_o/V_T)^2(\sin\beta_o + \cos^2\beta_o \ln[\tan(\beta_o/2+\pi/4)]}}$ $x_a = L(T-t_a)/\sqrt{2(t_a^2+2h/g)}$, $r = 2V_a\sqrt{2h/g}$ time from launch to apex $(V = V_a)$, $t_a = \sqrt{\dfrac{2h}{g}+\left(\dfrac{hV_a}{2V_T^2}\right)^2} - \left(\dfrac{hV_a}{2V_T^2}\right)^2$ Terminal velocity $V_T = \sqrt{2Mg/AC_D}$, time of flight, $T = t_a + 2h/(gt_a)$ Drag of air is included approximately. See case 16 without drag. β_o = launch angle. Magnitude of velocity = V.	
19. Dropped Particle Rebounds off Floor 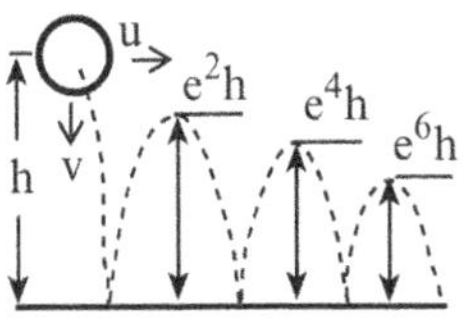	Velocity of initial impact $v = \sqrt{2gh}$ Time to initial impact $t = \sqrt{2h/g}$ Height of rebound $\dfrac{h_{i+1}}{h} = \sqrt{e} = \dfrac{\lvert v_{rebond}\rvert^{1/2}}{\lvert v_{income}\rvert^{1/2}}$ v = vertical velocity	Velocity is zero at initial height = h. h_i = max height after i^{th} impact. e = coefficient of restitution, dim'lss *Impulse* delivered to floor after drop from height h, $\int F\,dt = (1+e)M(2gh)^{1/2}$ Horizontal velocity u is constant.
20. Particle Falls from Height h and Impacts Spring and Rebounds 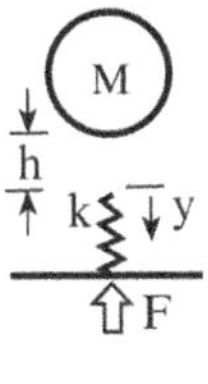 $y = 0$, $t = 0$ at first contact	$F = ky - Mg$ at $0 < t <= \pi\sqrt{M/k}$ $\dfrac{d^2y}{dt^2} = -\omega_n V_o \sin\omega_n t - g\cos\omega_n t$ $\dfrac{dy}{dt} = V_o\cos\omega_n t - \dfrac{g}{\omega_n}\sin\omega_n t$ $y = \dfrac{V_o}{\omega_n}\sin\omega_n t - \dfrac{g}{\omega_n^2}\cos\omega_n t + \dfrac{Mg}{k}$ velocity at initial contact, t=0 $V_o = \sqrt{2gh}$, case 10, $\omega_n = (k/M)^{1/2}$ h = initial height above spring. e = 1.	Maximum deformation and accel $y_{max} = \left(\dfrac{M}{k}\right)^{1/2}\left(V_o^2 + \dfrac{Mg^2}{k}\right)^{1/2} + \dfrac{Mg}{k}$ $\dfrac{y_{max}}{y_s} = 1 + \left(1 + 2\dfrac{h}{y_s}\right)^{1/2}$ $d^2y/dt^2 \rvert_{max} = \sqrt{V_o^2\omega_n^2 + g^2}$ $y_s = \dfrac{Mg}{k} = \dfrac{g}{\omega_n^2}$ = static deformation Max spring force $F_{max} = ky_{max}$ Final velocity of mass $= -V_o$
21 Mass with Vel. V_o Impacts Rigid Wall 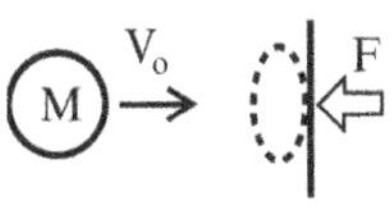 + velocity ⟶ V_o = initial velocity	Rebound velocity $v = \begin{cases} -V_o & \text{elastic impact} \\ 0 & \text{inelastic impact} \\ -eV_o & \text{general impact} \end{cases}$ Impulse on wall $\int F dt = \begin{cases} 2MV_o & \text{elastic impact} \\ MV_o & \text{inelastic impact} \\ M(1+e)V_o & \text{general impact} \end{cases}$	Mass rebounds at original velocity, after perfectly elastic impact, e=0. Mass lodges on wall after inelastic impact, e=1. Coefficient of restitution, $e = \dfrac{\lvert\text{rebound velocity}\rvert}{\lvert\text{incoming velocity}\rvert}$, $0 \le e \le 1\dfrac{1}{2}$ F = force on wall.

Table 2.3 Particle dynamics, continued, Collisions

Particle Motion	Force F, Acceleration a, Velocity v, Position x	Comments and Special Cases
22 Mass with Velocity V_o Impacts Spring and Rebounds 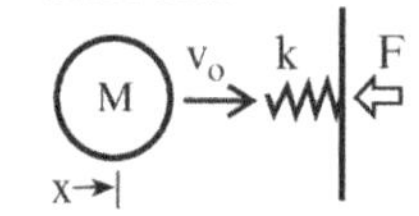 x=0, t=0 at first contact with spring.	$Md^2x/dt^2 + kx = 0$ F = kx = force on statonary wall $\frac{d^2x}{dt^2} = -V_o\sqrt{\frac{k}{M}}\sin\sqrt{\frac{k}{M}}t$ $\frac{dx}{dt} = V_o\cos\sqrt{\frac{k}{M}}t$ $x = V_o\sqrt{\frac{M}{k}}\sin\sqrt{\frac{k}{M}}t,\ 0\le t\le \pi\sqrt{M/k}$	max deceleration $a_{max} = (k/M)^{1/2}V_o$ Final velocity of mass $V_f = -V_o$ Max spring force $F_{max} = (kM)^{1/2}V_o$ Max deform. $x_{max} = (M/k)^{1/2}V_o$ F(t) = Force on wall = k x(t) Impulse = $2MV_o$
23. Mass with Velocity V_o Impacts Damped Spring and Rebounds 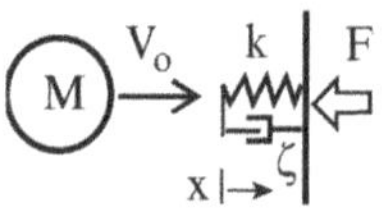 x=0 and t=0 at first contact with spring. Force on stationary wall $F = kx + 2M\zeta\omega_n dx/dt$ $F_{max} \approx kx_{max}$	$Md^2x/dt^2 = F = -kx - 2M\zeta\omega_n dx/dt,\quad \omega_n = (k/M)^{1/2}$ $x = \frac{V_o}{\omega_n(1-\zeta^2)^{1/2}}e^{-\zeta\omega_n t}\sin\omega_n(1-\zeta^2)^{1/2}t,\ 0\le t\le \frac{\cos^{-1}(2\zeta^2-1)}{\omega_n(1-\zeta^2)^{1/2}}$ x_{max} = maximum spring compression. dim'lss damping factor, $0\le\zeta\le 1$. $x_{max} = \frac{V_o}{\omega_n}Exp\left[\frac{-\zeta}{(1-\zeta^2)^{1/2}}\cos^{-1}\zeta\right]$, at time $t = \frac{1}{\omega_n(1-\zeta^2)^{1/2}}\cos^{-1}\zeta$ $\approx (V_o/\omega_n)(1-\pi\zeta/2)$, at time $t = \omega_n^{-1}(\pi/2-\zeta)$, for $\zeta<<1$, V_e = rebound velocity, $= -V_o Exp\left[\frac{-\zeta}{(1-\zeta^2)^{1/2}}\cos^{-1}[-(1-2\zeta^2)]\right][1-2\zeta^2\{1+(1-\zeta^2)^{1/2}\}]$ $\approx -V_o(1-\pi\zeta)$ at $t\approx(1/\omega_n)(\pi-2\zeta)$, a_{max} = maximum acceleration $\approx -V_o\omega_n(1-\pi\zeta/2),\ \zeta<<1$	
24. Mass M Strikes and Penetrates Plate 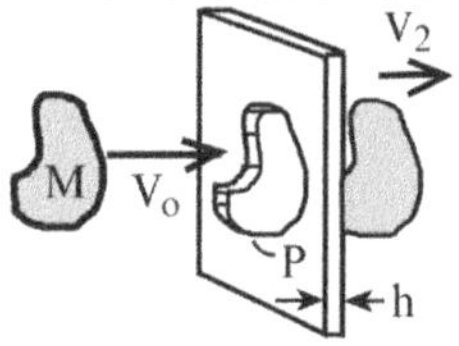	Velocity of particle after penetrating plate $V_2 = \sqrt{V_o^2 - 2P\tau h^2/M}$ if $h<h_{min}$, otherwise $V_2 = 0$. Minimum thickness no penetration $h_{min} = \sqrt{\frac{MV_o^2}{2P\tau}}$	Material / τ, MPa / lb/in^2 Aluminum2024 210 30600 Titanium 6-4 860 128000 Stainless steel 1300. 188500 Inconel® 625 2069. 298000 Graphite/epoxy 210.3 30500 τ = dynamic shear Refs [12, 13, 14] P = perimeter of body V_o = component of initial velocity perpendicular to face of plate
25. Mass M_1 with Velocity V_o Impacts Stationary Mass M_2 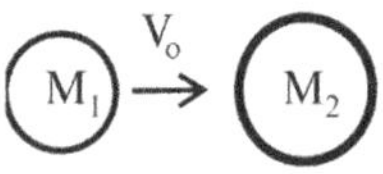 + velocity ⟶ Restitution coefficient $e = (V_1-V_2)/V_o$	general impact, $0\le e\le 1$: $V_1 = \frac{M_1 - eM_2}{M_1+M_2}V_o$, $V_2 = \frac{(1+e)M_1}{M_1+M_2}V_o$ elastic impact, e=1: $V_1 = \frac{M_1 - M_2}{M_1+M_2}V_o$, $V_2 = \frac{2M_1}{M_1+M_2}V_o$ inelastic impact, e = 0: $V_1 = \frac{M_1}{M_1+M_2}V_o$, $V_2 = V_1$ V_o = initial velocity of M_1. Initial velocity of M_2 is zero. Ref. [2], p. 158. V_1 = velocity of M_1 after impact. V_2 = velocity of M_2 after impact. $M_1V_o\vert_{before\ impact} = (M_1V_1 + M_2V_2)\vert_{after\ impact}$	

Table 2.3 Particle dynamics, continued, Collisions

Particle Motion	Position x, Velocity v, Acceleration a, and Force F	Comments and Special Cases
26. Moving Mass M_1 Strikes Stationary Mass M_2 and side distance d 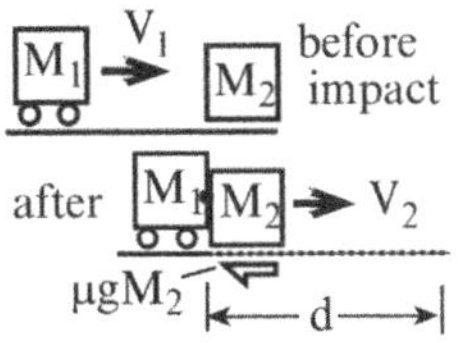 	V_1 = initial velocity of M_1. Initial vel M_2 is zero. After impact the two masses slide together distance d. Initial post impact velocity is V_2. $V_2 = \frac{M_1}{M_1+M_2} V_o$ slide distance after impact, $d = \frac{1}{2\mu g} \frac{M_1}{M_1+M_2} \frac{M_1}{M_2} V_o^2$	Ref. [15] Frictional force $\mu g M_2$ opposes the sliding. Dimensionless friction coefficient μ = friction force/(gM_2) $0.1 \le \mu \le 0.6$, typical
27. Mass with Vel. V_{1o} Impacts Mass Vel. V_{2o} 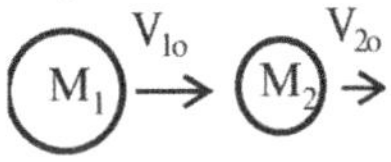 + velocity ⟶ Initial velocity $M_1 = V_{10}$, Initial velocity $M_2 = V_{20}$	After perfectly elastic impact (e=1) $V_1 = \frac{M_1 - M_2}{M_1+M_2} V_{1o} + \frac{2M_2}{M_1+M_2} V_{2o}$ $V_2 = \frac{2M_1}{M_1+M_2} V_{1o} + \frac{M_2 - M_1}{M_1+M_2} V_{2o}$ After inelastic impact (e=0) $V_1 = V_2 = \frac{M_1 V_{1o}}{M_1+M_2} + \frac{M_2 V_{2o}}{M_1+M_2}$	After general impact, 0<=e<=1 $V_1 = \frac{M_1 - eM_2}{M_1+M_2} V_{1o} + \frac{(1+e)M_2}{M_1+M_2} V_{2o}$ $V_2 = \frac{(1+e)M_2}{M_1+M_2} V_{1o} + \frac{M_2 - eM_1}{M_1+M_2} V_{2o}$ Restitution coefficient, Ref. [1], p.158 $e = (V_1 - V_2)/(V_{1o} - V_{2o})$, $M_1 V_{10} + M_2 V_{2o} = M_1 V_1 + M_2 V_2$
28. Mass M_1 Velocity V_1 Strikes and Scatters Stationary Mass M_2, 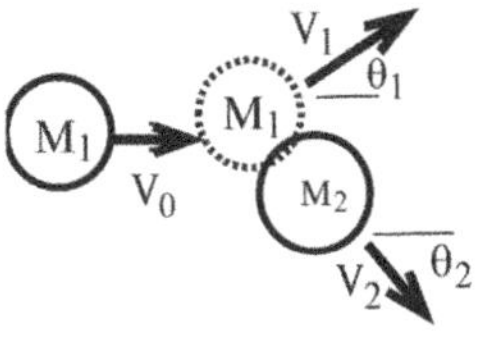 	$V_2^2 = \left(\frac{M_1}{M_2}\right)^2 \left(V_1^2 + V_0^2 - 2V_0 V_1 \cos\theta_1\right)$ $\frac{V_1}{V_0} = \frac{\cos\theta_1 \pm \sqrt{M_2^2/M_1^2 - \sin^2\theta_1}}{1 + M_2 M_1}$ $\frac{\sin\theta_1}{\sin\theta_2} = \frac{M_1 V_1}{M_2 V_2}$ $\sin\theta_1 = M_2/M_1$ is limit scatter angle	Elastic impact, e =1. Only the minus sign applies in 2nd expression unless θ_1 =0 and in that case only + sign applies, θ_2 =0 reduces to case 25. These expressions relate initial and final states given θ_1. Additional relationships needed to get final state from initial state. V_0 = initial velocity of M_1 V_1 = final velocity of M_1 V_2 = final velocity of M_2 Ref. [10]

Momentum, like velocity and force, is a vector. In constant-mass systems, the force vector **F** equals the rate of change of the momentum vector **v**:

$$\mathbf{F} = \frac{d(M\mathbf{v})}{dt} \tag{2.6}$$

This equation implies Newton's first law: in the absence of applied forces, the momentum vector (Eq. 2.5) is constant. This implies that the center of mass of a swarm of interacting particles has constant velocity in the absence of external forces.

2.2.2 *Variable Mass Systems*

The formula $F = Ma$ does not apply to systems that gain or lose mass because it does include momentum of exiting mass [1] (p. 168). Consider a boat with mass M that carries an inebriated physicist who tosses an empty wine bottle, mass dM, with velocity $\mathbf{u}$ relative to the boat with absolute velocity $\mathbf{v}$. Because the impulse provided to the bottle has the same magnitude but opposite sign as the impulse to the boat, momentum is conserved: $M\ d\mathbf{v} = \mathbf{u}\ dM$, where $d\mathbf{v}$ is the change in the boat's velocity. Dividing by dt and including external force $\mathbf{F}$ on mass M give Meshchersky's equation for a variable mass system [10, 11]:

$$M\frac{d\mathbf{v}}{dt} = \mathbf{F} + \frac{dM}{dt}\mathbf{u} \tag{2.7}$$

dM/dt is the rate of mass addition; it is positive if mass is added to the body and negative if mass is lost. This equation is applied to the rocket thrust provided in Table 2.4. Cases 6 and 7 of Table 2.3 present generalizations of this expression.

2.2.3 *Particle Trajectories*

The trajectory of a constant-mass particle responding to applied forces on it is determined by integrating Newton's second law (Eq. 2.4) in time from initial conditions. For motion of a particle moving along the x-axis from initial position x_1 with velocity v_1 at t_1,

$$\begin{aligned}&\text{Acceleration}\quad a = \frac{d^2x}{dt^2} = \frac{dv}{dt} = \frac{F}{M}, \qquad \text{velocity}\quad v = \frac{dx}{dt} = \int_{t_1}^{t_2} \frac{F}{M} dt + v_1\\ &\text{Displacement}\quad x = \int_{t_t}^{t_2} \left(\int_{t_1}^{t} \frac{F}{M} dt \right) dt + v_1(t_2 - t_1) + x_1\end{aligned} \tag{2.8}$$

The time history calculations integrate force over mass. If the force is gravity, $F/M = g$, the time to fall, without resistance, distance h without friction is $t = (2h/g)^{1/2}$, which is independent of mass. This was demonstrated by Galileo at Pisa and by Apollo 15 astronaut David Scott on the moon.

Numerical solutions are used when exact solutions are unavailable. Numerical particle trajectories replace the infinitesimal differential time step dt shown in Equation 2.8 with a finite time step Δt:

$$\begin{aligned}&\text{Acceleration:} \quad a_{n+1} = \frac{F_n}{M}\\ &\text{Velocity:} \quad v_{n+1} = \left(\frac{F_n}{M}\right)\Delta t + v_n\\ &\text{Displacement:} \quad x_{n+1} = v_n \Delta t + x_n\\ &\text{Time:} \quad t = n\Delta t \quad \text{and} \quad n = 0, 1, 2, 3, \ldots\end{aligned} \tag{2.9}$$

Numerical integration marches forward step-by-step in time from the initial condition, $n = 0, 1, 2$. This is Euler's explicit point slope method – all quantities are held constant during each time step. Accuracy increases as the time step decreases and the number of digits

Table 2.4 Rockets and orbits

Notation: A = area; a = acceleration or, major axis; b = minor axis; c = speed of sound; E = energy; e = 2.71828, or energy per unit mass; F = force; G = gravitational constant (case 9); g = acceleration due to gravity (Table 1.2); h = r dθ/dt = angular momentum per unit mass; H = height above earth surface; I_{sp} = specific impulse; K = GM (case 9); M = mass; **n** = unit normal vector; p = pressure; p_e = pressure of nozzle gas; p_{atm} = atmospheric pressure; r = radius from center; t = time; T = time period; v = velocity; **v** = velocity vector; v_e = exit velocity; ω = rotation, rad/s; ρ = density. Resistance of air is neglected. Consistent sets of units are given in Table 1.2 Refs [1, 5, 19–24]

Rocket Motor	Position x, Velocity v Acceleration a, and Thrust F	Comments and Special Cases
1. Rocket Motor Thrust 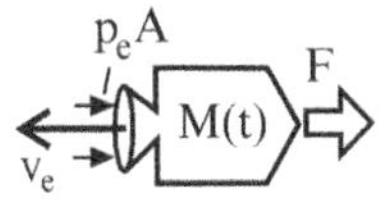 v_e = exhaust velocity relative to rocket	Mass rate fuel burn $= dM/dt$ Thrust $F = \frac{dM}{dt} v_e + (p_e - p_{atm})A$ Specific Impulse $I_{sp} = \frac{F}{g(dM/dt)}$ $= v_e / g$, if $p_e = p_{patm}$	$(p_e-p_{atm})A$ = net pressure force on nozzle. Ref. [1], p.172; Ref. [5] p. 241; [24]
2. Rocket Trajectory 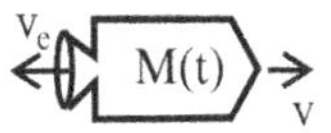	thrust, $I_{sp} g \frac{dM}{dt} = v_e \frac{dM}{dt} = M \frac{dv}{dt}$ acceleration, $\frac{dv}{dt} = I_{sp} g = \frac{I_{sp} g}{M} \frac{dM}{dt}$ velocity, $v - v_0 = I_{sp}\, g \log_e \frac{M_0}{M}$,	Rocket mass M(t) decreases at rate dM/dt as fuel is ejected from the nozzle at velocity v_e relative to rocket. Gravity and air resistance are neglected Ref. [5] p. 240–241.
3. Water Rocket 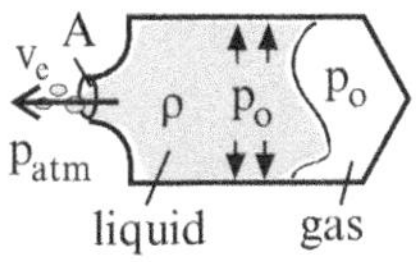	exit velocity, $v_e = \sqrt{2(p_o - p_{atm})/\rho}$ mass flow, $dM/dt = \sqrt{2\rho(p_o - p_{atm})}A$ thrust, $F = 2(p_o - p_{atm})A$ specific impulse, $I_{sp} = v_e / g$ p_o = gas pressure in motor; p_{atm} = atmospheric pressure;	Pressurized gas propels incompressible liquid through subsonic nozzle. (Adding divergent section does not increase subsonic nozzle flow.) ρ = liquid density, a constant A = nozzle exit area
4. Rocket Motor with Subsonic Gas Exhaust 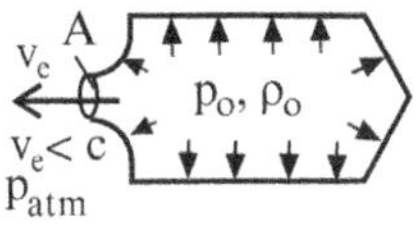 p_{atm} =atmospheric press. A – min exhaust area Compressible gas with subsonic exhaust v_e<c, through convergent nozzle. c = sound speed.	Exit Velocity $v_e = \left(\frac{2\gamma}{\gamma-1}\frac{p_o}{\rho_o}\right)^{1/2}\left[1 = \left(\frac{p_{atm}}{p_o}\right)^{(\gamma-1)/\gamma}\right]^{1/2}$ compressible ideal gas γ = ratio of specific heats; γ ≈ 1.4 for air Mass Flow $\frac{dM}{dt} = \left(\frac{2\gamma}{\gamma-1} p_o \rho_o\right)^{1/2} \left(\frac{p_{atm}}{p_o}\right)^{1/\gamma} \left[1 - \left(\frac{p_{atm}}{p_o}\right)^{(\gamma-1)/\gamma}\right]^{1/2} A$ Thrust $F = v_e \frac{dM}{dt} = p_o A \frac{2\gamma}{\gamma-1}\left(\frac{p_{atm}}{p_o}\right)^{1/\gamma}\left[1 - \left(\frac{p_{atm}}{p_o}\right)^{(\gamma-1)/\gamma}\right]$ nozzle subsonic for $p_o / p_{atm} < ((\gamma+1)/2)^{\gamma/(\gamma-1)}$ nozzle chokes at $p_o = \frac{1}{2} p_{atm} ((\gamma+1)/2)^{\gamma/(\gamma-1)}$, $I_{sp} = v_e / g$ Ref. [19]	

Table 2.4 Rockets and orbits, continued

Particle Motion	Force F, Velocity V, Accel. a	Comments and Special Cases
5. Rocket Motor with Supersonic exhaust, Mach 1 at throat, and divergent exhaust nozzle 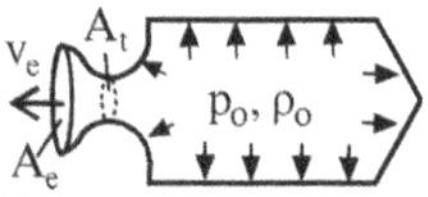 p_{atm} isentropic flow	Exit Velocity $v_e = \left(\frac{2\gamma}{\gamma-1}\frac{p_o}{\rho_o}\right)^{1/2}\left[1-\left(\frac{p_{atm}}{p_o}\right)^{(\gamma-1)/\gamma}\right]^{1/2}$ Mass Flow $dM/dt = (\gamma p_o \rho_o)^{1/2}[2/(\gamma+1)]^{(\gamma+1)/[2(\gamma-1)]}A_t$ Thrust $F = v_e\frac{dM}{dt} = p_o A_t\left(\frac{2\gamma^2}{\gamma-1}\left(\frac{2}{\gamma+1}\right)^{(\gamma+1)/(\gamma-1)}\left[1-\left(\frac{p_{atm}}{p_o}\right)^{(\gamma-1)/\gamma}\right]\right)^{1/2}$ supersonic exhaust $p_o > p_{choke} = p_0\left[(\gamma+1)/2\right]^{\gamma/(\gamma-1)}$, $I_{sp} = v_e/g$ Refs [19–21]	compressible ideal gas γ = ratio of specific heats; $\gamma = 1.4$ for air
6. Force on Variable Mass M(t) 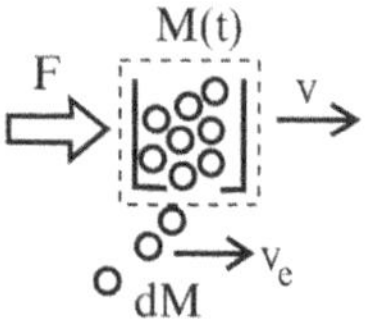	$F = \frac{dM}{dt}(v - v_e) + M\frac{dv}{dt}$ $a = \frac{F}{M} + \frac{v_e - v}{M}\frac{dM}{dt}$ v_e = velocity of mass increment v = velocity of mass M(t) Note system gains mass for dM/dt>0.	$v_e - v$ = velocity of dM relative to v For constant dM/dt: $M_2 = M_1 + (dM/dt)t_2$ and for F=0 and v_e=0 the velocity change between initial and final time, $v_2/v_1 = M_1/M_2$. Ref. [1], p. 168; Ref. [11] see Eq. 2.7.
7. Force on Deformable Variable mass Volume 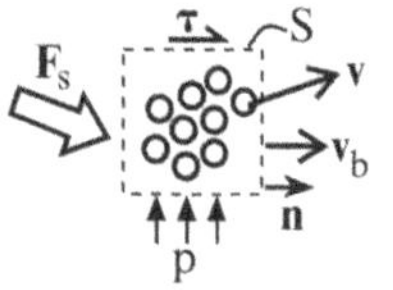 **n** = unit outward normal to control surface S.	$\mathbf{F}_s = -\int_S p\mathbf{n}dS + \int_S \boldsymbol{\tau}dS + \int_V \rho\,\mathbf{g}dV$ $= \frac{d}{dt}\int_V \rho\,\mathbf{v}dV - \int_V \rho\,\mathbf{v}(\mathbf{v} - \mathbf{v}_b)\bullet\mathbf{n}dV$ **F**s =mechanical force vector on volume V. p = pressure on S. ρ =fluid density. **τ** = shear stress. **v** =fluid velocity. **v**b =velocity of S.	Fluid of density ρ has velocity v. Fluid can pass through control surface S that bounds the control volume V. Left hand side is sum of external forces on control volume and right hand side is rate of change of momentum within control volume. Refs [1, 22]
8. Circumferential accel. for Arc of Radius R 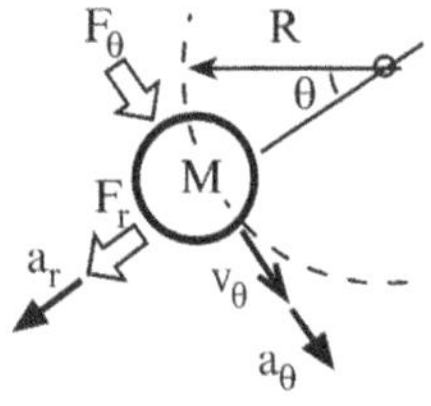	$F_r = -Ma_r,\ F_\theta = Ma_\theta$ Radial accel $a_r = -R(d\theta/dt)^2$ Circum. accl $a_\theta = R\frac{d^2\theta}{dt^2} = \frac{F_\theta}{M}$ $v_\theta = F_\theta t/M + \omega_o R,\ v_r = 0$ $\theta = \frac{F_\theta t^2}{2M} + \omega_o t + \theta_o$	See case 2 of Table 2.1. Angular momentum per unit mass $h = Rd\theta/dt$ Initial condition at t = 0 $d\theta/dt = \omega_o$ $\theta = \theta_o$
9. Gravity Attraction between Two Masses 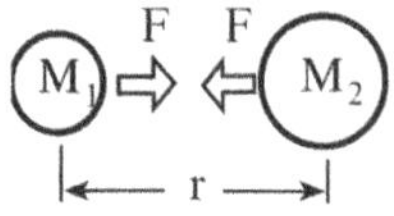	$F = \frac{GM_1M_2}{r^2} = \frac{KM_1}{r^2}$ Acceleration of M_1 due to M_2 $g = K/r^2 = GM_2/r^2$ $K = GM_2$, a constant.	$G = 6.673\times10^{-11}\ \text{N-m}^2/\text{kg}^2$ $= 3.439\times10^{-8}\ \text{ft}^4/\text{lb-s}^4$. $K_{earth} = GM_{earth}$ Ref. [2] $= 3.986\times10^{14}\ \text{m}^3/\text{s}^2$ $= 1.4076\times10^{16}\ \text{ft}^3/\text{s}^2$

Table 2.4 Rockets and orbits, continued

Orbit	Orbit Radius r, velocity V, period T	Comments and Special Cases
10. Earth Trajectory with Initial Velocity v_o and Angle β 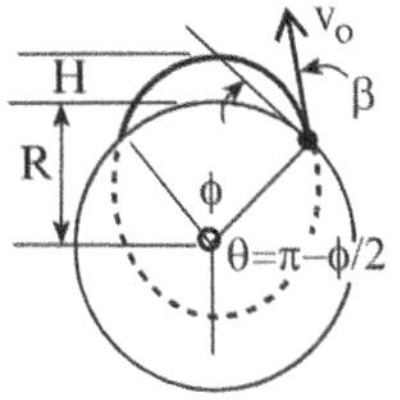 	Given v_o, R, and β, then height above earth is a function of φ, $\frac{H}{R} = \frac{\varepsilon}{1-\varepsilon}\left(1-\cos\frac{\phi}{2}\right)$ $\varepsilon^2 = \left(\frac{Rv_o^2}{K}-1\right)^2\cos^2\beta+\sin^2\beta$ $\tan\frac{\phi}{2} = \frac{(Rv_o^2/K)\sin\beta\cos\beta}{(Rv_o^2/K)\cos^2\beta-1}.$	Time of flight Ref. [5], p. 93 $t = \frac{2(H+R)^{2/3}}{K^{1/2}(1+\varepsilon)^{2/3}}$ $\left[\pi-\left\{2\tan^{-1}\left(\left(\frac{1-\varepsilon}{1+\varepsilon}\right)^{1/2}\tan\frac{\theta}{2}\right)\right.\right.$ $\left.\left.-\frac{\varepsilon\sqrt{1-\varepsilon^2}\sin\theta}{1+\varepsilon\cos\theta}\right\}\right]$
11. Satellite Circular Orbit 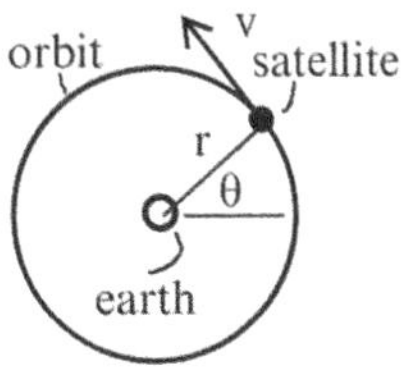 	$v=(K/r)^{1/2}$ $T = \frac{2\pi r^{3/2}}{K^{1/2}} = \frac{2\pi r}{v}$ r = constant Angular velocity $d\theta/dt = K^{1/2}/r^{3/2}$ Energy per unit mass, $e = -K/(2r)$ Potential energy = 0 at r = ∞ and it is negative for finite radius r.	Escape velocity $v_e = (2K/r)^{1/2}$. Satellite escapes orbit if $v>v_e$ Escape from surface of earth, $r_{earth} = 6.373\times10^6$ m $(20.90\times10^6$ ft) $v_e = 11185$ m/s (25000 mile/hr). Angular momentum / mass h =r dθ/dt = rv = constant
12. Elliptical Orbit 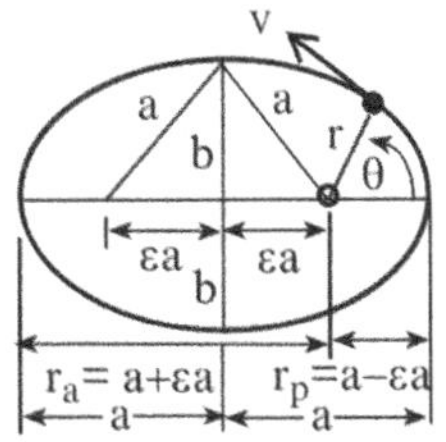 Detail: 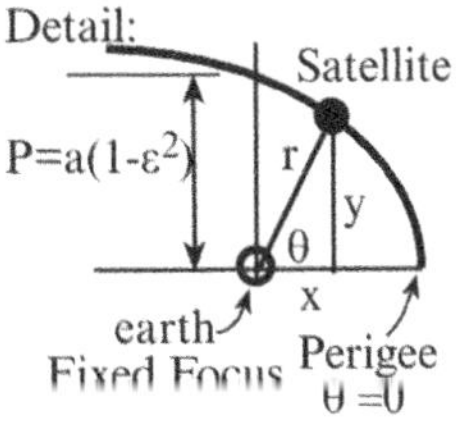 	Angular momentum per unit mass $h = r^2(d\theta/dt) = Ka(1-\varepsilon^2)^{1/2}$ Energy per unit mass $e = v^2/2 - K/r$ Both e and h are constant during orbit $r = \frac{a(1-\varepsilon^2)}{1+\varepsilon\cos\theta} = \frac{h^2/K}{1+\cos\theta\sqrt{1+2\varepsilon h^2/K^2}}$ $T = 2\pi\sqrt{a^3/K}$, for $0\le\varepsilon<1$ Eccentricity of orbit, $\varepsilon = \sqrt{1-b^2/a^2} = (r_a - r_p)/(r_a + r_p)$ $= \sqrt{1+\frac{2eh^2}{K^2}} = \begin{cases}\varepsilon = 0, \text{ circle} \\ 0<\varepsilon\le1, \text{ ellipse}\end{cases}$ $a = \frac{h^2}{K(1-\varepsilon^2)} = \frac{b}{(1-\varepsilon^2)^{1/2}} = \frac{K}{2e}$	Velocity, at apogee and perigee $v = \sqrt{2(e+K/r)} = \sqrt{K(2/r-1/a)}$ Apogee $v_a = \sqrt{K(1-\varepsilon)/[a(1+\varepsilon)]}$ Perigee $v_p = \sqrt{K(1+\varepsilon)/[a(1-\varepsilon)]}$ $dr/dt = \varepsilon h\sin\theta/[a(1-\varepsilon^2)]$ $d\theta/dt = h/r^2$ $dx/dt = -h\sin\theta/[a(1-\varepsilon^2)]$ $dy/dt = h(\varepsilon+\cos\theta)/[a(1-\varepsilon^2)]$ Earth is at fixed focus of ellipse. Potential energy = −K/r is zero at r=∞. Energy e<=0. Cartesian coordinates of satellite, x = r cos θ, y = r sin θ, $(x/a)^2 + (y/b)^2 = 1$ K from case 9. See Figs. 2.5, 2.6 Refs [1, 5, 23].

Table 2.4 Rockets and orbits, continued

Orbit	Orbit Radius r, Velocity V, Period T	Comments and Special Cases
13. Hyperbolic Orbit 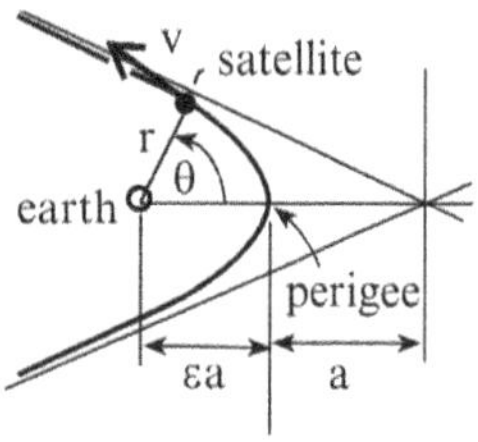	$r=\dfrac{a(\varepsilon^2-1)}{1+\varepsilon\cos\theta}=\dfrac{h^2/K}{1+\cos\theta\sqrt{1+2\dfrac{eh^2}{K^2}}}$ Angular momen. / unit mass $h=r^2\dfrac{d\theta}{dt}$ Energy per unit mass, $e=\dfrac{v^2}{2}-\dfrac{K}{r}>0$ $a=\dfrac{h^2}{K(1-\varepsilon^2)}=\dfrac{K}{2e}$ Both e and h are constant during orbit.	$v=\sqrt{2(e+K/r)}=\sqrt{K(2/r+1/a)}$ $\varepsilon=\sqrt{1+\dfrac{2eh^2}{K^2}}=\begin{cases}1, & \text{parabola}\\ \varepsilon>1, & \text{hyperbola}\end{cases}$ $d\theta/dt=h/r$ $v_{r=\infty}=h/a$ K from case 9 Satellite escapes to infinity, $T=\infty$ Refs [1, 5]. Figs 2.5, 2.6.
14. Two Bodies Spin about their CG 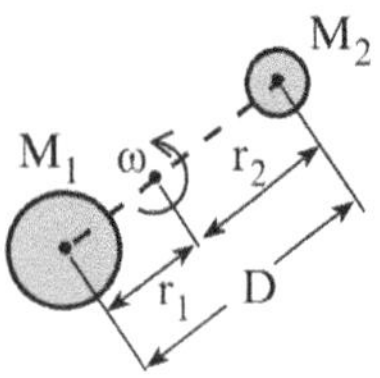	Distance between centers, $D^3=\dfrac{G(M_1+M_2)}{\omega^2}$ ω = angular velocity, rad/s. G = gravitational constant, case 9. Period, $T=\dfrac{2\pi}{\omega}=\left(\dfrac{(2\pi)^2D^3}{G(M_1+M_2)}\right)^{1/2}$	Center of mass and rotation. $r_1=D\dfrac{M_2}{M_1+M_2}$ $r_2=D\dfrac{M_1}{M_1+M_2}$ Gravitational attraction equals centrifugal force as bodies spin about each other Ref. [5], p. 56
15. Hohmann Elliptical Transfer Between Circular Orbits 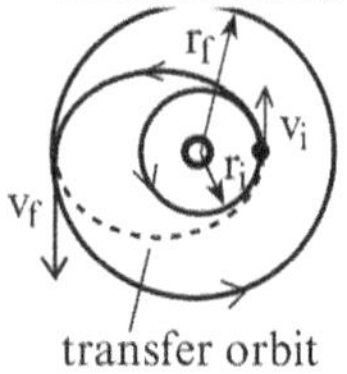	$v_i=\sqrt{\dfrac{K}{r_f}},\ v_f=\sqrt{\dfrac{K}{r_i}}$ Velocity change for transfer to larger circular orbit, $\Delta v_{total}=$ $v_f\left[\left(\dfrac{2r_i/r_f}{1+r_i/r_f}\right)^{1/2}\left(1-\dfrac{r_i}{r_f}\right)+\left(\dfrac{r_i}{r_f}\right)^{1/2}-1\right]$ v_i = initial velocity in circular orbit r_i v_f = final velocity in circular orbit r_f	The Hohmann orbit transfer between a smaller and a larger circular orbit is obtained by two velocity increases. One to enter an elliptical transfer orbit and one to leave it. From required velocity changes, thrust, burn time, and fuel mass are computed. See cases 11 and 12. Ref. [5], pp. 66–68
16. Orbit from Initial Velocity v at angle θ 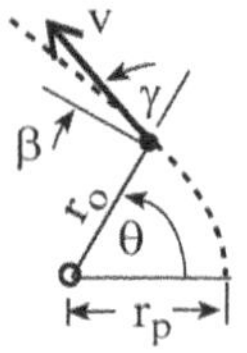	Given, r, β, v angular momentum per unit mass h is constant $h=r_o v\cos\theta$ $\tan\theta=\dfrac{(r_o v^2/K)\sin\theta\cos\theta}{(r_o v^2/K)\cos^2\theta-1}$ $\varepsilon=\left(\dfrac{r_o v^2}{K}-1\right)^2\cos^2\theta+\sin^2\theta$ $r_p=a(1-\varepsilon)=(vr_o\cos\theta)^2/[K(1+\varepsilon)]$ $\gamma=\pi/2-\beta$ = zenith angle	For $\beta=0$, $r_o=r_p$, $\varepsilon=\dfrac{r_o v_o^2}{K}-1$ $a=r_o/(1-\varepsilon)$ $b=r_o\sqrt{(1+\varepsilon)/(1-\varepsilon)}$ Ref. [5], p. 53

retained in the calculation increases. The central difference and the fourth-order (error of order Δt^4) Runge–Kutta explicit time-stepping methods for integrating Newton's second law are shown in Appendix B.

Example 2.2 Calculation of velocity

Black powder is an explosive invented in the seventh century that can develop 10,000 psi (68 MPa) pressure in a confined space. What is the muzzle velocity of a 1 in. (2.54 cm) long lead bullet (density $\rho = 0.5$ lb per cubic inch, 13.83 g/cc; Chapter 8) fired from a black powder musket with a 24 in. (60.96 cm) barrel?

Solution: The gun powder pressure p acts over the cross-sectional area A of the barrel and exerts force pA on the bullet. The acceleration of the bullet down the barrel presented in case 1 of Table 2.3 neglects friction and assumes constant pressure as the gas burns:

$$\text{Acceleration} \quad a = \frac{d^2x}{dt^2} = \frac{F}{M} = \frac{pA}{M} = \text{constant}$$

$$\text{Velocity} \quad v = \int_0^t a dt = \frac{dx}{dt} = \left(\frac{pA}{M}\right) t, \quad \text{with zero initial position at } t = 0$$

$$\text{Position} \quad x = \int_0^t v dt = \left(\frac{1}{2}\right)\left(\frac{pA}{M}\right) t^2, \quad \text{with zero initial position at } t = 0$$

The bullets exit the end of the barrel, $x = L$, at time $t_e = (2ML_{\text{barrel}}/(pA))^{1/2}$ with velocity

$$v_e = v(t_e) = \sqrt{\frac{2pAL_{\text{barrel}}}{M}} = \sqrt{\frac{2p}{\rho_{\text{bullet}}}\frac{L_{\text{barrel}}}{L_{\text{bullet}}}}$$

The bullet mass, $M = \rho(\pi D^2 4)L_{\text{bullet}}$, and the area of the barrel, $A = \pi D^2 4$, where D is the barrel diameter, have been substituted in the second expression.

Numerical results are computed with US and cgs units provided in cases 2 and 7 of Table 1.2. Conversion factors are shown in Table 1.4:

$$v_e = \sqrt{\frac{2 \times 10{,}000 \text{ lb/in.}^2}{0.5 \text{ lb/in.}^3/386.1 \text{ in./s}^2}\frac{24 \text{ in.}}{1 \text{ in.}}} = \sqrt{\frac{2 \times 68.947 \times 10^7 \text{ dyne/cm}^2}{13.84 \text{ g/cm}^3}\frac{60.96 \text{ cm}}{2.54 \text{ cm}}}$$

$$= 19{,}250 \text{ in./s} = 1600 \text{ ft/s} = 48{,}900 \text{ cm/s} = 489 \text{ m/s}$$

The exit velocity of the bullet is predicted to be 1600 ft/s (489 m/s). Measured bullet velocities are reduced by barrel friction.

2.2.4 *Work and Energy*

Work and energy are scalars. Work is the force on a particle times the displacement of the particle in the direction of the force. Work increases kinetic energy. This is shown by taking the vector dot product (Eq. 1.25) of Newton's second law (Eq. 2.4) with a vector increment of particle displacement dr and integrating along the path between points A and B [1] (p. 81):

$$\text{Work} = \int_A^B \mathbf{F} \bullet d\mathbf{r} = \int_A^B M\frac{d^2\mathbf{r}}{dt^2} \bullet d\mathbf{r} = \int_A^B M\left(\frac{1}{2}\frac{d}{dt}\left[\frac{d\mathbf{r}}{dt} \bullet \frac{d\mathbf{r}}{dt}\right]\right)dt = \frac{1}{2}M(V_B^2 - V_A^2)$$

$$\text{Potential energy} = -\text{work} = -\int_A^B \mathbf{F}\ d\mathbf{r} \tag{2.10}$$

$$\text{Kinetic energy} = \frac{1}{2}MV^2$$

$$\text{Power} = \mathbf{F} \bullet \left(\frac{d\mathbf{r}}{dt}\right)$$

The magnitude of velocity can be expressed as a vector dot product (Eq. 1.25), $V = |dr/dt| = (dr/dt \bullet dr\,/\,dt)^{1/2}$. Work can also produce other forms of energy, such as potential energy, which is stored work or heat, so a more general statement is that work on a system increases system energy.

Consider that gravity acts on mass M that is suspended on a spring with spring constant k. The difference in energy as the system moves from position A to B is equal to the work performed by gravity. The displacement x and height h are positive upward, and gravity acts downward:

$$\begin{aligned} \text{Work}|_A^B &= \mathbf{F} \bullet d\mathbf{r} = \text{Energy}_B - \text{energy}_A \\ Mg(h_A - h_B) &= \left(1/2MV_B^2 + 1/2kX_B + Mgh_B\right) - \left(1/2MV_A^2 + 1/2kX_A + Mgh_A\right) \end{aligned} \tag{2.11}$$

$(1/2)\ kX^2$ is the energy stored in the spring. Since Mgh is the potential energy of gravity, then this can be rewritten as conservation of total energy between positions A and B:

$$\begin{aligned} \text{Total energy}_A &= \text{Total energy}_B \\ 1/2MV_A^2 + Mgh_A + 1/2kX_A^2 &= 1/2MV_B^2 + Mgh_B + 1/2kX_B^2 \end{aligned} \tag{2.12}$$

Conservation of energy and *conservation of momentum* are major tools for determining particle trajectories (see Examples 2.2, 2.3, and 2.4). Hamilton's principle uses energy to develop the equations of motion [1, 5].

Example 2.3 Mass impacts spring

Force F acts on the mass M with velocity V_o (Figure 2.2a and b) as the mass strikes an uncompressed spring with spring coefficient k. What is the maximum compressive force in the spring?

Solution: Energy equation (Eq. 2.12) is applied between the initial contact and the maximum spring compression. The spring displacement is x. The energy stored in the spring when the mass comes to a stop is the sum of the work that the force F does on the spring, $F\,x_{max}$, as the spring compresses and the initial kinetic energy of the mass, $1/2MV_o^2$:

$$\int_0^{x_{max}} kx dx = \frac{1}{2}kx_{max}^2 = Fx_{max} + \frac{1}{2}MV_o^2$$

Solving for the maximum displacement,

$$x_{\max} = \frac{1}{k}\left(F + \sqrt{F^2 + kMV_o^2}\right)$$

$$F_{\max} = kx_{\max} = F + \sqrt{F^2 + kMV_o^2}$$

Initial kinetic energy increases both the deformation and maximum force.

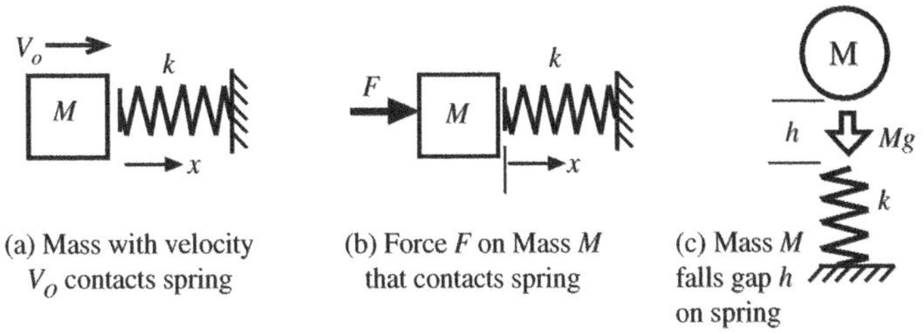

Figure 2.2 Three cases of mass impact against an uncompressed spring

Now, consider Figure 2.2c. A mass falls through gap h under gravity. Its velocity is $V_o = (2gh)1/2$ when it contacts and compresses a spring. This velocity is substituted into the previous equation, which is divided by the static deflection, Mg/k:

$$\frac{F_{\max}}{Mg} = \frac{x_{\max}}{\delta_s} = 1 + \sqrt{1 + \frac{2h}{\delta_s}}, \quad \text{where } \delta_s = \frac{Mg}{k} \tag{2.13}$$

Dynamic force amplification factor, on the right, is the ratio of the maximum spring force on the particle to the weight of the mass or equivalently the maximum spring displacement divided by the 1 g static spring displacement.

2.2.5 *Impulse*

Newton's second law (Eq. 2.4) is applied in differential form when force varies in time:

$$\mathbf{F}dt = Md\mathbf{v} \tag{2.14}$$

Force times dt is the increase in momentum. Integrating between t_1 and t_2 gives impulse

$$\text{Impulse} = \int_{t_1}^{t_2} \mathbf{F}dt = M(\mathbf{v}_2 - \mathbf{v}_1) \tag{2.15}$$

When the impulse is known, the details of the force time history are not required to compute changes in momentum. Impulses are applied in cases 25–28 of Table 2.3.

Collisions are short-term events dominated by particle interactions. If external forces, such as gravity or air resistance, have negligible effect during a collision, then the momentum (Eq. 2.5) and the velocity of the center of mass of the sum of the particles are constant during the collision.

Additional relations describing the impact are needed to uniquely define the post collision trajectories. The most common assumptions are:

1. During perfectly elastic collisions, energy is conserved. Restitution coefficient is unity. The post collision relative normal component of velocity, along a line between the centers of the particles, is the negative of its precollision value. No energy is lost.
2. After inelastic collisions, the particles stick together and/or depart with the same velocity. The restitution coefficient is zero. Energy is lost.
3. A common assumption is that tangent component velocities, perpendicular to a line between particle centers, do not change during the impact.

Restitution coefficient is the energy lost during the collision divided by the initial energy in a coordinate frame moving with the center of gravity:

$$e = |V_1 - V_2|/|V_{1o} - V_{2o}| \tag{2.16}$$

Particle 1 with mass M_1 has velocity V_{1o} before the collision and velocity V_1 after the collision. For Particle 2 with mass M_2, these are V_{2o} and V_2. No external forces are applied. Momentum (Eq. 2.5) is conserved, which implies the center of mass of the two particles has constant velocity, V_c, during the collision:

$$V_c = \frac{(M_1 V_1 - M_2 V_2)}{(M_1 + M_2)} = \frac{(M_1 V_{1o} - M_2 V_{2o})}{(M_1 + M_2)} \tag{2.17}$$

The particle velocities relative to the center of mass are $V_1 - V_c = M_2(V_1 - V_2)/(M_1 + M_2)$ and $V_2 - V_c = M_1(V_2 - V_1)/(M_1 + M_2)$. Using these, it is easy to show that the ratio of the total kinetic energy, measured relative to the moving center of mass, before the collision to after the collision is proportional to the square of the coefficient of restitution and independent of mass:

$$e^2 = \frac{\text{KE}_{\text{rel cg, after}}}{\text{KE}_{\text{rel cg, before}}} = \frac{M_1(V_1 - V_c)^2 + M_2(V_2 - V_c)^2}{M_1(V_{10} - V_c)^2 + M_2(V_{2o} - V_c)^2} = \frac{(V_1 - V_2)^2}{(V_{1o} - V_{2o})^2} \tag{2.18}$$

If $e = 1$, the collision is elastic, and kinetic energy is conserved, that is, constant, during the collision. The relative velocity $V_2 - V_1$ between the departing particles is maximized. If $e = 0$, the collision is inelastic, kinetic energy is lost during the collision, and both particles leave the collision at the same velocity, $V_1 = V_2$ (see cases 24–28 of Table 2.3).

Example 2.4 Two-car collision

Who has not been in a collision? Car 1 with velocity V_o and mass M_1 strikes stationary car 2 with mass M_2, which has its brakes on (Figure 2.3). The two cars then slide together distance d against road friction before stopping. The cars leave skid marks of length d. What was the initial velocity V_o of car 1?

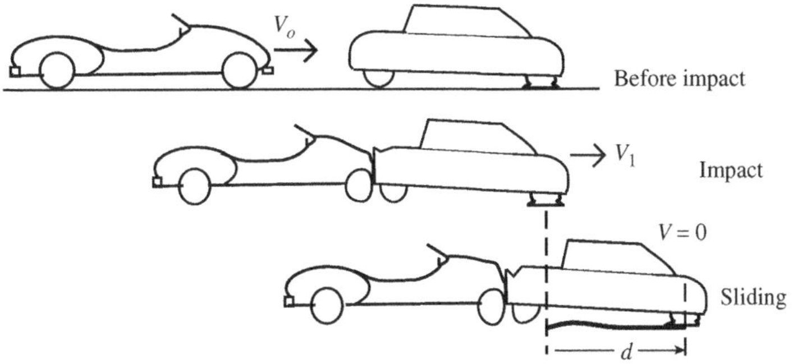

Figure 2.3 Car with velocity V_o strikes a stationary car. They slide together distance d before stopping.

Solution: The solution is in two steps: first the collision and second the slide. Momentum is conserved during the collision (Eq. 2.5). The momentum of the first car before the collision equals the momentum of the two cars just after the collision. This momentum equation is solved for the initial sliding velocity V_1:

$$M_1 V_o = (M_1 + M_2)V_1; \quad \text{hence,} \quad V_1 = \frac{V_o M_1}{(M_1 + M_2)}$$

The initial kinetic energy of the cars as they slide together equals the work done against the road friction force $F = \mu M_2 g$. This energy equation is solved for V_o:

$$\mu M_2 g d = {}^1\!/_2 M_1 V_1^2 = {}^1\!/_2 M_1 [V_o M_1/(M_1 + M_2)]^2; \text{ hence,}$$

$$V_o = \frac{M_1 + M_2}{M_1} \sqrt{2\mu g d \frac{M_2}{M_1}}$$

A dimensionless friction coefficient μ of 0.6 is often assumed for a dry road and 0.1 for wet road [15]. If both cars have wheels locked during the slide, then the sliding distance is shorter, and there are two sets of skid marks.

Example 2.5 Golf club

A golfer strikes a golf ball with a golf club. The face of the club flexes as the ball springs off the club. The United States Golf Association (USGA) recognized the potential for thin-faced metallic clubs to have coefficients of restitution (Eq. 2.18) as high as 0.9, whereas traditional solid persimmon wood clubs (*woods*) have a coefficient of restitution of 0.78. In 1981, the USGA limited the coefficient of restitution of driver clubs to 0.83. Calculate the initial speed of balls hit with clubs with coefficients of restitution of 0.78, 0.83, and 1. Assume club head speed for a good amateur golfer of 100 mph (147 ft/s, 44.81 m/s).

Solution: The initial velocities of the ball (M_2) and the club head (M_1) are $V_{2_o} = 0$ and $V_{1_o} = 147$ ft/s (44 m/s), respectively. Since the club mass is 5 to 10 times greater than the ball mass, ċase 25 of Table 2.3 shows that the speed of the ball off the club exceeds the club speed by one plus the restitution coefficient (e):

$$V_2 = \frac{(1+e)M_1 V_{1_o}}{M_1 + M_2} \approx (1+e)V_{1_o} = \begin{cases} 79.8 \text{ m/s} \ \ (262 \text{ ft/s}), & e = 0.78 \\ 82.0 \text{ m/s} \ \ (269 \text{ ft/s}), & e = 0.83 \\ 89.6 \text{ m/s} \ \ (294 \text{ ft/s}), & e = 1.0 \end{cases}$$

The maximum possible golf ball speed ($e = 1$) is twice the club head speed. There is significant advantage to a club with high coefficient of restitution.

2.2.6 *Armor*

The solid particle shown in Figure 2.4 strikes and penetrates a stationary plate. The work (Eq. 2.11) required to punch a hole through a plate is the shear force of deformation times the plate thickness. For moderate velocity impacts, shear force is modeled by the dynamic shear stress τ times the fragment perimeter P times the plate thickness h. The change in kinetic energy of the particle is equal to the work done by the particle as it penetrates the plate:

$$\tfrac{1}{2}MV_0^2\cos^2\theta - \tfrac{1}{2}MV_2^2\cos^2\theta = h^2 P\tau \tag{2.19}$$

M is the particle mass, and V_0 is its initial velocity with angle θ from normal to the plate face. V_2 is its exit velocity. The equation is solved for the final particle velocity and the minimum thickness $h_{\min}$ that stops the particle:

$$\begin{aligned} &V_2 \cos\theta - \sqrt{V_0^2\cos^2\theta - 2P\tau h^2/M}, \quad \text{if } h < h_{\min}; \ \text{otherwise,} \ V_2 = 0 \\ &\text{Thickness at threshold of penetration,} \ h_{\min} = \sqrt{\frac{MV_0^2\cos^2\theta}{2P\tau}} \end{aligned} \tag{2.20}$$

Dynamic shear stress for medium velocity impacts is indicated in case 24 of Table 2.3. The potential for penetration is greatest if the particle impacts normal to the plate with its

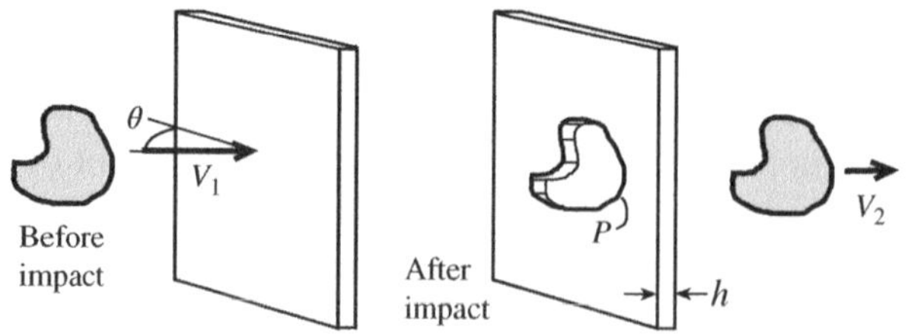

Figure 2.4 Irregular object with perimeter P and initial velocity V_o strikes and penetrates a stationary plate. V_2 is its exit velocity

smallest perimeter toward the plate. If the cutout piece of the plate and exiting particle leave together, then the final velocity is further reduced by the ratio $M/(M + M_{\text{plate}})$, where M is the particle mass and M_{plate} is the mass of the cut piece of the plate (case 25 of Table 2.3).

Rocket Thrust and Orbits (Table 2.4; [1, 2, 5, 19–24]). Static thrust of a rocket motor (Eq. 2.7; [1], p. 241; [21]) is the sum of (1) the change in momentum of fuel mass dM as it burns and accelerates through the rocket nozzle and (2) the net pressure force of exhaust gases on nozzle area A_e:

$$F = v_e \frac{dM}{dt} + (p_e - p_{\text{atm}})A_e \tag{2.21}$$

High gas pressure inside a solid rocket motor case, typically 35–70 atm, propels hot exhaust gases through the engine nozzle at sonic velocity [19, 20]. Once past the throat, the thrust and exhaust velocity are maximized by expanding exhaust gas supersonically through a divergent exhaust nozzle that is sized so that nozzle exit static pressure equals the surrounding atmospheric pressure, $p_e = p_{\text{atm}}$. Equation 2.21 shows that if fuel is burned at a constant rate dM/dt and exhausted at a constant velocity v_e relative to the rocket and the exhaust gas pressure relative to the atmosphere is constant, then the rocket thrust is constant and independent of the rocket velocity.

Specific impulse I_{sp} is the thrust divided by the initial weight of fuel times the time in seconds that the fuel burns [22]:

$$I_{\text{sp}} = \frac{\int_0^t F\,dt}{\int_0^t (g_c\ dM/dt)dt} \approx \frac{F}{(g_c\ dM/dt)} \approx \frac{v_e}{g_c} \tag{2.22}$$

I_{sp} has units of seconds; g_c is the acceleration due to gravity (Table 1.2). The specific impulse (I_{sp}) of ammonia perchlorate oxidizer burning asphalt or rubber is 170–210 s [21]. Oxygen burning hydrogen has a specific impulse of 450 s [20] owing to its high sonic velocity. The maximum specific impulse is limited to the sonic velocity of exhaust gases at the nozzle throat.

The differential equation of motion of a rocket with an optimum nozzle is obtained by applying Equations 2.7 and 2.21 with $p_e = p_{\text{atm}}$:

$$\frac{MdV}{dt} = v_e \left(\frac{dM}{dt}\right) \tag{2.23}$$

M is the mass of the rocket including unburned fuel, V is the rocket velocity vector, and v_e is the exhaust gas velocity vector relative to the rocket. The solution is provided in case 5 of Table 2.5. The vector dot product (Eq. 1.25) of this equation with the rocket velocity vector is Oberth's equation [26] for rocket kinetic energy, KE = ½ $MV \bullet V$:

$$\frac{d\text{KE}}{dM} = \mathbf{v}_e \bullet \mathbf{V} \tag{2.24}$$

The increase in rocket kinetic energy per mass of fuel burned is maximized by maximizing the exhaust velocity ($\mathbf{v}_e$) and minimizing turns.

Table 2.5 Rigid body rotation theory

Notation: $\mathbf{a}$ = acceleration; C = center of mass; $\mathbf{F}$ = force vector; F_x, F_y, F_z = forces on body; $\mathbf{H}$ = angular momentum vector; J = mass moment of inertia in x, y, z axes about body-fixed origin O′ Tables 1.6, 1.7; M, dm = mass; $\boldsymbol{M}$ = vector moment on body; O′ = body-fixed origin, not necessarily the center of mass; O = inertial (fixed) origin; R = radius from inertial origin O; r = radius from body-fixed origin O′; t = time; x, y, z = body-fixed coordinates with origin at O′; X, Y, Z = inertial coordinates rel. O; $\boldsymbol{\omega}$ = angular velocity vector; × = vector cross product Eq. 1.25. Consistent units are shown in Table 1.2.

Rotation Case	Velocity v and Acceleration a, Moment M and Force F
1. Velocity of Rigid Body Rotation 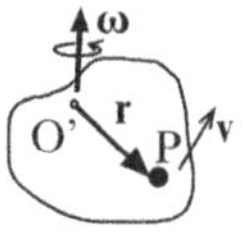	$\frac{d\mathbf{r}}{dt} = \mathbf{v} = \boldsymbol{\omega} \times \mathbf{r}$ = kinematic velocity due to rigid body rotation Velocity of point P in body with respect to body-fixed origin O' is cross production of fixed length vector **r** between O' and P with the body's angular velocity vector $\boldsymbol{\omega}$ through O'. Ref. [1]
2. Point on Rotating and Translating Rigid Body 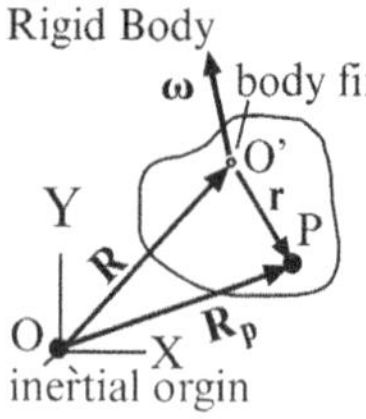	$\mathbf{R}_p = \mathbf{R} + \mathbf{r}$ **r** is fixed length vector Ref. [1] $\frac{d\mathbf{R}_p}{dt} = \frac{d\mathbf{R}}{dt} + \boldsymbol{\omega} \times \mathbf{r}$, note: $\frac{d(\boldsymbol{\omega} \times \mathbf{r})}{dt} = \frac{d\boldsymbol{\omega}}{dt} \times \mathbf{r} + \boldsymbol{\omega} \times (\boldsymbol{\omega} \times \mathbf{r})$ $\frac{d^2\mathbf{R}_p}{dt^2} = \frac{d^2\mathbf{R}}{dt^2} + \frac{d\boldsymbol{\omega}}{dt} \times \mathbf{r} + \boldsymbol{\omega} \times (\boldsymbol{\omega} \times \mathbf{r})$ R_p is differentiated in inertial X,Y, Z frame with origin at O to enable application of Newton's 2nd law in case 3.
3. Force and Moment in inertial coordinates on mass element dm 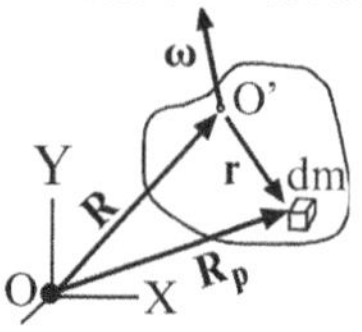	$dF = (dm)\frac{d^2\mathbf{R}_p}{dt^2} = \left(\frac{d^2\mathbf{R}}{dt^2} + \frac{d\boldsymbol{\omega}}{dt} \times \mathbf{r} + \boldsymbol{\omega} \times (\boldsymbol{\omega} \times \mathbf{r})\right) dm$ $d\boldsymbol{M} = \mathbf{r} \times dF = \mathbf{r} \times \left(\frac{d^2\mathbf{R}}{dt^2} + \frac{d\boldsymbol{\omega}}{dt} \times \mathbf{r} + \boldsymbol{\omega} \times (\boldsymbol{\omega} \times \mathbf{r})\right) dm$ Vector **r** from origin O' to dm. Integrating over body gives the force **F** at O′ and moment $\boldsymbol{M}$ acting about O′ on the rotating body, below.
4. Force and Moment on a Rotating and Translating Rigid Body 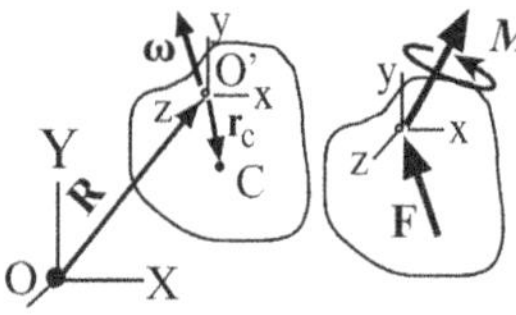	$\mathbf{F} = M\mathbf{a}_o + M\frac{d(\boldsymbol{\omega} \times \mathbf{r}_c)}{dt} = M(\mathbf{a}_o + \frac{d\boldsymbol{\omega}}{dt} \times \mathbf{r}_c + \boldsymbol{\omega} \times (\boldsymbol{\omega} \times \mathbf{r}_c)), \mathbf{a}_o = \frac{d^2\mathbf{R}}{dt^2}$ $\boldsymbol{M} = M\mathbf{r}_c \times \mathbf{a}_o + d\mathbf{H}/dt$, $\mathbf{r}_c, \mathbf{r}, \boldsymbol{\omega}$: instantaneous positions $\mathbf{H} = \int_M \boldsymbol{\omega} \times (\boldsymbol{\omega} \times \mathbf{r}) dm = [J]\,\boldsymbol{\omega}$, relative to inertial X,Y,Z $\mathbf{r}_c = \int_M \mathbf{r}\, dm / M$, $M = \int_M dm$. **F** and $\boldsymbol{M}$ written out below.

5. Euler's Equations. x,y,z are instantaneous positions of body-fixed coordinates in inertial X,Y,Z.

$F_x = M(a_{ox} - y_c d\omega_z/dt + z_c d\omega_y/dt - x_c\omega_y^2 - x_c\omega_z^2 + y_c\omega_x\omega_y + z_c\omega_x\omega_z)$, Fig 2.7. Ref. [7]

$F_y = M(a_{oy} + x_c d\omega_z/dt - z_c d\omega_x/dt - y_c\omega_x^2 - y_c\omega_z^2 + x_c\omega_x\omega_y + z_c\omega_y\omega_z)$, J about O'. a_o is accleration

$F_z = M(a_{oz} - x_c d\omega_y/dt + y_c d\omega_x/dt - z_c\omega_y^2 - z_c\omega_x^2 + y_c\omega_y\omega_z + x_c\omega_x\omega_z)$, of O' in inertial coordinates

$M_x = M(a_{oz}y_c - a_{oy}z_c) + J_x d\omega_x/dt - J_{xy}(d\omega_y/dt - \omega_x\omega_z) - J_{xz}(d\omega_z/dt + \omega_x\omega_y) + (J_z - J_y)\omega_y\omega_z - J_{yz}(\omega_y^2 - \omega_z^2)$

$M_y = M(a_{ox}z_c - a_{oz}x_c) + J_y d\omega_y/dt - J_{xy}(d\omega_x/dt + \omega_y\omega_z) - J_{yz}(d\omega_z/dt - \omega_x\omega_y) + (J_x - J_z)\omega_x\omega_z - J_{xz}(\omega_z^2 - \omega_x^2)$

$M_z = M(a_{oy}x_c - a_{ox}y_c) + J_z d\omega_z/dt - J_{xz}(d\omega_x/dt - \omega_y\omega_z) - J_{yz}(d\omega_y/dt + \omega_x\omega_z) + (J_y - J_x)\omega_x\omega_y - J_{xy}(\omega_x^2 - \omega_y^2)$

2.2.7 Gravitation and Orbits

If the integral of $\mathbf{F} \bullet d\mathbf{r}$ (the work of Eq. 2.11) is zero about any closed path, then the force is described as a conservative force because there is no net energy dissipation, or work, around the path. The most important conservative force is gravity. Newton's law of gravitational attraction between two particle masses, M_1 and M_2, with distance r between their centers is

$$F = -\frac{G\ M_1 M_2}{r^2} \tag{2.25}$$

G is the universal gravitational constant $G = 6.673 \times 10^{-11}\ N\text{-m}^2/\text{kg}^2 (3.439 \times 10^{-8}\ \text{ft}^4/\text{lb} - s^4)$ [27]. Consider that mass M_1 is fixed in space. The gravitational attraction it exerts on M_2 results in gravitational acceleration g of M_2 toward M_1:

$$g = \frac{F}{M_2} = -G\left(\frac{M_1}{r^2}\right) = -\frac{K}{r^2} \tag{2.26}$$

$K = gr^2 = GM_1$ is a measure of the gravitation force of a mass. The minus signs denote gravitation is an attractive force.

Gravitational attraction provides the centripetal acceleration required to maintain a satellite in orbit. The equations that describe radial and tangential orbital motions in the absence of atmospheric drag are obtained by setting the vector force required for radial acceleration shown in case 3 of Table 2.1 equal to the gravitation force (Eq. 2.25) and setting the tangential force to zero:

$$\text{Radial force:}\ \frac{d^2r}{dt^2} - r\left(\frac{d\theta}{dt}\right)^2 = -\frac{K}{r^2},\quad \text{tangential force:}\ r\frac{d^2\theta}{dt^2} + 2\frac{dr}{dt}\frac{d\theta}{dt} = 0 \tag{2.27}$$

The second equation is equivalent to $d(r^2 d\theta/dt)/dt = 0$, which implies Kepler's second law: the angular momentum of the satellite about a planet, $r^2 d\theta/dt$, is constant. By setting d^2r/dt^2 to zero in the first equation, the velocity of satellites in circular orbits with radius r is determined:

$$v_c = \frac{rd\theta}{dt} = \left(\frac{K}{r}\right)^{1/2} \tag{2.28}$$

For example, the earth has a mass of $5.98E + 24$ kg, a radius of 6,380,000 m, and $K = 3.99E + 14\ \text{m}^2/\text{s}^2$. The escape velocity for the earth is 11,200 m/s, and the velocity of a tree top orbit on the earth, neglecting air resistance, is $v_c = 7910\ \text{m/s}$ [23].

Circular orbits are produced if the satellite velocity vector is tangent to the earth with the velocity given by Equation 2.28. Otherwise, the orbit is elliptical, parabolic, or hyperbolic (see Figures 2.5 and 2.6, cases 11 through 13 of Table 2.4, and Refs [1, 5, 23, 24, 26]). The gravitational mass is at a focus of the elliptical orbit, which is Kepler's first law. The velocity required to escape to deep space, starting at radius r, is $2^{1/2}$ times the circular orbit velocity (Eq. 2.28); satellites in parabolic and hyperbolic orbits escape to infinity. Satellites in elliptical orbits have energies between the escape velocity and the circular orbit velocity.

Atmospheric drag is neglected in this analysis. The atmospheric drag slows satellites orbiting the earth, and their orbits decay inward. The International Space Station orbiting about 350 km (220 miles) above the earth loses about 65 m (200 ft) of altitude per day due

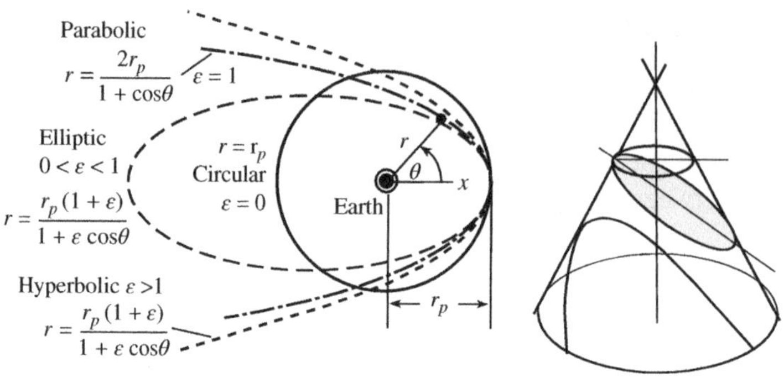

Figure 2.5 Circular, elliptical, parabolic, and hyperbolic orbits are conic sections

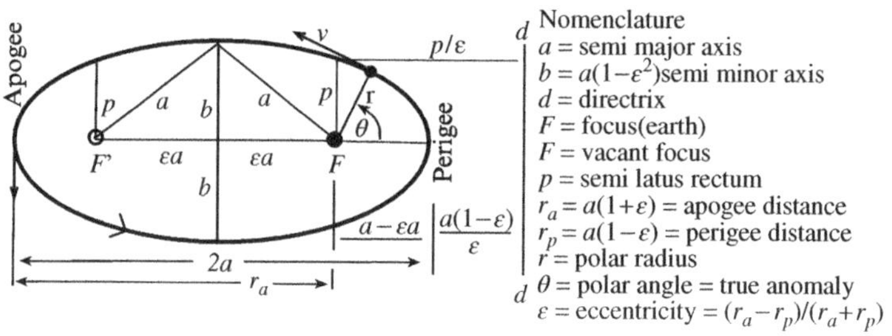

Figure 2.6 Elliptical orbit nomenclature

to atmospheric drag. The entry point to the top of the atmosphere is often considered to be 122 km (400,000 ft, 75 miles) above the earth where some limited aerodynamic control of vehicles becomes possible. Griffin and French [27] discuss atmospheric drag reentry calculations.

Example 2.6 Geosynchronous orbit

What orbit places a satellite at a fixed point in the sky?

Solution: Geosynchronous orbits pass over a given point once per sidereal day, 86,164 s (23.93 h), which is 4 min shorter than the 24 h solar day because of the earth's daily advance in its orbit about the sun (Figure 2.1c). Case 11 of Table 2.4 is solved for the radius of a circular geosynchronous orbit with the period of a sidereal day, $K = 3.986E14\ \mathrm{m}^3/\mathrm{s}^2 (1.408E16\mathrm{ft}^3/\mathrm{s}^2)$:

$$r = \left(\frac{TK^{1/2}}{4\pi^2}\right)^{1/3} = 42{,}136\ \mathrm{km}\,(26{,}198\ \mathrm{miles})$$

The geosynchronous orbit radius is 35,786 km (22,236 miles), about 6 diameters, above the surface of the earth. Round trip communication to a geosynchronous satellite takes about one-half second at the speed of light, 299E6 m/s (981E6 ft/s).

Geosynchronous satellites over the equator (0° inclination) are geostationary: they remain over the same point on the equator. It is not possible to have a geostationary satellite, say, above London because a focus of an earth orbit must be the center of the earth.

2.3 Rigid Body Rotation

2.3.1 Rigid Body Rotation Theory

Each point P in a rotating body (Fig. 2.7a) has an instantaneous velocity vector **v** that is the cross product of its position vector **r** with the angular velocity vector **ω**, which is along the axis of rotation:

$$\mathbf{v} = \boldsymbol{\omega} \times \mathbf{r}$$

Where × is the vector cross product (case 1 of Table 2.5, [1, 7]). Newton's second law (Eq. 2.4) is applied to rigid bodies by differentiating **v** in an inertial frame to obtain acceleration a and then integrating force $dF = a(dM)$ on mass element dM over the body. Force F equals mass times the acceleration of the center of mass of the body. Torque, or moment, on the body equals rate of change of angular momentum $\boldsymbol{T} = \mathbf{r} \times \mathbf{F} = d\mathbf{H}/dt$ (cases 4 and 5 of Table 2.5). Solutions for single-axis rotation are shown in Table 2.6 and multiaxis rotation in Table 2.7.

2.3.2 Single-Axis Rotation

Table 2.6 shows the rotation about a single fixed axis [1, 5, 16, 17]. Torque (or moment) applied about the axis of rotation (axle) causes angular acceleration of a rotor:

$$T = J\frac{d^2\theta}{dt^2} \tag{2.29}$$

θ is the angular position in radians. J is the polar mass moment of inertia of the rotor (Tables 1.6 and 1.7) about its axis of rotation between collinear bearings. Consistent sets of units

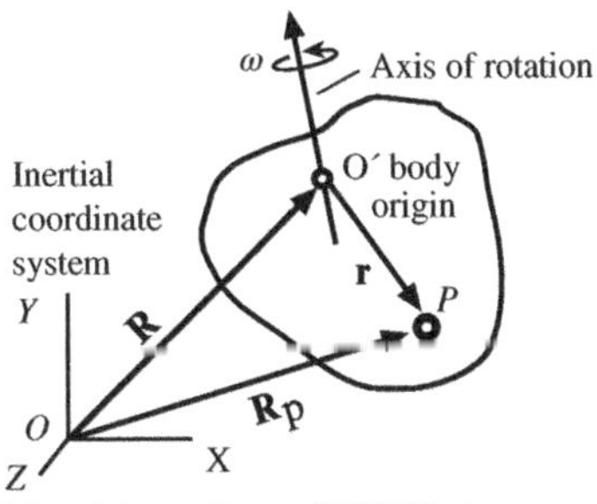

(a) Inertial coordinates X,Y,Z. Vector **R** to the body fixed origin O'.The Body rotates about the vector ω.

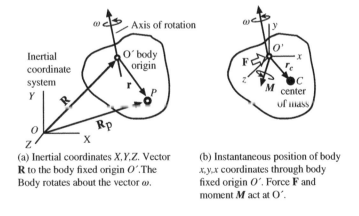

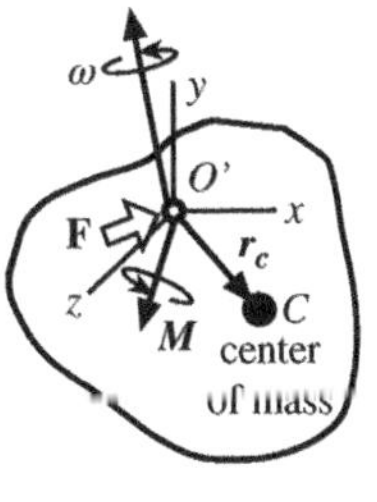

(b) Instantaneous position of body x,y,x coordinates through body fixed origin O'. Force **F** and moment $\boldsymbol{M}$ act at O´.

Figure 2.7 Inertial coordinate system x–y–z is instantaneously fixed in the rotating body on the axis of rotation

Table 2.6 Rotation about single axis

Notation: C = center of mass, F_x = force on shaft in lateral direction; F_y = force on shaft in vertical direction; g = acceleration due to gravity in −y direction; k = spring force/deflection; J = polar mass moment of inertial about axis of rotation Tables 1.6 and 1.7; M, m = mass; r, R = radius from pivot; t = time; T, M = torque or moment on body; x = lateral position with respect to pivot; y = vertical position with respect to pivot; θ = angular position, radian; dθ/dt = rotational velocity, radian/s; $d^2\theta/dt^2$ = angular acceleration, rad/s^2; ω = angular velocity, rad/s. Consistent sets of units are in Table 1.2. Bodies are rigid Refs [1, 5, 16, 17].

Rotational Motion	Torque T, Rotation Angle θ	Relationships at Time t
1. Torque on Rotor 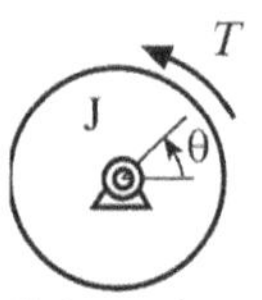 J about pivot	$T = J\frac{d^2\theta}{dt^2} = constant$ $\frac{d\theta}{dt} = \frac{T}{J}t$ $\theta = \frac{T}{2J}t^2$ Initial conditions at t=0: θ=dθ/dt=0.	$t = d\theta/dt \,/\, d^2\theta/dt^2$ $= \sqrt{2\theta / d^2\theta/dt^2}$ $\theta = (d\theta/dt)^2 / d^2\theta/dt^2$ $\frac{d\theta}{dt} = \sqrt{2\theta\frac{d^2\theta}{dt^2}}$ $\frac{d^2\theta}{dt^2} = \frac{1}{t}\frac{d\theta}{dt} = \frac{T}{J}$
2. Constant Torque on Rotor and initial velocity 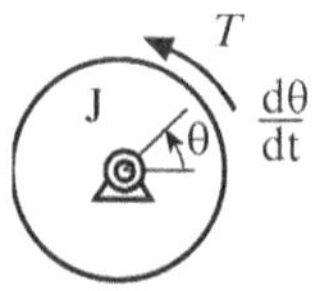 J about pivot	$T = J\frac{d^2\theta}{dt^2} = constant$ $\frac{d\theta}{dt} = \frac{T}{J}t + \left.\frac{d\theta}{dt}\right\|_{t=0}$ $\theta = \frac{T}{2J}t^2 + \left.\frac{d\theta}{dt}\right\|_{t=0} t$ at t=0: $\theta = 0, d\theta_o/dt$	$t = \frac{dt^2}{d^2\theta}\left(\frac{d\theta_o}{dt} + \sqrt{(\frac{d\theta_o}{dt})^2 + 2\theta\frac{d^2\theta}{dt^2}}\right)$ $= (d\theta/dt - d\theta_o/dt)/d^2\theta/dt^2$ $\theta = \frac{1}{2}\frac{d\theta}{dt}\frac{dt^2}{d^2\theta}(\frac{d\theta}{dt} - (\frac{d\theta_o}{dt})^2)$ $d\theta/dt = \sqrt{(d\theta_o/dt)^2 + 2\theta d^2\theta/dt^2}$
3. Linearly Increasing Torque on Rotor 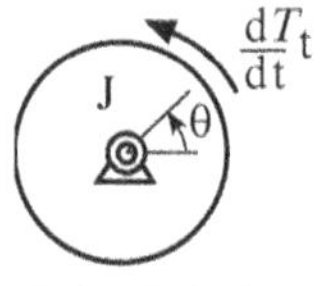 J about pivot	$T = \frac{dT}{dt}t = J\frac{d^2\theta}{dt^2}$ $\frac{d\theta}{dt} = \frac{dT}{dt}\frac{1}{2J}t^2$ $\theta = \frac{dT}{dt}\frac{1}{6J}t^3, \quad \frac{dT}{dt} = const$ at t=0: $\theta = d\theta/dt = 0$	$t = \left(\frac{6J\theta}{dT/dt}\right)^{1/3}$ $\theta = \frac{2^{1/2}}{3}\left(\frac{J(d\theta/dt)^3}{dT/dt}\right)^{1/3}$ $\frac{d\theta}{dt} = \left(\frac{6^2\theta^2}{J}\frac{dT}{dt}\right)^{1/3}$
4. Varying Torque Rotor 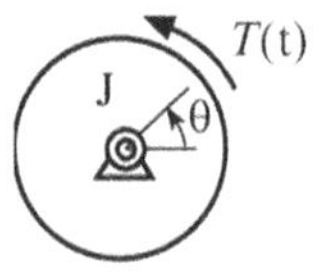 J about pivot	$\frac{d^2\theta}{dt^2} = \frac{T(t)}{J}, \quad T(t) = J\frac{d^2\theta}{dt^2}$ $\frac{d\theta}{dt} = \frac{1}{J}\int_0^t T(t)dt + d\theta_o/dt$ $\theta = \frac{1}{J}\int_0^t\left(\int_0^t T(t)dt\right)dt + td\theta_o/dt + \theta_o$ $T(t) = Jd^2\theta/dt^2$ At t=0: $\theta = \theta_o, d\theta/dt = d\theta_o/dt$.	Reduces to case 1 for constant torque. See Appendix B for integration.

Table 2.6 Rotation about single axis, continued

Rotational Motion	Torque T, Rotation Angle θ	Comments and Special Cases
5. Torque does Work on Rotor 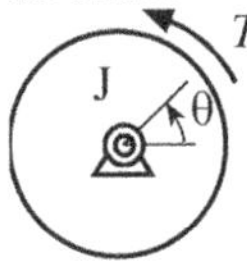	$\int_{\theta_1}^{\theta_2} T d\theta = \frac{1}{2}J\left(\left[\frac{d^2\theta_2}{dt^2}\right]^2 - \left[\frac{d^2\theta_1}{dt^2}\right]^2\right)$ Torque increases kinetic energy as body rotates from angle θ_1 to θ_2. J about pivot.	Work equals change in torsional kinetic energy.
6. Geared System 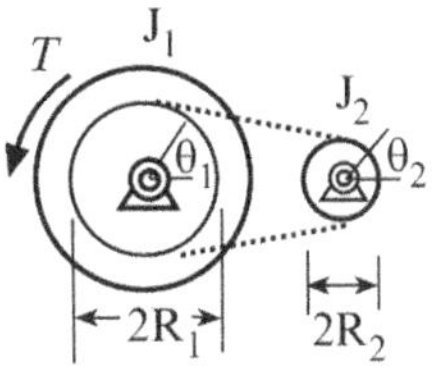	$d^2\theta_1/dt^2 = J_{eff}/T$, $T = J_{eff}d^2\theta_1/dt^2$ Effect. inertia $J_{eff} = J_1 + \left(\frac{R_2}{R_1}\right)^2 J_2$ Relationship between rotations, $\frac{\theta_2}{\theta_1} = \frac{R_1}{R_2}$	If wheel 2 is held stationary, then torque on wheel 2 due to torque applied at wheel 1 is $T_2 = \frac{R_2}{R_1}T_1$ J_1 and J_2 about respective pivots.
7. Rotating Imbalance m 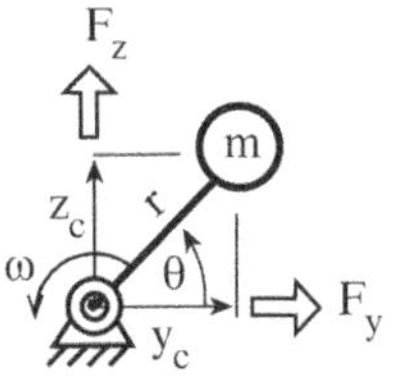	Force on pivot by imbalance m at r, $F_y = m y_c(d\theta/dt)^2$, $y_c = r\cos\theta$ $F_z = m z_c(d\theta/dt)^2$, $z_c = r\sin\theta$ $\lvert F\rvert = m r(d\theta/dt)^2$, $r = \sqrt{y_c^2 + z_c^2}$ $d\theta/dt = \text{constant}, d^2\theta/dt^2 = 0$	If $d^2\theta/dt^2 \neq 0$, $F_y = m y_c(d\theta/dt)^2 + m z_c d^2\theta/dt^2$ $F_z = m z_c(d\theta/dt)^2 - m y_c d^2\theta/dt^2$ $\lvert F\rvert = m r\sqrt{(d\theta/dt)^4 + (d^2\theta/dt^2)^2}$ also see case 9 of Table 2.7.
8. Rotating Shaft with Unbalanced Mass m 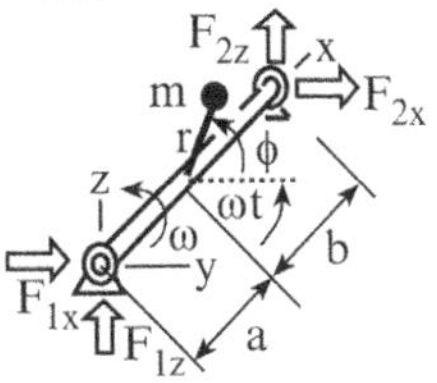	Forces on shaft by imbalance mass m at radius r and angle ϕ. $F_y = m r \omega^2 \cos(\omega t + \phi)$ $F_z = m r \omega^2 \sin(\omega t + \phi)$ Constant rotation speed ω, radian/s.	Forces on bearings applied by ends of rigid shaft $F_{1y} = F_y b/(a+b)$ $F_{1z} = F_z b/(a+b)$ $F_{2y} = F_y a/(a+b)$ $F_{2z} = F_z a/(a+b)$ also see case 9 of Table 2.7.
9. Inclined Disk on Rotating Shaft 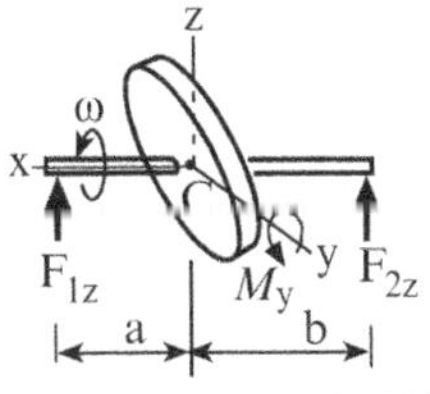	$M_y = J_{xz}\omega^2$ Moment applied by the shaft to the rotating body about y axis. Reactions at end of shaft. $F_z = M_y/(a+b)$ Constant angular velocity ω.	Jxy is the product of inertia about the coordinate system fixed to the disk and the shaft, Table 1.6. Center of gravity C is on shaft. Also see case 10.

Table 2.6 Rotation about single axis, continued

Rotation Motion	Torque T, Position θ, Force F	Comments and Special Cases
10. Body Rotating about x-axis 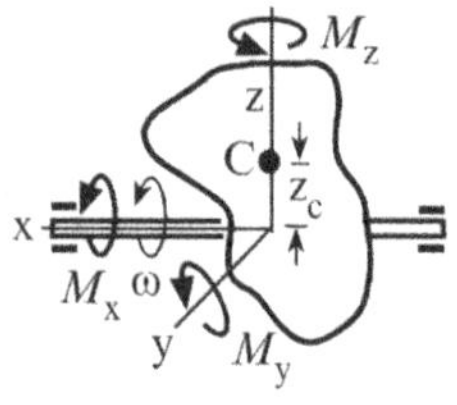	Forces and moments applied to the rotating body by the shaft. $M_x = J_x d\omega / dt$ $M_y = -J_{xy} d\omega / dt + J_{xz}\omega^2$ $M_z = -J_{xz} d\omega / dt - J_{xy}\omega^2$ $F_y = -M z_c d\omega / dt - M y_c \omega^2$ $F_z = M y_c d\omega / dt - M z_c \omega^2$ x-y-z fixed in body M=body mass.	The force and moments on the body are relative to the xyz axes that are fixed in the body. x-axis is along shaft. z-axis passes through center of mass that is located at y=yc and z=zc along z-axis. Jxy Jxz are about x,y,x axes, Table 1.6. ω = angular velocity, radian/ s. Reactions on shaft are negative of these. See case 5 of Table 2.5. Gravity load not included.
11. Spinning Governor 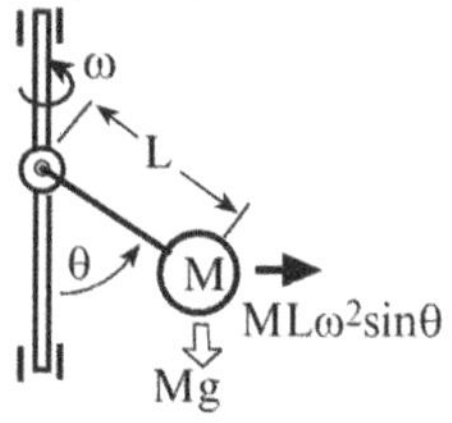	Equilibrium angles. Ref. [1], pp. 395–397 $\theta = \begin{cases} 0, & \text{stable for } \omega < \sqrt{\frac{g}{L}} \\ \text{arc cos}\frac{\omega^2 L}{g}, & \text{stable for } \omega > \sqrt{\frac{g}{L}} \end{cases}$ A mass on pivoted rod spins about a central axis. See case 6 of Table 2.7.	
12. Despinning a Satellite with Masses on Strings 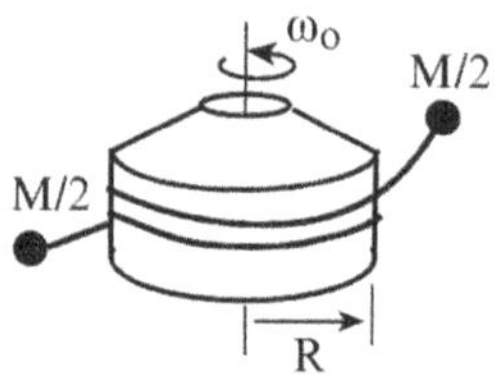	Angular velocity of satellite with despinning masses. Ref. [5], p. 211 $\frac{\omega}{\omega_o} = \frac{1 + \frac{J}{MR^2} - (\omega_o t)^2}{1 - \frac{J}{MR^2} + (\omega_o t)^2}$ Two masses M/2 are released at t=0 at radius R, from spinning satellite with initial angular velocity ω_o. The angular velocity of satellite decreases to zero at $t=[(1+J/MR^2)]^{1/2}/\omega_o$ as masses unwind. J = polar moment	
13. Ladder Slides between Wall and Floor 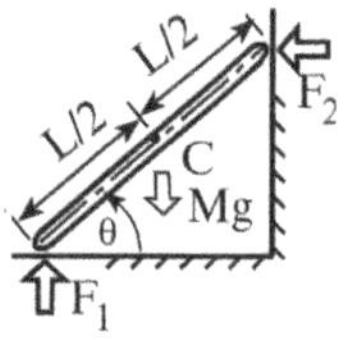	$\frac{d^2\theta}{dt^2} = -\frac{MgL}{J + ML^2/4}\cos\theta$ $\frac{d\theta}{dt} = -\sqrt{\frac{2MgL}{J + ML^2/4}(\sin\theta_o - \sin\theta)}$ θ_o = initial angle, $\theta_o / dt = 0$, J = moment of inertia about centroid C, Table 1.6 M= mass of bar, Ref. [17], p. 248	
14. Rotor Rolls down Inclined Plane 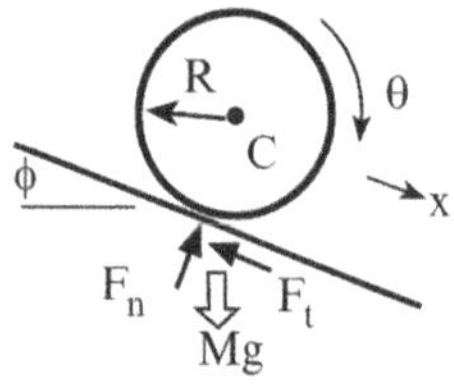	$\frac{d^2\theta}{dt^2} = \frac{1}{R}\frac{d^2x}{dt^2} = \frac{g\sin\phi}{1 + J/(MR^2)}$ $\frac{d\theta}{dt} = \frac{1}{R}\frac{dx}{dt} = \frac{g\sin\phi}{1 + J/(MR^2)}t + \left.\frac{d\theta}{dt}\right\vert_{t=0}$ $\theta = \frac{x}{R} = \frac{g\sin\phi}{1 + J/(MR^2)}\frac{t^2}{2} + \left.\frac{d\theta}{dt}\right\vert_0 t + \theta_o$	

Table 2.6 Rotation about single axis, continued

Rotational Motion	Torque T, Position θ, Force F	Comments and Special Cases
15. Pendulum 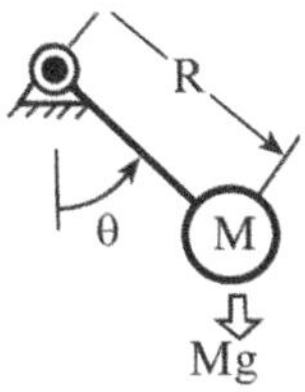	$\frac{d^2\theta}{dt^2} = -\frac{g}{R}\sin\theta$ $\frac{d\theta}{dt} = \sqrt{\frac{2R}{g}(\cos\theta - \cos\theta_o)}$ $\theta \approx \theta_o \sin\frac{2\pi t}{T}$, $\theta_o < 2$ θ_o = maximum amplitude, $\theta_o \ge \theta$ See Example 2.8. $d\theta/dt = 0$ at $\theta=\theta_o$	Period of free oscillation Ref. [16]. $T = 4\sqrt{\frac{R}{g}}K\left(\sin^2\frac{\theta_o}{2}\right)$, K = complete elliptic integral $\approx 2\pi\sqrt{\frac{R}{g}}\left(\frac{\sin\theta_o}{\theta_o}\right)^{-3/8}$, $\theta_o < 2$ $= 2\pi\sqrt{\frac{R}{g}}$, $\theta_o << 1$, case 1 Table 3.4
16. Body Pendulum 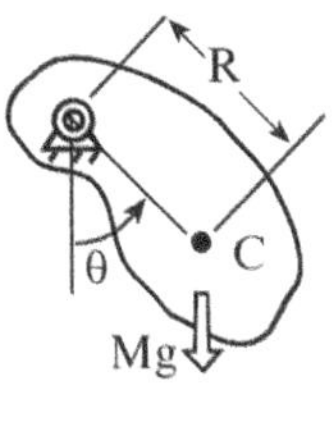	$\frac{d^2\theta}{dt^2} = -\frac{MRg}{J}\sin\theta$ $\frac{d\theta}{dt} = \sqrt{\frac{2MgR}{J}(\cos\theta - \cos\theta_o)}$ $\theta = \theta_o \cos\frac{2\pi t}{T}$ $d\theta/dt = 0$ at $\theta = \theta_o$	Period of free oscillation Ref. [16]. $T = 4\sqrt{\frac{J}{MgR}}K\left(\sin^2\frac{\theta_o}{2}\right)$, K = cplete ellticintral $T \approx 2\pi\sqrt{\frac{J}{MgR}}\left(\frac{\sin\pi_o}{\theta_o}\right)^{-3/8}$, $\theta_o < 2$ $\approx 2\pi\sqrt{\frac{J}{MgR}}$, $\theta_o < 1$, case 11; Table 3.4
17. Center of Percussion of Free Slender Bat 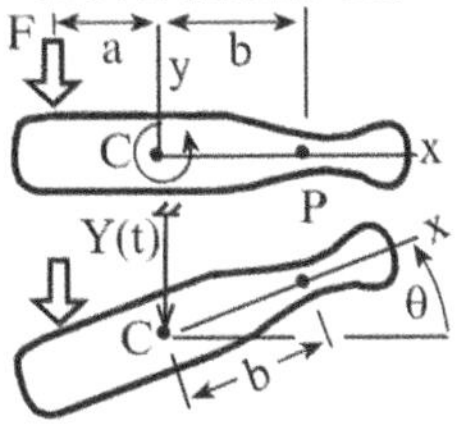	$M\frac{d^2Y(t)}{dt^2} = F$, $J\frac{d^2\theta}{dt^2} = Fa$ $Y(x,t) = Y(t) + x\theta$, $\frac{d^2Y(x,t)}{dt^2} = \frac{F}{M}\left(1 - \frac{aM}{J}x\right)$ $b = J/aM$, $d^2Y(b,t)/dt^2 = 0$ M = mass, J = mass moment inertia about C, Table 1.6.	The force F is applied at distance a outboard of center of mass C of free rigid bat. Instantaneous response is a combination of translation of center of mass and rotation about center of mass. Center of percussion (x=b) is point that has zero instantaneous displacement and acceleration. Ref. [17], p. 246.
18. Body Rotates about fixed Pivot 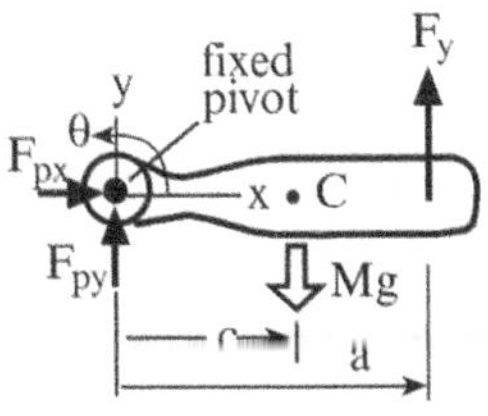	Angular Acceleration, $\frac{d^2\theta}{dt^2} = \frac{F_y a - Mgc}{J_z}$ Forces pvot: $F_{px} = Mc(\frac{d\theta}{dt})^2$, F_{py} $= F_y\left(\frac{Mca}{J_z} - 1\right) - Mg\left(\frac{Mc^2}{J_z} - 1\right)$ Out-of-Plane moments on body $M_x = J_{xz}\frac{d^2\theta}{dt^2} + J_{yz}(\frac{d\theta}{dt})^2$ $M_y = -J_{yz}\frac{d^2\theta}{dt^2} - J_{xz}(\frac{d\theta}{dt})^2$	Forces Fpx, Fpy and out-of-plane moments Mx, My are applied by pivot to body; forces and moments on pivot by body are negative of these. Jz is polar mass moment of inertia of body about pivot axis. C is center of mass. Fy induces no vertical force on pivot if a = J/z(Mc). J/z = Mass moments of inertia about pivot Table 1.6.

Table 2.6 Rotation about single axis, continued

Rotational Motion	Position θ, Force F, Torque T	Comments and Special Cases
19. Door Swings against Spring Stop 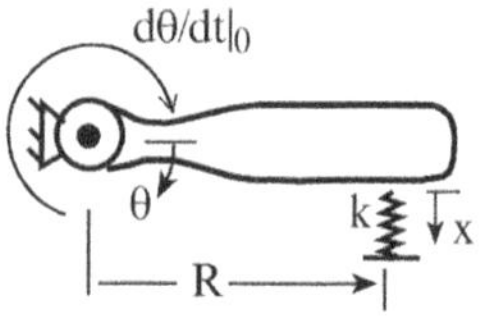 	spring torque on door $T = -kR\theta = -J\omega_n^2\theta$ $\frac{d\theta}{dt} = \left.\frac{d\theta}{dt}\right\|_{t=0}\cos\omega_n t, \quad 0 \le \omega_n t \le \pi$ $\theta(t) = (d\theta/dt\|_{t=0}/\omega_n)\sin\omega_n t$ $\omega_n = \sqrt{\frac{kR}{J}}$ = nat. freq., Hz At t=0: θ=0, dθ/dt>0.	$x \approx R\theta$ $x_{max} = \sqrt{\frac{J}{k}}\left.\frac{d\theta}{dt}\right\|_{t=0}$ $\theta_{max} = \frac{1}{\omega_n}\left.\frac{d\theta}{dt}\right\|_{t=0}$ $F_{max} = kx_{max} = \sqrt{Jk}\left.\frac{d\theta}{dt}\right\|_{t=0}$ $M_{max} = RF_{max}$
20. Door Swings into Force Stop 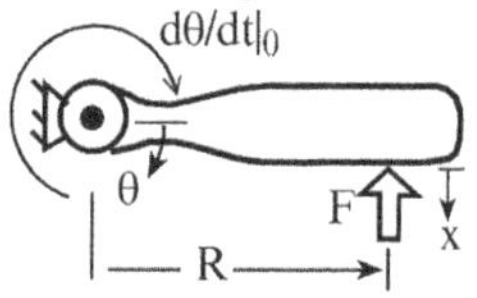 	$FR = -J\frac{d^2\theta}{dt^2}$ $\theta(t) = -[FR/(2J)]t^2 + d\theta/dt\|_{t=0}$ $\theta_{max} = \frac{J}{2FR}\left[\left.\frac{d\theta}{dt}\right\|_{t=0}\right]^2$ at $t = \frac{J}{FR}\left.\frac{d\theta}{dt}\right\|_{t=0}$	At t=0: θ=0, dθ/dt>0.
21. Door Swings against Spring Stop with Force 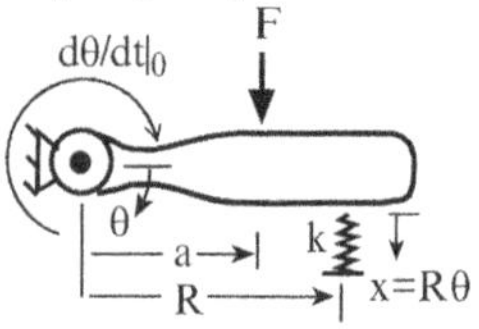 	$Fa - kR\theta = -J\frac{d^2\theta}{dt^2}$ = torque on door $\theta(t) = (d\theta/dt\|_{t=0}/\omega_n)\sin\omega_n t$ $-(Fa/kR^2)(1-\cos\omega_n t)$ $\omega_n = \sqrt{\frac{kR}{J}}$ = nat. freq., Hz At t=0: θ=0, dθ/dt>0.	$\theta_{max} = \frac{Fa}{kR^2} + \sqrt{\left(\frac{1}{\omega_n}\left.\frac{d\theta}{dt}\right\|_{t=0}\right)^2 + \left(\frac{Fa}{kR^2}\right)^2}$ $\pi/2 \le \omega_n t \le \pi$

are shown in Table 1.2. Torque $\boldsymbol{T}$, also called moment, is the vector cross product (Eq. 1.24) of force vector **F** applied to the rotor and the radius vector **r** from the axis of rotation to the point of force application:

$$\boldsymbol{T} = \mathbf{r} \times \mathbf{F} \tag{2.30}$$

If force F is applied perpendicularly to the radius r from the rotation axis, then the torque is $T = rF$.

Integrating Equation 2.29 with respect to time shows that the change in angular momentum

$$\text{Angular momentum} = J\left(\frac{d\theta}{dt}\right) \tag{2.31}$$

equals the torsional impulse applied between time t_1 and t_2:

$$\int_{t_1}^{t_2} T dt = J\left(\left.\frac{d\theta}{dt}\right|_{t_2} - \left.\frac{d\theta}{dt}\right|_{t_1}\right) \tag{2.32}$$

Table 2.7 Multiaxis rotation

Notation: C = center of mass; F = force; g = acceleration due to gravity (Table 1.2); J_x, J_y, J_z = mass moments of inertia about body-fixed x, y, z-axes, respectively; J_{xy}, etc. = mass products of inertia; J_1, J_2, J_3 = principal mass moments of inertia (Eq. 2.10); H = angular momentum; M = mass of body; M = moment; t = time; = spin angular velocity about 1- or x-axis, rad/s; θ = angular position; ω_x, ω_y, ω_z = angular velocity, rad/s. Positive rotations are defined using the right-hand rule. Overdot (.) denotes derivative with respect to time. x, y, z = body-fixed coordinate system; X, Y, Z = stationary coordinate system. Consistent units are shown in Table 1.3 Refs [1, 5, 18, 25].

Rotation	Angular Position, Velocity, Moments, and Comments
1. Rotor Reactions 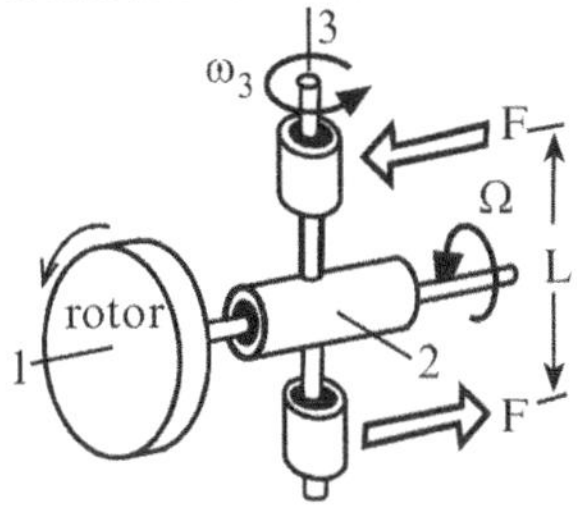	An axially symmetric rotor, $J_{12}=J_{13}=J_{23}=0$, $J_2=J_3<<J_1$, spins at constant angular velocity Ω about the 1-axis, the axis of symmetry. The 1, 2, and 3 axes are fixed to the rotor support assembly that spins about vertical axis with angular velocity ω_3. An external moment M_2=FL is applied about transverse 2 axis to counter the equal but opposite gyroscopic moment. J_1= mass moment of inertia of the rotor about the 1 axis. See Eg.2.39 Ref. [1], p. 439.
2. Body Rotates about Principal Axes with Moments 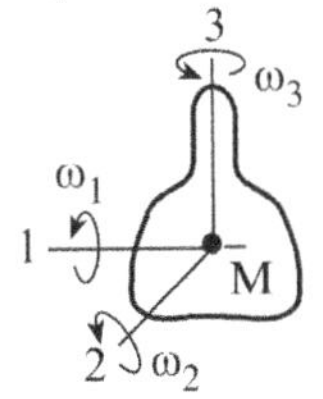	$M_1 = J_1 d\omega_1/dt + (J_3 - J_2)\omega_2\omega_3$ $M_2 = J_2 d\omega_2/dt + (J_1 - J_3)\omega_1\omega_3$ $M_3 = J_3 d\omega_3/dt + (J_2 - J_1)\omega_1\omega_2$ Moments M_1, M_2, M_3 exerted on the body about principal axes 1, 2, and 3. Origin is at the center of mass. See Table 2.5 for general case Ref. [1], p. 392.
3. Precession of Freely Spinning Axisymmetric Body 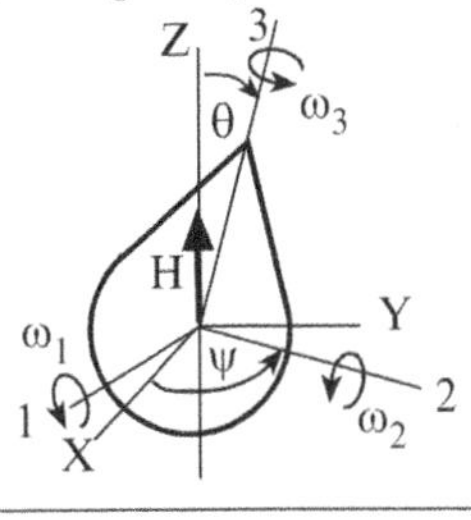	Angular velocity about body fixed x,y,z axes. No moments. ω_3 = constant $\omega_2 = A \sin \alpha t$ $\omega_1 = A \cos \alpha t$, $\alpha = \omega_3(J_3 - J_1)/J_1$ Euler angles θ and Ψ transform from fixed cartesian to body axis. Bank angle θ = constant. Body precesses at constant rate. $d\Psi/dt = \omega_3 J_3/(J_1 \cos\theta)$ Body is axisymmetric about x-axis. $J_1=J_2$ Ref. [5], pp. 114–117
4. Stability of Body Freely Rotating about Principal Axes 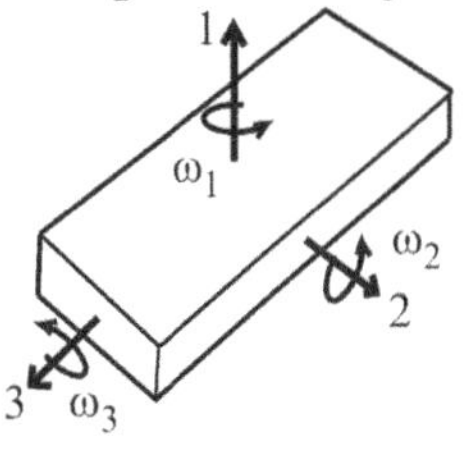	Axes 1, 2, and 3 are principal axes mass moments of inertia are J_1, J_2, and J_3. Their magnitudes are in a decreasing series. $J_1 > J_2 > J_3$ The stability of rotation about each of these principal axes is, 1. Rotation about the 1-axis, the max principal axis, is stable. 2. Rotation about the 2-axis, the intermediate axis, is unstable. 3. Rotation about the 3-axis is stable provided there is no internal energy dissipation. Stable rotation is continuing rotation without increase in wobble or change in axis of rotation. Rotation about the 2 axis will cause rotation into rotation about 1 axis Ref. [1], pp. 397–398.

Table 2.7 Multiaxis rotation, continued

<table>
<tr><th>Rotation</th><th>Angular Position, Velocity, and Moments</th></tr>
<tr><td>5. Aerodynamic Stability of Rotating Finless Projectile with Mass M, Velocity V, and Angular Velocity ω
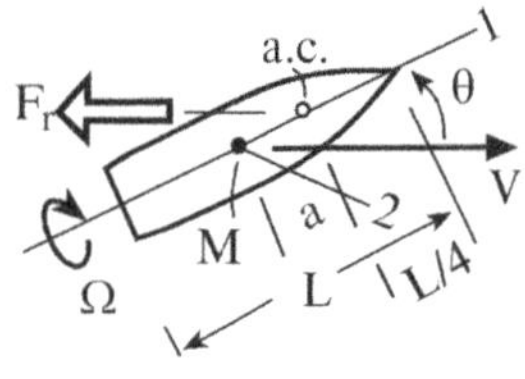
</td><td>In order to be stable so that the wobble angle θ does not grow in time, the spin angular velocity ω about longitudinal axis must be, Ref. [5], p. 142
$$\Omega \geq \frac{2}{J_2}\sqrt{F_r a J_1}$$
F_r = aerodynamic resistance = $(1/2)\rho V^2 c_f$, A is the cross sectional area, $c_f \sim 0.5$, and ρ = fluid density. R has units of force.
a = distance from center of mass to the aerodynamic center, a.c., located one quarter of the overall length aft of the nose.
J_1, J_2 = mass moments of inertia about the center of mass along longitudinal (1) and transverse (2) axes of the projectile, through C.</td></tr>
<tr><td>6. Spinning Pendulum
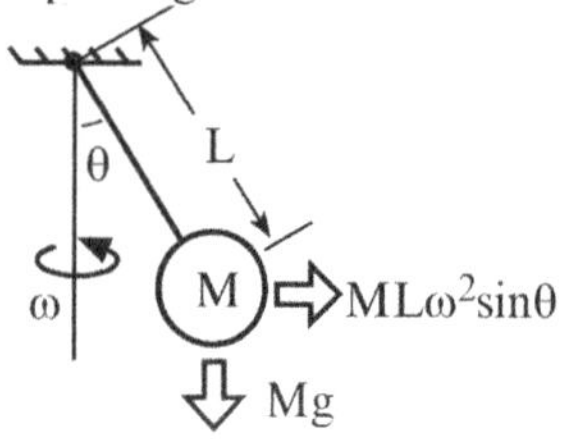
</td><td>A mass on a string or rod spins with angular velocity ω, radian/s, about a central axis. There are two equilibrium solutions. Natural frequencies are oscillations in θ about equilibrium solution.
Ref. [1], pp. 395–397
<table>
<tr><th>Equilibrium θ</th><th>Nat. freq., fn Hz</th><th>Stability Criteria</th></tr>
<tr><td>θ=0</td><td>$\frac{1}{2\pi}\sqrt{\frac{g}{L}-\omega^2}$</td><td>$\omega < \sqrt{\frac{g}{L}}$</td></tr>
<tr><td>$\theta = \arccos\frac{\omega^2 L}{g}$</td><td>$\frac{1}{2\pi}\sqrt{\omega^2 - \frac{g}{L}}$</td><td>$\omega > \sqrt{\frac{g}{L}}$</td></tr>
</table></td></tr>
<tr><td>7. Horizontal Gyro Fixed Point and Gravity
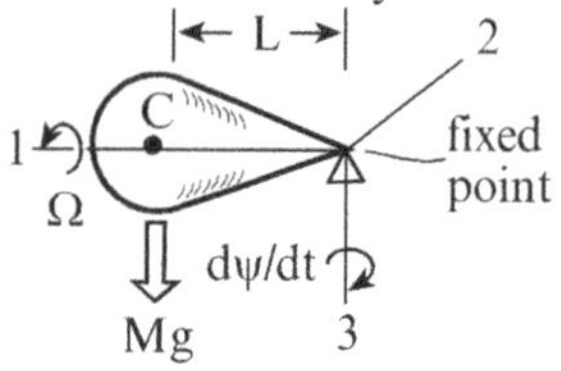
</td><td>Constant precession rate for a axisymmetric, $J_2=J_3$ top under gravity moment
$$\frac{d\Psi}{dt} = -\frac{MgL}{J_1\Omega}$$
Axisymmetric gyro topples for $\omega<2(J_2MgL)^{1/2}/J_1$.

Ref. [5], p. 141.</td></tr>
<tr><td>8. Spinning Gyro with Point Fixed and Gravity Moment
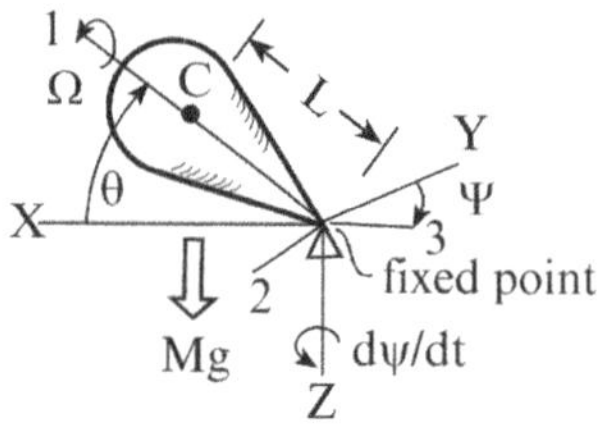
</td><td>The constant precession rate for a constant bank angle θ,
$$\frac{d\Psi}{dt} = -\frac{J_1\Omega}{2J_2\sin\theta}\left[1 \quad \sqrt{1-\frac{4J_2MgL\cos\theta}{J_1^2\Omega^2}}\right]$$
For stability, gyro is not toppled by gravity, spin velocity must be
$$\Omega > \frac{2\sqrt{J_2MgL\cos\theta}}{J_1}$$
Axisymmetric body, $J_2 = J_3$. J_1 = moment of inertia about 1 axis.

For vertical top θ = π/2 and cos π = 0. See case 8.
Ref. [2], pp. 439–440.</td></tr>
</table>

Table 2.7 Multiaxis rotation, continued

Rotation	Angular Position, Velocity, Moments, and Comments
9. Unbalanced Whirling disk On Rotating Shaft 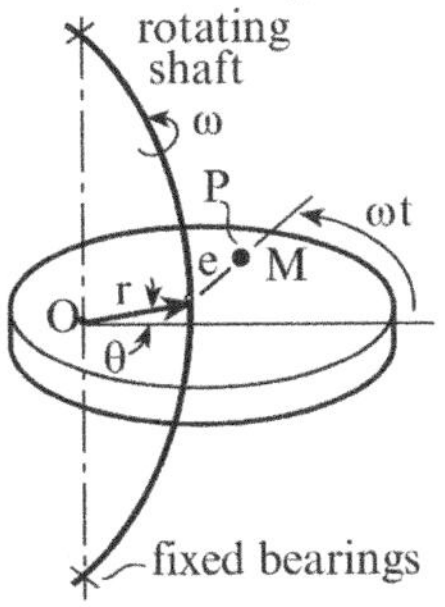 Disk with mass M spins at angular velocity ω radians/s on shaft through disk. **r** =unit vector in r direction **j** = unit tangential vector e = offset of M from shaft	Equations of motion. Ref. [18], pp. 59–61. $Md^2r/dt^2 + 2M\zeta\omega_n dr/dt + (\omega_n^2 - (d\omega/dt)^2)Mr = Me\omega^2\cos(\omega t - \phi)$ $Mrd^2\theta/dt^2 + (2M\zeta\omega_n + 2Mdr/dt)d\omega/dt = Me\omega^2\sin(\omega t - \phi)$ Center of mass of disk is distance e from shaft. Shaft deflects laterally distance r its equilibrium position on line between fixed bearings. Vector position **P** of center of mass M with respect to O, case 3 of Table 2.4, in r, θ cylindrical coordinates. $\mathbf{P} = [r + e\cos(\omega t - \theta)]\mathbf{r} + e\cos(\omega t - \theta)\mathbf{j}$ $d^2\mathbf{P}/dt^2 = [d^2r/dt^2 - r(d\theta/dt)^2 - e\omega^2\cos(\omega t - \theta)]\mathbf{r}$ $+ [rd^2\theta/dt^2 + 2(dr/dt)d\theta/dt - e\omega^2\sin(\omega t - \theta)]\mathbf{j}$ For synchronous whirling, $d^2\theta/dt^2 = d^2r/dt^2 = dr/dt = 0$, $\theta = \omega t - \phi$, $r = \frac{e(f/f_n)}{\sqrt{(1-(f/f_n)^2)^2 + (2\zeta f/f_n)^2}}$, $\tan\phi = \frac{2\zeta(f/f_n)}{1-(f/f_n)^2}$, ζ = dim'lss damping factor $f_n = (k/M)^{1/2}/(2\pi)$, k = shaft stiffness. Also see case 8 of Table 2.6.
10. Critical frequencies of Overhung Rotor 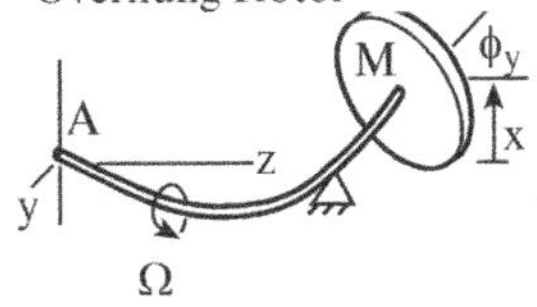 x = vertical rotor deflection y = lateral rotor deflection. ϕ_x = rotor angle about x axis ϕ_y = rotor angle about y axis k_{xx},k_{yy} = shaft spring stiffness for rotor displacements x, y J_{xx},J_{yy},J_{zz} = mass moments of inertia of rotor about x,y,z. $R_{\theta\theta}$ = *moment due to rotation* θ $R_{x\theta}$ = Force due to rotation θ	Equations of motion with gyroscopic terms. Ref. [25], pp. 61–64 $M\ddot{x} + k_{xx}x - k_{x\varphi_y}\varphi_y = 0,\ J_{zz}\ddot{\varphi}_x + (J_{zz} - J_{xx})\Omega\dot{\varphi}_x + k_{\varphi_x y}y + k_{\varphi_x\varphi_x}\varphi_x = 0,$ $M\ddot{y} + k_{yy}y + k_{y\varphi_x}\varphi_x = 0, J_{zz}\ddot{\varphi}_y - (J_{zz} - J_{xx})\Omega\dot{\varphi} - k_{\varphi_y x}x + k_{\varphi_y\varphi_y}\varphi_y = 0$ For symmetric shaft $k_{yy}=k_{xx}$, $k_{\varphi x\varphi x}= k_{\varphi y\varphi y}= k_{\varphi\varphi}$, $k_{\varphi xy}= k_{y\varphi x}= k_{\varphi x}$ Natural frequencies without rotation, Ω=0, $(2\pi f_n)^2 = \frac{Mk_{\varphi\varphi} + J_{zz}k_{xx} - \sqrt{(Mk_{\varphi\varphi} - J_{zz}k_{xx})^2 + 4MJ_{zz}k_{x\varphi}^2}}{2MJ_{zz}}$ Critical whirling frequencies for $\Omega = f_n$ and thin disk, $J_{zz}=J_{xx} + J_{yy}$ $(2\pi f_n)^2 = -\frac{1}{2}\left[\frac{k_{\varphi\varphi}}{J_{xx}} - \frac{k_{xx}}{M} - \sqrt{\left(\frac{k_{\varphi\varphi}}{J_{xx}} - \frac{k_{xx}}{M}\right)^2 - 4\frac{k_{x\varphi}^2}{J_{xx}M}}\right]$, fwd rotation $(2\pi f_n)^2 - \frac{1}{2}\left[\frac{k_{\varphi\varphi}}{3J_{xx}} + \frac{k_{xx}}{M} \pm \sqrt{\left(\frac{k_{\varphi\varphi}}{3J_{xx}} - \frac{k_{xx}}{M}\right)^2 + 4\frac{k_{x\varphi}^2}{J_{xx}M}}\right]$, aft rotation

Integrating Equation 2.32 with respect to an angular increment $d\theta$ shows that the angular work produces a change in angular kinetic energy $(1/2)J(d\theta/dt)^2$ of the rotating body:

$$\int_{t_1}^{t_2} T d\theta = \frac{J}{2}\int_{t_1}^{t_2} \frac{d}{dt}\left(\frac{d\theta}{dt}\right)^2 dt = \frac{1}{2}J\left[\left(\left.\frac{d\theta}{dt}\right|_{t_2}\right)^2 - \left(\left.\frac{d\theta}{dt}\right|_{t_1}\right)^2\right] \quad (2.33)$$

Kinetic energy for multiaxis rotation is given in Equation 2.37.

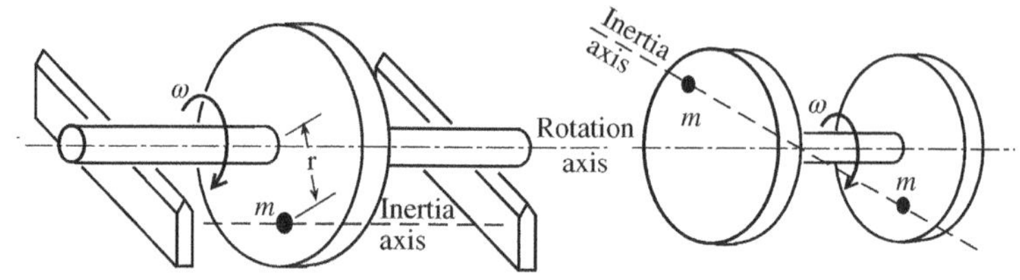

Figure 2.8 Static balancing a rotor by resting the shaft on knife-edges. Two-plane imbalance generates a transverse moment on the bearings. After Ref. [18]

Balancing is a procedure for adjusting the mass distribution of a rotor so the center of mass of the rotor approaches the axis of rotation. Rotor out of balance is expressed in gram-centimeters, gram-inches, or ounce-inches of mass off the axis of rotation. Figure 2.8 shows two out-of-balance cases: (1) the center of mass is off the rotation axis (cases 7 and 8 of Table 2.6), and (2) the mass product of inertia about the rotation axis is nonzero, which generates a moment on the rotor as the rotor spins (cases 9 and 10 of Table 2.6).

To static balance a rotor in a single plane, the rotor is decoupled from the drive mechanism, and, if possible, the shaft is placed on level knife-edges to minimize friction that impedes rotation as shown in Figure 2.8a. The unrestrained rotor oscillates with the unbalanced mass tending downward (cases 15 and 16 of Table 2.6). A small trial weight is added to the rotor 180° opposite the unbalance. The procedure is repeated incrementally until a satisfactory static balance is obtained.

Dynamic balancing uses vibration measurements. A small trial balance weight, such as a piece of metallic tape, is attached to the perimeter of the rotor. Vibration is measured while the rotor spins; an accelerometer on the bearing housing is often used. See case 9 of Table 2.7. The trial weight is then moved along the circumference. Vibration measurements are repeated at the same rotor speed. If the vibration decreases, then the balance weight is being moved in the proper direction.

Static balance does not eliminate the couple imbalance as shown in Figure 2.8b and case 9 of Table 2.6. Balance weights in two planes are required to simultaneously reduce the static and couple imbalances. Their locations could be approximately determined by statically balancing each end of a rotor, but this is very approximate. Two-plane balancing methods use multiple measurements to determine the phase and the locations of the two-plane balance weights as discussed in (Refs [17, 28–30]) and industry balancing standards ISO 21940 and API RP 684 and Mil-STD-167.

Critical Rotor Speeds are speeds of rotation corresponding to shaft natural frequencies. Deflections of an out-of-balance rotor increase as the rotor speed approaches rotor shaft natural frequencies (case 9 of Table 2.7). Critical speeds are shaft angular velocities that equal the natural frequencies of rotor deflection (Chapter 4). The large amplitude deflections at critical speeds are called synchronous rotor whirl case 10 of Table 2.7. Avoidance of critical speeds is fundamental to rotating machinery design and safe operation [28–30].

Example 2.7 Accelerating a rotor

A round steel disk on a rotating shaft is 1 m (1000 mm, 39.37 in.) in diameter and 10 mm (0.3937 in.) thick. The shaft is through the center of the disk, perpendicular to the face of the disk. What torque accelerates the disk from 0 to 110 revolutions per minute in 20 s?

Solution: The solution is presented in case 1 of Table 2.6. The mass and mass moment of inertia of the disk about an axis through its center are shown in case 13 of Table 1.7. J and M are computed as follows for a steel (density = 8 g/cc = 8×10^{-6} kg/mm^3 = 0.289 lb/in.^{3}; Chapter 8) disk:

$$M = \rho \pi R^2 h = (8 \times 10^{-6}\ \text{kg/mm}^3)\ 3.14159(500\ \text{mm})^2 10\ \text{mm} = 62.83\ \text{kg}\ \ (138.5\ \text{lb})$$

$$J = \left(\frac{M}{2}\right) R^2 = (62.83\ \text{kg}/2)\ \ (0.5\ \text{m})^2 = 7.853\ \text{kg–m}^2\ \ (186.3\ \text{lb–ft}^2)$$

The angular acceleration is the change in angular velocity, $110(2\pi)$ rad/60 s, divided by the time interval of 20 s. This is equal to 1.1152 rad/s^2. Case 1 of Table 2.6, or Equation 2.29, gives the required torque. In SI units (case 1 of Table 1.2),

$$T = J\ \frac{d^2\theta}{dt^2} = 7.853\ \text{kg–m}^2\ \ (1.1152\ \text{rad/s}^2) = 8.758\ \text{N–m}\ \ (6.460\ \text{ft–lb})$$

Example 2.8 Pendulum swings

The pendulum presented in Figure 2.9 swings freely about a fixed pivot. Describe its motions.

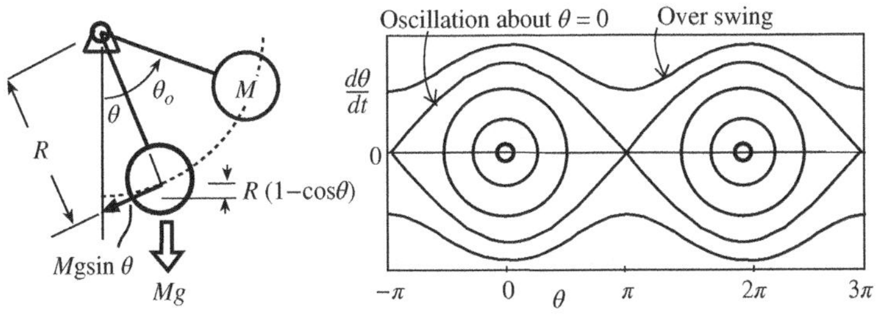

Figure 2.9 Freely swinging pendulum angular velocity versus angle

Solution: As the pendulum ball swings from θ to an angle θ_0, the change in potential energy equals the changes in kinetic energy. Potential energy of the pendulum ball is equal to its weight times its height. Its kinetic energy is mass times its velocity $MR(d\theta/dt)^2$. The sum of potential and kinetic energy is constant:

$$\text{Potential energy from } \theta \text{ to } \theta_0 + \text{kinetic energy from } \theta \text{ to } \theta_0 = 0$$

$$MgR(1 - \cos\theta) - MgR(1 - \cos\theta_0) + {}^1\!/_2 M\left(\frac{Rd\theta}{dt}\right)^2 - {}^1\!/_2 M\left(\frac{Rd\theta_0}{dt}\right)^2 = 0$$

$d\theta_0/dt$ is the angular velocity when $\theta = \theta_0$. Solving for $d\theta/dt$ gives two solutions for pendulum velocity that are plotted in Figure 2.9:

$$\frac{d\theta}{dt} = \begin{cases} \pm\left[\left(\frac{2g}{R}\right)(\cos\ \theta - \cos\theta_0)\right]^{1/2}, & |\theta| < |\theta_0| \leq \pi \\ \left[\left(\frac{d\theta_0}{dt}\right)^2 \pm \left(\frac{2g}{R}\right)(1+\cos\theta)\right]^{1/2}, & \left(\frac{d\theta_0}{dt}\right)^2 > 4\left(\frac{g}{R}\right) \end{cases} \tag{2.34}$$

In the first case, $|\theta| < \theta_0$ and the pendulum swings back and forth with a maximum amplitude $\theta_0 < \theta$; this case can be linearized for small angles (Section 3.6, Table 3.4). Exact, approximate, and numerical solutions for the natural period of the pendulum are given in cases 14 and 15 of Table 2.6 and Appendix B. In the second case, the pendulum has sufficient angular velocity $d\theta_0/dt$ to swing over the top ($\theta = n\pi,\ n = 1, 3, 5\ \ldots$) like an Olympic gymnast on the bar.

2.3.3 *Multiple-Axis Rotation*

Table 2.7 has solutions for multi-axis rotation. Euler's vector equations of motion for rigid body motion are expressed terms of the acceleration $\mathbf{a}_o$ of a body-fixed origin and angular velocity $\boldsymbol{\omega}$ about an axis through that origin. Both $\mathbf{a}_o$ and $\boldsymbol{\omega}$ must be measured in an inertial (fixed, nonrotating) frame for application of Newton's law ([1], p. 390; [7], p. 180; Table 2.5):

$$\mathbf{F} = M\mathbf{a}_o + M\frac{d(\boldsymbol{\omega}\times\mathbf{r}_c)}{dt} \tag{2.35a}$$

$$\boldsymbol{M} = M\mathbf{r}_c\mathbf{a}_o + \frac{d\mathbf{H}}{dt} \tag{2.35b}$$

$\mathbf{F}$ is the vector force on the body at the origin, and $\boldsymbol{M}$ is the vector moment about the origin. The time derivatives are calculated in an inertial frame. These six coupled, nonlinear equations are written out in case 5 of Table 2.5. They simplify if the body origin is either the center of mass ($r_c = 0$), in which case Equation 2.35a reduces to Newton's second law (Eq. 2.4), or a fixed point ($a_o = 0$), in which case Equation 2.35b has the same form as single-axis rotation (Eq. 2.29).

The body's angular momentum vector $\mathbf{H}$ about O' is the 3-by-3 inertia tensor $[J]$ (Table 1.6) times the angular velocity vector:

$$\mathbf{H} = [J]\boldsymbol{\omega} = \int_M \mathbf{r}\times(\boldsymbol{\omega}\times\mathbf{r})dm = \begin{bmatrix} J_{xx} & -J_{xy} & -J_{xz} \\ -J_{yx} & J_{yy} & -J_{yz} \\ -J_{zx} & -J_{zy} & J_{zz} \end{bmatrix} \begin{pmatrix} \omega_x \\ \omega_y \\ \omega_z \end{pmatrix}$$
$$= (J_{xx}\omega_x - J_{xy}\omega_y - J_{xz}\omega_z)\mathbf{i} + (-J_{yx}\omega_x + J_{yy}\omega_y - J_{yz}\omega_z)\mathbf{j}$$
$$+ (-J_{zx}\omega_x - J_{zy}\omega_y + J_{zz}\omega_z)\mathbf{k} \tag{2.36}$$

$[J]$ may change with time as the body rotates relative to the inertial coordinate frame. The rotational kinetic energy of the body about O' [1] (p. 314) is

$$\mathrm{KE} = \left(\frac{1}{2}\right)(J_{xx}\omega_x^2 + J_{yy}\omega_y^2 + J_{zzx}\omega_z^2 - 2J_{xy}\omega_x\omega_y - 2J_{yz}\omega_y\omega_z - 2J_{xz}\omega_x\omega_z) \tag{2.37}$$

The reader is cautioned that some authors [1], for example, define the mass products of inertia (J_{xy}, J_{xz}, J_{yz}) as the negative of the expressions shown in Equation 1.13 and Table 1.5.

2.3.3.1 Small Rotational Velocity

If the rotational velocities of a body are small, for example, small vibrations of an otherwise still body, then Equations 2.35 (case 5 of Table 2.5) are linearized by dropping terms proportional to ω_2. In matrix form, these six equations are

$$\begin{Bmatrix} F_x \\ F_y \\ F_z \\ M_x \\ M_y \\ M_z \end{Bmatrix} = \begin{bmatrix} M & 0 & 0 & 0 & Mz_c & -My_c \\ 0 & M & 0 & -Mz_c & 0 & Mx_c \\ 0 & 0 & M & My_c & -Mx_c & 0 \\ 0 & -Mz_c & My_c & J_x & -J_{xy} & -J_{xz} \\ Mz_c & 0 & -Mx_c & -J_{xy} & J_y & -J_{yz} \\ -My_c & Mx_c & 0 & -J_{xz} & -J_{yz} & J_z \end{bmatrix} \begin{Bmatrix} \ddot{x} \\ \ddot{y} \\ \ddot{z} \\ \dot{\omega}_x \\ \dot{\omega}_y \\ \dot{\omega}_z \end{Bmatrix} \tag{2.38}$$

These linear equations give the instantaneous moments (M_x, M_y, M_z) and force (F_x, F_y, F_z) in inertial Cartesian x, y, z coordinates based at the body origin. The overdot (.) denotes derivative with respect to time.

The right-hand side of Equation 2.38 shows the symmetric 6-by-6 fully coupled mass matrix. Off-diagonal terms couple degrees of freedom. Placing the body origin at the center of mass ($x_c = y_c = z_c = 0$) eliminates 12 of the 18 off-diagonal terms.

2.3.4 *Gyroscopic Effects*

Consider an axisymmetric rotor that spins about the 1-axis as shown in Figure 2.10a. Symmetry about the 1-axis, $J_{12} = J_{23} = J_{13} = 0$ and $J_2 = J_3$, simplifies Euler's moment equations shown in case 5 of Table 2.5 ([1], p. 439; [31], p. 239):

$$\begin{aligned} M_1 &= J_1\left(\frac{d\omega_1}{dt} + \frac{d\Omega}{dt}\right) \\ M_2 &= J_2\left(\frac{d\omega_2}{dt}\right) - (J_2 - J_1)\omega_1\omega_3 + J_1\Omega\omega_3 \\ M_3 &= J_2\left(\frac{d\omega_3}{dt}\right) + (J_2 - J_1)\omega_1\omega_2 - J_1\Omega\omega_2 \end{aligned} \tag{2.39}$$

M_1, M_2, and M_3 are moments applied to the rotor about the 1-, 2-, and 3-axes, respectively. The total spin about the 1-axis is the sum of the body rotational velocity Ω and ω_1.

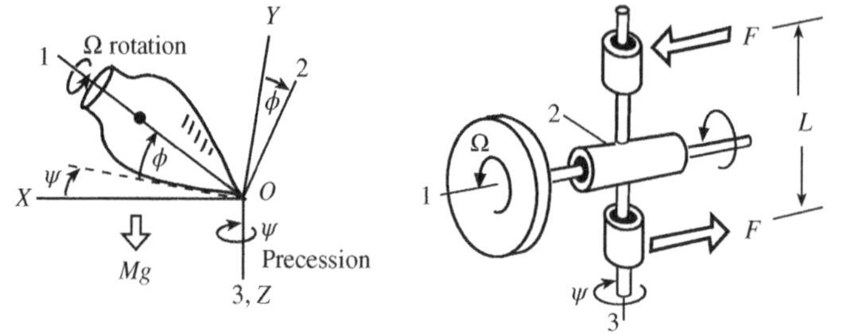

(a) Precession of spinning top (b) Rotation about the 1 and 3 axes

Figure 2.10 Spinning top (a) and constrained rotor precession (b)

For the constant speed rotor, $\omega_1 = d\omega_1/dt = M_1 = d\Omega/dt = 0$ and Equation 2.39 reduces to

$$M_2 = J_2\left(\frac{d\omega_2}{dt}\right) + J_1\Omega\omega_3$$
$$M_3 = J_2\left(\frac{d\omega_3}{dt}\right) - J_1\Omega\omega_2 \tag{2.40}$$

Asymmetric gyroscopic moments $J_1\Omega\omega_3$ and $-J_1\Omega\omega_2$ act perpendicular to the rotor axis. They couple pitch (rotation about 2-axis) and yaw (rotation about 3-axis) angular motions, which are called precession, a defining characteristic of gyroscopes and large spinning rotor disks [31] (p. 240). For example, in-plane rotation about the 3-axis generates an out-of-plane M_2 gyroscope moment about the 2-axis as shown in Figure 2.10b:

$$M_2 = J_1\Omega\omega_3 \tag{2.41}$$

Gyroscopic moments are applied to precession of wheel in the following example.

Example 2.9 Precession

A bicycle wheel is spun to angular velocity Ω. Its horizontal axle is supported only at its outboard tip as shown in Fig 2.11. Describe its subsequent motions using Equation 2.41.

Solution: The local body system is chosen so the 1-axis is along the axle, the horizontal plane is the 2-axis, and vertically downward through the support is 3-axis. x and y define the horizontal plane. Assume the 1-axis (the axle) remains in the horizontal plane; then the Euler angles ϕ and θ and their derivatives are zero. $\omega_3 = d\Psi/dt$ is the only nonzero angular velocity (see case 6 of Table 2.2). Gravity applies the moment $M_2 = -MgL$ to the wheel about the 2-axis, where L is the distance between the end of the axle and the center of gravity and M is the wheel mass. Equation 2.41 gives the following expression for precession, about the vertical axis:

$$-MgL = J_1\Omega\left(\frac{d\Psi}{dt}\right) \quad \text{hence} \quad \frac{d\Psi}{dt} = -\frac{MgL}{J_1\Omega} \tag{2.42}$$

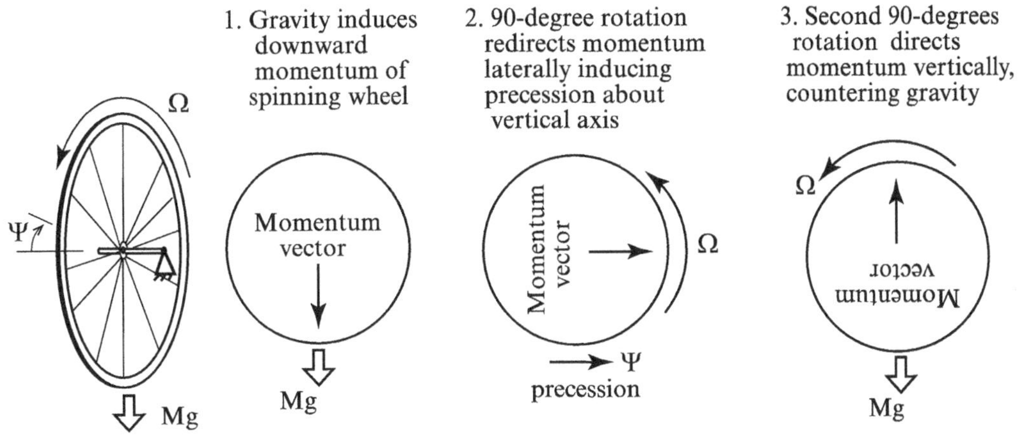

Figure 2.11 A spinning bicycle wheel is supported by a point at one end of its axle. The wheel precesses about this point. See case 7 of Table 2.7.

The wheel precesses about the vertical axis at a rate proportional to its gravity weight moment divided by its angular momentum (see cases 7 and 8 of Table 2.7).

Gravity produces downward acceleration and downward momentum. The rotation carries this momentum and redirects it, first horizontally inducing precession about the vertical axis and then vertically, where it counters gravity's downward pull.

References

[1] Greenwood, D. T., Principles of Dynamics, 2nd ed., Prentice Hall, Englewood Cliffs, N.J., 1988.

[2] Rektorys, K., Survey of Applicable Mathematics, The MIT Press, Cambridge, M.A., 1969, p. 325.

[3] McDougle, P., Vector Algebra, Wadsworth Publishing Co., Belmont, C.A., 1973.

[4] Boiffier, J-T, The Dynamics of Flight, The Equations, John Wiley, N.Y., 1998.

[5] Thomson, W. T., Introduction to Space Dynamics, Dover, N.Y., 1986.

[6] Hughes, W. F. and E. B. Gaylord, Basic Equations of Engineering Science, Schaum, New York, 1964, p. 136.

[7] Wells, D. A., 1967, Lagrangian Dynamics, Schaum's Outline Series, McGraw-Hill, pp. 180–182.

[8] Timoshenko, S. and D. H. Young, Advanced Dynamics, McGraw-Hill, 1948, pp. 29–31.

[9] Chudinov, P. S., 2001, The Motion of a Point Mass in a Medium with a Square Law of Drag, Journal of Applied Mathematics and Mechanics, vol. 65, pp. 421–426.

[10] Irodov, I. E., Fundamental Laws of Mechanics, English translation, Mir Publishers, 1980.

[11] Pinheiro, M. J., 2004, Some Remarks about Variable Mass Systems, European Journal of Physics, vol. 25, pp. L5–L7.

[12] Gunderson, C. O., Study to Improve Airframe Turbine Engine Rotor Blade Containment, US Federal Aviation Agency Report, FAA RD 77–44, 1977.

[13] Ludin, S. J., Engine Debris Fuselage Penetrations Testing Phase II, DOT/FAA/AR-01/27, II.

[14] Mueller, R. B. A Study of Improvements to the Existing Penetration Equations Including Structural Elements, NAWCD TM 8339, 2002.

[15] Brach, R. M., Mechanical Impact Dynamics, Rigid Body Collisions, Wiley-Interscience, revised ed., 1991.

[16] Parwani, R. R., An Approximate Expression for the Large Angle Period of a Simple Pendulum, European Journal of Physics, vol. 25, pp. 37–39, 2004.

[17] Den Hartog, J.P., Mechanics, Dover, 1961, reprint of 1948 edition.

[18] Thomson, W. T., Theory of Vibrations with Applications, Prentice Hall, 3rd ed., p. 109, 1988.

[19] Shapiro, A. S., The Dynamics and Thermodynamics of Compressible Fluid Flow, vol. 1, John Wiley, N.Y., pp. 100–101, 1953.

[20] Sutton, G. P., and O. Biblarz, Rocket Propulsion Elements, 7th ed., John Wiley, N.Y., 2001.
[21] Brown, C. D., Spacecraft Propulsion, AIAA, Washington, D.C., p. 17, 1996.
[22] Hansen, A. G. Fluid Mechanics, Wiley, N.Y., 1967.
[23] Logsdon, T., Orbital Mechanics, John Wiley, N.Y., 1998.
[24] Kaplan, M. H., Modern Spacecraft Dynamics & Control, John Wiley, N.Y., 1976.
[25] Dimentberg, F. M., Flexural Vibrations of Rotating Shafts, Butterworths, London, 1961.
[26] Ley, W., Rockets, Missiles, and Men in Space, The Viking Press, N.Y., p. 513, 1968.
[27] Griffin, M. D. and J. R. French, Space Vehicle Design, American Institute of Aeronautics and Astronautics, Washington, D.C. 1981.
[28] Rao, J. S., Rotor Dynamics, 2nd ed, Wiley, 1990.
[29] Ehrich, F.F., Handbook of Rotordynamics, McGraw-Hill, 1992.
[30] Wowk, V., Machinery Vibration: Balancing, McGraw-Hill, N.Y., 1994.
[31] Housner, G. W., and D. E. Hudson, Applied Mechanics Dynamics, 2nd ed., Van Nostrand Reinhold, 1959.

3

Natural Frequency of Spring–Mass Systems, Pendulums, Strings, and Membranes

This chapter presents the formulas for natural frequencies and mode shapes of spring–mass systems, strings, cables, and membranes for small elastic deformations. Forced vibration of these systems is shown in Table 7.1 of Chapter 7. Harmonic motion, discussed in the following section, is the basis for all vibration analysis.

3.1 Harmonic Motion

Harmonic motion is sinusoidal in time (Table 3.1). *Frequency* is the oscillations per unit time – the number of positive (or negative) peaks per unit time. Frequency in Hertz has units of 1/s; $2\pi f = \omega$ is the *circular frequency* in radians per second. The reciprocal relationship between frequency in Hertz and period T in seconds is $f = 1/T$.

Period T is the time for one sinusoidal cycle, $T = 1/f$. But integer-loving humans generally state frequency when frequency is greater than 1 Hz (1 cycle per second) and period (inverse of frequency) for frequencies is less than 1 Hz. For example, surfers and oceanographers say that ocean waves come in at about 8 s period, not that the wave frequency is 0.125 Hz.

Sinusoidal displacement in time (t) is $y(t) = Y \sin(2\pi ft)$ (Eq. 3.6). Velocity is the derivative of displacement with respect to time t. Acceleration is the derivative of velocity; it is proportional to frequency squared.

$$\begin{aligned}
\text{Displacement:} \quad & y(t) = Y \sin(2\pi f)t \\
\text{Velocity:} \quad & \frac{dy}{dt} = (2\pi f)Y \cos(2\pi f)t \\
\text{Acceleration:} \quad & \frac{d^2y}{dt^2} = -(2\pi f)^2 Y \sin(2\pi f)t
\end{aligned} \tag{3.1}$$

Formulas for Dynamics, Acoustics and Vibration, First Edition. Robert D. Blevins.
© 2016 John Wiley & Sons, Ltd. Published 2016 by John Wiley & Sons, Ltd.

Table 3.1 Harmonic motion

Notation: X_o = displacement amplitude, zero to peak displacement, largest displacement during a cycle; $\ddot{X}_o$ = acceleration amplitude, the maximum acceleration during harmonic motion; f = frequency, Hertz; V_o = maximum velocity during harmonic motion. See Equation 3.1 and Table 1.2.

Given	Frequency, Hz f	Max Displ't. X_o	Peak-to-Peak displacement	Max Velocity V_o	Max Accel. $\ddot{X}_o$
1. f, X_o	f	X_o	$2X_o$	$(2\pi f)X_o$	$(2\pi f)^2 X_o$
2. f, V_o	f	$V_o/(2\pi f)$	$2V_o/(2\pi f)$	V_o	$(2\pi f)V_o$
3. f, $\ddot{X}_o$	f	$\ddot{X}_o/(2\pi f)^2$	$2\ddot{X}_o/(2\pi f)^2$	$\ddot{X}_o/(2\pi f)$	$\ddot{X}_o$
4. X_o, $\ddot{X}_o$	$\sqrt{\ddot{X}_o/X_o}/2\pi$	X_o	$2X_o$	$\sqrt{X_o}\sqrt{\ddot{X}_o}$	$\ddot{X}_o$
5. X_o, V_o	$V_o/(2\pi X_o)$	X_o	$2X_o$	V_o	V_o^2/X_o
6. V_o, $\ddot{X}_o$	$\sqrt{\ddot{X}_o/V_o}/2\pi$	$V_o^2/\ddot{X}_o$	$2V_o^2/\ddot{X}_o$	V_o	$\ddot{X}_o$

Example: Case 1, 1000 Hz and 2 mm displacement amplitude. Velocity amplitude = $(2\pi f)X_o$ = 12.56 m/s.

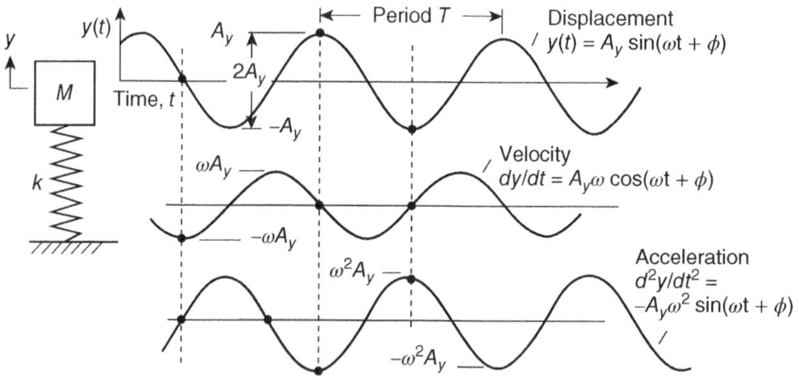

Figure 3.1 Sinusoidal motion during free vibration of a spring–mass system. Displacement and velocity are out of phase. The period $T = 2\pi/\omega = 1/f$

The force on an accelerating mass is determined with Newton's 2nd Law (Eq. 1.1). Y is a single amplitude and $2Y$ is a double amplitude. As shown in Figure 3.1, sinusoidal (harmonic) velocity and displacement are *out of phase*; velocity leads displacement by 90°. Displacement and acceleration are *in phase* but opposite in sign. The maximum displacement is Y, maximum acceleration is $(2\pi f)^2 Y$, maximum velocity is $(2\pi f)Y$, and frequency is f. If two of these are known, then formulas given in Table 3.1 compute the remainder. The *tripartite* plot, shown in Figure 3.2, uses these relationships to specify the vibration environment for machinery.

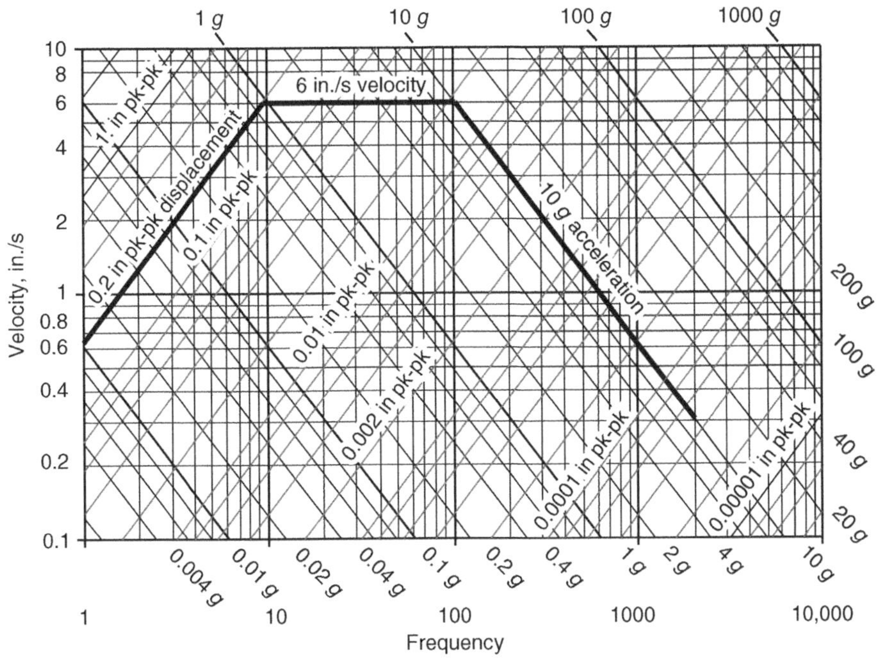

Figure 3.2 The three-axis plot of acceleration, velocity, and peak-to-peak displacement amplitudes for harmonic motion is used to specify a vibration environment envelope of a vibrating base

3.2 Spring Constants

Extensional spring constant, k, is the force required to extend the spring by a unit length; its units are force over length (Table 3.2). As shown in Figure 3.3, $k = F/x$, where F is the applied force and x is the resultant displacement. *Torsional spring constant*, $k = T/\theta$, is the moment required to produce one radian rotation of the spring, where T is the applied torque and θ is the resultant rotation. References [1–5] present the design and analysis of springs, flexures, and elastic structures.

Example 3.1 Elastic pad

A thin elastic pad is compressed between two rigid plates presented in case 10 of Table 3.1. Derive its spring constant from the theory of elasticity.

Solution: The stress–strain equations for normal stress in a material with *elastic modulus* E and *Poisson's ratio* ν (Chapter 8) are applied to the thin elastic pad in the y–z plane [2]:

$$\varepsilon_x = \frac{[\sigma_x - \nu(\sigma_y + \sigma_z)]}{E} \tag{3.2a}$$

$$\varepsilon_y = \frac{[\sigma_y - \nu(\sigma_x + \sigma_z)]}{E} \tag{3.2b}$$

$$\varepsilon_z = \frac{[\sigma_z - \nu(\sigma_x + \sigma_y)]}{E} \tag{3.2c}$$

Table 3.2 Spring stiffness

Notation: A = cross sectional area; c = distance from neutral axis to edge of cross section (Section 4.1); E = modulus of elasticity; F = applied load; $G = E/[2(1+\nu)]$ = shear modulus; I = area moment of inertia, Table 2.1; g = acceleration due to gravity, Table 1.2; $k = dF/d\delta$, force per unit length; $k = dM/d\theta$, moment per radian; L = length of beam or string; M = applied moment; P = mean tension; t = 2x(neutral axis to edge), thickness for symmetric sections; W = weight; θ = angle of rotation, radians; ν = Poisson's ratio; δ = deformation; σ = max stress Refs [1–8].

Extension Spring	Extension Spring Constant, k
1. Extension Spring 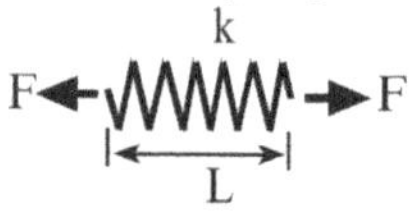	$k = \frac{dF}{dL}$ See case 28 for helical springs.
2. Two Springs in Series 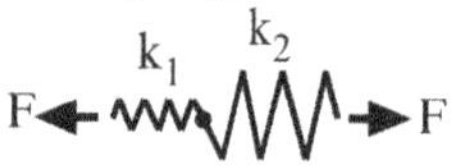	$\frac{k_1 k_2}{k_1 + k_2}$ Both springs have load F.
3. N Springs in Series 	$\frac{1}{\frac{1}{k_1}+\frac{1}{k_2}+\ldots+\frac{1}{k_N}} = \frac{k}{N}$, if $k_1 = k_2 = .. = k_N = k$ Each spring has load F.
4. Two Springs in Parallel 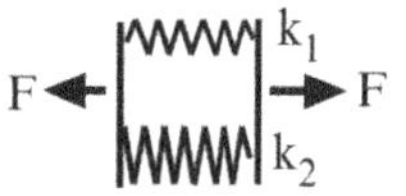	$k_1 + k_2$ Bars do not tilt. Springs have equal elongation.
5. N Springs in Parallel 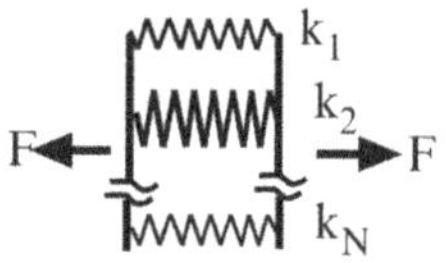	$k_1 + k_2 + \ldots + k_N$ Bars do not tilt. Springs have equal elongation.
6. Two Springs in Parallel, Bar Tilts 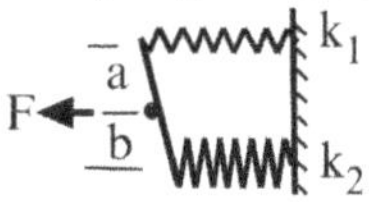	$\frac{k_1 k_2 (a+b)^2}{k_1 a^2 + k_2 b^2}$
7. Uniform Bar Stretches 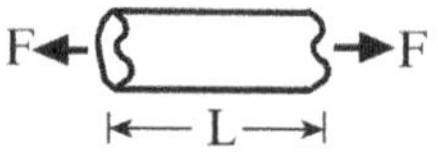	$\frac{EA}{L}$ E = modulus of elasticity, $\sigma = F/A$ A = cross sectional area. Section is uniform.
8. Stranded Cable or Rope 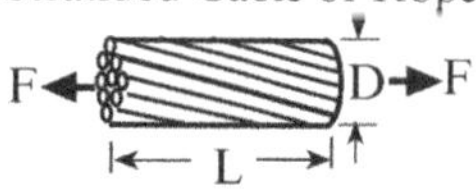	$\alpha \frac{E}{L} \frac{\pi D^2}{4}$ See Section 3.4. $\frac{\sigma}{\delta} = \alpha \frac{E}{L}$ E = modulus of elasticity of strands α = packing factor, typically 0.3 to 0.5.

Table 3.2 Spring stiffness, continued

Extension Spring	Extension Spring Constant, k
9. Tapered Circular Bar 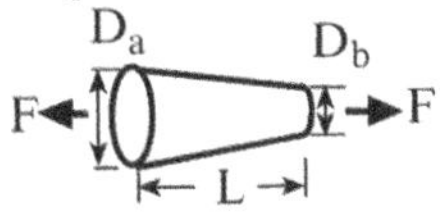	$\frac{\pi E D_a D_b}{4L}$, $\frac{\sigma}{\delta}=\frac{E D_a}{L D_b}$
10. Elastic Compression Pad 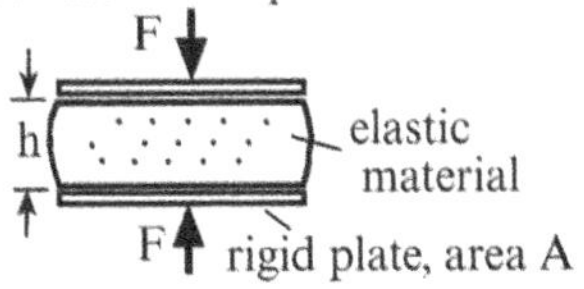	$k=\begin{cases}\frac{1-\nu}{(1+\nu)(1-2\nu)}\frac{EA}{h} & \text{material bonded to plate}\\ \frac{EA}{h} & \text{material slides on plate}\end{cases}$ E = modulus of elasticity, ν = Poisson's ratio.
11. Shear Pad 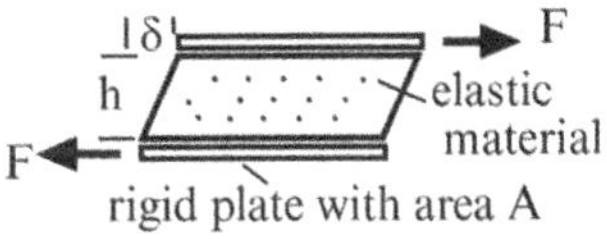	$\frac{GA}{h}$, $G=\frac{E}{2(1+\nu)}$ = shear modulus Elastic material bonded to plates. Shear stress =F/A.
12. Cantilever Beam 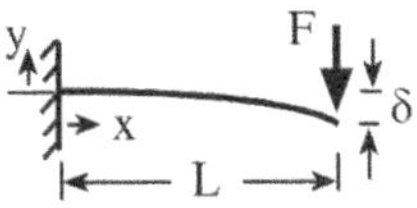	$\frac{EI}{L^3}$ $y(x)=-\frac{\delta}{2}\left(\frac{x}{L}\right)^2\left(3-\frac{x}{L}\right)$ maximum bending moment = FL $\sigma/\delta = 3Et/2L^2$, t/2 = distance from neutral axis edge
13. Pinned Beam Center Load 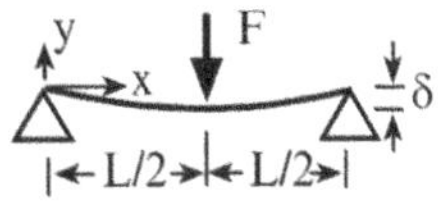	$\frac{48EI}{L^3}$ $y(x)=-\delta\left(3\frac{x}{L}-4\left(\frac{x}{L}\right)^3\right)$, $0\le x\le\frac{L}{2}$ Ends of beam rotate but do not translate. maximum bending moment σ/δ = FL/4,
14. Clamped Beam Center Load 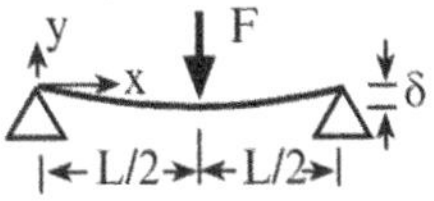	$\frac{192EI}{L^3}$ $y(x)=-\delta\left(\frac{x}{L}\right)^2\left(4\frac{x}{L}-3\right)$, $0\le x\le\frac{L}{2}$ Ends of beam do not translate or rotate. maximum bending moment = FL/8, $=12Et/L^2$
15. Pinned Beam Off Center Load 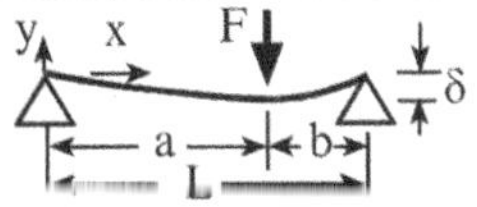	$\frac{3EI(a+b)}{a^2b^2}$ $y(x)=\delta\left(1-\left(\frac{b}{L}\right)^2-\left(\frac{x}{L}\right)^2\right)\frac{b}{L}\frac{L^4}{2a^2b^2}$ maximum bending moment = Fab/L,
16. Clamped-Pinned Beam Center F 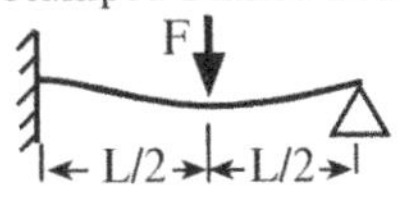	$\frac{768EI}{7L^3}$ maximum bending moment = 3FL/16

Table 3.2 Spring stiffness, continued

Extension Spring	Extension Spring Constant, k
17. Clamped Beam Off Center Load 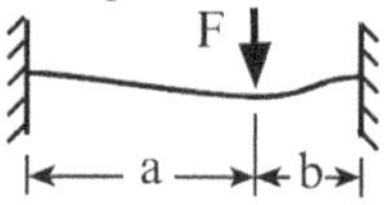	$\frac{3EI(a+b)^3}{a^3b^3}$, $k=\frac{192EI}{L^3}$ when $a=b=L/2$, max bending moment = 0.1481FL for a =L/3
18. Extended Cantilever with Support 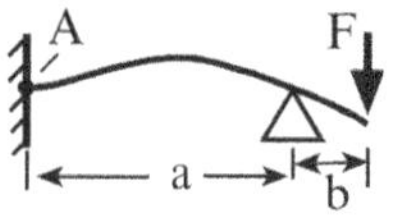	$\frac{3EI}{(a+b)b^2}$ A pinned $\frac{24EI}{(3a+8b)b^2}$ A clamped Ref. [10], p. 37
19. Clamped-Guided Beam End Load 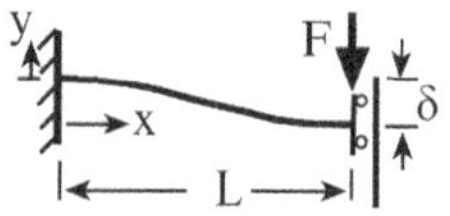	$\frac{12EI}{L^3}$ $y(x)=-\delta\left(\frac{x}{L}\right)^2\left(3-2\frac{x}{L}\right)$ maximum bending moment = FL/2
20. Multi Beam Support 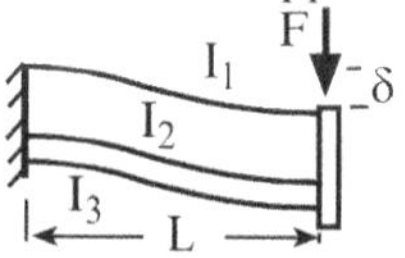	$\frac{12E}{L^3}(I_1+I_2+...+I_N)$ The beams are equal length and they are offset. The ends of the beams define two parallel planes. Overall bending has been neglected; offset between beams is large and beams do not buckle.
21. Tensioned String 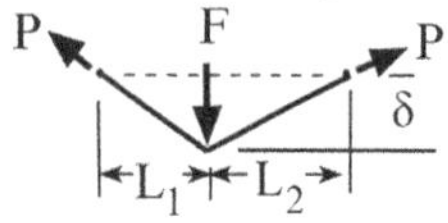	$k=\frac{F}{\delta}=\frac{P}{L_1}+\frac{P}{L_2}=\frac{P}{2L}$, if $L_1=L_2$ P = tension in string $\delta<<L$
22. Hanging Mass 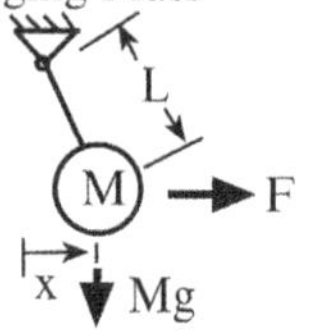	$k=\frac{dF}{dx}=Mg\frac{L^2}{(L^2-x^2)^{3/2}}\approx\frac{Mg}{L}$ for $x<<L$ $F=Mg\frac{x}{(L^2-x^2)^{1/2}}$
23. Weight Suspended by Two Equal Length Strings 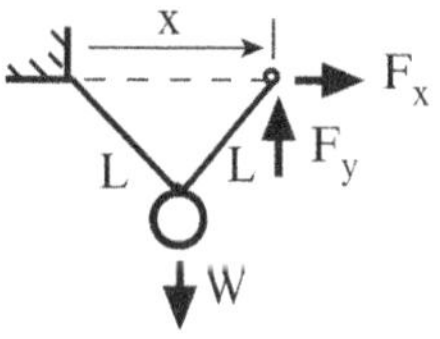	$k=\frac{dF_x}{dx}=\frac{4MgL^2}{(4L^2-x^2)^{3/2}}$ $F_x=\frac{Mg\,x}{(4L^2-x^2)^{1/2}}=\frac{Tx}{2L}$ $F_y=Mg/2$, L = length of each string, Mg = weight T = tension in each string $=2MgL/(4L^2-x^2)^{1/2}$

Table 3.2 Spring stiffness, continued

Spring	Extension or Torsion Spring Constants, k, k
24. Gas Spring 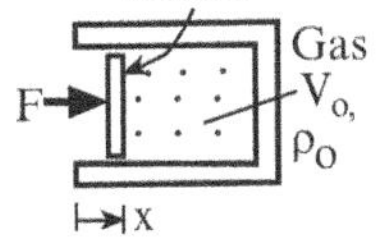 Mass of gas is trapped by piston	$k = \begin{cases} p_o A^2 / V_o & \text{isothermal} \\ c_o^2 \rho_o A^2 / V_o & \text{adiabatic, } c_o = \text{sound speed} \end{cases}$ pV^{κ} = constant. κ =1 for isothermal and γ for adiabatic. $k = \frac{dF}{dx} = \kappa \frac{p}{V} A^2 = \frac{\kappa p_o V_o^{\kappa}}{V^{1-\kappa}} A^2$ where Volume $V = V_o - Ax$ For small oscillations, $Ax << V_o$, $V < V_o$, $p < p_o$
25. Belleville Disk Spring 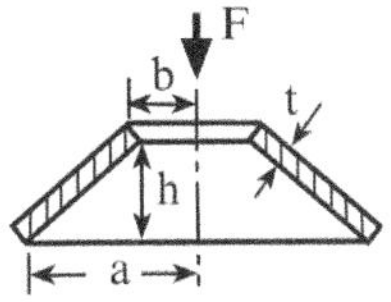 	$k = \frac{E}{(1-\nu^2)\beta a^2}\left[(h-\delta)(h-\delta/2)t - t^3\right]$ δ = deflection. Spring is nonlinear. Refs [1, 2] a/b 1.0 1.2 1.4 1.6 1.8 2.0 3.0 4.0 5.0 β 0 0.31 0.46 0.57 0.64 0.70 0.78 0.80 0.79
26. Triangle Form Cantilever 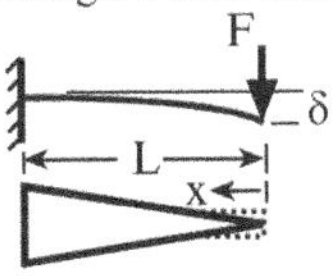 	$k = \frac{2EI_o}{L^3}$ $\quad y(x) = -\frac{FL}{2EI_o}(L-x)^2$ I_o = moment of inertia at base of triangle. Thickness is constant. maximum bending moment = FL. $\sigma/\delta = Et/L^2$
27. Frame with Lateral Load 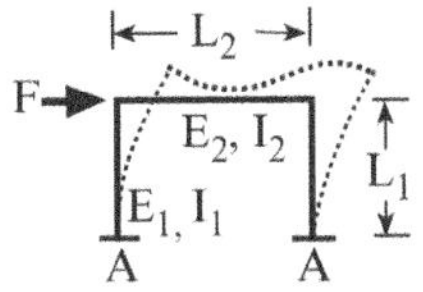 	$k = \frac{6EI_1}{L_1^3}\frac{4 + 2L_2E_1I_1/(3L_1E_2I_2)}{1 + 2L_2E_1I_1/(3L_1E_2I_2)}$, clamped feet at A $= \frac{6EI_1}{L_1^3}\frac{1}{1 + L_2E_1I_1/(2L_1E_2I_2)}$, pinned feet at A
28. Helical Spring Extension 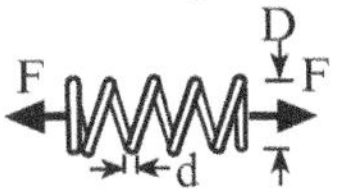 	$k = \frac{Gd^4}{8nD^3}$ circular wire, $k = \frac{Gd^4}{5.59nD^3}$ square wire $G = E/(2(1+\nu))$, n=number of active turns. n=4 shown. d = wire diameter. Refs [1–3], p.56. L >> R, max shear stress $= (8\,FD/\pi d^3)\left[1 + (5d/4D) + (7/8)(d/D)^2\right]$
29. Helical Spring in Bending 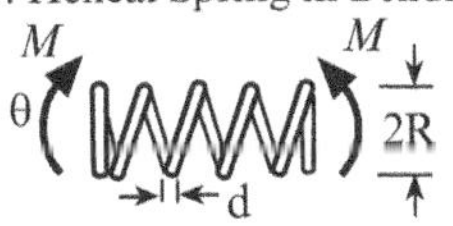 	$k = \frac{dM}{d\theta} = \frac{Ed^4}{64nR}\left(\frac{1}{2+\nu}\right)$ For L >> R, See Refs [1, 2] n= number of active turns. n=4 shown.
30. Torsion of Helical Spring 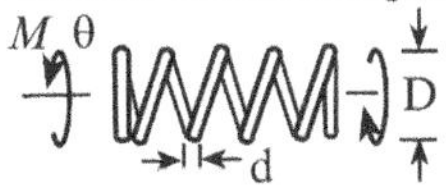 	$k = \frac{dM}{d\theta} = \frac{Ed^4}{10.8(2\pi)nD}$ For L >> R n = number of active turns. See Refs [1, 2]. n = 4 shown. Note k per radian. Ref. [3], p. 138.

Table 3.2 Spring stiffness, continued

Torsion Spring	Torsion Spring Constant, $k = dM/d\theta$
31. Spiral Torsion Spring 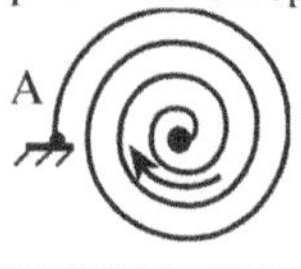	$\frac{EI}{L}$ if A is clamped Ref. [2] $0.8\frac{EI}{L}$ if A is pinned L = length of spiral, I = area mom. inertia, Table 1.5
32. Torsion of Uniform Shaft 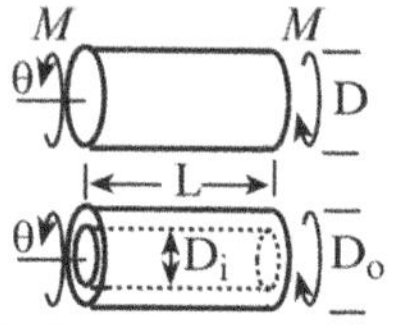	$\frac{GI_p}{L} = \frac{\pi G D^4}{32L}$ circular shaft, D = outside diameter $\frac{\pi G(D_o^4 - D_i^4)}{32L}$ hollow circular shaft $\frac{CG}{L}$ noncircular shaft, C given in Table 4.15.
33. Torsion of tapered Circular Shaft 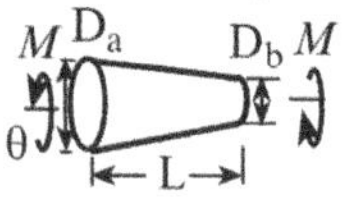	$\frac{3\pi}{32}\frac{D_b^4 G}{L}\frac{1}{\frac{D_b}{D_a} + \left(\frac{D_b}{D_a}\right)^2 + \left(\frac{D_b}{D_a}\right)^3}$
34. Torsion of Two Springs in Series 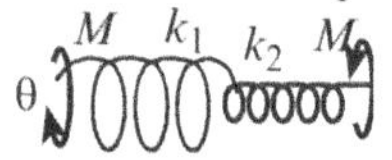	$k = \frac{k_1 k_2}{k_1 + k_2}$, N springs in series $k = \frac{1}{\frac{1}{k_1} + \frac{1}{k_2} + \ldots + \frac{1}{k_N}}$
35. Torsion of Two Parallel Springs 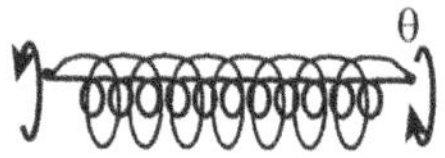	$k_1 + k_2$ For N springs in parallel $k = k_1 + k_2 + \ldots + k_N$ Springs have the same angular rotation.
36. Torsion of Two Geared Shafts 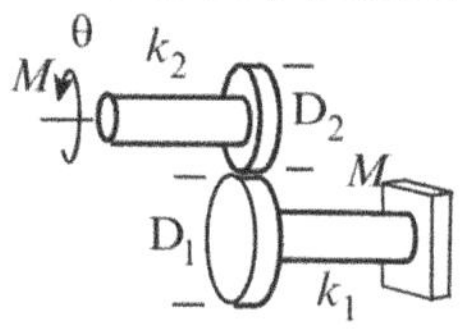	$\frac{k_1 k_2}{k_1 + n^2 k_2}$ where $n = \frac{D_1}{D_2} = \frac{\text{speed of shaft 2}}{\text{speed of shaft 1}}$ For N geared shafts in series, $\frac{1}{\frac{n_1^2}{k_1} + \frac{n_2^2}{k_2} + \ldots + \frac{n_N^2}{k_N}}$, $n_i = \frac{D_i}{D_{i+1}} = \frac{\text{speed of shaft i+1}}{\text{speed of shaft i}}$
37. Torsion of Hanging Mass 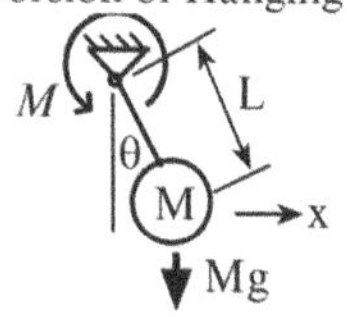	$\frac{dM}{d\theta} = MgL\cos\theta =$ torsional stiffness $\approx MgL$ for $\theta << 1$ radian $M = MgL\sin\theta$ $k = Mg/L =$ stiffness in x direction
38. Torsion Pinned Beam Off Center 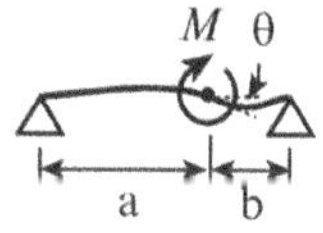	$\frac{EIL}{L^2 - 3ab}$, L = a + b, max moment = Ma/L, Mb/L $\frac{3EI}{b}$, if a=0 or b=0; $\frac{12EI}{L}$, if $a = b = \frac{L}{2}$

Table 3.2 Spring stiffness, continued

Torsion Spring	Torsion Spring Constant, $k = dM/d\theta$
39. Torsion Clamped Beam Off Ctr 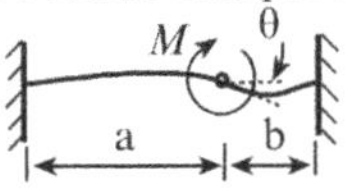	$\frac{EIL^3}{ab(L^2-3ab)}, \frac{16EI}{L}$ if $a=b=\frac{L}{2}$; $L=a+b$ max bend. moment $= M[4\frac{a}{L}-9(\frac{a}{L})^2+6(\frac{a}{L})^3]$
40. Torsion Clamped-Pinned Beam 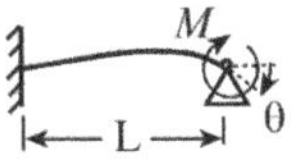	$\frac{4EI}{L}$ maximum bending moment = M
41. Torsion of Clamped-Free Beam 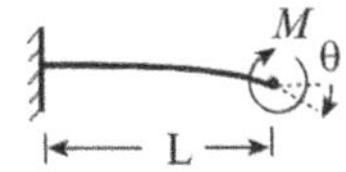	$\frac{EI}{L}$ maximum bending moment = M
42. End Moments on Pinned Beam 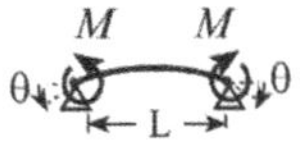	$\frac{2EI}{L}$ maximum bending moment = M
43. Bending of Circular Notch Bar 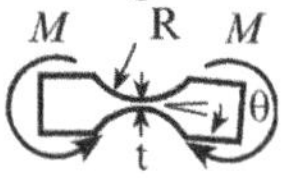	$\frac{2Ebt^{5/2}}{9\pi R^{1/2}}$ two notches in rectangular bar, width =b $\frac{Et^{7/2}}{20R^{1/2}}$ axisymmetric groove in circular bar, Ref. [5]
44. Torsion of Pipe Bend 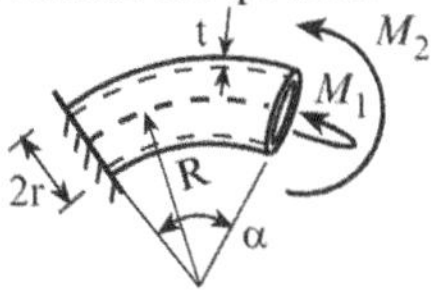	$\frac{EI}{R\alpha}\beta$ for both M_1 and M_2, $R/r>1.7$, $R\alpha>2r$ $\beta = \frac{tR}{1.65r^2}\left[1+6\frac{rP}{tE}\left(\frac{r}{t}\right)^{4/3}\left(\frac{R}{r}\right)^{1/3}\right]$, P = internal pressure if $\beta>1$ then $\beta=1$. ASME Code, Section III, NB-3686.1
45. 3-Bar Linkage Torsion Springs 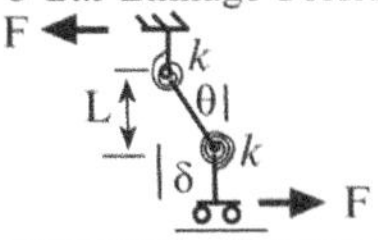	$k = \frac{2k}{L^2}$ k = torsion constant of each torsion spring.
46. Inclined Spring 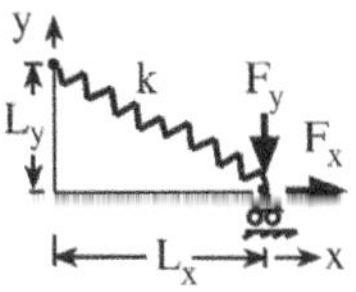	$k_x = \frac{dF_x}{dL_x} = k\left[1-\frac{L_o L_y^2}{(L_x^2+L_y^2)^{3/2}}\right]$, $L_o =$ unstressed length of spring $F_x = k\left[1-\frac{L_o}{(L_x^2+L_y^2)^{1/2}}\right]L_x$, $F_y = k\left[1-\frac{L_o}{(L_x^2+L_y^2)^{1/2}}\right]L_y$ Motion in y direction, substitute x for y in k_x formula.
47. Slender Ring, diametric Load 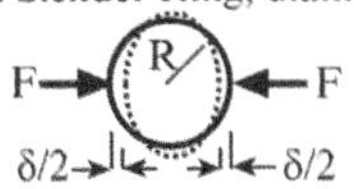	$k = \frac{F}{\delta} = 6.721\frac{EI}{R^3}$ Ring is slender beam radius R. I is moment of inertia of beam cross section. δ = diametric deformation. Ref. [1]

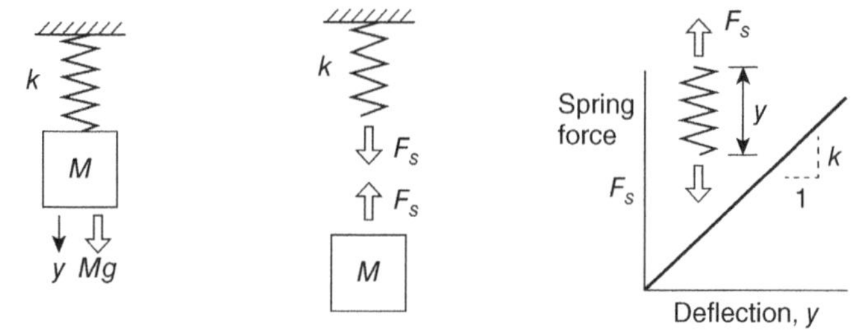

Figure 3.3 Mass with a linear extension spring

The out-of-plane stress $\sigma_x = F/A$ is the normal force F on the pad divided by the pad's area, A. Out-of-plane strain ε_x is the change in thickness divided by the pad thickness $\varepsilon_x = \delta_x/h$.

If the pad is lubricated so it can expand and contract freely, in-plane stresses are zero, $\sigma_y = \sigma_z = 0$, which is called plane stress, and Equation 3.2a reduces to $\varepsilon_x = \sigma_x/A$:

$$\varepsilon_x = \frac{\delta_x}{h} = \frac{\sigma_x}{E} = \frac{F/A}{E}, \quad \text{hence,} \quad F = \left(\frac{EA}{h}\right)\delta_x \tag{3.3}$$

Equation 3.2b and 3.2c give the in-plane strains $\varepsilon_y = \varepsilon_z = -\nu\sigma_x/E$. When the pad is in tension, its sides pull in; in compression, its sides bulge out owing to Poisson's ratio.

Now, consider that rigid plates are bonded to the top and bottom of the pad. They constrain in-plane displacements and strains to zero, $\varepsilon_y = \varepsilon_z = 0$, which is plain strain. In-plane stresses are also proportional to Poisson's ratio: $\sigma_y = \sigma_z = \nu\sigma_x/(1-\nu)$ (Eqs 3.2a, 3.2b, and 3.2c). The constrained-pad spring constant is found to be a function of Poisson's ratio:

$$k = \begin{cases} \dfrac{EA}{h}, & \text{unconstrained} \\ \dfrac{(EA/h)(1-\nu)}{(1+\nu)(1-2\nu)}, & \text{constrained} \end{cases} \tag{3.4}$$

In-plane constraint increases the pad's out-of-plane stiffness by a factor of 2.14 for a rubber pad with a Poisson's ratio of 0.4.

Example 3.2 Cantilever beam spring

A cantilevered steel plate 4 mm (0.175 in.) thick, 75 mm (3 in.) wide, and 300 mm (11.8 in.) tall is loaded on the free edge as shown in Figure 3.4(a). What is its spring rate? How far can it bend before yielding the mild steel? The modulus of elasticity is 191E9 Pa (27.9E6 psi) and the yield stress is 275 MPa (40,000 psi).

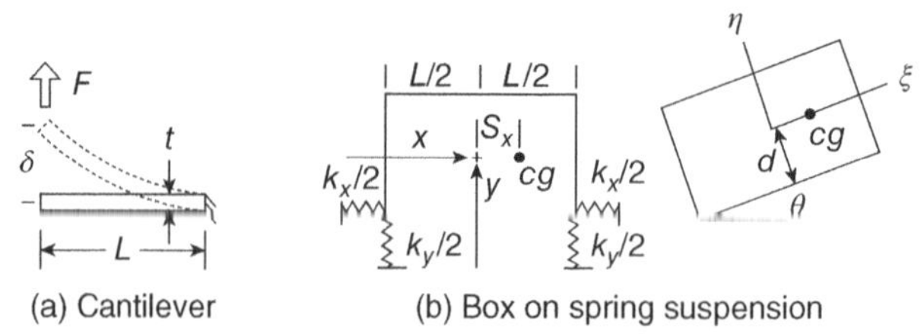

Figure 3.4 Cantilever plate spring and a box on a two-dimensional spring suspension

Solution: Case 12 of Table 3.2 shows the spring constant k. The moment of inertia I of the plate cross section is given in case 4 of Table 1.5:

$$I = \frac{1}{12}bt^3 = \left(\frac{1}{12}\right)\ 0.075\,\text{m}\ (0.004\,\text{m})^3 = 4\times 10^{-10}\ \text{m}^4(0.010\,\text{in.}^4)$$

$$k = 3\left(\frac{EI}{L^3}\right) = 3\ (191\times 10^9\ \text{Pa})\ 4\ \times\ 10^{-10}\ \text{m}^4/(0.3\,\text{m})^3 = 8489\ \text{N/m}\ \ (509\,\text{lb/in.})$$

Yielding is governed by the maximum tensile stress in the outer fiber. The load F that produces yield stress, 275 MPa, is obtained by solving tensile bending stress, $\sigma = My/I$, where $M = FL$ (Chapter 4, Eq. 4.2) for the load at yielding:

$$F = \left(\frac{\sigma I}{yL}\right) = 275\ \text{E6 Pa}\ \frac{(4E\times 10\,\text{m}^4)}{(0.002\,\text{m}\ 0.3\,\text{m})} = 183\,\text{N}\ (41.1\,\text{lb})$$

The corresponding tip deflection is $F/k = 183\text{N}/(8499\ \ \text{N/m}) = 0.0215\,\text{m}\ \ (0.846\,\text{in.})$.

3.3 Natural Frequencies of Spring–Mass Systems

3.3.1 Single-Degree-of-Freedom

The natural frequency of a single-degree-of-freedom spring–mass system (one mass in one direction) is determined by applying Newton's second law to the mass and solving the resultant linear ordinary differential equations of motion for the frequency of periodic free vibration solutions (Tables 3.3–3.5).

Figure 3.3 shows a simple spring–mass system. The mass M hangs from a spring with spring constant k. The spring force opposes the downward displacement y of the mass with force $F = -ky$. The acceleration of the mass M is d^2y/dt^2. Gravity acts downward with acceleration g. Substituting these into Newton's second law (Eq. 2.4) gives the *equation of motion* [7] (p. 18):

$$M\frac{d^2y}{dt^2} + ky = Mg \tag{3.5}$$

This linear equation can be decomposed into two equations: a homogeneous dynamic equation with an oscillatory solution and a static equation for the mean deformation. The total solution is the sum of the static and dynamic solutions:

$$\text{Static equation}\quad ky = Mg$$

$$\text{Static solution}\quad y = \frac{Mg}{k}$$

$$\text{Dynamic equation}\quad M\frac{d^2y}{dt^2} + ky = 0$$

$$\text{Dynamic solution}\quad y(t) = Y\sin\,\omega_n t$$

$$\text{Total solution}\quad y(t) = \frac{Mg}{k} + Y\sin\,\omega_n t \tag{3.6}$$

$\omega_n = (k/M)^{1/2}$ is the circular natural frequency, which is verified by substituting the dynamic solution into its equation of motion. Its units are radians per second.

Table 3.3 Natural frequency of spring–mass systems

Notation: E = modulus of elasticity; f = natural frequency, Hz; I = area moment of inertia, Table 2.1; k = deflection spring constant = (force/deformation), Table 3.1; L = length of beam or string; M = mass; T = tension in string; x = displacement; over bar (~) denotes mode shape.
Table 1.2 gives consistent sets of units. Springs are massless. Also see Table 4.9.

Spring Mass System	Natural Frequency f, Hz	Mode Shape and Comments
1. Mass, Spring 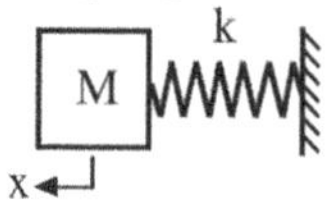	$\frac{1}{2\pi}\left(\frac{k}{M}\right)^{1/2}$	For a massive spring, Table 4.8. $\frac{1}{2\pi}\left(\frac{k}{M+M_s/3}\right)^{1/2}$ Ms = total mass of spring.
2. Two Equal Masses and Two Equal Springs 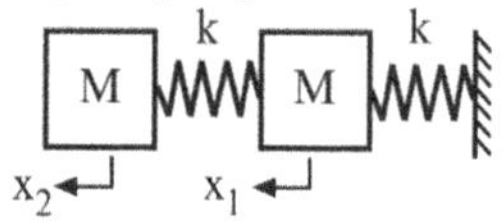	$\frac{(3-5^{1/2})^{1/2}}{2^{3/2}\pi}\left(\frac{k}{M}\right)^{1/2}$, $\frac{(3+5^{1/2})^{1/2}}{2^{3/2}\pi}\left(\frac{k}{M}\right)^{1/2}$	$\begin{bmatrix}\tilde{x}_1\\ \tilde{x}_2\end{bmatrix}=\begin{bmatrix}1\\ \frac{1+5^{1/2}}{2}\end{bmatrix},\begin{bmatrix}1\\ \frac{1-5^{1/2}}{2}\end{bmatrix}$
3. Two Unequal Masses and Two Unequal Springs 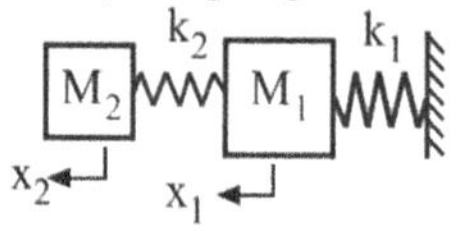	$\frac{1}{2^{3/2}\pi}\left(\frac{k_1+k_2}{M_1}+\frac{k_2}{M_2}\mp\left[\left(\frac{k_1+k_2}{M_1}+\frac{k_2}{M_2}\right)^2-\frac{4k_1k_2}{M_1M_2}\right]^{1/2}\right)^{1/2}$ $\begin{bmatrix}\tilde{x}_1\\ \tilde{x}_2\end{bmatrix}=\begin{bmatrix}1\\ 1+\frac{k_1}{k_2}-\frac{M_1}{k_2}(2\pi f_i)^2\end{bmatrix}$, $i=1,2$	Ref. [10]
4. Three Equal Masses and Three Equal Springs 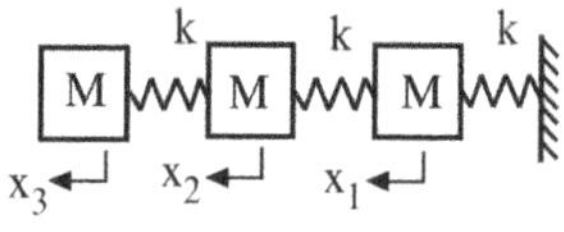	$0.0782\left(\frac{k}{M}\right)^{1/2}$, $0.1985\left(\frac{k}{M}\right)^{1/2}$, $0.2868\left(\frac{k}{M}\right)^{1/2}$ $\begin{bmatrix}\tilde{x}_1\\ \tilde{x}_2\\ \tilde{x}_3\end{bmatrix}=\begin{bmatrix}1\\ 1.802\\ 2.247\end{bmatrix},\begin{bmatrix}1\\ 0.445\\ -0.802\end{bmatrix},\begin{bmatrix}1\\ -1.247\\ 0.555\end{bmatrix}$	
5. N Equal Masses N Equal Springs 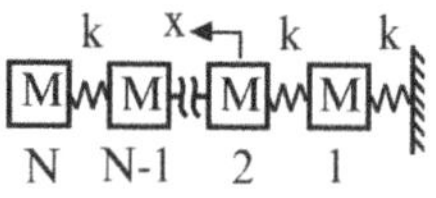	$\frac{1}{\pi}\sin\left[\frac{(2i-1)}{(2N+1)}\frac{\pi}{2}\right]\left(\frac{k}{M}\right)^{1/2}$ $i=1,2,\ldots,N$	Ref. [9], p. 343
6. Mass and Two Equal Springs 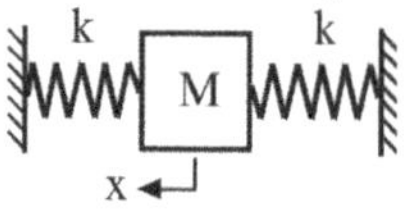	$\frac{1}{2\pi}\left(\frac{2k}{M}\right)^{1/2}$	
7. Mass and Two Unequal Springs 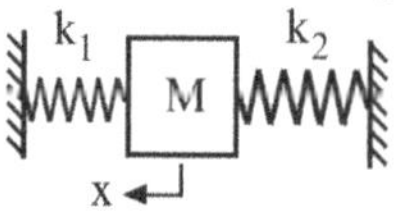	$\frac{1}{2\pi}\left(\frac{k_1+k_2}{M}\right)^{1/2}$	

Table 3.3 Natural frequency of spring–mass systems, continued

Spring Mass System	Natural Frequency f, Hz	Mode Shape and Comments
8. Two Equal Masses and Three Equal Springs 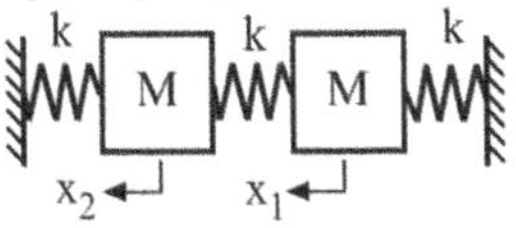	$\frac{1}{2\pi}\left(\frac{k}{M}\right)^{1/2}, \frac{1}{2\pi}\left(\frac{3k}{M}\right)^{1/2}$	$\begin{bmatrix}\tilde{x}_1\\ \tilde{x}_2\end{bmatrix}=\begin{bmatrix}1\\1\end{bmatrix},\begin{bmatrix}1\\-1\end{bmatrix}$
9. Two Unequal Masses, Three Unequal Springs 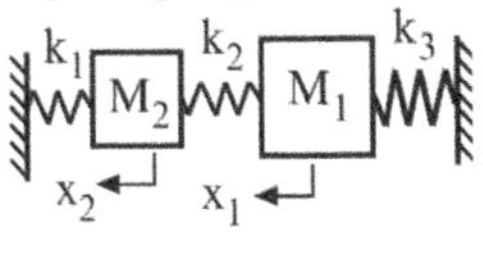	$\frac{1}{2^{3/2}\pi}\left(\frac{k_1+k_2}{M_1}+\frac{k_2+k_3}{M_2}\mp\left[\left(\frac{k_1+k_2}{M_1}+\frac{k_2+k_3}{M_2}\right)^2-\frac{4}{M_1M_2}(k_1k_2+k_2k_3+k_1k_3)\right]^{1/2}\right)^{1/2}$, $\begin{bmatrix}\tilde{x}_1\\ \tilde{x}_2\end{bmatrix}=\begin{bmatrix}1\\ 1+\frac{k_1}{k_2}-\frac{M_1}{k_2}(2\pi f_i)^2\end{bmatrix}$, $i=1,2$	
10. Three Equal Masses and Four Equal Springs 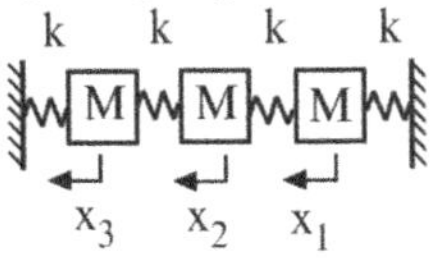	$\frac{(2-2^{1/2})}{2\pi}\left(\frac{k}{M}\right)^{1/2}, \frac{2^{1/2}}{2\pi}\left(\frac{k}{M}\right)^{1/2}, \frac{(2+2^{1/2})}{2\pi}\left(\frac{k}{M}\right)^{1/2}$ $\begin{bmatrix}\tilde{x}_1\\ \tilde{x}_2\\ \tilde{x}_3\end{bmatrix}=\begin{bmatrix}1\\2^{1/2}\\1\end{bmatrix},\begin{bmatrix}1\\0\\-1\end{bmatrix},\begin{bmatrix}1\\-2^{1/2}\\1\end{bmatrix}$	
11. N Equal Masses and N+1 Equal Springs 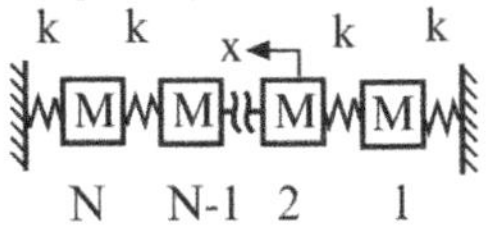	$\frac{1}{\pi}\sin\left[\frac{i}{(N+1)}\frac{\pi}{2}\right]\left(\frac{k}{M}\right)^{1/2}$ $i=1,2,\ldots,N$	Ref. [10], p. 349, 457
12. Two Equal Masses and Spring 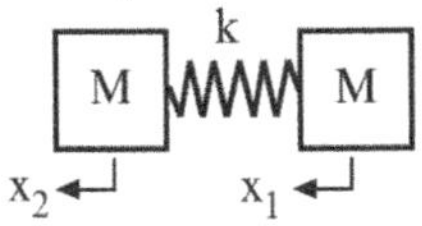	$0, \frac{1}{2\pi}\left(\frac{2k}{M}\right)^{1/2}$	$\begin{bmatrix}\tilde{x}_1\\ \tilde{x}_2\end{bmatrix}=\begin{bmatrix}1\\1\end{bmatrix},\begin{bmatrix}1\\-1\end{bmatrix}$ The zero frequency mode is a rigid body translation.
13.Two Unequal Masses, Spring 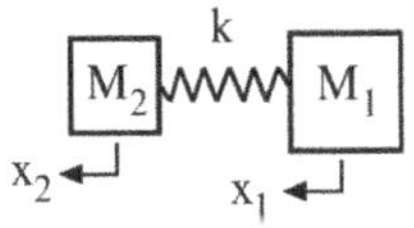	$0, \frac{1}{2\pi}\left(\frac{k(M_1+M_2)}{M_1M_2}\right)^{1/2}$	$\begin{bmatrix}\tilde{x}_1\\ \tilde{x}_2\end{bmatrix}=\begin{bmatrix}1\\1\end{bmatrix},\begin{bmatrix}1\\-M_1/M_2\end{bmatrix}$
14. Three Equal Masses and Two Equal Springs 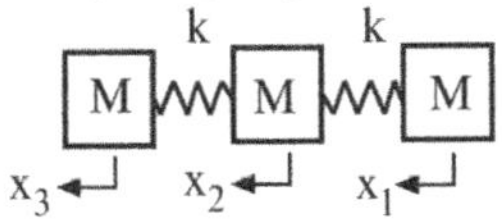	$0, \frac{1}{2\pi}\left(\frac{k}{M}\right)^{1/2}, \frac{1}{2\pi}\left(\frac{3k}{M}\right)^{1/2}$	$\begin{bmatrix}\tilde{x}_1\\ \tilde{x}_2\\ \tilde{x}_3\end{bmatrix}=\begin{bmatrix}1\\1\\1\end{bmatrix},\begin{bmatrix}1\\0\\-1\end{bmatrix},\begin{bmatrix}1\\-2\\1\end{bmatrix}$

Table 3.3 Natural frequency of spring–mass systems, continued

Spring Mass System	Natural Frequency f, Hz	Mode Shape and Comments
15. Three Unequal Masses and Two Unequal Springs 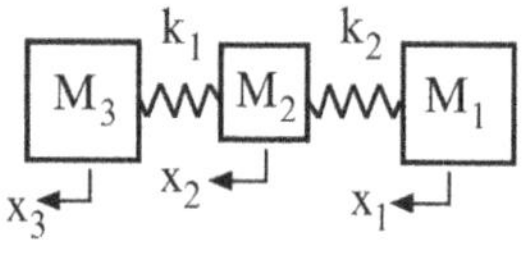 	$0,\ \frac{1}{2^{3/2}\pi}\left(\frac{k_1+k_2}{M_2}+\frac{k_1}{M_1}+\frac{k_2}{M_3}\mp\left[\left(\frac{k_1+k_2}{M_2}+\frac{k_1}{M_1}+\frac{k_2}{M_3}\right)^2 - 4k_1k_2\left(\frac{1}{M_1M_2}+\frac{1}{M_2M_3}+\frac{1}{M_1M_3}\right)\right]^{1/2}\right)^{1/2}$ $\begin{bmatrix}\tilde{x}_1\\ \tilde{x}_2\\ \tilde{x}_3\end{bmatrix}=\begin{bmatrix}1\\ 1-(M_1/k_1)(2\pi f_i)^2\\ (1-(M_1/k_1)(2\pi f_i)^2)/(1-(M_3/k_2)(2\pi f_i)^2)\end{bmatrix},\ i=1,2$	
16. Four Equal Masses and Three Equal Springs 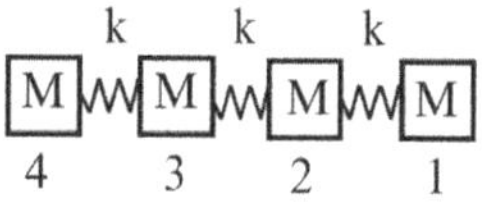 	$0,\ \frac{(2-2^{1/2})^{1/2}}{2\pi}\left(\frac{k}{M}\right)^{1/2},\ \frac{2^{1/2}}{2\pi}\left(\frac{k}{M}\right)^{1/2},\ \frac{(2+2^{1/2})^{1/2}}{2\pi}\left(\frac{k}{M}\right)^{1/2}$ $\begin{bmatrix}\tilde{x}_1\\ \tilde{x}_2\\ \tilde{x}_3\\ \tilde{x}_4\end{bmatrix}=\begin{bmatrix}1\\ 1\\ 1\\ 1\end{bmatrix},\begin{bmatrix}1\\ -1+2^{1/2}\\ 1-2^{1/2}\\ -1\end{bmatrix},\begin{bmatrix}1\\ -1\\ -1\\ 1\end{bmatrix},\begin{bmatrix}1\\ -1-2^{1/2}\\ 1+2^{1/2}\\ -1\end{bmatrix}$	
17. N Equal Masses N–1 Springs 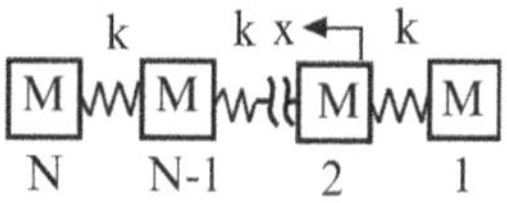 	$\frac{1}{\pi}\sin\left[\frac{i-1}{N}\frac{\pi}{2}\right]\left(\frac{k}{M}\right)^{1/2}$ $i=1,2,...,N$	Ref. [10], p. 247
18. Three Equal Masses and Six Equal Springs 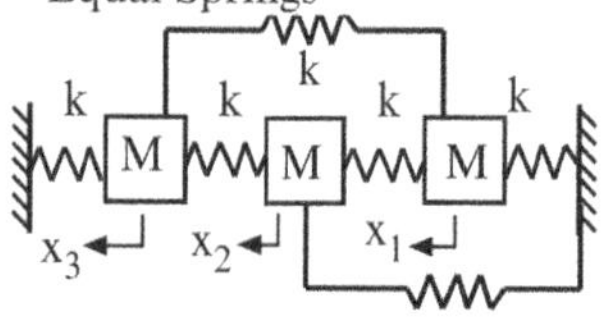 	$\frac{1}{2\pi}\left(\frac{k}{M}\right)^{1/2},\ \frac{1}{2\pi}\left(\frac{k}{M}\right)^{1/2},\ \frac{1}{2\pi}\left(\frac{k}{M}\right)^{1/2}$ $\begin{bmatrix}\tilde{x}_1\\ \tilde{x}_2\\ \tilde{x}_3\end{bmatrix}=\begin{bmatrix}1\\ 1\\ 1\end{bmatrix},\begin{bmatrix}1\\ 0\\ -1\end{bmatrix},\begin{bmatrix}1\\ -2\\ 1\end{bmatrix}$	Ref. [11] Note repeated natural frequencies.
19. Mass on Cantilever Beam 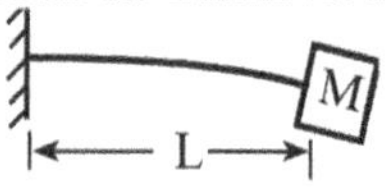 	$\frac{1}{2\pi}\left(\frac{3EI}{ML^3}\right)^{1/2}$ Also see Table 4.8	Mass is compact. Mass of beam is neglected. Spring constants for other beams are given in Table 3.1.
20. Mass on Pinned-Pinned Beam 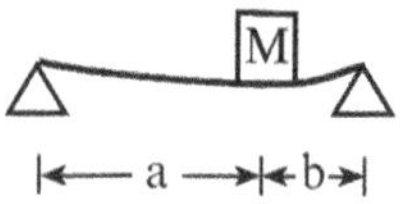 	$\frac{1}{2\pi}\left(\frac{3(a+b)EI}{Ma^2b^2}\right)^{1/2}$ Also see Table 4.8	If a = b = L/2, $f=\frac{2}{\pi}\left(\frac{3EI}{ML^3}\right)^{1/2}$
21. Mass on Clamped Beam 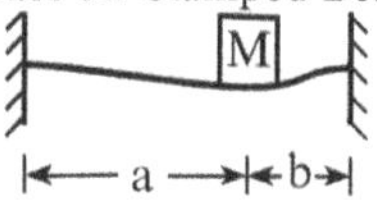 	$\frac{1}{2\pi}\left(\frac{3(a+b)^3EI}{Ma^3b^3}\right)^{1/2}$ Also see Table 4.8	If a = b = L/2, $f=\frac{4}{\pi}\left(\frac{3EI}{ML^3}\right)^{1/2}$

Table 3.3 Natural frequency of spring–mass systems, continued

Spring Mass System	Natural Frequency f, Hz	Mode Shape and Comments
22. Center Mass on Clamped-Pinned Beam 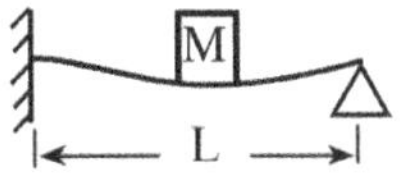	$\frac{1}{2\pi}\left(\frac{768EI}{7ML^3}\right)^{1/2}$ Also see Table 4.8	Other beam spring constants are given in Table 3.1. Mass is compact. Mass of beam is neglected.
23. Mass Multiple Beam Support 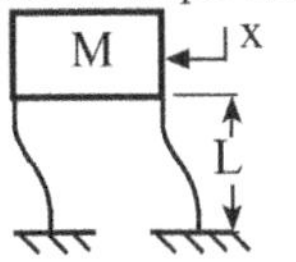	$\frac{1}{\pi}\left(\frac{3EI_s}{ML^3}\right)^{1/2}$	$I_s = \sum I_i$ Beams are off set. Their ends define two parallel panes.
24. Two Unequal Masses on Multiple Beam Supports 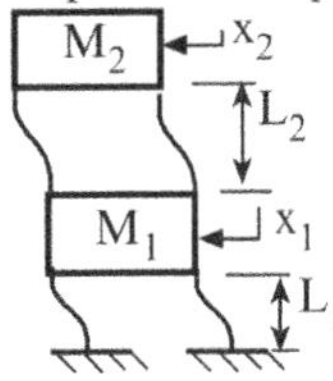	See case 3 with $k_1 = 12I_s / L_1^3$ $k_2 = 12I_s / L_2^3$	Other cases of multiple beam supports with clamped-free boundary conditions are given by cases 1 through 5 where k is given by case 18 of Table 3.1. Mass of beams is neglected.
25. Two Unequal Masses on Free Multiple Beam Supports 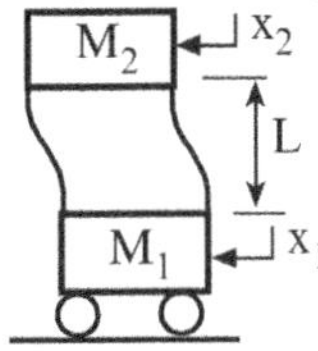	See case 13 with $k = 12I_s / L^3$	Other cases of multiple beam supports with free boundary conditions are given by cases 13 through 16 where k is given by case 18 of Table 3.1. Mass of beams is neglected.
26. Mass on Clamped Multi Beam Supports 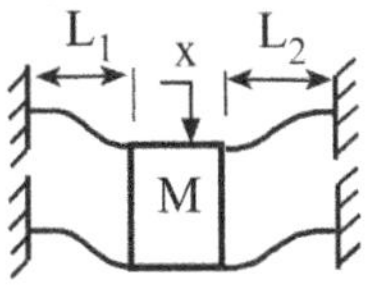	See case 7 with $k_1 = 12I_s / L_1^3$ $k_2 = 12I_s / L_2^3$	Other cases of multiple beam supports with clamped boundary conditions are in cases 7 through 11 where k is case 18 of Table 3.1. Mass of beams is neglected.
27. Mass Centered on String 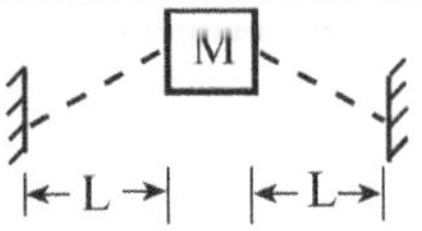	$\frac{1}{2\pi}\left(\frac{2T}{ML}\right)^{1/2}$ T = tension in string	Other cases of tensioned spring supports are in cases 7 through 11 where $k = \frac{\text{string tension}}{\text{string span}}$
28. Mass Off Center on String 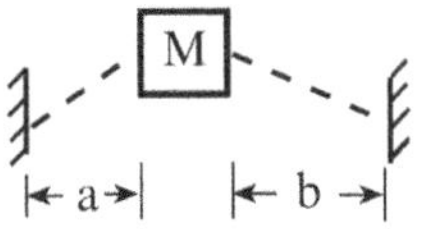	$\frac{1}{2\pi}\left(\frac{T(a+b)}{Mab}\right)^{1/2}$ T = tension in string	Other cases of tensioned spring supports are in cases 7 through 11 where $k = \frac{\text{string tension}}{\text{string span}}$

Table 3.4 Natural frequency of torsion spring and mass system

Notation: C = center of mass; E = modulus of elasticity; f = frequency Hz; I = area; m = moment of inertia (Table 1.5); J = mass moment of inertia about centroid C (Table 1.6); k = spring constant (Table 3.2); k = torsion spring constant, moment/rad (Table 3.2); L = length; M = applied moment; θ = angle of rotation radians; Table 1.2 shows the consistent units for use in the formulas. Also see Tables 3.5 and 4.16.

Torsion Spring, Rigid Body	Natural Frequency f, Hz	Mode Shape and Comments
1. Body with Torsion Spring 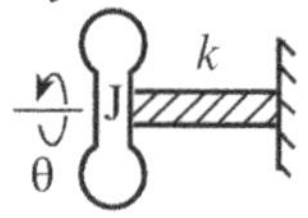	$\frac{1}{2\pi}\left(\frac{k}{J}\right)^{1/2}$	This and other torsion spring systems can be obtained from translation spring and mass systems, Table 3.3, substituting k for k and J for M.
2. Two Equal Bodies, Two Springs 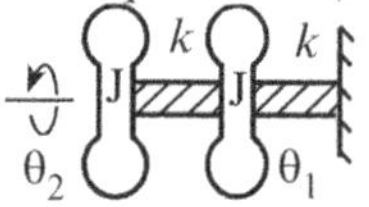	$\frac{(3-5^{1/2})^{1/2}}{2^{3/2}\pi}\left(\frac{k}{J}\right)^{1/2}$, $\frac{(3+5^{1/2})^{1/2}}{2^{3/2}\pi}\left(\frac{k}{J}\right)^{1/2}$	$\begin{bmatrix}\tilde{\theta}_1\\ \tilde{\theta}_2\end{bmatrix}=\begin{bmatrix}1\\ \frac{1+5^{1/2}}{2}\end{bmatrix},\begin{bmatrix}1\\ \frac{1-5^{1/2}}{2}\end{bmatrix}$
3. Two Unequal Bodies and Two Unequal Springs 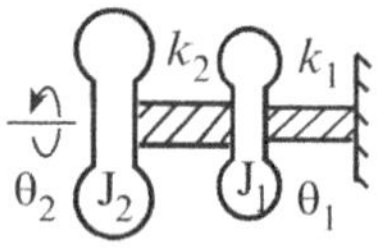	$f_i=\frac{1}{2^{3/2}\pi}\left(\frac{k_1}{J_1}+\frac{k_2}{J_1}+\frac{k_2}{J_2}\mp\left[\left(\frac{k_1}{J_1}+\frac{k_2}{J_1}+\frac{k_2}{J_2}\right)^2-\frac{4k_1k_2}{J_1J_2}\right]^{1/2}\right)^{1/2}$ $\begin{bmatrix}\tilde{\theta}_1\\ \tilde{\theta}_2\end{bmatrix}=\begin{bmatrix}1\\ 1+\frac{k_1}{k_2}-\frac{J_1}{k_2}(2\pi f_i)^2\end{bmatrix}$, $i=1,2$	
4. Three Equal Bodies and Three Equal Torsion Springs 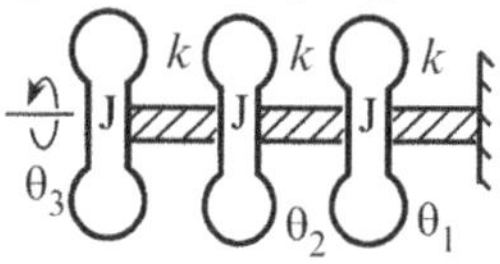	$0.07083\left(\frac{k}{J}\right)^{1/2}$, $0.1985\left(\frac{k}{J}\right)^{1/2}$, $0.2868\left(\frac{k}{J}\right)^{1/2}$ $\begin{bmatrix}\tilde{\theta}_1\\ \tilde{\theta}_2\\ \tilde{\theta}_3\end{bmatrix}=\begin{bmatrix}1\\ 1.802\\ 2.247\end{bmatrix},\begin{bmatrix}1\\ 0.445\\ -0.802\end{bmatrix},\begin{bmatrix}1\\ -1.247\\ 0.555\end{bmatrix}$	
5. N Equal Masses and N Equal Torsion Springs 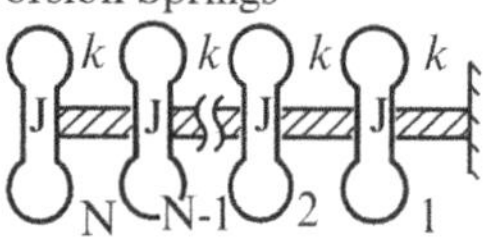	$\frac{1}{\pi}\sin\left[\frac{(2i-1)}{(2N+1)}\frac{\pi}{2}\right]\left(\frac{k}{J}\right)^{1/2}$ $i=1,2,\ldots,N$	Ref. [5], p. 343
6. Body and Two Equal Springs 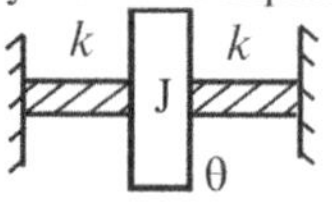	$\frac{1}{2\pi}\left(\frac{2k}{J}\right)^{1/2}$	Other torsion spring systems can be obtained from spring mass systems, Table 3.2, substituting k for k and J for M.
7. Two Equal Bodies and Three Equal Torsion Springs 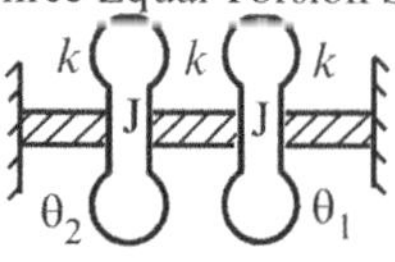	$\frac{1}{2\pi}\left(\frac{k}{J}\right)^{1/2}$, $\frac{1}{2\pi}\left(\frac{3k}{J}\right)^{1/2}$	$\begin{bmatrix}\tilde{\theta}_1\\ \tilde{\theta}_2\end{bmatrix}=\begin{bmatrix}1\\ 1\end{bmatrix},\begin{bmatrix}1\\ -1\end{bmatrix}$

Table 3.4 Natural frequency of torsion spring systems, continued

Torsion Spring, Rigid Body	Natural Frequency f, Hz	Mode Shape and Comments
8. Two Unequal Bodies and Three Unequal Torsion Springs. 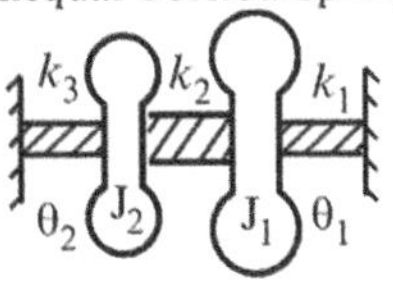	$\frac{1}{2^{3/2}\pi}\left(\frac{k_1+k_2}{J_1}+\frac{k_2+k_3}{J_2}\mp\left[\left(\frac{k_1+k_2}{J_1}+\frac{k_2+k_3}{J_2}\right)^2-\frac{4}{J_1J_2}(k_1k_2+k_2k_3+k_1k_3)\right]^{1/2}\right)^{1/2}$, $\begin{bmatrix}\tilde{\theta}_1\\ \tilde{\theta}_2\end{bmatrix}=\begin{bmatrix}1\\ 1+\frac{k_1}{k_2}-\frac{J_1}{k_2}(2\pi f_i)^2\end{bmatrix}$, i = 1,2	
9. Two Equal Bodies and Spring 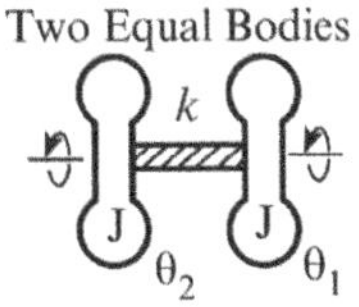	$0,\ \frac{1}{2\pi}\left(\frac{2k}{J}\right)^{1/2}$ $\begin{bmatrix}\tilde{\theta}_1\\ \tilde{\theta}_2\end{bmatrix}=\begin{bmatrix}1\\ 1\end{bmatrix},\begin{bmatrix}1\\ -1\end{bmatrix}$	This and other torsion spring systems can be obtained from translation spring and mass systems, Table 3.2, substituting k for k and J for M.
10. Two Unequal Bodies and Spring 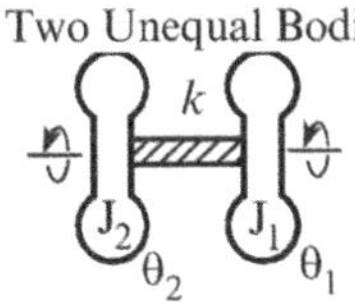	$0,\ \frac{1}{2\pi}\left(\frac{k(J_1+J_2)}{J_1J_2}\right)^{1/2}$	$\begin{bmatrix}\tilde{\theta}_1\\ \tilde{\theta}_2\end{bmatrix}=\begin{bmatrix}1\\ 1\end{bmatrix},\begin{bmatrix}1\\ -\frac{J_1}{J_2}\end{bmatrix}$
11. Three Unequal Bodies and Two Unequal Torsion Springs 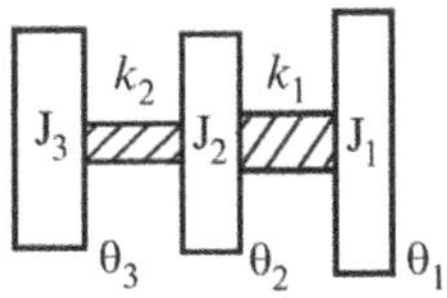	$0,\frac{1}{2^{3/2}\pi}\left(\frac{k_1+k_2}{J_2}+\frac{k_1}{J_1}+\frac{k_2}{J_3}\mp\left[\left(\frac{k_1+k_2}{J_2}+\frac{k_1}{J_1}+\frac{k_2}{J_3}\right)^2-4k_1k_2\left(\frac{1}{J_1J_2J_3}(J_1+J_2+J_3)\right)\right]^{1/2}\right)^{1/2}$, $\begin{bmatrix}\tilde{\theta}_1\\ \tilde{\theta}_2\\ \tilde{\theta}_3\end{bmatrix}=\begin{bmatrix}1\\ 1-\frac{J_1}{k_1}(2\pi f_i)^2\\ \frac{1-(J_1/k_1)(2\pi f_i)^2}{1-(J_3/k_2)(2\pi f_i)^2}\end{bmatrix}$, i = 1,2,3, Ref. [11], p.430	
12. One Body and Three Spring Geared System 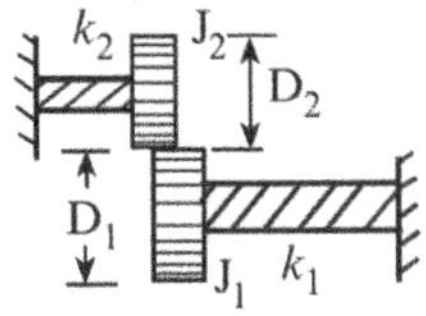	$\frac{1}{2\pi}\left(\frac{k_1+k_{eq}}{J}\right)^{1/2}$ where $k_{eq}=\frac{k_2k_3}{n^2k_2+k_3}$	$n=\frac{D_3}{D_2}$ $=\frac{\text{speed of shaft 2}}{\text{speed of shaft 3}}$ Gears are massless.
13. Two Massive Gears and Two Torsion Springs 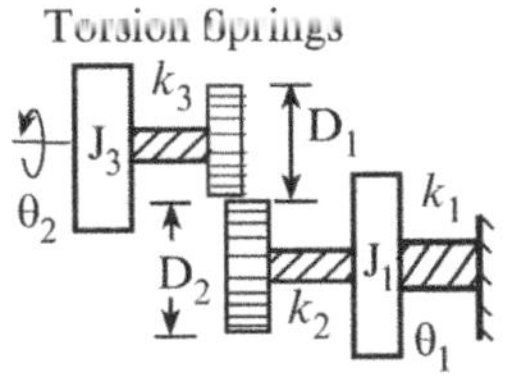	$\frac{1}{2\pi}\left(\frac{(k_1+n^2k_2)}{J_1+n^2J_2}\right)^{1/2}$ where $n=\frac{D_1}{D_2}$	The speed ratio of gears and the transmitted moment is proportional to the inverse with the ratio of their diameters.

Table 3.4 Natural frequency of torsion spring systems, continued

Torsion Spring, Rigid Body	Natural Frequency f, Hz	Mode Shape and Comments
14. Two Unequal bodies and Three Unequal Springs, $n=D_2/D_3$ 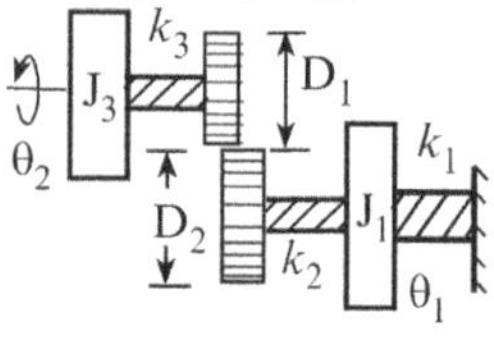	$\frac{1}{2^{3/2}\pi}\left\{\frac{k_1}{J_1}+\frac{n^2 k_{eq}}{J_1}+\frac{k_{eq}}{J_3}\mp\left[\left(\frac{k_1}{J_1}+\frac{n^2 k_{eq}}{J_1}+\frac{k_{eq}}{J_3}\right)^2-\frac{4k_1 k_{eq}}{J_1 J_3}\right]^{1/2}\right\}^{1/2}$ $\begin{bmatrix}\tilde{\theta}_1\\ \tilde{\theta}_2\end{bmatrix}=\begin{bmatrix}1\\ -n^2-\frac{k_1}{k_{eq}}+\frac{J_1}{k_{eq}}(2\pi f_i)^2\end{bmatrix}$, $i=1,2$ $\quad k_{eq}=\frac{k_2 k_3}{k_2+n^2 k_3}$	
15. Two Unequal Bodes and Two Unequal Geared Springs 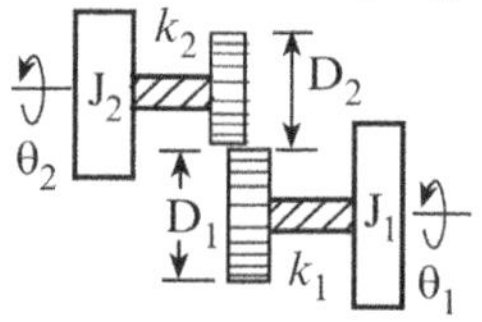	$0,\ \frac{1}{2\pi}\left(\frac{k_1 k_2 (J_1+n^2 J_2)}{J_1 J_2 (k_1+n^2 k_2)}\right)^{1/2}$ Gears are massless. Ref. [11], p. 430	$\begin{bmatrix}\tilde{\theta}_1\\ \tilde{\theta}_2\end{bmatrix}=\begin{bmatrix}1\\ -n\end{bmatrix},\begin{bmatrix}1\\ \frac{J_1}{J_2}\end{bmatrix}$ $n=\frac{D_1}{D_2}=\frac{\text{speed of shaft 2}}{\text{speed of shaft 1}}$
16. Mass on Rigid Lever Spring 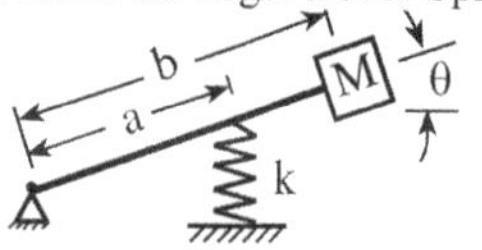	$\frac{1}{2\pi}\left(\frac{ka^2}{Mb^2}\right)^{1/2}$ Lever is massless.	k = extensional (not torsional) spring constant, Table 3.2.
17. Mass on Lever. Two Springs 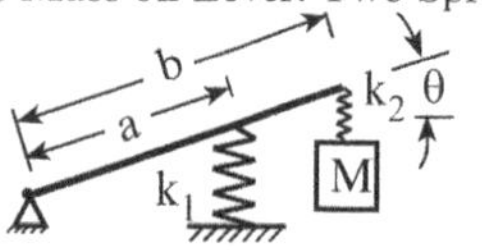	$\frac{1}{2\pi}\left(\frac{k_1 k_2 a}{M(k_1 a+k_2 b)}\right)^{1/2}$ Lever is massless.	
18. Pivoted Body with Spring 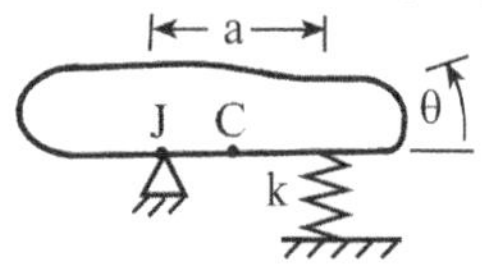	$\frac{1}{2\pi}\left(\frac{k}{J}\right)^{1/2}$ J is polar mass moment of inertia about the pivot.	Solution only applies if center of gravity C has same elevation as the pivot. See case 21 of Table 3.5 for gravity effect if the center of gravity of body is not same elevation as pivot.
19. Pivoted Body and Springs 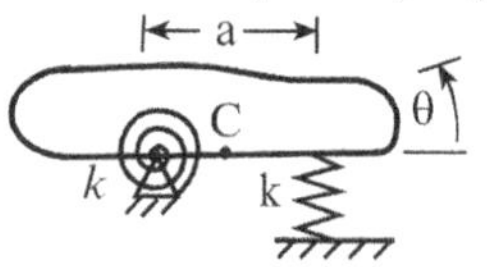	$\frac{1}{2\pi}\left(\frac{k+a^2 k}{J}\right)^{1/2}$ J is polar mass moment of inertia about the pivot.	Solution only applies if center of gravity C has same elevation as the pivot. See case 21 of Table 3.5 for gravity effect if the center of gravity of body is not same elevation as pivot.
20. Symmetric Body Held by Two Tensioned Strings 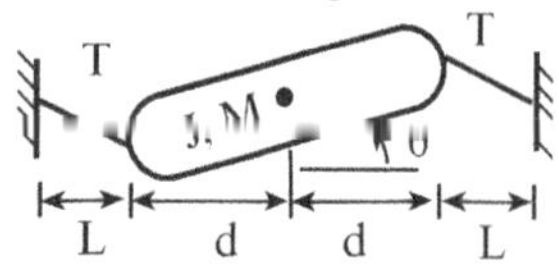	$\frac{1}{2\pi}\left(\frac{2d^2 T}{JL}\right)^{1/2}$ torsion shown $\frac{1}{2\pi}\left(\frac{2T}{ML}\right)^{1/2}$ translation	Body is symmetric left to right. J is the polar mass moment of inertia about the center of gravity C. $J = Md^2/3$ if body is a rod. T = tension in the strings.

Table 3.4 Natural frequency of torsion spring systems, continued

Torsion Spring, Rigid Body	Natural Frequency f, Hz	Mode Shape and Comments
21. Symmetric Body Supported on Two Equal Springs 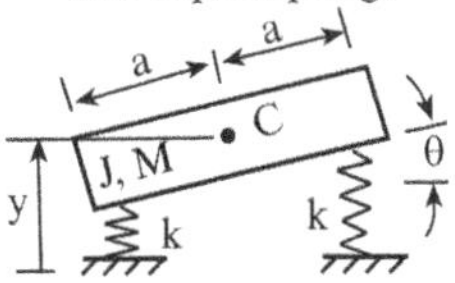	$\frac{1}{2\pi}\left(\frac{2k}{M}\right)^{1/2}, \frac{1}{2\pi}\left(\frac{2a^2k}{J}\right)^{1/2}$ First mode is vertical translation and second mode is torsion.	$\begin{bmatrix}\tilde{y}\\ \tilde{\theta}\end{bmatrix}=\begin{bmatrix}1\\0\end{bmatrix},\begin{bmatrix}0\\1\end{bmatrix}$ J is polar mass moment of inertia taken about center of gravity C. Also see Example 3.4.
22. Body Supported on Two Springs 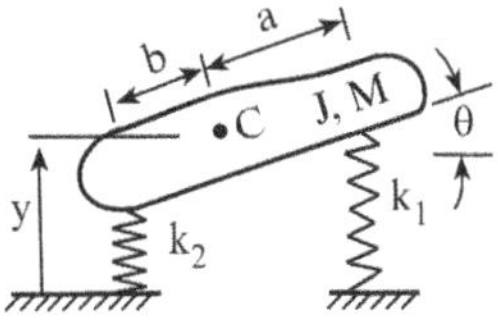	$f_i=\frac{1}{2^{3/2}\pi}\left\{\frac{k_1+k_2}{M}+\frac{a^2k_1+b^2k_2}{J}\mp\left[\left(\frac{k_1+k_2}{M}+\frac{a^2k_1+b^2k_2}{J}\right)^2-\frac{4k_1k_2(a+b)^2}{JM}\right]^{1/2}\right\}^{1/2}$, $\begin{bmatrix}\tilde{y}\\ \tilde{\theta}\end{bmatrix}=\begin{bmatrix}1\\ \dfrac{k_1+k_2-M(2\pi f_i)^2}{ak_1-bk_2}\end{bmatrix}, i=1,2$	
23. Body with Torsion, Extension Springs, Off-Center Mass 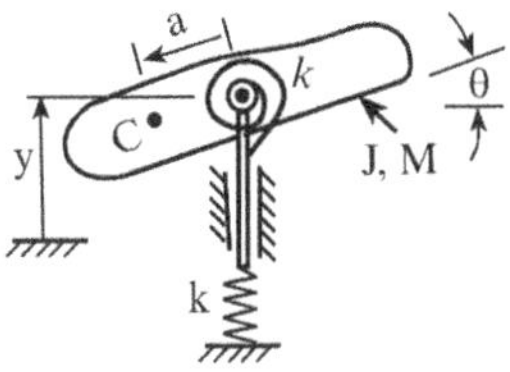	$f_i=\frac{\left(f_y^2+f_\theta^2\mp\left[\left(f_y^2+f_\theta^2\right)^2-4f_y^2f_\theta^2(1-Ma^2/J)\right]^{1/2}\right)^{1/2}}{2^{1/2}(1-Ma^2/J)^{1/2}}$ $\begin{bmatrix}\tilde{y}\\ \tilde{\theta}\end{bmatrix}=\begin{bmatrix}1\\ \frac{1}{a}\left(1-k/\left[M(2\pi f_i)^2\right]\right)\end{bmatrix}$, $i=1,2$, J about pivot. $J>=Ma^2$ Uncoupled frequencies (a=0), $f_y=\frac{1}{2\pi}\left(\frac{k}{M}\right)^{1/2}, f_\theta=\frac{1}{2\pi}\left(\frac{k}{J}\right)^{1/2}$	
24. Roller with Extension Spring 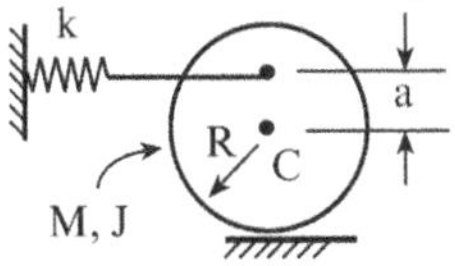	$\frac{1}{2\pi}\left(\frac{k(R+a)^2}{MR^2+J}\right)^{1/2}$ J taken about center of curvature, C.	Roller does not slip.
25. Two Pullleys and Two Springs 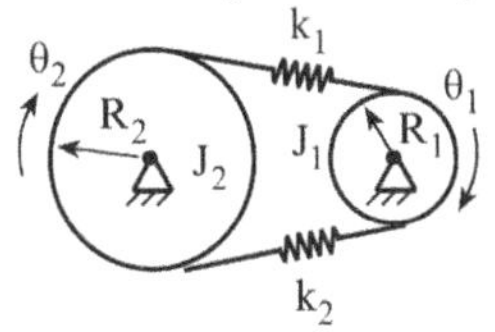	$0, \frac{1}{2\pi}\left(\frac{(k_1+k_2)(R_1^2J_2+R_2^2J_1)}{J_1J_2}\right)^{1/2}$ $\begin{bmatrix}\tilde{\theta}_1\\ \tilde{\theta}_2\end{bmatrix}=\begin{bmatrix}1\\ \frac{R_1}{R_2}\end{bmatrix},\begin{bmatrix}1\\ -\frac{R_2J_1}{R_1J_2}\end{bmatrix}$	J_1 and J_2 are about pivots.
26. Mass, Spring supported Pulley 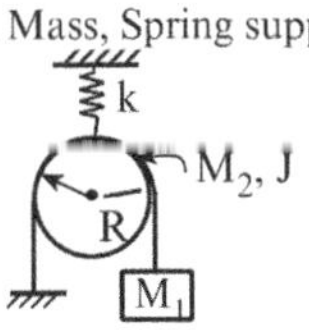	$\frac{1}{2\pi}\left(\frac{k}{M_2+4M_1+J/R^2}\right)^{1/2}$	J taken about center of pulley wheel. String does not slip with respect to the pulley.

Table 3.5 Natural frequency of pendulum systems

Notation: C = center of mass; f = frequency, Hz; g = acceleration due to gravity, Table 1.3, Section 1.4; J = mass moment of inertia about C unless otherwise noted, Table 1.6; k = spring constant, Table 3.2; k = torsion spring constant = (moment/rotation), Table 3.2; M = point mass; L = length of rod or string from pivot or center of mass to pivot; r, R = radius of curvature. Table 1.2 has consistent sets of units. Formulas are valid for small angles, $\theta \ll 1$.

Pendulum System	Natural Frequency f, Hz	Mode Shape and Comments
1. Mass on a Pivoted Rod 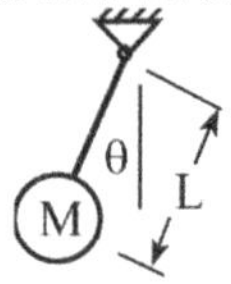	$\frac{1}{2\pi}\left(\frac{g}{L}\right)^{1/2}$	Valid for small angles θ<<1. See case 14 for hanging rigid body. See Example 2.8 for non linear solution for larger amplitudes. Case 6 of Table 2.7 has rotating pendulum.
2. Double Pendulum 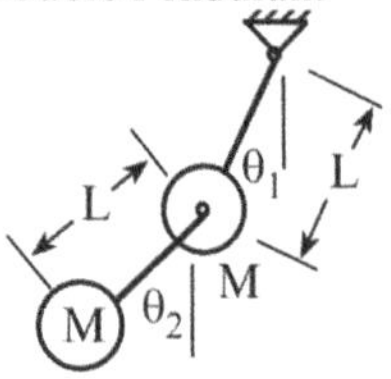	$f_i = \frac{(2\mp 2^{1/2})^{1/2}}{2\pi}\left(\frac{g}{L}\right)^{1/2}, i=1,2$	$\begin{bmatrix}\tilde{\theta}_1\\ \tilde{\theta}_2\end{bmatrix} = \begin{bmatrix}1\\ 2^{1/2}\end{bmatrix}, \begin{bmatrix}1\\ -2^{1/2}\end{bmatrix}$
	For three equal masses and three equal length rods, $f_1 = 0.1026\left(\frac{g}{L}\right)^{1/2}, f_2 = 0.2411\left(\frac{g}{L}\right)^{1/2}, f_3 = 0.3992\left(\frac{g}{L}\right)^{1/2}$	
3. Two Unequal Masses, Unequal Length Rods 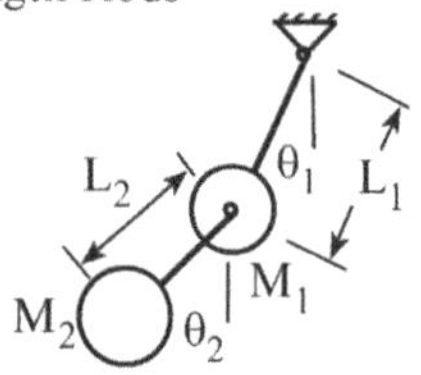	$f_i = \frac{g^{1/2}}{2^{3/2}\pi}\left(\beta \mp \left[\beta^2 - \frac{4}{L_1 L_2}\left(1+\frac{M_2}{M_1}\right)\right]^{1/2}\right)^{1/2}, i=1,2$	$\begin{bmatrix}\tilde{\theta}_1\\ \tilde{\theta}_2\end{bmatrix} = \begin{bmatrix}1\\ \frac{L_1(2\pi f_i)^2}{g - L_2(2\pi f_i)^2}\end{bmatrix}, i=1,2 \quad \beta = \left(1+\frac{M_2}{M_1}\right)\left(\frac{1}{L_1}+\frac{1}{L_2}\right)$
4. Two Mass Spring Coupled 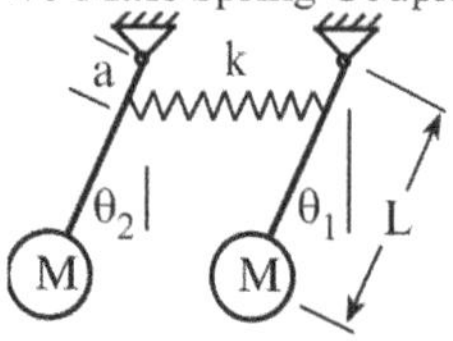	$\frac{1}{2\pi}\left(\frac{g}{L}\right)^{1/2}$, $\frac{1}{2}\left(\frac{g}{L}+\frac{2ka^2}{ML^2}\right)^{1/2}$	$\begin{bmatrix}\tilde{\theta}_1\\ \tilde{\theta}_2\end{bmatrix} = \begin{bmatrix}1\\ 1\end{bmatrix}, \begin{bmatrix}1\\ -1\end{bmatrix}$
5. Torsion and Deflection Springs 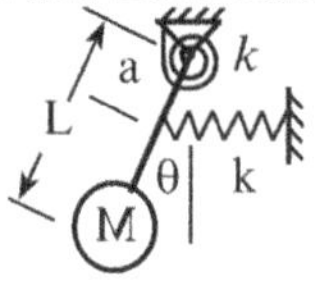	$\frac{1}{2\pi}\left(\frac{g}{L}+\frac{ka^2+k}{ML^2}\right)^{1/2}$	
6. Inverted Pendulum with Torsion and Deflection Springs 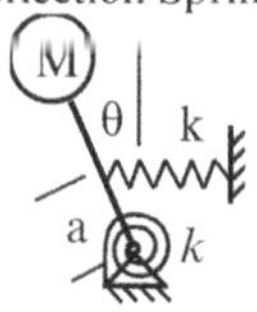	$\frac{1}{2\pi}\left(\frac{ka^2+k}{ML^2}-\frac{g}{L}\right)^{1/2}$	Pendulum is stable only if $ka^2 + k > MgL$; L = pendulum arm length

Table 3.5 Natural frequency of pendulum systems, continued

Pendulum System	Natural Frequency f, Hz	Mode Shape and Comments
7. Counter Balanced Pendulum 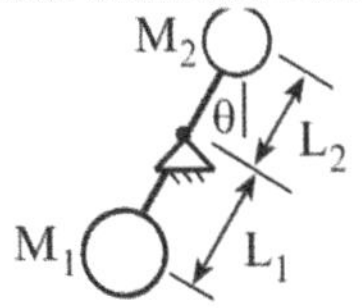	$\frac{g^{1/2}}{2\pi}\left(\frac{M_1L_1 - M_2L_2}{M_1L_1^2 + M_2L_2^2}\right)^{1/2}$	Stable if $M_1L_1 > M_2L_2$
8. Submerged Pendulum 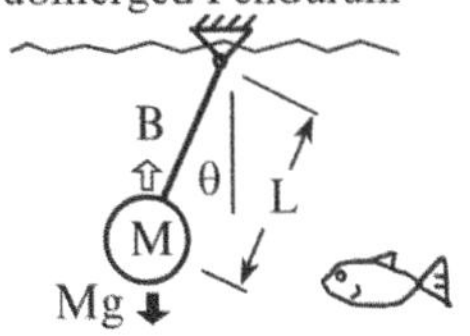	$\frac{1}{2\pi}\left(\frac{Mg - B}{(M + M_{add})L}\right)^{1/2}$ B = buoyancy of mass M M_{add} = entrained fluid added mass, Chapter 6	
9. Compound Pendulum 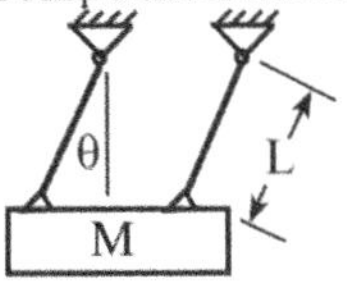	$\frac{1}{2\pi}\left(\frac{g}{L}\right)^{1/2}$ Body M remains parallel to a line between pivots	
10. Stiffened Compound Pendulum 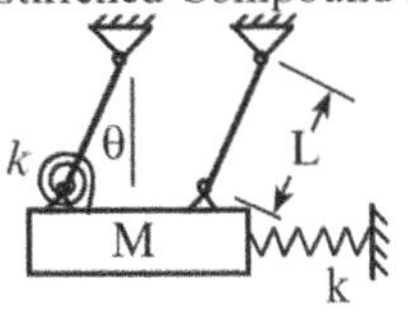	$\frac{1}{2\pi}\left(\frac{1}{M}\left\{\frac{Mg}{L} + k + \frac{k}{L^2}\right\}\right)^{1/2}$ *k* = torsion spring.	
11. Submerged Stiffened Compound Pendulum 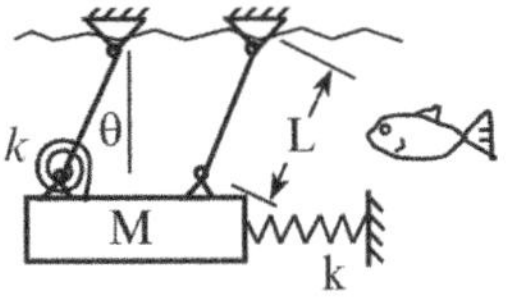	$\frac{1}{2\pi}\left(\frac{1}{M + M_{add}}\left\{\frac{Mg - B}{L} + k + \frac{k}{L^2}\right\}\right)^{1/2}$ B = buoyancy of mass M M_{add} = entrained fluid added mass *k* = torsion spring.	
12. Skate Board in Trough 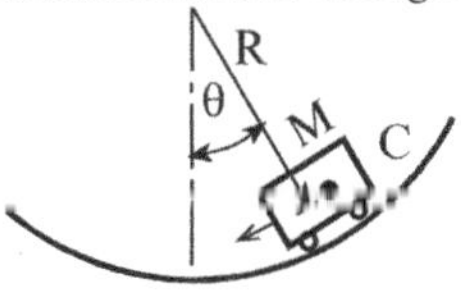	$\frac{1}{2\pi}\left(\frac{g}{R}\right)^{1/2}$	R to center of mass. Also see case 27.
13 Pendulum Damper, Rotor 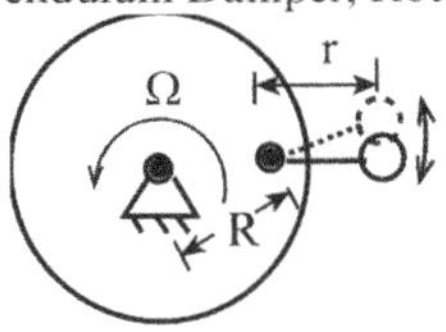	$\frac{\Omega}{2\pi}\left(1 + \frac{r}{R}\right)^{1/2}$	Ref. [6], p. 251. Pivoting damper is tuned to expend energy in vibration mode of interest.

Table 3.5 Natural frequency of pendulum systems, continued

Pendulum System	Natural Frequency f, Hz	Mode Shape and Comments
14. Two Unequal Masses, Rollers 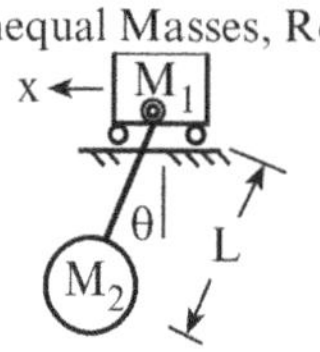	$0,\ \frac{1}{2\pi}\left[\frac{g}{L}\left(1+\frac{M_2}{M_1}\right)\right]^{1/2}$	$\begin{bmatrix}\tilde{x}_1\\ \tilde{\theta}_2\end{bmatrix}=\begin{bmatrix}1\\ 0\end{bmatrix},\begin{bmatrix}1\\ -\left(1+\frac{M_1}{M_2}\right)\frac{1}{L}\end{bmatrix}$ Zero frequency mode is rigid body translation.
15. Mass on Rollers and Pendulum 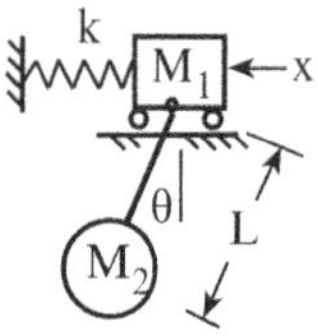	$f_i=\frac{1}{2^{3/2}\pi}\left[\beta\mp\left(\beta^2-\frac{4kg}{M_1L}\right)^{1/2}\right]^{1/2},\ \beta=\left(1+\frac{M_2}{M_1}\right)\frac{g}{L}+\frac{k}{M_1}$ $\begin{bmatrix}\tilde{x}\\ \tilde{\theta}\end{bmatrix}=\begin{bmatrix}1\\ \frac{k}{M_2L(2\pi f_i)^2}-\left(1+\frac{M_1}{M_2}\right)\frac{1}{L}\end{bmatrix},\ i=1,2$	
16.Two Masses, Torsion Spring 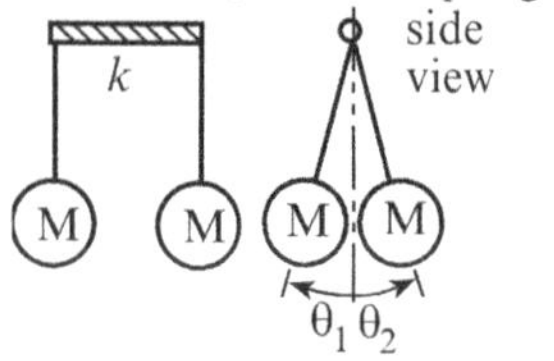	$\frac{1}{2\pi}\left(\frac{g}{L}\right)^{1/2},$ $\frac{1}{2\pi}\left(\frac{g}{L}+\frac{2k}{ML^2}\right)^{1/2}$	$\begin{bmatrix}\theta_1\\ \theta_2\end{bmatrix}=\begin{bmatrix}1\\ 1\end{bmatrix},\begin{bmatrix}1\\ -1\end{bmatrix}$
17. Mass, Inclined Axis Rotation 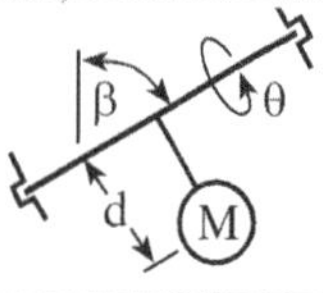	$\frac{1}{2\pi}\left(\frac{g\sin\beta}{d}\right)^{1/2}$	Also see case 26.
18. Hanging Rigid Body 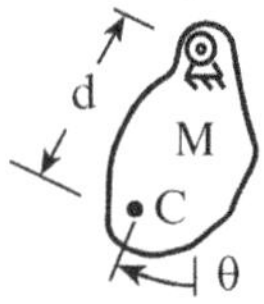	$\frac{1}{2\pi}\left(\frac{Mgd}{J}\right)^{1/2}$	J is polar mass moment of inertia about pivot. C is location of center of mass.
19. Two Pinned Hanging Bodies 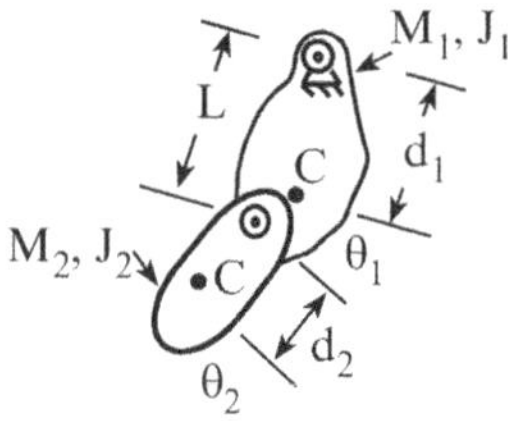	$f_i=\frac{g^{1/2}}{2^{3/2}\pi\alpha^{1/2}}\left[\beta\mp\left(\beta^2-4\alpha\gamma\right)^{1/2}\right]^{1/2},$ J about pivot $\begin{bmatrix}\tilde{\theta}_1\\ \tilde{\theta}_2\end{bmatrix}=\begin{bmatrix}1\\ \frac{M_1gd_1+M_2gL}{M_2d_2L(2\pi f_i)^2}-\frac{J_1'}{M_2d_2L}\end{bmatrix},\ i=1,2,$ $J_1'=J_1+M_1d_1^2+M_2L^2,\ J_2'=J_2+M_2d_2^2,\ \alpha=J_1'J_2'-M_2^2d_2^2L^2$ $\beta=J_1'M_2d_2+J_2'(M_1d_1+M_2L),\ \gamma=M_2d_2(M_1d_1+M_2L)$	

Table 3.5 Natural frequency of pendulum systems, continued

Pendulum System	Natural Frequency f, Hz	Mode Shape and Comments
20. Two Spring Coupled Bodies 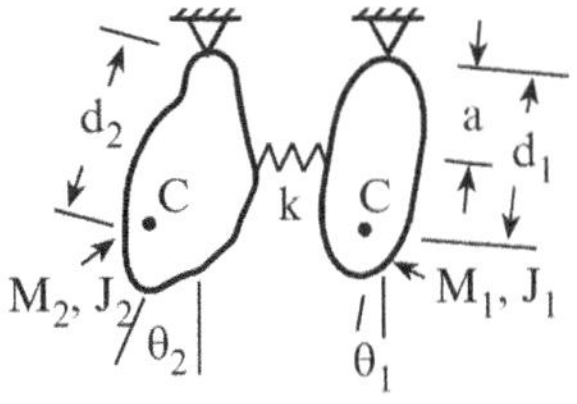	$f_i = \frac{g^{1/2}}{2^{3/2}\pi}\left[\beta \mp \left(\beta^2 - 4\gamma\right)^{1/2}\right]^{1/2}, \ \beta = \frac{ka^2}{J_1} + \frac{ka^2}{J_2} + \frac{gM_1d_1}{J_1} + \frac{gM_2d_2}{J_2}$ $\gamma = \frac{g^2d_1d_2M_1M_2}{J_1J_2} + \frac{ka^2gM_1d_1}{J_1J_2} + \frac{ka^2gM_2d_2}{J_1J_2}$ $\begin{bmatrix}\tilde{\theta}_1 \\ \tilde{\theta}_2\end{bmatrix} = \begin{bmatrix}1 \\ 1 + \frac{gM_1d_1}{ka^2} - \frac{J_1}{ka^2}(2\pi f_i)^2\end{bmatrix}, \ i = 1, 2,$	
21. Two Hanging Bodies, Torsion Bar 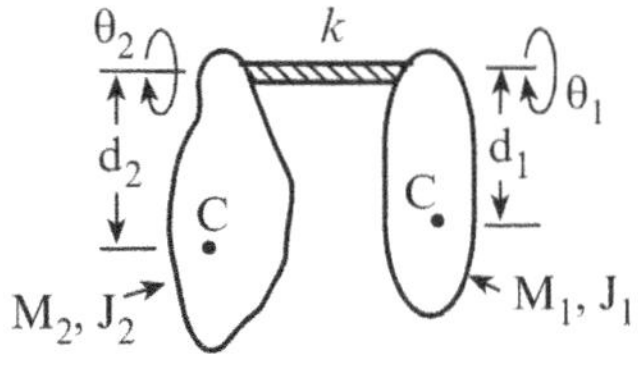	$\frac{1}{2^{3/2}\pi}\left[\beta \mp \left(\beta^2 - 4\gamma\right)^{1/2}\right]^{1/2}, \ \beta = \frac{k}{J_1} + \frac{k}{J_2} + \frac{gM_1d_1}{J_1} + \frac{gM_2d_2}{J_2}$ $\gamma = \frac{g^2d_1d_2M_1M_2}{J_1J_2} + \frac{kgM_1d_1}{J_1J_2} + \frac{kgd_2M_2}{J_1J_2}$ $\begin{bmatrix}\tilde{\theta}_1 \\ \tilde{\theta}_2\end{bmatrix} = \begin{bmatrix}1 \\ 1 + \frac{gM_1d_1}{k} - \frac{J_1}{k}(2\pi f_i)^2\end{bmatrix}, \ i = 1, 2,$	
22. Rigid Body Inverted 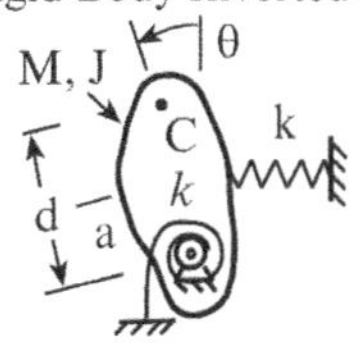	$\frac{1}{2\pi}\left(\frac{ka^2 + k}{J} - \frac{Mgd}{J}\right)^{1/2}$ J about pivot Stable only if $ka^2 + k > Mgd$	
23. Body Suspended on String 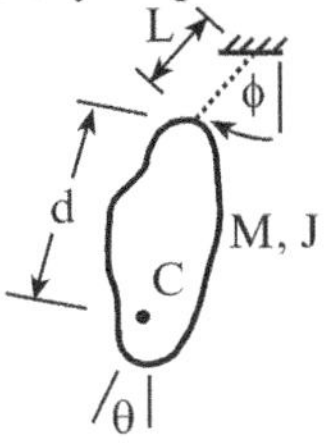	$\frac{g^{1/2}}{2^{3/2}\pi J^{1/2}}\left[Md + \frac{J + Md^2}{L} \mp \left(\left[Md + \frac{J + Md^2}{L}\right]^2 - \frac{4JMd}{L}\right)^{1/2}\right]^{1/2}$ $\begin{bmatrix}\tilde{\Phi}_1 \\ \tilde{\theta}_2\end{bmatrix} = \begin{bmatrix}1 \\ \frac{g}{d(2\pi f_i)^2} - \frac{L}{d}\end{bmatrix}, \ i = 1, 2$ J about center of gravity C.	
24. Torsion of Body Suspended on Two Equal Length Strings 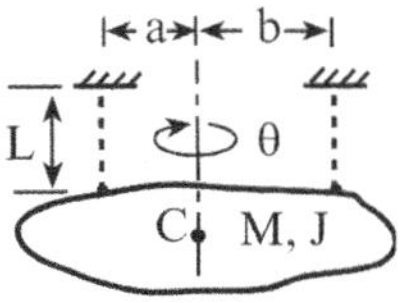	$\frac{1}{2\pi}\left(\frac{Mgab}{JL}\right)^{1/2}$	J about center of mass C. Strings are vertical. Rotation is about center of mass.

Table 3.5 Natural frequency of pendulum systems, continued

Pendulum System	Natural Frequency f, Hz	Mode Shape and Comments
25. Torsion of Body Suspended on N Equal Length Strings 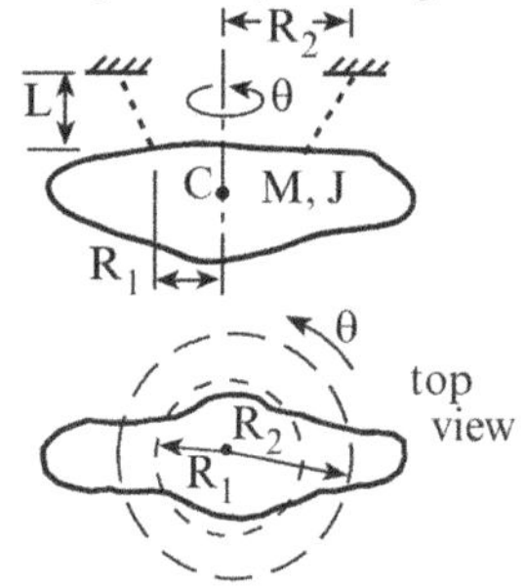	$\frac{1}{2\pi}\left(\frac{MgR_1R_2}{JL}\right)^{1/2}$	J about center of mass C. Rotation is about center of mass. Strings are splayed between circles of radius R_1 on body and R_2 at attachment so that each string bears the same load. Result is independent of number of strings.
26. Rigid Body with Inclined Axis 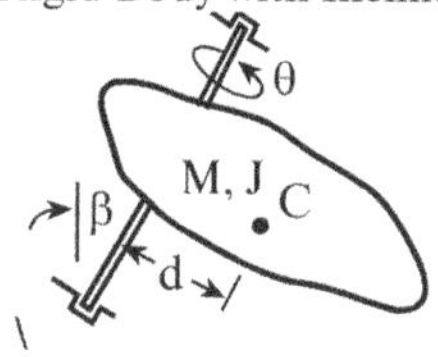	$\frac{1}{2\pi}\left(\frac{Mgd\sin\beta}{J}\right)^{1/2}$	J is polar mass moment of inertia, Table 1.6, about fixed axis of rotation. Also see case 17.
27. Ball or Cylinder in a Trough 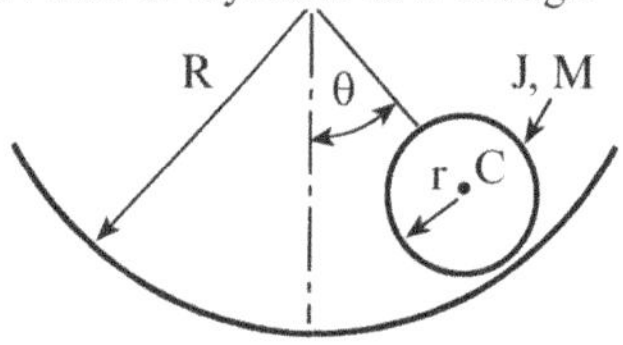	$\frac{1}{2\pi}\left(\frac{Mg}{(R-r)(M+J/r^2)}\right)^{1/2}$	J about center of mass and curvature, C. Body does not slip. If body slips freely see case 12.
28. Body Balanced on Pivot, Spring 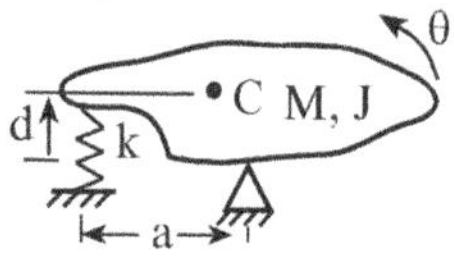	$\frac{1}{2\pi}\left(\frac{ka^2-Mgd}{J+Md^2}\right)^{1/2}$ d is distance of center of gravity C above pivot. J about C.	Body is stable only if $ka^2>Mgd$. Body is always stable if C is below pivot so that d is negative. If pivot is adjusted so that d=0 then mass moment of inertia is $J=ka^2/(2\pi f)^2$.
29. Body Balanced on an Arc 	$\frac{1}{2\pi}\left(\frac{(r-d)Mg}{J+Md^2}\right)^{1/2}$ d is distance that center of gravity C is above pivot.	J about center of mass, C. Body does not slip. Body is stable only if r > d. Body is always stable if C is below pivot so that d is negative.
30. Rounded Body on Plane 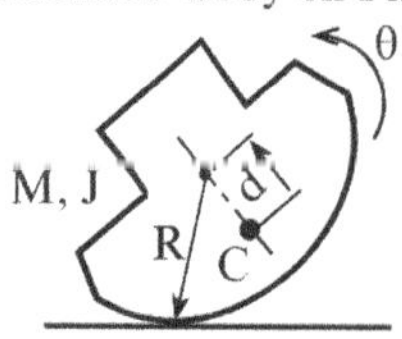	$\frac{1}{2\pi}\left(\frac{Mgd}{J+M(R-d)^2}\right)^{1/2}$ d is positive upward from center of mass C to center of curvature.	J about center of mass, C. Body does not slip. Body is stable only if d>0. Natural frequency falls to zero if d=0 and body continues to roll.

The *natural frequency* f_n in units of vibration cycles per second, *Hertz*, is $1/(2\pi)$ times ω_n:

$$f_n = \frac{\omega}{2\pi} = \frac{1}{2\pi}\sqrt{\frac{k}{M}}, \quad \text{Hz} \tag{3.7}$$

The natural frequency is proportional to the square root of stiffness k divided by mass M (Appendix A, Eq. A.7). Natural frequency increases with stiffness and decreases with increasing mass:

$$\text{Natural frequency} \sim \sqrt{\frac{\text{stiffness}}{\text{mass}}}$$

Mean static deformation due to gravity has no effect on the natural frequencies and mode shapes of linear systems.

3.3.2 Two-Degree-of-Freedom System

Consider the two-mass system shown in case 2 of Table 3.3. Their axial displacements are x_1 and x_2, respectively. Newton's second law (Eq. 2.4) gives an equation of motion for each mass:

$$\begin{aligned} M\ \ddot{x}_1 &= -kx_1 + k(x_2 - x_1) \\ M\ \ddot{x}_2 &= -k(x_2 - x_1) \end{aligned} \tag{3.8}$$

The over dots $(\cdot)$ denote derivative with respect to time. In matrix form,

$$\begin{bmatrix} M & 0 \\ 0 & M \end{bmatrix} \begin{Bmatrix} \ddot{x}_1 \\ \ddot{x}_2 \end{Bmatrix} + \begin{bmatrix} 2k & -k \\ -k & k \end{bmatrix} \begin{Bmatrix} x_1 \\ x_2 \end{Bmatrix} = 0 \tag{3.9}$$

The trial solutions for the free vibration are sinusoidal and in phase:

$$x_1(t) = \tilde{x}_1 \sin \omega t, \quad x_2(t) = \tilde{x}_2 \sin \omega t \tag{3.10}$$

The over tilde $(\sim)$ denotes a mode shape that is independent of time.

Substituting Equation 3.10 into Equation 3.9 gives the matrix equation for the free vibrations of the two-mass system:

$$\begin{bmatrix} -\omega^2 M + 2k & -k \\ -k & -\omega^2 M + k \end{bmatrix} \begin{Bmatrix} \tilde{x}_1 \\ \tilde{x}_2 \end{Bmatrix} = 0 \tag{3.11}$$

This is an *eigenvalue* problem. Nonzero solutions exist only if the determinant $(|\,|)$ of the matrix on the left-hand side is zero [10]. This produces a *characteristic frequency polynomial* for *natural frequencies*. There are two solutions:

$$\begin{vmatrix} -\omega^2 M + 2k & -k \\ -k & -\omega^2 M + k \end{vmatrix} = \omega^4 - \frac{3k}{M}\omega^2 + \frac{k^2}{M^2} = 0 \tag{3.12}$$

$$f_i = \frac{\omega_i}{2\pi} = \frac{1}{2\pi}\left(\frac{3 \mp 5^{1/2}}{2}\right)^{1/2} \left(\frac{k}{M}\right)^{1/2} \quad i = 1, 2 \tag{3.13}$$

The lowest is the *fundamental frequency*. The *natural frequencies* are substituted back into the equation of motion (Eq. 3.11) to obtain the associated *mode shapes*:

$$\begin{Bmatrix} \tilde{x}_1 \\ \tilde{x}_2 \end{Bmatrix} = \begin{pmatrix} 1 \\ (1+5^{1/2})/2 \end{pmatrix}, \begin{pmatrix} 1 \\ (1-5^{1/2})/2 \end{pmatrix} \quad (3.14)$$

Mode shapes, also called *eigenvectors*, are the *relative* displacements of the masses as they vibrate freely at their natural frequencies. Mode shapes are independent of time and amplitude. In the first mode of vibration, the two masses move in the same direction with the same amplitude; they move in opposite directions in the second mode.

Displacement in a mode is equal to its mode shape times the modal amplitude. Total displacement is the sum of the modal displacements, as discussed in Chapter 7. The load in the springs during vibration is the spring constant times spring displacement.

Multi-degree-of-freedom system equations of motion are placed in *matrix form*. Each matrix row is the ordinary differential equation of motion of a mass degree of freedom. (A mass element can have one through six degrees of freedom.)

$$[M]\{\ddot{x}\} + [K]\{x\} = 0 \quad (3.15)$$

The *mass matrix* $[M]$ and the *stiffness matrix* $[K]$ are N by N square symmetric matrices [8]. $\{x\}$ is the N by 1 *displacement vector*, a vertical column of the 1-to-N displacements of the N masses. The free vibration of the system can be expressed compactly with *complex notation*: i is the imaginary mathematical constant $(-1)^{1/2}$, and $e = 2.71828$ is the base of the natural logarithm:

$$\{x(t)\} = \{\tilde{x}\}e^{i\omega t} = \{\tilde{x}\}(\cos\omega t + i\sin\omega t) \quad (3.16)$$

The over tilde (~) denotes that $\tilde{x}$ is independent of time. Substituting this solution form into the previous equation of motion produces an *eigenvalue problem* for the circular natural frequencies ω and their *mode shapes* $\{\tilde{x}\}$:

$$(-\omega^2[M] + [K])\{\tilde{x}\} = 0 \quad (3.17)$$

This equation has nontrivial solutions (solutions other than zero) only if the determinant of the matrix $|-\omega^2[M] + [K]| = 0$ [12]. If the mass and stiffness matrices are self-adjoint, the usual case, then this is a characteristic polynomial whose solution is a sequence of N unique positive *eigenvalues* (natural frequencies) ω_i, $i = 1, 2, 3, \ldots, N$, which are inserted back into Equation 3.17 to obtain their associated *eigenvectors* (mode shapes $\tilde{x}_i$, $i = 1, 2, 3, \ldots, N$) [13].

Eigenvectors are *orthogonal* with respect to the mass and stiffness [10]:

$$\{\tilde{x}_s\}^T[M]\{\tilde{x}_r\} = \{\tilde{x}_s\}^T[K]\{\tilde{x}_r\} = 0 \text{ if } \omega_r \neq \omega_s \quad (3.18)$$

For example, the two mode shapes in Equation 3.14 are orthogonal to the mass and stiffness matrices in Equation 3.9:

$$\begin{bmatrix} 1 & \frac{1+5^{1/2}}{2} \end{bmatrix} \begin{bmatrix} 1 & 0 \\ 0 & 1 \end{bmatrix} \begin{bmatrix} 1 \\ (1-5^{1/2})/2 \end{bmatrix} = \begin{bmatrix} 1 & \frac{1+5^{1/2}}{2} \end{bmatrix} \begin{bmatrix} 2 & -1 \\ -1 & 1 \end{bmatrix} \begin{bmatrix} 1 \\ (1-5^{1/2})/2 \end{bmatrix} = 0 \quad (3.19)$$

Exact and approximate methods for eigenvalue matrix solution are discussed in Appendix A. Numerical solution of Equation 3.17 is the basis for finite element modal vibration analysis ([14] in Chapter 2). Chapter 7 discusses the forced vibration of these systems.

Example 3.3 Natural frequency of a helical spring mass system

A 2.2 kg (4.85 lb) mass is supported on rollers with a helical spring shown in Figure 3.5a. The spring is made from 1 mm (0.03937 in.) steel wire. It has a coil diameter of 12 mm (0.4724 in.) and four active coils. Determine the natural frequency of the spring–mass system.

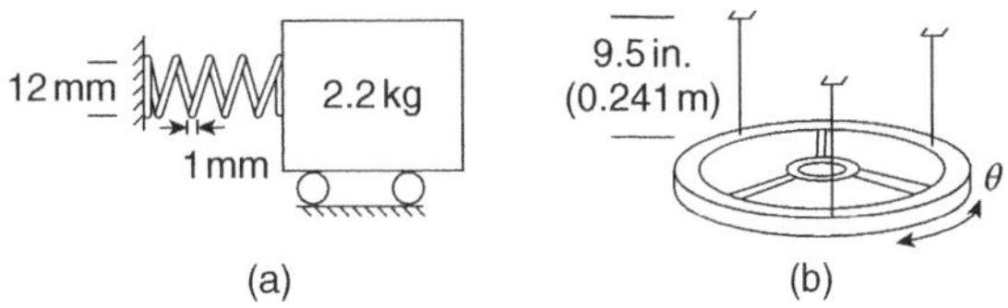

Figure 3.5 Examples of a spring–mass system and a torsional pendulum

Solution: The elastic modulus of carbon steel at room temperature is $E = 20$ E10 Pa (29 E6 psi), and its Poisson's ratio is $\nu = 0.3$ (Chapter 8), so the shear modulus is $G = E/(2(1 + \nu)) = 7.69$E10 Pa (11.15 E6 psi). The spring stiffness is computed with case 28 of Table 3.2 with $n = 4$, $d = 0.001$ m, and $R = 0.006$ m:

$$k = \frac{Gd^4}{64nR^3} = \frac{7.69\text{E}10\ \text{Pa}(0.001\ \text{m})^4}{64 \times 4(0.006\ \text{m})^3} = 1391\ \text{N/m}\ (7.94\ \text{lb/in})$$

The spring natural frequency is given in case 1 of Table 3.3. In SI units (case 1 of Table 1.2) and in US customary units (case 7 of Table 1.2),

$$f = \frac{1}{2\pi}\sqrt{\frac{k}{M}} = \frac{1}{2\pi}\sqrt{\frac{1391\ \text{N/m}}{2.2\ \text{kg}}} = \frac{1}{2\pi}\sqrt{\frac{7.94\,\text{lb/in.}}{4.85\,\text{lb/in./386.1 in./s}^2}} = 4.00\ \text{Hz}$$

The spring–mass of the spring is negligible compared to the 2.2 kg mass (see Section 3.4).

Lagrange's equations often provide an easier and less error-prone means of deriving equations of motion for complex systems when expressions for kinetic energy of mass and potential energy of conservative restoring forces are available. The kinetic energy of mass, here given the symbol T, and the elastic potential energy U are written in terms of generalized coordinates q_i, $i = 1, 2, \ldots, N$ and its derivative with respect to time. The Lagrange's equations of motion are [9, 13]

$$L = T - U, \quad \frac{d}{dt}\left(\frac{\partial L}{\partial \ddot{q}_i}\right) - \frac{\partial L}{\partial q_i} = Q_i \tag{3.20}$$

The generalized forces Q_i are such that the work done on the system to produce an increment in the ith coordinate is $\delta W = Q_i \delta_{q_i}$. The following example will provide clarification.

Example 3.4 Elastic suspension

Figure 3.4b shows a two-dimensional suspension of a rigid package. x and y are the inertial horizontal and vertical positions of reference point O that is fixed in the body at the same height as the center of mass C. Find its modes and natural frequencies.

Solution: The body fixed ξ, η coordinate system has origin at O. The coordinate transform given in case 1 of Table 2.1 for small tilt angle θ is applied to obtain the inertial coordinates X,Y to each point on the body:

$$X(t) = x(t) + \xi - \eta\theta(t), \quad Y(t) = y(t) + \xi + \eta\theta(t)$$

$$\ddot{X}(t) = \ddot{x}(t) + \xi - \eta\ddot{\theta}(t), \quad \ddot{Y}(t) = \ddot{y}(t) + \ddot{\eta} + \eta\ddot{\theta}(t)$$

The kinetic energy T of the body is the integral of the kinetic energy of body elements:

$$T = \frac{1}{2}\int_V \rho(\dot{X}^2 + \dot{Y}^2) d\xi d\eta dz = \frac{1}{2}M(\dot{x}^2 + \dot{y}^2) + \frac{1}{2}J\dot{\theta}^2 + S_x\dot{y}\dot{\theta}, \quad \text{where}$$

$$M = \int_V \rho d\xi d\eta dz, \quad J = \int_V \rho(\xi^2 + \eta^2) d\xi d\eta dz, \quad S_x = \int_V \rho\xi d\xi d\eta dz$$

The body has mass M and density ρ. The potential energy stored in a spring is $kX^2/2$, where k is a spring constant and X is its axial deformation. Springs are attached to two lower corners at $x + \theta d$, $y \pm \theta L/2$. The total stored potential energy U in the four springs is

$$U = \left(\frac{1}{2}\right) k_x(x^2 + 2xd\theta + \theta^2 d^2) + \left(\frac{1}{2}\right) k_y \left(\frac{y^2 + \theta^2 L^2}{4}\right)$$

The two equations of motion, one for each mass, are obtained by substituting T and U into Equation 3.20 with $q_i = x, y, \theta$, $i = 1, 2, 3$:

$$M\ddot{x} + k_x x + dk_x\theta = F_x, \quad M\ddot{y} + S_x\ddot{\theta} + k_y y = F_y$$

$$J\ddot{\theta} + S_x\ddot{y} + dk_x x + \left(\frac{d^2 k_x + k_y L^2}{4}\right)\theta = M_\theta$$

F_x, F_y and M_θ are the forces in the horizontal and vertical direction and the moment about the origin. Setting these forces to zero for free vibration leads to equations that have a solution of the form $X\ Exp(i\omega t)$ and a sixth-order polynomial for the circular natural frequencies ω. For $S_x = 0$, the center of mass is centered above the springs, and the vertical motions and horizontal motions uncouple. But the rotational mode and the vertical motions are coupled by the offset d of the center of mass above the springs:

$$f_y = \frac{1}{2\pi}\sqrt{\frac{k_y}{M}}, \quad \text{Hz},$$

$$f_{x-\theta,i} = \frac{1}{4\pi}(\omega_a^2 + \omega_b^2 + \omega_c^2) \pm \frac{1}{2}\sqrt{(\omega_a^2 + \omega_b^2 + \omega_c^2)^2 - 4\omega_a^2\omega_c^2}, \quad i = 1, 2 \tag{3.21}$$

Here, $\omega_a{}^2 = k_x/M,\ \ \omega_b{}^2 = d^2/J,\ \ \omega_c{}^2 = L^2k_y/4J.$

3.4 Modeling Discrete Systems with Springs and Masses

3.4.1 Springs with Mass

Mass of the springs is neglected as shown in column 2 of Table 3.3. Spring–mass can be included approximately in the computation of the fundamental natural frequency by using an assumed mode shape (Raleigh technique; case 9 of Table 4.9 and Appendix A) or exactly by continuous system analysis (Table 4.13).

The accuracy of the massless-spring model is assessed with formulas from case 1 of Table 3.2, case 8 of Table 4.8, and case 12 of Table 4.13, which is the exact solution. See Fig. 3.6, the spring has mass M_s. M is the supported mass:

$$f_n \text{ (Hz)} = \begin{cases} \left(\frac{1}{2\pi}\right)\sqrt{\frac{k}{M}} & \text{massless spring, } M_s = 0, \text{ case 1 of Table 3.2} \\ \left(\frac{1}{2\pi}\right)\sqrt{\frac{k}{(M+M_s/3)}} & \text{massive spring, case 8 of Table 4.8} \\ \left(\frac{\lambda}{2\pi}\right)\sqrt{\frac{k}{M}} & \text{continuous spring, case 12 of Table 4.13} \end{cases} \tag{3.22}$$

The natural frequencies are a function of the ratio of discrete mass to spring–mass, M/M_s.

	Values of $2\pi f(M/k)^{1/2}$					
Spring–mass/discrete mass, M_s/M	0	0.01	0.1	1	10	100
Massless spring, $M_s = 0$	1	1	1	1	1	1
Raleigh solution	1	0.998	0.837	0.866	0.4804	0.1707
Exact solution n/a	0.994	0.983	0.860	0.4519	0.157	

Including the mass of the spring–mass lowers the predicted natural frequency. The most conservative (lowest) estimate is to lump the spring–mass in the mass M.

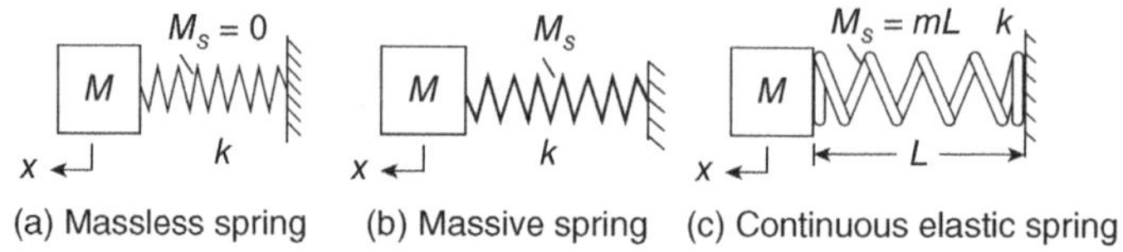

Figure 3.6 Three models for a spring in a spring–mass system

3.4.2 Bellows

Gerlach [6] modeled axial vibrations of bellows with $2N_c - 1$ discrete masses at each convolution crown and root. The N_c convolution crowns are connected by $2N_c$ axial springs (case 11 of Table 3.3 and Figure 3.7). The predicted bellows natural frequencies for axial vibration are

$$f_i = \frac{2}{2\pi} \sin\left(\frac{i\pi}{4N_c}\right) \sqrt{\frac{k_{c/2}}{m_c + m_a}}, \quad i = 1, 2, 3, \ldots \tag{3.23}$$

The half-convolution axial stiffness is $k_{c/2} = k/2N$. The bellows overall axial stiffness k, mass of half a crown or root m_c, and its mass of internal entrained fluid m_a are shown in Figure 3.6 [7, 8].

Finite element modeling discretizes an elastic structure into spring–mass elements. The accuracy of a finite element model increases with the number of elements. Consider longitudinal vibrations of continuous uniform elastic beam that has length L, area A, density ρ, and modulus of elasticity E. The left end of the beam is fixed and the right end is free. The beam is sectioned into N elements, each with length L/N, mass $m = \rho AL/N$, and axial stiffness (case 7 of Table 3.2), $k = EA(N/L)$. The discrete spring–mass model's natural frequencies are given in case 5 of Table 3.3. The exact beam result is shown in case 2 of Table 4.14:

$$\text{Finite element}: \quad f_i = \frac{N}{\pi} \sin\left(\frac{2i-1}{2N+1}\frac{\pi}{2}\right) \sqrt{\frac{E}{\rho L^2}}, \quad \text{Hz}, \quad i = 1, 2, 3, \ldots$$

$$\text{Exact beam}: \quad f_i = \frac{(2i-1)}{4L} \sqrt{\frac{E}{\rho}}, \quad \text{Hz}, \quad i = 1, 2, 3, \ldots \tag{3.24}$$

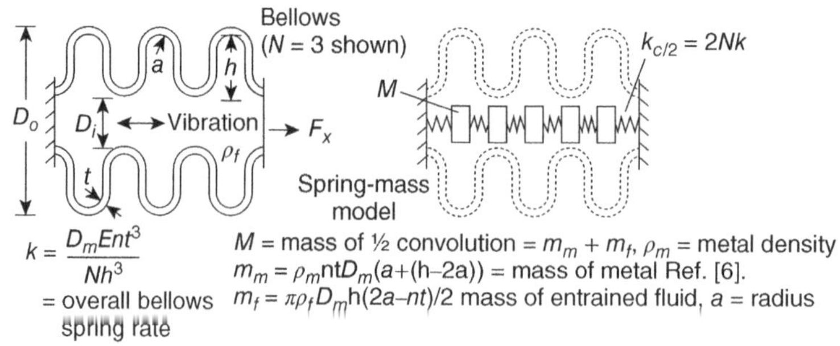

Figure 3.7 Bellows modeling

The first three natural frequencies are plotted in Figure 3.8 as a function of the number of elements. The finite element result approaches the exact result from below as the number of elements increases. At least three elements are required to represent a half wave in the mode shape (see Appendix D).

3.5 Pendulum Natural Frequencies

Pendulums swing about a fixed pivot under the influence of gravity (Table 3.4 and Figure 3.9). The restoring torque of gravity about the pivot is $-Mg\,L\ \sin\ \theta$, where M is the mass, L is the length of the pendulum, and θ is the angle of rotation from vertical radians. Newton's second law (Eq. 2.4) gives the pendulum's equation of motion:

$$J\frac{d^2\theta}{dt^2} + Mg\ \ L\sin\theta = 0 \tag{3.25}$$

where J is the polar moment of inertia of the mass about the pivot (Tables 1.5 and 1.6) and L is the distance from the pivot to point mass M. This equation is nonlinear owing to the $\sin\theta$ term, but it has an exact nonlinear periodic solution given in case 14 of Table 2.6 and Example 2.8.

Equation 3.25 is linearized for small angular rotations, much less than 1 rad (57.29°) by setting $\sin\theta \sim \theta$:

$$ML^2\frac{d^2\theta}{dt^2} + MgL\theta = 0 \tag{3.26}$$

Following Equations 3.5 through 3.7, this linear equation of motion is solved for the pendulum natural frequency:

$$f_n = \frac{1}{2\pi}\sqrt{\frac{g}{L}}\quad \text{Hz} \tag{3.27}$$

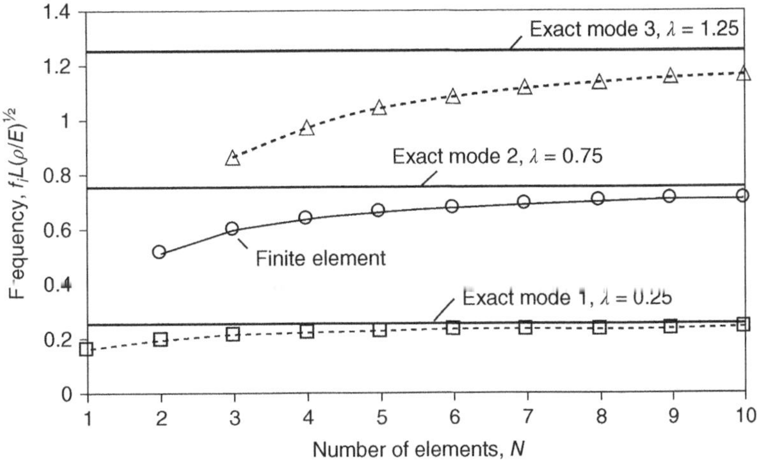

Figure 3.8 Spring–mass discretized longitudinal beam natural frequencies as a function of number of elements

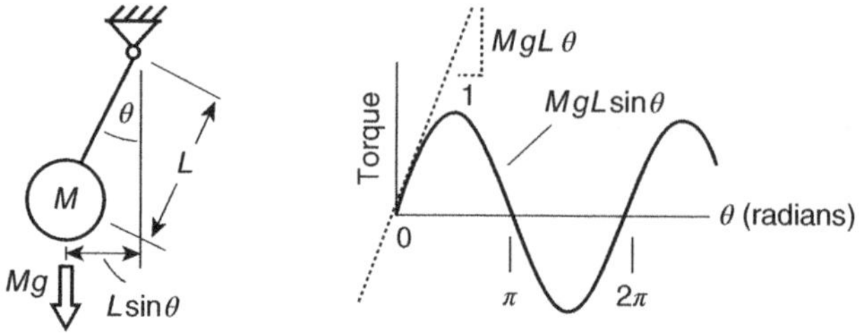

Figure 3.9 Linearization of the gravity-induced torque on the pendulum for small deflections

The pendulum natural frequency is independent of the mass of the pendulum bob and inversely proportional to the square root of arm length. A 2 s period clock pendulum is 0.99 m (39.1 in.) long, and its escapement ticktocks once per second at the extremes of the pendulum swing.

3.5.1 Mass Properties from Frequency Measurement

Experimental measurements of natural period during a free vibration experiment can be used to compute mass properties. Consider the system shown in Figure 3.3. Its period is given by Equation 3.7, which can be solved for mass M as a function of period T and spring stiffness k:

$$M = \frac{T^2 k}{(2\pi)^2} \tag{3.28}$$

If the average elapsed time for several periods of the system shown in Figure 3.3 is timed with a stopwatch, the system mass can be calculated for the known spring constant k and measured period T.

The spring constant k can be eliminated from the previous calculation if a known mass m is added to the system. The ratio of the period, $T_1 = (2\pi\ \ k)^2/M$, without added mass to that with the additional mass, $T_2 = (2\pi\ \ k)^2/(M+m)$, is $T_1{}^2/T_2{}^2 = M/(M+m)$. This is solved for given the initial mass M in terms of the added mass m and the period ratio T_2/T_1:

$$M = \frac{m}{(T_2/T_1)^2 - 1} \tag{3.29}$$

The accuracy of the calculation of M depends on the precision of the measurement of T_2/T_1; a task is made easier if the added mass m is comparable in size to M. A 1% error in the T_2/T_1 measurement produces a 20% error in M if m is 10% of M but only a 1% error if m equals M.

The **tilt table** (case 28 of Table 3.5) is a teeter-totter used to measure the polar moment of inertia of bodies about a horizontal axis. The center of gravity of the body is positioned directly above the pivot. A known soft spring is then attached between the table and ground at distance "a" from the pivot. The table is set into motion and its period T is timed. Case 21 of Table 3.4 gives the natural period T, which is solved for the polar mass moment:

$$J_{\text{table+protype}} = \frac{(2\pi)^2 k a^2}{T^2} \tag{3.30}$$

The polar mass moment of inertia J is calculated from the measured period.

Example 3.5 Pendulum period

Figure 3.6b shows a 3.5 lb (1.588 kg) circular flywheel suspended by three vertical wires, 9.5 in. (0.2413 m) long, that are spaced at equal intervals, so each wire bears the same load. The wheel is given a torsional impulse and allowed to oscillate freely. Its natural period is measured by a stopwatch to be 0.714 s. Calculate the polar mass moment of inertia of the flywheel from the measured period.

Solution: The polar mass moment of inertia is computed by solving the formula of the natural frequency of case 25 of Table 3.5, ($R_1 = R_2 = R$), for the polar mass moment of inertia. J is calculated with in-lb-s units (case 7 of Table 1.2):

$$J = \frac{M\, gR^2}{(2\pi f)^2 L} = \frac{2.5\ \text{lb}\ (4\ \text{in.})^2}{(2\pi \times 1.4\ \text{Hz})^2 9.5\ \text{in.}} = 0.0544\ \text{lb-in.-s}^2 = 21.01\ \text{lb-in.}^2\,(0.009074\ \text{kg-m}^2)$$

In these units, (lb-s^2/in.) is a unit of mass. Multiplying it by $g_c = 386.1$ in./s^2 gives $J = 21.01$ lb-in.^{2} where lb is now a unit of force.

3.6 Tensioned Strings, Cables, and Chain Natural Frequencies

Strings, cables, and chains have no bending rigidity (Table 3.6); tension generates their lateral stiffness (case 21 of Table 3.2). *Cables* and *strings* can be a single strand or woven from many fibers. Cables stretch under load, whereas *chains* are inextensionable. The term *string* is used for light straight tensioned members, whereas the heavier *cables* sag into catenaries under lateral load of gravity or fluid drag.

3.6.1 Equation of Motion

Figure 3.10a shows a uniform string of length L stretched between two tie-downs. The partial differential equation of motion for lateral displacements $y(x,t)$ of the string is the *one-dimensional wave equation* [11] (pp. 244–245):

$$\frac{m\partial^2 y(x,t)}{\partial t^2} - \frac{P\partial^2 y(x,t)}{\partial x^2} = F(x,t) \tag{3.31}$$

P is the mean tension in the string, which is considered constant, and m is the mass per unit length of cable. $F(x,t)$ is the exciting force per unit length. Solutions to Equation 3.31 are a function of distance x along the cable multiplied by a function of time t: $y(x,t) = y(x,t) = \tilde{y}(x)T(t)$. Substituting into the previous equation and dividing through by $\tilde{y}(x)T(t)$ shows

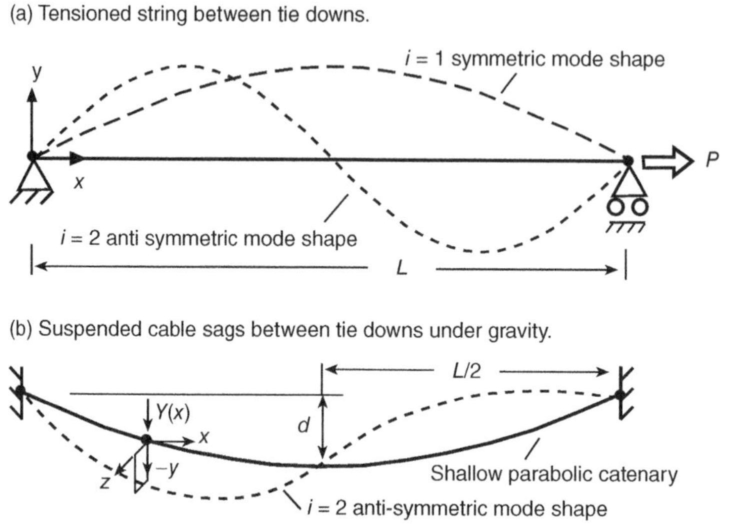

Figure 3.10 Tensioned cables

that the left side of the equation is only a function of time. Since it equals a function of the independent space variable x on the right side, both sides of the equation must be equal to the same constant, chosen to be $-\omega^2$:

$$\frac{T''(t)}{T(t)} = \left(\frac{P}{m}\right)\frac{\tilde{y}''(x)}{\tilde{y}(x)} = -\omega^2 = \text{constant, hence,} \tag{3.32a}$$

$$T''(t) + \omega^2 T(t) = 0 \tag{3.32b}$$

$$\left(\frac{P}{m}\right)\tilde{y}''(x) + \omega^2\tilde{y}(x) = 0 \tag{3.32c}$$

The primes (′) denote derivative with respect to x. The temporal solution to Equation 3.32b is $T(t) = \sin(\omega t)$ or $\cos(\omega t)$.

The solution to Equation 3.32c is a sinusoidal mode shape, $\tilde{y}(x) = \sin[\omega x(m/P_o)^{1/2}]$. The natural frequencies are determined by applying the boundary conditions of zero displacement at the ends of the cable, $y(0, t) = y(L, t) = 0$ so $\tilde{y}(L) = \sin[\omega L(m/P)^{1/2}] = 0$, which holds only if $\omega L(m/P)^{1/2}$ is an integer multiple of π : $= \omega = i\pi(P/m)^{1/2}$, $i = 1, 2, 3, \ldots$. These are the circular *natural frequencies*. The assembled solution to Equation 3.32 describes the *free vibrations* of the string in its ith mode:

$$y(x, t) = Y_i \sin\left(\frac{i\pi x}{L}\right)\sin(\omega_i t)$$

$$f_i = \frac{\omega_i}{2\pi} = \frac{i}{2L}\sqrt{\frac{P}{m}} \quad \text{Hz}, \quad i = 1, 2, 3, \ldots \tag{3.33}$$

Table 1.2 shows the consistent sets of units for this equation. Like all one-dimensional continuous systems, the cable has a semi-infinite series of unique positive natural frequencies beginning with the fundamental mode and orthogonal mode shapes. The total displacement is the sum of the modal displacements.

Cable modulus, $E = L(dP/dL)$, is the proportionality constant in the relationship between the incremental change in tension dP and incremental change in cable length dL, where A is the reference cross-sectional area. If the cable is a solid rod, then E is the modulus of elasticity. The cable modulus of a cable woven from fibers is a function of the modulus of the fibers, the packing factor, the orientation of the fibers, and whether the ends of the cable are constrained against rotation. The cable modulus of helically woven cables is often 30–50% of the modulus of their component fibers [27].

3.6.2 *Cable Sag*

The massive structural cables and chains used in civil engineering sag under the force of gravity into catenaries [14, 28, 29] (see Figure 3.10b). For small sags, less than about one eighth of the span, the sag is obtained by dropping the time-dependent term from Equation 3.32 and adding in a gravity load milligram per unit length; m is the mass per unit length:

$$P\frac{d^2Y(x)}{dx^2} = -mg, \quad \text{which has solution} \quad Y(x) = -\frac{mgx}{2P}\left(1 - \frac{x}{L}\right) \tag{3.34}$$

The sag-induced horizontal component of cable tension is $P = mgL^2/(8d)$, where d is the midspan sag $d = mgL/(8P)$. The developed length of the sagged cable, $L_c = L[1 + (8/3)(d/L)^2 - (32/5)(d/L)^4 +] \cdots$, is larger than the span L between tie-downs.

Sag increases the cable fundamental natural frequency of the fundamental symmetric in-plane (vertical) mode because the cable stretches in this mode. Out-of-plane modes do not stretch the cable to first order, and they are independent of cable modulus. Solutions are provided in Table 3.6.

Cables that are tied together are cable truss networks that support sails, covers, and bridges. A cable truss consists of two uniform pretensioned cables anchored at both ends and separated by a series of lightweight vertical spacers [28, 35, 36].

The mode shapes and natural frequencies of a straight string with an *attached viscous damper* can be analyzed exactly [37]. As a nominally straight string vibrates (Eq. 3.34), it stretches slightly, $dL \approx (i\pi Yi)^2/(2L)$, to accommodate its deflected shape. The increase in length is small and second order in the vibration amplitude Y for a nominally straight string, but it produces small oscillations in tension at twice the vibration frequency (also see Ref. [38]).

Example 3.6 Tensioned string natural frequency

A dulcimer is a folk stringed instrument with tensioned 72 cm (28.35 in.) long steel strings. One string is 0.34 mm (0.0133 in.) in diameter. What tension is required to achieve a natural frequency of 523 Hz, which is C5 on the musical scale?

Solution: The mass per unit length of the steel string is density (8 g/cc) times its cross-sectional area. Using gram-centimeter-seconds (cgs) units (case 5 of Table 1.2), the

Table 3.6 Natural frequencies of strings, cables, and chains

Notation: A = cross sectional area of cable; d = length or midspan sag, case 7; E = extensional modulus, Chapter 8; g = accelerations due to gravity, Table 1.2; J_0 = Bessel function of 1st kind; L = length between supports; L_c = length of unloaded cable; m = mass per unit length of cable; P = mean tension in cable. w = vertical load per unit length, mg for horizontal cable; $Y_0()$ = Bessel function of 2nd kind; x = distance from support; λ = dimensionless frequency parameter. Table 1.2 has consistent sets of units.

Cable Geometry	Natural Frequency f, Hz	Mode Shape and Comments
1. Straight String or Cable 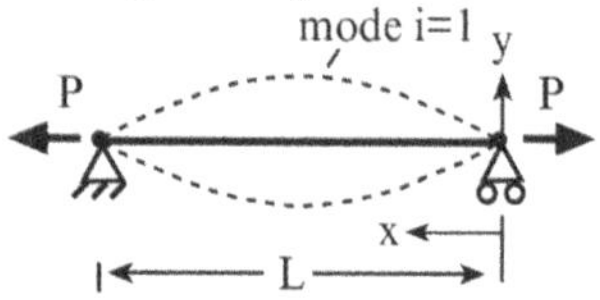	$\frac{i}{2L}\left(\frac{P}{m}\right)^{1/2}$, $i=1,2,3...$ m = mass per unit length of cable	$\tilde{y}(x)=\sin\left(\frac{i\pi x}{L}\right)$ See Fig. 3.10a. P = horizontal component of tension in cable. Ref. [9]
2. Inclined Suspended Cable Shallow Parabolic Sag 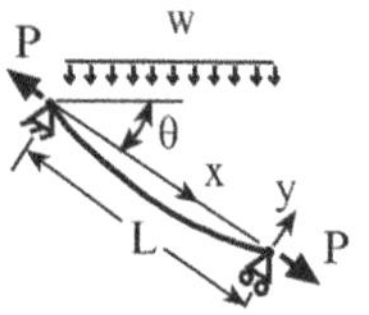	Out-of-plane modes are given by case 1. In- plane modes given in case 7.	Shallow parabolic sag is identical to case 7 with substitution w cos θ for w. Refs [28–31]
3. Hanging Chain 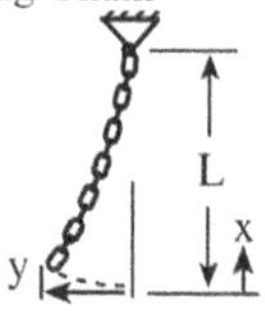	$\frac{\lambda_i}{2\pi}\left(\frac{g}{L}\right)^{1/2}$ i = 1,2,3.. λ_i = 1.2026, 2.7602, 4.3266, 5.8955... [$J_o(2\lambda_i) = 0$].	$\tilde{y}(x)=J_o\left(2\lambda_i(x/L)^{1/2}\right)$ $i=1,2,3....$ Chain hangs under own weight. Ref. [32]
4. Hanging Chain with End Mass 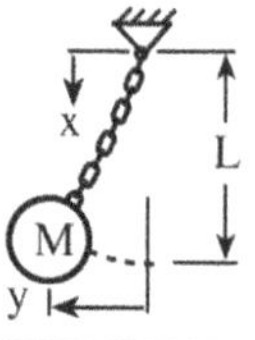	$\frac{1}{2\pi}\left(\frac{g}{L}\frac{M+M_c/2}{M+M_c/3}\right)^{1/2}$ M = mass of mass M M_c = mass of chain	$\tilde{y}(x)\frac{x}{L}$ Fundamental mode Also see Table 3.5. Approximate. Ref. [33] gives exact transcendental equation for λ.
5. Hanging Chain with End Constraint 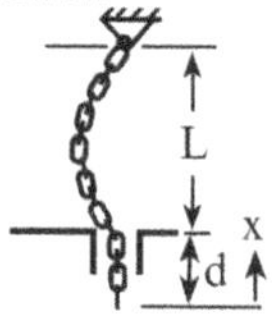	$\frac{\lambda_i}{2\pi}\left(\frac{g}{L}\right)^{1/2}$ d+L = total length of chain Ref. [21]	See table below. Values of λ. See case 6 for formulas for λ.

λ	d / L				
	0	0.01	0.1	1.0	10
λ_1	1.203	1.648	2.108	3.787	10.18
λ_2	2.760	3.410	4.268	7.582	20.35
λ_3	4.327	5.160	6.418	11.37	30.53

Values of λ. See case 6 for formulas for λ.

Table 3.6 Natural frequencies of strings, cables, and chains, continued

Cable Geometry	Natural Frequency f, Hz	Mode Shape and Comments
6. Chain with Linearly Varying Tension due to Traction w 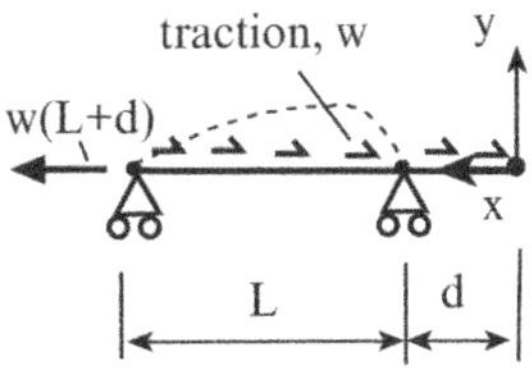 w = traction on cable = mg for vertically hanging chain or cable. Tension is zero at x = 0. m = mass per unit length of cable	$\frac{\lambda_i}{2\pi}\left(\frac{w}{mL}\right)^{1/2}$ $i = 1,2,3...$ λ tabulated in case 5. λ_i are found from the solutions of $J_0\left[2\lambda\left(\frac{d}{L}\right)^{1/2}\right]Y_0\left[2\lambda\left(\frac{d+L}{L}\right)^{1/2}\right] = J_0\left[2\lambda\left(\frac{d+L}{L}\right)^{1/2}\right]Y_0\left[2\lambda\left(\frac{d}{L}\right)^{1/2}\right]$	Mode shape over $d < x \le L + d$ $\tilde{y}(x) = J_0\left[2\lambda_i\left(\frac{x}{L}\right)^{1/2}\right] - \frac{J_0\left[2\lambda_i\left(\frac{d}{L}\right)^{1/2}\right]}{Y_0\left[2\lambda_i\left(\frac{d}{L}\right)^{1/2}\right]}Y_0\left[2\lambda_i\left(\frac{x}{L}\right)^{1/2}\right]$ Cases 5 and 6 are equivalent. Ref. [34]
7a. Suspended Cable with Shallow Parabolic Sag under Lateral Load w 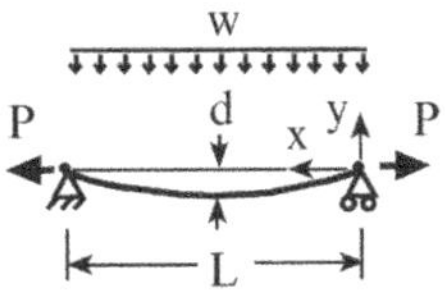 z = out of plane Mean sag $\bar{y}(x) = -\frac{wL^2}{2P}\left[\frac{x}{L} - \left(\frac{x}{L}\right)^2\right]$ Maximum sag $d = \frac{wL^2}{8P}$ at $x = \frac{L}{2}$ Formulas valid for d<L/8 Refs [14 ,28, 29]	Out-of-plane modes (z) $\frac{i}{2L}\left(\frac{P}{m}\right)^{1/2}$ $i = 1,2,3...$	$\tilde{z}(x) = \sin\left(\frac{i\pi x}{L}\right)$ See Fig. 10b. P = horizontal component of tension in cable. Ref. [9]
	In-plane anti-symmetric modes (x-y) $\frac{i}{L}\left(\frac{P}{m}\right)^{1/2}$ $i = 1,2,3...$	$\tilde{y}(x) = \sin\left(\frac{2i\pi x}{L}\right)$ See Fig. 3.10b. $\lvert\tilde{x}(x)\rvert / \lvert\tilde{y}(x)\rvert =$ of order d/L $\int_0^L \tilde{y}(x)dx = 0$, no cable stretch.
	In-plane symmetric modes $\frac{\lambda_i}{2L}\left(\frac{P}{m}\right)^{1/2}$ $i = 1,2,3...$ λ in 7b, below.	$\tilde{y}(x) = 1 - \tan(\pi\lambda/2)\sin(\pi\lambda/L) - \cos(\pi\lambda x/L)$ $\lvert\tilde{x}(x)\rvert / \lvert\tilde{y}(x)\rvert =$ of order d/L $\int_0^L \tilde{y}(x)dx \neq 0$, cable stretch

7b. λ for in-plane symmetric modes are solutions of transcendental equation in terms of stiffness α.

$$\tan\frac{\pi\lambda}{2} = \frac{\pi\lambda}{2} - \frac{4}{\alpha^2}\left(\frac{\pi\lambda}{2}\right)^3, \text{where } \alpha^2 = \left(\frac{8d}{L}\right)^2\frac{EA}{T_o}\frac{L}{L_e}, \quad \frac{L_e}{L} = 1 + \frac{8}{3}\left(\frac{d}{L}\right)^2 - \frac{32}{5}\left(\frac{d}{L}\right)^4 + ..$$

For first mode

α^2	0	2	3	6	10	14	20	24	35	40	60	80	100	150	> 1000
λ_1	1.0	1.079	1.115	1.220	1.345	1.484	1.609	1.701	1.921	2.008	2.291	2.480	2.597	2.727	2.86

L_e = virtual cable lengths, . L = distance between supports, L_c >L Refs [14 ,28, 29]

mass per unit length of the string is calculated:

$$m = \rho A = \rho\left(\frac{\pi}{4}\right)D^2 = 8\ \text{g/cm}^3\left(\frac{\pi}{4}\right)(0.034\ \text{cm})^2$$
$$= 0.00726\ \text{g/cm}\ (0.00004065\ \text{lb/in.})$$

The natural frequencies are calculated from case 1 of Table 3.6 using a tension $P = 100\,N = 10{,}000{,}000$ dyne (22.25 lb) and consistent units shown in Table 1.2:

$$f = \frac{i}{2L}\sqrt{\frac{P}{m}} = \frac{1}{2\ (72\ \text{cm})}\sqrt{\frac{10{,}000{,}000\ \text{dyne}}{0.00726\ \text{g/cm}}} = \frac{1}{2\ (28.35\ \text{in.})}\sqrt{\frac{22.25\ \text{lb} \times 386.1\ \text{in./s}^2}{0.00004065\ \text{lb/in.}}}$$
$$= 257.7\ \text{Hz}$$

To achieve 523 Hz in the first mode, the frequency must be increased by the factor $523/257.7 = 2.03$, which requires an increase in tension by a factor $2.03^2 = 4.12$, $P = 412\,N$. Section 6.1 and Refs [39–41] discuss music and vibration.

3.7 Membrane Natural Frequencies

3.7.1 Flat Membranes

Table 3.7 shows the natural frequencies and mode shapes of flat and curved membranes. Membranes are thin surfaces that have mass and support tension but not bending. The mean tension provides the linear restoring force during out-of-plane vibration (Figure 3.11).

The equation of motion of small out-of-plane displacement (w) of a tensioned flat membrane is the *two-dimensional wave equation* [13]:

$$T\left(\frac{\partial^2 w}{\partial x^2} + \frac{\partial^2 w}{\partial y^2}\right) = \gamma\frac{\partial^2 w}{\partial t^2} - q \quad \text{Cartesian } x\text{–}y \text{ coordinates} \tag{3.35a}$$

$$T\left(\frac{\partial^2 w}{\partial r^2} + \frac{1}{r}\frac{\partial w}{\partial r} + \frac{1}{r^2}\frac{\partial^2 w}{\partial \theta^2}\right) = \gamma\frac{\partial^2 w}{\partial t^2} - q \quad \text{cylindrical coordinates} \tag{3.35b}$$

The tension per unit length of edge is T and γ is the mass per unit area of membrane. The mean load per unit area q is applied perpendicular to the membrane surface. Shear is not supported and the uniform membrane tension is independent of small displacements.

The natural frequencies of out-of-plane vibration increase with membrane tension:

$$f_{ij} = \frac{\lambda_{ij}}{2}\sqrt{\frac{T}{\gamma A}} \quad \text{Hz} \quad i = 1, 2, 3, \ldots \quad j = 1, 2, 3, \ldots \tag{3.36}$$

The area of the membrane is A. λ_{ij} is the dimensionless natural frequency parameter which, like the mode shapes, is enumerated by two indices (i and j) that correspond to half waves in the two coordinate directions as shown in Figure 3.12. The mode shapes are orthogonal over the membrane.

Table 3.7 Natural frequencies of membranes

Notation: a,b = lengths; i, j = integers, A = area of membrane; f = natural frequency, Hz; $J_i()$ = Bessel function of first kind and i order; T = tension per unit length of edge; β, θ = angle; i = mass per unit area of membrane; λ_{ij} = dimensionless natural frequency parameter. $\tilde{z}$ = mode shape for vibration out-of-plane of the membrane. Consistent units are in Table 1.2. Membrane edges are fixed except case 14.

$$\text{Natural Frequency, } f_{ij} = \frac{\lambda_{ij}}{2}\left(\frac{T}{\gamma A}\right)^{1/2}, \quad \text{Hz}$$

<table>
<tr><th>Flat Membrane Shape</th><th>Natural Frequency Parameter λ_{ij}, Area A, and Mode Shape $\tilde{z}$</th></tr>
<tr><td>1. Rectangular Membrane
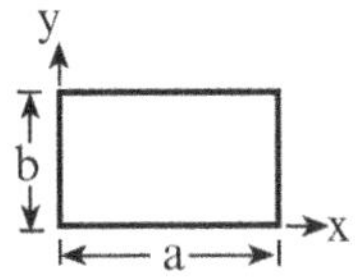
</td><td>$\lambda_{ij} = \left(i^2 \frac{b}{a} + j^2 \frac{a}{b}\right)^{1/2}$, $i = 1,2,3..,\ j = 1,2,3...$ Refs [15, 20]
$A = ab$
$\tilde{z} = \sin\frac{i\pi x}{a}\sin\frac{j\pi y}{b}$</td></tr>
<tr><td>2. Isosceles Right Triangle
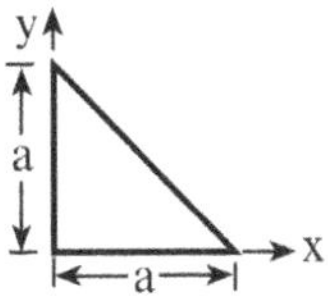
</td><td>$\lambda_{ij} = \begin{cases} 1.5811, 2.2361, 2.5499, 2.9155.. \\ \left(i^2 + j^2\right)^{1/2} / 2^{1/2},\ i,j = (1,2), (2,1), (3,1), (3,2), (2,3)..\ (i \neq j) \end{cases}$
$A = a^2/2$ Ref. [20], p. 318
$\tilde{z} = \sin\frac{i\pi x}{a}\sin\frac{j\pi y}{b} \pm \sin\frac{i\pi x}{a}\sin\frac{j\pi y}{b}$, − if i + j = even; + if i + j = odd</td></tr>
<tr><td>3. Isosceles Triangle
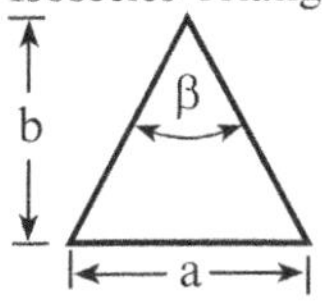
</td><td>

β, deg.	20	30	40	50	60	70	80	90
a/b	0.352	0.535	0.727	0.983	1.155	1.400	1.628	2.00
λ	1.782	1.631	1.564	1.528	1.520	1.527	1.548	1.581

A= ab/2
Fundamental mode, Ref. [15]</td></tr>
<tr><td>4. Parallelogram
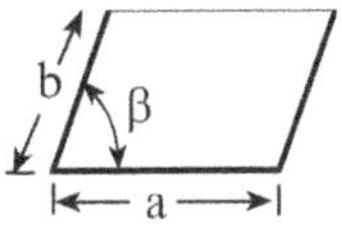
</td><td>Fundamental model, λ A=ab sin β Ref. [16]

b/a	β, degrees			
	90	75	60	45
1	1.414	1.429	1.478	1.579
1.5	1.472	1.489	1.546	1.662
2	1.581	1.602	1.670	1.810
3	1.826	1.853	1.943	2.127

</td></tr>
<tr><td>5. L Shape
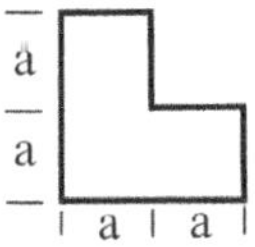
</td><td>

Mode	1	2	3	4	5
λ	1.717	2.150	2.449	2.999	3.124

$A = 3a^2$
Ref. [17]</td></tr>
<tr><td>6. H Shape
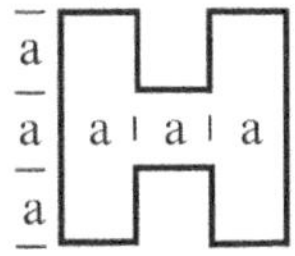
</td><td>

Mode	1	2	3
λ	6.556	12.06	16.63

$A = 7a^2$
Ref. [17]</td></tr>
</table>

Table 3.7 Natural frequencies of membranes, continued

$$\text{Natural Frequency, } f_{ij} = \frac{\lambda_{ij}}{2}\left(\frac{T}{\gamma A}\right)^{1/2}, \quad \text{Hz}$$

Flat Membrane Shape	Natural Frequency Parameter λ_{ij}, Area A, and Mode Shape $\tilde{z}$

7. Regular Polygon with N Equal Sides

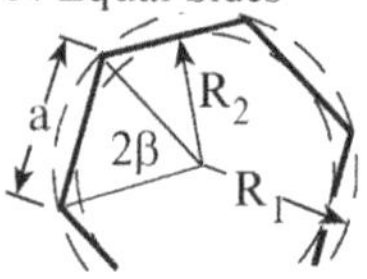

N sides	3	4	5	6	7	8	∞
λ	1.520	1.414	1.385	1.372	1.366	1.362	1.357

$A = (N/4)a^2 \cot\beta$, N = number of sides

$R_1 = a/(2\sin\beta)$, $R_2 = a/(2\tan\beta)$, $\beta = \pi/N$, radians Ref. [18]

Fundamental mode

8. Circle

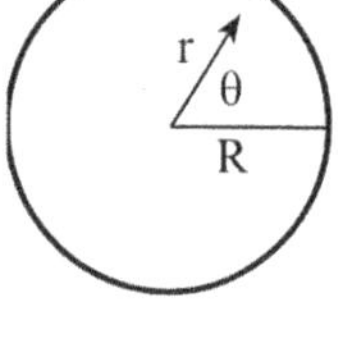

Natural frequency parameter λ_{ij}

j	number of modal diameters, i			
	0	1	2	3
1	1.357	2.162	2.897	3.600
2	3.114	3.958	4.749	5.507
3	4.882	5.740	6.556	7.343
4	6.653	7.517	8.348	9.153

$\lambda_{ij} \approx \pi^{1/2}(j+i/2-1/4)$, i,j>3, Ref. [19]

$\tilde{z}_{ij} = J_i(\pi^{1/2}\lambda_{ij} r/R)\cos i\theta$

Eqn for λ: $J_i(\pi^{1/2}\lambda_{ij}) = 0$

i = number of nodal diameters

j = number nodal radii

$A = \pi R^2$ Refs [13, 20]

9. Sector of a Circle

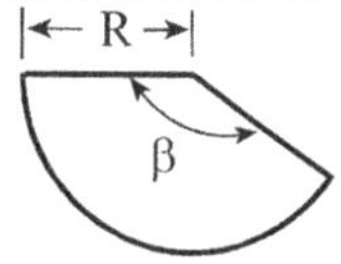

Fundamental mode

β, deg.	20	30	45	60	90	180	360
λ	1.724	1.618	1.514	1.470	1.448	1.529	1.357

$A = R^2\beta/2$, β in radians

Ref. [20], pp. 332–334

10. Ellipse

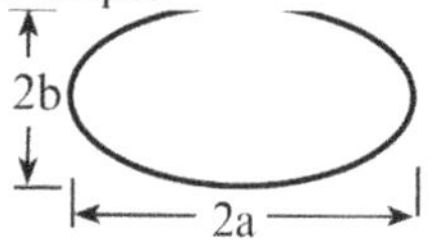

Fundamental mode $A = \pi ab$

$$\lambda = 2.405\left[\frac{\frac{b}{a}+\frac{a}{b}}{2\pi}\right]^{1/2}$$

Ref. [21]

11. Annulus

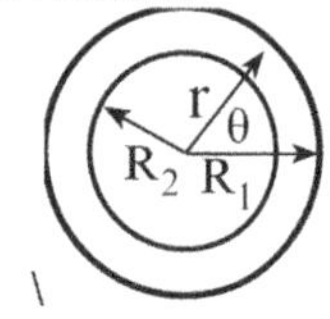

For narrow annuli R_2/R_1`>0.5 See Example 3.7

$$\lambda_{ij} = \frac{1}{\pi^{1/2}}\left[4i^2\frac{R_1-R_2}{R_1+R_2} + j^2\pi^2\frac{R_1+R_2}{R_1-R_2}\right]^{1/2} \quad i = 1,2,3..\ \ j = 0,1,2..$$

$$\tilde{z}_{ij} = \cos i\theta \sin\frac{j\pi(r-R_2)}{R_1-R_2}, \qquad A = \pi(R_1^2 - R_2^2)$$

12. Parabolic Section

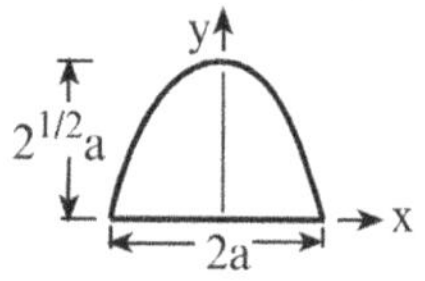

Fundamental mode Ref. [21]

$$\lambda = \frac{4.6701}{\pi}$$

$$\text{outline: } x^2 = a^2\left(1 - \frac{y}{\sqrt{2}a}\right), \quad A = 4\frac{\sqrt{2}}{3}a^2$$

13. Any Flat Membrane

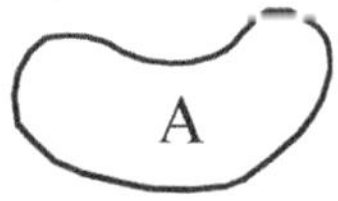

Lower bound on λ for ith mode

$\lambda_i = 1.357\, i^{1/2}$ $i = 1,2,3..$

Area = A Refs [22, 23]

Table 3.7 Natural frequencies of membranes, continued

Curved Membrane	Natural Frequency f, Hz, and Mode Shape $\tilde{z}$
14. Pressurized Ring, 2-D Cylinder Section i = 2 shown 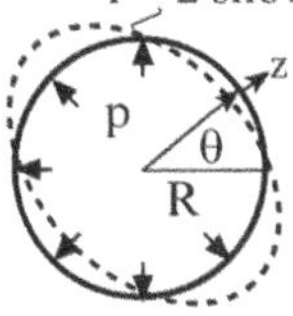	$f_i = \frac{1}{2\pi R}\left(\frac{i^2(i^2-1)}{i^2+1}\right)^{1/2}\left[\frac{T}{m}\right]^{1/2}$, Hz $i = 1,2,3..$ $\tilde{z} = \cos i\theta$, $i = 2,3,4..$ $p = T/R$ = internal pressure, R = equilibrium radius T = circumferential tension per unit depth = pR m = mass per unit length of ring Ref. [24]
15. Axially Tensioned Cylinder j=1 mode 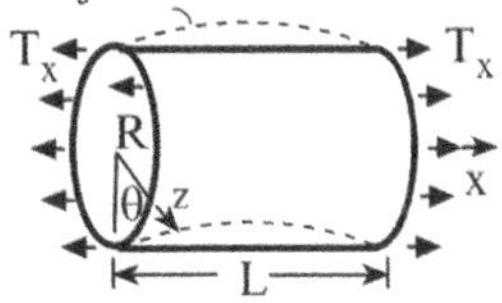	$f_i = \frac{j}{2L}\left[\frac{T_x}{\gamma}\right]^{1/2}$ Hz $j = 1,2,3..$ Ref. [25] $\tilde{z}(x) = \sin j\pi x/L$ T_x = axial tension per unit length of edge = pR/2 if end cap and internal pressure p γ = membrane mass per unit area. Edges at x=0, L fixed in z direction. Section remains circular.
16. Pressurized Cylindrical Membrane with Axial Tension Edges at x=0,L fixed in z j=1 mode 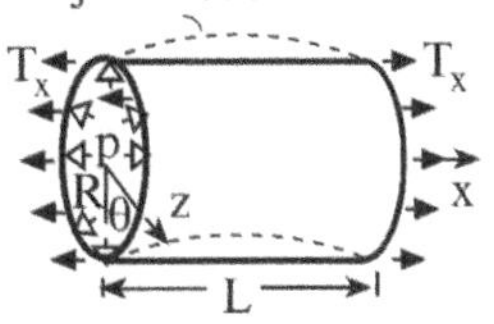	$f_{ij} = \frac{1}{2\pi}\left[\frac{T_x}{\gamma}\left(\frac{j\pi}{L}\right)^2 + \frac{T_\theta}{\lambda R^2}\frac{i^2(i^2-1)}{i^2+1}\right]^{1/2}$, Hz $i = 1,2,3..$, $j = 1,2,3..$ $\tilde{z}_{ij} = \sin(j\pi x/L)\cos i\theta$, $i = 0,1,2,3..$, $j = 1,2,3..$, ends are fixed T_θ = circumferential tension per unit edge, constant = pR for internal pressure p T_x = axial tension per unit length of edge = pR / 2 for axial load due to internal pressure p γ = membrane mass per unit area Ref. [26]
17. Pressurized Spherical Membrane 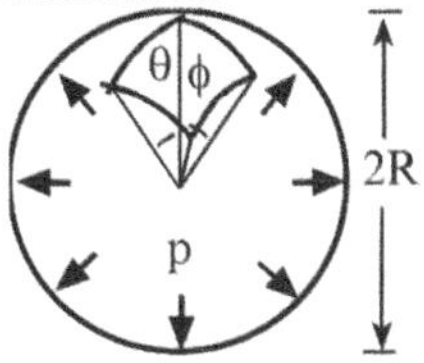	$f_{ij} = \frac{1}{2\pi}\left(i^2 + j^2 + 2\right)^{1/2}\left[\frac{T}{\gamma}\right]^{1/2}$, Hz $i = 0,1,2..$, $j = 0,1,2..$ $\tilde{z}_{ij} = \cos j\phi \cos i\theta$ $p = T/R$ = internal pressure, R = equilibrium radius T = uniform tension per unit edge = pR for interal pressure p γ = membrane mass per unit area author's result
18. Flat Membrane - Plate Analog 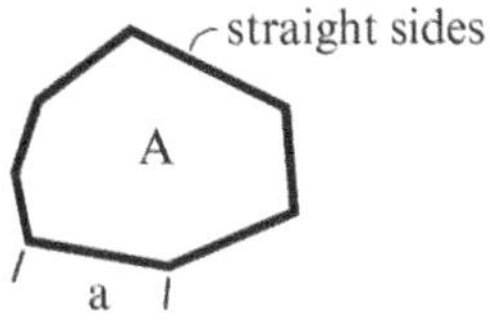	The natural frequencies of a flat membrane with straight sides and a similarly shaped flat plate with simply supported edges are related. $\frac{\lambda_{ij}\vert_{plate}}{\pi a} = \frac{\lambda_{ij}\vert_{membrane}}{A^{1/2}}$ A= area of membrane, a = length of reference side. Refs [15, 22, 23]. f_{ij} membrane given by Eq. 3.36, f_{ij} plate given by Eq. 5.15.

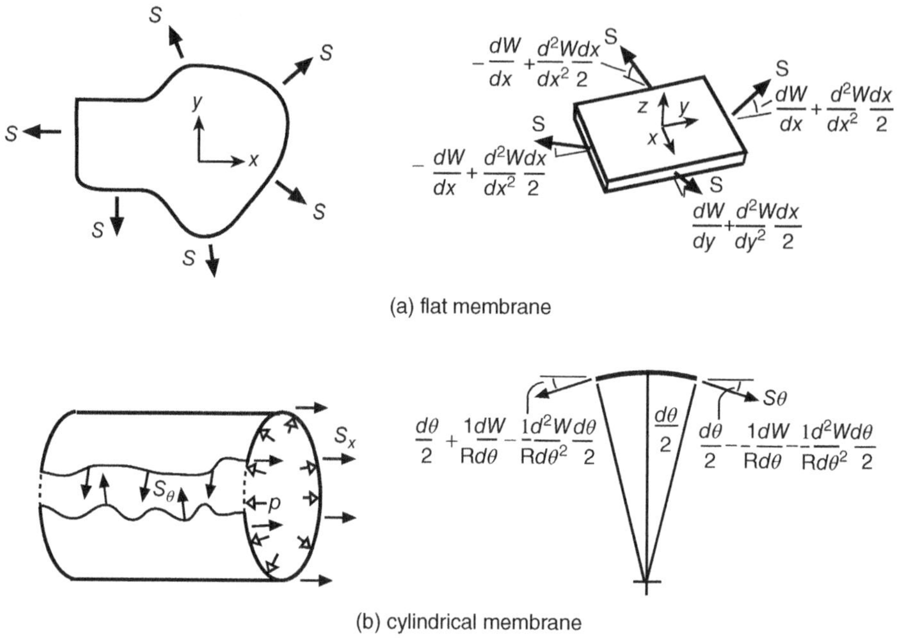

Figure 3.11 Flat and cylindrical tensioned membranes

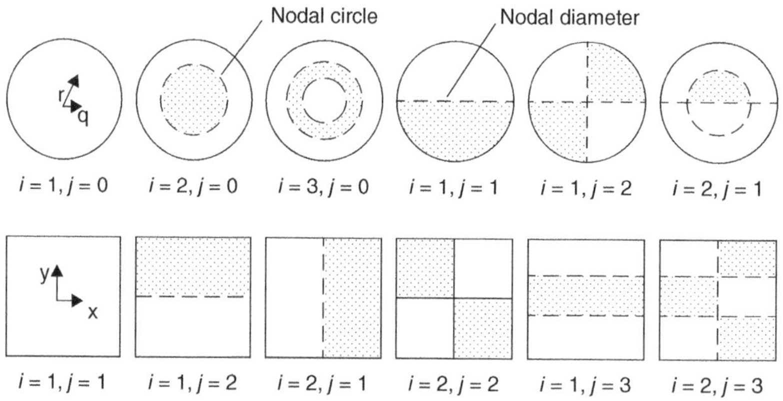

Figure 3.12 Out-of-plane modal deformation patterns of flat membranes

Membrane analogy is a relationship between the natural frequencies of polygonal membranes and similarly shaped simply supported plates (case 18 of Table 3.7) because they have the same mode shapes. Tensioned membrane solutions can be used to include the effect of in-plane loads on natural frequencies of plates and shells. Consider a round membrane shown in case 8 of Table 3.7 and the round plate in case 2 of Table 5.7. The natural frequency with in-plane loads is the sum of the squares of circular membrane natural frequency and the plate bending natural frequency without in-plane loads. This sums the stiffness terms:

$$f_{ij}^2\Big|_{\substack{\text{round plate}\\ \text{in-plane load}}} = f_{ij}^2\Big|_{\substack{\text{round plate}\\ \text{no in-plane load}}} + f_{ij}^2\Big|_{\substack{\text{round}\\ \text{tensioned}\\ \text{membrane}}}$$

$$f_{ij}\Big|_{\substack{\text{round plate}\\ \text{in-plane load}}} = \frac{1}{2\pi a^2}\sqrt{\frac{Eh^3}{12\gamma(1-\nu^2)}}\left(\lambda_{ij}^4\Big|_{\substack{\text{round}\\ \text{plate}}} + \frac{12(1-\nu^2)\pi a^2 T}{Eh^3}\lambda_{ij}^2\Big|_{\substack{\text{round}\\ \text{tensioned}\\ \text{membrane}}}\right)^{1/2} \tag{3.37}$$

This solution is identical to case 14 of Table 5.2.

3.7.2 Curved Membranes

Unlike flat membranes where the boundaries provide tension, the tension in curved membranes (Table 3.7 cases 14–17) is the result of internal pressure. Equilibrium between internal pressure, tension, and radius, $p = 2T/R$ for the spherical membrane and $p = T/R$ for the cylindrical membrane, results in a zero-frequency mode ($i = 1$) where the membrane expands radially (w) without limit. The pressure model for a bubble in chapter 6, section 6.4, produces a nonzero-frequency axisymmetric mode of spherical membranes.

Example 3.7 Annular membrane natural frequency

Derive the approximate natural frequencies of the flat annular membrane given in case 11 of Table 3.7.

Solution: The annular membrane mode shape is assumed to be the product of circumferential and radial sinusoidal components:

$$w(r,\theta) = \sin\left[\frac{j\pi(r-R_2)}{(R_1-R_2)}\right]\cos i\theta \sin\omega t \quad i = 0,1,2,3,\ldots \quad j = 1,2,3,\ldots \tag{3.38}$$

The index i is the number of circumferential waves and j is the number of radial half waves. The radius r varies between the inner fixed radius R_1 and the outer fixed radius R_2. θ is the circumferential angle. This mode shape is substituted into Equation 3.35b, which is then multiplied by the mode shape and integrated over the area of the membrane (Galerkin technique; Appendix A):

$$T\left(\frac{\pi}{R_1-R_2}\right)^2\int_{R_1}^{R_2}\sin^2 u(r)dr + T\left(\frac{j\pi}{R_1-R_2}\right)\int_{R_1}^{R_2}\frac{1}{r}\sin u(r)\cos u(r)dr$$
$$+ Ti^2\int_{R_1}^{R_2}\frac{1}{r^2}\sin^2 u(r)dr = {}^2\int_{R_1}^{R_2}\sin^2 u(r)dr, \quad u(r) = j\pi\left(\frac{r-R_2}{R_1-R_2}\right)$$

The average radius, $1/r \approx 2/(R_1 + R_2)$, is substituted into the integrals. The result is solved for the approximate natural frequencies in hertz of a flat narrow annular membrane:

$$f_{ij} = \frac{1}{2\pi^{1/2}} \left[4i^2 \frac{R_1 - R_2}{R_1 + R_2} + j^2\pi^2 \frac{R_1 + R_2}{R_1 - R_2} \right]^{1/2} \left(\frac{T}{\gamma A} \right)^{1/2}, \quad \text{Hz} \qquad (3.39)$$

The membrane area is $A = \pi(R_1{}^2 - R_2{}^2)$. The membrane analogy shown in case 18 of Table 3.7 converts this to the natural frequency of an analogous thin flat plate with simply supported edges.

References

[1] Roark, J. J., and W. C. Young, Formulas for Stress and Strain, 4th ed., McGraw-Hill, N.Y., 1975.
[2] Timoshenko, S., and J. N. Goodier, Theory of Elasticity, 2nd ed., McGraw-Hill, p. 7, 1951.
[3] Wahl, A. M., Mechanical Springs, 2nd ed., McGraw-Hill, N.Y., 1963.
[4] SAE Spring Committee, Spring Design Manual, AE-11, Society of Automotive Engineers, Warrendale, PA, USA, 1990.
[5] Smith, S. T., Flexures, Elements of Elastic Mechanisms, CRC Press, New York, p. 181, 212, 2000.
[6] Gerlach, C. R., Flow-Induced Vibrations of Metal Bellows, Journal of Engineering for Industry, vol. 91, pp. 1196–1202, 1969.
[7] Assessment of Flexible Lines for Flow-Induced Vibration, 1990, NASA30M02540, Rev. D, Huntsville, Alabama.
[8] Standards of the Expansion Joint Manufacturers Association, 7th ed., Expansion Joint Manufacturers Association, White Plains, NY, USA, 1998.
[9] Thomson, W. T., Theory of Vibration with Applications, 3rd ed., Prentice Hall, Englewood Cliffs, N.J., 1988.
[10] Brelawa, R. L., Rotary Wing Structural Dynamics and Aeroelasticity, 2nd ed., American Institute of Aeronautics and Astronautics, Reston, VA, 2006.
[11] Seto, W. W., Theory and Problems of Mechanical Vibrations, Schaum's Outline Series, McGraw-Hill, N.Y., 1964.
[12] Bellman, R., Introduction to Matrix Analysis, 2nd ed., McGraw-Hill, N.Y., pp. 32–42, 1970.
[13] Meirovitch, L., Analytical Methods in Vibrations, Macmillan, N.Y., pp. 141–142, 166–178, 1967.
[14] Irvine, H. M. and T. K. Caughey, The Linear Theory of Free Vibrations of a Suspended Cable, Proceedings of Royal Society of London, Series A, vol. 342, pp. 299–315, 1974.
[15] Conway, H. D. and K. A. Karnham, Free Flexural Vibration of Triangular, Rhombic, Parallelogram Plates and Some Analogies, International Journal of Mechanical Sciences, vol. 7, pp. 811–817, 1965.
[16] Durvasula, S., Natural Frequencies and Mode Shapes of Skew Membranes, Journal of Acoustical Society of America, vol. 44, pp. 1636–1646, 1968.
[17] Milsted, M. G., and J. R. Hutchinson, Use of Trigonometric Terms in the Finite Element with Application to Vibrating Membranes, Journal of Sound and Vibration, vol. 32, pp. 327–346, 1974.
[18] Shahady, P. A., R. Pasarelli, and P. A. A. Laura, Application of Complex Variable Theory to the Determination of the Fundamental Frequency of Vibrating Plates, Journal of Acoustical Society of America, vol. 42, pp. 806–809, 1967.
[19] Abramowitz, A. and I. Stegun, Handbook of Mathematical Functions, National Bureau of Standards, Applied Math Series 55, U.S. Government Printing Office, p. 371, 1964.
[20] Rayleigh, J. W. S., The Theory of Sound, Volume 1, Dover Publications, New York, 1945, pp. 332–345, 543–545. (first edition published in 1894.)
[21] Mazumdar, J., Transverse Vibration of Membranes of Arbitrary Shape by the Method of Constant Deflection Contours, Journal of Sound and Vibration, vol. 27, pp. 47–57, 1973.
[22] Pnueli, D., Lower Bound to the nth Eigenvalue of the Helmholtz Equation over Two Dimensional Regions of Arbitrary Shape, Journal of Applied Mechanics, vol. 36, pp. 630–631, 1969.
[23] Pnueli, D., Lower Bounds to the Gravest and All Higher Frequencies of Homogeneous Vibrating Plates of Arbitrary Shape, Journal of Applied Mechanics, vol. 42, pp. 815–820, 1975.
[24] Koga, T. and S. Morimatsu, Bifurcation Buckling of Circular Cylindrical Shells Under Uniform External Pressure, AIAA Journal, vol. 27, pp. 242–248, 1989.
[25] Sodel, W., Vibration of Shells and Plates, 3rd ed. Revised, p. 320, 2004.
[26] Armenakas, A. E., Influence of Initial Stress on the Vibrations of Simply Supported Cylindrical Shells, AIAA Journal, vol. 2, pp. 1607, 1964.

[27] Costello, G. A., and J. W. Phillips, Effective Modulus of Twisted Wire Cables, ASCE Journal Engineering Mechanics Division, vol. 102, pp. 171–181, 1976.
[28] Irvine, M., Cable Structures, Dover, N.Y., 1981.
[29] Ramberg, S. E. and O. M. Griffin, Free Vibrations of Taut and Slack Marine Cables, ASCE Journal of the Structural Division, vol. 103, pp. 2079–2092, 1977. Also see discussion by A. S. Richardson, 1978, Journal of the Structural Division, vol. 104, pp. 1036–1041 and closure pp. 1926–1927.
[30] Ramberg, S. E. and C. L. Bartholomew, 1982, Vibrations of Inclined Slack Cables, ASCE Journal of the Structural Division, vol. 108, pp. 1662–1664.
[31] Wilson, A. J., and R. J. Wheen, Inclined Cables Under Load Design Expressions, ASCE Journal of Structural Division, vol. 103, 1977, pp. 1061–1078.
[32] Lamb, H., The Dynamical Theory of Sound, Dover Press, New York, 1980. pp. 84–86. First published in 1925.
[33] Sujith, R. I., and D. H. Hodges, 1995, Exact Solution for the Free Vibration of a Hanging Chord with a Tip Mass, Journal of Sound and Vibration, vol. 179, pp. 359–361.
[34] Huang, T. and D. W. Doreing, Frequencies of a Hanging Chain, Journal of American Acoustical Society, vol. 45, 1969, pp. 1046–1049.
[35] Irvine, H. M., Statics and Dynamics of Cable Trusses, ASCE Journal Engineering Mechanics Division, vol. 101, 1975, pp. 429–446.
[36] Troitsky, M. S., Cable-Stayed Bridges, 2nd ed., Van Nostrand Reinhold, N.Y., 1988.
[37] Main, J. A. and N. P. Jones, Free Vibrations of Taut Cable with Attached Damper, Part 1: Linear Viscous Damper, Journal of Engineering Mechanics, vol. 128, pp. 1062–1071, 2002.
[38] Bolwell, J. E., The Flexible String's Neglected Term, Journal of Sound and Vibration, vol. 206, pp. 618–623, 2004.
[39] Helmholtz, H. L. F., On the Sensations of Tone, Dover, N.Y., 1954. Reprint of 1885 edition, 2004.
[40] Bendale, A. H., Fundamentals of Musical Acoustics, Dover, N.Y., 1990.
[41] Woods, A., The Physics of Music, W.H. Freeman and Co., San Francisco, 1978.

4

Natural Frequency of Beams

Beams are slender structural elements that have bending stiffness. Bernoulli-Euler-Timoshenko beam theory postulates that plane cross sections of slender beams remain plane and normal to the longitudinal fibers during bending, and stress varies linearly over the cross section, which provides simple elegant solutions for the beam natural frequencies that are in this chapter.

4.1 Beam Bending Theory

4.1.1 Stress, Strain, and Deformation

Plane cross sections remain plane as a beam. Transversely displaces $Y(x, t)$ and bends into an arc with curvature $\partial^2 Y/\partial x^2$, Table 4.1 [1–6]. Each beam element extends or compresses as shown in Figure 4.1 in proportion to its distance from the neutral axis. Beam curvature is proportional to the bending moment M_z about an axis (z) perpendicular to the beam longitudinal axis. Equilibrium requires that the moment gradient along the span equals the shear force. The shear gradient equals the lateral load F per unit length (Fig. 4.1):

$$M_z = EI_z \frac{\partial^2 Y(x,t)}{\partial x^2}, \quad V_y = \frac{\partial M_z}{\partial x}, \quad -F = \frac{\partial V_y}{\partial x} = \frac{\partial^2 M_z}{\partial x^2} = EI_z \frac{\partial^4 Y}{\partial x^4} \tag{4.1}$$

$Y(x, t)$ = transverse displacement of the neutral axis (z) of the beam, positive up.
c = transverse coordinate in cross section from the local neutral axis.
I_z = area moment of inertia of cross section about z-axis through the centroid (Table 1.5).
R = radius *of curvature*, a function of x. $R = -1/(\partial^2 Y/\partial^2 x)$ for small displacements, positive convex in $-y$ direction-*a* valley.
E = modulus of elasticity.
M_z = moment about the transverse z-axis (Fig. 4.1), positive about z on $+x$ face.
V_y = transverse shear force, positive in $+y$ direction on $+x$ face.
F = transverse load on beam per unit length, here positive in the $+y$ direction.
x = global longitudinal distance along an undeformed beam axis.
y = transverse axis, transverse to x and z, $z = x \times y$.
z = z axis is the neutral axis, normal to bending arc and through the centroid.

Formulas for Dynamics, Acoustics and Vibration, First Edition. Robert D. Blevins.
© 2016 John Wiley & Sons, Ltd. Published 2016 by John Wiley & Sons, Ltd.

Table 4.1 Beam bending theory

Notation: A = cross section area; E = modulus of elasticity; I = area moment of inertia of beam cross section, Table 2.1; J = polar mass moment of inertia, Table 2.2; K = shear factor Table 8.14; k = extension spring constant, force/displacement, Table 3.1; k = torsion spring constant, moment/angle; M = mass; M_z = bending moment about z (neutral) axis; P = axial load, positive tensile; R = radius of curvature can be + or −, V_x shear force; x = axial coordinate; Y = lateral displacement, positive up. Refs [1–6]. See Fig. 4.1.

Stress	Strain	Loads and Curvature
$\sigma_x = -Ey\frac{\partial^2 Y}{\partial x^2} = -\frac{M_z y}{I}$	$\varepsilon_{xx} = -y\frac{\partial^2 Y}{\partial x^2}$	$M_z = -\int_A y\sigma_x dA = EI_z\frac{\partial^2 Y}{\partial x^2}$
$\sigma_{xy}\|_{centroid} = -\frac{EI_z}{KA}\frac{\partial^3 Y}{\partial x^3}$	$\varepsilon_{yy} = \varepsilon_{zz} = \nu y\frac{\partial^2 Y}{\partial x^2}$	$V_x = \frac{\partial M_z}{\partial x} = EI_z\frac{\partial^3 Y}{\partial x^3} = \int_A \sigma_{xy} dA = \sigma_{xy}\|_{avg} A$
$\sigma_{xy}\|_{average} = -\frac{EI_z}{A}\frac{\partial^3 Y}{\partial x^3}$	$\varepsilon_{xy} = \varepsilon_{xz} = \varepsilon_{yz} = 0$	$\frac{1}{R} = \frac{\partial^2 Y(x,t)}{\partial x^2} = \frac{M_z}{EI_z} = -\frac{\sigma_x}{Ey}$
$\sigma_{yy} = \sigma_{zz} = \sigma_{yz} = \sigma_{xz} = 0$		$F = \partial V_x / \partial x$

Potential Energy $= \frac{1}{2}\int_0^L EI\left(\partial^2 Y/\partial x^2\right)^2 dx$, Kinetic Energy $= \frac{1}{2}\int_0^L m(\partial Y/\partial t)^2 dx$, per unit Lenth

Boundary Condition	Symbol	Deformation B.C.	Beam Load at Boundary
1. Pinned, also called Simply-Supported	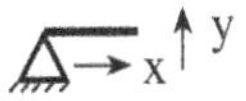	$Y = \frac{\partial^2 Y}{\partial x^2} = 0$	$M_z = 0,\ V_x = EI\frac{\partial^3 Y}{\partial x^3}$
2. Clamped, also called Fixed		$Y = \frac{\partial Y}{\partial x} = 0$	$M_z = EI\frac{\partial^2 Y}{\partial x^2},\ V_x = EI\frac{\partial^3 Y}{\partial x^3}$
3. Guided, also called Sliding- Support	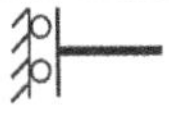	$\frac{\partial Y}{\partial x} = 0$	$M_z = EI\frac{\partial^2 Y}{\partial x^2},\ V_x = EI\frac{\partial^3 Y}{\partial x^3}$
4. Free		$\frac{\partial^2 Y}{\partial x^2} = \frac{\partial^3 Y}{\partial x^3} = 0$	$M_z = 0,\ V_x = 0$
5. Axial load	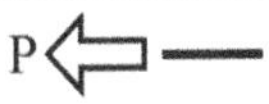	not applicable	$V_x = EI\frac{\partial^3 Y}{\partial x^3} - P\frac{\partial Y}{\partial x}$
6. Extensional spring	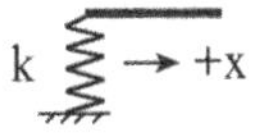	$EI\frac{\partial^3 Y}{\partial x^3} = -kY$ (a)	$V_x = -kY$ (a)
7. Torsional Spring	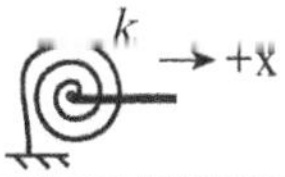	$EI\frac{\partial^2 Y}{\partial x^2} = k\frac{\partial Y}{\partial x}$ (a)	$M_z - k\frac{\partial Y}{\partial x}$ (a)
8. Concentrated Mass	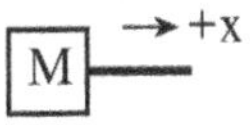	$EI\frac{\partial^3 Y}{\partial x^3} = -M\frac{\partial^2 Y}{\partial t^2}$ (a)	$V_x = -M\frac{\partial^2 Y}{\partial t^2}$ (a)
9. Rotational Inertia	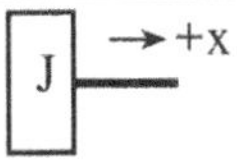	$EI\frac{\partial^2 Y}{\partial x^2} = -J\frac{\partial^3 Y}{\partial x \partial t^2}$ (a)	$M_z = -J\frac{\partial^3 Y}{\partial x \partial t^2}$ (a), $V_x = 0$

(a) The right hand side of these equations changes sign if the direction of increasing x is towards boundary rather than away from boundary as shown.

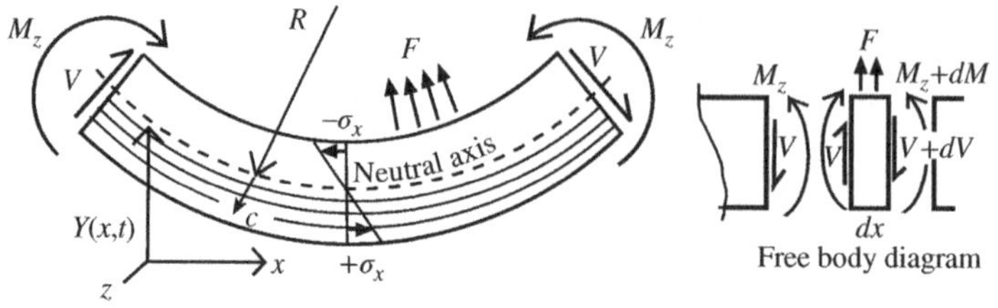

Figure 4.1 Coordinates for stress and bending deformation of a beam. σ_x is the axial beam bending stress and R is the radius of curvature

The neutral axis, which passes through the centroid of the cross section of a homogeneous beam, neither lengthens nor shortens as the beam bends.

Longitudinal stress σ_{xx} and strain ε_{xx} in a beam are proportional to the moment M_z [1, p. 95]:

$$\sigma_{xx} = E\varepsilon_{xx} = \frac{M_z c}{I_z} = Ec\frac{\partial^2 Y(x,t)}{\partial x^2} \tag{4.2}$$

Coordinate c is the transverse distance from the neutral axis to a longitudinal fiber in the cross section, positive in negative y direction. Moment M is positive about the z-axis defined by the right-hand rule: $z = x \times y$ on the $(+x)$ cut face and counterclockwise on the negative $(-x)$ face. Equilibrium requires that the moment M and shear V change sign (direction) across an axial element of the beam [1, p. 75–78] (also see the discussion of shear stress in Section 4.4.4). The strain energy per unit length in a bending beam is equal to $M^2/(2EI)$ [4, p. 235].

4.1.2 Sandwich Beams

Sandwich beams in Figure 4.2 consists of two identical thin and stiff face sheets that are bonded to a lightweight core spacer. Case 44 of Table 1.5 shows formulas for equivalent stiffness in terms of h, the distance between face sheets:

$$EI|_{\text{equiv.}} = \frac{1}{2}Etbh^2\left[1 + \frac{1}{3}\left(\frac{t}{h}\right)^2\right] \approx \frac{1}{2}Etbh^2, \text{ for } t \ll h \tag{4.3}$$

If a solid homogeneous beam is replaced with a sandwich beam with the (1) same depth, (2) same material for face sheets, and (3) massless core (rarely the case), then the ratio of

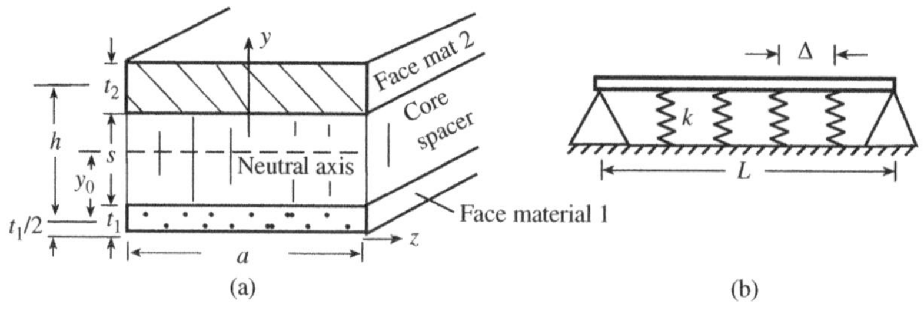

Figure 4.2 Cross section of an asymmetric sandwich beam and beam on an elastic foundation, (a) sandwich beam and (b) beam on elastic foundation

beam stiffness to beam mass increases by a factor of 3, and the natural frequencies will increase by $3^{1/2}$. See example 4.2.

The neutral axis position (y_0) and the equivalent beam stiffness about the neutral axis are integral over the cross section for a general multilayered sandwich:

$$y_0 = \frac{\int_A yE(y,z)dydz}{\int_A E(y,z)dydz} = \frac{\sum_i y_{ci}E_iA_i}{\sum_i E_iA_i} \quad \text{discretized section} \tag{4.4a}$$

$$EI|_{\text{equiv.}} = \int_A (y-y_0)^2 E(y,z)dydz = \sum_i E_iI_i \quad \text{discretized section, } I_i \text{ about } y_0 \tag{4.4b}$$

$E(y,z)$ is the modulus of elasticity of material on the cross section at point y,z. See case 45 of Table 1.5.

4.1.3 Beam Equation of Motion

This is Newton's second law (Eq. 2.4) applied to a small longitudinal element of the beam. The change in shear force per unit length of beam equals the mass per unit length m times its lateral acceleration, $\partial^2 Y/\partial t^2$ (Eq. 4.1):

$$\frac{EI_z\partial^4 Y}{\partial x^4} + \frac{m\partial^2 Y}{\partial t^2} = 0 \tag{4.5}$$

This fourth-order partial differential equation of motion is generalized to include spanwise varying properties, axial tension P, rotary inertia, and the transverse external load F per unit length in the $+y$ direction:

$$\frac{\partial^2}{\partial x^2}\left(EI_z\frac{\partial^2 Y}{\partial x^2}\right) - \frac{\partial}{\partial x}\left(P\frac{\partial Y}{\partial x}\right) - \rho I_z\frac{\partial^4 Y}{\partial x^2\partial t^2} + m\frac{\partial^2 Y}{\partial t^2} = F(x,t) \tag{4.6}$$

See Refs [2, 3 and 5]; and Figure 4.1.

4.1.4 Boundary Conditions and Modal Solution

The four classical boundary conditions on the ends of a beam are (1) free boundary, which bears no load; (2) pinned, which allows rotation but not displacement; (3) guided, also called sliding, which allows displacement but not rotation; and (4) clamped, which allows neither rotation nor displacement [2, p. 161; 3]:

$$\begin{array}{ll} \text{Free boundary} & \dfrac{\partial^2 Y}{\partial x^2} = \dfrac{\partial^3 Y}{\partial x^3} = 0 \text{ which implies } M = V = 0 \\ \text{Pinned boundary} & \dfrac{\partial^2 Y}{\partial x^2} = \dfrac{\partial^3 Y}{\partial x^3} = 0 \text{ which implies } M = 0 \\ \text{Clamped boundary} & Y = \dfrac{\partial Y}{\partial x} = 0 \\ \text{Guided boundary} & \dfrac{\partial Y}{\partial x} = 0 \end{array} \tag{4.7}$$

Clamped, pinned, and guided boundaries apply moments(M) (Eq. 4.1) and/or shear(V) (Eq. 4.24) to the ends of the beam; free boundary does not. An ideal clamped boundary is difficult to achieve in practice because of the high support stiffness, about 10 times the beam stiffness (Table 4.10), required to prevent rotation. Similarly, the frictionless rotation at a pin can only be approximated. Most real beam supports are intermediate between clamped and pinned. Other boundary conditions are given in Table 4.1. Solutions are provided in Sections 4.2 and 4.3.

The *separation-of-variable modal* solution to the unforced ($F(x,t)=0$) equation of motion is a function of distance x along the beam axis times a periodic function of time [2, 3, 5, 6, 7]:

$$Y(x,t)=\tilde{y}(x)T(t) \tag{4.8}$$

Substituting this into Equation 4.5 and dividing by $\tilde{y}(x)T(t)$ leads to functions of space (x) on the right-hand side of the equal sign and a functions of time (t) on the left:

$$\frac{1}{T(t)}\frac{\partial^2 T(t)}{\partial}=\frac{1}{\tilde{y}(x)}\frac{EI}{m}\frac{\partial^4\tilde{y}(x)}{\partial x^4}=-\omega^2=\text{constant} \tag{4.9}$$

Since both sides are equal and independent, they must equal the same constant. Choosing this to be $-\omega^2$ produces two ordinary differential equations, one in time and one in space:

$$\frac{d^2T(t)}{dt^2}+\omega^2T(t)=0 \tag{4.10a}$$

$$\frac{d^4\tilde{y}(x)}{dx^4}-\omega^2\left(\frac{m}{EI_z}\right)\tilde{y}(x)=0 \tag{4.10b}$$

Solutions to Equation 4.10a are ω: $T(t)=\sin(\omega t)$ and $\cos(\omega t)$. They oscillate in time with circular frequency ω.

Solutions of Equation 4.10b are sums of sine, cosine, sinh, cosh, and exponential functions of $\lambda x/L$. These are called *mode shapes*, $\tilde{y}(\lambda x/L)$. Discrete values of the dimensionless *natural frequency parameter* λ satisfy the *boundary conditions* (Table 4.1) and determine natural frequencies. For example, sinusoidal mode shapes, $\sin i\pi x/L$ where $\lambda=[\omega L^2(m/EI)^{1/2}]^{1/2}=i\pi,\ i=1,2,3,\ \ldots$ satisfy both Equation 4.10b and the boundary conditions of zero displacement and zero curvature boundary conditions at the ends, $x=0$, L, of a pinned–pinned beam Eq. 4.7.

Natural frequencies of beams are functions of the eigenvalues and the beam geometry and material properties:

$$f_i=\frac{\omega_i}{2\pi}=\frac{\lambda_i^2}{2\pi L^2}\sqrt{\frac{EI_z}{m}},\ \text{Hz},\quad i=1,2,3,\ \ldots \tag{4.11}$$

Beam natural frequencies are an increasing semi-infinite series beginning with the fundamental mode, $i=1$. Each successive mode has approximately one more half wave in its mode shape than its predecessor. E is the modulus of elasticity, I_z is the area moment of inertia of the cross section about the neutral axis (Table 1.5), L is the span, and m is the mass per unit length including any attached or nonstructural mass. The total beam mass is in mL. Consistent units required for this equation are shown in Table 1.2. The dimensionless *natural frequency parameter*, denoted by the symbol λ_i, depends on the boundary conditions shown in Table 4.2. (Note that some other authors use the symbol βL, kL, BL, or Ω for this parameter.)

Table 4.2 Natural frequencies of single-span beams

Notation: E = modulus of elasticity; f = natural frequency, Hz; i = 1,2,3.. modal index; I = area moment of inertia, Table 1.5, about z axis; L = length between supports; m = mass per unit length of beam, m = ρA plus nonstructural mass per unit length where ρ = beam material density and A is area of cross section; x = axial coordinate; y = transverse coordinate, in direction of displacement; z = neutral axis through centroid; λ, σ = nondimensional parameters, case 11; prime (') = derivative with respect to λx/L. Table 1.2 has consistent sets of units. Dashed line (---) in 1st column shows mode shape of first mode; see Fig. 4.5. Boundary conditions are defined in Table 4.1. Ref. [9].

$$\text{Natural Frequency,}\quad f_i = \frac{\lambda_i^2}{2\pi L^2}\sqrt{\frac{EI}{m}},\quad i=1,2,3\ldots\ \text{Hertz}$$

Boundary Conditions	λ_i, i=1,2,3.	Mode Shape, $\tilde{y}_i(\lambda_i x / L)$	σ_i, i=1,2,3.
1. Free-Free 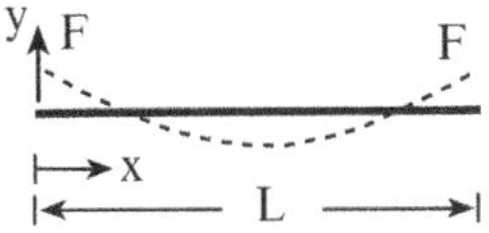	4.73004074 7.85320462 10.9956078 14.1371635 17.2787597 (2i+1)π/2,i>5	$\cosh\frac{\lambda_i x}{L} + \cos\frac{\lambda_i x}{L} - \sigma_i\left(\sinh\frac{\lambda_i x}{L} + \sin\frac{\lambda_i x}{L}\right)$ $\approx 2.16[1 - \frac{3}{2}\sin\frac{\pi x}{L}]$, for i=1	0.982502215 1.000777312 0.999966450 1.000001450 ≈ 1 for i >5
2. Free-Sliding 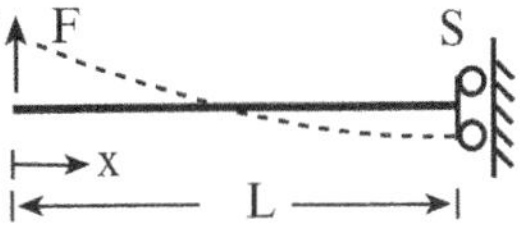	2.36502037 5.49780392 8.63937983 11.78097245 14.92256510 (4i–1)π/4,i>5	$\cosh\frac{\lambda_i x}{L} + \cos\frac{\lambda_i x}{L} - \sigma_i\left(\sinh\frac{\lambda_i x}{L} + \sin\frac{\lambda_i x}{L}\right)$	0.982502207 0.999966450 0.999999933 0.999999993 ≈ 1 for i >5,
3. Clamped-Free 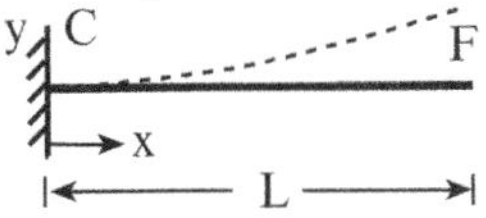	1.87510407 4.69409113 7.85475744 10.99554073 14.13716839 (2i–1)π/2,i>5	$\cosh\frac{\lambda_i x}{L} - \cos\frac{\lambda_i x}{L} - \sigma_i\left(\sinh\frac{\lambda_i x}{L} - \sin\frac{\lambda_i x}{L}\right)$ $\approx 2.1[1 - \cos(\pi x / 2L)]$, i=1	0.734095514 1.018467319 0.999224497 1.000033553 ≈ 1 for i >5
4. Free-Pinned 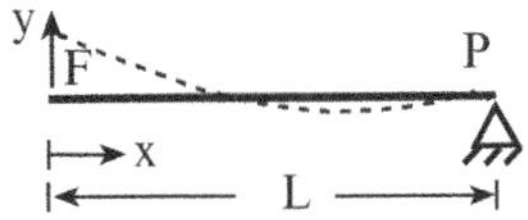	3.92660231 7.06858275 10.21017612 13.35176878 16.49336143 (4i+1)π/4,i>5	$\cosh\frac{\lambda_i x}{L} + \cos\frac{\lambda_i x}{L} - \sigma_i\left(\sinh\frac{\lambda_i x}{L} + \sin\frac{\lambda_i x}{L}\right)$	1.000777304 1.000001445 1.000000000 1.000000000 ≈1 for i >5
5. Pinned-Pinned 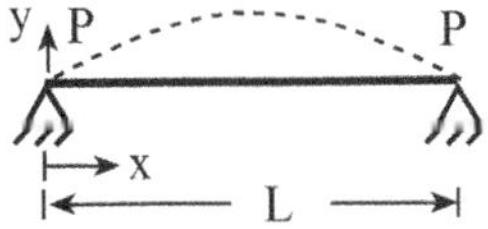	iπ, i=1,2,3..	$\sin\frac{i\pi x}{L}$	
6. Clamped-Pinned 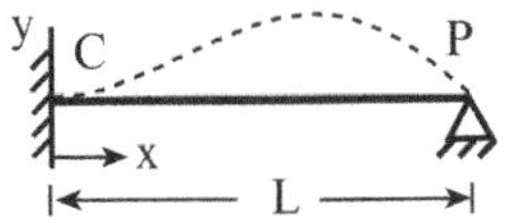	3.92660231 7.06858275 10.21017612 13.35176878 16.49336143 (4i+1)π/4,i>5	$\cosh\frac{\lambda_i x}{L} - \cos\frac{\lambda_i x}{L} - \sigma_i\left(\sinh\frac{\lambda_i x}{L} - \sin\frac{\lambda_i x}{L}\right)$ $\approx 105^{1/2}\left(\frac{x^2}{L^2} - \frac{x^3}{L^3}\right)$, for i=1	1.000777304 1.000001445 1.000000000 ≈ 1 for i >3

Table 4.2 Natural frequencies of single-span beams, continued

Boundary Conditions	λ_i, i=1,2,3..	Mode Shape, $\tilde{y}_i(\lambda_i x / L)$	σ_i, i=1,2,3..
7. Clamped-Clamped 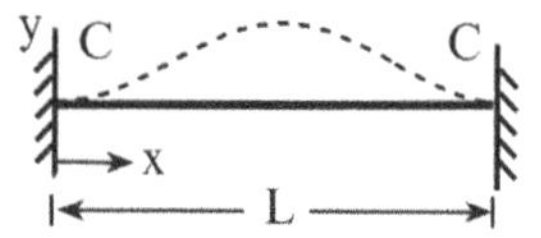	4.73004074 7.85320462 10.9956079 14.1371655 17.2787597 (2i+1)π/2, i>5	$\cosh\frac{\lambda_i x}{L} - \cos\frac{\lambda_i x}{L} - \sigma_i\left(\sinh\frac{\lambda_i x}{L} - \sin\frac{\lambda_i x}{L}\right)$ $\approx\sqrt{\frac{2}{3}}\left(1-\cos\frac{2\pi x}{L}\right)$, for i=1	0.982502207 1.000777312 0.999966450 1.000001450 ≈ 1 for i >5
8. Clamped-Sliding 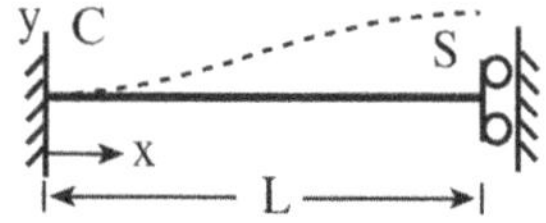	2.36502037 5.49780392 8.63937983 11.78097245 14.92256510 (4i-1)π/4, i>5	$\cosh\frac{\lambda_i x}{L} - \cos\frac{\lambda_i x}{L} - \sigma_i\left(\sinh\frac{\lambda_i x}{L} - \sin\frac{\lambda_i x}{L}\right)$ $\approx\sqrt{\frac{2}{3}}\left(1-\cos\frac{\pi x}{L}\right)$, i=1	0.982502207 0.999966450 0.999999933 0.999999993 ≈ 1 for i >5
9. Sliding-Pinned 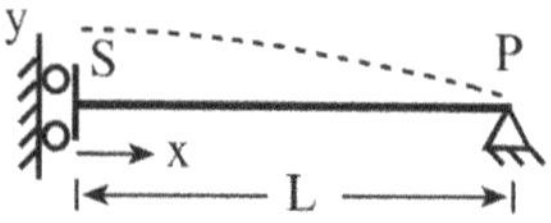	$\frac{\pi}{2}, \frac{3\pi}{2}, ..$ $(2i-1)\frac{\pi}{2}$ i=1,2,3..	$\cos\frac{(2i-1)\pi x}{2L}$	
10. Sliding-Sliding 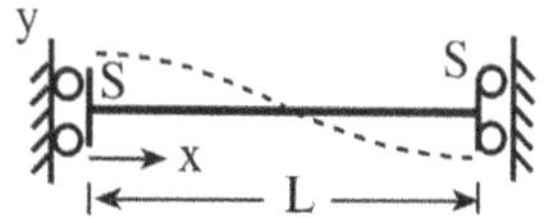	iπ, i=1,2,3..	$\cos\frac{i\pi x}{L}$	
11. Boundary Condition	Transcendental Equation for λ	Relationship for Mode Shapes	Formula for σ
11 a. Free-Free and Clamped-Clamped	cosλ coshλ=1 Two zero freq. modes	$\tilde{y}_i\|_{free-free} = \tilde{y}_i''\|_{clamp-clamp}$ $\tilde{y}_i'\|_{free-free} = \tilde{y}_i'''\|_{clamp-clamp}$ $\tilde{y}_i''\|_{free-free} = \tilde{y}_i\|_{clamp-clamp}$ $\tilde{y}_i'''\|_{free-free} = \tilde{y}_i'\|_{clamp-clamp}$	$\frac{\cosh\lambda - \cos\lambda}{\sinh\lambda - \sin\lambda}$
11 b. Free-Sliding and Clamped-Sliding	tanλ +tanh λ =0 One zero frequency mode	$\tilde{y}_i\|_{free-slide} = \tilde{y}_i''\|_{clamp-slide}$ $\tilde{y}_i'\|_{free-slide} = \tilde{y}_i'''\|_{clamp-slide}$ $\tilde{y}_i''\|_{free-slide} = \tilde{y}_i\|_{clamp-slide}$ $\tilde{y}_i'''\|_{free-slide} = \tilde{y}_i'\|_{clamp-slide}$	$\frac{\sinh\lambda - \sin\lambda}{\cosh\lambda + \cos\lambda}$
11 c. Free-Pinned and Clamped-Pinned	tanλ –tanh λ=0 One zero frequency mode	$\tilde{y}_i\|_{free-pin} = \tilde{y}_i''\|_{clamp-pin}$ $\tilde{y}_i'\|_{free-pin} = \tilde{y}_i'''\|_{clamp-pin}$ $\tilde{y}_i''\|_{free-pin} = \tilde{y}_i\|_{clamp-pin}$ $y_i'''\|_{free-pin} = y_i\|_{clamp-pin}$	$\frac{\cosh\lambda - \cos\lambda}{\sinh\lambda - \sin\lambda}$

4.1.5 Beams on Elastic Foundations

A massless elastic foundation, shown in Figure 4.2b, increases the natural frequencies of the attached beam [6, 8]:

$$f_i = \left[\left(f_i|_{Ef=0} \right)^2 + \frac{E_f}{4\pi^2 m} \right]^{1/2} \quad i = 1, 2, 3, \ldots \text{ Hz} \tag{4.12}$$

The foundation modulus E_f is the ratio of the load per unit spanwise length applied to the elastic foundation to the deformation this load produces. For example, the natural frequencies of the pinned–pinned beam (case 5 of Table 4.2, Eq. 4.11) on an elastic foundation are

$$f_i = \left[\frac{i^4 \pi^4}{4\pi^2 L^2} \frac{EI}{m} + \frac{k}{4\pi^2 m \Delta} \right]^{1/2} \quad i = 1, 2, 3, \ldots \text{ Hz} \tag{4.13}$$

The foundation modulus E_f has units of force per unit area. $E_f = k\Delta$ for the foundation shown in Figure 4.2b where k is the spring constant (Table 3.2) of discrete springs separated by distance Δ.

The foundation is massless in the earlier equations; case 8 of Table 4.8 implies that the effective beam mass per unit length increases by one-third of the foundation mass per unit length, provided the foundation natural frequencies substantially exceed the beam natural frequencies.

4.1.6 Simplification for Tubes

The tubular cross section's (case 27 of Table 1.5) ratio of moment of inertia, I, to the mass per unit length, m, is nearly independent of the tube wall. If nonstructural mass, such as internal fluid, is negligible, then the natural frequency of tubular beams is approximately proportional to the tube diameter:

$$\frac{I}{m} = \frac{I}{(\rho A)} = \frac{(D_o^2 + D_i^2)}{(16\rho)} = \frac{D_{\text{avg}}^2}{(8\rho)} \tag{4.14}$$

D_o is the tube outside diameter, D_i is the inside diameter, $D_{\text{avg}} = (D_o + D_i)/2$ is the average diameter, A is the cross-sectional area, and ρ is the tube material density. Substituting Equation 4.14 into Equation 4.11 gives a simplification for tubes:

$$f_i = \frac{\lambda_i^2}{2\pi L^2} \sqrt{\frac{EI}{m}} = \frac{\lambda_i^2}{8\pi L^2} \sqrt{\frac{E(D_o^2 + D_i^2)}{\rho}} \approx \frac{\lambda_i^2 D_{\text{avg}}}{8\pi L^2} \sqrt{\frac{2E}{\rho}}, \text{ Hz} \tag{4.15}$$

Further, since the ratio of material modulus of elasticity to material density, E/ρ, is approximately constant for engineering metals such as aluminum, steel, and titanium, the natural frequencies of tubular beams are nearly independent of the material.

4.2 Natural Frequencies and Mode Shapes of Single-Span and Multiple-Span Beams

4.2.1 Single-Span Beams

Natural frequencies and mode shapes of uniform single-span beams are given in Table 4.2 for the ten combinations of the four classical boundary conditions (Table 4.1). The single-span beam mode shapes are plotted in Figure 4.3 and tabularized in Table 4.3.

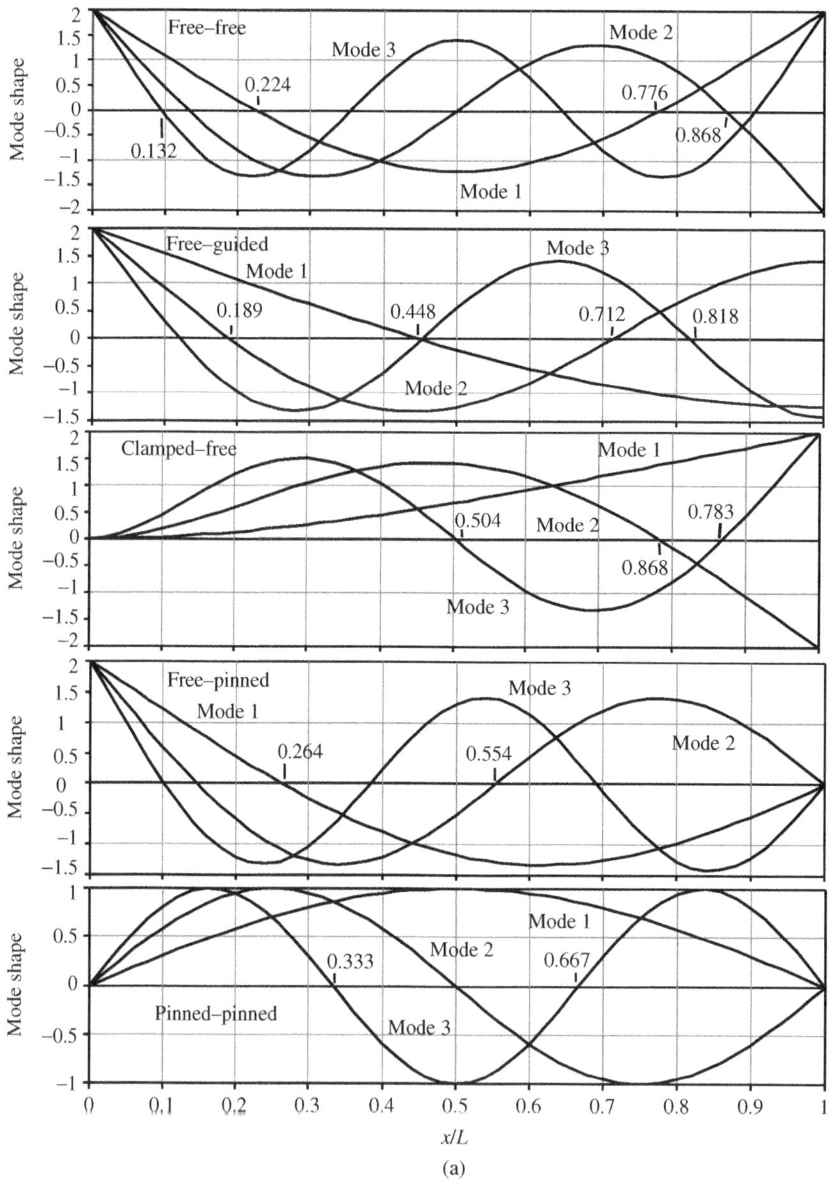

Figure 4.3 Mode shapes of single-span beams (Table 4.2)

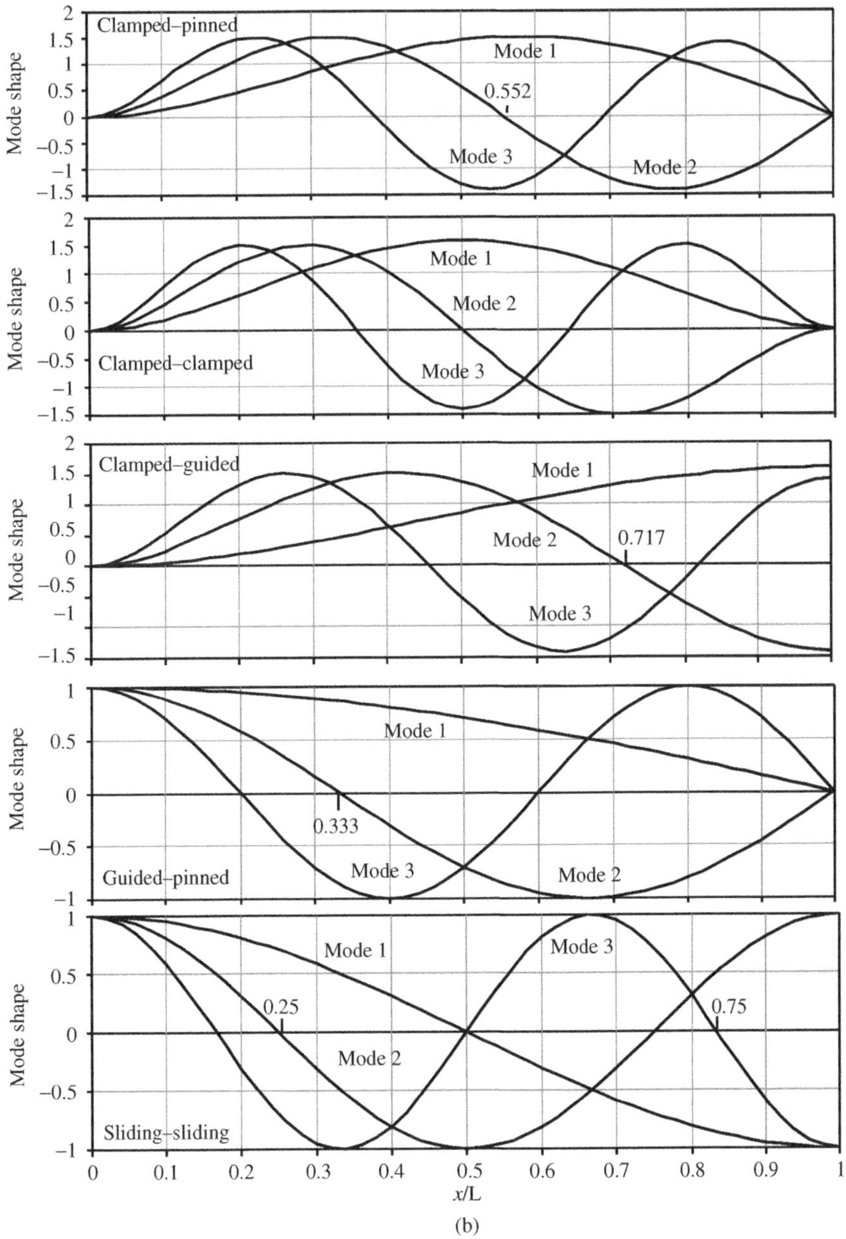

Figure 4.3 (*continued*)

Table 4.3a Beam mode shapes and their derivatives (Ref. [9])

Clamped-Clamped Beam. Also Free-Free Beam with case 11a of Table 4.2

	First Mode				*Second Mode*			
$\frac{x}{L}$	$\tilde{y}_1$	$\frac{L}{\lambda_1}\frac{d\tilde{y}_1}{dx}$	$\frac{L^2}{\lambda_1^2}\frac{d^2\tilde{y}_1}{dx^2}$	$\frac{L^3}{\lambda_1^3}\frac{d^3\tilde{y}_1}{dx^3}$	$\tilde{y}_2$	$\frac{L}{\lambda_2}\frac{d\tilde{y}_2}{dx}$	$\frac{L^2}{\lambda_2^2}\frac{d^2\tilde{y}_2}{dx^2}$	$\frac{L^3}{\lambda_2^3}\frac{d^3\tilde{y}_2}{dx^3}$
0.00	0.00000	0.00000	2.00000	-1.96500	0.00000	0.00000	2.00000	-2.00155
0.02	0.00867	0.18041	1.81412	-1.96473	0.02338	0.28944	1.68568	-2.00031
0.04	0.03358	0.34324	1.62832	-1.96285	0.08834	0.52955	1.37202	-1.99203
0.06	0.07306	0.48850	1.44284	-1.95792	0.18715	0.72055	1.06060	-1.97079
0.08	0.12545	0.61624	1.25802	-1.94862	0.31214	0.86296	0.75386	-1.93187
0.10	0.18910	0.72655	1.07433	-1.93383	0.45574	0.95776	0.45486	-1.87177
0.12	0.26237	0.81956	0.89234	-1.91254	0.61058	1.00643	0.16712	-1.78813
0.14	0.34363	0.89546	0.71270	-1.88393	0.76958	1.01105	-0.10555	-1.67975
0.16	0.43126	0.95450	0.53615	-1.84732	0.92601	0.97427	-0.35923	-1.54651
0.18	0.52370	0.99702	0.36346	-1.80219	1.07363	0.89940	-0.59009	-1.38932
0.20	0.61939	1.02342	0.19545	-1.74814	1.20675	0.79029	-0.79450	-1.21002
0.22	0.71684	1.03418	0.03299	-1.68494	1.32032	0.65138	-0.96917	-1.01128
0.24	0.81458	1.02986	-0.12305	-1.61250	1.41006	0.48755	-1.11133	-0.79652
0.26	0.91124	1.01113	-0.27180	-1.53085	1.47245	0.30410	-1.21875	-0.56977
0.28	1.00546	0.97870	-0.41240	-1.44017	1.50485	0.10661	-1.28992	-0.33555
0.30	1.09600	0.93338	-0.54401	-1.34074	1.50550	-0.09916	-1.32402	-0.09872
0.32	1.18168	0.87608	-0.66581	-1.23296	1.47357	-0.30736	-1.32106	0.13566
0.34	1.26141	0.80774	-0.77704	-1.11735	1.40913	-0.51224	-1.28181	0.36247
0.36	1.33419	0.72942	-0.87699	-0.99452	1.31313	-0.70820	-1.20786	0.57665
0.38	1.39913	0.64219	-0.96500	-0.86516	1.18740	-0.88997	-1.10158	0.77340
0.40	1.45545	0.54723	-1.04049	-0.73007	1.03456	-1.05270	-0.96606	0.94823
0.42	1.50246	0.44574	-1.10297	-0.59008	0.85794	-1.19210	-0.80507	1.09714
0.44	1.53962	0.33897	-1.15201	-0.44611	0.66151	-1.30449	-0.62295	1.21670
0.46	1.56647	0.22821	-1.18728	-0.29911	0.44974	-1.38693	-0.42455	1.30414
0.48	1.58271	0.11478	-1.20854	-0.15007	0.22752	-1.43727	-0.21508	1.35742
0.50	1.58815	0.00000	-1.21564	0.00000	0.00000	-1.45420	0.00000	1.37533
0.52	1.58271	-0.11478	-1.20854	0.15007	-0.22752	-1.43727	0.21508	1.35742
0.54	1.56647	-0.22821	-1.18728	0.29911	-0.44974	-1.38693	0.42455	1.30414
0.56	1.53962	-0.33897	-1.15201	0.44611	-0.66151	-1.30449	0.62295	1.21670
0.58	1.50246	-0.44574	-1.10297	0.59008	-0.85794	-1.19210	0.80507	1.09714
0.60	1.45545	-0.54723	-1.04049	0.73007	-1.03456	-1.05270	0.96606	0.94823
0.62	1.39913	-0.64219	-0.96500	0.86516	-1.18740	-0.88997	1.10158	0.77340
0.64	1.33419	-0.72942	-0.87699	0.99452	-1.31313	-0.70820	1.20786	0.57665
0.66	1.26141	-0.80774	-0.77704	1.11735	-1.40913	-0.51224	1.28181	0.36247
0.68	1.18168	-0.87608	-0.66581	1.23296	-1.47357	-0.30736	1.32106	0.13566
0.70	1.09600	-0.93338	-0.54401	1.34074	-1.50550	-0.09916	1.32402	-0.09872
0.72	1.00546	-0.97870	-0.41240	1.44017	-1.50485	0.10661	1.28992	-0.33555
0.74	0.91124	-1.01113	-0.27180	1.53085	-1.47245	0.30410	1.21875	-0.56977
0.76	0.81458	-1.02986	-0.12305	1.61250	-1.41006	0.48755	1.11133	-0.79652
0.78	0.71684	-1.03418	0.03299	1.68494	-1.32032	0.65138	0.96917	-1.01128
0.80	0.61939	-1.02342	0.19545	1.74814	-1.20675	0.79029	0.79450	-1.21002
0.82	0.52370	-0.99702	0.36346	1.80219	-1.07363	0.89940	0.59009	-1.38932
0.84	0.43126	-0.95450	0.53615	1.84732	-0.92601	0.97427	0.35923	-1.54651
0.86	0.34363	-0.89546	0.71270	1.88393	-0.76958	1.01105	0.10555	-1.67975
0.88	0.26237	-0.81956	0.89234	1.91254	-0.61058	1.00643	-0.16712	-1.78813
0.90	0.18910	-0.72655	1.07433	1.93383	-0.45574	0.95776	-0.45486	-1.87177
0.92	0.12545	-0.61624	1.25802	1.94862	-0.31214	0.86296	-0.75386	-1.93187
0.94	0.07306	-0.48850	1.44284	1.95792	-0.18715	0.72055	-1.06060	-1.97079
0.96	0.03358	-0.34324	1.62832	1.96285	-0.08834	0.52955	-1.37202	-1.99203
0.98	0.00867	-0.18041	1.81412	1.96473	-0.02338	0.28944	-1.68568	-2.00031
1.00	0.00000	0.00000	2.00000	1.96500	0.00000	0.00000	-2.00000	-2.00155
|Max|	1.58815	1.03418	2.00000	1.96500	1.50550	1.45420	2.00000	2.00155

Table 4.3b Beam mode shapes and their derivatives, continued

Clamped-Clamped Beam. Also Free-Free Beam with case 11a of Table 4.2

	Third Mode				*Fourth Mode*			
$\frac{x}{L}$	$\tilde{y}_3$	$\frac{L}{\lambda_3}\frac{d\tilde{y}_3}{dx}$	$\frac{L^2}{\lambda_3^2}\frac{d^2\tilde{y}_3}{dx^2}$	$\frac{L^3}{\lambda_3^3}\frac{d^3\tilde{y}_1}{dx^3}$	$\tilde{y}_4$	$\frac{L}{\lambda_4}\frac{d\tilde{y}_4}{dx}$	$\frac{L^2}{\lambda_4^2}\frac{d^2\tilde{y}_4}{dx^2}$	$\frac{L^3}{\lambda_4^3}\frac{d^3\tilde{y}_4}{dx^3}$
0.00	0.00000	0.00000	2.00000	-1.99993	0.00000	0.00000	2.00000	-2.00000
0.02	0.04482	0.39147	1.56038	-1.99658	0.07241	0.48557	1.43502	-1.99300
0.04	0.16510	0.68646	1.12323	-1.97469	0.25958	0.81207	0.87658	-1.94824
0.06	0.33975	0.88609	0.69428	-1.91998	0.51697	0.98325	0.33937	-1.83960
0.08	0.54803	0.99303	0.28189	-1.82280	0.80177	1.00789	-0.15633	-1.65333
0.10	0.77005	1.01202	-0.10393	-1.67794	1.07449	0.90088	-0.58802	-1.38736
0.12	0.98720	0.95005	-0.45253	-1.48447	1.30078	0.68345	-0.93412	-1.05012
0.14	1.18265	0.81648	-0.75348	-1.24534	1.45308	0.38242	-1.17673	-0.65879
0.16	1.34190	0.62284	-0.99738	-0.96697	1.51208	0.02893	-1.30380	-0.23724
0.18	1.45317	0.38256	-1.17658	-0.65867	1.46765	-0.34350	-1.31068	0.18649
0.20	1.50783	0.11049	-1.28573	-0.33199	1.31923	-0.70122	-1.20092	0.58286
0.22	1.50059	-0.17760	-1.32221	-0.00003	1.07549	-1.01271	-0.98634	0.92349
0.24	1.42971	-0.46574	-1.28637	0.32333	0.75348	-1.25091	-0.68630	1.18364
0.26	1.29690	-0.73832	-1.18164	0.62424	0.37700	-1.39515	-0.32640	1.34442
0.28	1.10719	-0.98086	-1.01443	0.88956	-0.02537	-1.43265	0.06348	1.39439
0.30	0.86863	-1.18058	-0.79386	1.10762	-0.42268	-1.35944	0.45136	1.33056
0.32	0.59186	-1.32695	-0.53144	1.26880	-0.78413	-1.18058	0.80569	1.15876
0.34	0.28949	-1.41222	-0.24051	1.36606	-1.08158	-0.90972	1.09776	0.89319
0.36	-0.02444	-1.43171	0.06439	1.39528	-1.29186	-0.56793	1.30395	0.55537
0.38	-0.33527	-1.38399	0.36811	1.35553	-1.39857	-0.18205	1.40755	0.17245
0.40	-0.62836	-1.27099	0.65569	1.24912	-1.39351	0.21752	1.40010	-0.22494
0.42	-0.88987	-1.09783	0.91301	1.08148	-1.27726	0.59923	1.28198	-0.60506
0.44	-1.10739	-0.87257	1.12747	0.86096	-1.05919	0.93288	1.06244	-0.93759
0.46	-1.27060	-0.60585	1.28859	0.59841	-0.75676	1.19208	0.75879	-1.19604
0.48	-1.37174	-0.31031	1.38852	0.30668	-0.39407	1.35629	0.39504	-1.35983
0.50	-1.40600	0.00000	1.42238	0.00000	0.00000	1.41251	0.00000	-1.41592
0.52	-1.37174	0.31031	1.38852	-0.30668	0.39407	1.35629	-0.39504	-1.35983
0.54	-1.27060	0.60585	1.28859	-0.59841	0.75676	1.19208	-0.75879	-1.19604
0.56	-1.10739	0.87257	1.12747	-0.86096	1.05919	0.93288	-1.06244	-0.93759
0.58	-0.88987	1.09783	0.91301	-1.08148	1.27726	0.59923	-1.28198	-0.60506
0.60	-0.62836	1.27099	0.65569	-1.24912	1.39351	0.21752	-1.40010	-0.22494
0.62	-0.33527	1.38399	0.36811	-1.35553	1.39857	-0.18205	-1.40755	0.17245
0.64	-0.02444	1.43171	0.06439	-1.39528	1.29186	-0.56793	-1.30395	0.55537
0.66	0.28949	1.41222	-0.24051	-1.36606	1.08158	-0.90972	-1.09776	0.89319
0.68	0.59186	1.32695	-0.53144	-1.26880	0.78413	-1.18058	-0.80569	1.15875
0.70	0.86863	1.18058	-0.79386	-1.10762	0.42268	-1.35944	-0.45136	1.33056
0.72	1.10719	0.98086	-1.01443	-0.88956	0.02537	-1.43265	-0.06348	1.39438
0.74	1.29690	0.73832	-1.18164	-0.62424	-0.37701	-1.39515	0.32639	1.34442
0.76	1.42971	0.46574	-1.28637	-0.32333	-0.75348	-1.25091	0.68630	1.18364
0.78	1.50059	0.17760	-1.32221	0.00003	-1.07550	-1.01271	0.98634	0.92348
0.80	1.50783	-0.11049	-1.28573	0.33199	-1.31923	-0.70122	1.20092	0.58286
0.82	1.45317	-0.38256	-1.17658	0.65867	1.46766	-0.34351	1.31067	0.18649
0.84	1.34190	-0.62284	-0.99738	0.96697	-1.51209	0.02893	1.30379	-0.23725
0.86	1.18266	-0.81648	-0.75348	1.24534	-1.45309	0.38241	1.17672	-0.65880
0.88	0.98720	-0.95005	-0.45253	1.48447	-1.30079	0.68343	0.93411	-1.05013
0.90	0.77005	-1.01202	-0.10393	1.67794	-1.07451	0.90087	0.58800	-1.38738
0.92	0.54803	-0.99303	0.28190	1.82280	-0.80179	1.00786	0.15631	-1.65335
0.94	0.33975	-0.88609	0.69428	1.91998	-0.51700	0.98322	-0.33940	-1.83963
0.96	0.16511	-0.68646	1.12323	1.97469	-0.25962	0.81203	-0.87662	-1.94828
0.98	0.04482	-0.39147	1.56038	1.99659	-0.07246	0.48552	-1.43507	-1.99305
1.00	0.00000	0.00000	2.00000	1.99994	-0.00007	-0.00007	-2.00007	-2.00007
|Max|	1.50783	1.43171	2.00000	1.99994	1.51208	1.41251	2.00007	2.00007

Table 4.3c Beam mode shapes and their derivatives, continued

Clamped-Free Beam

	First Mode				*Second Mode*			
$\frac{x}{L}$	$\tilde{y}_1$	$\frac{L}{\lambda_1}\frac{d\tilde{y}_1}{dx}$	$\frac{L^2}{\lambda_1^2}\frac{d^2\tilde{y}_1}{dx^2}$	$\frac{L^3}{\lambda_1^3}\frac{d^3\tilde{y}_1}{dx^3}$	$\tilde{y}_2$	$\frac{L}{\lambda_2}\frac{d\tilde{y}_2}{dx}$	$\frac{L^2}{\lambda_2^2}\frac{d^2\tilde{y}_2}{dx^2}$	$\frac{L^3}{\lambda_2^3}\frac{d^3\tilde{y}_2}{dx^3}$
0.00	0.00000	0.00000	2.00000	-1.46819	0.00000	0.00000	2.00000	-2.03693
0.02	0.00139	0.07397	1.94494	-1.46817	0.00853	0.17879	1.80878	-2.03667
0.04	0.00552	0.14588	1.88988	-1.46805	0.03301	0.33962	1.61764	-2.03483
0.06	0.01231	0.21572	1.83483	-1.46773	0.07174	0.48253	1.42680	-2.03002
0.08	0.02168	0.28350	1.77980	-1.46710	0.12305	0.60755	1.23661	-2.02097
0.10	0.03355	0.34921	1.72480	-1.46607	0.18526	0.71475	1.04750	-2.00658
0.12	0.04784	0.41287	1.66985	-1.46455	0.25670	0.80428	0.86004	-1.98590
0.14	0.06449	0.47446	1.61496	-1.46245	0.33573	0.87631	0.67485	-1.95814
0.16	0.08340	0.53400	1.56016	-1.45969	0.42070	0.93108	0.49261	-1.92267
0.18	0.10451	0.59148	1.50549	-1.45617	0.51002	0.96892	0.31409	-1.87901
0.20	0.12774	0.64692	1.45096	-1.45182	0.60211	0.99020	0.14007	-1.82682
0.22	0.15301	0.70031	1.39660	-1.44656	0.69544	0.99539	-0.02864	-1.76592
0.24	0.18024	0.75167	1.34247	-1.44032	0.78852	0.98501	-0.19123	-1.69625
0.26	0.20936	0.80100	1.28859	-1.43302	0.87992	0.95970	-0.34687	-1.61792
0.28	0.24030	0.84832	1.23500	-1.42459	0.96827	0.92013	-0.49475	-1.53113
0.30	0.27297	0.89364	1.18175	-1.41497	1.05227	0.86707	-0.63410	-1.43625
0.32	0.30730	0.93696	1.12889	-1.40410	1.13068	0.80136	-0.76419	-1.33373
0.34	0.34322	0.97832	1.07646	-1.39191	1.20236	0.72389	-0.88431	-1.22416
0.36	0.38065	1.01771	1.02451	-1.37834	1.26626	0.63565	-0.99384	-1.10821
0.38	0.41953	1.05516	0.97309	-1.36334	1.32141	0.53764	-1.09222	-0.98667
0.40	0.45977	1.09070	0.92227	-1.34685	1.36694	0.43093	-1.17895	-0.86040
0.42	0.50131	1.12435	0.87209	-1.32884	1.40209	0.31665	-1.25365	-0.73034
0.44	0.54408	1.15612	0.82262	-1.30924	1.42619	0.19593	-1.31600	-0.59749
0.46	0.58800	1.18606	0.77391	-1.28801	1.43871	0.06995	-1.36578	-0.46291
0.48	0.63301	1.21418	0.72604	-1.26512	1.43920	-0.06012	-1.40289	-0.32772
0.50	0.67905	1.24052	0.67905	-1.24052	1.42733	-0.19307	-1.42733	-0.19307
0.52	0.72604	1.26512	0.63301	-1.21418	1.40289	-0.32772	-1.43920	-0.06012
0.54	0.77391	1.28801	0.58800	-1.18606	1.36578	-0.46291	-1.43871	0.06995
0.56	0.82262	1.30924	0.54408	-1.15612	1.31600	-0.59749	-1.42619	0.19593
0.58	0.87209	1.32884	0.50131	-1.12435	1.25365	-0.73034	-1.40209	0.31665
0.60	0.92227	1.34685	0.45977	-1.09070	1.17895	-0.86040	-1.36694	0.43093
0.62	0.97309	1.36334	0.41953	-1.05516	1.09222	-0.98667	-1.32141	0.53764
0.64	1.02451	1.37834	0.38065	-1.01771	0.99384	-1.10821	-1.26626	0.63565
0.66	1.07646	1.39191	0.34322	-0.97832	0.88431	-1.22416	-1.20236	0.72389
0.68	1.12889	1.40410	0.30730	-0.93696	0.76419	-1.33373	-1.13068	0.80136
0.70	1.18175	1.41497	0.27297	-0.89364	0.63410	-1.43625	-1.05227	0.86707
0.72	1.23500	1.42459	0.24030	-0.84832	0.49475	-1.53113	-0.96827	0.92013
0.74	1.28859	1.43302	0.20936	-0.80100	0.34687	-1.61792	-0.87992	0.95970
0.76	1.34247	1.44032	0.18024	-0.75167	0.19123	-1.69625	-0.78852	0.98501
0.78	1.39660	1.44656	0.15301	-0.70031	0.02864	-1.76592	-0.69544	0.99539
0.80	1.45096	1.45182	0.12774	-0.64692	-0.14007	-1.82682	-0.60211	0.99020
0.82	1.50549	1.45617	0.10451	-0.59148	-0.31409	-1.87901	-0.51002	0.96892
0.84	1.56016	1.45969	0.08340	-0.53400	-0.49261	-1.92267	-0.42070	0.93108
0.86	1.61496	1.46245	0.06449	-0.47446	-0.67485	-1.95814	-0.33573	0.87631
0.88	1.66985	1.46455	0.04784	-0.41287	-0.86004	-1.98590	-0.25670	0.80428
0.90	1.72480	1.46607	0.03355	-0.34921	-1.04750	-2.00658	-0.18526	0.71475
0.92	1.77980	1.46710	0.02168	-0.28350	-1.23661	-2.02097	-0.12305	0.60755
0.94	1.83483	1.46773	0.01231	-0.21572	-1.42680	-2.03002	-0.07174	0.48253
0.96	1.88988	1.46805	0.00552	-0.14588	-1.61764	-2.03483	-0.03301	0.33962
0.98	1.94494	1.46817	0.00139	-0.07397	-1.80878	-2.03667	-0.00853	0.17879
1.00	2.00000	1.46819	0.00000	0.00000	-2.00000	-2.03693	0.00000	0.00000
|Max|	2.00000	1.46819	2.00000	1.46819	2.00000	2.03693	2.00000	2.03693

Table 4.3d Beam mode shapes and their derivatives, continued

Clamped-Free Beam

	Third Mode				*Fourth Mode*			
$\frac{x}{L}$	$\tilde{y}_3$	$\frac{L}{\lambda_3}\frac{d\tilde{y}_3}{dx}$	$\frac{L^2}{\lambda_3^2}\frac{d^2\tilde{y}_3}{dx^2}$	$\frac{L^3}{\lambda_3^3}\frac{d^3\tilde{y}_1}{dx^3}$	$\tilde{y}_4$	$\frac{L}{\lambda_4}\frac{d\tilde{y}_4}{dx}$	$\frac{L^2}{\lambda_4^2}\frac{d^2\tilde{y}_4}{dx^2}$	$\frac{L^3}{\lambda_4^3}\frac{d^3\tilde{y}_4}{dx^3}$
0.00	0.00000	0.00000	2.00000	-1.99845	0.00000	0.00000	2.00000	-2.00007
0.02	0.02339	0.28953	1.68610	-1.99721	0.04482	0.39147	1.56035	-1.99672
0.04	0.08839	0.52979	1.37287	-1.98892	0.16510	0.68645	1.12317	-1.97482
0.06	0.18727	0.72099	1.06188	-1.96766	0.33974	0.88606	0.69420	-1.92012
0.08	0.31237	0.86367	0.75558	-1.92871	0.54801	0.99298	0.28179	-1.82294
0.10	0.45614	0.95879	0.45702	-1.86854	0.77002	1.01194	-0.10407	-1.67809
0.12	0.61120	1.00785	0.16973	-1.78480	0.98714	0.94994	-0.45270	-1.48463
0.14	0.77049	1.01291	-0.10245	-1.67629	1.18256	0.81633	-0.75368	-1.24552
0.16	0.92728	0.97665	-0.35563	-1.54286	1.34177	0.62264	-0.99762	-0.96717
0.18	1.07535	0.90237	-0.58594	-1.38540	1.45299	0.38230	-1.17687	-0.65891
0.20	1.20901	0.79394	-0.78975	-1.20575	1.50758	0.11017	-1.28608	-0.33228
0.22	1.32324	0.65580	-0.96375	-1.00656	1.50027	-0.17801	-1.32262	-0.00038
0.24	1.41376	0.49285	-1.10515	-0.79124	1.42928	-0.46625	-1.28688	0.32290
0.26	1.47707	0.31040	-1.21172	-0.56380	1.29634	-0.73895	-1.18226	0.62370
0.28	1.51055	0.11405	-1.28189	-0.32872	1.10648	-0.98164	-1.01518	0.88888
0.30	1.51248	-0.09041	-1.31485	-0.09085	0.86774	-1.18153	-0.79478	1.10676
0.32	1.48203	-0.29711	-1.31055	0.14479	0.59073	-1.32813	-0.53258	1.26772
0.34	1.41932	-0.50026	-1.26974	0.37310	0.28808	-1.41368	-0.24191	1.36469
0.36	1.32534	-0.69422	-1.19397	0.58908	-0.02621	-1.43351	0.06264	1.39357
0.38	1.20196	-0.87368	-1.08556	0.78797	-0.33748	-1.38622	0.36594	1.35339
0.40	1.05185	-1.03374	-0.94753	0.96533	-0.63112	-1.27376	0.65299	1.24643
0.42	0.87841	-1.17003	-0.78359	1.11722	-0.89330	-1.10126	0.90964	1.07812
0.44	0.68568	-1.27881	-0.59802	1.24030	-1.11166	-0.87683	1.12327	0.85675
0.46	0.47822	-1.35704	-0.39555	1.33189	-1.27592	-0.61115	1.28336	0.59315
0.48	0.26103	-1.40247	-0.18131	1.39005	-1.37836	-0.31690	1.38199	0.30011
0.50	0.03938	-1.41367	0.03938	1.41367	-1.41424	-0.00819	1.41424	-0.00819
0.52	-0.18131	-1.39005	0.26103	1.40247	-1.38199	0.30011	1.37836	-0.31690
0.54	-0.39555	-1.33189	0.47822	1.35704	-1.28336	0.59315	1.27592	-0.61115
0.56	-0.59802	-1.24030	0.68568	1.27881	-1.12327	0.85675	1.11166	-0.87683
0.58	-0.78359	-1.11722	0.87841	1.17003	-0.90964	1.07812	0.89330	-1.10126
0.60	-0.94753	-0.96533	1.05185	1.03374	-0.65299	1.24643	0.63112	-1.27376
0.62	-1.08556	-0.78797	1.20196	0.87368	-0.36594	1.35339	0.33748	-1.38622
0.64	-1.19397	-0.58908	1.32534	0.69422	-0.06264	1.39357	0.02621	-1.43351
0.66	-1.26974	-0.37310	1.41931	0.50026	0.24191	1.36469	-0.28808	-1.41368
0.68	-1.31055	-0.14479	1.48203	0.29711	0.53258	1.26772	-0.59073	-1.32813
0.70	-1.31485	0.09085	1.51248	0.09041	0.79478	1.10676	-0.86774	-1.18153
0.72	-1.28189	0.32872	1.51055	-0.11405	1.01518	0.88888	-1.10648	-0.98164
0.74	-1.21172	0.56380	1.47707	-0.31040	1.18226	0.62370	-1.29634	-0.73895
0.76	-1.10515	0.79124	1.41376	-0.49285	1.28688	0.32290	-1.42928	-0.46624
0.78	-0.96375	1.00656	1.32324	-0.65580	1.32262	-0.00038	-1.50027	-0.17801
0.80	-0.78975	1.20575	1.20901	-0.79394	1.28608	-0.33228	-1.50758	0.11017
0.82	-0.58594	1.38540	1.07535	-0.90237	1.17687	-0.65890	-1.45299	0.38230
0.84	0.35563	1.54286	0.92728	0.97665	0.99763	-0.96717	-1.34177	0.62264
0.86	-0.10245	1.67629	0.77049	-1.01291	0.75369	-1.24552	-1.18256	0.81633
0.88	0.16973	1.78480	0.61120	-1.00785	0.45270	-1.48463	-0.98714	0.94994
0.90	0.45701	1.86854	0.45614	-0.95879	0.10407	-1.67809	-0.77001	1.01194
0.92	0.75558	1.92871	0.31237	-0.86367	-0.28178	-1.82293	-0.54801	0.99298
0.94	1.06188	1.96766	0.18727	-0.72099	-0.69419	-1.92012	-0.33973	0.88606
0.96	1.37287	1.98892	0.08839	-0.52979	-1.12317	-1.97482	-0.16510	0.68645
0.98	1.68610	1.99721	0.02339	-0.28953	-1.56034	-1.99671	-0.04481	0.39147
1.00	2.00000	1.99845	0.00000	0.00000	-1.99999	-2.00006	0.00001	0.00001
|Max|	2.00000	1.99845	2.00000	1.99845	1.99999	2.00006	2.00000	2.00007

Table 4.3e Beam mode shapes and their derivatives, continued

Clamped-Pinned Beam. Also Free-Pinned Beam with case 11c of Table 4.2

	First Mode				*Second Mode*			
$\frac{x}{L}$	$\tilde{y}_1$	$\frac{L}{\lambda_1}\frac{d\tilde{y}_1}{dx}$	$\frac{L^2}{\lambda_1^2}\frac{d^2\tilde{y}_1}{dx^2}$	$\frac{L^3}{\lambda_1^3}\frac{d^3\tilde{y}_1}{dx^3}$	$\tilde{y}_2$	$\frac{L}{\lambda_2}\frac{d\tilde{y}_2}{dx}$	$\frac{L^2}{\lambda_2^2}\frac{d^2\tilde{y}_2}{dx^2}$	$\frac{L^3}{\lambda_2^3}\frac{d^3\tilde{y}_2}{dx^3}$
0.00	0.00000	0.00000	2.00000	-1.99845	0.00000	0.00000	2.00000	0.00000
0.02	0.02339	0.28953	1.68610	-1.99721	0.04482	0.39147	1.56035	0.02339
0.04	0.08839	0.52979	1.37287	-1.98892	0.16510	0.68645	1.12317	0.08839
0.06	0.18727	0.72099	1.06188	-1.96766	0.33974	0.88606	0.69420	0.18727
0.08	0.31237	0.86367	0.75558	-1.92871	0.54801	0.99298	0.28179	0.31237
0.10	0.45614	0.95879	0.45702	-1.86854	0.77002	1.01194	-0.10407	0.45614
0.12	0.61120	1.00785	0.16973	-1.78480	0.98714	0.94994	-0.45270	0.61120
0.14	0.77049	1.01291	-0.10245	-1.67629	1.18256	0.81633	-0.75368	0.77049
0.16	0.92728	0.97665	-0.35563	-1.54286	1.34177	0.62264	-0.99762	0.92728
0.18	1.07535	0.90237	-0.58594	-1.38540	1.45299	0.38230	-1.17687	1.07535
0.20	1.20901	0.79394	-0.78975	-1.20575	1.50758	0.11017	-1.28608	1.20901
0.22	1.32324	0.65580	-0.96375	-1.00656	1.50027	-0.17801	-1.32262	1.32324
0.24	1.41376	0.49285	-1.10515	-0.79124	1.42928	-0.46625	-1.28688	1.41376
0.26	1.47707	0.31040	-1.21172	-0.56380	1.29634	-0.73895	-1.18226	1.47707
0.28	1.51055	0.11405	-1.28189	-0.32872	1.10648	-0.98164	-1.01518	1.51055
0.30	1.51248	-0.09041	-1.31485	-0.09085	0.86774	-1.18153	-0.79478	1.51248
0.32	1.48203	-0.29711	-1.31055	0.14479	0.59073	-1.32813	-0.53258	1.48203
0.34	1.41932	-0.50026	-1.26974	0.37310	0.28808	-1.41368	-0.24191	1.41932
0.36	1.32534	-0.69422	-1.19397	0.58908	-0.02621	-1.43351	0.06264	1.32534
0.38	1.20196	-0.87368	-1.08556	0.78797	-0.33748	-1.38622	0.36594	1.20196
0.40	1.05185	-1.03374	-0.94753	0.96533	-0.63112	-1.27376	0.65299	1.05185
0.42	0.87841	-1.17003	-0.78359	1.11722	-0.89330	-1.10126	0.90964	0.87841
0.44	0.68568	-1.27881	-0.59802	1.24030	-1.11166	-0.87683	1.12327	0.68568
0.46	0.47822	-1.35704	-0.39555	1.33189	-1.27592	-0.61115	1.28336	0.47822
0.48	0.26103	-1.40247	-0.18131	1.39005	-1.37836	-0.31690	1.38199	0.26103
0.50	0.03938	-1.41367	0.03938	1.41367	-1.41424	-0.00819	1.41424	0.03938
0.52	-0.18131	-1.39005	0.26103	1.40247	-1.38199	0.30011	1.37836	-0.18131
0.54	-0.39555	-1.33189	0.47822	1.35704	-1.28336	0.59315	1.27592	-0.39555
0.56	-0.59802	-1.24030	0.68568	1.27881	-1.12327	0.85675	1.11166	-0.59802
0.58	-0.78359	-1.11722	0.87841	1.17003	-0.90964	1.07812	0.89330	-0.78359
0.60	-0.94753	-0.96533	1.05185	1.03374	-0.65299	1.24643	0.63112	-0.94753
0.62	-1.08556	-0.78797	1.20196	0.87368	-0.36594	1.35339	0.33748	-1.08556
0.64	-1.19397	-0.58908	1.32534	0.69422	-0.06264	1.39357	0.02621	-1.19397
0.66	-1.26974	-0.37310	1.41931	0.50026	0.24191	1.36469	-0.28808	-1.26974
0.68	-1.31055	-0.14479	1.48203	0.29711	0.53258	1.26772	-0.59073	-1.31055
0.70	-1.31485	0.09085	1.51248	0.09041	0.79478	1.10676	-0.86774	-1.31485
0.72	-1.28189	0.32872	1.51055	-0.11405	1.01518	0.88888	-1.10648	-1.28189
0.74	-1.21172	0.56380	1.47707	-0.31040	1.18226	0.62370	-1.29634	-1.21172
0.76	-1.10515	0.79124	1.41376	-0.49285	1.28688	0.32290	-1.42928	-1.10515
0.78	-0.96375	1.00656	1.32324	-0.65580	1.32262	-0.00038	-1.50027	-0.96375
0.80	-0.78975	1.20575	1.20901	-0.79394	1.28608	-0.33228	-1.50758	-0.78975
0.82	-0.58594	1.38540	1.07535	-0.90237	1.17687	-0.65890	-1.45299	-0.58594
0.84	-0.35563	1.54286	0.92728	-0.97665	0.99763	-0.96717	-1.34177	-0.35563
0.86	-0.10245	1.67629	0.77049	-1.01291	0.75369	-1.24552	-1.18256	-0.10245
0.88	0.16973	1.78480	0.61120	-1.00785	0.45270	-1.48463	-0.98714	0.16973
0.90	0.45701	1.86854	0.45614	-0.95879	0.10407	-1.67809	-0.77001	0.45701
0.92	0.75558	1.92871	0.31237	-0.86367	-0.28178	-1.82293	-0.54801	0.75558
0.94	1.06188	1.96766	0.18727	-0.72099	-0.69419	-1.92012	-0.33973	1.06188
0.96	1.37287	1.98892	0.08839	-0.52979	-1.12317	-1.97482	-0.16510	1.37287
0.98	1.68610	1.99721	0.02339	-0.28953	-1.56034	-1.99671	-0.04481	1.68610
1.00	2.00000	1.99845	0.00000	0.00000	-1.99999	-2.00006	0.00001	2.00000
|Max|	2.00000	1.99845	2.00000	1.41367	1.50758	1.39357	2.00000	1.39357

Table 4.3f Beam mode shapes and their derivatives, continued

Clamped-Pinned Beam. Also Free-Pinned Beam with case 11c of Table 4.2

	Third Mode				*Fourth Mode*			
$\frac{x}{L}$	$\tilde{y}_3$	$\frac{L}{\lambda_3}\frac{d\tilde{y}_3}{dx}$	$\frac{L^2}{\lambda_3^2}\frac{d^2\tilde{y}_3}{dx^2}$	$\frac{L^3}{\lambda_3^3}\frac{d^3\tilde{y}_1}{dx^3}$	$\tilde{y}_4$	$\frac{L}{\lambda_4}\frac{d\tilde{y}_4}{dx}$	$\frac{L^2}{\lambda_4^2}\frac{d^2\tilde{y}_4}{dx^2}$	$\frac{L^3}{\lambda_4^3}\frac{d^3\tilde{y}_4}{dx^3}$
0.00	0.00000	0.00000	2.00000	-2.00000	0.00000	0.00000	2.00000	-2.00000
0.02	0.03886	0.36671	1.59173	-1.99731	0.06496	0.46278	1.46633	-1.99408
0.04	0.14410	0.65019	1.18531	-1.97961	0.23451	0.78357	0.93791	-1.95600
0.06	0.29879	0.85122	0.78508	-1.93509	0.47105	0.96521	0.42662	-1.86287
0.08	0.48626	0.97168	0.39742	-1.85535	0.73820	1.01441	-0.05091	-1.70171
0.10	0.69037	1.01491	0.03009	-1.73537	1.00204	0.94270	-0.47581	-1.46893
0.12	0.89584	0.98593	-0.30845	-1.57331	1.23237	0.76665	-0.82947	-1.16955
0.14	1.08857	0.89148	-0.60968	-1.37037	1.40407	0.50751	-1.09559	-0.81599
0.16	1.25604	0.74002	-0.86560	-1.13046	1.49825	0.19041	-1.26206	-0.42659
0.18	1.38759	0.54152	-1.06927	-0.85985	1.50306	-0.15704	-1.32223	-0.02380
0.20	1.47476	0.30725	-1.21523	-0.56678	1.41422	-0.50624	-1.27577	0.36779
0.22	1.51147	0.04939	-1.29988	-0.26098	1.23501	-0.82944	-1.12901	0.72343
0.24	1.49419	-0.21934	-1.32168	0.04683	0.97582	-1.10140	-0.89465	1.02023
0.26	1.42202	-0.48616	-1.28137	0.34551	0.65324	-1.30107	-0.59110	1.23893
0.28	1.29662	-0.73864	-1.18195	0.62397	0.28879	-1.41295	-0.24121	1.36537
0.30	1.12212	-0.96520	-1.02863	0.87171	-0.09274	-1.42807	0.12917	1.39164
0.32	0.90489	-1.15556	-0.82867	1.07934	-0.46510	-1.34455	0.49299	1.31666
0.34	0.65324	-1.30107	-0.59110	1.23893	-0.80250	-1.16772	0.82386	1.14636
0.36	0.37703	-1.39512	-0.32637	1.34445	-1.08149	-0.90963	1.09785	0.89328
0.38	0.08727	-1.43330	-0.04597	1.39199	-1.28266	-0.58823	1.29518	0.57571
0.40	-0.20439	-1.41364	0.23807	1.37996	-1.39201	-0.22602	1.40159	0.21644
0.42	-0.48616	-1.33665	0.51362	1.30919	-1.40200	0.15152	1.40934	-0.15886
0.44	-0.74658	-1.20525	0.76897	1.18287	-1.31209	0.51780	1.31771	-0.52342
0.46	-0.97504	-1.02471	0.99329	1.00646	-1.12877	0.84697	1.13308	-0.85127
0.48	-1.16223	-0.80235	1.17711	0.78747	-0.86513	1.11580	0.86843	-1.11909
0.50	-1.30050	-0.54726	1.31263	0.53513	-0.53994	1.30530	0.54246	-1.30782
0.52	-1.38422	-0.26994	1.39411	0.26005	-0.17628	1.40210	0.17821	-1.40403
0.54	-1.41001	0.01818	1.41807	-0.02625	0.20000	1.39937	-0.19853	-1.40084
0.56	-1.37687	0.30521	1.38344	-0.31179	0.56222	1.29734	-0.56109	-1.29847
0.58	-1.28624	0.57929	1.29160	-0.58465	0.88466	1.10326	-0.88379	-1.10413
0.60	-1.14194	0.82907	1.14631	-0.83344	1.14445	0.83092	-1.14379	-0.83159
0.62	-0.95000	1.04421	0.95356	-1.04778	1.32317	0.49963	-1.32266	-0.50014
0.64	-0.71844	1.21582	0.72135	-1.21873	1.40813	0.13289	-1.40774	-0.13328
0.66	-0.45690	1.33678	0.45927	-1.33915	1.39330	-0.24329	-1.39301	0.24300
0.68	-0.17628	1.40210	0.17821	-1.40403	1.27973	-0.60226	-1.27950	0.60203
0.70	0.11175	1.40907	-0.11017	-1.41064	1.07546	-0.91855	-1.07529	0.91837
0.72	0.39519	1.35742	-0.39391	-1.35870	0.79497	-1.16974	-0.79484	1.16960
0.74	0.66228	1.24931	-0.66123	-1.25036	0.45814	-1.33802	-0.45804	1.33792
0.76	0.90188	1.08924	-0.90103	-1.09010	0.08884	-1.41146	-0.08876	1.41138
0.78	1.10404	0.88388	-1.10335	-0.88457	-0.28675	-1.38486	0.28681	1.38480
0.80	1.26036	0.64176	-1.25979	-0.64232	-0.64202	-1.26010	0.64206	1.26005
0.82	1.36432	0.37294	-1.36386	-0.37340	-0.95177	-1.04601	0.95180	1.04598
0.84	1.41161	0.08861	-1.41123	0.08899	1.19405	-0.75779	1.19407	0.75776
0.86	1.40026	-0.19942	-1.39995	0.19911	-1.35169	-0.41585	1.35171	0.41583
0.88	1.33073	-0.47917	-1.33048	0.47892	-1.41351	-0.04443	1.41352	0.04441
0.90	1.20592	-0.73903	-1.20571	0.73882	-1.37513	0.33014	1.37514	-0.33015
0.92	1.03100	-0.96818	-1.03083	0.96801	-1.23928	0.68130	1.23929	-0.68131
0.94	0.81325	-1.15711	-0.81311	1.15697	-1.01558	0.98417	1.01559	-0.98417
0.96	0.56171	-1.29796	-0.56160	1.29785	-0.71989	1.21727	0.71990	-1.21728
0.98	0.28683	-1.38488	-0.28674	1.38479	-0.37317	1.36409	0.37317	-1.36409
1.00	0.00004	-1.41425	0.00004	1.41418	0.00000	1.41421	0.00000	-1.41422
|Max|	1.51147	1.43330	2.00000	2.00000	1.50306	1.41421	2.00000	2.00000

4.2.2 *Orthogonality, Normalization, and Maximum Values*

Beam mode shapes and their derivatives are functions of the dimensionless spanwise coordinate $\lambda_i x/L$:

$$\begin{aligned}\frac{d\tilde{y}_i}{dx} &= \frac{(\lambda_i/L)d\tilde{y}_i}{d(\lambda_i x/L)}\\ \frac{d^{n+4}\tilde{y}_i}{d(\lambda_i x/L)^{n+4}} &= \frac{d^n\tilde{y}_i}{d(\lambda_i x/L)^n} = \frac{(\lambda_i/L)^4 d^n\tilde{y}_i}{dx^n}\end{aligned} \tag{4.16}$$

Each mode is orthogonal over the span to the other modes. Modes can be normalized to an arbitrary scalar value. The mode shapes of beams in Tables 4.2 and 4.3 are normalized as follows:

$$\begin{aligned}&\int_0^L \tilde{y}_i(x)\tilde{y}_j(x)dx = 0, \qquad i \neq j\\ &\int_0^L \tilde{y}_i^2(x)dx = \begin{cases} L/2, & \text{for } P-P, S-P, S-S\\ L, & \text{other beams in Table 4.2}\end{cases}\end{aligned} \tag{4.17}$$

P is pinned and S is sliding (also called guided) boundary condition. Free–free, sliding–sliding, free–sliding, and pinned–free boundary conditions allow one or more rigid body translation and rotation modes with zero natural frequency. Integrals of beam mode shapes are discussed in Appendix D and Felgar [9].

Maximum values of beam mode shapes and the second derivative of the beam mode shapes, used in stress Equation 4.2, are

$$\begin{aligned}\tilde{y}_i(x)|_{\max} &= \begin{cases} 2, & F-F, F-S, C-F, F-P\\ 1.51, & CP, CC\ CS \text{ for } i>1;\ 1.588, CC\ CS \text{ for } i=1\\ 1, & P-P, S-P, S-S\end{cases}\\ \left.\frac{d^2\tilde{y}_i(x)}{dx^2}\right|_{\max} &= \frac{\lambda_i^2}{L^2}\times\begin{cases} 2, & C-F, C-P, C-C, C-S\\ 1.51, & FP, FF\ FS \text{ for } i>1;\ 1.588, FF\ FS \text{ for } i=1\\ 1, & P-P, S-P, S-S\end{cases}\end{aligned} \tag{4.18}$$

Errors as small as 10^{-6} in the dimensionless parameters λ or σ (Table 4.2) can produce significant computational errors in beam mode shapes above the fourth mode [10]. By expanding a clamped–free beam mode shape in a series of exponentials and sines about $x = 0$, Dowell [11] showed that as λ_i becomes large, the higher mode shapes are

$$\tilde{y}_i\left(\frac{\lambda_i x}{L}\right) \cong \frac{\sin \lambda_i x}{L} - \frac{\cos \lambda_i x}{L} + e^{-\lambda_i x/L} + (-1)^{i+1} e^{-\lambda_i(1-x/L)} \tag{4.19}$$

This suggests that beam mode shapes can be approximated with sines and cosines except those adjacent to a clamped end.

4.2.3 *Beams Stress*

During vibration in a single mode (Eq. 4.8), the beam displacement is equal to the mode shape times a function of time $T(t)$. Beam bending stress (Eq. 4.2) is proportional to the second derivative of the mode shape, quantity in parenthis below, which is a function of the dimensionless parameter $\lambda_i x/L$:

$$\sigma_{xx} = Ec\frac{\partial^2 Y(x,t)}{\partial x^2} = Ec\frac{\lambda^2}{L^2}\left(\frac{d^2\tilde{y}(\lambda x/L)}{d(\lambda x/L)^2}\right)T(t) \tag{4.20}$$

Maximum stress occurs at the edge of the beam cross section (maximum c) where the beam curvature is greatest. The maximum value of beam mode curvature over the span is in Table 4.3 and Eq. 4.18.

4.2.4 *Two-Span Beams*

Table 4.4 [12] has formulas for the natural frequencies and made shapes of two span beams. They have a single intermediate pinned support. They reduce to a single-span beam (Table 4.2) when the intermediate support approaches one end of the beam as shown in Figure 4.4.

4.2.5 *Multispan Beams*

Table 4.5 [13, 14] has formulas for the natural frequencies and mode shapes of multi-span beams. These consist of N equal length spans with length L and intermediate pinned supports. (Clamped spans are independent.) $N = 1$ is a single-span beam and Tables 4.2 and 4.5 are identical for this case. Miles [15] has shown that the natural frequencies of multi-span beams with pinned intermediate supports and clamped or pinned ends fall only in frequency passbands: $\pi < \lambda < (3/2)\pi$, $2\pi < \lambda < (5/2)\pi$, and $n\pi < \lambda < (n + 1/2)\pi$, $n = 1, 2, \ldots$. Inspection of Table 4.5c–e verifies that there are no values of λ between $(3/2)\pi$ and 2π and between $(5/2)\pi$ and 3π. Frequency passbands also occur with periodically supported plates [16, 17].

Figure 4.5 [14] gives the fundamental natural frequency of multispan beams with clamped extreme ends, variable length outermost spans, and pinned intermediate supports.

Example 4.1 Natural frequency of a tube

The $D_o = 1$ in. (25 mm) outside diameter of the steel tube shown in Figure 4.6 spans 102 in. (259 mm) between clamped supports. The tube wall thickness is 0.089 in. (2.26 mm), and the tube carried a water–steam mixture with a density of 10 pounds per cubic feet (160.1 kg/m^3). The tube temperature is 500 F (260 °C). What is the tube's fundamental natural frequency? Use Equation 4.11 and case 7 of Table 4.2.

Solution

The modulus of elasticity of steel at 500 F (260 °C) is estimated to be $E = 26{,}400{,}000$ psi (182.0E9 Pa) (Chapter 8). The moment of inertia and area are given in case 27 of Table 1.5:

$$I = \left(\frac{\pi}{64}\right)(D_o^4 - D_i^4) = \frac{\pi}{64}[(1\,\text{in.})^4 - (1\,\text{in.} - 2(0.089\,\text{in.}))^4] = 0.02668\,\text{in.}^4(11,100\,\text{mm}^4)$$

$$A = \frac{\pi}{4}(D_o^2 - D_i^2) = \frac{\pi}{4}[(1\text{ in.})^2 - (0.822\text{ in.})^2] = 0.2547/N^2(164.3\text{ mm}^2)$$

Table 4.4 Natural frequencies of two-span beams

Notation: a = length of first span; E = modulus of elasticity; I = area moment of inertia about neutral axis (Table 1.5); L = beam length; m = mass per unit length. Table 1.2 has consistent sets of units. Ref. [12].

1. Pinned-Pinned-Pinned

Also see Figure 4.4

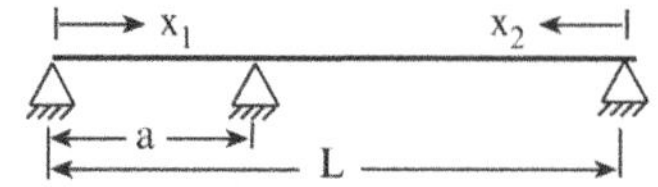

Natural Frequency

$$f_i = \frac{\lambda_i^2}{2\pi L^2}\sqrt{\frac{EI}{m}} \quad \text{Hz} \quad i = 1, 2, 3$$

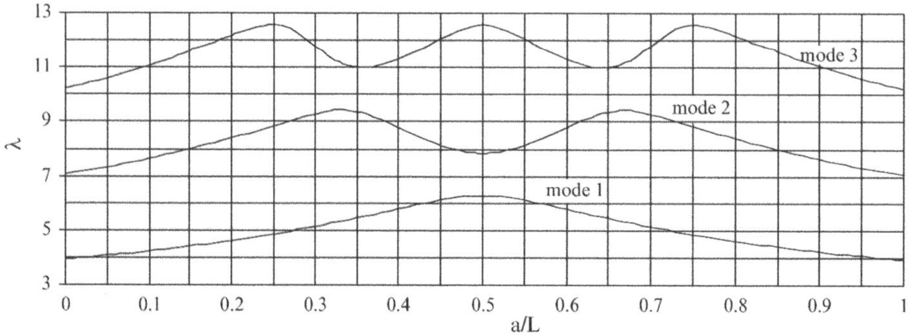

Mode shape $\tilde{y}(x) = \begin{cases} \sin\lambda x_1/L + \sigma_1 \sinh\lambda x_1/L & x_1 \le a \\ \sigma_2(\sin\lambda x_2/L + \sigma_3 \sinh\lambda x_2/L) & x_2 \le L-a \end{cases}$

where $\sigma_1 = \frac{-\sin\lambda a/L}{\sinh\lambda a/L}, \sigma_2 = \frac{\sin\lambda a/L}{\sin\lambda(1-a/L)}, \sigma_3 = \frac{-\sin\lambda(L-a/L)}{\sinh\lambda(L-a/L)}$

Transcendental eqn λ: $\cosh\lambda \ \sin\lambda - \cosh\lambda(1-2a/L)\ \sin\lambda - 2\ \sin\lambda\ a/L\ \sin\lambda(1-a/L)\ \sinh\lambda = 0$

2. Pinned-Pinned-Free

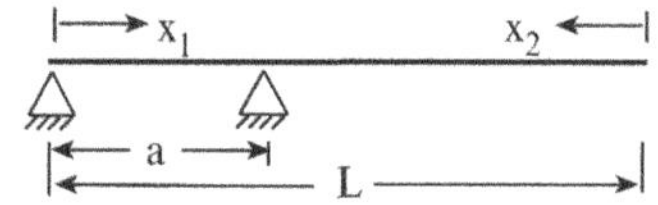

Natural Frequency

$$f_i = \frac{\lambda_i^2}{2\pi L^2}\sqrt{\frac{EI}{m}} \quad \text{Hz} \quad i = 1,2,3$$

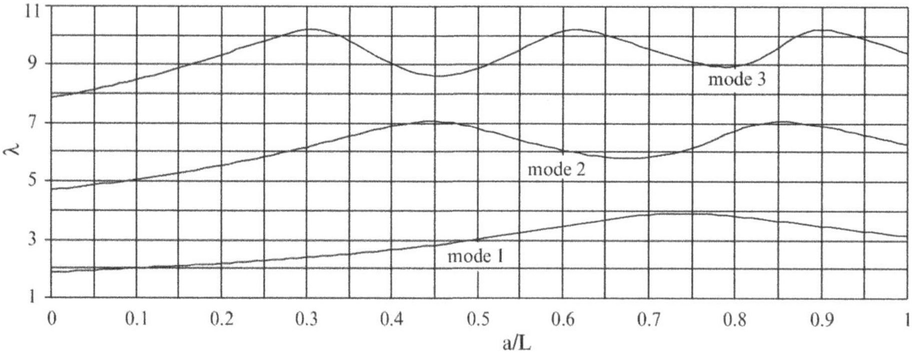

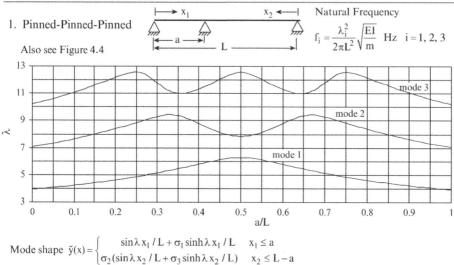

Mode shape $\tilde{y}(x) = \begin{cases} \sin\lambda x_1/L + \sigma_1 \sinh\lambda x_1/L & x_1 \le a \\ \sigma_2[\sin\lambda x_2/L + \sin\lambda x_2/L + \sigma_3(\cos\lambda x_2/L + \cosh\lambda x_2/L)] & x_2 \le L-a \end{cases}$

where $\sigma_1 = \frac{-\sin\lambda a/L}{\sinh\lambda a/L}, \quad \sigma_3 = -\frac{\sin\lambda(L-a)/L + \sinh\lambda(L-a)/L}{\cos\lambda(L-a)/L + \cosh\lambda(L-a)/L}$

$$\sigma_2 = \frac{(\sin\lambda a/L\cosh\lambda a/L - \cos\lambda a/L\sinh(1-a/L)[\cosh\lambda(1-a/L) + \cos\lambda(1-a/L)]}{2\sinh\lambda a/L[1+\cos\lambda(1-a/L)\cosh\lambda(L-a/L)]}$$

Transcendental equation for λ: $-\cos\lambda(1-2a/L)\ \cosh\lambda + \cos\lambda\ \cosh\lambda(1-2a/L) + 2\ \sin\lambda\ \sinh\lambda - \sin\lambda(1-2\,a/L)\ \sinh\lambda + 4\ \sin\lambda a/L\ \sinh\lambda a/L - \sin\lambda\sinh\lambda(1-2\,a/L) = 0$

Table 4.4 Natural frequencies of two-span beams, continued

3. Clamped-Pinned-Pinned

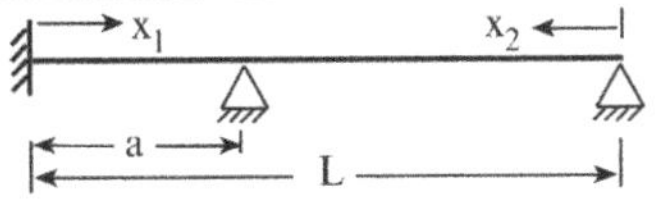

Natural Frequency

$$f_i = \frac{\lambda_i^2}{2\pi L^2}\sqrt{\frac{EI}{m}}, \text{ Hz} \quad i = 1, 2, 3$$

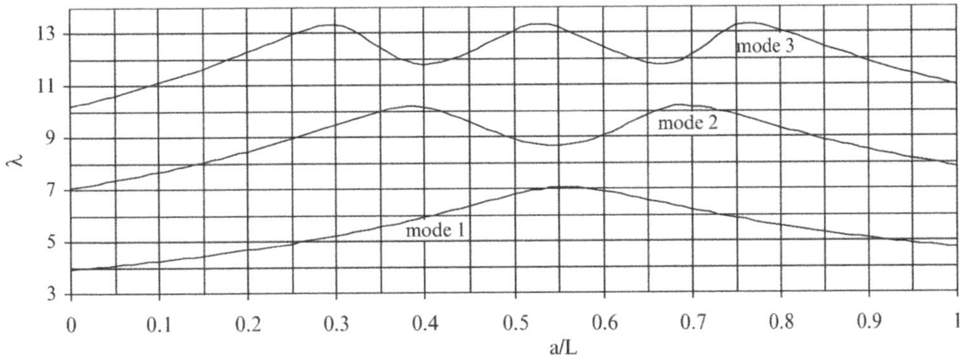

Mode shape $\tilde{y}(x) = \begin{cases} \sin\lambda x_1/L - \sin\lambda x_1/L + \sigma_1(\cos\lambda x_1/L - \cosh\lambda x_1/L) & x_1 \le a \\ \sigma_2\ (\sin\lambda x_2/L + \sigma_1 \sinh\lambda x_2/L) & x_2 \le L - a \end{cases}$

where $\sigma_1 = \dfrac{-\sinh\lambda a/L - \sin\lambda a/L}{\cos\lambda a/L - \cosh\lambda a/L}, \quad \sigma_3 = -\dfrac{\sin\lambda(1-a/L)}{\sinh\lambda(1-a/L)}$

$$\sigma_2 = \frac{-2(1-\cos\lambda a/L\cosh\lambda a/L)}{(\cos\lambda a/L - \cosh\lambda a/L)[\cos\lambda(1-a/L)\sinh\lambda(1-a/L) - \sin\lambda(1-a/L)\cosh\lambda(1-a/L)]}$$

Transcendental eqn. for λ: $-\cos\lambda(1-2a/L)\cosh\lambda + \cos\lambda\cos\lambda(1-2a/L) + 2\sin\lambda\ \sinh\lambda$
$+\sin\lambda(1-2a/L)\ \sinh\lambda + \sin\lambda\sinh\lambda(1-2a/L) - 4\sin\lambda(1-a/L)\sinh\lambda(1-a/L) = 0$

4. Clamped-Pinned Clamp

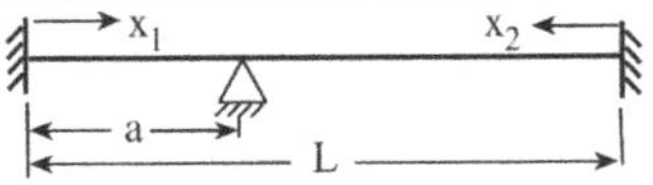

Natural Frequency

$$f_i = \frac{\lambda_i^2}{2\pi L^2}\sqrt{\frac{EI}{m}}, \text{ Hz} \quad i = 1,2,3$$

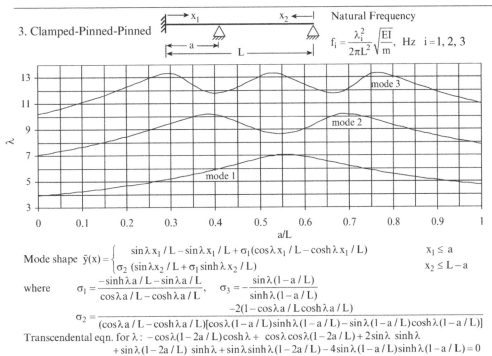

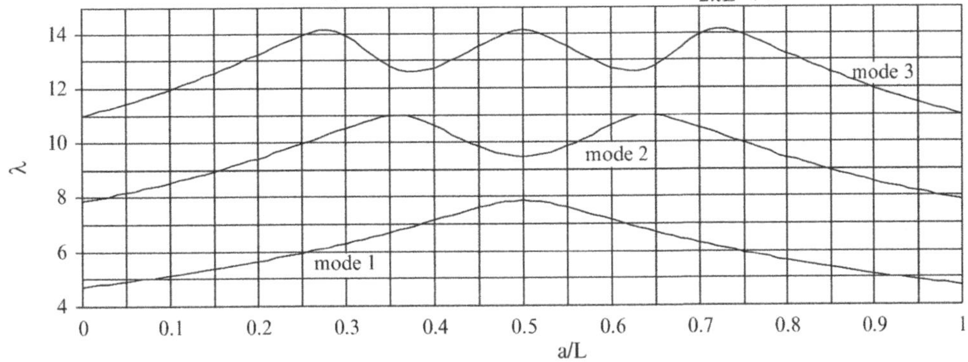

Mode shape $y(x) = \begin{cases} \sin\lambda x_1/L - \sinh\lambda x_1/L + \sigma_1(\cos\lambda x_1/L - \cosh\lambda x_1/L) & x_1 \le a \\ \sigma_2\ [\sin\lambda x_2/L - \sinh\lambda x_2/L + \sigma_3(\cos\lambda x_2/L - \cosh\lambda x_2/L)] & x_2 \le L - a \end{cases}$

where $\sigma_1 = \dfrac{\sinh\lambda a/L - \sin\lambda a/L}{\cos\lambda a/L - \cosh\lambda a/L}, \quad \sigma_3 = \dfrac{\sinh\lambda(1-a/L) - \sin(1-\lambda a/L)}{\cos\lambda(1-a/L) - \cosh\lambda(1-a/L)}$

$$\sigma_2 = \frac{(1-\cos\lambda a/L\cosh\lambda a/L)[\cosh\lambda(1-a/L) - \cos\lambda(1-a/L)]}{(\cos\lambda a/L - \cosh\lambda a/L)[1-\cos\lambda(1-a/L)\cosh\lambda(1-a/L)]}$$

Eqn. for λ: $\cosh\lambda\sin\lambda + \cosh\lambda(1-2a/L)\sin\lambda - 2\cosh\lambda a/L\sin\lambda a/L - 2\ \cosh\lambda(1-a/L)\ \sin\lambda(1-a/L) -$
$\cos\lambda a/L\ \sinh\lambda a/L - \cos\lambda(1-2a/L)\sinh\lambda + 2\ \cos\lambda a/L\ \sinh\lambda a/L + 2\sin\lambda(1-2\,a/L)\ \sinh\lambda(1-2\,a/L) = 0$

Table 4.4 Natural frequencies of two-span beams, continued

5. Clamped-Pinned-Free

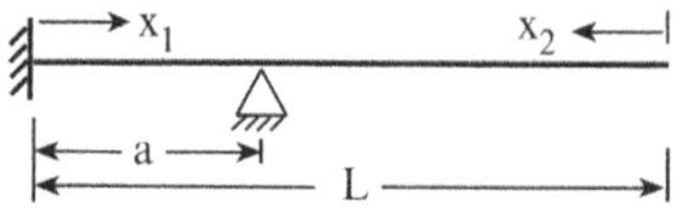

Natural Frequency

$$f_i = \frac{\lambda_i^2}{2\pi L^2}\sqrt{\frac{EI}{m}}, \quad \text{Hz} \quad i = 1, 2, 3$$

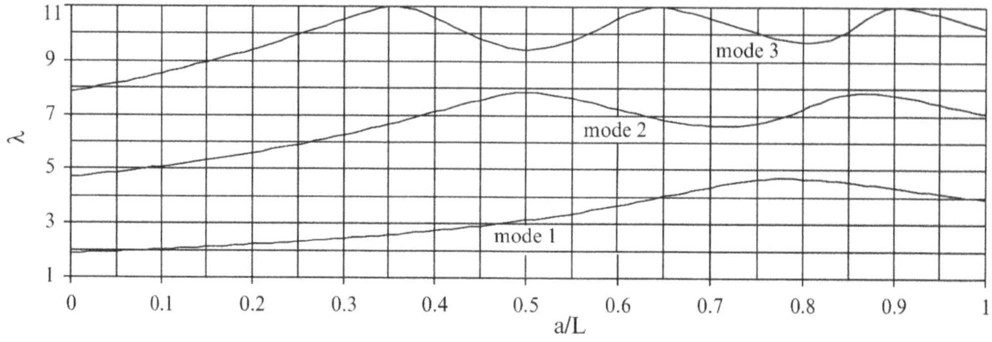

Mode shape
$$\tilde{y}(x) = \begin{cases} \sin\lambda x_1/L - \sinh\lambda x_1/L + \sigma_1(\cos\lambda x_1/L - \cosh\lambda x_1/L) & x_1 \le a \\ \sigma_2\,[\sin\lambda x_2/L + \sinh\lambda x_2/L + \sigma_3(\cos\lambda x_2/L + \cosh\lambda x_2/L)] & x_2 \le L - a \end{cases}$$

where
$$\sigma_1 = \frac{\sinh\lambda a/L - \sin\lambda a/L}{\cos\lambda a/L - \cosh\lambda a/L}, \quad \sigma_3 = \frac{\sinh\lambda(1-a/L) - \sin(1-\lambda a/L)}{\cos\lambda(1-a/L) + \cosh\lambda(1-a/L)}$$

$$\sigma_2 = \frac{(1-\cos\lambda a/L\cosh\lambda a/L)[\cos\lambda(1-a/L) + \cosh\lambda(1-a/L)]}{(\cosh\lambda a/L - \cos\lambda a/L)[1+\cos\lambda(1-a/L)\cosh\lambda(1-a/L)]}$$

Tran eqn:
$$-\cosh\lambda\sin\lambda - \cosh\lambda(1-2a/L)\sin\lambda - 2\cosh\lambda a/L\sin\lambda a/L + 2\cosh\lambda(1-a/L)\sin\lambda(1-a/L) + \cos\lambda\sinh\lambda + \cos\lambda(1-2a/L)\sinh\lambda + 2\cos\lambda a/L\sinh\lambda a/L - 2\cos\lambda(1-a/L)\sinh\lambda(1-a/L) = 0$$

6. Free-Pinned-Free

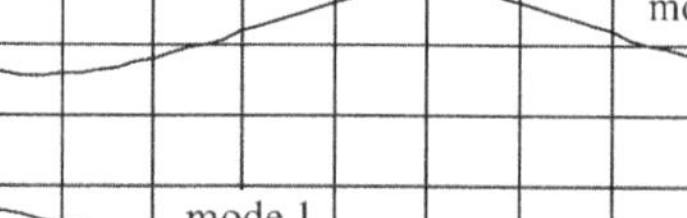

Natural Frequency

$$f_i = \frac{\lambda_i^2}{2\pi L^2}\sqrt{\frac{EI}{m}}, \quad \text{Hz} \quad i = 1,2,3$$

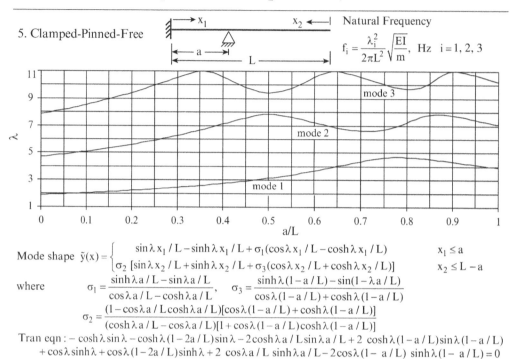

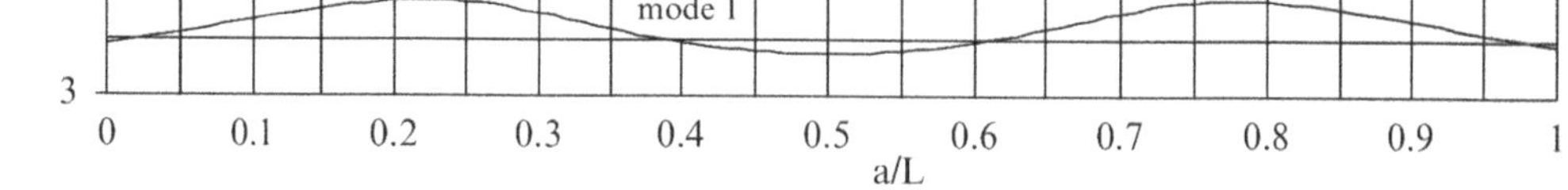

Mode shape
$$\tilde{y}(x) = \begin{cases} \sin\lambda x_1/L + \sinh\lambda x_1/L + \sigma_1(\cos\lambda x_1/L + \cosh\lambda x_1/L) & x_1 \le a \\ \sigma_2\,[\sin\lambda x_2/L + \sinh\lambda x_2/L + \sigma_3(\cos\lambda x_2/L + \cosh\lambda x_2/L)] & x_2 \le L - a \end{cases}$$

where
$$\sigma_1 = \frac{-\sin\lambda a/L - \sinh\lambda a/L}{\cos\lambda a/L + \cosh\lambda a/L}, \quad \sigma_3 = \frac{-\sinh\lambda(1-a/L) - \sin(1-\lambda a/L)}{\cos\lambda(1-a/L) + \cosh\lambda(1-a/L)}$$

$$\sigma_2 = \frac{(1+\cos\lambda a/L\cosh\lambda a/L)[\cos\lambda(1-a/L) + \cosh\lambda(1-a/L)]}{(\cos\lambda a/L + \cosh\lambda a/L)[1+\cos\lambda(1-a/L)\cosh\lambda(1-a/L)]}$$

Tran eq λ:
$$\cosh\lambda\sin\lambda + \cosh\lambda(1-2a/L)\sin\lambda + 2\cosh\lambda a/L\sin\lambda a/L + 2\cosh\lambda(1-a/L)\sin\lambda(1-a/L) - \cos\lambda\sinh\lambda - \cos\lambda(1-2a/L)\sinh\lambda - 2\cos\lambda a/L\sinh\lambda a/L - 2\cos\lambda(1-a/L)\sinh\lambda(1-a/L) = 0$$

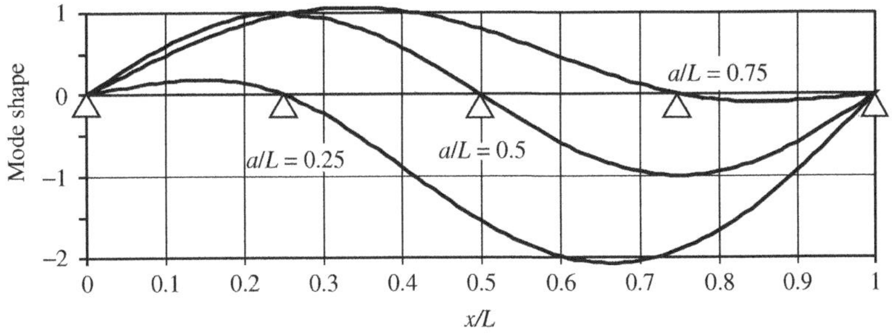

Figure 4.4 Fundamental mode shape of a pinned–pinned beam as a function of position of intermediate support (case 1 of Table 4.4)

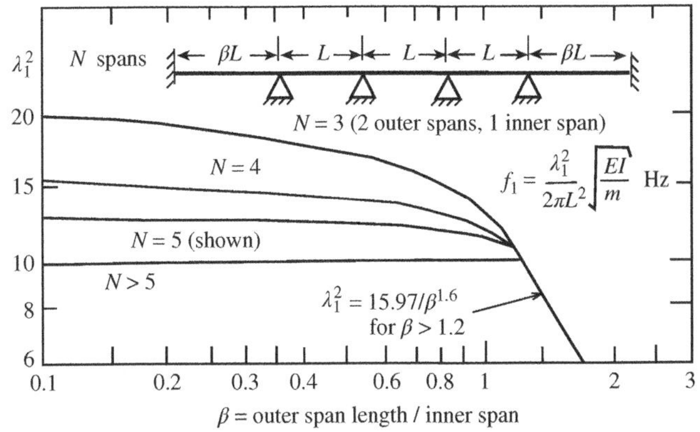

Figure 4.5 Fundamental natural frequency of an N-span beam with ends clamped, variable length outer spans, and intermediate pinned supports spaced at length L [14]

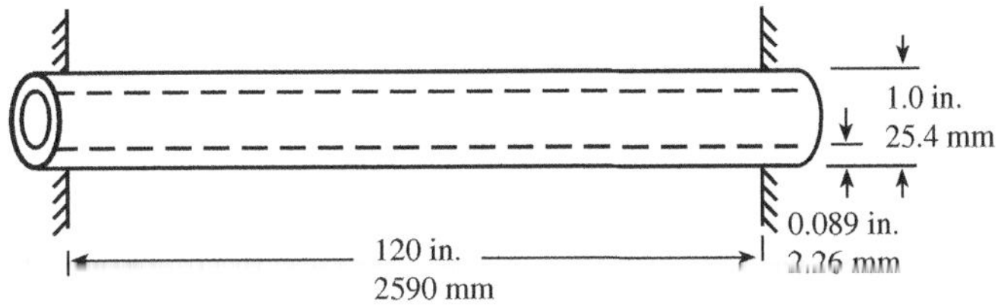

Figure 4.6 Example of a clamped–clamped tube

Table 4.5 Natural frequencies of multi-span beams with pinned intermediate supports

Notation: Span = beam segment of length L; E = modulus of elasticity; fi = natural frequency in ith mode in units of Hertz; I = area moment of inertia of beam cross section about neutral axis, Table 1.5; m = mass per unit length of beam, Table 1.5. Consistent sets of units are in Table 1.2. Refs [13, 14].

a. Free-Free Ends

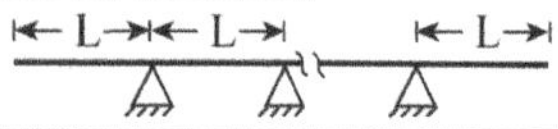

$$f_i = \frac{\lambda_i^2}{2\pi L^2}\sqrt{\frac{EI}{m}}$$

Num. of Spans	λ=λ_i (Number of Spans)				
	Mode Number, i				
	1	2	3	4	5
1	4.730	7.853	11.00	14.14	17.28
2	1.875	3.927	4.694	7.069	7.855
3	1.412	1.648	3.580	4.273	4.707
4	1.506	1.571	3.413	3.928	4.438
5	1.530	1.548	3.324	3.710	4.144
6	1.537	1.542	3.270	3.568	3.927
7	1.538	1.540	3.237	3.471	3.770
8	1.539	1.539	3.215	3.404	3.653
9	1.539	1.539	3.200	3.353	3.564
10	1.539	1.539	3.189	3.316	3.496
12	1.539	1.539	3.174	3.265	3.400
15	1.539	1.539	3.162	3.222	3.313

b. Free-Pinned Ends

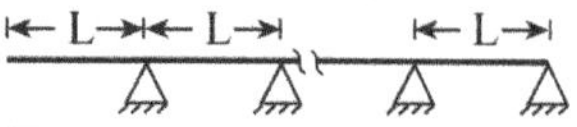

$$f_i = \frac{\lambda_i^2}{2\pi L^2}\sqrt{\frac{EI}{m}}$$

Num. of Spans	λ=λ_i (Number of Spans)				
	Mode Number, i				
	1	2	3	4	5
1	3.927	7.069	10.21	13.35	16.49
2	1.505	3.412	4.431	6.541	7.574
3	1.536	3.270	3.927	4.580	6.410
4	1.539	3.215	3.653	4.200	4.640
5	1.539	3.188	3.496	3.926	4.358
6	1.539	3.173	3.400	3.742	4.112
7	1.539	3.166	3.337	3.611	3.927
8	1.539	3.160	3.294	3.519	3.786
9	1.539	3.156	3.263	3.449	3.679
10	1.539	3.153	3.241	3.397	3.595
12	1.539	3.150	3.212	3.325	3.475
15	1.539	3.147	3.187	3.263	3.367

c. Clamped-Free Ends

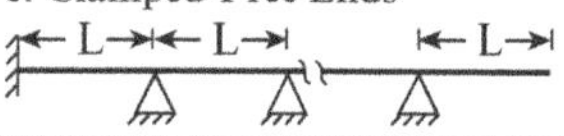

$$f_i = \frac{\lambda_i^2}{2\pi L^2}\sqrt{\frac{EI}{m}}$$

Num. of Spans	λ=λ_i (Number of Spans)				
	Mode Number, i				
	1	2	3	4	5
1	1.875	4.694	7.855	11.00	14.14
2	1.570	3.923	4.707	7.058	7.842
3	1.541	3.570	4.283	4.720	6.707
4	1.539	3.403	3.928	4.450	4.723
5	1.539	3.316	3.706	4.148	4.538
6	1.539	3.265	3.563	3.927	4.292
7	1.539	3.233	3.466	3.767	4.086
8	1.539	3.213	3.399	3.649	3.926
9	1.539	3.198	3.349	3.560	3.802
10	1.539	3.187	3.312	3.492	3.703
12	1.539	3.173	3.263	3.397	3.559
15	1.539	3.162	3.221	3.311	3.427

d. Pinned-Pinned Ends

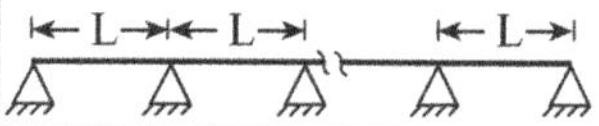

$$f_i = \frac{\lambda_i^2}{2\pi L^2}\sqrt{\frac{EI}{m}}$$

Num. of Spans	λ=λ_i (Number of Spans)				
	Mode Number, i				
	1	2	3	4	5
1	3.142	6.283	9.425	12.57	15.71
2	3.142	3.927	6.283	7.069	9.424
3	3.142	3.557	4.297	4.713	6.707
4	3.142	3.393	3.928	4.463	6.283
5	3.142	3.310	3.700	4.152	4.550
6	3.142	3.260	3.557	3.927	4.293
7	3.142	3.230	3.460	3.764	4.089
8	3.142	3.210	3.394	3.645	3.926
9	3.142	3.196	3.344	3.557	3.800
10	3.142	3.186	3.309	3.488	3.700
12	3.142	3.173	3.261	3.393	3.557
15	3.142	3.161	3.219	3.309	3.424

Table 4.5 Natural frequencies of multi-span beams with pinned intermediate supports, continued

e. Clamped-Pinned Ends $f_i = \frac{\lambda_i^2}{2\pi L^2}\sqrt{\frac{EI}{m}}$

Num. of Spans	$\lambda=\lambda_i$ (Number of Spans) Mode Number, i				
	1	2	3	4	5
1	3.927	7.069	10.21	13.35	16.49
2	3.393	4.463	6.545	7.591	9.687
3	3.261	3.927	4.600	6.410	7.070
4	3.210	3.645	4.207	4.655	6.357
5	3.186	3.488	3.926	4.366	4.682
6	3.173	3.393	3.738	4.115	4.463
7	3.164	3.331	3.607	3.927	4.247
8	3.159	3.290	3.514	3.784	4.069
9	3.156	3.260	3.444	3.675	3.927
10	3.153	3.239	3.393	3.592	3.813
12	3.149	3.210	3.322	3.472	3.645
15	3.147	3.186	3.261	3.364	3.489

f. Clamped-Clamped Ends $f_i = \frac{\lambda_i^2}{2\pi L^2}\sqrt{\frac{EI}{m}}$

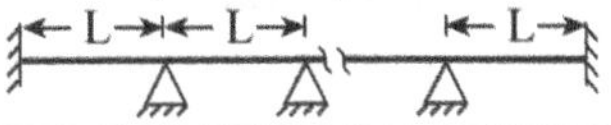

Num. of Spans	$\lambda=\lambda_i$ (Number of Spans) Mode Number, i				
	1	2	3	4	5
1	4.730	7.853	11.00	14.14	17.28
2	3.927	4.730	7.068	7.853	10.21
3	3.557	4.297	4.730	6.707	7.430
4	3.393	3.928	4.463	4.730	6.545
5	3.310	3.700	4.152	4.550	4.730
6	3.260	3.557	3.927	4.298	4.602
7	3.230	3.460	3.764	4.089	4.394
8	3.210	3.394	3.645	3.926	4.208
9	3.196	3.344	3.557	3.800	4.053
10	3.186	3.309	3.488	3.700	3.927
12	3.173	3.261	3.393	3.557	3.738
15	3.161	3.219	3.309	3.424	3.557

The mass per unit length (m) of the tube is the sum of the mass of the tube metal (ρ_{steel} = 0.289 lb/in.^{3}, 8 gm/cc) and the mass of the internal steam:

$$m_{steel} = \rho_{steel}\left(\frac{\pi}{4}\right)(D_o^2 - D_i^2)$$

$$= 0.289\,\text{lb/in.}^3(\pi/4)[1\,\text{in.}^2 - (1\,\text{in.} - 2\times 0.089\,\text{in.})^2] = 0.0736\,\text{lb/in.}(1.314\,\text{gm/mm})$$

$$m_{steam} = \rho_{steam}\left(\frac{\pi}{4}\right)D_i^2$$

$$= \left(\frac{10\ \text{lb}}{1728\ \text{in.}^3}\right)(1\ \text{in.} - 2\times 0.089\ \text{in.})^2 = 0.00307\ \text{lb/in.}\ (0.0548\ \text{gm/mm})$$

D_i = 0.0822 in. (20.88 mm) is the inside diameter of the tube. The total mass per unit length, m = 0.0767 lb/in. (1.37 gm/mm; Table 1.5), is the sum of the steel and steam contributions. I, E, m, and L are converted into the consistent units of case 6 for US customary units. Case 1 of Table 1.3 is used for SI units. These are substituted into the formula in case 7 of Table 1.3a with $\lambda = 4.73$:

$$f_1 = \frac{\lambda_1^2}{2\pi L^2}\sqrt{\frac{EI}{m}} = \frac{\dfrac{4.73^2}{2\pi(102\ \text{in.})^2}\sqrt{\dfrac{26400000\ \text{lb/in.}^2\times 0.02688\ \text{in.}^4}{0.0767\ \text{lb/in./}386.1\ \text{in./s}^2}}}{\dfrac{4.73^2}{2\pi(2.59\ \text{m})^2}\sqrt{\dfrac{182\times 10^9\ \text{Pa}(1.11\times 10^{-8}\ \text{m}^4)}{1.37\ \text{kg/m}}}} = 20.46\ \text{Hz}$$

The result has units of 1/s, which is Hertz.

4.3 Axially Loaded Beam Natural Frequency

4.3.1 Uniform Axial Load

Shaker's [18] formula for natural frequencies of the pinned–pinned beam bearing tensile axial load P (Figure 4.7a; Table 4.6) is

$$f_i = \frac{(i\pi)^2}{2\pi L^2}\left(1 + \frac{PL^2}{(i\pi)^2 EI}\right)^{1/2}\left(\frac{EI}{m}\right)^{1/2} \qquad i = 1, 2, 3 \ldots, \text{ Hz} \tag{4.21}$$

P is the mean axial load; it is positive for tensile loads. E is the modulus of elasticity, L is the span, and I is the area moment of inertia (Table 1.5). This equation is identical to the Rayleigh approximation (Eq. A.14). It also applies to guided–guided beam when P is replaced with $4P$.

Tensile axial load ($P > 0$) increases the natural frequencies of lateral vibration. Compressive axial load decreases the lateral vibration natural frequencies and leads to buckling. Buckling loads for beams are given in Table 4.6 [6, 18, 19, 20]. The pinned beam buckles in ith mode shape when $P = -(i\pi)^2 EI/L^2$.

The mode shape of vibration and the buckled shape of a pinned beam are identical and the same as the vibration mode shape of a tensioned string (Table 3.6). These three shapes are similar for other beam boundary conditions. As a result, compressive axial load can be modeled as a component of beam bending stiffness that is proportional to $-P$. This leads to an approximate expression for natural frequency of the ith mode for beams with axial load in terms of the compressive load that buckles the beam in that mode, P_{bi}:

$$\frac{f_i|_{P\neq 0}}{f_i|_{P=0}} = \left(1 + \frac{P}{|P_{bi}|}\right)^{1/2} \tag{4.22}$$

This is exact for beams with sinusoidal mode shapes and within a few percent of the exact solutions of Shaker [18] and Bokaian [19] for most other beams supported at both ends. P_b is the buckling load shown in Table 4.6. An exact solution form for compressive loads on single-span beams with spring-supported ends is given by Maurizi and Belles [5]. Zuo [20] discusses follower loads.

As a tensile axial load increases beyond the absolute value of buckling load, the tension stiffness dominates the bending stiffness, and the beam natural frequencies and mode

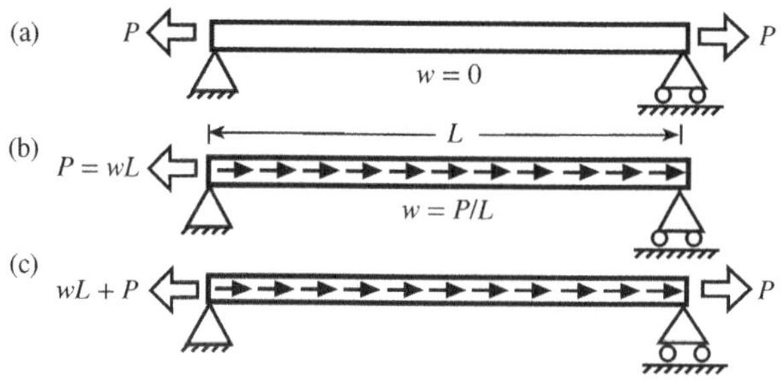

Figure 4.7 Axially loaded beams with pinned ends. w = traction force/length, (a) uniform axial load, (b) traction reacted at one end, and (c) traction reacted at both ends

Table 4.6 Buckling load and mode for single span beams

Notation: E = modulus of elasticity; I = area moment of inertia, Table 1.5; I = mode number; L = span of beam; P = axial load, positive tensile; x = axial distance along beam from left hand end. Refs [6, 18, 19, 20]. See Fig. 4.7a. Results for higher modes are approximate.

Boundary Conditions	Buckling Load P_b (a)	Static Buckled Shape (a)	Higher Mode Buckling Load P_i, i=2,3,4...	Buckled Shape
1. Free-Free	$\pi^2 EI/L^2$	$\sin \pi x/L$	$i^2\pi^2 EI/L^2$	
2. Free-Sliding	$\pi^2 EI/(4L^2)$	$\sin \pi x/(2L)$	$(2i-1)^2\pi^2 EI/(4L^2)$	
3. Clamped-Free	$\pi^2 EI/(4L^2)$	$1-\cos \pi x/(2L)$	$(2i-1)^2\pi^2 EI/(4L^2)$	
4. Free-Pinned	$\pi^2 EI/L^2$	$\sin \pi x/L$	$i^2\pi^2 EI/L^2$	
5. Pinned-Pinned	$\pi^2 EI/L^2$	$\sin \pi x/L$	$i^2\pi^2 EI/L^2$	
6. Clamped-Pinned	$2.05\pi^2 EI/L^2$	--	$(2i+1)^2\pi^2 EI/(4L^2)$	
7. ClampedClamped	$4\pi^2 EI/L^2$	$1-\cos 2\pi x/L$	$(i+1)^2\pi^2 EI/L^2$	
8. Clamped-Sliding	$\pi^2 EI/L^2$	$1-\cos \pi x/L$	$i^2\pi^2 EI/L^2$	
9. Sliding-Pinned	$\pi^2 EI/(4L^2)$	$\cos \pi x/(2L)$	$(2i-1)^2\pi^2 EI/(4L^2)$	
10. Sliding-Sliding	$\pi^2 EI/L^2$	$\cos \pi x/L$	$i^2\pi^2 EI/L^2$	
11. Follower Force on cantilever beam	Follower force does not buckle cantilever beam. Ref. [32] (a) Buckling load and shape for first buckling mode.			

shapes approach those of a tensioned string. This can be seen in Equation 4.22 by letting $P \gg EI/L^2$ and comparing with case 1 of Table 3.6. Conversely, if the absolute value of the axial load in the beam is much less than the buckling load (Table 4.6), then the axial load does not significantly change the natural frequencies and mode shapes.

4.3.2 Linearly Varying Axial Load

Table 4.7 [21] and Figures 4.8 and 4.9 presented the natural frequencies of beams with an axial traction (force per unit length) applied to each spanwise segment of the beam. Traction produces a static axial load that varies linearly along the span. Traction can be a gravity load on a vertical pipe or a centrifugal load on a spinning propeller.

The natural frequencies of the beam with uniform traction loads scale with the dimensionless traction parameter wL^3/EI (equivalent to $P_t L^2/EI$), the end reaction P_1, and the boundary conditions [21]. Note $P > 0$ is a tensile load:

$$\lambda_i = \lambda_i\left(\frac{wL^3}{EI}, \frac{P_1}{wL}, \text{boundary conditions}\right) \tag{4.23}$$

Table 4.7 Values of λ_i for beams with axial traction

Figs. 4.7b, 4.8, Ref. [21]. Tensile Loading. Values of λ. $f_i = (\lambda_i^2/2\pi L^2)(EI/m)^{1/2}$, Hz

b.c. Case	mode i	wL^3/EI = 0	200	400	600	800	1000
Case 11	1	3.1416	5.4652	6.3076	6.8804	7.3261	7.6957
Case 11	2	6.2832	8.4652	9.5456	10.3113	10.9171	11.4243
Case 11	3	9.4248	11.303	12.4565	13.3145	14.0089	14.5980
Case 12	1	0	4.1293	4.9077	5.4300	5.8342	6.1684
Case 12	2	3.9266	6.6210	7.6994	8.4435	9.0269	9.5131
Case 12	3	7.0686	9.2779	10.4467	11.2868	11.9574	12.522
Case 13	1	3.9266	6.0973	6.9535	7.5387	7.9947	8.3732
Case 13	2	7.0686	9.1389	10.2306	11.0098	11.6285	12.147
Case 13	3	10.2102	11.9999	13.1454	14.0087	14.7115	15.3097
Case 14	1	1.5708	4.2112	4.9728	5.4867	5.8856	6.2160
Case 14	2	4.7124	7.0946	8.1293	8.8488	9.4147	9.8873
Case 14	3	7.8540	9.9110	11.0588	11.8913	12.5578	13.119
Case 21	1	3.9266	5.6429	6.4353	6.9875	7.4212	7.7828
Case 21	2	7.0686	8.7817	9.7720	10.4969	11.0790	11.571
Case 21	3	10.210	11.744	12.7812	13.5811	14.2404	14.8056
Case 22	1	1.8751	4.2155	4.9762	5.4905	5.8898	6.2205
Case 22	2	4.6941	6.8154	7.8375	8.5586	9.1288	9.6062
Case 22	3	7.8548	9.6222	10.6904	11.4853	12.1295	12.6768
Case 23	1	4.7300	6.3202	7.1121	7.6704	8.1109	8.4790
Case 23	2	7.8532	9.4945	10.4851	11.2186	11.8105	12.3115
Case 23	3	10.996	12.468	13.4947	14.2966	14.9617	15.3141
Case 24	1	2.3650	4.3116	5.0489	5.5526	5.9453	6.2714
Case 24	2	5.4978	7.3379	8.3004	8.9896	9.5382	9.9992
Case 24	3	8.6394	10.293	11.3331	12.1155	12.7524	13.2945

Compressive Loading, w≤0, Values of λ.							Buckling
wL^3/EI		0	-5	-10	-15	-18	wL^3/EI
Case 11	1	3.1416	2.9168	2.6120	2.1089	1.3368	-18.569
Case 11	2	6.2832	6.1810	6.0728	5.9581	5.8857	-86.431
Case 11	3	9.4248	9.3577	9.2889	9.2185	9.1753	-196.29
wL^3/EI		0	-5	-10	-20	-30	
Case 13	1	3.9266	3.7660	3.5762	3.0347	0.5371	-30.009
Case 13	2	7.0686	6.9801	6.8877	6.6899	6.4717	-112.09
Case 13	3	10.2102	10.1489	10.0805	9.9575	9.8229	-234.99
wL^3/EI		0	-1	-2	-3	-3.4	
Case 14	1	1.5708	1.4442	1.2700	0.9580	0.6068	-3.4766
Case 14	2	4.7124	4.6844	4.6558	4.6266	4.6148	-44.138
Case 14	3	7.8540	7.8378	7.8214	7.8050	7.7983	-129.26
wL^3/EI		0	-20	-30	-40	-50	
Case 21	1	3.9266	3,5017	3.2039	2.7753	1.8629	-52.501
Case 21	2	7.0686	6.7895	6.6339	6.4648	6.2794	-129.05
Case 21	3	10.210	10.004	9.8951	9.7821	9.6645	-276.24
wL^3/EI		0	-2	-4	-6	-7.5	
Case 22	1	1.8751	1.7424	1.5694	1.3059	0.8550	-7.8373
Case 22	2	4.6941	4.6517	4.6081	4.5631	4.5285	-55.977
Case 22	3	7.8548	7.8289	7.8027	7.7762	7.7562	-148.51
wL^3/EI		0	-20	-40	-60	-74.5	
Case 23	1	4.7300	4.4011	3.9537	3.2127	0.9902	-74.629
Case 23	2	7.8532	7.6018	7.3177	6.9900	6.7157	-157.03
Case 23	3	10.996	10.8040	10.600	10.3820	10.2139	-325.51
wL^3/EI		0	-5	-10	-15	-18.5	
Case 24	1	2.3650	2.1919	1.9630	1.6013	0.9335	-18.956
Case 24	2	5.4978	5.4160	5.3300	5.2390	5.1721	-81.887
Case 24	3	8.6394	8.5812	8.5216	8.4605	8.4169	-189.22

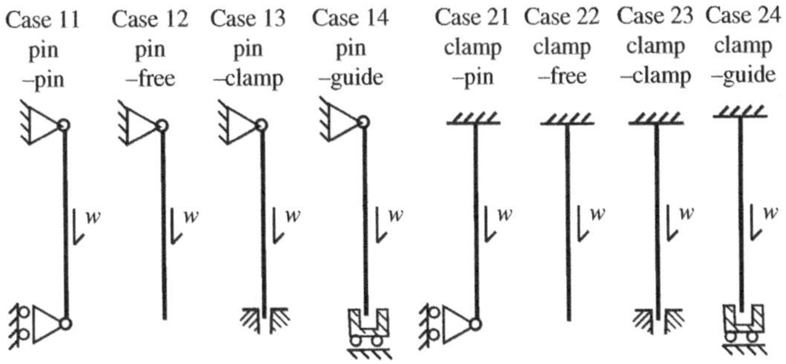

Figure 4.8 End conditions for beams with traction load ([6]; Table 4.7)

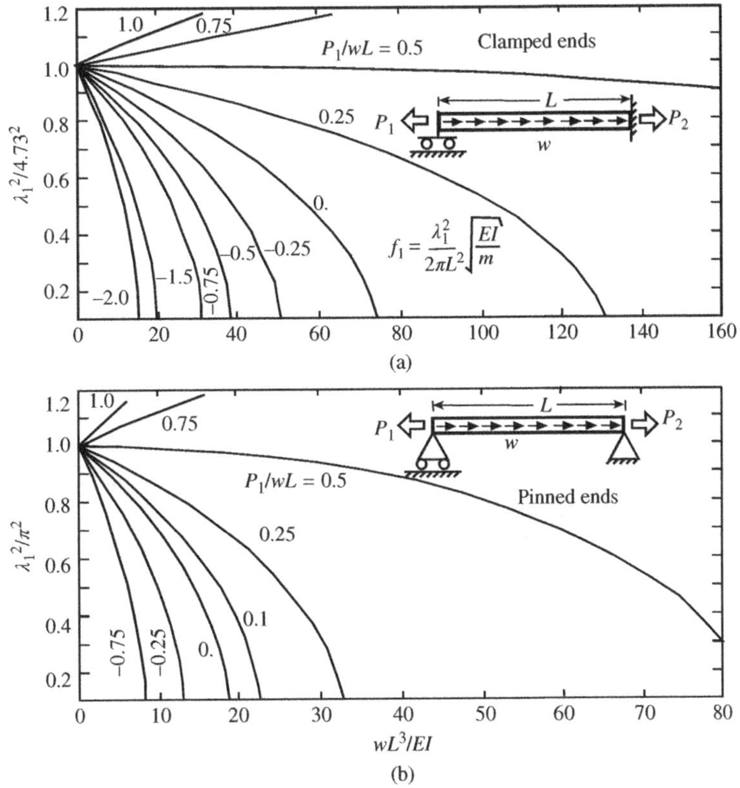

Figure 4.9 Effect of axial loads on natural frequency parameter (Eqs. 4.11, 4.21) of clamped–clamped beam (a) and pinned–pinned beam (b) with axial tractions reacted at both ends [21]

The natural frequencies increase with positive (tensile) traction loads. Tractions that put the beam in compression eventually buckle the beam as load increases and the fundamental natural frequency approaches zero. Table 4.7 provides the fundamental natural frequency parameters of beams with pinned and clamped ends that react the total traction load $P = wL$ on one end (Fig. 4.7b). Higher modes are given in Ref. [22]. Figure 4.9 presents the natural frequency of beams that have loads on both ends. Here, the total traction load $wL = P_1 - P_2$ is the sum of the reactions P_1 and P_2 at the ends of the beam (Fig. 4.7c).

As the traction parameter wL^3/EI becomes large, on the order of 1000, the tension-induced stiffness in the beam dominates the bending stiffness, and the beams with one end free can be modeled by hanging chains (Table 3.6). Consider a long slender beam that hangs vertically under gravity from its top (case 12 of Fig. 4.8). Traction is the weight per unit length, $w = mg$, where m is the mass per unit length and g is the acceleration of gravity. Equation 4.11 is rearranged to incorporate this traction with the substitution $m = w/g$:

$$f_i = \frac{\lambda_i^2}{2\pi L^2}\sqrt{\frac{EI}{m}} = \frac{\lambda_i^2}{2\pi}\left(\frac{EI}{wL^3}\right)^{1/2}\sqrt{\frac{g}{L}}, \text{ Hz}, \quad i = 1, 2, \ldots$$

$$\text{where} \quad \lim_{wL^3/EI\to\infty} \lambda_i^2\left(\frac{EI}{wL^3}\right)^{1/2} = \begin{cases} 1.20, & i = 1 \\ 2.86, & i = 2 \end{cases} \tag{4.24}$$

Comparing this result with case 3 of Table 3.6, it can be seen that the lower mode natural frequencies of long, heavy, and hanging beams are accurately predicted by hanging chains (Figure 4.10).

4.4 Beams with Masses, Tapered Beams, Beams with Spring Supports, and Shear Beams

4.4.1 *Beams with Masses*

Table 4.8 [2, 23–30] shows the fundamental natural frequencies of slender uniform beams with concentrated masses. These approximate formulas were developed with the Rayleigh technique (Appendix A). They neglect the rotary inertia of the masses. They are generally within about 1% of the exact solutions (also see Ref. [31] and Appendix A).

4.4.2 *Tapered and Stepped Beams*

Table 4.9 [32–37] has formulas for the natural frequencies and mode shapes of tapered and stepped beams. Tapered beams are described by the cross section at maximum width, the taper in two perpendicular planes, and the boundary conditions Eq. 4.7. The properties at the widest point on the beam are the reference values used in computation of natural frequency. The tapered beam formulas in Table 4.9 apply exactly to simple cross sections including square, rectangular, diamond, circular, and elliptical, and approximately to other cross sections [32–34]. Tapering reduces stiffness and mass. If the beam tapers from a clamped boundary toward a less restrictive boundary, then tapering increases the natural frequency of the fundamental mode. Most formulas in Table 4.9 are exact. Some were found by Kirchhoff. Additional solutions for tapered cantilever beams can be found in Refs [38–39]. Gorman [7] has extensive solutions for stepped beams.

Table 4.8 Natural frequencies of beams with concentrated masses

Notation: A = cross sectional area of beam, Table 2.1; E = modulus of elasticity; f_1 = fundamental natural frequency, Hz; I = area moment of inertia of beam cross section, Table 1.5, about z axis; M_p = mass; M_b = mass of beam; L = length of beam between supports; x = axial coordinate; y = transverse coordinate, direction of displacement for cases 1 through 6 and 9. Table 1.2 has consistent units. Also see Tables 3.2 and 3.3. Refs [2, 23–30].

Boundary Conditions	Natural Frequency, f_1, Hz	Mode Shape, $\tilde{y}, \tilde{x}$, Comment
1. Cantilever Beam with Mass 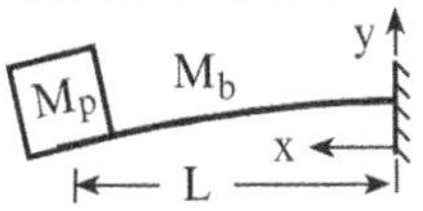	$\frac{1}{2\pi}\sqrt{\frac{3EI}{L^3(M_p+0.24M_b)}}$ Ref. [2], p. 28. Refs [23–27]	$\tilde{y}(x)=\left(\frac{x}{L}\right)^2\left(3-\frac{x}{L}\right)$, $0 \le x \le L/2$ For tip mass rotational mass moment inertial J, then use $0.24\,M_b+3J/L^3$
2. End Masses Free-Free beam 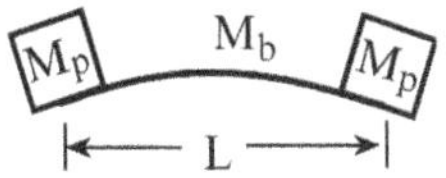	$\frac{\pi}{2}\left(1+\frac{5.45}{1-77.4M_p^2/M_b^2}\right)^{1/2}\sqrt{\frac{EI}{L^3M_b}}$ Ref. [28]. See this reference for higher modes.	
3. Ctr. Mass, Pinned-Pinned 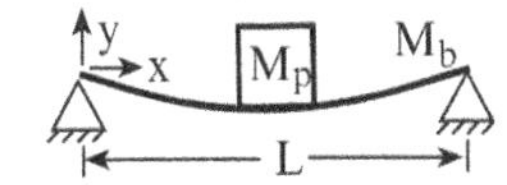	$\frac{2}{\pi}\sqrt{\frac{3EI}{L^3(M_p+0.486M_b)}}$	$\tilde{y}(x)=\frac{x}{L}\left(3\frac{x}{L}-4\left(\frac{x}{L}\right)^2\right)$, $0 \le x \le L/2$ Ref. [2], p. 26. Also see Refs [28–30].
4. Off Center Mass, Pinned-Pinned Beam 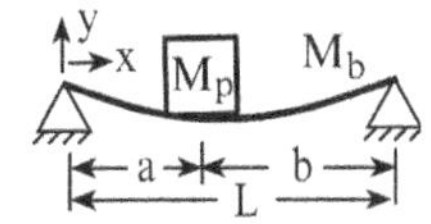	$\frac{1}{2\pi}\sqrt{\frac{3EI(a+b)}{a^2b^2(M_p+\alpha M_b)}}$ $\alpha=[2a^4+12a^3b+23a^2b^2+12ab^3+2b^4]/(105a^2b^2)$ $\tilde{y}(x)=[1-(b/L)^2-(x/L)^2](x/L)(b/L),\ 0 \le x \le a,$ $=[1-a^2/L^2-(1-x/L)^2](1-x/L)(a/L),\ a \le x \le L,$	Ref. [2], p. 33 and author's result
5. Ctr. Mass, Clamp-Clamped 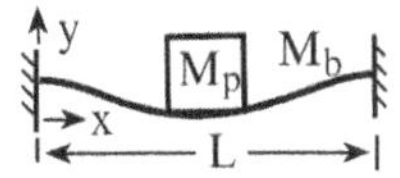	$\frac{4}{\pi}\sqrt{\frac{3EI}{L^3(M_p+0.37M_b)}}$	$\tilde{y}(x)=\left(\frac{x}{L}\right)^2\left(3-4\frac{x}{L}\right)$, $0 \le x \le L/2$ Ref. [30]
6. Off Center Mass, Clamped-Clamped Beam 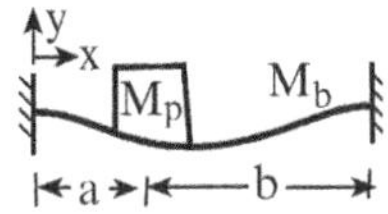	$\frac{1}{2\pi}\sqrt{\frac{3EIL^3}{a^3b^3(M_p+\alpha M_b)}}$, $L=a+b$, $\alpha=(3a^2+7ab+3b^2)L^2/(140a^2b^2)$ $\tilde{y}(x)=[3x/b+x/a-3L/b](x/a)^2$, $0 \le x \le a$, author's result $=[(3b/a+1)(L/b-x/b)/L-3L/a](L/b-x/b)^2$, $a \le x \le L$	
7. Mass Longitudinal Beam 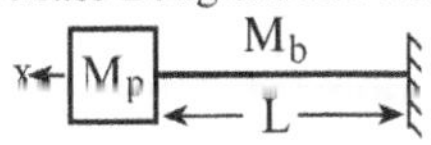	$\frac{1}{2\pi}\sqrt{\frac{EA}{L(M_p+M_b/3)}}$	$\tilde{x}(x)=x$ Note: Cases 7 and 8 are only case of longitudinal motion in this table.
8. Mass on Massive Spring 	$\frac{1}{2\pi}\sqrt{\frac{k}{M_p+M_s/3}}$	$\tilde{x}(x)=x$ k = spring constant. M_s = mass of spring
9. Triangle Cantilever tip mass 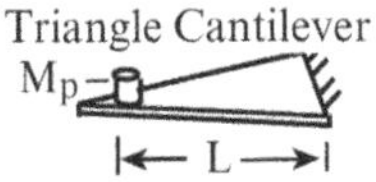	$\frac{1}{2\pi}\sqrt{\frac{2EI_o}{L^3(M_p+M_b/30)}}$	$\tilde{y}(x)\approx(1-x/L)^2$, author's result I_o = moment of inertial at fixed end.

Table 4.9 Natural frequencies of tapered and stepped beams

Notation: A_0 = area of widest section, b_0h_0, for rectangular sections, Table 1.5; E = modulus of elasticity; f_i = natural frequency, Hz; i = 1,2,3.. modal index; I_0 = area moment of inertia of widest section, $(1/12)b_0h_0^3$, for rectangular sections, Table 1.5; L = span of beam; ρ = material density. Table 1.2 has consistent units. Refs [7, 32–37].

$$\text{Natural Frequency, } f_i = \frac{\lambda_i^2}{2\pi L^2}\sqrt{\frac{EI_0}{\rho A_0}}, \quad i=1,2,3\ldots \text{ Hertz}$$

Tapered Beam	Frequency Parameter, λ_i, and Mode Shape $\tilde{y}_i(x)$

1. Tapered Cantilever Plate

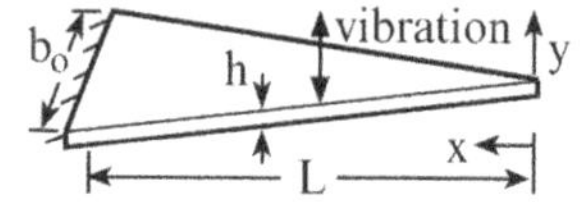

$\lambda_1 = 2.675$ Ref. [32]

$\tilde{y}_1(x) \approx (1 - x/L)^2$, approximate mode shape. See Table 5.3, case 11.

$A_o = b_o h, \; I_0 = (1/12)b_o h^3$

2. Cantilevered Wedge

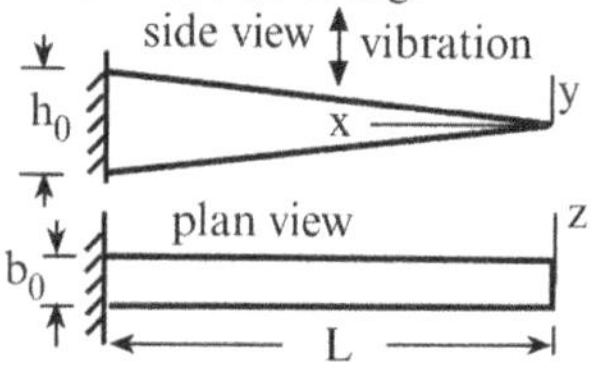

$\lambda_i = 2.3054, 3.8996, 5.4790, 7.0543, 10.200, \ldots$ Ref. [33]

$\tilde{y}_i(x) = x^{-1/2}\left(I_1[2\lambda_i]J_1[2\lambda_i(x/L)^{1/2}] - J_1[2\lambda_i]I_1[2\lambda_i(x/L)^{1/2}]\right)$

Transcendental equation for λ,

$J_1[2\lambda]I_2[2\lambda] + J_2[2\lambda]I_1[2\lambda] = 0$

Profile; $y = \pm(h_0/2)(x/L), \;\; z = \pm b_0/2$

3. Cantilevered Pyramid

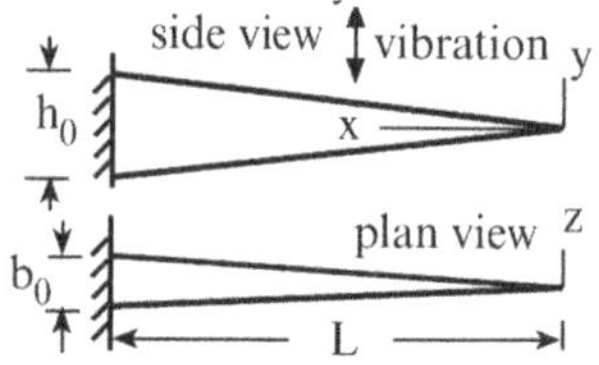

$\lambda_i = 29528, 4.5984, 6.2011, 9.3710, 10.051, ..$ Ref. [33]

$\tilde{y}_i(x) = x^{-1}\left(I_2[2\lambda_i]J_2[2\lambda_i(x/L)^{1/2}] - J_2[2\lambda_i]I_2[2\lambda_i(x/L)^{1/2}]\right)$

Transcendental equation for λ,

$J_2[2\lambda]I_3[2\lambda] + J_3[2\lambda]I_2[2\lambda] = 0$,

4. Tapered Cantilever Beam

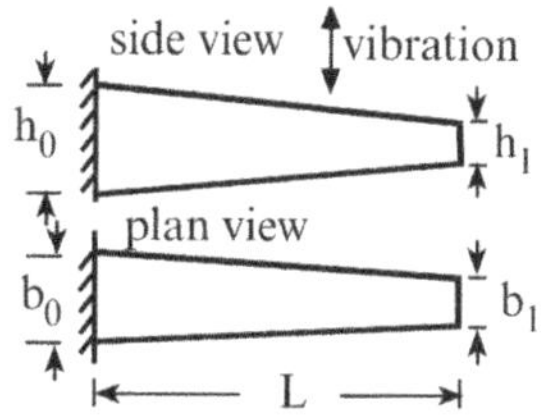

Values of λ_1. First mode. Refs [32, 34].

b_o/b_1	h_o/h_1, height taper							
	1	1.2	1.4	2	3	4	5	∞
1	1.875	1.895	1.912	1.995	2.007	2.043	2.072	2.352
1.2	1.927	1.947	1.964	2.006	2.056	2.093	2.120	2.401
1.4	1.973	1.992	2.009	2.054	2.099	2.135	2.162	2.442
2	2.077	2.095	2.211	2.151	2.198	2.232	2.259	2.539
3	2.192	2.209	2.225	2.262	2.308	2.341	2.366	2.646
4	2.268	2.285	2.300	2.336	2.381	2.413	2.437	2.718
5	2.323	2.340	2.355	2.391	2.434	2.465	2.489	2.769
∞	2.675	2.703	2.727	2.788	2.861	2.913	2.941	2.953

5. Cantilever Cone

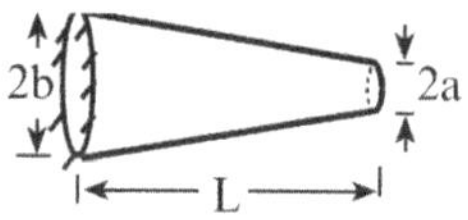

$\lambda_1^2 = 8.72\left(\frac{1 - 0.016(a/b)}{1 + 5.053(a/b)}\right)^{1/2}$ Ref. [35]

$A_0 = \pi b^2, \; I_0 = \pi b^4/4, \quad b \ge a.$

fundamental mode. Approximate solution.

Table 4.9 Natural frequencies of tapered and stepped beams, continued

Additional Notation: $J_k()$ = Bessel function of 1st kind and kth order; $I_k()$ = modified Bessel function of kth order; sym. = mode is symmetric about mid-span; asym. = mode is asymmetric about mid-span, Fig. 4.11.

$$\text{Natural Frequency,}\quad f_i = \frac{\lambda_i^2}{2\pi L^2}\sqrt{\frac{EI_0}{\rho A_0}},\quad i=1,2,3\ldots\ \text{Hertz}$$

Tapered Beam	Frequency Parameter, λ_i and Mode Shape $\tilde{y}_i(x)$

6. Clamped-Clamped Taper

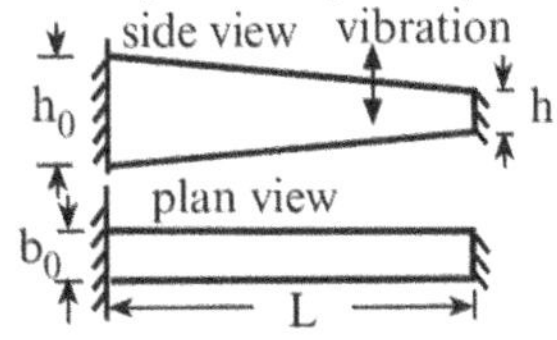

Values of λ_i Ref. [36]

h_0/h_1	1	2	3	4	10
i=1	4.730	4.040	3.742	3.565	3.144
i=2	7.853	6.708	6.200	5.901	5.194
i=3	10.990	9.394	8.667	8.250	7.262

7. Pinned-Pinned Single Taper

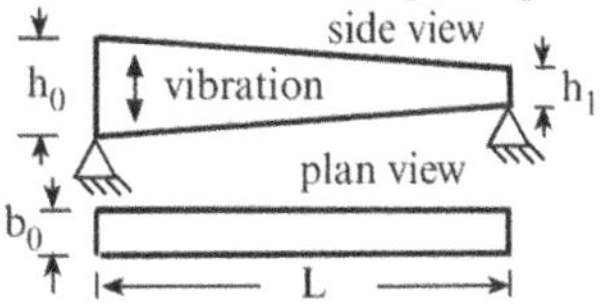

Values of λ_i Ref. [37]

h_0/h_1	1	2	3	4	10
i=1	3.142	2.669	2.449	2.313	1.972
i=2	6.283	5.385	4.993	4.773	4.260
i=3	9.425	8.062	7.483	7.123	6.326

8. Cantilever Power law Taper

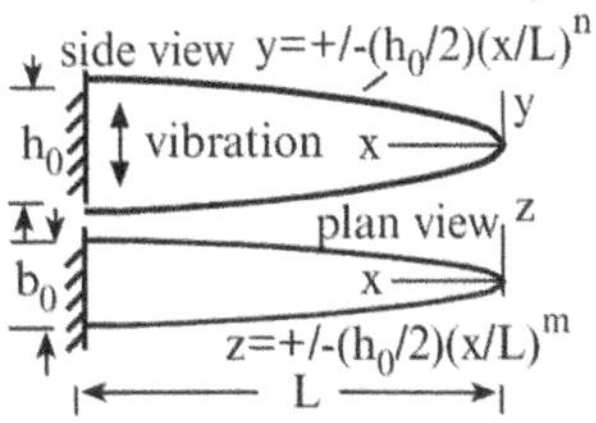

First mode. Values of λ_1 Ref. [32]

m	Exponent in power law taper, n					
	0	0.2	0.4	0.6	0.8	1.0
0	1.875	1.996	2.099	2.186	2.255	2.305
0.5	2.290	2.389	2.473	2.543	2.597	2.634
1.0	2.675	2.758	2.828	2.885	2.927	2.953

9. Cantilever with Sqrt Taper

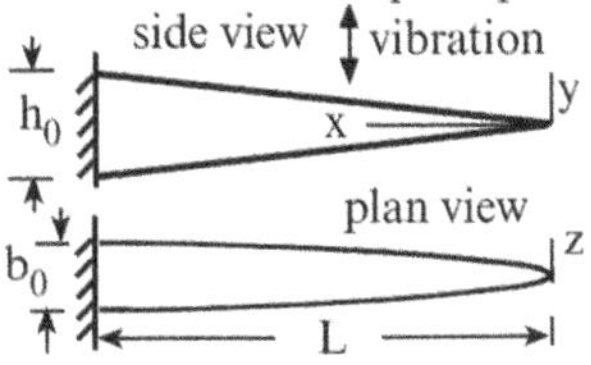

λ_i =2.6338, 4.2534, 5.8438,7.4252, 9.0027 Ref. [33]

$$\tilde{y}_i(x) = x^{-3/4}\left(I_{3/2}[2\lambda_i]J_{3/2}[2\lambda_i(x/L)^{1/2}] - J_{3/2}[2_i]I_{3/2}[2\lambda_i(x/L)^{1/2}]\right)$$

Eqn for λ: $J_{3/2}[2\lambda]I_{5/2}[2\lambda] + J_{5/2}[2\lambda]I_{3/2}[2\lambda] = 0$

Profiles: $y = \pm(h_0/2)(x/L)$, $z = \pm(b_0/2)(x/L)^{1/2}$

10. Free-Free Double Wedge

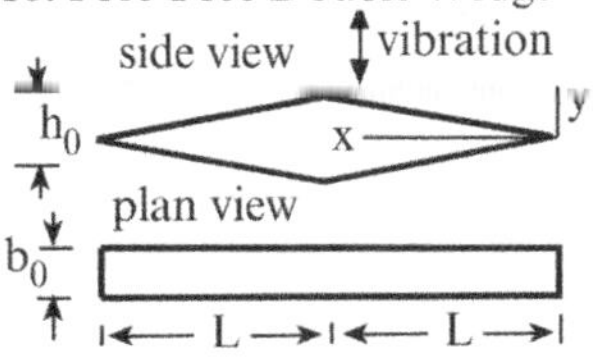

Symmetric modes Ref. [33]

λ_i = 2.5678, 4.2086, 5.8099, $\tilde{y}_i(x) = x^{-1/2}J_1[2\lambda_i(x/L)^{1/2}]$, $J_2[2\lambda]=0$

Asymmetric modes

λ_i =3.3019, 4.9630, 6.5730..

$$\tilde{y}_i(x) = x^{-1/2}\left(I_1[2\lambda_i]J_1[2\lambda_i(x/L)^{1/2}] - J_1[2\lambda_i]I_1[2\lambda_i(x/L)^{1/2}]\right)$$

Eqn for λ: $2\lambda J_1[2\lambda]I_1[2\lambda] + 2(J_0[2\lambda]I_1[2\lambda] - J_1[2\lambda]I_0[2\lambda]) = 0$

Table 4.9 Natural frequencies of tapered and stepped beams, continued

Boundary Conditions	Natural Frequency and Mode Shape $\tilde{y}_i(x)$

11. Free-Free Double Pyramid

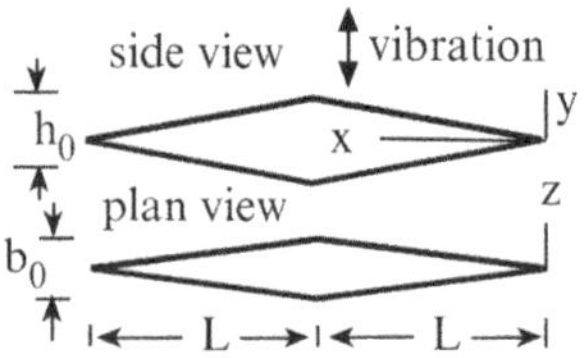

Natural Frequency, $f_i = \dfrac{\lambda_i^2}{2\pi L^2}\sqrt{\dfrac{EI_0}{\rho A_0}}$, i=1,2,3... Hertz

Symmetric modes Ref. [33]

λ_i =3.1901, 6.5076, 8.1117.., $\tilde{y}_i(x) = x^{-1} J_2[2\lambda_i (x/L)^{1/2}]$, $J_3[2\lambda] = 0$

Asymmetric modes

λ_i =3.9251, 5.6348, 7.2705, ..

$$\tilde{y}_i(x) = x^{-1}\left(I_2[2\lambda_i] J_2[2\lambda_i (x/L)^{1/2}] - J_2[2\lambda_i] I_2[2\lambda_i (x/L)^{1/2}]\right)$$

Eqn for λ: $2\lambda J_2[2\lambda] I_2[2\lambda] + 3(J_1[2\lambda] I_2[2\lambda] - J_2[2\lambda] I_1[2\lambda]) = 0$

12, Stepped Cantilever Beam Exact Solution

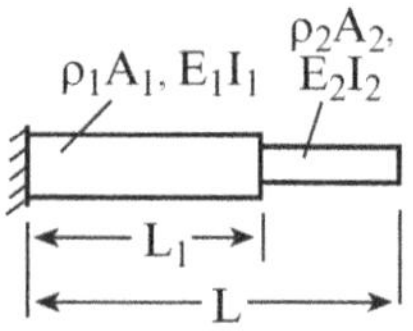

$$f_1 = \frac{\lambda_1^2}{2\pi L^2}\sqrt{\frac{E_1I_1}{\rho_1A_1}}, \text{ Hz}$$

First mode. Values of λ_1 exact solution Ref. [7]

E_2I_2/E_1I_1=0.1296

ρ_2A_2/ρ_1A_1	L_1/L 0.0	0.15	0.35	0.5	0.65	0.75	0.85	1.0
0.1296	1.875	2.146	2.571	2.764	2.571	2.233	2.146	1.875
0.2401	1.607	1.840	2.208	2.374	2.390	2.256	2.101	1.875
0.4096	1.406	1.610	1.934	2.149	2.198	2.135	2.039	1.875
1.0	1.313	1.471	1.685	1.804	1.861	1.872	1.875	1.875

E_2I_2/E_1I_1=0.2401

	0.0	0.15	0.35	0.5	0.65	0.75	0.85	1.0
0.1296	2.500	2.718	2.903	2.865	2.687	2.235	2.146	1.875
0.2401	2.188	2.388	2.565	2.544	2.403	2.287	2.101	1.875
0.4096	1.875	2.039	2.216	2.267	2.216	2.138	2.039	1.875
0.6561	1.667	1.695	1.817	1.966	2.028	2.005	1.962	1.875
1.0	1.500	1.471	1.685	1.804	1.861	1.872	1.875	1.875

E_2I_2/E_1I_1=0.6561

ρ_2A_2/ρ_1A_1	L_1/L 0.0	0.15	0.35	0.5	0.65	0.75	0.85	1.0
0.1296	2.812	2.943	3.019	2.883	2.579	2.355	2.146	1.875
0.2401	2.401	2.523	2.607	2.556	2.405	2.258	2.101	1.875
0.4096	2.109	2.208	2.289	2.289	2.219	2.138	2.039	1.875
0.6561	1.875	1.962	2.099	2.057	2.039	2.007	1.962	1.875
1.0	1.687	1.776	1.817	1.863	1.873	1.871	1.875	1.875

13. Stepped Cantilever Beam Approximate solution

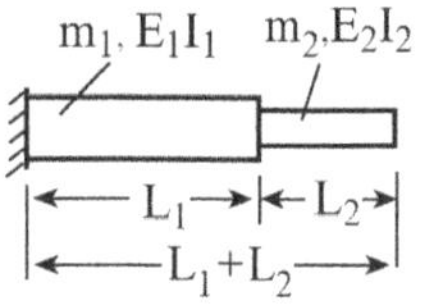

Natural frequency, $f_1 = \dfrac{\lambda_1^2}{2\pi L_1^2}\sqrt{\dfrac{E_1I_1}{m_1}}$, Hz

$$\lambda_1^2 = \frac{8^{1/2}}{\left[1 + \frac{4}{3}\frac{L_2}{L1}\left(1 + 2\frac{m_2}{m_1}\right) + 6\frac{L_2^2}{L_1^2}\frac{m_2}{m_1} + 4\frac{L_2^3}{L_1^3}\frac{m_2}{m_1} + \frac{L_2^4}{L_1^4}\frac{m_2}{m_1}\frac{E_1I_1}{E_2I_2}\right]^{1/2}}$$

Author's result for fundamental mode using static deformation.
m = mass per unit length: $m_1 = \rho_1A_1$, $m_2 = \rho_2A_2$
Reduces to case 3 of Table 4.2, for $E_1I_1 = E_2I_2$ and $m_1 = m_2$.
Exact solutions in case 12.

4.4.3 Spring-Supported Beams

Table 4.10 [40–45] has the natural frequencies of beams with springs. These are slender, uniform, single-span beams with massless rotational and extensional spring boundary conditions, given in Table 3.2 (Table 4.10; [40–45]). As the spring constants approach limit cases of either zero or infinity, the beam natural frequencies approach those given in Tables 4.2 and 4.4. As the springs become soft relative to the beam stiffness, the beam–spring system approaches a spring-supported rigid body (Fig. 4.10; Table 3.3). Additional results are provided in Refs [41].

4.4.4 Shear Beams

Transverse deformation of real beams is the sum of shear and bending deformations [1, 2, 3, 47, 50, 51]. Shear deformations are the result of shearing of the cross section under the transverse shear force V which, is proportional to the slope times the shear modulus G:

$$V = GA\frac{\partial Y}{\partial x} \tag{4.25}$$

Average shear stress over the section is just the shear force V divided by the section area A. Maximum shear stress is at the centroid of the cross section. Shear stress falls to zero at the edges of the section. The ratio of the average shear stress to the maximum shear stress is denoted by the dimensionless shear coefficient K. Table 4.11 [47–51] has shear coefficients.

$$K = \frac{\sigma_{xy}|_{avg}}{\sigma_{xy}|_{max}} \quad \text{where} \quad \sigma_{xy}|_{max} = G\frac{\partial Y}{\partial x} \quad \text{and} \quad \sigma_{xy}|_{avg} = KG\frac{\partial Y}{\partial x} \tag{4.26}$$

Timoshenko found $K = 3/4$ for a circular section and $K = 2/3$ for a rectangular cross section [1, p. 118 and 120, 2, p. 346]. Stevens [49] found that both $K = 5/(6 - \nu)$ and $K = \pi^2/12$ have merit for rectangular sections. In general, K is a function of both Poisson's ratio and section shape. The shear stress is related to the shear strain by $\partial Y/\partial x = \varepsilon_{xy,y=0} = \tau_{xy,y=0}/G$. [Ref 1, p. 171].

The equation of motion for shear beams, neglecting any flexural deformation, is obtained by setting the derivative of transverse shear force, $V_y = KAG\partial Y/\partial x$ (Eq. 4.25), equal to the mass times its transverse acceleration of a beam element dx:

$$\frac{m\partial^2 Y(x,t)}{\partial t^2} = KAG\frac{\partial^2 Y(x,t)}{\partial x^2} \tag{4.27}$$

This is the *wave equation*, a second-order equation, whereas beam bending is a fourth-order equation (Eq. 4.5). Solutions are shown in Table 4.12 (Also see discussion that follows of buildings' natural frequencies). At a fixed end, $Y(x, t) = 0$, and at a free end, there is no stress, $\partial Y(x, t)/\partial x = 0$. The shear modulus $G = E/[2(1 + \nu)]$ where E is the modulus of elasticity and ν is Poisson's ratio. Eq. 4.30 and cases 23 to 26 of Table 3.3.

Table 4.10 Natural frequencies of beams with spring supports

Notation: E = modulus of elasticity of beam material; f_i = natural frequency, Hz; k = extensional spring constant (force/length), Table 3.1; *k* = torsional spring constant (moment/angle), Table 3.1; i = 1,2,3.. modal index; I = area moment of inertia of beam, Table 2.1; L = span of beam; m = mass per unit length of beam including any non-structural mass; M = concentrated mass. Table 1.2 has consistent sets of units. Boundary conditions are defined in Table 4.1.

$$\text{Natural Frequency,}\quad f_1 = \frac{\lambda^2}{2\pi L^2}\sqrt{\frac{EI}{m}},\quad i = 1,2,3\ldots\ \text{Hertz}$$

Boundary Conditions | **Dim'less Frequency Parameter, $\lambda_1(k_1L/EI,\ k_2L/EI)$**

1. Pinned-Pinned Beam with Unequal Torsion Springs at Pinned Joints

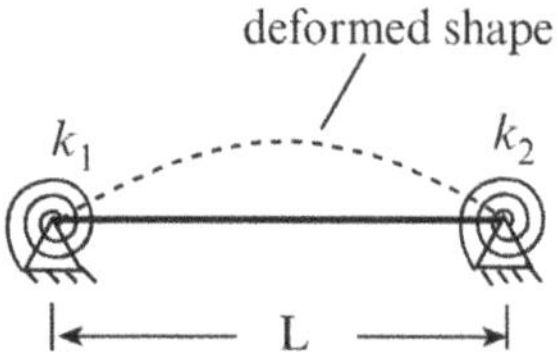

First mode Values of λ_1 Ref. [6]. Also see Ref. [6].

k_2L/EI	Torsion spring stiffness, k_1L/EI						
	0	0.01	0.1	1	10.	100	∞
0	3.142	3.143	3.157	3.273	3.664	3.889	3.926
0.01	3.143	3.144	3.158	3.274	3.780	3.890	3.927
0.1	3.157	3.158	3.172	3.288	3.678	3.902	3.939
1	3.273	3.274	3.285	3.398	3.780	4.004	4.041
10	3.664	3.666	3.678	3.780	4.155	4.390	4.430
100	3.889	3.890	3.902	4.004	4.390	4.641	4.685
∞	3.926	3.927	3.939	4.041	4.430	4.685	4.712

2. Cantilever Beam with Torsion Spring at Pined End

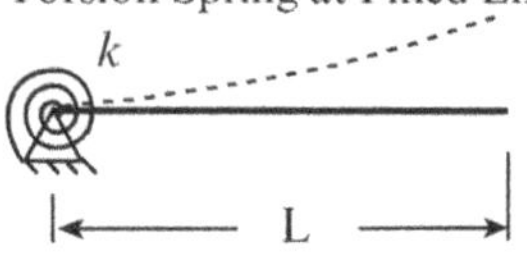

Values of λ_i, i=1,2,3 Ref. [41]

i	Torsion spring stiffness, kL/EI						
	0	0.01	0.1	1	10.	100	∞
1	0	0.4159	0.7357	1.248	1.723	1.857	1.875
2	3.927	3.928	3.928	4.031	4.400	4.650	4.694
3	7.069	7.069	7.076	7.134	7.451	7.783	7.855

3. Beam with Equal Torsion Springs at Pinned Ends and a Central Point Mass

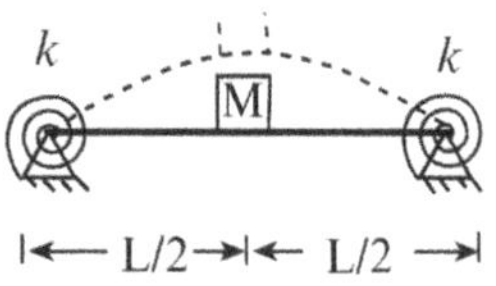

First mode values of λ_1 Ref. [42]

$\frac{kL}{2EI}$	Point Mass / Beam mass, M/mL				
	0	0.2	1	5	10
0	3.142	2.887	2.384	1.719	1.463
0.2	3.258	2.992	2.467	1.779	1.512
1	3.577	3.277	2.692	1.937	1.646
5	4.156	3.783	3.081	2.205	1.872
10	4.374	3.969	3.220	2.299	1.952
∞	4.730	4.250	3.440	2.466	2.072

4. Cantilever Beam with Torsion Spring at One End Point Mass at Other End

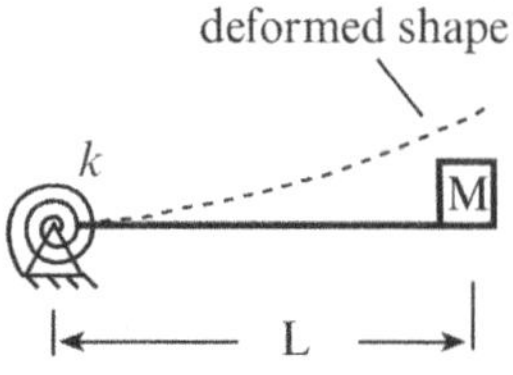

Note that $\lambda_1 \to 0$ as $k \to 0$

First mode, values of λ_1 Ref. [43]

$\frac{kL}{EI}$	Point Mass / Beam mass, M/mL					
	0	0.01	0.1	1	10.	100
0.01	0.4159	0.4129	0.3895	0.2941	0.1762	0.0998
0.1	0.7359	0.7303	0.6887	0.5194	0.3111	0.1762
1	1.2479	1.2381	1.1642	0.8705	0.5194	0.2941
10	1.7227	1.7071	1.5912	1.1642	0.6887	0.3895
100	1.8568	1.8388	1.7071	1.2381	0.7303	0.4129
∞	1.8751	1.861	1.723	1.247	0.735	0.416

Table 4.10 Natural frequencies of beams with spring supports, continued

$$\text{Natural Frequency, } f_i = \frac{\lambda_i^2}{2\pi L^2}\sqrt{\frac{EI}{m}}, \quad i = 1,2,3\ldots \text{ Hertz}$$

Boundary Conditions	Dimensionless Frequency Parameter, λ

5. Pinned-Free Beam with Torsion Spring at One End Translations Spring at Other

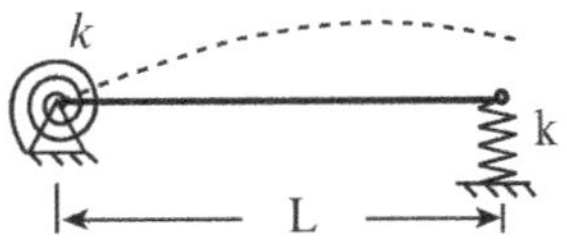

First mode, values of λ_1 Ref. [44]

$\frac{kL}{EI}$	Extension spring stiffness, kL^3/EI						
	0	0.01	0.1	1	10.	100	∞
0	0	0.4162	0.7397	1.3098	2.2313	2.9886	3.1416
0.01	0.4159	0.4948	0.7577	1.3134	2.2326	2.9901	3.1432
0.1	0.7359	0.7541	0.8782	1.3437	2.2434	3.0030	3.1572
1	1.2479	1.2520	1.2870	1.5358	2.3265	3.1084	3.2733
10	1.7227	1.7245	1.7406	1.8793	2.5388	3.4412	3.6646
100	1.8568	1.8583	1.8720	1.9939	2.6262	3.6133	3.8892
∞	1.8751	1.8766	1.8900	2.0100	2.6389	3.6405	3.9266

6. Beam on Equal Extension Springs Pinned to Ends

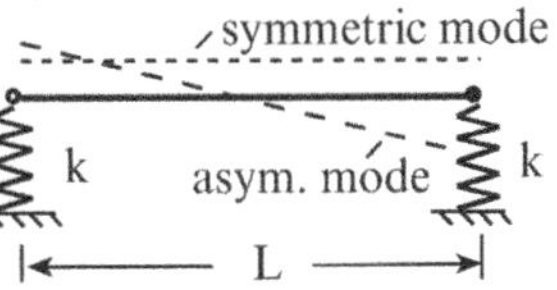

See Figure 4.11

Values of λ_i, i=1,2 author's result

Mode, i	Extension spring stiffness, kL^3/EI							
	0	0.01	0.1	1	10.	100	1000	∞
Sym, i=1	0	0.3760	0.6685	1.184	2.032	2.877	3.111	π
Asym, i=2	0	0.4948	0.8799	1.564	2.766	4.664	6.037	2π

7. Cantilever Beam Rotation and Translation Restraints Other End.

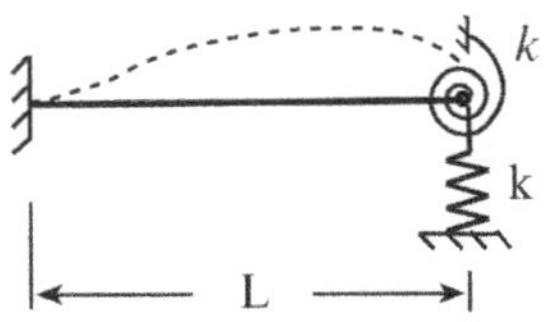

First mode, values of λ_1 Ref. [45]

$\frac{kL}{EI}$	Extension spring stiffness, kL^3/EI					
	0	1	10	100	1000	∞
0	1.8752	2.0100	2.6389	3.6405	3.8978	3.9266
1	2.0539	2.1491	2.6663	3.6818	4.0042	4.0410
10	2.2911	2.3470	2.7147	3.7888	4.3562	4.4300
100	2.3564	2.4037	2.7309	3.8403	4.4845	4.685
1000	2.3642	2.4104	2.7330	3.8475	4.6205	4.725
∞	2.3650	2.4112	2.7332	3.8483	4.6247	4.7300

8. Cantilever with Translational Spring at Part Span

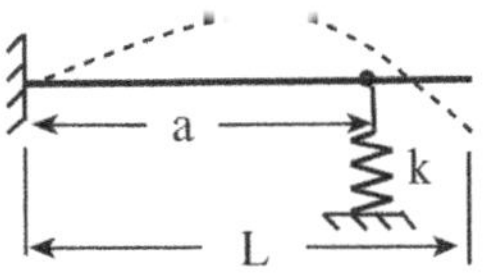

also see [46]

First mode, values of λ_1 Ref. [45]

$\frac{kL^3}{EI}$	a/L					
	0	0.2	0.4	0.6	0.8	1.0
0	1.8751	1.8751	1.8751	1.8751	1,8751	1.8751
1	1.8751	1.8757	1.8830	1.9065	1.9503	2.0100
10	1.8751	1.8812	1.9456	2.1303	2.4029	2.6389
100	1.8751	1.9273	2.2791	2.7366	3.8271	3.6404
1000	1.8751	2.0938	2.6539	3.5723	4.6790	3.8978
∞	1.8751	2.2160	2.7462	3.6830	4.6825	3.9266

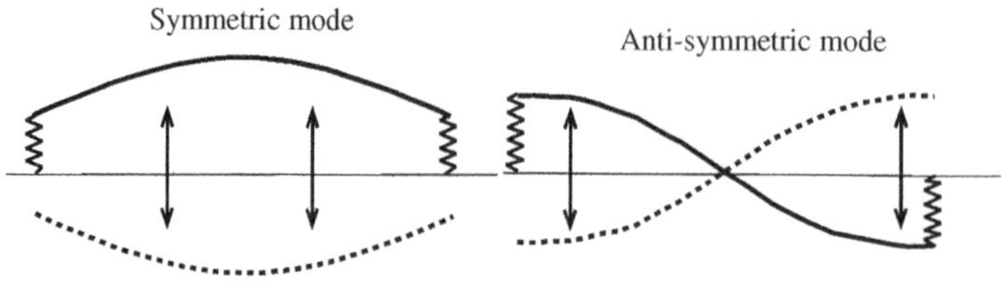

Figure 4.10 Mode shape of beams with spring-supported ends

4.4.5 *Effect of Shearing Force on the Deflections of Beams*

For beams with sinusoidal mode shapes, the ratio of shear deformation to flexural deformation is proportional to the radius of gyration of the cross section $r = (I/A)^{1/2}$ to the span L:

$$\frac{\text{Shear deformation}}{\text{Flexural deformation}} = i\frac{r}{L}, \qquad \text{mode } i = 1, 2, 3, \ldots \tag{4.28}$$

This is also the ratio of the modal stiffness ($m(2\pi f_i)^2$) of a shear beam (case 3 of Table 4.12) to that of a flexure beam (case 5 of Table 4.3a). Shear beam natural frequencies are proportional to the mode number divided by the length, whereas transverse bending mode natural frequencies are proportional to the square of the mode number divided by the square of the length (Eq. 4.11). Thus, shear deformation becomes increasingly important in higher bending modes.

4.4.6 *Rotary Inertia*

Figure 4.11 shows that beam elements rotate to conform to the curved shape. Like shear deformation, rotary inertia of rotating beam elements is neglected in the slender beam theory (Section 4.1). Shear deformation and rotary inertia both lower the natural frequency from that predicted by the flexural beam theory [2, 3, 50, 47, 51], and they become increasingly important in the higher modes of short beams.

Goodman and Sutherland's [52], also see Ref. [2], p. 436, [53], natural frequencies of a pinned–pinned beam include the effects of flexure deformation, rotary inertia, and shear deformation. i is the mode number:

$$\begin{aligned}\frac{f_i|_{\text{rotary+shear}}}{f_i|_{\text{flexure only}}} &= \frac{L}{ri}(B - (B^2 - D)^{1/2})^{1/2}, \quad i = 1, 2, 3, \ldots \\ &\approx 1 - \left(\frac{i^2\pi^2}{2}\right)\left(\frac{r}{L}\right)^2\left[1 + \frac{E}{(KG)}\right], \quad \text{for } \frac{ir}{L} \ll 1\end{aligned} \tag{4.29}$$

$$\text{where } B = \frac{1}{(2\pi^2)} + \left(\frac{D}{2}\right)\left[\pi^2 + \left\{\frac{L}{(ir)}\right\}^2\right], \quad D = \frac{KG}{(\pi^4 E)}$$

Table 4.11 Section shear coefficients

Notation: A = area of cross section; K = average shear stress over cross section/maximum shear stress, dimensionless; ν = Poisson's ratio; --- = axis perpendicular to shear load, bending neutral axis. Refs [2, 47–51].

Section	Shear Coefficient, K	Section	Shear Coefficient, K
1. Circle Diameter D 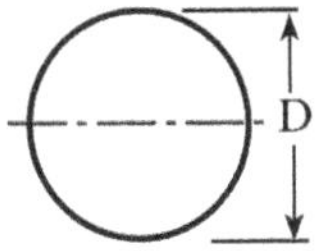	$\frac{6(1+\nu)^2}{7+12\nu+4\nu^2}$ Ref. [47] $= 0.9260$ for $\nu = 0.3$ 0.75, Ref. [1], p.122 $A = \pi D^2/4$	7. Rectangle 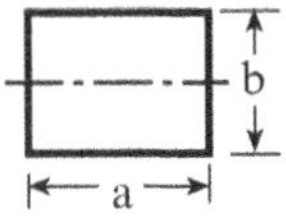	b/a=2 1 1/2 ¼ .8331 .8295 .7962 .6308 $\nu = 0.25$ Ref. [47]. 0.833, Ref. [2], p. 436 $5(1+\nu)/(6+5\nu)$, Ref. [48] A = ab Also see Ref. [49]
2. Annulus 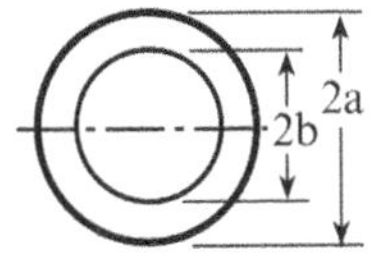	$\frac{6(1+m^2)^2(1+\nu)^2}{7+34m^2+7m^4+F_2}$ $A = \pi(a^2-b^2)$, $m = b/a$ $F_2(m)$ in note (a) Ref. [50]	8.Thin Hollow Square 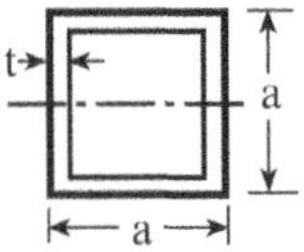	$K = 5/12$ for $\nu = 0.0$ $A = 4at$ Ref. [51] for cases 8 to 12 with ν=0 Ref. [50], p. 87.
3. Thin Annulus, a>>t 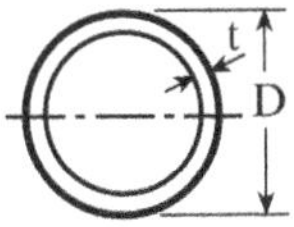	$\frac{1+\nu}{2+\nu}$ $K = 0.5652$ for $\nu = 0.3$ $A = \pi Dt$ Ref. [50]	9. Thin-Walled Box 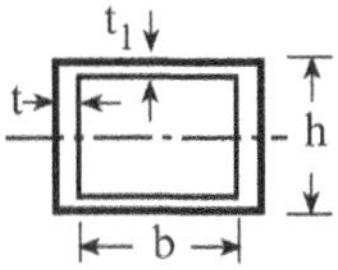	$\frac{10(1+3m)^2}{F_9(m,n)}$ for $\nu = 0$ $F_9(m,n)$ is in note (a) $m = bt_1/ht$, $n = b/h$ $A = 2(bt_1+ht)$
4. Semi Circle 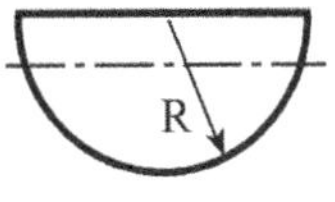	$\frac{1+\nu}{1.305+1.273\nu}$ $A = \pi R^2/2$ Ref. [51]	10 Thin Wall I Sec. 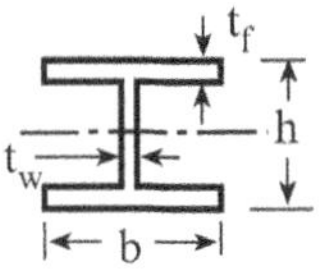	$\frac{10(1+3m)^2}{F_9(m,n)}$ for $\nu = 0$ $F_9(m,n)$ is in note (a) $m = 2bt_f/ht_w$, $n = b/h$ $A = 2bt_f + ht_w$
5. Ellipse 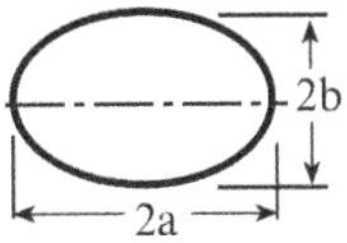	$\frac{6a^2(3a^2+b^2)(1+\nu)^2}{20a^4+8a^2b^2+F_5}$ $A = \pi ab$, $m = b/a$ $F_5(m)$ in foot note (a) Ref. [50]	11. Spars and Web 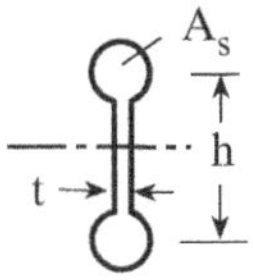	$\frac{10(1+3m)^2}{F_9(m,n)}$ for $\nu = 0$ $F_9(m,n)$ in (a) with n=0 $m = 2A_s/ht$. $A=2A_s+ht$ A_s = area of one spar
6. Square 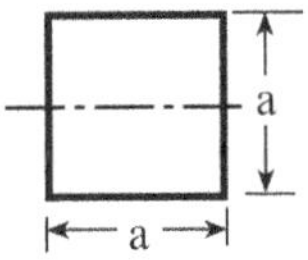	0.8333, $\nu = 0$ 0.8295, ν= 0.25 0.8228, $\nu = 0.5$, Ref. [47] $5(1+\nu)/(6+5\nu)$, Ref. [48] $A = a^2$	12. Thin Walled T 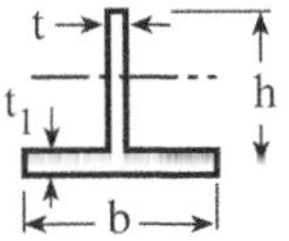	$\frac{10(1+4m)^2}{F_{12}(m,n)}$ for $\nu = 0$ $F_{12}(m,n)$ is in note (a) $m = bt_1/ht$, $n = b/h$ $A = bt_1 + ht$

(a) $F_2 = \nu(12+48m^2+12m^4)+\nu^2(4+16m^2+4m^4)$, $F_5 = \nu(37+10m^2+m^4)+\nu^2(17+2m^2-3m^4)$, Ref. [51]

$F_9(m,n) = 12+72m+150m^2+90m^3+30n^2(m+m^2)$, $F_{12}(m,n) = 12+96m+276m^2+192m^3+30n^2(m+m^2)$

Table 4.12 Natural frequencies of shear beams

Notation: A = cross sectional area, G = E/[2(1+ν)]; shear modulus; E = modulus of elasticity; K = shear coefficient, Table 4.11; k = translational spring constant, Table 3.1; i = 1,2,3.. ; modal index; L = span of beam; M = concentrated mass; m = ρA = mass per unit length of beam, plus any non-structural mass; x = axial coordinate; y = transverse coordinate; ρ = beam material density; λ = dimensionless natural frequency parameter. Table 1.2 has consistent units. Beams deform only in shear.

<table>
<tr><td colspan="3">$$f_i = \frac{\lambda_i}{2\pi L}\sqrt{\frac{KAG}{m}} = \frac{\lambda_i}{2\pi L}\sqrt{\frac{KG}{\rho}},\ i = 1,2,3..\ \text{Hz}$$</td></tr>
<tr><td>Shear Beam</td><td>λ_i, i=1,2,3..</td><td>Mode Shape, $\tilde{y}(x/L)$ and Remarks</td></tr>
<tr><td>1. Free-Free
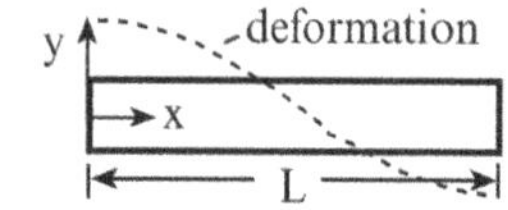
</td><td>$i\pi$
i = 1,2,3..</td><td>$\cos\frac{i\pi x}{L}$</td></tr>
<tr><td>2. Fixed-Free
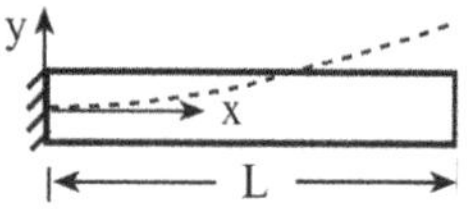
</td><td>$\frac{(2i-1)\pi}{2} = \frac{\pi}{2}, \frac{3\pi}{2}, \frac{5\pi}{2}...$
i = 1,2,3..</td><td>$\cos\frac{(2i-1)\pi x}{L}$</td></tr>
<tr><td>3. Fixed-Fixed
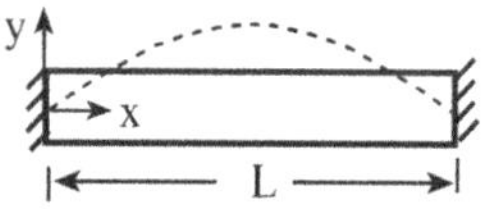
</td><td>$i\pi$
i = 1,2,3..</td><td>$\sin\frac{i\pi x}{L}$</td></tr>
<tr><td>4. Spring-Free
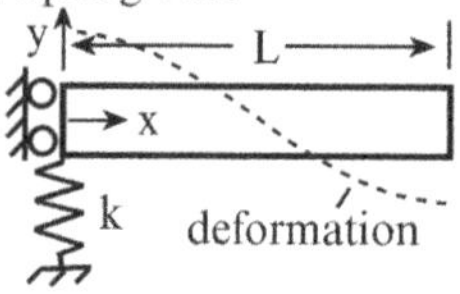
</td><td>λ determined by solutions of $\tan\lambda = (kL/KAG)/\lambda$
See table at right for λ_1.
$\lambda_1 = (kL/KAG)^{1/2}$
for KAG>>kL</td><td>$\cos\lambda_i \cos(\lambda_i x/L) + \sin(\lambda_i x/L)$
<table>
<tr><td>kL/KAG</td><td>λ_i</td><td>kL/KAG</td><td>λ_i</td></tr>
<tr><td>0</td><td>0</td><td>1</td><td>0.860</td></tr>
<tr><td>0.02</td><td>0.141</td><td>5</td><td>1.313</td></tr>
<tr><td>0.1</td><td>0.311</td><td>∞</td><td>π/2</td></tr>
</table></td></tr>
<tr><td>5. Fixed-Spring
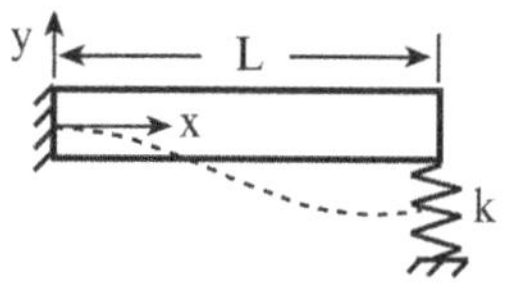
</td><td>λ determined by solutions of $\cot\lambda = -(kL/KAG)/\lambda$
See table at right for λ_1.
$\lambda_1 = \pi/2 + 2kL/(\pi KAG)$
for KAG>>kL</td><td>$\sin\lambda_i x/L$
<table>
<tr><td>kL/KAG</td><td>λ_i</td><td>kL/KAG</td><td>λ_i</td></tr>
<tr><td>0</td><td>π/2</td><td>2</td><td>2.289</td></tr>
<tr><td>0.2</td><td>1.688</td><td>10</td><td>2.863</td></tr>
<tr><td>1.0</td><td>2.029</td><td>∞</td><td>π</td></tr>
</table></td></tr>
<tr><td>6. Fixed-Mass
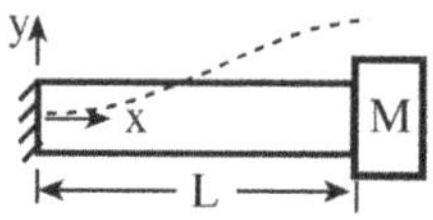
</td><td>λ determined by solutions of $\cot\lambda = (M/mL)\lambda$
$\lambda_1 = (M/mL)^{1/2}$
for mL>>M</td><td>$\sin\lambda_i x/L$
λ_i given in case 4 after substituting mL/M for kL/KAG.</td></tr>
<tr><td>7. Free-Mass
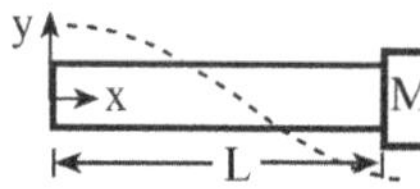
</td><td>λ determined by solutions of $\tan\lambda = -(M/mL)\lambda$
$\lambda_1 = \pi/2 + 2mL/(\pi M)$
for M>>mL</td><td>$\cos\lambda_i x/L$
λ_i given in case 5 after substituting mL/M for kL/KAG.
Also see case 8, Table 4.8</td></tr>
</table>

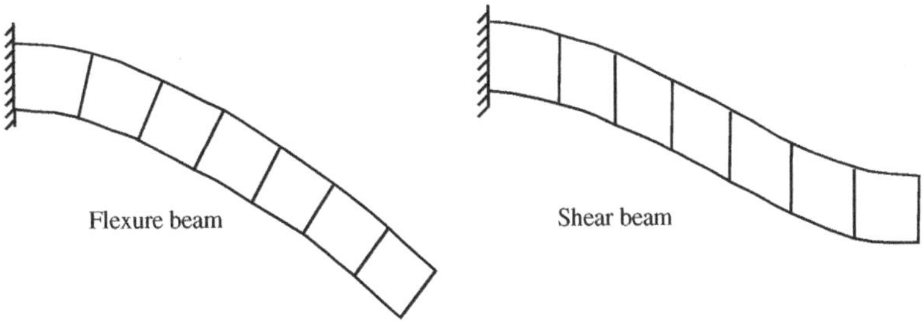

Figure 4.11 First-mode deflection of cantilever flexure and shear beams

For $K = 0.85$ and $\nu = 0.3$, the reduction in natural frequency is as follows:

$L/(ir)$	1	2	3	4	6	8	10	20
$f_i\|_{r+s}/f_i\|_{\text{flex}}$	0.177	0.334	0.463	0.565	0.707	0.795	0.851	0.954

These factors also apply to guided–hinged beams, guided–guided beams, infinitely long beams, and periodically supported beams with sinusoidal mode shapes [47]. The flexure-only frequency ($f_{i-\text{flex}}$) is given in case 4 of Table 4.3. I is the moment of inertia, A is the cross-sectional area, i is the mode number ($i = 1, 2, 3$), and $r = (I/A)^{1/2}$ is the radius of gyration (Eq. 1.6). Kang [54] gives the frequency equation of a clamped beam. Additional frequency factors for shear are given by Picket [55] and in ASTM Standard E1876 [56].

For slender beams ($L > 20(I_z/A)^{1/2}$), the reduction in natural frequency due to both shear deformation and rotary inertia is less than 5%. The reduction due to shear is about three times greater than that due to rotary inertia. Equation 4.28 is plotted in Figure 4.12, which

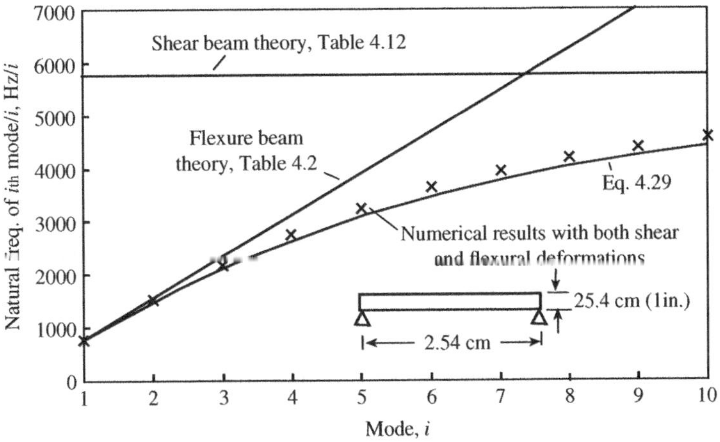

Figure 4.12 Natural frequencies of transverse modes of a pinned–pinned round steel rod compared to flexural and shear beam theories

shows the increasing importance of shear deformations in higher modes (also see Refs [2, 3, 50]; Example 4.2).

4.4.7 Multistory Buildings

The natural frequencies and mode shapes of most multistory buildings are dominated by shear deformations between floors. The floors shear parallel to each other, like a deck of cards, as the vertical support beams flex laterally ([57]; cases 23–25 of Table 3.3; case 12 of Table 4.17; Fig. 4.11). Equation 4.29 suggests that the shear beam model ($f \sim 1/\text{height}$; Table 4.12) is more appropriate for a building than a flexure beam model ($f \sim 1/\text{height}^2$; Table 4.2).

Semiempirical formulas have been developed for the fundamental lateral natural frequency of multistory buildings:

$$f_1\ (\text{Hz}) = \begin{cases} \dfrac{10}{N} & N = \text{number of storys}, \quad \text{Ref. [67]} \\ \dfrac{c}{H} & c = 46\ \text{m/s}\ (151\ \text{ft/s}), \text{Ref. [67]} \\ \dfrac{1}{CH^{3/4}} & C = 0.035,\ 0.030,\ \text{or}\ 0.020\ \text{s/ft}^{3/4}, \quad \text{Ref. [68]} \\ \dfrac{cD^{1/2}}{H} & c = 11\ \text{m}^{1/2}/\text{s} \quad (20\ \text{ft}^{1/2}/s), \quad \text{Ref. [69]} \end{cases} \tag{4.30}$$

D is the building width in the direction of vibration and H is the building height. In the third formula, $C = 0.085,\ 0.073$, or $0.049\ \text{s/m}^{3/4}$ for steel frame, reinforced concrete, and other constructions, respectively. The first torsion mode natural frequency about the vertical axis is approximately c/H (Hz) where $c = 72$ ft and H is the height in feet [67]. Buildings often have overall average volumetric densities on the order of 12 lb/ft^3 (192 kg/m^3).

Example 4.2 Sandwich beam

Figure 4.13 shows a 914.4 mm (36 in.) long aluminum honeycomb beam that is suspended by long strings. The beam vibrates in its free–free modes perpendicular to the plane of its face sheets. Determine its natural frequencies. The beam weighs 1.71 kg (3.77 lb) [56]. Its face sheets are 1.625 mm (0.064 in.) thick.

Solution

Neglecting the stiffness of the flexible core spacer, the moment of inertia of the cross section is entirely due to the face sheets, and it is computed from Equation 4.3 with $h = 1.625\ \text{mm} + 50.8\ \text{mm} = 54.525\ \text{mm}$:

$$I = \frac{1}{2}tbh^2 = (1/2)1.625\ \text{mm} \times 152.4\ \text{mm}(52.425\ \text{mm})^2 = 340,318\ \text{mm}^4(0.8179\ \text{in.}^4)$$

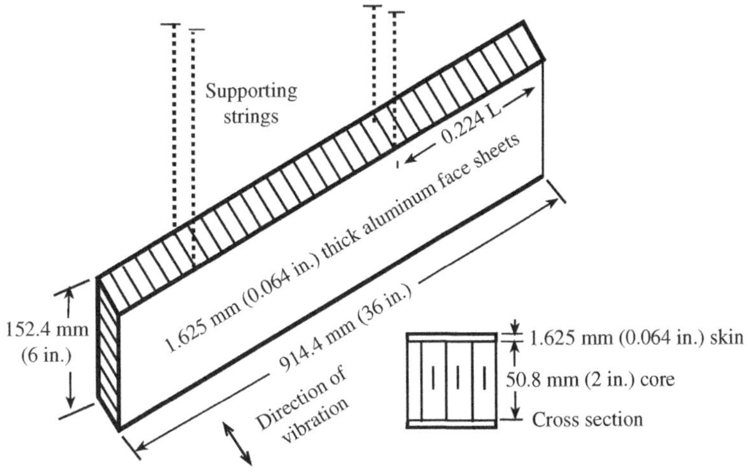

Figure 4.13 An aluminum honeycomb beam suspended on strings

The mass per unit length of the beam is its total mass divided by its span: $m = 1.71$ kg/914.4 mm = 0.00187 kg/mm (0.1047 lb/in.). The string supports do not offer significant resistance to the free bending vibrations as they are located at the nodes of the first free–free mode (0.224 L and 0.776 L; Fig. 4.3a, top). Case 3 of Table 1.2 offers a consistent set of units with millimeter as a unit of length. In this system, 10^4 kg is the unit of mass so $m = 1.87\text{E} - 7$ (10^4 Kg/mm). The modulus of elasticity of aluminum is $E = 6894.4\,\text{dN/mm}^2$ (10E6 psi). The fundamental natural frequency is then computed in these consistent units and US customary units from case 7 of Table 1.2:

$$f_1 = \frac{\lambda_i^2}{2\pi L^2}\sqrt{\frac{EI}{m}} = \begin{cases} \dfrac{4.73004^2}{2\pi(914.4\text{ mm})^2}\sqrt{\dfrac{6894.9\text{ dN/mm}^2 \times 340,450\text{ mm}^4}{1.87\text{E}-7\ (10^4\text{kg/mm})}} \\ \dfrac{4.73004^2}{2\pi(36\text{ in.})^2}\sqrt{\dfrac{10\times 10^6\text{ lb/in.}^2 \times 08179\text{ in.}^4}{0.1047\text{ lb/in./386.1 in./s}^2}} \end{cases} = 477.1\text{ Hz}$$

Clary [56] measured the natural frequencies of this beam.

Mode, i	1	2	3	4
f_i, calculation, Table 4.2 (Hz)	477.1	1315.	2578.	4261.
f_i, calc. with shear and rot., Ref. [70] (Hz)	416	882	1370	1833
f_i, measurement, Ref. [70] (Hz)	411	887	1340	1825

The flexure-only natural frequencies (Table 4.2) compare well with the measurements in the lower modes. In the higher modes, the shear deformations become increasingly important as the beam wavelengths become comparable to the beam width and the flexure theory overpredicts the measurements. Shear deformation of sandwich beams is further considered in Refs [71, 72].

4.5 Torsional and Longitudinal Beam Natural Frequencies

4.5.1 Longitudinal Vibration of Beams and Springs

Table 4.13 presents natural frequencies and mode shapes of beams stretching and contracting longitudinally along the beam axis. Owing to the high axial stiffness, longitudinal mode natural frequencies of slender beams ordinarily have much higher natural frequencies than their transverse bending natural frequencies (Table 4.2) (also see cases 1–17 of Table 3.3).

During axial vibration, longitudinal deformation $X(x,t)$ and axial stress σ_x are uniform over a cross section perpendicular to the beam axis. The relationships between strain and stress and deformation are

$$\begin{aligned} &\varepsilon_x = \frac{\partial X}{\partial x} \quad \varepsilon_y = \varepsilon_z = \frac{-\nu \partial X}{\partial x} \quad \varepsilon_{xy} = \varepsilon_{xz} = \varepsilon_{yz} = 0 \\ &\sigma_x = \frac{E\partial X}{\partial x} \quad \sigma_y = \sigma_z = \sigma_{xy} = \sigma_{yz} = 0 \end{aligned} \tag{4.31}$$

ε_x, ε_y, and ε_z are normal strains in the x, y, and longitudinal directions, respectively. Poisson's effect creates the lateral contractions in the lateral y and z directions even though the stresses in these directions, σ_y and σ_z, are zero. Shear stress and strains, denoted by the double subscripts in Equation 4.31, are zero. The equation of motion for free longitudinal motions of a uniform, elastic beam is the *wave equation*:

$$m\frac{\partial^2 X(x,t)}{\partial t^2} = EA\frac{\partial^2 X(x,t)}{\partial x^2} \tag{4.32}$$

Boundary conditions include $X = 0$ for a fixed end, $\partial X/\partial x = 0$ for a free end, $kX = -EA\ \partial X/\partial x$ for a spring end, and $-M\partial^2 X/\partial x^2 = EA\ \partial X/\partial x$ for a tip mass M. A is the cross-sectional area and E is the beam modulus of elasticity. The signs in these last two equations are reversed if the boundaries are applied to the beam face in the $(-x)$ direction.

Example 4.3 Longitudinal vibration

Consider the uniform beam, L long, with one end fixed and a mass on the other end in case 6 of Table 4.13. Determine its *longitudinal* natural frequencies and mode shapes.

Solution

The deformation in the x direction is sinusoidal in space and time:

$$X(x,t) = c_1 \sin\left(\frac{\lambda x}{L}\right) \sin \omega t$$

Substituting this equation into Equation 4.32 shows that the circular natural frequencies ω are related to the dimensionless frequency parameter λ:

$$\omega = \frac{\lambda}{L}\sqrt{\frac{EA}{m}}$$

Table 4.13 Natural frequencies of longitudinal vibration of beams and springs

Notation: A = cross sectional area; E = modulus of elasticity; k = translational spring constant, Table 3.2; f = natural frequency, Hz; i = 1,2,3.. modal index; L = span of beam; M = concentrated mass; m = mass per unit length of beam, mL = mass of beam, ρA plus any non-structural mass per unit length; x = axial coordinate, in the direction of deformation; ρ = beam material mass density; λ = dimensionless natural frequency parameter. Table 1.2 has consistent sets of units. Also see Example 4.3.

$$\text{Natural Frequency,}\quad f_i = \frac{\lambda_i}{2\pi L}\sqrt{\frac{EA}{m}} = \frac{\lambda_i}{2\pi L}\sqrt{\frac{E}{\rho}},\quad i=1,2,3\ldots\ \text{Hertz}$$

Axial Beam Motion	λ_i, i=1,,2,3	Mode Shape, $\tilde{x}_i(x/L)$, Remarks
1. Free-Free Beam 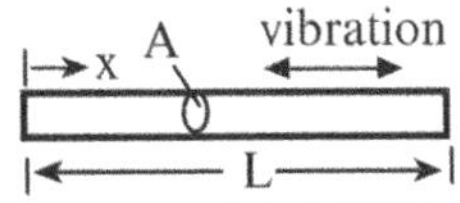	$i\pi$ i=1,2,3..	$\cos\frac{i\pi x}{L}$
2. Fixed-Free Beam 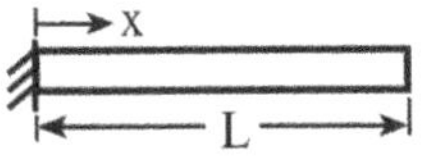	$\frac{(2i-1)\pi}{2} = \frac{\pi}{2}, \frac{3\pi}{2}, \frac{5\pi}{2}..$ i = 1,2,3..	$\sin\frac{(2i-1)\pi x}{2L}$
3.Fixed-Fixed Beam 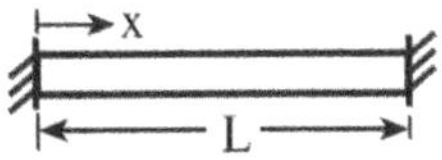	$i\pi$ i =1,2,3..	$\sin\frac{i\pi x}{L}$
4. Fixed Spring-Free Beam 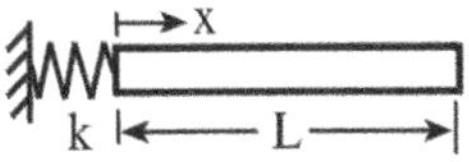	λ solutions of $\tan\lambda = (kL/EA)/\lambda$ See table at right for λ_1. $\lambda_1 \approx (kL/EA)^{1/2}$, EA>>kL	$\cos\lambda_i \cos\lambda_i x/L + \sin\lambda_i x/L$ kL/EA — λ_i — kL/EA — λ_i 0 — 0 — 1 — 0.860 0.02 — 0.141 — 5 — 1.313 0.1 — 0.311 — ∞ — π/2
5. Spring Beam-Fixed 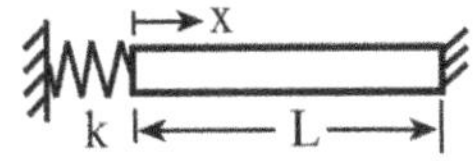	λ solutions of $\cot\lambda = -(kL/EA)/\lambda$ See table at right for λ_1. $\lambda_1 \approx \pi/2 + 2kL/\pi EA$ for EA >>kL	$\sin\lambda_i x/L$ kL/EA — λ_i — kL/EA — λ_i 0 — π/2 — 2 — 2.289 0.02 — 1.688 — 10 — 2.863 0.1 — 2.029 — ∞ — π
6. Fixed-Beam Mass 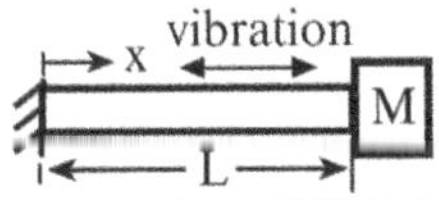	λ solutions of $\cot\lambda = (M/mL)\lambda$ $\lambda_1 \approx (mL/M)^{1/2}$, M>>mL	$\sin\lambda_i x/L$ λ_1 given in case 4, after substituting mL/M for kL/EA
7. Free Beam –Mass 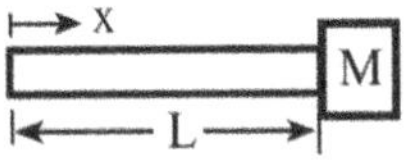	λ solutions of $\tan\lambda = -(M/mL)\lambda$ $\lambda_1 \approx \pi/2 + 2mL/(\pi M)$ for M >>mL	$\cos\lambda_i x/L$ λ_1 given in case 5, after substituting mL/M for kL/EA
8. Beam with Two Masses 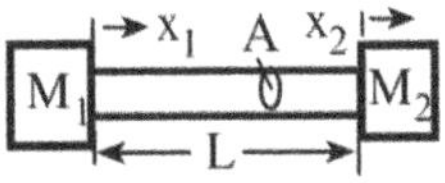	λ determined by solutions of transcendental equation. $\tan\lambda = \frac{\lambda(M_1+M_2)(mL)}{\lambda^2 M_1 M_2 - (mL)^2}, \quad \frac{\tilde{x}_1}{\tilde{x}_2} = -\lambda\frac{M_1}{mL}$ $\lambda_1 \approx [mL(M_1+M_2)/(M_1 M_2)]^{1/2}$ for M_1, M_2>>mL	

Table 4.13 Natural frequencies of longitudinal vibration, continued

Additional Notation: f = natural frequency, Hz; i = 1,2,3.. modal index; k = translational spring constant, case 24 of Table 3.2, force/ length; L = span of spring; M = concentrated mass; M_s = total mass of spring or bellows, x = axial coordinate, in the direction of deformation; Table 1.2 has consistent units. Author's results.

$$\text{Natural Frequency, } f_i = \frac{\lambda_i}{2\pi}\sqrt{\frac{k}{M_s}}, \quad i = 1,2,3\ldots \text{ Hertz}$$

Description	λ_i, i=1,2,3..	Mode Shape, $\tilde{x}_i(x/L)$ and Remarks
9. Free-Free Spring 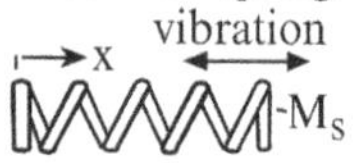	$i\pi$ i=1,2,3..	$\cos\frac{i\pi x}{L}$
10. Fixed-Free Spring 	$\frac{(2i-1)\pi}{2} = \frac{\pi}{2}, \frac{3\pi}{2}, \frac{5\pi}{2}..$ $i = 1,2,3..$	$\sin\frac{(2i-1)\pi x}{2L}$
11. Fixed-Fixed Spring 	$i\pi$ i = 1,2,3..	$\sin\frac{i\pi x}{L}$
12. Fixed Spring Mass 	λ determined by solutions of $\cot\lambda = (M/M_s)\lambda$ $\lambda_1 \approx (M_s/M)^{1/2}$	$\sin\lambda_i x/L$ λ_1 given in case 4, insert M_s/M for kL/EA.
13. Free Spring Mass 	λ determined by solutions of $\tan = -(M/M_s)\lambda$ $\lambda_1 \approx \pi/2 + 2M_s/\pi M,\ M >> M_s$	$\cos\lambda_i x/L$ λ_1 given in case 5, insert M_s/M for kL/EA.
14. Mass-Spring-Mass 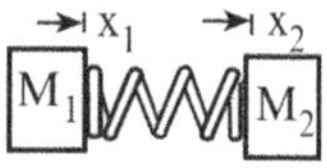	λ determined by $\tan\lambda = \frac{\lambda(M_1+M_2)(M_s)}{\lambda^2 M_1 M_2 - (M_s)^2}$ $\lambda_1 \approx [M_s(M_1+M_2)/(M_1 M_2)]^{1/2}$ for $M_1, M_2 >> M_s$	$\frac{\tilde{x}_1}{\tilde{x}_2} = -\lambda\frac{M_1}{M_s}$
15. Bellows Axial Modes 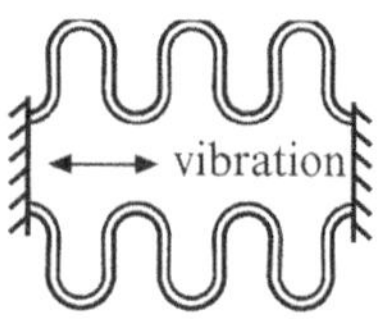	$\lambda_i = \begin{cases} i\pi, & \text{continuous model} \\ 4N_c \sin\left[\frac{i\pi}{4N_c}\right], & \text{discrete model} \end{cases}$ $i = 1,2,3..(2N_c - 1)$ N_c = number of convolutions N_c = 3 shown at left. Eq. 3.23	Bellows stiffness k in case 46 Table 3.1. Bellows mass M_s, including fluid added mass in convolutions, from Table 3.1. Discrete model is equivalent to $f_i = \frac{2}{\pi}\sin\left[\frac{i\pi}{4N_c}\right]\sqrt{\frac{k_{c/2}}{m_{c/2}}}$, Hz where $k = 2N_c k_{c/2}$, $M_s = 2N_c m_{c/2}$ discrete model is preferred. Ch. 3, Ref. [13].

The boundary conditions are fixed at $x = 0$ and Newton's second law (Eq. 2.4) for mass M at $x = L$:

$$X(0) = 0, \quad M\frac{\partial^2 X(L)}{\partial t^2} = -EA\frac{\partial X(L)}{\partial x}$$

Substituting in the trial mode shape gives a linear homogeneous equation:

$$c_1\left[-M\omega^2 \sin\lambda + \left(\frac{EA\lambda}{L}\right)\cos\lambda\right] = 0$$

The term in brackets must be zero for nonzero solution. Inserting the previous equation for ω produces a transcendental equation for the dimensionless natural frequency parameter λ of nontrivial solutions:

$$\tan\lambda = \frac{(mL/M)}{\lambda}, \text{ or its inverse, } \cot\lambda = \frac{M}{mL}\lambda$$

There are multiple solutions for λ given mL/M. The first of these is found by expanding the cotangent that is a series about zero, $\cot\lambda = 1/\lambda - \lambda/3..$; substituting the first two terms into the second equation; and solving. This gives $\lambda_1 = (M/mL + 1/3)^{-1/2}$, which fortuitously is within 10% of the exact solution over the full range of the ratio M/mL:

mL/M	0	0.01	0.1	1	10	100	∞
Exact λ_1	0	0.0994	0.311	0.860	1.429	1.555	$\pi/2$
$1/(M/mL + 1/3)^{1/2}$	0	0.0998	.311	0.866	1.519	1.706	$3^{1/2}$

4.5.2 *Torsional Vibration of Beams and Shafts*

Table 4.14 [1, 59, 73–76] presents the torsion constants of cross sections. Additional values for structural sections are provided in Ref. [77]. Table 4.15 shows the torsional natural frequencies of shafts and beams. Table 3.4 in Chapter 3 provides solutions for torsional systems consisting of rotary inertia mass, such as heavy disks, connected by massless torsional springs.

4.5.3 *Circular Cross Section*

Exact solutions exist for torsion circular cross sections such as tube and round rods [73]. Their circumferential shear stress and strain increase linearly with distance from the center of the circular shaft; all other stresses and strain components are zero. The torque (moment) borne by the cross section is proportional to the shear modulus, $G = E/[2(1 + \nu)]$ where E is the modulus of elasticity and ν is Poisson's ratio, and the polar moment of inertia of a circular cross section:

$$M = GI_p\frac{\partial\theta}{\partial x} \tag{4.33}$$

x is the distance along the center of the shaft, and the polar area moment of inertia of the shaft is $I_p = \pi R^4/2$ for a circular rod of radius R. θ is the angle of twist about the centroid (center of mass). The maximum shear stress $\tau = 2M/\pi R^3$ is on the outer radius R (see Fig 4.14).

Table 4.14 Torsion constants of beam cross sections

Notation: $C = M/(G\, d\theta/dx)$. M = moment; $d\theta/dx$ = twist gradient; G = shear modulus. I_p = polar moment of inertia. Refs [59, 73–76].

Section	Torsion Constant, C	Section	Torsion Constant, C
1. Circle with Radius R 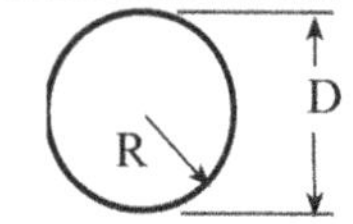 	$\frac{1}{2}\pi R^4 = \frac{1}{32}\pi D^4$ For circles, C is polar area moment of inertia, I_p=C, Table 1.5 Ref. [73]	8. Equilateral Triangle 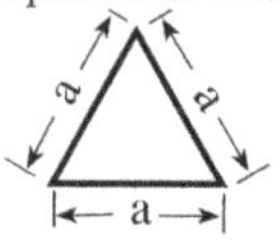 	$\frac{3^{1/2}}{80}a^4$ $I_p = 0.03a^4$ based on case 6
2. Tube 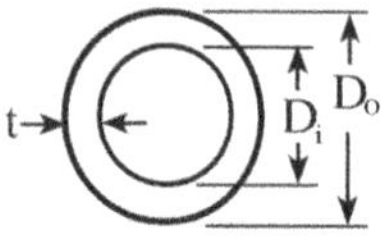 	$\frac{1}{32}\pi(D_o^4 - D_i^4) \approx \pi R^3 t$ where $R = (D_o + D_i)/4$ Ref. [73]	9. Hexagon 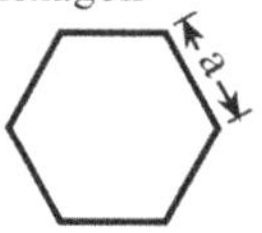 	$1.05a^4$ $I_p = 1.082a^4$ based on case 6.
3. Ellipse 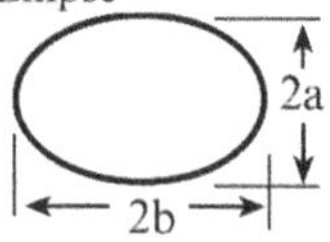 	$\frac{\pi a^3 b^3}{a^2 + b^2}$ Ref. [73], p.264 $I_p = \frac{\pi}{4}ab(a^2 + b^2)$	10. Thin Open Section with Uniform Wall 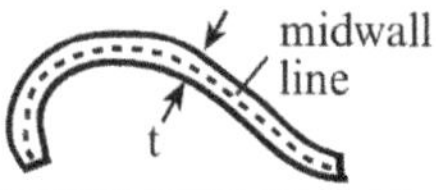 	$\frac{1}{3}St^3$ S = length of midwall based on case 5
4. Square 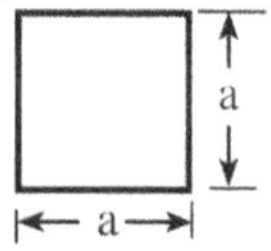 	$0.1406a^4$ $I_p = \frac{a^4}{6}$ Ref. [73], p. 277	11. Thin Flanged I beam 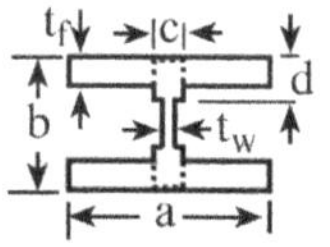 	$\frac{2}{3}(a-c)t_f^3 + \frac{1}{3}(b-d)t_w^3 +$ $[0.46 - 0.5(\frac{d}{c} - 1.15)^2]d^3c$ Table1.5 for I_p. Ref. [59]
5. Rectangle 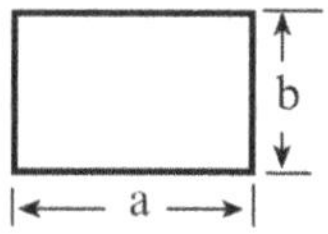 	$\frac{ca^3b^3}{a^2+b^2}$, $I_p = \frac{ab^3 + a^3b}{12}$ a/b 1 2 4 ∞ c 0.281 0.286 0.299 0.33 $c = ab^3/3$ for $a >> b$ Refs [73, 74]	12. Thin-Walled Section with Rectangles 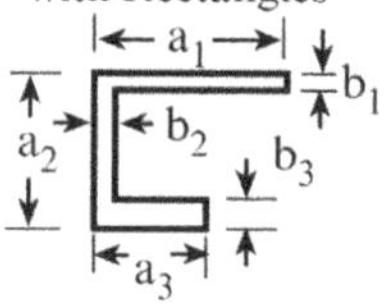 	$\sum_i \frac{c_i a_i b_i^3}{1 + b_i^2/a_i^2}$ $a_i \ge b_i$ c_i is given in case 5. $\frac{1}{3}\sum_i a_i b_i^3$ for $a_i >> b_i$ Refs [1, 59, 75]
6. Compact Section 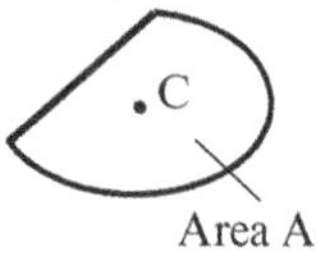 	$\frac{A^4}{40I_p}$ (approx., Ref. [76]) A = area of cross section I_p = polar area moment of inertia about centroid C, Table 1.5	13. Hollow Rectangle 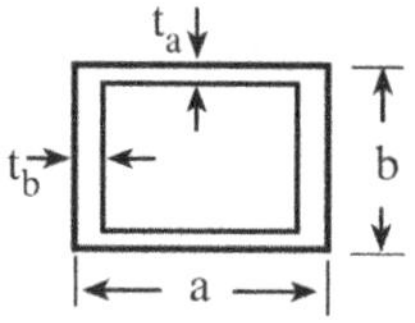 	$\frac{2t_a t_b (a - t_a)^2 (b - t_b)^2}{a t_a + b t_b - t_a^2 - t_b^2}$ $\frac{2ta^2b^2}{a+b}$, $t = t_a = t_b$ $I_p = (a+b)^3 t$, Ref. [74]
7. ElongatedSymmetric Section 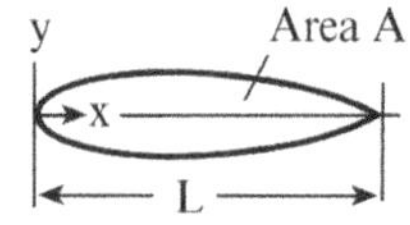 symmetric about x	$\frac{4I_x}{1 + 16I_x/(AL^2)}$ I_x = area moment of inertia about axis of symmetry, x-axis, Table 1.5 Ref. [75] A = cross sectional area	14. Thin Walled Closed Section 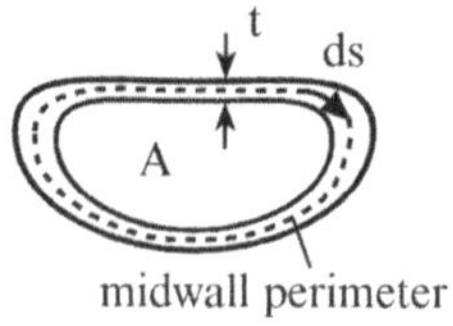 	$4A^2 / \int_0^S (1/t)ds$ $= 4A^2 t/S$, t = constant A = area enclosed by midwall perimeter S = length of midwall perimeter. Ref. [76]

Table 4.15 Natural frequencies of shafts in torsion

Notation: C = torsion constant of shaft cross section, Table 4.14; G = shear modulus, $E/[2(1+\nu)]$ where E = modulus of elasticity and ν = Poisson's ratio; I_p = polar moment of inertia of cross section about torsion axis, Eq. 1.9, Table 1.5; J = rotor mass moment of inertia about shaft axis, Eq. 1.13, Table 1.6; k = torsion spring constant, Table 3.2; I = 1,2,3.. modal index; L = span of shaft; x = axial coordinate; ρ = mass density; θ = angle of twist. Table 1.2 has consistent units. Also see Table 3.4.

$$\text{Natural Frequency, } f_i = \frac{\lambda_i}{2\pi L}\sqrt{\frac{CG}{\rho I_p}} = \frac{\lambda_i}{2\pi L}\sqrt{\frac{G}{\rho}}\Big|_{\text{circular section}, C=I_p}, \text{ Hz, } i=1,2,3..$$

Torsion Shaft	λ_i, i=1,2,3..	Mode Shape, $\tilde{\theta}(x/L)$
1. Free-Free Shaft 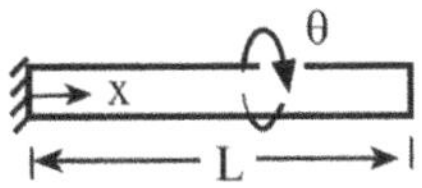	$i\pi$ $i = 0,1,2,3..$	$\cos\frac{i\pi x}{L}$
2. Fixed-Free Shaft 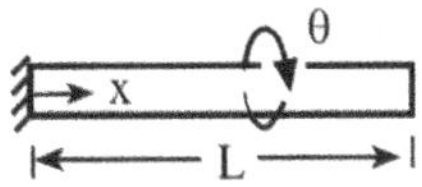	$\frac{(2i-1)\pi}{2} = \frac{\pi}{2}, \frac{3\pi}{2}, \frac{5\pi}{2}..$	$\sin\frac{i\pi x}{L}$
3. Fixed-Fixed Shaft 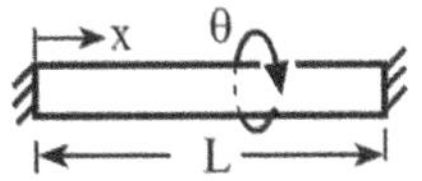	$i\pi$ $i = 0,1,2,3..$	$\sin\frac{i\pi x}{L}$
4. Fixed-Free Shaft 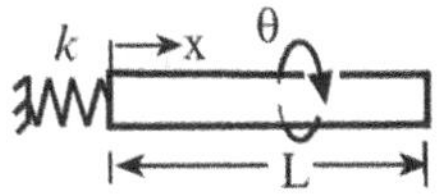	λ determined by $\tan\lambda = (kL/CG)\lambda$ See table at right for λ_1.	$\cos\lambda_i \cos(\lambda_i x/L) + \sin\lambda_i x/L$ kL/CG 0 .02 0.1 1 5 ∞ λ_1 0 .141 .311 .86 1.31 π
5. Fixed-Spring Shaft 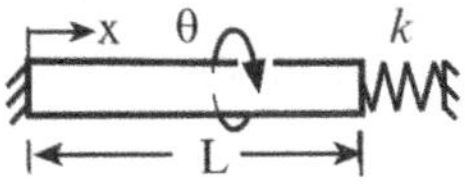	λ determined by $\cot\lambda = -(kL/CG)/\lambda$ See table at right for λ_1.	$\sin\lambda_i x/L$ kL/CG 0 .02 0.1 2 10 ∞ λ_1 $\pi/2$ 1.69 2.03 2.29 2.86 π
6. Fixed – Rotor Shaft 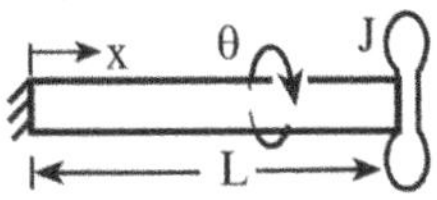	λ determined by $\cot\lambda = (J/\rho I_p L)\lambda$ See case 1, Table 3.3	$\sin\frac{i\pi x}{L}$ λ_1 given in case 4 after substituting $\rho I_p L/J$ for kL/CG.
7. Free-Rotor Shaft 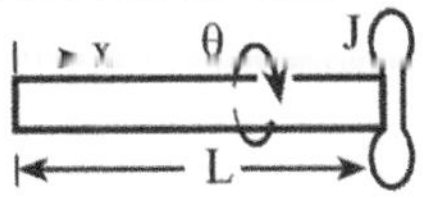	λ determined by $\tan\lambda = -(J/\rho I_p L)\lambda$ There is a zero freuency mode.	$\cos\lambda_i x/L$ λ_1 given in case 5 after substituting $\rho I_p L/J$ for kL/CG.
8. Rotors on Shaft Ends 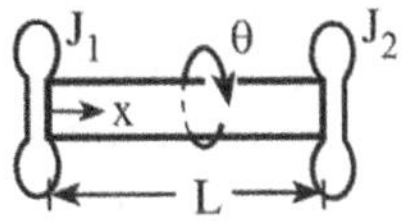	λ determined by $\tan\lambda = -\frac{J_1+J_2}{\rho I_p L}\frac{\lambda}{(\lambda/\rho I_p L)^2 J_1 J_2 - 1}$, $\frac{\tilde{\theta}_1}{\tilde{\theta}_2} = -\lambda\frac{J_1}{\rho I_p L}\sin\lambda + \cos\lambda$ Also see case 10 of Table 3.3 Ref. [2], p. 404	

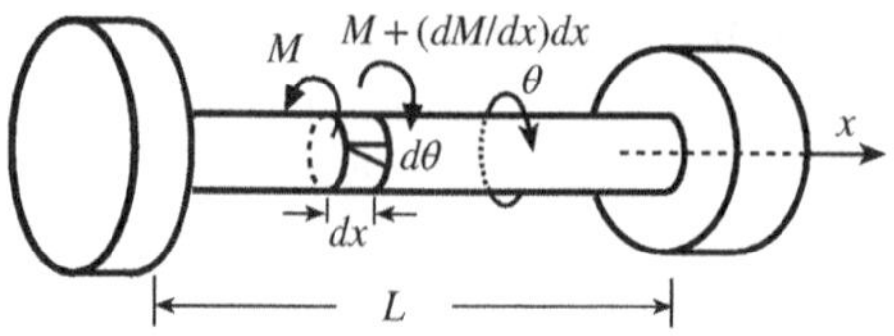

Figure 4.14 Torsional deformation of a shaft with rigid end disks (case 8 of Table 4.15)

4.5.4 Noncircular Cross Sections

Shafts with noncircular cross sections distort in plane (plane sections do not remain plane) during torsion, which reduces the torsional moment below that predicted by Equation 4.33 [59, 74, 76]. The angular moment during torsion of a noncircular section is

$$M = GC\, \partial\theta/\partial x \tag{4.34}$$

The torsion constant C has units of length to the fourth power. C is less than the polar area moment of inertia. Values of the torsion constant C for various cross sections are given in Table 4.14.

Thin-walled open sections warp during torsion. Warping is the axial distortion of the cross section out of its plane. Gere [78] implies that the effect of warping on the stiffness of thin open cross section is negligible if

$$\frac{1}{i}\frac{t}{D}\frac{L}{D} > 10 \tag{4.35}$$

where $i = 1, 2, 3, \ldots$ is the mode number, t is a typical wall thickness, D is the diameter of the cross section, and L is its length. If the center of mass does not lie on the axis of rotation (shear center) or bearings do not constrain the rotation to be about the center of mass of the shaft (see Section 2.5), then the torsional vibration will be coupled with bending [78, 79].

This equation of motion of free torsional vibration is derived by setting the incremental moment (Eq. 4.39) acting on a small length dx of shaft equal to the angular momentum of that incremental length (Eq. 2.29) and neglecting warping [2, 73]:

$$\rho I_p \frac{\partial^2\theta(x,t)}{\partial t^2} = CG\frac{\partial^2\theta(x,t)}{\partial x^2} \tag{4.36}$$

Boundary conditions include

$$\text{Free, } \frac{\partial\theta}{\partial x} = 0; \quad \text{fixed, } \theta = 0; \quad \textit{spring, } k\theta = -GC\frac{\partial\theta}{\partial x}; \quad \text{inertial mass, } -J\frac{\partial^2\theta}{\partial t^2} = GC\frac{\partial\theta}{\partial x} \tag{4.37}$$

The negative signs in the equations for the spring and inertial mass apply for boundaries on the end of the shaft facing the $+x$ direction; they are replaced with (+) signs for boundaries in the $-x$ direction. The solutions to Equation 4.36 and the boundary conditions of Equation 4.37 are given in Table 4.15. These solutions assume that the center of mass of the shaft, the shear center, and the centroid of end conditions are all on the x-axis and there is no warping. Additional solutions are in Table 3-1 and Appendix A.

4.6 Wave Propagation in Beams

Lateral vibration of strings (Eq. 3.31, beam torsion vibration (E.q. 4.35), shear beam vibration (Eq. 4.27), longitudinal vibration of beams (Eq. 4.32) and acoustic pressure ducts (Eq. 6.14) are described by the *wave equation*. The wave equation has *propagating wave* solutions that propagate energy longitudinally. Their wave shape $Y(x, t)$ holds constant with the wave parameter $kx - \omega t$ as the wave propagates in time (t) and space (x).

$$Y(x,t) = Y_o \cos(kx - \omega t) \tag{4.38}$$

Y_o = wave displacement amplitude, $\omega = 2\pi f =$ wave circular frequency in radians per second, $f =$ wave frequency in Hertz, $k = \omega/c = 2\pi f/c = 2\pi/\lambda =$ wave number, $\lambda =$ wave length, the distance between wave crests. The propagation *wave speed c* is independent of frequency, which is called *non dispersive* propagation. Wave speed is found by inserting Eq 4.38 into the wave equation. The results are as follows

Case	Lateral Vibration of String	Longitudinal Vibration of Rod	Torsion Vibration of Rod	Lateral Vib. of Shear Beam	Acoustic Waves
Wave speed, c	$(P/m)^{1/2}$	$(E/\rho)^{1/2}$	$(CG/\rho I_p)^{1/2}$	$(KAG/m)^{1/2}$	$(B/\rho)^{1/2}$

E is modulus of elasticity, P is axial tension, K is shear coefficient and m is mass per unit length. See Sections 3.4 and 6.2. Their standing wave (modal) solutions in Tables 4.12, 4.13, and 4.15 are the same when adjusted for wave speed.

The equation of motion of flexure waves in beams is not the wave equation but it does support wave propagation. Substituting Equation 4.38 into the equation of motion of lateral deformation of a slender uniform beam, Eq. 4.5, results in the wave speed of flexure waves.

$$c\big|_{\substack{\text{flexure}\\ \text{beam}}} = \frac{2\pi}{\lambda}\sqrt{\frac{EI}{m}} = (2\pi f)^{1/2}\left(\frac{EI}{m}\right)^{1/4} \tag{4.39}$$

The two expressions are equivalent through the relationship $\lambda = c/f$. Flexure waves are a function of the wave frequency: they are *dispersive*. Beam flexure wave speed increases without limit as frequency increase and wavelengths decrease.

Equation 4.6 generalizes this result by including rotary inertia and tensile axial load P.

$$c\big|_{\substack{\text{flex+shear}\\ \text{+axial load}}} = \left(\frac{\left(\frac{2\pi}{\lambda}\right)^2 \frac{EI}{m} + \frac{P}{m}}{1 + \left(\frac{2\pi}{\lambda}\right)^2 \frac{I}{A}}\right)^{1/2} = \begin{cases} (P/m)^{1/2}, & \lambda >> I/A, E = 0 \\ 2\pi/\lambda)(EI/m)^{1/2}, & \lambda >> I/A \\ (E/\rho)^{1/2}, & \lambda << I/A \end{cases} \tag{4.40}$$

Limit cases are on the right hand side. ρ is material density and A is cross sectional area. As the wavelength becomes short, on the order of the cross section radius of gyration $r = I/A = \rho I/m$, Eq. 1.6, the flexure wave speed is limited by rotary inertia to $(E/\rho)^{1/2}$, which is the longitudinal wave speed. A summary of wave propagation speed in structures and fluids is in Table 6.1.

Example 4.4 Wave Propagation Speed

Compute the longitudinal, shear, torsion and flexure wave speeds for the tube in Example 4.1.

Solution

Parameters for the 25 mm outside diameter tube in Example 4.1 are as follows.

$I = 0.02668$ in.^{4}(11100 mm^4), $A = 0.2547$ in^2, $m = 0.0736$ lb/in. (1.314 gm/mm), $E = 26400000$ psi (182.0E9 Pa), $\rho = 0.289$ lb/in^3 (8 gm/cc)

The longitudinal wave speed and the flexure wave speed, at 20 Hz, are computed.

$$\text{longitudinal wave speed} = \left(\frac{E}{\rho}\right)^{1/2} = \begin{cases} \left[26.4\times10^6\ \text{psi}/\left(0.289\,\text{lb/in}^3/386.1\,\text{in/s}^2\right)\right]^{1/2} \\ [182\times10^9\ \text{Pa}/(8000\ \text{kg/m}^3)]^{1/2} \end{cases}$$

$$= 4770\ \text{m/s}\ (188\times10^3\,\text{in./s};\ 15{,}650\ \text{ft}/s)$$

$$\begin{array}{c}\text{transverse}\\ \text{wave speed}\end{array} = (2\pi f)^{1/2}\left(\frac{EI}{m}\right)^{1/4}$$

$$= \begin{cases} (2\pi 20\,\text{Hz})^{1/2}[26.4\times10^6\,\text{psi}\times0.02668\,\text{in}^4/(0.0736\,\text{lb/in}/386.1\,\text{in/s}^2)]^{1/4} \\ (2\pi 20\,\text{Hz})^{1/2}[182\times10^9\,\text{Pa}\times1.11\times10^{-8}\,\text{m}^4/(1.314\,\text{kg/m})]^{1/4} \end{cases}$$

$$= 70.2\,\text{m/s}\,(2760\,\text{in./s};\ 230\,\text{ft/s})$$

The transverse (flexure) wave speed in beams is much less that the longitudinal wave speed at low frequencies. The ratio of the flexure wave speed to the longitudinal wave speed for a uniform beam is $2\pi r_g/\lambda$ where the radius of gyration of the section, $r_g = (I/A)^{1/2}$.

4.7 Curved Beams, Rings, and Frames

The bending mode shapes of curved beam vibration modes are inherently more complex than the mode shapes of straight beams because curvature couples longitudinal and transverse motions. As a result, the bending natural modes of vibration of curved beams are two- and three-dimensional (Figure 4.14).

4.7.1 Complete Rings

Table 4.16 [2, 63–66, 80, 81] shows the natural frequencies and mode shapes of complete circular slender rings. Shear deformation is neglected and the cross sections are assumed to be symmetric about the x- and y-axes through the cross section (Fig. 4.15) so that the

Table 4.16 Natural frequencies of circular rings

Notation: C = torsion constant, Table 4.14; E = modulus of elasticity; G = E/[2(1+ν)], shear modulus; f = natural frequency, Hz; i = 1,2,3.. modal index; I_x, I_y = area moment of inertia of cross section about x and y axes, Table 1.5; IP = polar moment of inertia of cross section about z axis; K = shear coefficient, Table 4.11; R = radius to centroidal axis of ring; m = mass per unit circumferential length, including non-structural mass; x = coordinate in plane of ring toward center of ring; y = coordinate transverse to plane of ring; z = coordinate in circumferential direction, tangent to ring; c = angular position along ring; ρ = material mass density; ν = Poisson's ratio; s = angular twist of ring about z axis. Overbar (~) denotes mode shape. Table 1.2 has consistent sets of units. See Fig. 4.15. Refs [2, 63–66, 80, 81].

Description of Ring	Natural Frequency, f_i, Hz	Mode Shape
1. Extension Modes of Ring 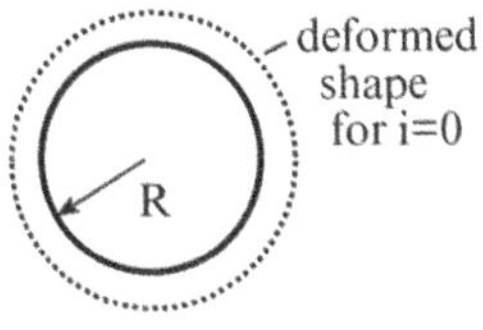 	$\frac{1}{2\pi R}\left(\frac{E}{\rho}\right)^{1/2}$, i = 0 (shown) $\frac{(1+i^2)^{1/2}}{2\pi R}\left(\frac{E}{\rho}\right)^{1/2}$, i = 0,1,2,3.. Extension modes of slender rings have higher natural frequencies than flexure modes.	$\begin{bmatrix}\tilde{x}\\ \tilde{y}\\ \tilde{z}\\ \tilde{\theta}\end{bmatrix}_i = \begin{bmatrix}\cos i\alpha\\ 0\\ -i\sin i\alpha\\ 0\end{bmatrix}$ Ref. [80]
2. Torsion Modes of Ring 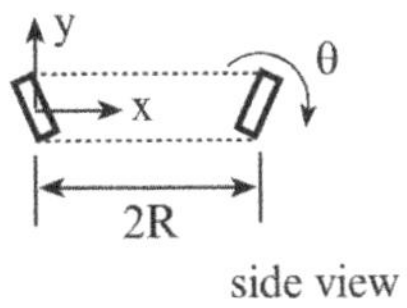 	$\frac{1}{2\pi R}\left(\frac{i^2CG+I_xE}{\rho I_{zz}}\right)^{1/2}$, i = 0,1,2,3..	$\begin{bmatrix}\tilde{x}\\ \tilde{y}\\ \tilde{z}\\ \tilde{\theta}\end{bmatrix}_i = \begin{bmatrix}0\\ \varepsilon\\ 0\\ \cos i\alpha\end{bmatrix}$ $\varepsilon \ll 1$; ε = 0 for i = 0 Refs [2, 80]
	For circular cross sections, $\frac{(i^2+\nu+1)^{1/2}}{2\pi R}\left(\frac{G}{\rho}\right)^{1/2}$, i = 0,1,2,3..	
3. In-Plane Flexure Modes 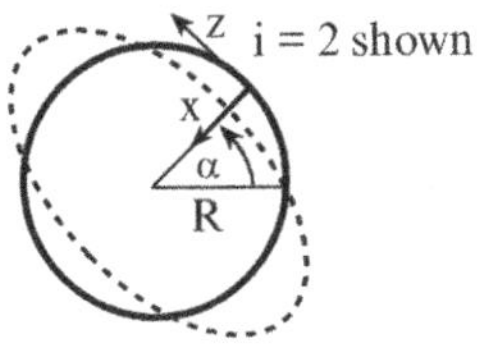 	$\frac{i(i^2-1)}{2\pi R^2(i^2+1)^{1/2}}\left(\frac{EI_y}{m}\right)^{1/2}$ i = 1,2,3.. i = 1 is a rigid body mode. y is out of plane.	$\begin{bmatrix}\tilde{x}\\ \tilde{y}\\ \tilde{z}\\ \tilde{\theta}\end{bmatrix}_i = \begin{bmatrix}i\cos i\alpha\\ 0\\ \sin i\alpha\\ 0\end{bmatrix}$ Refs [2, 80]
4. Out-of-Plane Flexure Modes 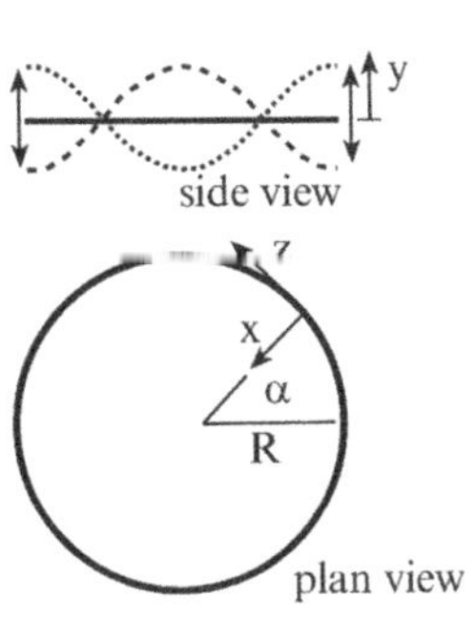 	$\frac{i(i^2-1)}{2\pi R^2}\left(\frac{EI_x}{m[i^2+EI_x/(GC)]}\right)^{1/2}$ i = 1,2,3,.. Refs [2, 80]	$\begin{bmatrix}\tilde{x}\\ \tilde{y}\\ \tilde{z}\\ \tilde{\theta}\end{bmatrix}_i = \begin{bmatrix}0\\ \sin i\alpha\\ 0\\ -\frac{i^2}{R}\frac{1+CG/(EI_x)}{1+i^2CG/(EI_x)}\sin i\alpha\end{bmatrix}$
	For circular cross sections, $\frac{i(i^2-1)}{2\pi R^2(i^2+1+\nu)^{1/2}}\left(\frac{EI}{m}\right)^{1/2}$ i = 1,2,3..	$\begin{bmatrix}\tilde{x}\\ \tilde{y}\\ \tilde{z}\\ \tilde{\theta}\end{bmatrix}_i = \begin{bmatrix}0\\ \sin i\alpha\\ 0\\ -\frac{i^2}{R}\frac{2+\nu}{i^2+1+\nu}\sin i\alpha\end{bmatrix}$

Table 4.16 Natural frequencies of circular rings, continued

Description of Ring	Natural Frequency, f_i, Hz	Mode Shape
5. In-Plane Mode of Ring on Massless Elastic Foundation 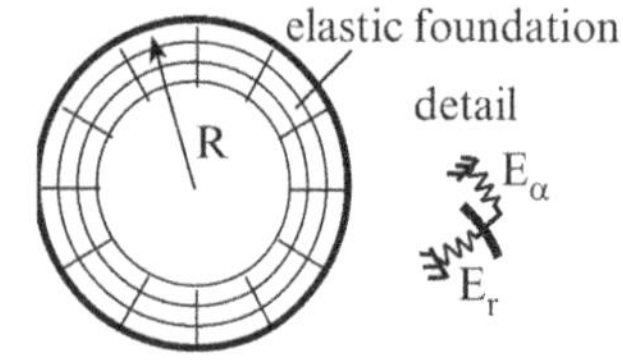 	$\frac{1}{2\pi}\sqrt{\frac{i^2(i^2-1)^2}{R^4(i^2+1)}\frac{EI_y}{m}+\frac{i^2E_r+E_\alpha}{(i^2+1)m}}$ E = modulus of elasticity of ring E_r= radial foundation modulus E_α= circumferential foundation modulus	$\begin{bmatrix}\tilde{x}\\ \tilde{y}\\ \tilde{z}\\ \tilde{\theta}\end{bmatrix}_i=\begin{bmatrix}\cos i\alpha\\ 0\\ -i\sin i\alpha\\ 0\end{bmatrix}$ Ref. [81]
6. In-Plane Mode of Rotating Ring 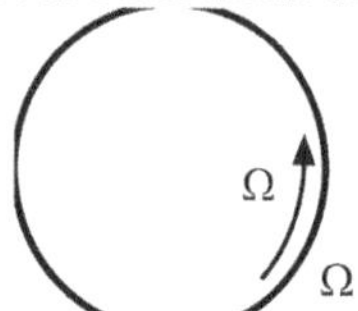 Ω = rotation speed, radian/s	$\frac{1}{2\pi}\left[\frac{2i}{i^2+1}\Omega\pm\sqrt{(2\pi f_i)^2+\frac{i^2(i^2-1)^2}{(i^2+1)^2}\Omega^2}\right]$ $i=1,2,3,4..$ where f_i is natural frequency of case 3 or case 5. + = fwd rotating mode; – = backward rotating mode. Refs [63, 64]	$\begin{bmatrix}\tilde{x}\\ \tilde{y}\\ \tilde{z}\\ \tilde{\theta}\end{bmatrix}_i=\begin{bmatrix}\cos(i\alpha\pm 2\pi f_i t)\\ 0\\ -i\sin(i\alpha\pm 2\pi f_i t)\\ 0\end{bmatrix}$ These are forward and backward traveling waves.
7. In-Plane Mode of Thick Ring 	$\frac{f_i\vert_{case\ 3}}{\left[1+i^2\gamma+\frac{I_y}{AR}\frac{(i^2-1)^2}{(i^2+1)(1+i^2\gamma)}\right]^{1/2}}$ where $\gamma=\frac{I_y}{AR^2}\frac{E}{G}\frac{1}{K}$, K shear coef., Table 4.11	$\begin{bmatrix}\tilde{x}\\ \tilde{y}\\ \tilde{z}\\ \tilde{\theta}\end{bmatrix}_i=\begin{bmatrix}\cos i\alpha\\ 0\\ -i\sin i\alpha\\ 0\end{bmatrix}$ Ref. [65]
8. In-Plane Mode of Elliptical Ring 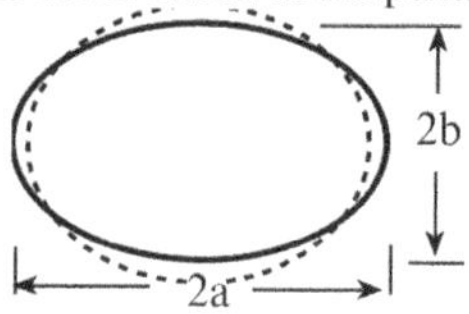 	fundamental mode $\frac{\lambda_1^2}{2\pi^2}\left(\frac{EI_y}{m}\right)^{1/2}$, Hz a/b 1.0 1.1 1.2 1.4 1.7 2.0 2.5 3.0 λ_1 1.638 1.558 1.481 1.342 1.167 1.028 0.8528 0.7271	Ref. [66]
9. In-Plane Flexure Modes Arc with circumferentially Sliding/Pinned Ends i=2 anti-symmetric mode 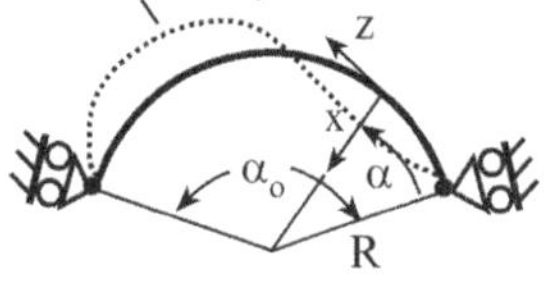 $\tilde{z}=d^2\tilde{z}/d\alpha^2=0, \alpha=0,\alpha_o$ $\tilde{x}\neq 0$ at $\alpha=0,\alpha_o$	$\frac{i^2\pi^2}{2\pi(\alpha_o R)^2}\left[\frac{\left\{1-\left(\frac{\alpha_o}{i\pi}\right)^2\right\}^2}{1+\left(\frac{\alpha_o}{i\pi}\right)^2}\right]^{1/2}\left(\frac{EI_y}{m}\right)^{1/2}$, Hz $i=1,2,3,.....$ Based on case 3 of Table 4.16	

Table 4.16 Natural frequencies of circular rings, continued

Description of Ring	Natural Frequency, f_i, Hz
10. Ring on 2i Equally Spaced Supports 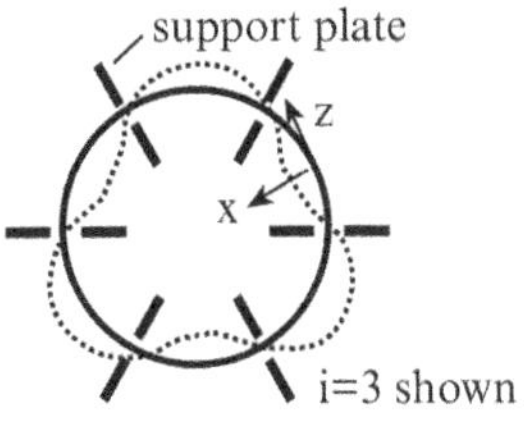 y out-of-plane	For rings with 2i equally spaced supports, i=2,3,4,.., 1) If support plate restricts out-of-plane (y) displacement but not rotation, fundamental out-of-plane mode is given by *ith* out-of-plane flexural mode, case 4 2) If supports restricts *both* out-of plane (y) and tangential disp. (z), th fundamental in-plane flexure mode is given by the 2i mode (case 3). 3) If the supports restricts the radial displacement (x) but not the tangential displacement (z), then the fundamental inplane mode is t the i mode, case 3. Author's result
11. Helix with Multiple, Equally Spaced Supports 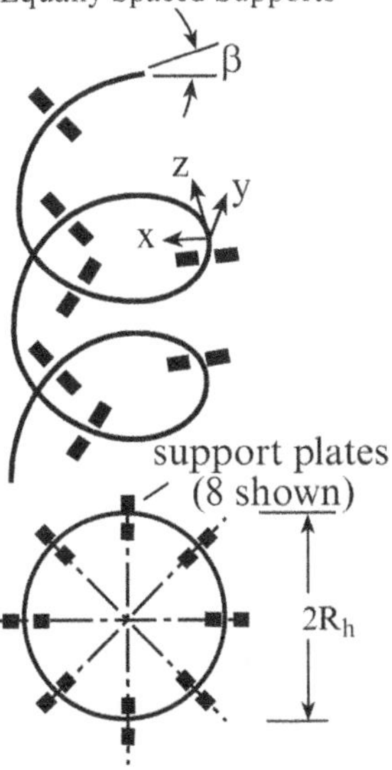	The effective radius of the helix is $R = R_h(1+\beta^2)^{1/2}$ β = helix angle, radians 1) If support plates restrict the out-of-plane (y) displacement at the supports, the fundamental out-of-plane mode of a helix with 2i equally spaced supports per turn is given by ith out-of-plane flexural model of case 4. 2) If the supports restricts both out-of-plane motion (y) and the tangential displacement (z), then the fundamental in-plane mode is given by the i/2 mode of case 3. 3) For odd number of supports use i = (number of supports per turn)/2. 4) If the supports prevent both in and out-of-plane displacement and twist, use case 1 with i=1 for the fundamental in-plane mode.
12. In plane Flexure Modes Pressurized Ring i = 2 shown 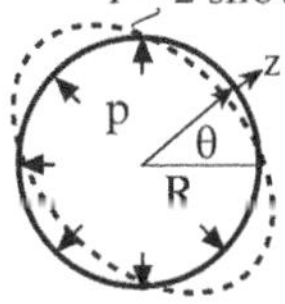	$f_i = \frac{i(i^2-1)}{2\pi R^2(i^2+1)^{1/2}}\left(\frac{EI_y}{m}\right)^{1/2} + \frac{1}{2\pi R}\left(\frac{i^2(i^2-1)}{i^2+1}\right)^{1/2}\left[\frac{pR}{m}\right]^{1/2}$ $i = 1,2,3..$ Mode shape given in case 3. Also see case 14 of Tabl‹3.7 '. i =1 is a rigid body mode. Ref. [82]. p = internal pressure positive outward.

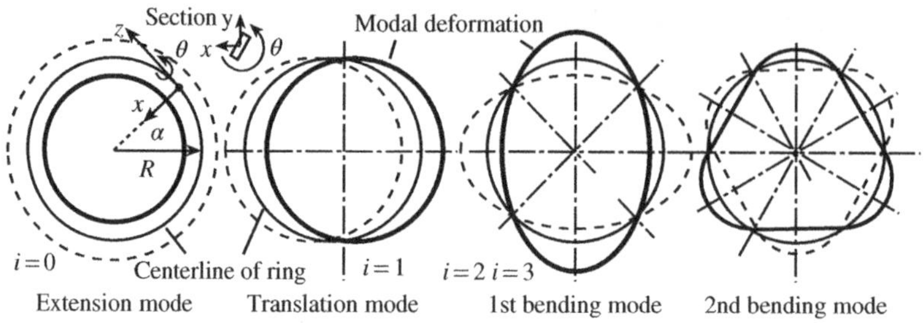

Figure 4.15 In-plane modes of a complete ring

torsional and flexural modes are uncoupled. It is also assumed that the cross sections do not warp in torsional modes.

The ring modes are classified by their mode shape [2, 80]:

Case 1: *Extensional modes*, the ring stretches circumferentially
Case 2: *Torsional modes*, twist about the ring's axis
Case 3: *In-plane flexural modes*, bending in the plane of the ring
Case 4: *Out-of-plane flexural modes*, out-of-plane bending

The mode number i is the number of complete modal waves around the circumference of the ring. Figure 4.15 shows that the number of equally spaced vibration nodes is equal to $2i$. $i = 0$ is a radial expansion for the extension mode and uniform twist for the torsion mode (case 1 of Table 4.16). The mode $i = 1$ is a rigid body translation and $i = 2, 3, 4, \ldots$ are flexural modes (cases 3 and 4 of Table 4.16). The flexural modes are inextensional.

Low-frequency, in-plane inextensional, flexural modes bend but do not extend (stretch) along the axis of the arc, which requires that radial deformations (X) and circumferential deformations (Z) are related by $X + \partial Z/\partial\theta = 0$ where θ is the circumferential angle [78]. For example, if the circumferential deformation is $\sin(i\theta)$, then the radial deformation will be $i\ \cos(i\theta)$.

If a ring has a circular cross section, then the area moments of inertia about two perpendicular axes are equal $I_x = I_y$, and the torsion constant is $C = 2\ I_x$, which simplifies the formulas for natural frequency (cases 2 and 4 of Table 4.16). The in-plane and out-of-plane natural frequencies of a ring with a circular cross section are within 2.96% for the fundamental bending mode, $i = 2$, and 1.5% for the $i = 3$ mode.

The ring on a massless elastic foundation (case 5 of Table 4.16) is very similar to the straight beam on an elastic foundation (Fig. 4.2b and Eq. 4.14). A massless elastic foundation raises the ring natural frequencies. The foundation modulus E_r is defined as the ratio of the load per unit circumference on the foundation to the radial deformation it produces, and E_θ is the ratio of the circumferential shear load per unit length of the foundation to the circumferential displacement it produces. This is a model for a tire.

4.7.2 *Stress and Strain of Arcs*

The stress–strain–deformation relationships for slender circular rings and circular arcs with radius R are as follows [78]. For *extension modes*, the radial inward deflection is X. The circumferential strain (ε), stress (σ), and load (N) are

$$\varepsilon_z = -\frac{X}{R},\ \sigma_{zz} = -\frac{EX}{R},\ N_z = \int_A \sigma_{zz} dA = -\frac{EA}{R} \tag{4.41}$$

For *torsional* modes of rings with a circular cross section,

$$\begin{aligned}
&\varepsilon_z = \frac{\theta y}{R},\ \sigma_{zz} = \frac{E\theta y}{R},\ \varepsilon_{z\theta} = \left(\frac{r}{R}\right)\frac{\partial\theta}{\partial\alpha},\ \varepsilon_{z\theta} = \frac{G(r/R)\partial\theta}{\partial\alpha}\\
&M_z = \int_A \sigma_{zz} y dA = \frac{EI_x\theta}{R},\ M_\theta = \int_A \sigma_{r\theta} r dA = \frac{(GC/R)\ \partial\theta}{\partial\alpha¶}
\end{aligned} \tag{4.42}$$

θ is the torsional deformation and C is the torsion constant (Table 4.14). y is the out of plane within the cross section, α is the circumferential coordinate (Fig. 4.16), and M denotes the bending moment about the subscripted axis. For *in-plane* flexural modes,

$$\varepsilon_z = \left(\frac{x}{R^2}\right)\frac{\partial^2\theta}{\partial\alpha^2},\ \sigma_{zz} = E\varepsilon_z,\ M_z = \int_A \sigma_{zz} x dA = \left(\frac{EI_y}{R^2}\right)\frac{\partial^2 X}{\partial\alpha^2} \tag{4.43}$$

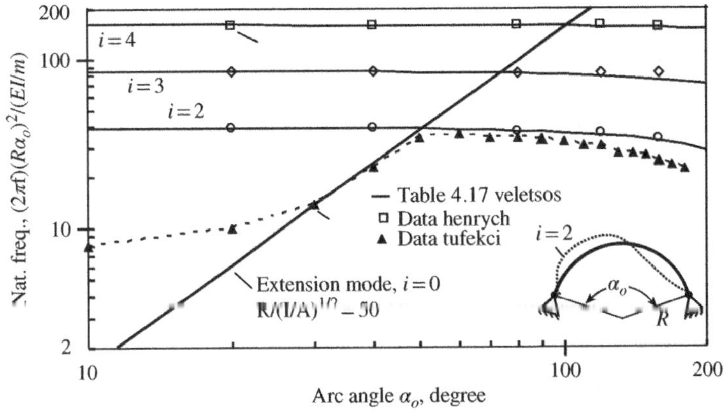

Figure 4.16 In-plane natural frequency of a circular arc with pinned end with circumferential constraint. Cases 1 and 2 of Table 4.17 in comparison with Henrych [83] and Tufekci [84]

For *out-of-plane* flexural modes, with a circular cross section where r is radius from the centroid of the section,

$$
\begin{aligned}
\varepsilon_z &= \frac{\theta y}{R} - \left(\frac{y}{R^2}\right)\frac{\partial^2\theta}{\partial\alpha^2}, \quad \sigma_{zz} = E\varepsilon_z, \\
\varepsilon_{z\theta} &= \left(\frac{r}{R}\right)\left[\frac{\partial\theta}{\partial\alpha} + \left(\frac{1}{R}\right)\frac{\partial Y}{\partial\alpha}\right], \quad \sigma_{z\theta} = G\varepsilon_z, \\
M_z &= \int_A \sigma_{zz} yA = \left(\frac{EI_y}{R}\right)\left(\theta - \left(\frac{1}{R}\right)\frac{\partial^2 Y}{\partial\alpha^2}\right) \\
M_\theta &= \int_A \sigma_{r\theta} rdA = \left(\frac{GC}{R}\right)\left\{\frac{\partial\theta}{\alpha} + \left(\frac{1}{R}\right)\frac{\partial Y}{\partial\alpha}\right\}
\end{aligned} \tag{4.44}
$$

These relationships can be applied mode by mode to the X, Y, and θ deformations in Table 4.16 to determine stress, strain, and internal loads in the ring. Results for a unit modal deformation can be factored to a known deformation.

4.7.3 *Supported Rings and Helices*

Cases 10 and 11 of Table 4.16 were adapted from the exact solutions for complete rings. For ring in-plane flexural modes, the $2i$ equally spaced nodes, without transverse displacement, can be considered to be pinned supports that suppress transverse motion but allow axial motion. The *multiply-supported helix*, which arises in heat exchangers where a tube is coiled through drilled plates arranged in radial pattern about a cylindrical core, is shown in case 11. An effective radius of the helix $R(1+\beta^2)^{1/2}$, where β is the helix angle (case 7 of Table 2.1) and R is the plan view ring radius, can be used to correct the ring analogy for the developed length of the helix.

The angle between nodes in a vibrating ring is $\pi/(2i)$ radians where $i = 1, 2, 3, \ldots$ is the mode number. An arc with radius R subtends angle β_o. Scaling the ring so nodes are at the ends of the arc gives $\beta_o/k = 2\pi/(2i)$, $k = 1, 2, 3, \ldots$, or $i = (\pi/\beta_o)k$, where i is no longer an integer and k is the arc mode number. The resultant arc solution is presented in case 9 of Table 4.16 and case 3 of Table 4.17. The analysis is exact if β_o is an even integer submultiple of π. Note that there is tangential movement at the ends of these arcs, just as there is Tangential movement of nodes with flexural modes of complete rings.

4.7.4 *Circular Arcs, Arches, and Bends*

Table 4.17 (Refs [60–62]) shows the natural frequencies and mode shapes of homogeneous, slender, circular arcs and bends. Shear deformation, rotary inertia, and warping are neglected. The torsional and flexural deformations are not coupled, which is the case if the cross section is symmetric about the x- and y-axes.

The notation for displacements of arcs is identical to that of complete rings (Figure 4.15): X is a radial displacement, positive inward; Y is an out-of-plane displacement; Z is a tangential displacement; and θ is a rotation about the z-axis. The boundary conditions for

Table 4.17 Natural frequencies of arcs, bends, and frames

Notation: C = torsion constant, Table 4.15; E = modulus of elasticity; G = E/[2(1+ν)], shear modulus; f = natural frequency, Hz; i = 1,2,3.. modal index; I_x, I_y = area moment of inertia of cross section about x and y axes, Table 1.5; I_{zz} = polar moment of inertia of cross section about z axis; L = length between supports; m = mass per unit circumferential length including non-structural mass; R = radius to center line of arc; x = coordinate in plane of arc toward center of ring; y = coordinate transverse to plane of arc, out-of-plane of paper; z = coordinate in circumferential direction, tangent to arc; α = angular position; α_o = subtended angle, radian; ρ = material mass density; ν = Poisson's ratio; s = angular twist of ring about z axis. X, Y, Z = modal displacements. Table 1.2 has consistent units. Refs [63–66].

Description	Natural Frequency, f_i, Hz	Mode Shape
1. In-Plane Extension Mode ends prevent circ. motion 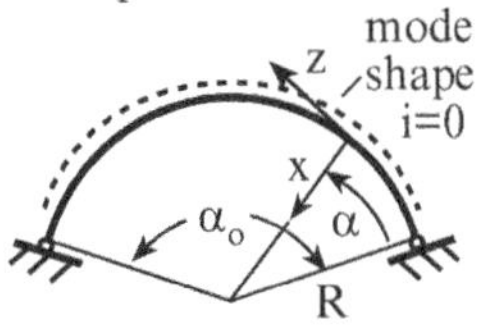 $Z(0)=Z(\alpha_o)=0$	$\frac{1}{2R}\left(\frac{E}{\rho}\right)^{1/2}$, i = 0 (ends slide radially) $\frac{1}{2\alpha_o R}\left[\alpha_o^2+(i\pi)^2\right]^{1/2}\left(\frac{E}{\rho}\right)^{1/2}$, i = 1,2,3,. Modes dominated by longitudinal stretching (extension) along axis of arc	$\begin{bmatrix} X \\ Y \\ Z \\ \theta \end{bmatrix}_i = \begin{bmatrix} \cos i\pi\alpha/\alpha_o \\ 0 \\ -(i\pi/\alpha_o)\sin i\pi\alpha/\alpha_o \\ 0 \end{bmatrix}$ Based on case 1 of Table 4 16 and case 3 of Table 4.13. Ref. [58]
2. In-Plane Flexure Mode Pinned-Pinned Arc 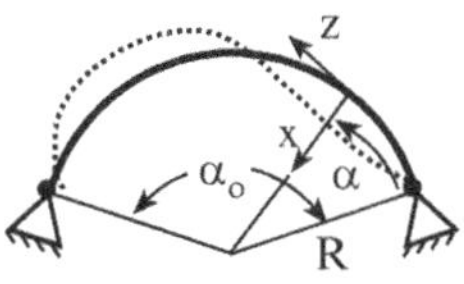 i=2 anti-symmetric mode	i=2,4,6,. X anti-symmetric, Z symmetric $\frac{i^2\pi^2}{2\pi(\alpha_o R)^2}\left[\frac{\left\{1-\left(\frac{\alpha_o}{i\pi}\right)^2\right\}^2}{1+3\left(\frac{\alpha_o}{i\pi}\right)^2}\right]^{1/2}\left(\frac{EI_y}{m}\right)^{1/2}$	$\begin{bmatrix} X \\ Y \\ Z \\ \theta \end{bmatrix}_i = \begin{bmatrix} \sin i\pi\alpha/\alpha_o \\ 0 \\ \frac{\alpha_o}{i\pi}(1-\cos\frac{i\pi\alpha}{\alpha_o}) \\ 0 \end{bmatrix}$ Ref. [59]
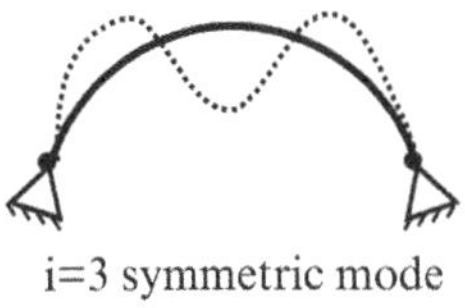 i=3 symmetric mode $Z = d^2X/d\alpha^2 = 0,\ \alpha = 0,\alpha_o$ $X = 0$ at $\alpha = 0,\alpha_o$ Ends prevent displacement	i=3,5,7,. X symmetric, Z anti-symmetric, $\frac{i^2\pi^2}{2\pi(\alpha_o R)^2}\left[\frac{\left\{1-\left(\frac{\alpha_o}{i\pi}\right)^2\right\}^2}{1+\frac{1}{i^2}+2\left(\frac{\alpha_o}{i\pi}\right)^2}\right]^{1/2}\left(\frac{EI_y}{m}\right)^{1/2}$ Use case 1 for i = 1.	$\begin{bmatrix} X \\ Y \\ Z \\ \theta \end{bmatrix}_i = \begin{bmatrix} \sin\frac{i\pi\alpha}{\alpha_o}-\frac{1}{i}\sin\frac{\pi\alpha}{\alpha_o} \\ 0 \\ \frac{-\alpha_o}{i\pi}(\cos\frac{i\pi\alpha}{\alpha_o}-\cos\frac{\pi\alpha}{\alpha_o}) \\ 0 \end{bmatrix}$ Ref. [58]
3. In-Plane Flexure Modes with circumferentially Sliding/Pinned Ends i=2 anti-symmetric mode 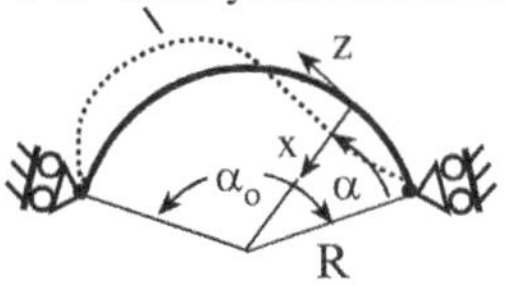	$\frac{i^2\pi^2}{2\pi(\alpha_o R)^2}\left[\frac{\left\{1-\left(\frac{\alpha_o}{i\pi}\right)^2\right\}^2}{1+\left(\frac{\alpha_o}{i\pi}\right)^2}\right]^{1/2}\left(\frac{EI_y}{m}\right)^{1/2}$ i = 1,2,3,..... Ends allow circumferential motion. Based on case 3 of Table 4.16	$\begin{bmatrix} X \\ Y \\ Z \\ \theta \end{bmatrix}_i = \begin{bmatrix} i\cos i\alpha \\ 0 \\ \sin i\alpha \\ 0 \end{bmatrix}$ $Z = d^2X/d\alpha^2 = 0, \alpha = 0,\alpha_o$ $X \neq 0$ at $\alpha = 0,\alpha_o$

Table 4.17 Natural frequencies of arcs, bends, and frames, continued

Description	Natural Frequency, f_i, Hz
4. In-Plane Flexure Mode of Clamped-Clamped Arc i=2 anti-symmetric mode 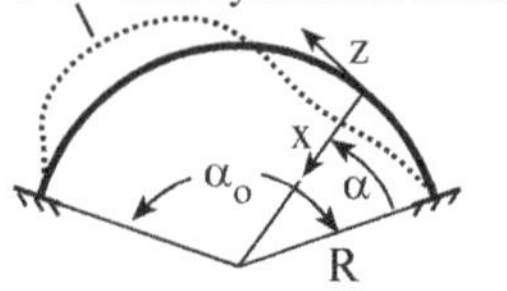 $Z = X = dX/d\alpha = 0$ at $\alpha = 0, \alpha_o$ Ends have circum. restraint. λ_i, σ_i, and Ψ_i are given in case 7 of Table 4.2 with $\Psi_i = \tilde{y}_i$, and substituting "α/α_o" for "x/L". Use case 1 for i =1.	Modes with X anti-symmetric, Z symmetric about midplane Ref. [58] $\frac{\lambda_i^2}{2\pi(\alpha_o R)^2}\left[\frac{1-2\sigma_i^2\left(1-\frac{2}{\sigma_i\lambda_i}\right)\left(\frac{\alpha_o}{\lambda_i}\right)^2+\left(\frac{\alpha_o}{\lambda_i}\right)^4}{1+5\sigma_i^2\left(1-\frac{2}{\sigma_i\lambda_i}\right)\left(\frac{\alpha_o}{\lambda_i}\right)^2}\right]^{1/2}\left(\frac{EI_y}{m}\right)^{1/2}, \quad i=2,4,6,..$ $X_i = \Psi_i,\ Z_i = (\alpha_o/\lambda_i)[(1/\lambda_i^3)d^3\Psi_i/d(\alpha/\alpha_o)^3 + 2\sigma_i],\ Y = \theta = 0$ $\lambda_2 = 7.853, \sigma_2 = 1.001, \lambda_4 = 14.137, \sigma_4 = 1.000$ Modes with X symmetric, Z anti-symmetric about midplane. $\frac{\lambda_i^2}{2\pi(\alpha_o R)^2}\left[\frac{1-1.82\sigma_i^2\left(\frac{\alpha_o}{\lambda_i}\right)^2}{1+\left(\frac{\lambda_1}{\lambda_i}\right)^2+2\left(1.63-\frac{5}{\lambda_i}\right)\left(\frac{\alpha_o}{\lambda_i}\right)^2}\right]^{1/2}\left(\frac{EI_y}{m}\right)^{1/2}, \quad i=3,5,7..$ $X_i = \Psi_i - (\sigma_i/\sigma_1)(\lambda_i/\lambda_1)\Psi_1,\ Y = \theta = 0$ $\lambda_1 = 4.730, \sigma_1 = 0.9825,\ \lambda_3 = 10.995, \sigma_3 = 1.000,\ \lambda_5 = 17.278, \sigma_5 = 1.000$
5. In-Plane Flexure Mode of Clamped-Pinned Arc i=2 anti-symmetric mode 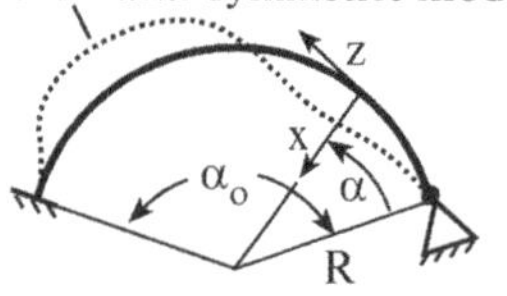 $Z = X = dX/d\alpha = 0$ at $\alpha = \alpha_o$ Ends restrict circumferential motion. λ_i, σ_i, and Ψ_i are given in case 6 of Table 4.2 with $\Psi_i = \tilde{y}_i$, and "α/α_o" for "x/L". Use case 1 for i =1.	Modes with X anti-symmetric, Z symmetric about midplane. $\frac{\lambda_i^2}{2\pi(\alpha_o R)^2}\left[\frac{1-\sigma_i^2\left(1-\frac{1}{\sigma_i\lambda_i}\right)\left(\frac{\alpha_o}{\lambda_i}\right)^2+\left(\frac{\alpha_o}{\lambda_i}\right)^4}{1+5\sigma_i^2\left(1-\frac{1}{\sigma_i\lambda_i}\right)\left(\frac{\alpha_o}{\lambda_i}\right)^2}\right]^{1/2}\left(\frac{EI_y}{m}\right)^{1/2}, \quad i=2,4,6,..$ $X_i = \Psi_i,\ Z_i = (\alpha_o/\lambda_i)[(1/\lambda_i^3)d^3\Psi_i/d(\alpha/\alpha_o)^3 + 2\sigma_i],\ Y = \theta = 0$ $\lambda_2 = 7.068, \sigma_2 = 1.001, \lambda_4 = 13.351, \sigma_4 = 1.000.$ based on Ref. [58]. Modes with X symmetric, Z anti-symmetric about midplane. $\frac{\lambda_i^2}{2\pi(\alpha_o R)^2}\left[\frac{1-2\sigma_i^2\left(\frac{\alpha_o}{\lambda_i}\right)^2}{1+\left(\frac{\lambda_1}{\lambda_i}\right)^2+2\left(1.31-\frac{4}{\lambda_i}\right)\left(\frac{\alpha_o}{\lambda_i}\right)^2}\right]^{1/2}\left(\frac{EI_y}{m}\right)^{1/2}, \quad i=3,5,7..$ $X_i = \Psi_i - (\sigma_i/\sigma_1)(\lambda_i/\lambda_1)\Psi_1,\ Y = \theta = 0$ $\lambda_1 = 3.927, \lambda_3 = 10.21, \sigma_1 = \sigma_3 = 1.000, \lambda_5 = 16.491, \sigma_5 = 1.000$

Table 4.17 Natural frequencies of arcs, bends, and frames, continued

Description	Natural Frequency, f_i, Hz	Mode Shape
6. Out-of-Plane Flexure Mode of Pinned-Pinned Arc First Mode 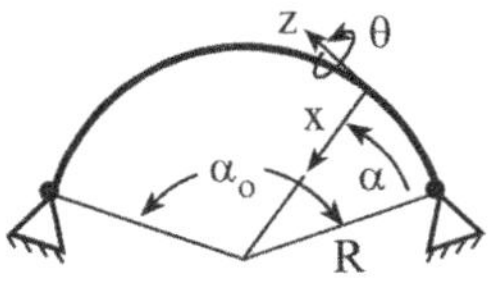 	$\frac{\pi^2}{2\pi(\alpha_o R)^2}\left[\frac{\frac{GC}{EI_x}\left\{1-\left(\frac{\alpha_o}{\pi}\right)^2\right\}^2}{\left(\frac{\alpha_o}{\pi}\right)^2+\frac{GC}{EI_x}}\right]^{1/2}\left(\frac{EI_x}{m}\right)^{1/2}$ $0<EI_x/CG<2$ Ref. [60]	$\begin{bmatrix} X \\ Y \\ Z \\ \theta \end{bmatrix}_i = \begin{bmatrix} 0 \\ \sin \pi\alpha/\alpha_o \\ 0 \\ \frac{\frac{-1}{R}\left(1+\frac{GC}{EI_x}\right)}{\left(\frac{\alpha_o}{\pi}\right)^2+\frac{GC}{EI_x}}\sin\frac{\pi\alpha}{\alpha_o} \end{bmatrix}$
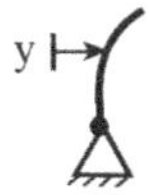 side view Fundamental mode only $\theta = 0$ at $\alpha = 0, \alpha_o$ $X = \frac{d^2X}{d\alpha^2} = 0$ at $\alpha = 0, \alpha_o$	For circular sections $GC/EI_x=1/(1+\nu)$, $\frac{\pi^2}{2\pi(\alpha_o R)^2}\left[\frac{\left\{1-\left(\frac{\alpha_o}{\pi}\right)^2\right\}^2}{1+(1+\nu)\left(\frac{\alpha_o}{\pi}\right)^2}\right]^{1/2}\left(\frac{EI_x}{m}\right)^{1/2}$	$\begin{bmatrix} X \\ Y \\ Z \\ \theta \end{bmatrix}_i = \begin{bmatrix} 0 \\ \sin \pi\alpha/\alpha_o \\ 0 \\ \frac{-\frac{(2+\nu)}{R}\sin\frac{\pi\alpha}{\alpha_o}}{1+(1+\nu)\left(\frac{\alpha_o}{\pi}\right)^2} \end{bmatrix}$
7. Out-of-Plane Flexure Mode of Clamped-Clamped Arc, First Mode. 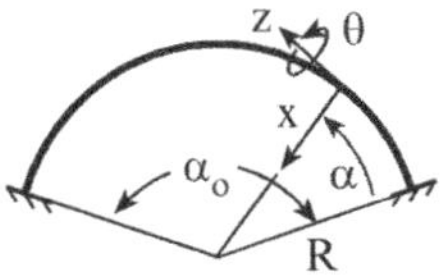  side view Fundamental Mode	$\frac{\pi^2}{2\pi(\alpha_o R)^2}\left[\frac{3.586\left(\frac{\alpha_o}{\pi}\right)^2+1.246\frac{GC}{EI_x}\beta}{\left(\frac{\alpha_o}{\pi}\right)^2+1.246\frac{GC}{EI_x}}\right]^{1/2}\left(\frac{EI_x}{m}\right)^{1/2}$ $0<EI_x/GC<2$. For circular cross sections, $GC/EI_x=1/(1+\nu)$. Ref. [60]	$\begin{bmatrix} X \\ Y \\ Z \\ \theta \end{bmatrix}_1 = \begin{bmatrix} 0 \\ \Psi_1(\alpha/\alpha_o) \\ 0 \\ \frac{-(1+GC/EI_x)(\Psi_1/R)}{0.8025(\alpha_o/\pi)^2+GC/EI_x} \end{bmatrix}$, $\beta=(\alpha_o/\pi)^4-2.49(\alpha_o/\pi)^2+5.139$ Ψ_1 is given in case 7 of Table 4.3 with $\Psi_1=Y_1$, and $\alpha/\alpha_o=x/L$.
8. Out-of-Plane Flexure Mode of Clamped-Pinned Arc, First Mode. 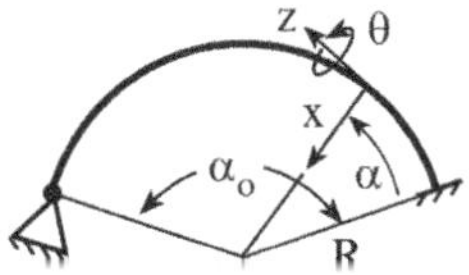  side view Fundamental Mode	$\frac{\pi^2}{2\pi(\alpha_o R)^2}\left[\frac{1.080\left(\frac{\alpha_o}{\pi}\right)^2+1.166\frac{GC}{EI_x}\beta}{\left(\frac{\alpha_o}{\pi}\right)^2+1.166\frac{GC}{EI_x}}\right]^{1/2}\left(\frac{EI_x}{m}\right)^{1/2}$ $0<EI_x/GC<2$. For circular cross sections, $GC/EI_x=1/(1+\nu)$. Ref. [60]	$\begin{bmatrix} X \\ Y \\ Z \\ \theta \end{bmatrix}_1 = \begin{bmatrix} 0 \\ \Psi_1(\alpha/\alpha_o) \\ 0 \\ \frac{-(1+GC/EI_x)(\Psi_1/R)}{0.8576(\alpha_o/\pi)^2+GC/EI_x} \end{bmatrix}$, $\beta=(\alpha_o/\pi)^4-2.32(\alpha_o/\pi)^2+2.44$ Ψ_1 is given in case 6 of Table 4.2 with $\Psi_1=Y_1$, and $\alpha/\alpha_o=x/L$.

Table 4.17 Natural frequencies of arcs, bends, and frames, continued

Description

Natural Frequency $f_i = \frac{\lambda^2}{2\pi R^2}\left(\frac{EI_y}{m}\right)^{1/2}$, Hz

9. Right Angle Bend, out-of-plane modes

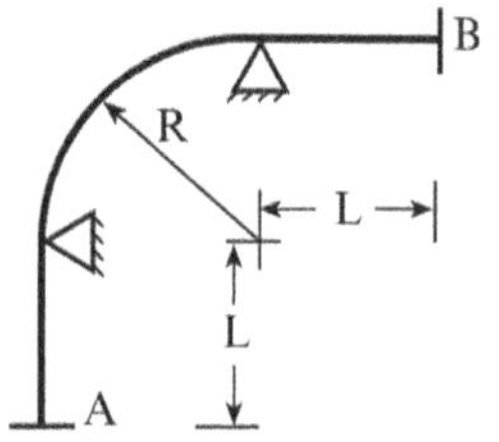

Clamped or pinned ends
P = pinned end in 3 axes.
C = clamped end at A or B
Intermediate constraints are pinned. y out-of-plane
$I_x=I_y$, $\nu=0.3$.

Values of λ^2 Ref. [61]

B.C. A,B	Mode i	L/R 0	0.2	0.4	0.6	0.8	1.0	1.5	2.0
In –plane modes									
P-P	1	23.6	19.7	18.0	16.8	14.1	11.0	5.2	3.2
P-P	2	43.2	38.4	36.0	30.0	19.1	12.5	7.6	2.7
C-P	1	22.6	19.8	18.3	17.3	15.0	14.5	5.6	4.6
C-P	2	43.2	38.9	36.9	31.5	21.2	16.0	8.0	4.4
C-C	1	22.6	20.1	18.9	17.7	16.6	14.1	5.7	4.6
C-C	2	43.2	39.2	36.9	31.5	21.2	16.0	8.0	4.4
Out-of-plane modes									
P-P	1	10.8	9.1	8.25	6.7	6.85	6.3	4.3	2.5
P-P	2	30.9	26.6	24.6	22.0	18.2	13.9	6.8	3.8
C-P	1	10.8	9.3	8.7	8.1	7.3	6.7	4.8	3.0
C-P	2	30.9	27.0	25.2	22.9	19.4	14.9	6.8	5.35
C-C	1	10.8	9.5	8.8	8.3	7.9	7.1	5.7	4.3
C-C	2	30.9	27.4	26.0	24.3	21.9	16.5	9.7	5.8

10. U-Bend, Overhanging

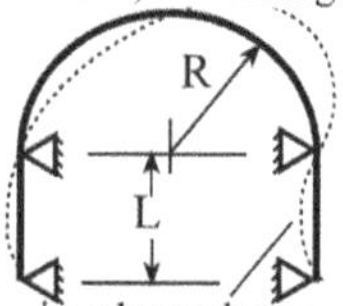

in-plane shape

$I_x = I_y$, $\nu = 0.3$

L/R	0	0.2	0.4	0.6	0.8	1.0	1.5	2.0	mode type
λ_3^2	2.29	4.02	3.75	3.55	3.39	3.25	2.92	2.39	1st in-plane mode
λ_1^2	2.13	1.41	1.35	1.29	1.24	1.19	1.09	0.99	1st out-of-plane mode

Pinned in 3 axes at A and B. y out-of-plane. author's result.
Clamped at A and B for L = 0

11. UBend with Intermediate Support

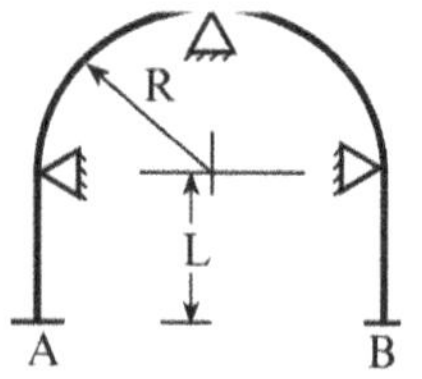

Clamped or pinned ends

P = pinned end
C = clamped end
y out-of-plane
$I_x=I_y$, $\nu=0.3$

Values of λ^2 Ref. [61]

B.C. A,B	Mode	L/R 0	0.2	0.4	0.6	0.8	1.0	1.5	2.0
In –plane modes									
P-P	1	4.4	4.0	3.8	3.6	3.5	3.3	2.9	2.45
P-P	2	17.9	16.8	16.3	15.7	14.7	11.6	6.5	3.6
C-P	1	4.4	4.0	3.85	3.65	3.55	3.45	3.10	2.55
C-P	2	17.9	16.8	16.3	15.7	14.7	11.6	6.5	4.0
C-C	1	4.4	4.0	3.9	3.7	3.68	3.5	3.2	2.9
C-C	2	17.9	16.9	16.4	15.9	15.4	14.4	6.5	4.0
Out-of-plane modes									
P-P	1	6.80	6.20	5.90	5.50	5.30	5.0	4.27	3.13
P-P	2	10.8	10.0	9.40	8.95	8.66	8.21	5.90	3.70
C-P	1	6.80	6.3	6.0	5.7	5.4	5.2	4.45	3.25
C-P	2	10.8	10.0	9.5	9.20	8.8	8.4	6.50	4.85
C-C	1	6.80	6.2	5.90	5.45	5.3	5.0	4.25	3.1
C-C	2	10.8	9.95	9.4	8.95	8.65	8.2	5.9	3.7

Table 4.17 Natural frequencies of arcs, bends, and frames, continued

Description	Natural Frequency, f_i, Hz	Mode Shape, Y(x)
12. Frame with Rigid Cap 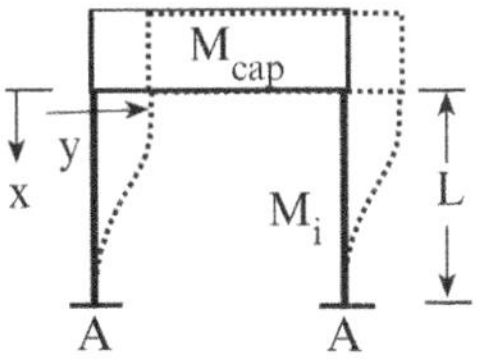 Lateral Mode. Clamped or pinned feet at A. Two legs shown.	Pinned feet at A $$\frac{1}{2}\pi\left[\frac{3\sum_i E_iI_i}{L^3(M_{cap}+0.48\sum_i M_i)}\right]^{1/2}$$	$$\frac{x^3}{6L^3}-\frac{x}{2L}+\frac{1}{2}$$ M_{cap} = total mass of cap M_i = total mass of each leg
	Clamped feet at A $$\frac{1}{2\pi}\left[\frac{12\sum_i E_iI_i}{L^3(M_{cap}+0.37\sum_i M_i)}\right]^{1/2}$$	$$\left(1-\frac{2x}{L}\right)^3-3\left(1-\frac{2x}{L}\right)+2$$ Summation over number of legs. E_i, M_i, I_i = properties of each leg
13. Frame with Rigid Legs 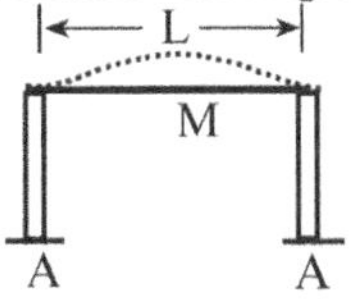	$$\frac{4.7304^2}{2\pi}\left[\frac{EI}{ML^3}\right]^{1/2}$$ M = mass of cap beam Legs either pinned or clamped at A. E, I, and M, are properties of cap beam.	See case 7 of Table 4.2 Vertical symmetric mode.
14. Lateral Asymmetric Frame Mode 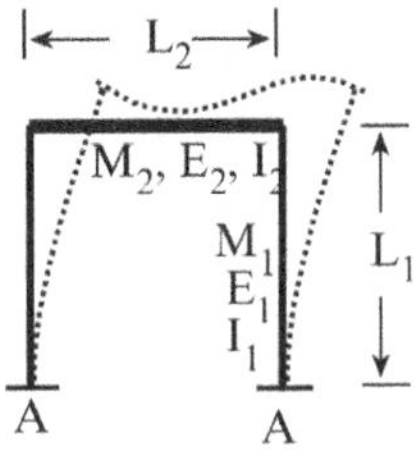	Pinned feet at A $$\frac{1}{2\pi}\left[\frac{6E_1I_1}{L_1^3M_1}\right]^{1/2}\left[\frac{2L_1E_2I_2}{2L_1E_2I_2+L_2E_1I_1}\right]^{1/2}\left[\frac{1}{1+0.97M_2/M_1}\right]^{1/2}$$ Each leg has properties M_1, E_1, I_1, L_1. Cap properties are M_2, E_2, I_2, L_2.	
Clamped or pinned feet A.	Clamped feet at A $$\frac{1}{2\pi}\left[\frac{6E_1I_1}{L_1^3M_1}\right]^{1/2}\left[\frac{12L_1E_2I_2+2L_2E_1I_1}{3L_1E_2I_2+2L_2E_1I_1}\right]^{1/2}\left[\frac{1}{1+0.74M_2/M_1}\right]^{1/2}$$ Author's result based on Table 3.2. See Ref. [62] for tabulated solution.	
15. Vertical Symmetric Frame Mode 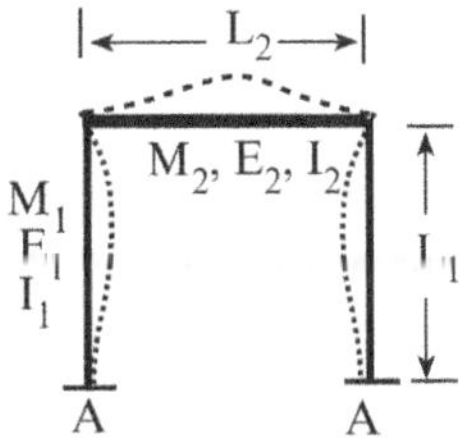	Pinned feet at A $$\frac{1}{2\pi}\left[\frac{E_2I_2}{L_2^3M_2}\right]^{1/2}\left[\frac{384}{5-\frac{4}{[1+2L_1E_2I_2/(3L_2E_1I_1)]}}\right]^{1/2}\left[\frac{1}{1+\frac{M_1}{M_2}}\right]^{1/2}$$ Each leg has properties E_1, I_1, L_1. Cap properties are E_2, I_2, L_2.	
Clamped or pinned feet A	Clamped feet at A $$\frac{1}{2\pi}\left[\frac{E_2I_2}{L_2^3M_2}\right]^{1/2}\left[\frac{384}{5-\frac{4}{1+L_1E_2I_2/(2L_2E_1I_1)}}\right]^{1/2}\left[\frac{1}{1+\frac{M_1}{M_2}}\right]^{1/2}$$ Author's result. See Ref. [62] for tabulated solution.	

in-plane modes, whose motion lies in the plane of the arc, so $Y = 0$, with pinned ends and axial constraint are as follows:

$$\begin{aligned} \text{Pinned – pinned:}\ X = \frac{\partial^2 X}{\partial \alpha^2} &= 0 \quad \text{at} \quad \alpha = 0, \alpha_o \text{ without axial constraint} \\ Z = X = \frac{\partial^2 X}{\partial \alpha^2} &= 0 \text{ at } \alpha = 0, \alpha_o \text{ with axial constraint} \\ \text{Clamped – clamped:}\ Z = X = \frac{\partial X}{\partial \alpha} &= 0 \text{ at } \alpha = 0, \alpha_o \end{aligned} \tag{4.45}$$

Clamped ends do not allow axial motion at the ends of the arc. The boundary conditions for *out-of-plane modes*, that is, modes whose motion is dominantly out of plane of the arc, $Z = 0 = X \approx 0$, are as follows:

$$\begin{aligned} \text{Pinned – pinned:}\ Z = Y = \frac{\partial^2 Y}{\partial z^2} = \frac{\partial^2 \theta}{\partial \alpha^2} &= 0 \quad \text{at } \alpha = 0, \alpha_o \\ \text{Clamped – clamped:}\ Z = Y = \frac{\partial Y}{\partial z} = \frac{\partial \theta}{\partial \alpha} &= 0 \quad \text{at } \alpha = 0, \alpha_o \end{aligned} \tag{4.46}$$

The stress and strain relationships of Equations 4.41–4.464.43 apply to complete rings and arcs. Henrych [83] provides a detailed analysis.

The solutions in cases 7–9 in Table 4.17 for bends with pinned supports are valid for symmetric sections, $I_x = I_y$. The out-of-plane solutions are only valid for $GC/EI_y = 1/(1 + \nu)$, which applies to circular cross sections such as tubes. The natural frequencies of bends with pinned supports are largely a function of the span of the adjacent straight sections if the straight spans are longer than the arc length.

The Rayleigh technique (Appendix A) was used to derive the results in Table 4.17 for arcs with axial constraints using curved beam. The mode shapes are adapted from straight beams, and they do not always satisfy all the boundary conditions. Nevertheless, the formulas in the table are generally within a few percent of the exact result for the lower modes of slender arcs with the arc span between 0 and 180 degrees. Comparison with the numerical methods of Henrych [83] and Tufekci and Araci [84] is made in Figure 4.16 (also see Ref. [85]).

4.7.5 *Lowest Frequency In-Plane Natural Frequency of an Arc*

The lowest in-plane natural frequency of an arc depends on circumferential restraint. If the ends restrain circumferential displacement (*Z*) along the axis of the arc, as in cases 1, 2, 4, and 5 in Table 4.17, then the arc must extend along its own axis to deform radially in the fundamental $i = 1$ half-sine mode shape which is similar to the first longitudinal mode of a fixed–fixed rod (case 3 of Table 4.13) since stiffness is provided by axial elongation (case 7 of Table 3.2). On the other hand, the $i = 2$ full-sine wave mode is an inextensional (no stretching) mode and the stiffness is by bending. Thus, the lowest in-plane natural frequency of a slender arc with axially restrained ends and low bending stiffness is the $i = 2$ flexural mode rather than the $i = 0$ extensional mode as shown in Figure 4.17 [83, 84].

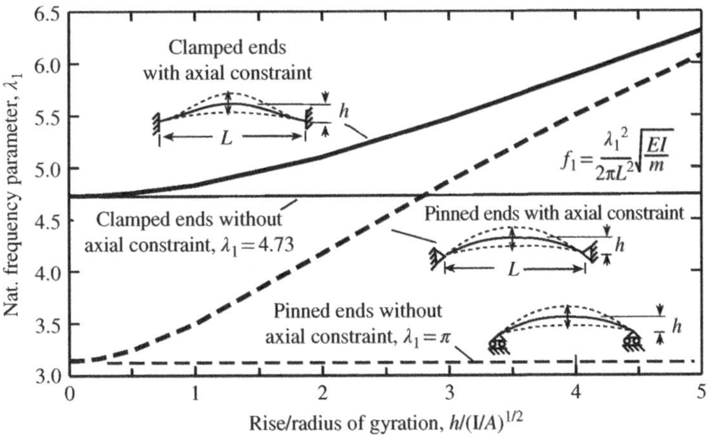

Figure 4.17 Effect of arc rise and axial restraint on transverse in-plane mode of slightly curved beams with pinned and clamped ends (Eq. 4.47)

4.7.6 *Shallow Arc*

A straight beam is curved into a shallow arc, as shown in Figure 4.17 with rise h and in Figure 4.16 with radius R. The rise of the arc, h, above the ends is related to the arc angle by $h = R(1 - \cos\,\theta_o/2)$. The arc radius is $R = (h^2 + L^2/4)/(2h)$. Reissner [86] gives the fundamental natural frequency of the fundamental sinusoidal mode of a slightly curved pinned–pinned beam whose ends are constrained not to move axially:

$$f_1 = \frac{\pi^2}{2\pi L^2}\left[1 + \frac{1}{2}\left(\frac{h}{r_g}\right)^2\right]^{1/2}\left(\frac{EI_y}{m}\right)^{1/2} = \begin{cases} \dfrac{\pi^2}{2\pi L^2}\left(\dfrac{EI_y}{m}\right)^{1/2}, & \text{if } h \ll r_g, \text{ straight beam} \\ \dfrac{1}{2}\dfrac{\pi}{2^{3/2}}\left(\dfrac{EA}{m}\right)^{1/2}, & \text{if } h \gg r_g, \text{ constrained arc} \end{cases} \tag{4.47}$$

As arcs become flat, their natural frequencies and mode shapes approach those of straight beams (Table 4.2). In particular, if the rise of an arc, h, is a small fraction of the radius of gyration r_g (Eq. 1.6) of the arc's cross section, $r_g = (I_y/A)^{1/2}$ where I_y is the moment of inertial and A is the cross section area:

$$\frac{h}{2}\left(\frac{A}{I_y}\right)^{1/2} \sim 1.74\frac{h}{t} \ll 1 \tag{4.48}$$

then the natural frequency of an arc can be predicted using straight beam formulas (Table 4.2). For a rectangular section, r_g is equal to 0.288 times its thickness, so if the rise is less than about one-half times the beam thickness t, the natural frequency of rectangular section arcs can be predicted from the straight beam theory. If the rise is more than 0.288 times the rectangular section thickness, then the in-plane mode natural frequency of the slightly curved beam with axial constraints is greater than the straight beam.

4.7.7 *Portal Frames*

A portal frame consists of two parallel leg beams and a cap beam. If these beams are slender, then their deformation is due to bending, not extension.

The fundamental asymmetric (lateral) in-plane bending modes of a portal frame are given in cases 12 and 14 of Table 4.17. The fundamental in-plane symmetric (vertical) mode is provided in cases 13 and 15. These approximate solutions are based on deformed shape of the frame under static load (Table 3.2). They are in reasonably good agreement with the tabulated exact solutions of Rieger and McCallion [62]. For example, for fixed ends and $L_1/L_2 = E_1I_1/E_2I_2 = (M_1/M_2)^{1/2} = 1.5$, they give $2\pi f(M_1L_1{}^3/E_1I_1)^{1/2} = 1.32^2$ for pinned feet and 1.98^2 for clamped feet as compared with 1.29^2 and 1.89^2 in Table 4.17. Numerical analysis of frames is discussed in Refs [87, 88].

Example 4.5 Frame natural frequency

Figure 4.18b,c shows a pin-footed portal frame and circular arc, both 1.20 m (47.24 in.) high. They are fabricated from 2 cm (0.787 in.) outside diameter of the aluminum tubing with 2 mm (0.0787 in.) wall thickness. They support fabric enclosures. Compute the fundamental natural frequencies of their fundamental in-plane transverse modes. The density and elastic modulus of aluminum are 2.77 gm/cc (0.1 lb/ sq. in.) and 6.89E11 dyne/cm^2 (10E6 lb/sq. in).

Solution

The flexure modes scale with the mass per unit length and moment of inertia of the cross section of the aluminum tube. These are computed using case 30 of Table 1.5 and using an average radius of $R = (2\ \text{cm}^{-2} \times 0.2\ \text{cm})/2 = 0.8$ cm:

$$m = \rho\pi A = \rho 2\pi Rt = (2.77\ \text{gm/cm}^3)2\pi(0.8\ \text{cm})(0.2\ \text{cm}) = 2.78\ \text{gm/cm}$$

$$I = R^3 t = \pi(0.8\ \text{cm})^3(0.2\ \text{cm}) = 0.321\ \text{cm}^4$$

The arc subtends 225 degrees, $\beta_o = 3.926$ radians with a radius of 0.894 meters (35.2 in.). The formula for the fundamental lateral mode is given in case 2 of Table 4.17 with $i = 2$ so

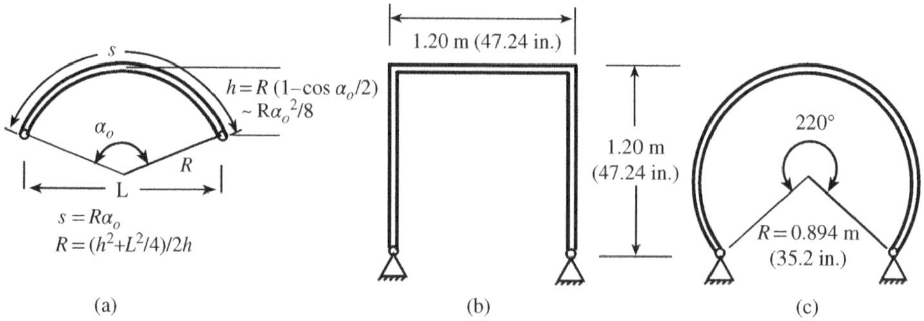

Figure 4.18 Pin-ended frame and circular arc

$\beta_o/(i\pi) = 0.6248$. In the dyne–cm–sec units of case 4 of Table 1.3, the natural frequency is computed as follows:

$$f_2 = \frac{i^2\pi^2}{2\pi(\alpha_o R)^2}\left[\frac{\{1-(\alpha_o/(i\pi))^2\}^2}{1+i^{-2}+2(\alpha_o/(i\pi))^2}\right]^{1/2}\left(\frac{EI_y}{m}\right)^{1/2}$$

$$= \frac{2^2\pi^2}{2\pi(3.926\times 89.4\text{ cm})^2}\left[\frac{\{1-0.6248^2\}^2}{1+2^{-2}+2\times 0.6248^2}\right]^{1/2}\left(\frac{(6.89\text{E}11\,\text{dyne/cm}^2)0.321\text{ cm}^4}{2.78\text{ g/cm}}\right)^{1/2}$$

$$= \frac{2\pi}{123.2\text{E}3\text{ cm}^2}[0.3002]^{1/2}(8.05E10\text{ cm}^4/\text{s}^2) = 7.93\text{ Hz}$$

The lateral natural frequency of the frame is calculated from case 14 of Table 4.17. With identical legs and cap, $E_1 = E_2 = E$; $I_1 = I_2 = I$, calculated previously; $L_1 = L_2 = L = 120$ cm; and $M_1 = M_2 = M = mL = 2.78$ gm/ cm (120 cm) = 333.6 gm:

$$f_1 = \frac{1}{2\pi}\left[\frac{EI}{L^3M}\right]^{1/2}\left[\frac{6}{1+\frac{1}{2}}\right]^{1/2}\frac{1}{[1+0.97]^{1/2}} = \frac{1.42}{2\pi}\left[\frac{(6.89\text{E}11\text{ dyne/cm}^2)\,0.321\text{ cm}^4}{(120\text{ cm})^3\,333.6\text{ g}}\right]^{1/2}$$

$$= 4.42\text{ Hz}$$

The natural frequency of the portal frame is about half that of a comparable arc. The mass of the two structures is about the same but the arc is stiffer.

References

[1] Timoshenko, S., Strength of Materials, 3th ed., 1955. Reprint by Robert Krieger, Malabar Florida.

[2] Weaver, W., S. P. Timoshenko, and D. H. Young, Vibration Problems in Engineering, 5th ed., John Wiley, New York, p. 417, 435, 483, 1990.

[3] Han, S. M., Benaroya, H. and Wei, T. Dynamics of Transversely Vibrating Beams using Four Engineering Theories, Journal of Sound and Vibration, vol. 225, pp. 935–989, 1999.

[4] Harris, C. O., Introduction to Stress Analysis, Macmillian Co., New York, 1959.

[5] Dimarogonas, A., Vibration for Engineers, 2nd ed., Prentice-Hall, New Jersey, 1996.

[6] Maurizi, M. J., and P. M. Belles. General Equation of Frequencies for Vibrating Uniform One-Span Beam Under Compressive Axial Loads, Journal of Sound and Vibration, vol. 145, pp. 345–347, 1991.

[7] Gorman, D. J., Free Vibration Analysis of Beams and Shafts, John Wiley, New York, p. 450, 1975.

[8] Stafford, J. W., Natural Frequencies of Beams and Plates on Elastic Foundations with Constant Modulus, Journal of the Franklin Institute, vol. 284, pp. 262–264, 1967.

[9] Young, D., and R. P. Felgar, Tables of Characteristic Functions Representing Normal Modes of Vibration of a Beam, University of Texas Publication No. 1913, July 1, 1949. Also see Felgar, R.P., Formulas for Integrals Containing Characteristic Functions of a Vibrating Beam, University of Texas Circular No. 14, Bureau of Engineering Research, Austin, TX, 1950.

[10] Chang, T. C., and R. R. Craig, Normal Modes of Uniform Beams, Journal Engineering Mechanics Division, American Society of Civil Engineers, vol. 95, pp. 1027–1031, 1969.

[11] Dowell, E. H., On Asymptotic Approximations to Beam Model Shapes, Journal of Applied Mechanics, vol. 51, p. 439, 1984.

[12] Gorman, D. J., Free Lateral Vibration Analysis of Double-Span Uniform Beams, International Journal of Mechanical Sciences, vol. 16, pp. 345–351, 1974.

[13] Gorman, D. J., and R. K. Sharma, Vibration Frequencies and Modal Shapes for Multi-Span Beams with Uniformly Spaced Supports, Ottawa University, Ontario, Canada, Report No. Conf-740330-1, 1974.

[14] Franklin, R. E., B. M. H. Soper, and R. H. Whittle, Avoiding Vibration-Induced Tube Failures in Shell and Tube Heat Exchangers, BNES Vibration in Nuclear Plant Conference, Paper 3:1, Keswick, U.K., 1978.
[15] Miles, J. W., Vibrations of Beams on Many Supports, Journal Engineering Mechanics Division, ASCE, paper 863, pp. 1–6, 1956.
[16] Sen Gupta, S., Natural Flexural Waves and the Normal Modes of Periodically-Supported Beams and Plates, Journal of Sound and Vibration, vol. 13, pp. 89–101, 1970.
[17] Yuan, J., and S. M. Dickinson, On the Determination of Phase Constants for the Study of Free Vibration of Periodic Structures, Journal of Sound and Vibration, vol. 179, pp. 369–383, 1995.
[18] Shaker, F. J., Effect of Axial Load on Mode Shapes and Frequencies of Beams, NASA TN D-8109, 1975.
[19] Bokaian, A., Natural Frequencies of Beams under Tensile Axial Loads, Journal of Sound and Vibration, vol. 142, pp. 481–498, 1990.
[20] Zuo, H. and K. D. Helmstad, Conditions for Bifurcation of a Cantilever Beam, Journal of Sound and Vibration, vol. 203, pp. 899–902, 1997.
[21] Huang, T., and D.W. Dareing, Buckling and Natural Frequencies of Long Vertical Pipes, American Society of Civil Engineers, Journal of the Applied Mechanics Division, vol. 95, pp. 167–181, 1969.
[22] Laird, W. M., and G. Fauconneau, Upper and Lower Bounds for the Eigenvalues of Vibrating Beams with Linearly Varying Axial Load, NASA Report CR-653, University of Pittsburgh, 1966.
[23] Stephens, N. G., Vibration of a Cantilever Beam Carrying a Heavy Tip Body, by Dunkerley's Method, Journal of Sound and Vibration, vol. 70, pp. 463–465, 1980.
[24] Laura, P. A. A., J. L. Pombo, and E. A. Susemihl, A Note on the Vibrations of a Clamped-Free Beam with a Mass at the Free End, Journal of Sound and Vibration, vol. 37, pp. 161–168, 1974. Also, vol. 108, 1986, pp. 123–131.
[25] Bhat, B. R., and H. Wagner, Natural Frequencies of a Uniform Cantilever with a Tip Mass Slender in the Axial Direction, Journal of Sound and Vibration, vol. 45, pp. 304–307, 1976.
[26] Joshi, A., Constant Frequency Solutions of a Uniform Cantilever Beam with Variable Tip Mass and Corrector Spring, Journal of Sound and Vibration, vol. 179, pp. 165–169, 1995.
[27] Gurgoze, M., 1984, A Note on the Vibrations of Restrained Beams and Rods with Point Masses, Journal of Sound and Vibration, vol. 96, pp. 461–468. Also vol. 100, pp. 588–589.
[28] Haener, J., Formulas for the Frequencies Including Higher Frequencies of Uniform Cantilever and Free-Free Beams with Additional Masses at the Ends, Journal of Applied Mechanics, vol. 25, p. 412, 1958.
[29] Chang, C. H. Free Vibrations of a Simply Supported Beam Carrying a Rigid Mass at the Middle, Journal of Sound and Vibration, vol. 237, pp. 733–774, 2000.
[30] Baker, W. E., Vibration of Uniform Beams with Central Masses, Journal of Applied Mechanics, vol. 31, pp. 35–37, 1964.
[31] Chen, D.-W., The Exact Solution for Free Vibration of Uniform Beam Carrying Multiple Two-Degree-of-Freedom Spring-Mass Systems, Journal of Sound and Vibration, vol. 295, pp. 342–360, 2006.
[32] Wang, H. C., and W. J. Worley, Tables of Natural Frequencies and Nodes for Transverse Vibration of Tapered Beams, NASA Report NASA-CR-443, University of Illinois, April 1966.
[33] Cranch, E. T., and A.A. Alder, Bending Vibrations of Variable Section Beams, Journal of Applied Mechanics, vol. 23, pp. 103–108, 1956.
[34] Mabie, H. H., and C.B. Rodgers, Transverse Vibrations of Double-Tapered Cantilever Beams, Journal of the Acoustical Society of America, vol. 44, pp. 1739–1741, 1968.
[35] Brock, J. E., Dunkerley-Mikhlin Estimates of Gravest Frequency of a Vibrating System, Journal of Applied Mechanics, vol. 43, pp. 345–347, 1976.
[36] Mabie, H. H., and C. B. Rodgers, Transverse Vibrations of Tapered Cantilever Beams with End Support, Journal of the Acoustical Society of America, vol. 51, pp. 1771–1774, 1972.
[37] Conway, H. D., and J. F. Dubil, Vibration Frequencies of Truncated-Cone and Wedge Beams, Journal of Applied Mechanics Division, vol. 32, pp. 932–934, 1965.
[38] Aucielo, N. M., Free Vibrations of a Linear Taped Cantilever beam with Constraining Springs and Tip Mass, Journal of Sound and Vibration, vol. 192, 905–911, 1996.
[39] Wu, J. -S., and C. -T. Chen, An Exact Solution for the Natural Frequencies and Mode Shapes of Immersed Wedge Beam with tip mass, Journal of Sound and Vibration, vol. 286, pp. 549–568, 2005.
[40] Hibbeler, R. C., Free Vibration of a Beam Supported by Unsymmetrical Spring-Hinges, Journal of Applied Mechanics, vol. 42, pp. 501–502, as corrected, 1975.
[41] Chun, K. R., Free Vibration of a Beam with One End Spring-Hinged and the Other Free End, Journal of Applied Mechanics, vol. 39, 1972, pp. 1154–1155.
[42] Hess, M. S., Vibration Frequency for a Uniform Beam with Central Mass and Elastic Supports, Journal of Applied Mechanics, vol. 31, pp. 556–558, 1964.

[43] Lee, W. T., Vibration Frequency for a Uniform Beam with One End Spring-Hinged and Carrying a Mass at the Other Free End, Journal of Applied Mechanics, vol. 95, p. 813–815, 1973.
[44] Maurizi, M. J., R. E. Rossi, and J. A. Reyes, Vibration Frequencies for a Uniform Beam with One End Spring-Hinged and Subjected to a Translational Restraint at the Other End, Journal of Sound and Vibration, vol. 48, pp. 565–568, 1976.
[45] Lau, J. H., Vibration Frequencies and Mode Shapes for a Constrained Cantilever, Journal of Applied Mechanics, vol. 51, pp. 182–187, 1984.
[46] Liu, W. W. H. and K. S. Chen, Effects of Lateral Support on the Fundamental natural Frequencies and Buckling Coefficients, Journal of Sound and Vibration, vol. 129, pp. 155–160, 1989.
[47] Gruttmann, F. and W. Wagner, Shear Correction factors in Timoshenko's Beam Theory for Arbitrary Shaped Cross-Sections, Computational Mechanics, vol. 37, pp. 199–207, 2001.
[48] Mendez-Sanchez, R.A., A. Morales, and J. Flores, Experimental Check on the Accuracy of Timoshenko Beam Theory, Journal of Sound and Vibration, vol. 279, 508–512, 2005.
[49] Stephen, N. G. and S. Puchegger, The On the valid range of Timoshenko Beam Theory, Journal of Sound and Vibration, vol. 297, pp. 1082–1087, 2006.
[50] Hutchinson, J.R., Shear Coefficients for Timoshenko Beam Theory, Journal of Applied Mechanics, vol. 68, pp. 87–92, 2001. Also see discussion and closure vol. 68, pp. 959–961.
[51] Cowper, G.R., The Shear Coefficient in Timoshenko's Beam Theorem, Journal of Applied Mechanics, vol. 33, pp. 335–340, 1966. Note: values only applicable to Poisson's ratio = 0 per Ref. [43].
[52] Goodman, L. E., and J. G. Sutherland, Discussion of Natural Frequencies of Continuous Beams of Uniform Span Length, Journal of Applied Mechanics, vol. 18, pp. 217–219, 1951.
[53] Hurty, W. C., and M. F. Rubenstein, On the Effect of Rotary Inertia and Shear in Beam Vibration, Journal of the Franklin Institute, vol. 278, pp. 124–132, 1964.
[54] Kang, J.-H., An Exact Frequency Equation for Timoshenko Beam Clamped at Both ends, Journal of Sound and Vibration, vol. 333, pp. 3332–3337, 2014.
[55] Pickett, G., Equations for Computing Elastic Constants from Vibration of Prisms and Cylinders, American Society for Testing Materials, Proceedings, vol. 45, pp. 846–863, 1945.
[56] ASTM Standard E1876-01, Standard Test Method for Dynamic Young's Modulus, Shear Modulus, and Poisson's Ratio by Impulse Excitation of Vibration, 2001.
[57] Rutenburg, A., Approximate Natural Frequencies for Coupled Shear Walls, Earthquake Engineering and Structural Division, vol. 4, pp. 95–100, 1975.
[58] Veletsos, A. S., et al., Free In-Plane Vibration of Circular Arches, ASCE Journal of the Engineering Mechanics Division, vol. 98, pp. 311–329, 1972.
[59] Kraus, M. and R. Kindmann, 2009, St. Venants Torsion Constant of Hot Rolled Steel Profiles and Position of the Shear Centre, Institute for Steel and Composite Structures, University of Bochum, Germany.
[60] Culver, C. G., Natural Frequencies of Horizontally Curved Beams, ASCE Journal of the Structural Division, vol. 93, pp. 189–203, 1967.
[61] Lee, L. S. S., Vibration of an Intermediately Supported U-Bend Tube, Journal of Engineering for Industry, vol. 97, pp. 23–32, 1975.
[62] Rieger, N. F., and H. McCallion,The Natural Frequencies of Portal Frames, International Journal of Mechanical Engineering Science, vol. 7, pp. 253–261, 1965.
[63] Endo, M., K. Hatamura, M. Sakata, and O. Taniguchi, Flexural Vibration of a Thin Rotating Ring, Journal of Sound and Vibration, vol. 92, pp. 261–272, 1984.
[64] Kim, W., and J. Chung, Free Non-Linear Vibration of a Rotating Thin Ring with the In-Plane and Out-of-Plane Motions, Journal of Sound and Vibration, vol. 258, pp. 167–178, 2002.
[65] Kirkhope, J., Simple Frequency Expression for the In-Plane Vibration of Thick Circular Rings, Journal of Acoustical Society of America, vol. 59, pp. 86–88, 1976.
[66] Sato, K., Free Flexural Vibrations of an Elliptical Ring in its Plane, Journal Acoustical Society of America, vol. 57, pp. 113–115, 1975.
[67] Ellis, B. R., An Assessment of the Fundamental Natural Frequency of Buildings, Proceeding of the Institution of Civil Engineers, part 2, vol. 69, pp 763–776, 1980.
[68] Uniform Building Code, 1997, Section 1630.2.2.
[69] Housner, G.W., and A. G. Brody, Natural Periods of Vibration of Buildings, Journal of the Engineering Mechanics Division, American Society of Civil Engineers, vol. 89, 1963, pp. 31–65.
[70] Clary, R. R., and S. A. Leadbetter, An Analytical and Experimental Investigation of the Natural Frequencies of Uniform Rectangular-Cross-Section Free-Free Sandwich Beams, NASA Technical Notes, NASA TN-D-1967, 1963.
[71] Murty, A. V. K., and R. P. Shimpi, Vibration of Laminated Beams, Journal of Sound and Vibration, vol. 36, pp. 273–284, 1974.

[72] Frostig, Y. and M. Baruch, Free Vibrations of Sandwich Beams, Journal of Sound and Vibration, vol. 176, pp. 195–208, 1994.

[73] Timoshenko, S., and J. N. Goodier, Theory of Elasticity, McGraw-Hill, p. 264, 266, 277, 287, 1951.

[74] Francu, J., P. Novackova, and P. Janicek, 2012, Torsion of a Non-Circular Bar, Engineering Mechanics, vol. 19, pp. 45–60.

[75] Roark, R. J. Formulas for Stress and Strain, 4th ed., McGraw-Hill, New York, 1965.

[76] Trayer, G. W. and H. W. March, The Torsion of Members Having Section Common in Aircraft Construction, National Advisory Committee on Aeronautics, report 334, 1930.

[77] Seaburg, P. A. and C. J. Carter, 2003, Torsional Analysis of Structural Steel Members, American Institute of Steel Construction.

[78] Gere, J. M., Torsional Vibration of Beams of Thin-Walled Open Section, Journal of Applied Mechanics Division, American Society of Civil Engineers, vol. 21, pp. 381–387, 1954.

[79] Carr, J. B., The Torsional Vibration of Uniform Thin-Walled Beams of Open Section, Aeronautical Journal (London), vol. 73, pp. 672–674, 1969.

[80] Love, A. E., A Treatise on the Mathematical Theory of Elasticity, 4th ed., (first published in 1927 by the Cambridge University Press), Dover Press, New York, 1944, pp. 446–454.

[81] Huang, S. C. and W. Soedel, Effects of Rotating Rings on Elastic Foundations, Journal of Sound and Vibration, vol. 115, pp 253–274, 1987.

[82] Koga, T. and T. Kodama, Bifurcation Buckling and Free Vibrations of Cylindrical Shells under Pressure, International Journal Pressure Vessels and Piping, vol. 45, pp. 223–235, 1991.

[83] Henrych, J., The Dynamics of Arches and Frames, Elsevier, New York, p. 186, 1981.

[84] Tufekci, E., Exact Solution of Free In-Plane Vibration of Shallow Circular Arches, International Journal of Structural Stability and Dynamics, vol. 1, pp 409–428, 2001.

[85] Auciello, N. M. and M. A. Rosa, Free Vibrations of Circular Arches: A Review, Journal of Sound and Vibration, vol. 176, pp.433–458, 1994.

[86] Reissner, E., Note on the Problem of Vibrations of Slightly Curved Bars, Journal of Applied Mechanics, pp. 195–196, 1954.

[87] Morales, C. A., Portal Frame Inertia and Stiffness Matrices, Journal of Sound and vibration, vol 283, pp. 1205–1215, 2005.

[88] Kolousek, V., Dynamics of Engineering Structures, Butterworth, London, 1973.

5

Natural Frequency of Plates and Shells

This chapter presents formulas and data for plate and shell natural frequencies and mode shapes. Plates are thin, flat elastic sheets that have mass, thickness, and bending stiffness. The natural frequencies of flat plates increase in proportion to their thickness. A curved plate is a shell. The curvature of a shell surface couples its radial and circumferential deformations. Shell modes are generally three-dimensional.

5.1 Plate Flexure Theory

5.1.1 Stress and Strain

A plate bends out of plane in response to loads on its surface [1, 2]. The plate theory (Table 5.1) is similar to the beam theory (Table 4.1) in that normals to the mid surface of the undeformed plate (halfway between the top and bottom surfaces) are assumed to remain straight and normal to the deformed mid surface. Bending stresses and strains are proportional to distance from the mid surface. The maximum stress is on the surface of the plate (see reviews in Refs [1–6]). If the displacement in a vibration mode is known, by measurement or forced vibration calculation (Chapter 7), the stress and strains, moments, and shears in the plate can be calculated from the plate theory (Table 5.1) (see Example 5.1).

5.1.2 Boundary Conditions

Three classical boundary conditions on the edges of thin plates with out-of-plane deformation W are [2, 6, 7]:

1. *Free edge* (F), no loads: $M = V = 0$
2. *Simply supported edge* (S), no moment or displacement: $\mathrm{W} = M = 0$
3. *Clamped edge* (C), no rotation or displacement: $W = \partial W/\partial n = 0$

Formulas for Dynamics, Acoustics and Vibration, First Edition. Robert D. Blevins.
© 2016 John Wiley & Sons, Ltd. Published 2016 by John Wiley & Sons, Ltd.

Table 5.1 Plate bending stress and strain theory

Notation: h = plate thickness; r, θ = polar coordinates in plane of plate, radius and angle; x, y = orthogonal in-plane coordinates; z = out-of-plane coordinate by right hand rule; W = out-of-plane deformation, in z direction; E = modulus of elasticity; σ = stress, force/area; ε = strain, unitless; ν = Poisson's ratio; G = E/[2(1 + ν)], shear modulus. Refs [1, 2].

Description	Cartesian Coordinates x,y	Polar coordinates r,θ
1. Strain 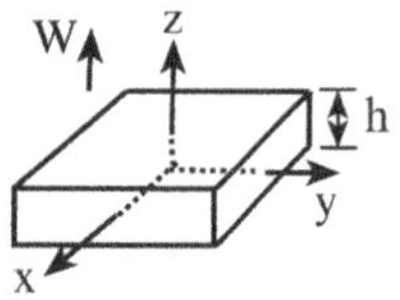	$\varepsilon_x = -z\frac{\partial^2 W}{\partial x^2}$, $\varepsilon_y = -z\frac{\partial^2 W}{\partial y^2}$, $\varepsilon_{xy} = -2z\frac{\partial^2 W}{\partial x\partial y}$, $\varepsilon_{xz} = \varepsilon_{yz} = \varepsilon_{zz} = 0$	$\varepsilon_r = -z\frac{\partial^2 W}{\partial r^2}$, $\varepsilon_\theta = -\frac{z}{r}\frac{\partial W}{\partial r} - \frac{z}{r^2}\frac{\partial^2 W}{\partial r^2}$, $\varepsilon_{r\theta} = -\frac{z}{r}\frac{\partial^2 W}{\partial r\partial\theta} - rz\frac{\partial}{\partial r}\left(\frac{1}{r^2}\frac{\partial W}{\partial\theta}\right)$, $\varepsilon_{rz} = \varepsilon_{\theta z} = \varepsilon_{zz} = 0$
2. Stress 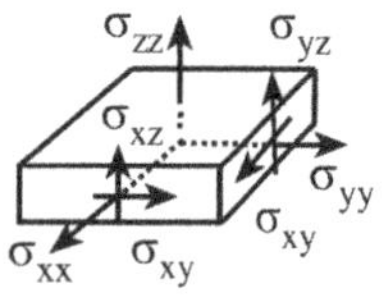	$\sigma_{xx} = \frac{E}{1-\nu^2}(\varepsilon_x + \nu\varepsilon_y)$, $\sigma_{yy} = \frac{E}{1-\nu^2}(\varepsilon_y + \nu\varepsilon_x)$, $\sigma_{xy} = G\varepsilon_{xy}$, $\sigma_{xz} = \sigma_{yz} = \sigma_{zz} = 0$	$\sigma_{rr} = \frac{E}{1-\nu^2}(\varepsilon_r + \nu\varepsilon_\theta)$, $\sigma_{\theta\theta} = \frac{E}{1-\nu^2}(\varepsilon_\theta + \nu\varepsilon_r)$, $\sigma_{r\theta} = G\varepsilon_{r\theta}$, $\sigma_{rz} = \sigma_{\theta z} = \sigma_{zz} = 0$
3.Moment Resultants 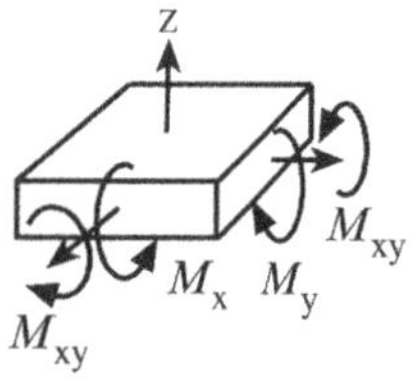 Note some authors (Ref. [2]) define M_{xy} with the opposite sign	$M_x = \int_{-h/2}^{h/2}\sigma_{xx}z\,dz$ $= -\frac{Eh^3}{12(1-\nu^2)}\left(\frac{\partial^2 W}{\partial x^2} + \nu\frac{\partial^2 W}{\partial y^2}\right)$ $M_y = \int_{-h/2}^{h/2}\sigma_{yy}z\,dz$ $= -\frac{Eh^3}{12(1-\nu^2)}\left(\frac{\partial^2 W}{\partial y^2} + \nu\frac{\partial^2 W}{\partial x^2}\right)$ $M_{xy} = \int_{-h/2}^{h/2}\sigma_{xy}z\,dz$ $= -\frac{Eh^3}{12(1+\nu)}\frac{\partial^2 W}{\partial x\partial y}$	$M_r = \int_{-h/2}^{h/2}\sigma_{rr}z\,dz$ $= -\frac{Eh^3}{12(1-\nu^2)}\left[\frac{\partial^2 W}{\partial r^2} + \frac{\nu}{r}\left(\frac{\partial W}{\partial r} + \frac{1}{r}\frac{\partial^2 W}{\partial\theta^2}\right)\right]$ $M_\theta = \int_{-h/2}^{h/2}\sigma_{\theta\theta}z\,dz$ $= -\frac{Eh^3}{12(1-\nu^2)}\left(\frac{1}{r}\frac{\partial W}{\partial r} + \frac{1}{r^2}\frac{\partial^2 W}{\partial\theta^2} + \nu\frac{\partial^2 W}{\partial r^2}\right)$ $M_{r\theta} = \int_{-h/2}^{h/2}\sigma_{r\theta}z\,dz$ $= -\frac{Eh^3}{12(1+\nu)}\frac{}{r}\left(\frac{1}{r}\frac{\partial W}{\partial\theta}\right)$
4. Shear Resultants 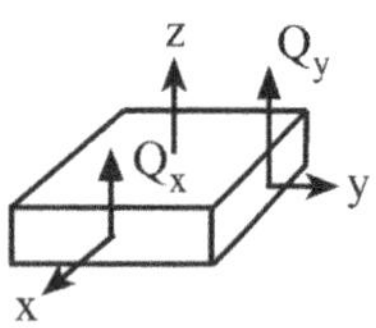	$Q_x = \int_{-h/2}^{h/2}\sigma_{xz}\,dz$ $= -\frac{Eh^3}{12(1-\nu^2)}\frac{\partial}{\partial x}\left(\frac{\partial^2 W}{\partial x^2} + \frac{\partial^2 W}{\partial y^2}\right)$ $Q_y = \int_{-h/2}^{h/2}\sigma_{yz}\,dz$ $= -\frac{Eh^3}{12(1-\nu^2)}\frac{\partial}{\partial y}\left(\frac{\partial^2 W}{\partial x^2} + \frac{\partial^2 W}{\partial y^2}\right)$	$Q_r = \int_{-h/2}^{h/2}\sigma_{rz}\,dz =$ $-\frac{Eh^3}{12(1-\nu^2)}\frac{}{r}\left(\frac{\partial^2 W}{\partial r^2} + \frac{1}{r}\frac{\partial W}{\partial r} + \frac{1}{r^2}\frac{\partial^2 W}{\partial\theta^2}\right)$ $Q_\theta = \int_{-h/2}^{h/2}\sigma_{\theta z}\,dz =$ $-\frac{Eh^3}{12(1-\nu^2)r}\frac{\partial}{\partial\theta}\left(\frac{\partial^2 W}{\partial r^2} + \frac{1}{r}\frac{\partial W}{\partial r} + \frac{1}{r^2}\frac{\partial^2 W}{\partial\theta^2}\right)$
5. Edge Reaction 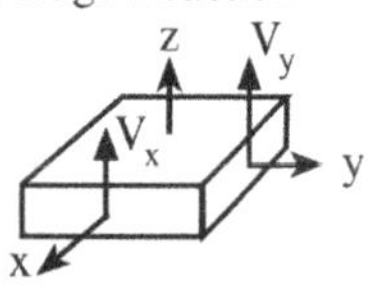	$V_x = Q_x + \frac{\partial M_{xy}}{\partial x}$ $V_y = Q_y + \frac{\partial M_{xy}}{\partial y}$	$V_r = Q_r + \frac{\partial M_{r\theta}}{r\partial\theta}$

These boundary conditions are applied to an edge where the in-plane coordinate n is normal to the edge of the plate, and the coordinate t is tangential to the edge. Out-of-plane shear Q is defined in Table 5.1. The moment M is defined in Table 5.2. $V = Q + \partial M_{nt}/\partial t$ is the out-of-plane Kelvin–Kirchhoff edge reaction See Refs [1–6].

The boundaries of many plates in engineering practice are intermediate between clamped and simply supported. The clamped boundary condition suppresses both edge rotation and displacement, and it is independent of Poisson's ratio. Poisson's ratio enters the boundary conditions for curved simply supported edges and for free edges through M and Q (Table 5.1; [3, p. 182]; [4, p. 108]; [6, 7]). The natural frequency parameter (Eq. 5.14) of a simply supported round plate increases with Poisson's ratio ν. The natural frequency parameter of an S–F–S–F square plate decreases with increasing Poisson's ratio. Maximum possible variation in Poisson's ratio for an isentropic material is 0 to 0.5.

Poisson's ratio ν	0.0	0.3	0.5
S–S round plate [26] λ_{ij}^2	4.443	4.935	5.213
S–F–S–F square plate [10] λ_{ij}^2	9.870	9.631	9.079

5.1.3 Plate Equation of Motion

The equations of motion of small out-of-plane deformations W of a flat, uniform, elastic plate in x–y Cartesian (Eq. 5.1) and r-θ polar (Eq. 5.2) coordinates are [1, p. 337], [2]

$$\frac{Eh^3}{12(1-\nu^2)}\left(\frac{\partial^4 W}{\partial x^4}+2\frac{\partial^4 W}{\partial x^2\partial y^2}+\frac{\partial^4 W}{\partial y^4}\right)+\gamma\frac{\partial^2 W}{\partial t^2}=q+N_x\frac{\partial^2 W}{\partial x^2}+2N_{xy}\frac{\partial^2 W}{\partial x\partial y}+N_y\frac{\partial^2 W}{\partial y^2} \tag{5.1}$$

$$\frac{Eh^3}{12(1-\nu^2)}\left(\frac{\partial^4 W}{\partial r^4}+\frac{2}{r^2}\frac{\partial^4 W}{\partial r^2\partial \theta^2}+\frac{1}{r^4}\frac{\partial^4 W}{\partial \theta^4}\right)+\gamma\frac{\partial^2 W}{\partial t^2}=q+N_r\left(\frac{\partial^2 W}{\partial r^2}+\frac{1}{r}\frac{\partial W}{\partial r}+\frac{1}{r^2}\frac{\partial^2 W}{\partial \theta^2}\right) \tag{5.2}$$

where E = modulus of elasticity of the plate material, ν = Poisson's ratio, h = plate thickness, and t = time. γ = mass per unit area of the plate, which is equal to ρh, where ρ is the material density and h is the plate thickness, plus any nonstructural mass per unit area, and q is the out-of-plane pressure load in the $+z$ direction. N_x and N_y are the normal in-plane mean loads per unit length of edge, positive outward. N_r is the outward mean radial load.

Modal solutions to the homogeneous equation of motion, Equation 5.1 without loads, are a spatial mode shape, $\widetilde{w}(x,y)$, times a function of time, $T(t)$ [1–6, 8]:

$$W(x,y,t)=\widetilde{w}(x,y)T(t)=\widetilde{x}(x)\widetilde{y}(y)T(t) \tag{5.3}$$

This *separation-of-variables* solution is substituted into Equation 5.1, which is then divided by $\widetilde{x}(y)\widetilde{y}(y)T(t)$. Neglecting in-plane loads results in

$$\frac{Eh^3}{12(1-\nu)\gamma}\frac{1}{\widetilde{w}(x,y)}\left(\frac{d^4\widetilde{w}(x,y}{dx^4}+2\frac{d^4\widetilde{w}(x,y)}{dx^2dy^2}+\frac{d^4\widetilde{w}(x,y)}{dy^4}\right)=\frac{1}{T(t)}\frac{d^2T(t)}{dt^2}=-\omega^2 \tag{5.4}$$

The left-hand side of this equation is only a function of space, x and y. Since it equals a function of time, t, on the right-hand side, both sides of the equation must be equal to the same constant. Choosing this to be $-\omega^2$, the partial differential equation bifurcates into two ordinary differential equations, one in time and one in space:

$$\frac{d^2T(t)}{dt^2} + \omega^2 T(t) = 0 \tag{5.5a}$$

$$\frac{d^4\tilde{w}(x,y)}{dx^4} + \frac{d^2\tilde{w}(x,y)}{2dx^2dy^2} + \frac{d^4\tilde{w}(x,y}{dy^4} - \omega^2\frac{12(1-\nu^2)\gamma}{Eh^3}\tilde{w}(x,y) = 0 \tag{5.5b}$$

Equation 5.5a's solutions, $T(t) = \sin(\omega t)$ and $\cos(\omega t)$, are sinusoidal in time at the natural frequency ω. Equation 5.5b is a *spatial eigenvalue equation* that is solved for the *circular natural frequencies* and mode shapes. The mode shapes of rectangular plates are combinations of sinusoidal and hyperbolic functions. Mode shapes of circular plates are Bessel functions ([1–6, 9–11] and Eq. 5.15).

5.1.4 Simply Supported Rectangular Plate

The modal solution to Eq 5.5b for a simply supported rectangular plate, case 2 of Table 5.2, with length a and width b has sinusoidal mode shapes, as shown in Figure 5.1 [10, 11]:

$$\tilde{w}(x,y) = \sin\left(\frac{i\pi x}{a}\right)\sin\left(\frac{j\pi y}{b}\right), \quad i = 1,2,3,\ \dots, \quad j = 1,2,3,\ \dots \tag{5.6}$$

The two integer modal indices i and j describe the number of half waves in the mode shape along the x- and y-coordinates, respectively. Increasing mode number corresponds to increasing number of waves in the mode shape as shown in Figure 5.1.

Substituting Equation 5.6 into Equation 5.5b gives the associated *natural frequency* of a simply supported rectangular plate in the i–jth vibration mode in Hertz, $f_{ij} = \omega_{ij}/2\pi$:

$$f_{ij} = \frac{1}{2\pi}\left[\left(\frac{i\pi}{a}\right)^2 + \left(\frac{j\pi}{b}\right)^2\right]\sqrt{\frac{Eh^3}{12\gamma(1-\nu^2)}}, \text{ Hz}, \quad i = 1,2,3,\ \dots \quad j = 1,2,3\ \dots \tag{5.7}$$

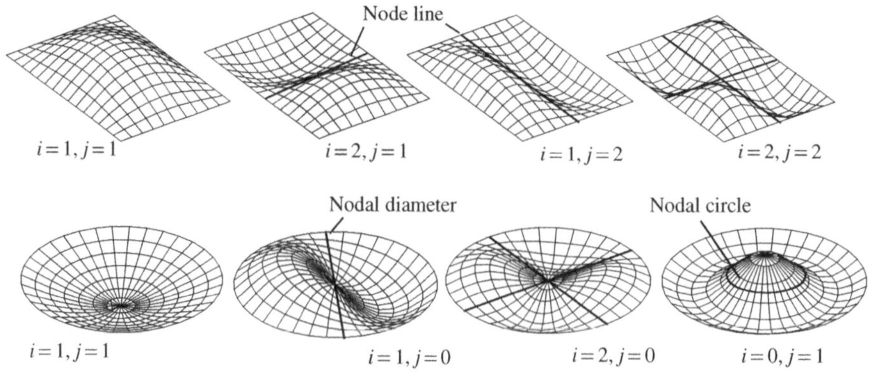

Figure 5.1 First six modes of simply supported rectangular and circular plates. Natural frequencies increase left to right in each row

The natural frequencies lie in doubly semiinfinite series beginning with the fundamental (lowest) natural frequency at $i = j = 1$.

5.1.5 Plates on Elastic Foundations

Elastic foundations have a foundation modulus, E_f that is the force per unit area per unit radial deformation. A massless foundation increases natural frequencies [6, 12]:

$$f_{ij} = \left[\left(f_{ij}\Big|_{Ef=0} \right)^2 + \frac{E_f}{4\pi^2\gamma} \right]^{1/2}, \quad i = 1, 2, 3, \ldots, \quad j = 1, 2, 3, \ldots \tag{5.8}$$

The term in parentheses (…) is the natural frequency of the plate *without* the foundation. For example, the natural frequencies of the simply supported plate with the spring foundation shown in Figure 4.2b are

$$f_{ij} = \left[\frac{1}{4\pi^2} \left\{ \left(\frac{i\pi}{a} \right)^2 + \left(\frac{j\pi}{b} \right)^2 \right\}^2 \frac{Eh^3}{12\gamma(1-\nu^2)} + \frac{k}{4\pi^2\gamma\,\Delta^2} \right]^{1/2}, \quad \begin{array}{l} i = 1, 2, 3, \ldots \\ j = 1, 2, 3, \ldots \end{array} \tag{5.9}$$

Case 8 of Table 4.8 and Soedel [6] (p. 48) suggest that the foundation mass can be approximately included by increasing the plate mass per area γ by one-third of the foundation mass per area.

5.1.6 Sandwich Plates

An equivalent stiffness for sandwich plates can be defined by assuming that normals to the mid plane remain normal during deformation as described in Section 4.1, Equations 4.4 and 4.5, and Figure 4.2. The stiffness $Eh^3/12$ in the plate natural frequency formula (Eqs 5.7 and 5.14) is replaced by the equivalent sandwich stiffness:

$$\frac{Eh^3}{12}\Big|_{\text{plate}} = \frac{1}{b} EI\Big|_{\text{equiv. beam}} \tag{5.10}$$

I/*b* is the sandwich beam moment of inertia *I* divided by the width *b*, cases 44 and 45 of Table 1.5. Also see case 3 of Table 5.5 and Equations 4.4 and 4.5. Perforated Plates are discussed in section 5.2.9.

5.1.7 Thick Plates and Shear Deformation

The thin plate bending theory (Table 5.1) neglects the out-of-plane shear deformations ε_{xz} and ε_{yz} (ε_{rz} and $\varepsilon_{\theta z}$ in polar coordinates where z is the out-of-plane coordinate) and the associated out-of-plane shear stresses σ_{yz} and σ_{yz} (σ_{rz} and $\sigma_{\theta z}$ in polar coordinates). However, equilibrium of the plate element in response to out-of-plane loads requires that the integrals of these shear stresses over the thickness (Q in case 4 of Table 5.1) cannot be zero. Out-of-plane shear stresses have a parabolic distribution through the thickness, maximum at the mid surface and zero at the surfaces of the plate. The Mindlin plate theory [13, 14] incorporates shear deformations and rotary inertia associated with local rotation of plate elements. Rotary inertia and shear are relatively insignificant for the fundamental mode of thin plates whose thicknesses are 10% or less of their typical spans (see Refs [13–23] and Section 4.4).

5.1.8 *Membrane Analogy and In-Plane Loads*

A mode shape that satisfies the flat membrane equation (Eq. 3.36) also satisfies the plate equation (Eqs 5.1, 5.2) with simply supported boundaries. The *membrane analogy* (look ahead to case 35 of Table 5.3) relates the natural frequencies of plates with simply supported straight edges and natural frequencies of flat membranes with supported edges [2]. Membrane solutions are also useful for estimating the effect of in-plane loads on natural frequency.

Consider steady in-plane edge load of N per unit length of edge applied perpendicularly to the edge in the plane of the plate to the edge. N is positive outward. An approximate expression for plate natural frequency of a rectangular plate with in-plane loads is obtained summing the plate bending stiffness and the membrane stiffness (chapter 3, case 1 of Table 3.7):

$$f_{ij}^2|_{\text{plate},N>0} = f_{ij}^2|_{\text{plate},N=0} + f_{ij}^2|_{\text{membrane}} \tag{5.11}$$

An approximate expression for effect of in-plane loads on rectangular plates is given in case 35 of Table 5.3.

$$f_{ij}^2|_{N_x,N_y} = f_{ij}^2|_{N_x=N_y=0} + \frac{N_x J_i}{4\gamma a^2} + \frac{N_x J_j}{4\gamma b^2} \tag{5.12}$$

N_x is the outward (tensile) load per unit length of edge in the x direction and applied to sides of length a, while N_y is the outward (tensile) load per unit length of edge in the y direction and applied to sides of length b. This result is the exact solution to the simply supported rectangular plate equation of motion (Eq. 5.2) where $J_i = i\pi$ and $J_j = j\pi$ [1]. The J parameters for other edge boundaries are given in cases 1 through 19 of Table 5.3. Tensile in-plane edge loads N_x, $N_y > 0$ increase natural frequency; very large values reduce to a tensioned membrane (case 1 of Table 3.7). Sufficient compressive loads, $N < 0$, will buckle the plate as the lowest natural frequency goes to zero.

5.1.9 *Orthogonality*

The mode shapes of most thin uniform-thickness plates can be shown to be *orthogonal* over the plate area A:

$$\int_A \widetilde{w}_{ij}\widetilde{w}_{rs}\, dA = 0 \quad \text{if } i \neq r \text{ and/or } j \neq s \tag{5.13}$$

provided that the natural frequencies of the modes are well separated and either the edges of the plates are clamped or the portion of the edges which are not clamped are straight and simply supported. Orthogonality of the mode shapes with free edges or curved plates with simply supported edges cannot be shown rigorously owing to the appearance of Poisson's ratio, but the lack of proof of orthogonality has not handicapped approximate modal analysis of plates.

5.2 Plate Natural Frequencies and Mode Shapes

5.2.1 Plate Natural Frequencies

Plate natural frequencies are expressed in terms of a dimensionless *natural frequency parameter* λ, the plate material density ρ or, equivalently, the plate mass per unit area $\gamma = \rho h$, and modulus of elasticity E:

$$f_{ij} = \frac{\lambda_{ij}^2}{2\pi a^2}\sqrt{\frac{Eh^3}{12\gamma(1-\nu^2)}}, \text{ Hertz}, \quad i = 1,2,3,\ \ldots\ , \quad j = 1,2,3,\ \ldots \tag{5.14}$$

Plate thickness is h and ν is Poisson's ratio. The natural frequencies of plates increase linearly with thickness, and they decrease inversely with the square of the plate length a. If the factor inside the square root can be simplified for homogeneous plates.

$$\frac{Eh^3}{12\gamma(1-\nu^2)} = \frac{Eh^2}{12\rho(1-\nu^2)}$$

The mass per unit area of homogeneous plates is $\gamma = \rho h$, where ρ is the material density.

The dimensionless frequency parameter λ is a function of the boundary conditions, plate geometry, and Poisson's ratio ν for free boundary conditions and curved simply supported edges (Section 5.1):

$$\lambda = \lambda(\text{boundary conditons, geomentry, } \nu)$$

Sections 5.1.3 and 5.1.4 has the derivation for a simply supported rectangular plate. Values of λ are provided in Tables 5.3 through 5.5 for other plate geometries. General reviews of plate vibrations are given by Leissa [1], Timoshenko [2], Reddy [4], Rao [5], Soedel [6], and Gorman [8].

5.2.2 Circular and Annular Plates

Table 5.2 [24–45] contains formulas for the natural frequencies and mode shapes of small out-of-plane bending vibrations of round, annular, and elliptical plates. The mode shapes for uniform, isentropic, constant-thickness, round, and annular plates (Eq. 5.2b) with axisymmetric boundary conditions are sums of Bessel functions [1, 3, 9]:

$$\widetilde{w}_{ij}(r,\theta) = \begin{cases} \left[J_i\left(\lambda_{ij}r/a\right) + b_i I_i(\lambda_{ij}r/a)\right]\cos i\theta, & \text{circular plate} \\ \left[J_i(\lambda_{ij}r/a) + b_i Y_i(\lambda_{ij}r/a) + c_i I_i(\lambda r/a) + d_i K_i(\lambda_{ij}r/a)\right]\cos i\theta, & \text{annular} \end{cases} \tag{5.15}$$

b_i, c_i, d_i = constants set by boundary conditions
λ = dimensionless natural frequency parameter
$J_i()$, $Y_i()$ = Bessel functions of first and second kinds and ith order
$I_i()$, $K_i()$ = modified Bessel functions of first and second kind and ith order

Table 5.2 Natural frequencies of round, annular, and elliptical plates

Notation: a = radius, inner radius for annular plates; b = outer radius; E = modulus of elasticity; f_{in} = natural frequency, Hz; i = 0, 1, 2, 3, … modal index, number of nodal diameters; j = 0, 1, 2, … modal index, number of nodal circles; h = plate thickness; I() = modified Bessel function of first kind; J() = Bessel function of first kind; K() = modified Bessel function of 2nd kind, Ref. [6]; N = force per unit length of edge, positive outward from center; γ = mass per unit area of plate (h where ρ = plate material density) plus non-structural mass; ν = Poisson's ratio. S = simply supported; C = clamped edge, F = free edge. Consistent units are in Table 1.2.

Natural Frequency, $f_{ij} = \dfrac{\lambda_{ij}^2}{2\pi a^2}\sqrt{\dfrac{Eh^3}{12\gamma(1-\nu^2)}} = \dfrac{\lambda_{ij}^2}{2\pi a^2}\sqrt{\dfrac{Eh^2}{12\rho(1-\nu^2)}}$, i = 0,1,2,3… j = 0,1,2,3… Hertz

Boundary Conditions	λ_{ij}^2	Mode Shape and Remarks

1. Round Plate with Free Edge

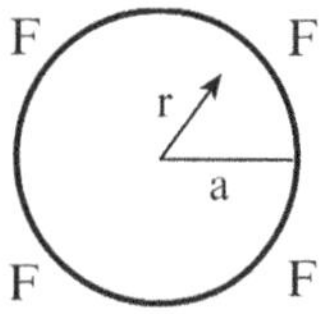

λ_{ij}^2

j \ i	0	1	2	3
0	0	0	5.262	12.24
1	9.063	20.51	35.24	52.92
2	38.51	59.86	84.37	111.2
3	87.81	119.0	153.3	190.7

Transcendental equation for λ is given in Ref. [25]

$\lambda_{41}^2 = 21.6$ for $\nu = 0.33$

Radii of nodal circles, r/a. Ref. [24]

j \ i	0	1	2	3
1	0.680	0.781	0.822	0.847
2	0.841	0.871	0.890	0.925
	0.391	0.497	0.562	0.605
3	0.893	0.932	0.936	0.939
	0.591	0.643	0.678	0.704
	0.257	0.351	0.414	0.440

i = number of nodal diameters
j = number of nodal circles

2. Round Plate Simply Supported

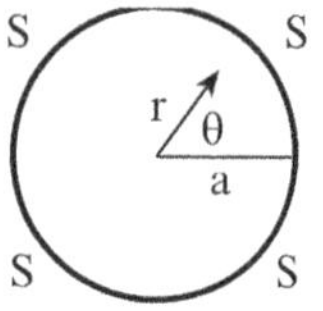

λ_{ij}^2

j \ i	0	1	2	3
0	4.935	13.89	25.61	39.96
1	29.76	48.51	70.14	94.35
2	74.16	102.8	134.3	168.7
3	138.3	176.8	218.2	262.5

ν=0.3. Transcendental equation for λ, Ref. [26]

$$\frac{J_{i+1}(\lambda)}{J_i(\lambda)} + \frac{I_{i+1}(\lambda)}{I_i(\lambda)} = \frac{2\lambda}{1-\nu}$$

Radii of nodal circles, r/a. Ref. [11]

j \ i	0	1	2	3
0	1	1	1	1
1	1	1	1	1
	0.441	0.550	0.613	0.605
2	1	1	1	0.939
	0.644	0.692	0.726	0.704
	0.279	0.378	0.443	0.440

$$\tilde{w}(r,\theta) = \left[J_i\left(\frac{\lambda_{ij} r}{a}\right) - \frac{J_i(\lambda_{ij})}{I_i(\lambda_{ij})} I_i\left(\frac{\lambda_{ij} r}{a}\right)\right]\cos i\theta$$

3. Round Plate with Clamped Edge

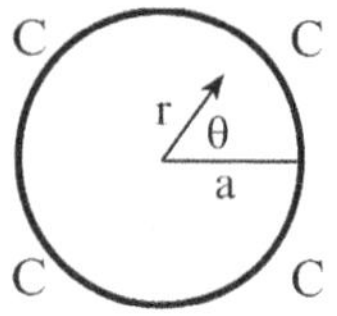

λ_{ij}^2 Refs [1, 27]

j \ i	0	1	2	3
0	10.22	21.26	34.88	51.04
1	39.77	60.82	84.58	111.0
2	89.10	120.1	153.8	190.3
3	158.2	199.1	242.7	289.2

λ is independent of ν. Ref. [22]. Transcen, equation for λ.

$$J_i(\lambda)I_{i+1}(\lambda) + I_i(\lambda)J_{i+1}(\lambda) = 0$$

Radii of nodal circles, r/a. Ref. [11]

j \ i	0	1	2	3
0	1	1	1	1
1	1	1	1	1
	0.379	0.490	0.559	0.606
2	1	1	1	1
	0.583	0.640	0.679	0.708
	0.255	0.350	0.414	0.462

$$\tilde{w}(r,\theta) = \left[J_i\left(\frac{\lambda_{ij} r}{a}\right) - \frac{J_i(\lambda_{ij})}{I_i(\lambda_{ij})} I_i\left(\frac{\lambda_{ij} r}{a}\right)\right]\cos i\theta$$

Table 5.2 Natural frequencies of round, annular, and elliptical plates, continued

Natural Frequency, $f = \frac{\lambda^2}{2\pi a^2}\sqrt{\frac{Eh^3}{12\gamma(1-\nu^2)}}$, Hertz

Boundary Conditions	λ^2, Mode Shape and Remark

4. Clamped or Free on Part of Edge

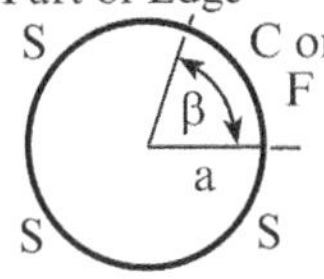

Fundamental mode — Refs [28,29]

β, deg	0	45	90	135	180	225	270	315	360
λ_{01}, C	4.935	5.85	6.34	6.87	7.49	8.198	9.039	9.827	10.22
λ_{01}, F	4.935	4.95	4.55	2.97	1.753	1.081	0.565	0.253	0.

ν=0.3. Approximate transcendental equation for λ for S-C

$$2\lambda[J_1(\lambda)/J_0(\lambda)+I_1(\lambda)/I_0(\lambda)]^{-1} = 1-\nu+1/\log_e(\sin\beta/2)$$

5. Free Edge, Clamped at Center

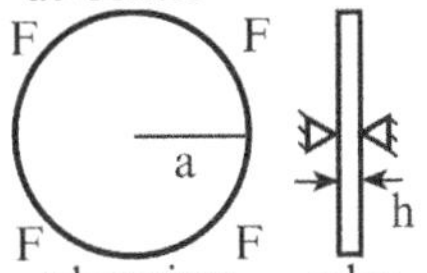

plan view edge

Ref. [24]

j	0	1	2	3	4	5	6
λ^2	3.752	20.91	61.2	120.6	199.9	298.2	416.6

Fundamental polar symmetric modes, i=0, j=1.
ν=1/3. In the higher modes values of λ are separated by π, $\lambda_{j+1} = \lambda_j + \pi$.

6. Simply Supported Clamped at Center

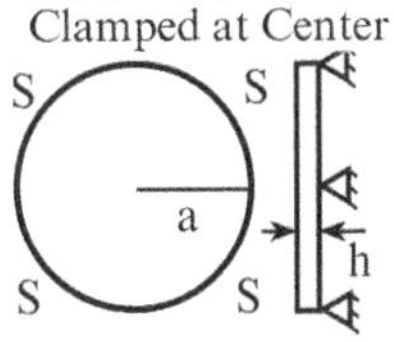

ν=0.3. Fundamental polar symmetric mode, i=0. Ref. [25]

j	0	1	2	3
λ^2	14.77	49.46	103.8	177.8

Transcendental equation for λ for polar symmetric modes.

$$0 = (1-\nu)\{[I_0(\lambda)-J_0(\lambda)][Y_1(\lambda)+(2/\pi)K_1(\lambda)] + [J_1(\lambda)+I_1(\lambda)][Y_0(\lambda)+(2/\pi)K_0(\lambda)]\} - 2\lambda[I_0(\lambda)Y_0(\lambda)+(2/\pi)J_0(\lambda)K_0(\lambda)]$$

7. Clamped Edge, Clamped at Center

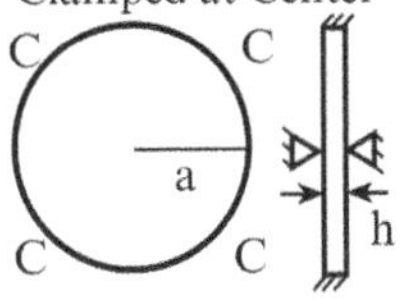

Polar symmetric modes, i=0 Ref. [30]

j	0	1	2	3	4
λ^2	22.74	61.95	121.2	200.1	298.8

Transcendental equation for polar symmetric modes.

$$[J_0(\lambda)-I_0(\lambda)][Y_1(\lambda)+(2/\pi)K_1(\pi)] - [J_1(\lambda)+I_1(\lambda)][Y_0(\lambda)+(2/\pi)K_0(\lambda)] = 0$$

8. Simply Supported at an Arbitrary Radius

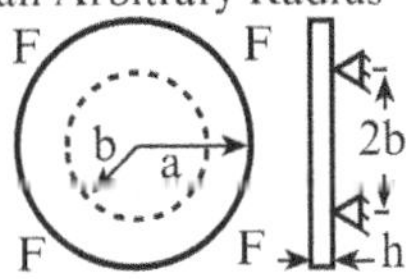

Fundamental mode Ref. [31]

b/a	0	0.2	0.4	0.6	0.8	1.0
λ^2	3.75	4.5	6.7	8.8	7.5	5.0

ν = 1/3

9. Elliptical Plate

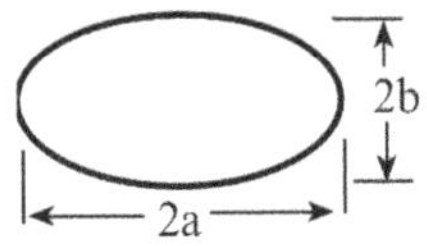

Also see appendix A

a/b	1	1.1	1.2	1.4	1.7	2.0	2.5	3.0	5.0
λ^2 Free, ν=.3	2.32	2.19	2.06	1.81	1.51	1.29	1.04	0.89	0.52
λ^2 Simple, ν=.25	4.865	4.45	4.157	3.77	3.46	3.29	3.13	3.03	2.85
λ^2 Clamped	10.2	9.35	8.73	7.92	7.28	6.94	6.66	6.52	6.35

Fund. mode for free, clamped and simply supported edge. Refs [32–35]

Table 5.2 Natural frequencies of round, annular, and elliptical plates, continued

Natural Frequency, $f_{ij} = \dfrac{\lambda_{ij}^2}{2\pi a^2}\sqrt{\dfrac{Eh^3}{12\gamma(1-\nu^2)}}$, Hertz

Boundary Conditions | **λ^2, Mode Shape and Remarks**

---

10. Simply-Supported with Torsional Springs

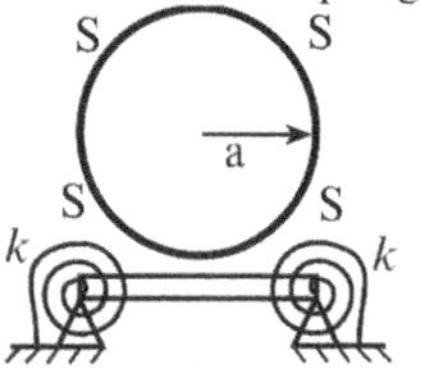

Ref. [36], also see Refs [37–39]

$Eh^3/[12ka(1-\nu^2)]$	0	0.001	0.01	0.1	1.0	∞
λ^2, i=0, j=0	4.93	6.05	8.76	10.0	10.2	10.2
λ^2, i=0, j=1	29.7	30.8	35.2	39.1	39.7	39.7
λ^2, i=1, j=0	13.9	15.0	18.6	20.9	21.2	21.2

k = torsional spring constant/ unit length of edge, Table 3.1.

---

11. Point Mass at Center

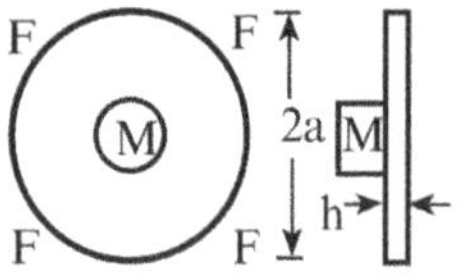

Ref. [40]

$M/(\gamma\pi a^2)$	0	0.05	0.1	0.2	0.4	0.6	1.0	1.4	∞
λ^2	9.0	8.2	7.5	6.8	5.9	5.5	4.9	4.7	3.7

ν=0.3, Polar symmetric mode, i=0, j=1
M = mass at center of plate

---

12. Simply-Supported, Point Mass at Center

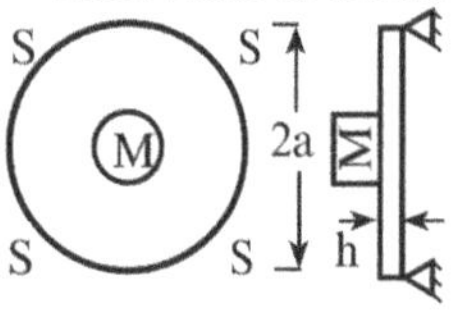

first symmetric mode, i = 0, j = 1 ν=0.3 Ref. [39]

$M/(\gamma\pi a^2)$	0	0.05	0.1	0.2	0.4	0.6	1.0	1.4	2.0
λ^2	4.95	4.57	4.27	3.82	3.22	2.84	2.36	2.06	1.77

$$\lambda_{01}^2 \approx 4.452\left(\frac{1+0.743\nu-0.19\nu^2}{1-0.0814\nu+3.32M/(\gamma\pi a^2)}\right)^{1/2}$$ adapted from Ref. [39]

---

13. Clamped Edge Point Mass at Center

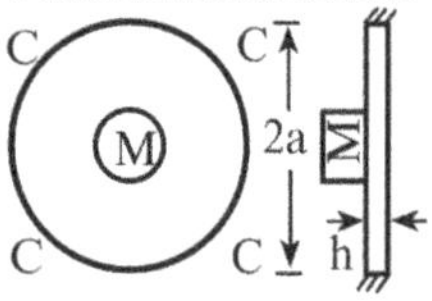

$M/(\gamma\pi a^2)$	0	0.05	0.1	0.2	0.4	0.6	1.0	1.4	Ref. [41]
λ^2	102	9.0	8.1	6.9	5.4	4.75	3.8	3.3	

Polar symmetric mode, i=0, j=1. Approximate formula from Ref. [39]

$$\lambda_{01}^2 = \left(\frac{320}{3[1+5M/(\gamma\pi a^2)]}\right)^{1/2}$$

---

14. Radial Edge Load on Round Plate

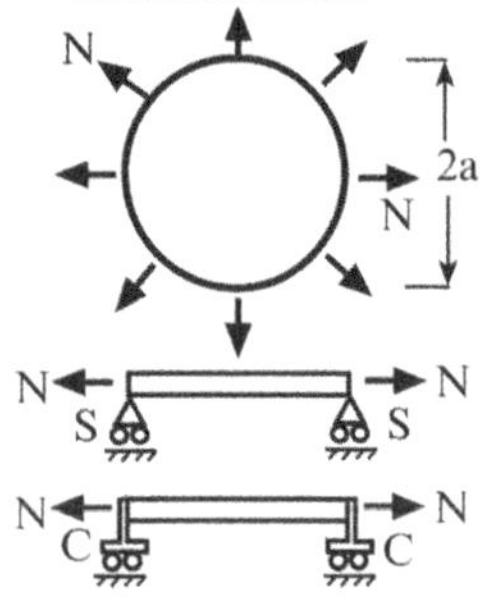

N = force per unit edge

Simply Supported Edge, Ref. [42]

$\dfrac{12(1-\nu^2)Na^2}{4.2Eh^3}$	λ^2 i=0 j=0	λ^2 i=1 j=1
-1.0	0	10.95
-0.5	3.46	12.23
-0.25	4.27	12.86
0.0	4.94	13.47
0.25	5.52	13.98
0.5	6.05	14.55
1.0	6.99	15.57
1.5	7.81	16.55
2.0	8.55	17.47

Clamped Edge, Ref. [42]

$\dfrac{12(1-\nu^2)Na^2}{14.68Eh^3}$	λ^2 i=0 j=0	λ^2 i=1 j=1
-1.0	0	14.31
-0.5	7.28	17.94
-0.25	8.91	19.61
0.0	10.21	21.25
0.25	11.39	22.81
0.5	12.44	24.00
1.0	14.30	26.41
1.5	15.92	28.51
2.0	17.37	30.61

Table 5.2 Natural frequencies of round, annular, and elliptical plates, continued

Additional Notation: a = outside radius; b = inside radius.

Natural Frequency, $f_{ij} = \frac{\lambda_{ij}^2}{2\pi a^2}\sqrt{\frac{Eh^3}{12\gamma(1-\nu^2)}}$, i =0,1,2,3... j =0,1,2,3... Hertz

Boundary | λ_{ij}^2 Refs [43, 44]

15. Free-Free Annular Plate

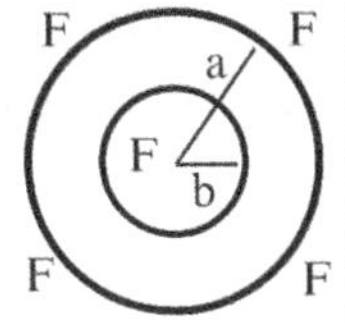

λ^2, $\nu = 0.3$

i	j	b/a 0.1	0.3	0.5	0.7
2	0	5.30	4.91	4.28	3.57
0	1	8.77	8.36	9.32	13.2
3	0	12.4	12.3	11.4	9.86
1	1	20.5	18.3	17.2	22.0
2	1	34.9	33.0	31.1	37.8

16. Free-Simple Annular Plate

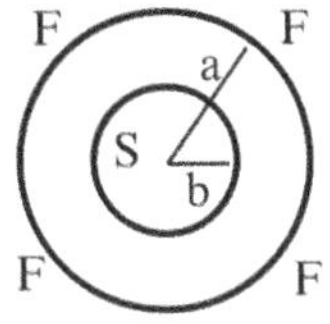

λ^2, $\nu = 0.3$

i	j	b/a 0.1	0.3	0.5	0.7
0	0	3.45	3.42	4.11	6.18
1	0	2.30	3.32	4.86	8.34
2	0	5.42	6.08	7.98	13.4
3	0	12.4	12.6	14.0	20.5
0	1	20.8	31.6	61.0	170

17. Simple-Free Annular Plate

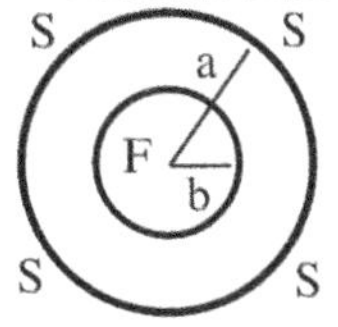

i	j	λ^2 b/a ν=0.3 0.1	0.3	0.5	0.7
2	0	4.86	4.66	5.07	6.94
0	0	13.9	12.8	11.6	13.3
3	0	25.4	24.1	22.3	24.3
1	0	40.0	38.8	35.7	37.2
2	1	29.4	36.9	65.8	175.

18. Free-Clamped Annular Plate

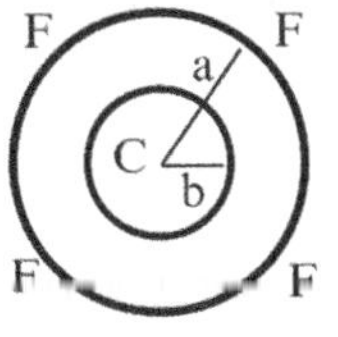

i	j	λ^2 b/a ν=0.3 0.1	0.3	0.5	0.7
0	0	4.23	6.66	13.0	37.0
1	0	3.14	6.33	13.3	37.5
2	0	5.62	7.95	14.7	39.3
3	0	12.4	13.3	18.5	42.6
0	1	25.3	42.6	85.1	239

19. Clamped-Free Annular Plate

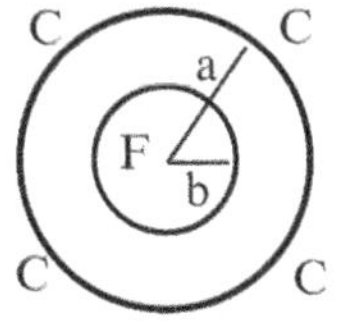

i	j	λ^2 b/a ν=0.3 0.1	0.3	0.5	0.7
0	0	10.2	11.4	17.7	43.1
1	0	21.1	19.5	22.1	45.3
2	0	34.5	32.5	32.0	51.5
3	0	51.0	49.1	45.8	61.3
0	1	39.5	51.7	93.8	253

20. SimpleSimple Annular Plate

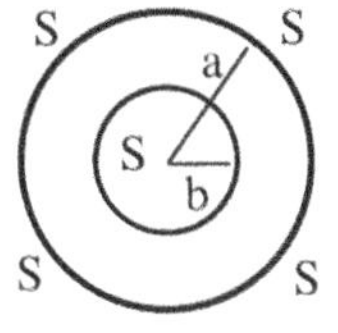

λ^2

i	j	b/a ν = 0.3 0.1	0.3	0.5	0.7
0	0	14.5	21.1	40.0	110
1	0	16.7	23.3	41.8	112
2	0	25.9	30.2	47.1	116
3	0	40.0	42.0	56.0	122
0	1	51.7	81.8	159.	439

21. Simple-Clamp Annular Plate

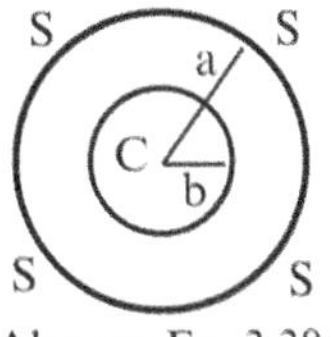

Also see Eq. 3.38.

λ^2, ν=0.3

i	j	b/a 0.1	0.3	0.5	0.7
0	0	17.8	29.9	59.8	168
1	0	19.0	31.4	61.0	170
2	0	26.0	36.2	64.6	172
3	0	40.0	45.4	71.0	177
0	1	60.1	100	198	552

22. Clamp-Simple Annular Plate

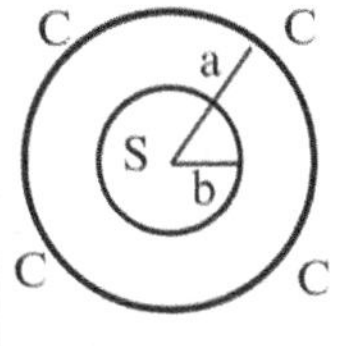

i	j	λ^2 b/a ν=0.3 0.1	0.3	0.5	0.7
0	0	22.6	33.7	63.9	175
1	0	25.1	35.8	65.4	175
2	0	35.4	42.8	70.0	178
3	0	51.0	54.7	78.1	185
0	1	65.6	104.	202.	558

23. Clamp-Clamp Annular Plate

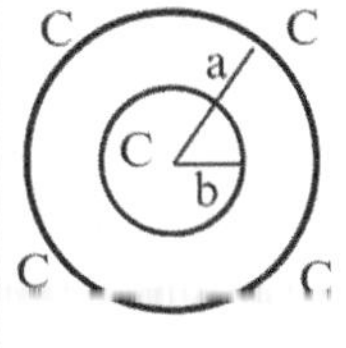

i	j	λ^2 b/a 0.1	0.3	0.5	0.7
0	0	27.3	45.2	89.2	248
1	0	28.4	45.6	90.2	249
2	0	36.7	51.0	93.3	251
3	0	51.2	60.0	99.0	256
0	1	75.3	125.	246.	686

24. Sector with Simply-Supported Straight Edges C, S, or F

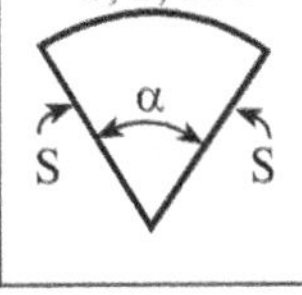

If the straight edges of a circular or annular plate are simply supported, then the previous natural frequencies apply with the following choice of the modal index i, i = 360/(2α) α is angle in degrees. See Ref. [45]

The radial distance r is from the center toward the plate radius a. θ is circumferential angle, radians, and f_i is the ith natural frequency in hertz. The Bessel functions $Y_i(\lambda r/a)$ and $K_i(\lambda r/a)$ approach infinity at the center of the plate as r approaches zero [9]; these terms are not included in circular plates, but they are retained for annular plates [2, 3]. The circumferential dependence $\cos i\theta$ is equally well met by $\sin i\theta$.

There are two orthogonal vibration modes with equal natural frequency for each value of the index i for a circular symmetric plate: one proportional to $\cos i\theta$ and one proportional to $\sin i\theta$. As shown in Figure 5.1, the modal index $i = 0, \ 1, \ 2, \ 3, \ldots$ is the number of nodal diameters in the mode shape, and the modal index $j = 0, \ 1, \ 2, \ 3, \ldots$ is the number of nodal rings in the mode shape not counting nodal rings enforced by boundary conditions. Boundary conditions on the edges determine λ_{ij}, b_i, c_i, and d_i. Consider a round plate with a clamped edge: $W = 0$ and $\partial W/\partial r = 0$ at $r = a$. Applying the first condition to the general circular plate mode shape (Eq. 5.16) implies $J_i(\lambda) + b_i I_i(\lambda) = 0$ so $b_i = -J_i(\lambda)/_i I_i(\lambda)$. The second condition then implies

$$J_i'(\lambda) - \left[\frac{J_i(\lambda)}{I_i(\lambda)}\right] I_i'(\lambda) = 0 \tag{5.16}$$

This transcendental equation is solved for a semiinfinite series of dimensionless natural frequency parameters λ_{ij}, $j = 0, \ 1, \ 2, \ 3, \ \ldots$ for each $i = 0, \ 1, \ 2, \ 3, \ \ldots$ Bessel function identities ([9], articles 9.1.27, 9.6.26) show that Equation 5.16 is identical to the corresponding equation in case 3 of Table 5.2.

5.2.3 *Sectorial and Circular Orthotropic Plates*

A sectorial plate is a pie-shaped slice of a circular or annular plate. Its natural frequency and mode shapes can be adapted from the higher mode natural frequencies of the complete plate, provided that the straight sides are simply supported and the included angle α of the sector is an integer submultiple of 360°, $\alpha = 360°/j$, where $j = 2, 3, 4, \ \ldots$ (see case 24 of Table 5.2). The natural frequencies of sectorial plates with clamped straight edges are provided in Ref. [42]. Variable thickness circular plates are discussed in Refs [46–51]. The thick plate theory includes shear deformation and rotary inertia [11, 18–21], which decrease natural frequency. Spinning circular disks are provided in Refs [52–54].

Cylindrical orthotropic circular and annular plates have different radial and circumferential material properties, as, for example, a plate constructed by winding fibers on a rotating mandrel. This results in a singularity/discontinuity at the center of the plate. The natural frequencies of cylindrical orthotropic plates are discussed in Refs [55–59].

5.2.4 *Rectangular Plates*

Table 5.3 [10, 60, 61, 63–84] presents natural frequencies of rectangular plates with length a and width b. Leissa [1] and Gorman [8, 63] review rectangular plate vibration. Leissa [10] used a series to represent the mode shape of rectangular plate and the Rayleigh technique

(Appendix A) to generate the subtables in cases 1, 3, 4, and 9 through 21 of Table 5.3:

$$\widetilde{w}_{ij}(x,y) = \sum_n \sum_m a_{nm}^{ij} \frac{X_n(x/a)}{Y_m(y/b)} \tag{5.17}$$

Waller [64] found that the mode shape of the i,j modes of a free rectangular plate was well represented by the sum of two terms:

$$\widetilde{w}_{ij}(x,y)|_{\text{free}} \approx \cos\left(\frac{i\pi x}{a}\right)\cos\left(\frac{j\pi y}{a}\right) \pm \cos\left(\frac{j\pi x}{a}\right)\cos\left(\frac{i\pi y}{a}\right) \tag{5.18}$$

Also see Li and Yu [62] and Gorman [63].

Rectangular plate mode shapes can also be represented approximately by the product of the beam mode shapes (Table 4.2) and denoted here by $X_i(x/a)$ and $Y_j(b/y)$ that satisfy boundary conditions [10, 61, 62]:

$$\widetilde{w}_{ij}(x,y) \approx X_i\left(\frac{x}{a}\right) Y_j\left(\frac{y}{b}\right) \tag{5.19}$$

The natural frequency parameter λ_{ij} (Eq. 5.14) is a function of dimensionless parameters G, H, and J given in cases 1–21 of Table 5.3, according to the edge boundary conditions:

$$\lambda_{ij}^2 \approx \pi^2 \left\{ G_i^4 + G_j^4\left(\frac{a}{b}\right)^4 + 2\left(\frac{a}{b}\right)^2 (J_iJ_j + 2\nu(H_iH_j - J_iJ_j)) \right\}^{1/2}$$

$$= \pi^2(i^2 + \left(j^2\frac{a^2}{b^2}\right), \text{ four simply supported sides, } i=1,2,3,\ \ldots\ ,\ j=1,2,3,\ \ldots \text{ (Eq. 5.7)} \tag{5.20}$$

If there are no free edges, then $J = H$ and the Poisson's ratio dependence disappears from the equation. The result is exact for plates with opposite sides and is simply supported ([1]; Eq. 5.6). Equation 5.19 also allows the beam stress theory (Chapter 4) to be applied to plate stress.

5.2.5 *Parallelogram, Triangular and Point-Supported Plates*

A point support allows rotation of the plate about the point but prevents out-of-plane deformation. Point supported plates are in cases 22 to 28 of Table 5.3. Point supports have been used to model vertical column support of building floor slabs [69]. The natural frequency of parallelogram plates on point supports is discussed in Ref. [66].

Parallelogram, rectangular, and triangular plates are provided in Table 5.4 [85–98].

5.2.6 *Rectangular Orthotropic Plates and Grillages*

Two material constants (E and ν) describe the stress–strain behavior of the isotropic plates considered thus far. An orthotropic plate has directional material properties, symmetric

Table 5.3 Natural frequency of rectangular plates

Notation: a = length of plate; b = width of plate; h = thickness of plate; i = number of half-waves in mode shape along horizontal (x) axis; j = number of half-waves in mode shape along vertical (y) axis; E = modulus of elasticity; f = natural frequency, Hz; G, H, J = dimensionless mode shape parameters. S = simply supported edge, C = clamped edge; F = free edge. Consistent units in Table 1.2. Notation continues. [10, 60, 61, 63–84].

Natural Frequency, $f_{ij} = \frac{\lambda_{ij}^2}{2\pi a^2}\left[\frac{Eh^3}{12\gamma(1-\nu^2)}\right]^{1/2} = \frac{\lambda_{ij}^2}{2\pi a^2}\left[\frac{Eh^2}{12\rho(1-\nu^2)}\right]^{1/2}$ Hz, $i = 1,2,3..; j = 1,2,3...$

Boundary Conditions | **Natural Frequency Parameter, λ_{ij}^2 and (ij)** Refs [10,61, 60]

1. Free Edges

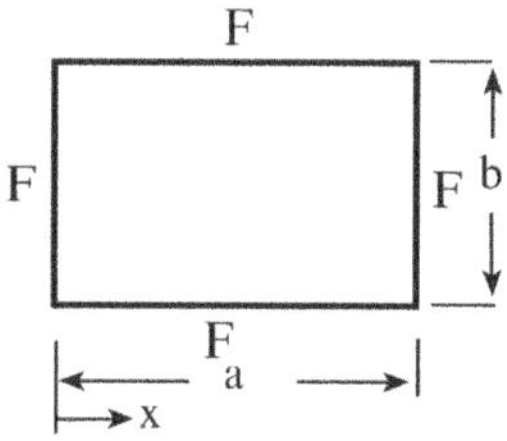

λ^2 and (ij) ν=0.3

a/b	Mode Sequence				
	1	2	3	4	5
0.4	3.463 (13)	5.288 (22)	9.622 (14)	11.44 (23)	18.79 (15)
2/3	8.946 (22)	9.602 (13)	20.74 (23)	22.35 (31)	25.87 (14)
1.0	13.49 (22)	19.79 (13)	24.43 (31)	35.02 (32)	35.02 (23)
1.5	20.13 (22)	21.60 (31)	46.65 (32)	50.29 (13)	50.20 (41)
2.5	21.64 (31)	33.05 (22)	60.14 (41)	71.48 (32)	117.5 (51)

$\lambda_{ij}^2 \approx \pi^2\left\{G_i^4 + (a/b)^4 G_j^4 + 2(a/b)^2[J_iJ_j + 2\nu(H_iH_j - J_iJ_j)]\right\}^{1/2}$

$G_1 = 0, G_2 = 0, G_3 = 1.506, G_k = k - 3/2$, k=4,5.Also see Refs [63, 64]

$H_1 = 0, H_2 = 1.248, H_3 = 1.506, H_k = (k-3/2)\left(1 - 2[(k-3/2)\pi]^{-1}\right), k = 4,5.$

$J_1 = 0, J_2 = 1.216, J_3 = 5.017, J_k = (k-3/2)^2(1 + 6[(k-3/2)\pi]^{-1}), k = 4,5..$

2. Simply Supported Edges

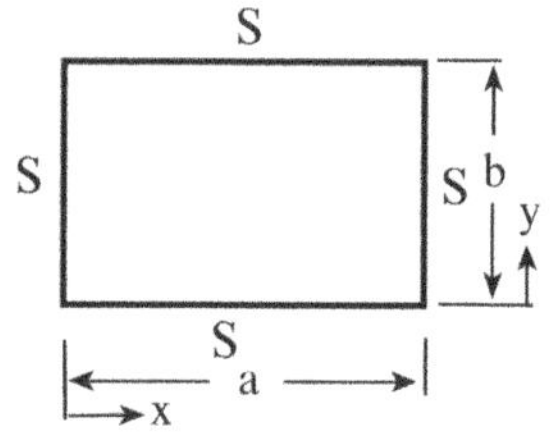

λ^2 and (ij) λ indep. of ν

a/b	Mode Sequence				
	1	2	3	4	5
0.4	11.45 (11)	16.19 (12)	24.08 (13)	35.14 (14)	41.06 (21)
2/3	14.26 (11)	27.42 (12)	43.86 (21)	49.35 (13)	57.02(22)
1.0	19.74 (11)	49.35 (21)	49.35 (12)	78.96 (22)	98.99(70)
1.5	32.08 (11)	61.69 (21)	98.70 (12)	111.0 (31)	128.3(22)
2.5	71.56 (11)	101.2 (21)	150.5 (31)	219.6 (41)	256.6(12)

$\lambda_{ij}^2 = \pi^2\{i^2 + j^2(a/b)^2\}$, $\tilde{w}_{ij} = \sin(i\pi x/a)\sin(j\pi y/b)$,

$H_j = J_j = j^2$, $H_i = J_i = i^2, G_i = i, G_j = j$. See Section 5.2.

3. Clamped Edges

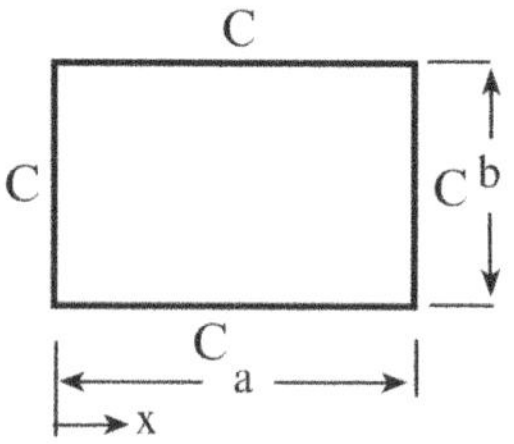

λ^2 and (ij) λ indep. of ν

a/b	Mode Sequence				
	1	2	3	4	5
0.4	23.65 (11)	27.82 (12)	35.45(13)	46.70 (14)	61.55 (15)
2/3	27.01 (11)	41.72 (12)	66.14 (21)	66.55 (13)	79.85 (22)
1.0	35.99 (11)	73.41 (21)	73.41(12)	108.3 (22)	131.6 (31)
1.5	60.77 (11)	93.86 (21)	148.8 (12)	149.7 (31)	179.7 (22)
2.5	147.8 (11)	173.9 (21)	221.5 (31)	291.9 (41)	384.7 (51)

$\lambda_{ij}^2 \approx \pi^2\left\{G_i^4 + (a/b)^4 G_j^4 + 2(a/b)^2 J_iJ_j\right\}^{1/2}$

$G_1 = 1.506,\ G_2 = 2.5,\ G_k = k + 1/2,\ k = 3,4...,$

$H_k = J_k, J_1 = 1.248,\ J_2 = 4.658, J_3 = 10.02,\ J_k = (k+1/2)^2(1-2[(k+1/2)p]^{-1}), k = 4,5..$

Table 5.3 Natural frequency of rectangular plates, continued

Notation continued: $\gamma = \rho h + \gamma_{ns}$ where ρ = plate material density and γ_{ns} = density of non-structural mass per unit area, if any; ν = Poisson's ratio; λ = dimensionless natural frequency parameter. $\tilde{y}_j(y/b)$, $\tilde{w}_{ij}(x,y)$ = mode shapes for out-of-plane deformation. Consistent sets of units are given in Table 1.2.

$$\text{Natural Frequency, } f_{ij} = \frac{\lambda_{ij}^2}{2\pi a^2}\left[\frac{Eh^3}{12\gamma(1-\nu^2)}\right]^{1/2} \text{ Hz, } i = 1,2,3..;\ j = 1,2,3...$$

Boundary Conditions	Natural Frequency Parameter, λ_{ij}^2 and (ij) Refs [10, 61, 60]

4. Simply Supported-Free-Simply Supported-Free

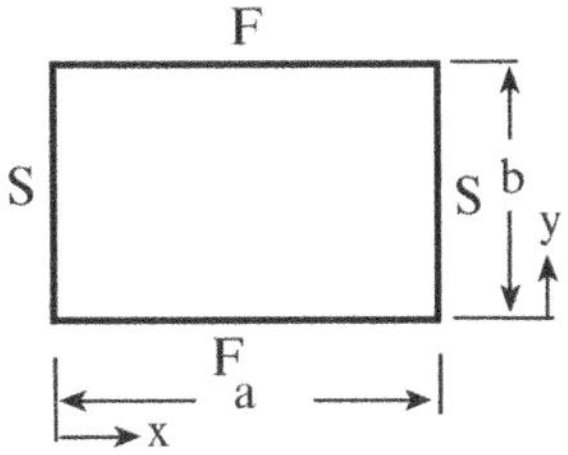

λ^2 and (ij) $\nu=0.3$

a/b	Mode Sequence				
	1	2	3	4	5
0.4	9.760 (11)	11.04 (12)	15.06 (13)	21.71 (14)	31.18 (15)
2/3	9.698 (11)	12.98 (12)	22.95 (13)	39.11 (21)	40.36 (14)
1.0	9.631 (11)	16.14 (12)	36.73 (13)	38.95 (21)	46.74 (22)
1.5	9.558 (11)	21.62 (12)	38.72 (21)	54.84 (22)	65.79 (13)
2.5	9.484 (11)	33.62 (12)	38.36 (21)	75.20 (22)	86.97 (31)

$\lambda_{ij}^2 \approx \pi^2\{i^4 + (a/b)^4 G_j^4 + 2i^2(a/b)^2[J_j + 2\nu(H_j - J_j)]\}^{1/2}$

$\tilde{w}_{ij} = \sin(i\pi x/a)\tilde{y}_j(y/b)$, $\tilde{y}_j$ from case 1 of Table 4.3.

G_j, H_j, J_j, from case 1.

5. Simply Supported-Free Simply Supported-Simply Supported

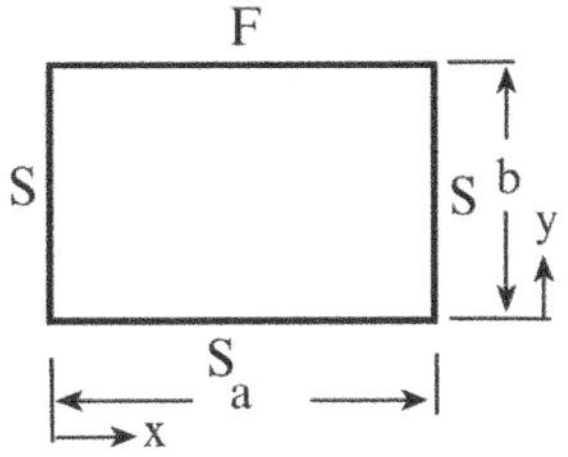

λ^2 and (ij) $\nu=0.3$

a/b	Mode Sequence				
	1	2	3	4	5
0.4	10.13 (11)	13.06 (12)	18.84 (13)	27.56(14)	39.34 (15)
2/3	10.67 (11)	18.30 (12)	33.70 (13)	40.13 (21)	48.41 (22)
1.0	11.68 (11)	27.76 (12)	41.20 (21)	59.17 (22)	61.86 (13)
1.5	13.71 (11)	43.57 (21)	47.86 (12)	81.48 (22)	92.69 (31)
2.5	18.80 (11)	50.54 (21)	100.2 (31)	100.2 (12)	147.6 (22)

$\lambda_{ij}^2 \approx \pi^2\{i^4 + (a/b)^4 G_j^4 + 2i^2(a/b)^2[J_j + 2\nu(H_j - J_j)]\}^{1/2}$

$\tilde{w}_{ij} = \sin(i\pi x/a)\tilde{y}_j(y/b)$, $\tilde{y}_j$ from case 4 of Table 4.3.

G_j, H_j, J_j, from case 9.

6. Simply Supported-Free Simply Supported-Clamped

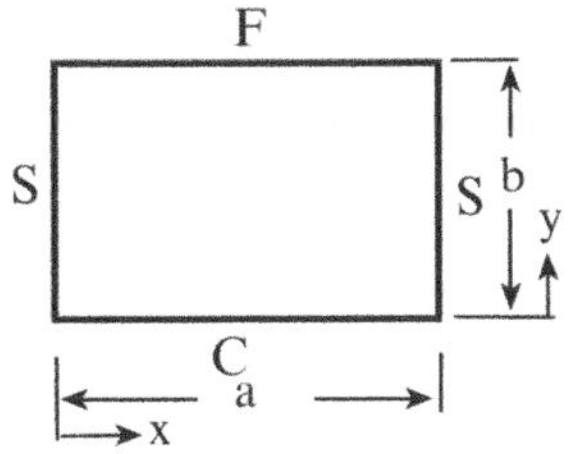

λ^2 and (ij) $\nu=0.3$

a/b	Mode Sequence				
	1	2	3	4	5
0.4	10.19 (11)	13.60 (12)	20.10 (13)	29.62 (14)	39.64 (21)
2/3	10.98 (11)	20.34 (12)	37.96 (13)	40.27 (21)	49.73 (22)
1.0	12.69 (11)	33.07 (12)	41.70 (21)	63.01 (22)	72.40 (13)
1.5	16.82 (11)	45.30 (21)	61.02 (12)	92.31 (22)	93.83 (31)
2.5	30.63 (11)	58.08 (21)	105.5 (31)	149.5 (12)	173.1 (41)

$\lambda_{ij}^2 \approx \pi^2\{i^4 + (a/b)^4 G_j^4 + 2i^2(a/b)^2[J_j + 2\nu(H_j - J_j)]\}^{1/2}$

$\tilde{w}_{ij} = \sin(i\pi x/a)\tilde{y}_j(y/b)$, $\tilde{y}_j$ from case 3 of Table 4.3.

G_j, H_j, J_j, from case 15.

Table 5.3 Natural frequency of rectangular plates, continued

Natural Frequency, $f_{ij} = \dfrac{\lambda_{ij}^2}{2\pi a^2}\left[\dfrac{Eh^3}{12\gamma(1-\nu^2)}\right]^{1/2}$ Hz, $i = 1,2,3..; j = 1,2,3...$

Boundary Conditions	Natural Frequency Parameter, λ^2 and (ij) Refs [10, 61, 60]

7. Simply Supported-Simply Supported-Simply Supported-Clamped

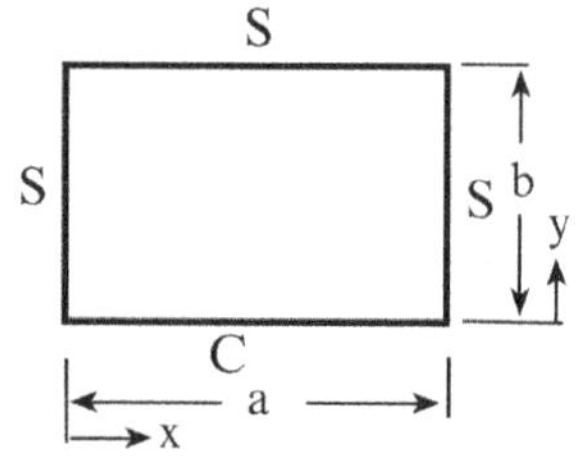

λ^2 and (ij)

a/b	Mode Sequence				
	1	2	3	4	5
0.4	11.75 (11)	17.19 (12)	25.92 (13)	37.83 (14)	41.21 (21)
2/3	15.58 (11)	31.07 (12)	44.56 (21)	55.39 (13)	59.46 (22)
1.0	23.65 (11)	51.67 (21)	58.65 (12)	86.13 (22)	100.3 (31)
1.5	42.53 (11)	69.00 (21)	116.3 (31)	121.0 (12)	147.6 (22)
2.5	103.9 (11)	128.3 (21)	172.4 (31)	237.3 (41)	320.8 (12)

$\lambda_{ij}^2 \approx \pi^2\left\{i^4 + (a/b)^4 G_j^4 + 2i^2(a/b)^2 J_j\right\}^{1/2}$

$\tilde{w}_{ij} = \sin(i\pi x/a)\tilde{y}_j(y/b)$, $\tilde{y}_j$ from case 6 of Table 4.3.

G_j, J_j from case 20.

8. Simply Supported-Clamp Simply Supported-Clamp

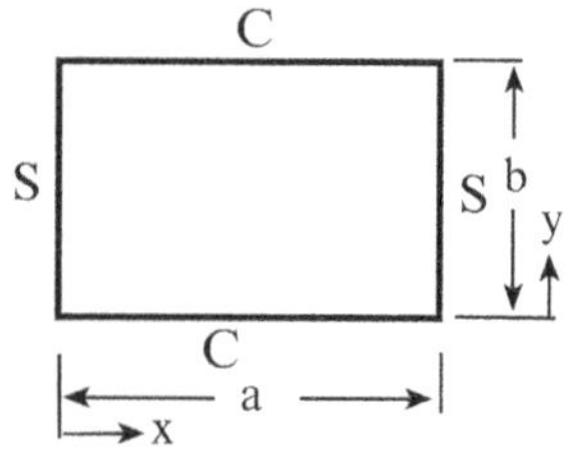

λ^2 and (ij)

a/b	Mode Sequence				
	1	2	3	4	5
0.4	12.13 (11)	18.36 (12)	27.97 (13)	40.75 (14)	41.38 (21)
2/3	17.37 (11)	35.34 (12)	45.43 (21)	62.05 (13)	62.31 (22)
1.0	28.95 (11)	54.74 (21)	69.12 (12)	94.59 (22)	102.2 (31)
1.5	56.35 (11)	78.98 (21)	123.2 (31)	146.3 (12)	170.1 (22)
2.5	145.5 (11)	164.7 (21)	202.2 (31)	261.1 (41)	342.1 (51)

$\lambda_{ij}^2 \approx \pi^2\left\{i^4 + (a/b)^4 G_j^4 + 2i^2(a/b)^2 J_j\right\}^{1/2}$

$\tilde{w}_{ij} = \sin(i\pi x/a)\tilde{y}_j(y/b)$, $\tilde{y}_j$ from case 7 of Table 4.3.

G_j, J_j, from case 3

9. Simply Supported-Free-Free-Simply Supported

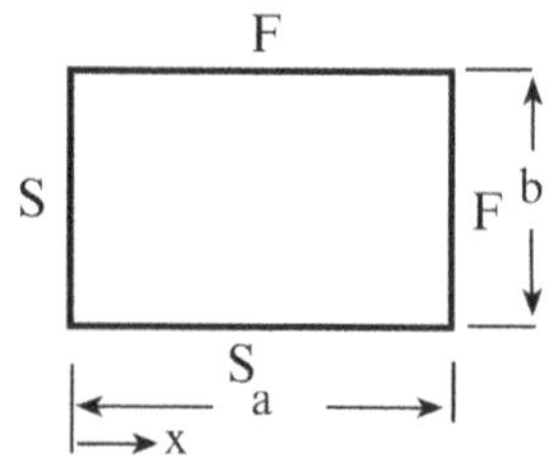

λ^2 and (ij) ν=0.3

a/b	Mode Sequence				
	1	2	3	4	5
0.4	1.320 (11)	4.743 (12)	10.36 (13)	15.87 (21)	18.93 (14)
2/3	2.234 (11)	9.575 (12)	16.76 (21)	24.66 (13)	27.06 (22)
1.0	3.369 (11)	17.41 (12)	19.37 (21)	28.29 (22)	51.32 (13)
1.5	5.026 (11)	21.54 (21)	37.72 (12)	55.49 (31)	60.88 (22)
2.5	8.251 (11)	29.65 (21)	64.76 (31)	99.21 (12)	118.3 (41)

$\lambda_{ij}^2 \approx \pi^2\left\{G_i^4 + (a/b)^4 G_j^4 + 2(a/b)^2[J_i J_j + 2\nu(H_i H_j - J_i J_j)]\right\}^{1/2}$

$G_1=0$, $G_2=1.25$, $G_3=2.25$, $G_k = k-1/2$, k=4,5..

$H_1=0$, $H_2=1.165$, $H_3=4.346$, $H_k = (k-3/4)^2(1-2[(k-3/4)\pi]^{-1})$, k=4,5,..

$J_1=0.3040$, $J_2=2.756$, $J_3=7.211$, $J_k = (k-3/4)^2(1+3[(k-3/4)\pi]^{-1})$, k=4,5,

Table 5.3 Natural frequency of rectangular plates, continued

$$\text{Natural Frequency, } f_{ij} = \frac{\lambda_{ij}^2}{2a^2}\left[\frac{Eh^3}{12(1-\nu^2)}\right]^{1/2} \text{ Hz, } i=1,2,3..;\ j=1,2,3...$$

Boundary Conditions	Natural Frequency Parameter, λ^2 and (ij) Refs [10,61, 60]

10. Simply Supported-Free-Free-Free

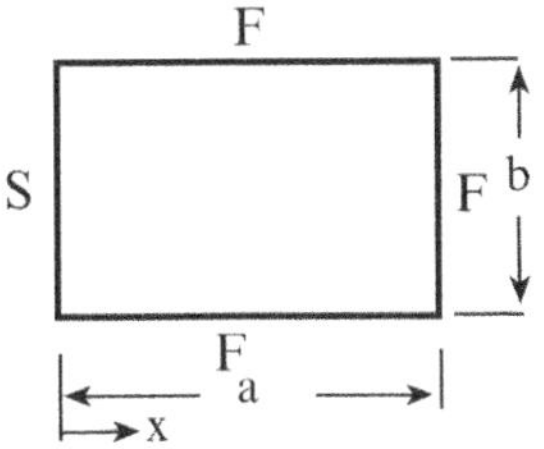

a/b	Mode Sequence λ^2 and (ij) ν=0.3				
	1	2	3	4	5
0.4	2.692 (12)	6.503 (13)	12.64 (14)	15.34 (21)	17.51 (22)
2/3	4.481 (12)	13.01 (13)	15.67 (21)	20.37 (22)	30.55 (14)
1.0	6.648 (12)	15.02 (21)	25.49 (22)	26.13 (13)	48.71 (31)
1.5	9.850 (12)	15.01 (21)	34.03 (22)	48.33 (31)	55.07 (13)
2.5	14.94 (21)	16.24 (12)	48.84 (31)	52.09 (22)	97.23 (32)

$\lambda_{ij}^2 \approx \pi^2\{G_i^4 + (a/b)^4 G_j^4 + 2(a/b)^2[J_iJ_j + 2\nu(H_iH_j - J_iJ_j)]\}^{1/2}$

G_i, H_i, $J_i$from case 9. G_j, H_j, J_j from case 1.

11. Clamped-Free-Free-Free

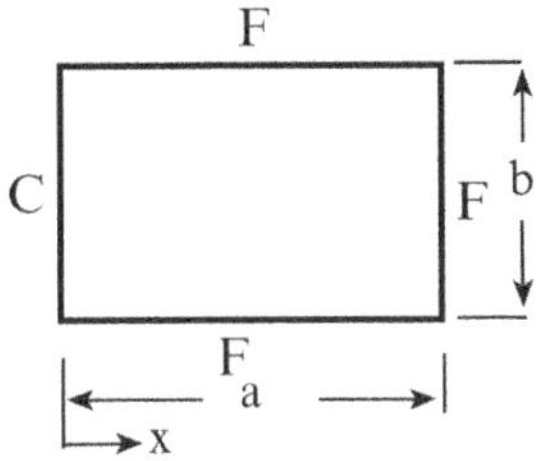

a/b	Mode Sequence λ^2 and (ij) ν=0.3				
	1	2	3	4	5
0.4	3.511 (11)	4.786 (12)	8.115 (13)	13.88 (14)	21.64 (21)
2/3	3.502 (11)	6.406 (12)	15.54 (13)	20.04 (21)	26.07 (22)
1.0	3.492 (11)	8.525 (12)	21.43 (21)	27.33 (13)	31.11 (22)
1.5	3.477 (11)	11.68 (12)	21.62 (21)	39.49 (22)	53.88 (13)
2.5	3.456 (11)	17.99 (12)	21.56 (21)	57.46 (22)	60.58 (31)

$\lambda_{ij}^2 \approx \pi^2\{G_i^4 + (a/b)^4 G_j^4 + 2(a/b)^2[J_iJ_j + 2\nu(H_iH_j - J_iJ_j)]\}^{1/2}$

G_i, H_i, J_i,from case 15.G_j. H_j, J_j from case 1. Note $\lambda^2_{11}=1.875^2$

12. Clamped-Free-Simply Supported-Free

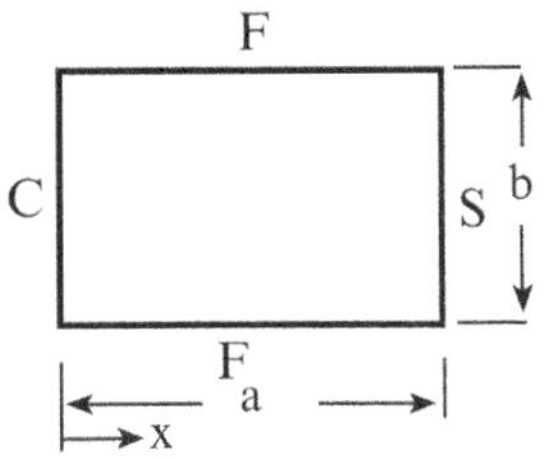

a/b	Mode Sequence λ^2 and (ij) ν=0.3				
	1	2	3	4	5
0.4	15.38 (11)	16.37 (12)	9.622 (13)	11.44 (14)	34.51 (15)
2/3	15.34 (11)	17.95 (12)	20.74 (13)	22.35 (14)	49.84 (21)
1.0	15.29 (11)	20.67 (12)	24.43 (13)	35.02 (21)	56.62 (22)
1.5	15.22 (11)	25.71 (12)	46.65 (21)	50.29 (22)	68.13 (13)
2.5	15.13 (11)	37.29 (12)	60.14 (21)	71.48 (22)	103.1 (31)

$\lambda_{ij}^2 \approx \pi^2\{G_i^4 + (a/b)^4 G_j^4 + 2(a/b)^2[J_iJ_j + 2\nu(H_iH_j - J_iJ_j)]\}^{1/2}$

G_i, H_i, J_i, from case 20. G_j, H_j, J_j from case 1.

13. Clamped-Free-Free-Simply Supported

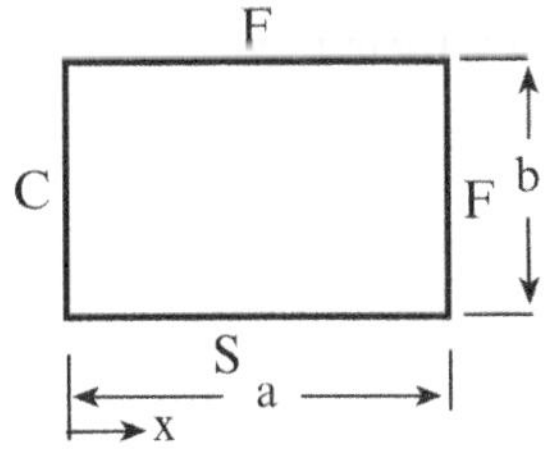

a/b	Mode Sequence λ^2 and (ij) ν=0.3				
	1	2	3	4	5
0.4	3.854 (11)	6.420 (12)	11.58 (13)	19.77 (14)	22.52 (21)
2/3	4.425 (11)	10.91 (12)	22.96 (21)	25.70 (13)	32.43 (22)
1.0	5.364 (11)	19.17 (12)	24.77 (21)	43.19 (22)	53.00 (13)
1.5	6.931 (11)	27.29 (21)	38.59 (12)	64.25 (22)	67.47 (31)
2.5	10.10 (11)	35.16 (21)	74.99 (31)	99.93 (12)	127.7 (22)

$\lambda_{ij}^2 \approx \pi^2\{G_i^4 + (a/b)^4 G_j^4 + 2(a/b)^2[J_iJ_j + 2\nu(H_iH_j - J_iJ_j)]\}^{1/2}$

G_i, H_i, J_i,from case 15. G_j, H_j, J_j from case 9.

Table 5.3 Natural frequency of rectangular plates, continued

$$\text{Natural Frequency, } f_{ij} = \frac{\lambda_{ij}^2}{2a^2}\left[\frac{Eh^3}{12(1-\nu^2)}\right]^{1/2} \text{ Hz, } i=1,2,3..;\ j=1,2,3...$$

Boundary Conditions	Natural Frequency Parameter, λ^2 and (ij) Refs [10,61,60]

14. Clamp-Free-Clamp-Free

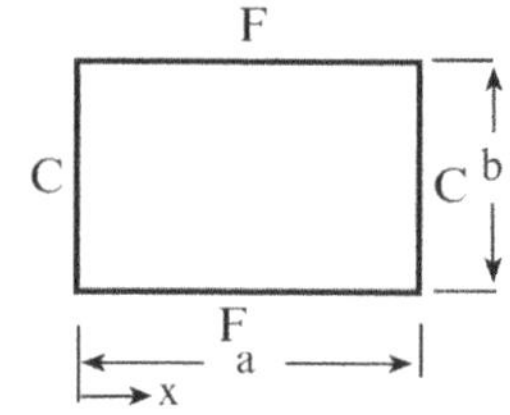

a/b	Mode Sequence	λ^2 and (ij)			ν=0.3
	1	2	3	4	5
0.4	22.35 (11)	23.09 (12)	25.67 (13)	30.63 (14)	38.69 (15)
2/3	22.31 (11)	24.31 (12)	31.70 (13)	46.82 (14)	61.57 (21)
1.0	22.27 (11)	26.53 (12)	43.66 (13)	61.47 (21)	67.55(22)
1.5	22.21 (11)	30.90 (21)	61.30 (21)	70.96 (13)	74.26 (22)
2.5	22.13 (11)	41.69 (21)	61.00 (21)	92.38 (22)	119.9 (31)

$$\lambda_{ij}^2 \approx \pi^2\{G_i^4 + (a/b)^4 G_j^4 + 2(a/b)^2[J_iJ_j + 2\nu(H_iH_j - J_iJ_j)]\}^{1/2}$$

G_i, H_i, J_i, from case 3. G_j, H_j, J_j from case 1.

15. Clamp-Free-Free-Clamp

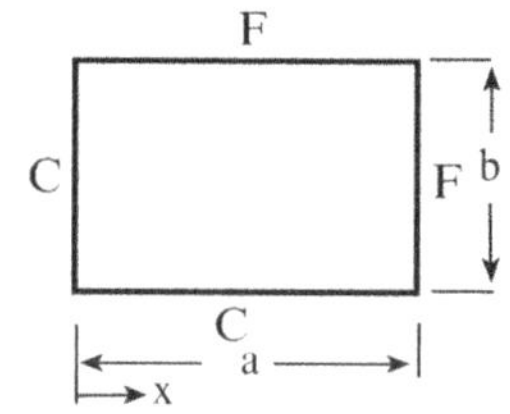

a/b	Mode Sequence	λ^2 and (ij)			ν=0.3
	1	2	3	4	5
0.4	3.986 (11)	7.155 (12)	13.10 (13)	21.84 (14)	22.90 (21)
2/3	4.985 (11)	13.29 (12)	23.38 (21)	30.26 (13)	34.24 (22)
1.0	6.942 (11)	24.03 (21)	26.68 (12)	47.78 (22)	63.04 (13)
1.5	11.22 (11)	29.90 (21)	52.62 (12)	68.09 (31)	77.04 (22)
2.5	24.91 (11)	44.72 (21)	81.88 (31)	136.5 (41)	143.1 (12)

$$\lambda_{ij}^2 \approx \pi^2\{G_i^4 + (a/b)^4 G_j^4 + 2(a/b)^2[J_iJ_j + 2\nu(H_iH_j - J_iJ_j)]\}^{1/2}$$

$G_1=0.597$, $G_2=1.494$, $G_3=2.50$, $G_k = k-1/2$, $k=4,5..$
$H_1=-0.087$, $H_2=1.347$, $H_3 = 4.658$, $H_k = (k-1/2)^2(1-2[(k-1/2)\pi]^{-1})$,
$J_1=0.471$, $J_2= 3.284$, $J_3 = 7.842$, $J_k = (k-1/2)^2(1+2[(k-1/2)\pi]^{-1})$,

16. Clamped-Free-Clamped-Simply Supported

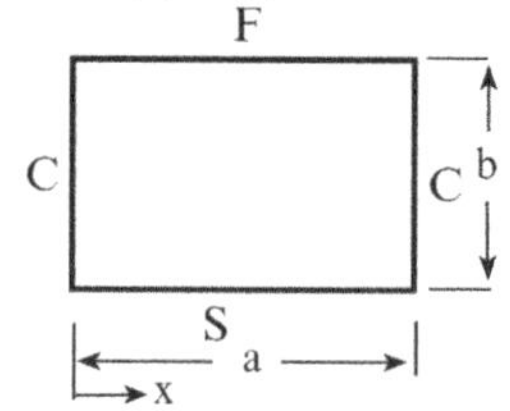

a/b	Mode Sequence	λ^2 and (ij)			ν=0.3
	1	2	3	4	5
0.4	22.54 (11)	24.30 (12)	28.34 (13)	35.35 (14)	45.71 (15)
2/3	22.86 (11)	27.97 (12)	40.68 (13)	62.31 (21)	62.70 (14)
1.0	23.46 (11)	35.61 (12)	63.13 (21)	66.81 (13)	77.50 (22)
1.5	24.78 (11)	53.73 (12)	64.96 (21)	97.26 (22)	124.5 (31)
2.5	28.56 (11)	70.56 (21)	114.0 (12)	130.8 (31)	159.5 (22)

$$\lambda_{ij}^2 \approx \pi^2\{G_i^4 + (a/b)^4 G_j^4 + 2(a/b)^2[J_iJ_j + 2\nu(H_iH_j - J_iJ_j)]\}^{1/2}$$

G_i, H_i, J_i from case 3. G_j, H_j, J_j from case 9.

17. Clamped-Free-Simply Supported-Simply Supported

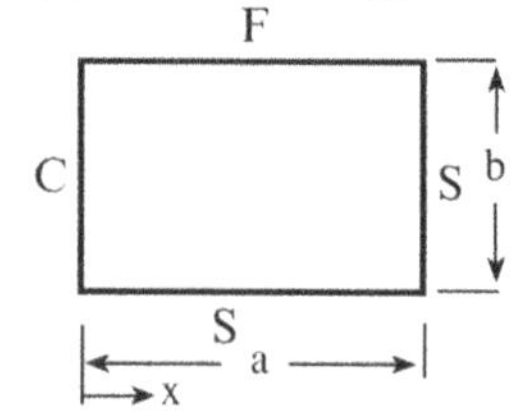

a/b	Mode Sequence	λ^2 and (ij)			ν=0.3
	1	2	3	4	5
0.4	15.65 (11)	17.95 (12)	22.90 (13)	30.89 (14)	42.11 (15)
2/3	16.07 (11)	22.45 (12)	36.70 (13)	50.70 (21)	57.91 (22)
1.0	16.87 (11)	31.14 (12)	51.63 (21)	64.04 (13)	67.65 (22)
1.5	18.54 (11)	50.44 (12)	53.72 (21)	88.80 (22)	108.2 (31)
2.5	23.07 (11)	59.97 (21)	112.0 (12)	115.1 (31)	153.2 (22)

$$\lambda_{ij}^2 \approx \pi^2\{G_i^4 + (a/b)^4 G_j^4 + 2(a/b)^2[J_iJ_j + 2\nu(H_iH_j - J_iJ_j)]\}^{1/2}$$

G_i, H_i, J_i, from case 20. G_j, H_j, J_j from case 9.

Table 5.3 Natural frequency of rectangular plates, continued

$$\text{Natural Frequency, } f_{ij} = \frac{\lambda_{ij}^2}{2\pi a^2}\left[\frac{Eh^3}{12\gamma(1-\nu^2)}\right]^{1/2} \text{ Hz, } i=1,2,3..;\ j=1,2,3...$$

Boundary Conditions	Natural Frequency Parameter, λ^2 and (ij) Refs [10,61,60]

18. Clamped-Free-Simply Supported-Clamped

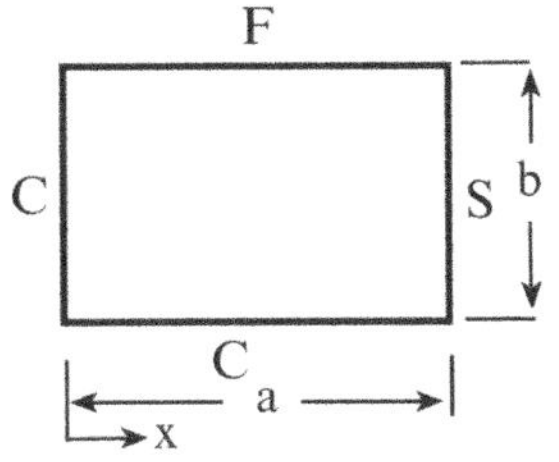

a/b	Mode Sequence 1	2	3	4	5
	λ^2 and (ij)				ν=0.3
0.4	15.70 (11)	18.37 (12)	23.99 (13)	32.18 (14)	44.86 (15)
2/3	16.29 (11)	24.20 (12)	40.70 (13)	50.82 (21)	59.07 (22)
1.0	17.62 (11)	36.05 (12)	52.07 (21)	71.19 (22)	74.35 (13)
1.5	21.04 (11)	55.18 (21)	63.18 (12)	99.01 (22)	109.2 (31)
2.5	33.58 (11)	66.61 (21)	119.9 (31)	150.8 (12)	187.6 (22)

$\lambda_{ij}^2 \approx \pi^2\{G_i^4 + (a/b)^4 G_j^4 + 2(a/b)^2[J_iJ_j + 2\nu(H_iH_j - J_iJ_j)]\}^{1/2}$

G_i, H_i, J_i, from case 20. G_j, H_j, J_j, from case 15.

19. Clamped-Free-Clamped-Clamped

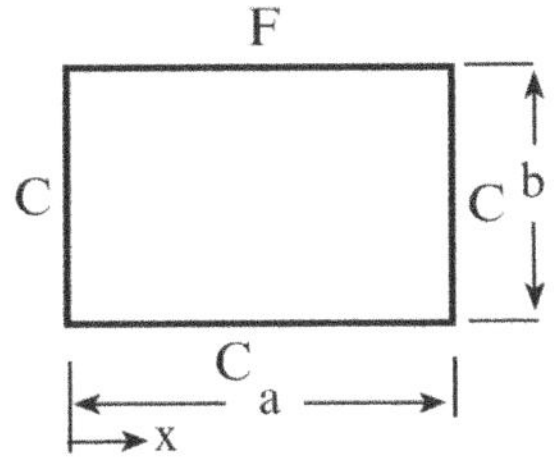

a/b	Mode Sequence 1	2	3	4	5
	λ^2 and (ij)				ν=0.3
0.4	22.58 (11)	24.62 (12)	29.24 (13)	37.06 (14)	48.28 (15)
2/3	23.02 (11)	29.43 (12)	44.36 (13)	62.42 (21)	68.89 (14)
1.0	24.02 (11)	40.04 (12)	63.49 (21)	76.76 (13)	80.71 (22)
1.5	26.73 (11)	65.92 (12)	66.22 (21)	106.8 (22)	125.4 (31)
2.5	37.66 (11)	76.41 (21)	135.2 (31)	152.5 (12)	193.0 (22)

$\lambda_{ij}^2 \approx \pi^2\{G_i^4 + (a/b)^4 G_j^4 + 2(a/b)^2[J_iJ_j + 2\nu(H_iH_j - J_iJ_j)]\}^{1/2}$

G_i, H_i, J_i, from case 3. G_j, H_j, J_j, from case 15

20. Clamp-Simply Supported Simply Supported-Clamped

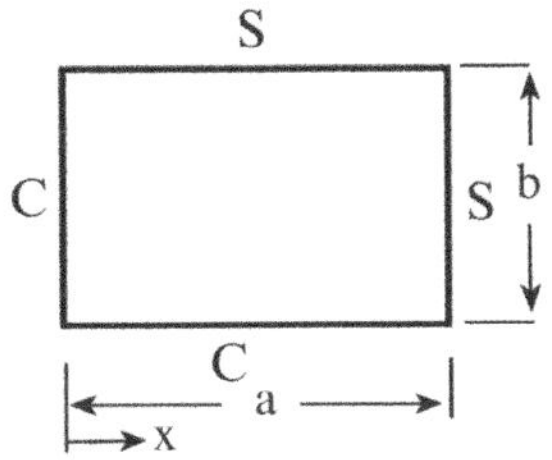

a/b	Mode Sequence 1	2	3	4	5
	λ^2 and (ij)				
0.4	16.85 (11)	21.36 (12)	29.24 (13)	40.51 (14)	51.46 (21)
2/3	19.95 (11)	34.02 (12)	54.37 (21)	57.52 (13)	67.81 (22)
1.0	27.06 (11)	60.54 (21)	60.79 (12)	92.86 (22)	114.6 (13)
1.5	44.89 (11)	76.55 (21)	122.3 (12)	129.3 (31)	152.6 (22)
2.5	105.3 (11)	133.5 (21)	182.7 (31)	253.2 (41)	321.6 (12)

$\lambda_{ij}^2 \approx \pi^2\{G_i^4 + (a/b)^4 G_j^4 + 2(a/b)^2 J_iJ_j\}^{1/2}$

$G_k = k+1/4$, $k=1,2,3..$, $H_k=J_k$;

$J_1 = 1.165$, $J_2=4.346$, $J_3=9.528$, $J_k = (k+1/4)^2(1-[(k+1/4)\pi]^{-1})$, $k=4,5$.

21. Clamp-Simply Supported Clamped-Clamped

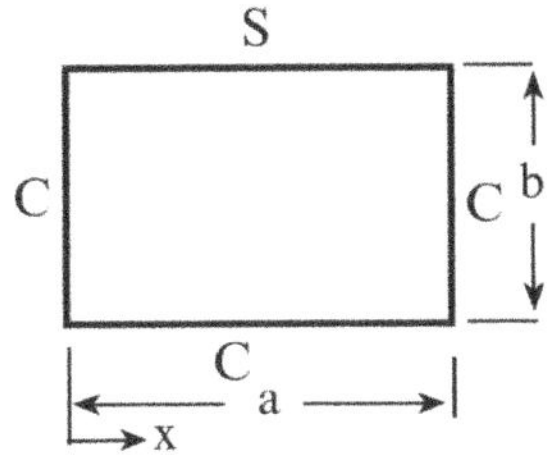

a/b	Mode Sequence 1	2	3	4	5
	λ^2 and (ij)				
0.4	23.44 (11)	27.02 (12)	33.80 (13)	44.13 (14)	58.03 (15)
2/3	25.86 (11)	38.10 (12)	60.33 (13)	65.62 (21)	77.56 (22)
1.0	31.83 (11)	63.35 (12)	71.08 (21)	100.8 (22)	116.4 (13)
1.5	48.17 (11)	85.51 (21)	124.0 (12)	144.0 (31)	158.4 (22)
2.5	107.1 (11)	139.7 (21)	194.4 (31)	270.5 (41)	322.6 (12)

$\lambda_{ij}^2 \approx \pi^2\{G_i^4 + (a/b)^4 G_j^4 + 2(a/b)^2 J_iJ_j\}^{1/2}$

G_i, H_i, J_i, from case 3. G_j, J_j, from case 20.

Table 5.3 Natural frequency of rectangular plates, continued

$$\text{Natural Frequency, } f_{ij} = \frac{\lambda_{ij}^2}{2\pi a^2}\left[\frac{Eh^3}{12\gamma(1-\nu^2)}\right]^{1/2} \text{ Hz, } i=1,2,3..;\ j=1,2,3...$$

Boundary Conditions | **Natural Frequency Parameter, λ^2**

---

22. Corner Point Supports

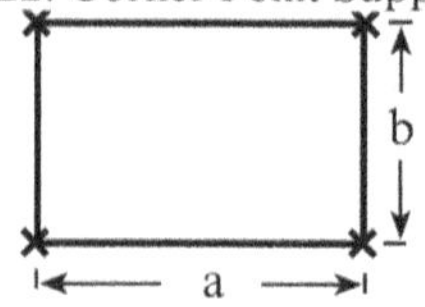

ν=0.3 Ref. [65]

a/b	1	1.5	2.0	2.5
λ_1^2	7.12	8.92	9.29	9.39
λ_2^2	15.8	21.5	27.5	35.5

Point supports allow rotation but not out-of-plane displacement.

---

23. Square Plate with Four Point Supports

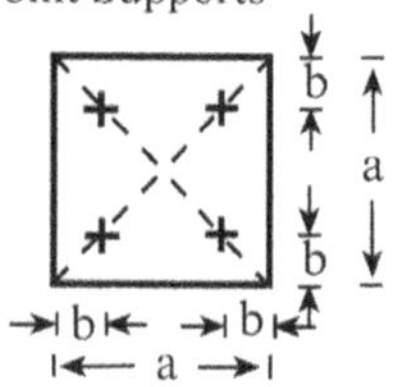

ν=0.3 Ref. [66]

b/a	0	0.1	0.2	0.3	0.4	0.5
λ_1^2	7.14	12.89	19.69	19.31	13.35	11.34
λ_2^2	15.79	19.69	23.13	19.72	14.06	13.47
λ_3^2	19.69	23.97	32.56	24.30	16.83	19.69

b/a=0.5 gives a single point support at center of plate.

---

24. Square Plate with n Equally Spaced Supports on Edges

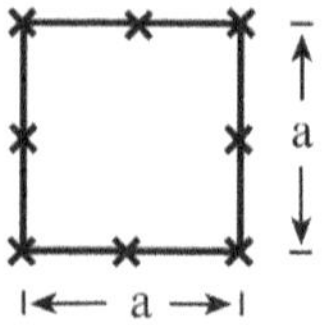

Ref. [67]

N	2	3	5	7	9	∞
λ_1^2	7.12	18.20	19.64	19.71	19.73	19.74

n = ∞ is a simply supported plate, case 2.

n = 3 shown

---

25. Square Plate with Point Supports at Midpoints of Sides

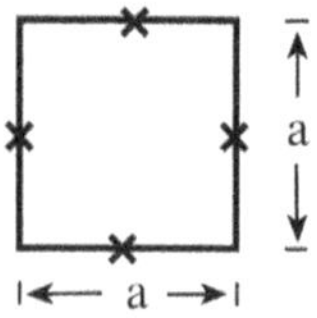

λ_1^2 = 13.5 Ref. [68]

ν = 0.3

Point supports allow rotation by not out-of-plane displacement.

---

26. N-Bay Point Supported Plate

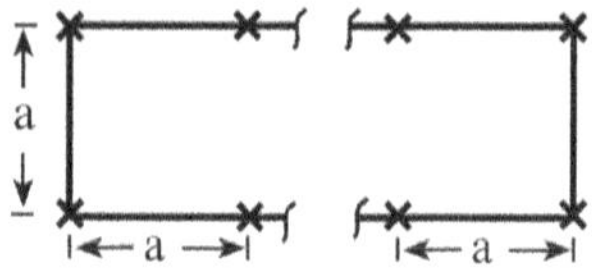

Ref. [69]

Num of Bays	1	2	3	4	5
λ_1^2	7.18	16.27	24.41	33.02	41.41
λ_2^2	16.30	16.76	25.41	33.41	41.86
λ_3^2	16.30	33.28	28.39	37.20	45.43

1 bay = square plate, case 22.

Table 5.3 Natural frequency of rectangular plates, continued

Natural Frequency, $f_{ij} = \dfrac{\lambda_{ij}^2}{2\pi a^2}\left[\dfrac{Eh^3}{12\gamma(1-\nu^2)}\right]^{1/2}$ Hz, $i=1,2,3..;\ j=1,2,3...$

<table>
<tr><th>Boundary Conditions</th><th>Natural Frequency Parameter, λ^2</th></tr>
<tr><td>27. Simply Supported Edges Central Point Support
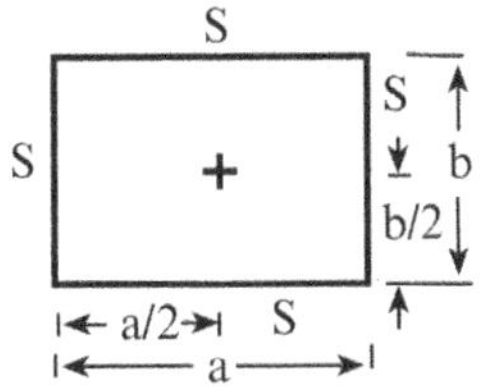</td><td>ν=0.3 Ref. [70]

a/b	1.0	1.5	2.0
λ_1^2	52.6	53.1	91.1

Also see Ref. [65]</td></tr>
<tr><td>28. Square Plate, Two Supported Sides and Single Corner Point Support
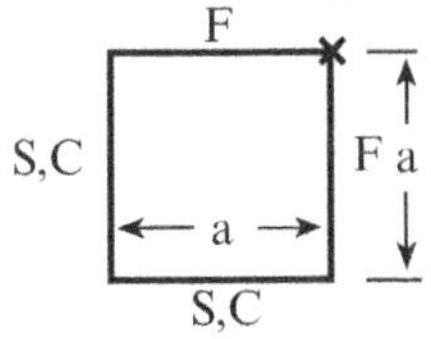</td><td>λ_1^2 = 9.00 for simply supported edges ν=0.3 Ref. [71]

λ_1^2 = 13.7 for clamped edges

Point supports allow rotation by not out-of-plane displacement.</td></tr>
<tr><td>29. Simply Supported Square Plate with Round Opening
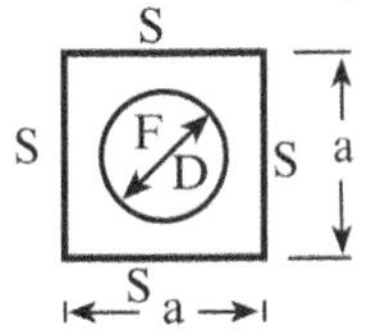</td><td>ν=0.3 Ref. [72]

2R/a	0	0.05	0.10	0.15	0.20	0.25	0.30
λ_1^2	19.9	19.75	19.5	19.4	19.3	19.35	19.5

Hole is centered in plate. Hole has free edges.</td></tr>
<tr><td>30. Clamped Square Plate with Round Opening
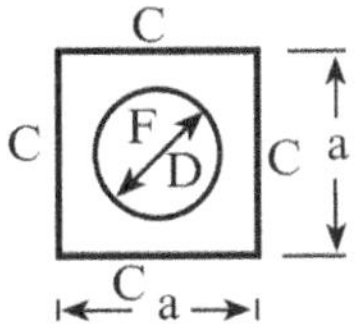</td><td>ν=0.3 Refs [72, 73]

2R/a	0	0.05	0.10	0.15	0.20	0.25
λ_1^2	36.0	35.5	35.1	35.2	35.7	36.7

Hole is centered in plate. Hole has free edges.</td></tr>
<tr><td>31. Supported Square Plate with Square Opening
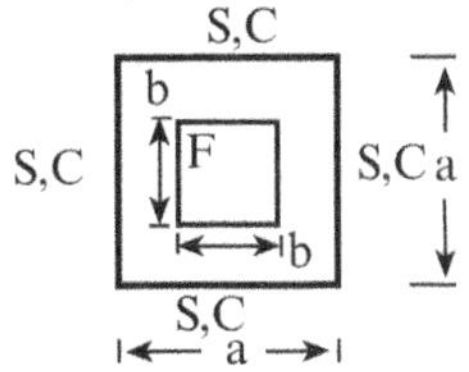</td><td>Ref. [74]

b/a	0	1/6	1/3	1/2
λ_1^2 simply supported edges S	19.63	19.48	21.45	26.05
λ_1^2 clamped edges C	34.85	35.8	43.25	62.40

ν = 0.0; λ_1^2 is 7 % to 10% lower for ν = 0.3

Opening is centered in plate and has free edges.</td></tr>
</table>

Table 5.3 Natural frequency of rectangular plates, continued

Natural Frequency, $f_{ij} = \frac{\lambda^2}{2a^2}\left[\frac{Eh^2}{12\rho(1-\nu^2)}\right]^{1/2}$ Hz, $i = 1,2,3..;\ j = 1,2,3...$

Boundary Conditions | **Natural Frequency Parameter, λ^2**

---

32. Tapered, Simply Supported

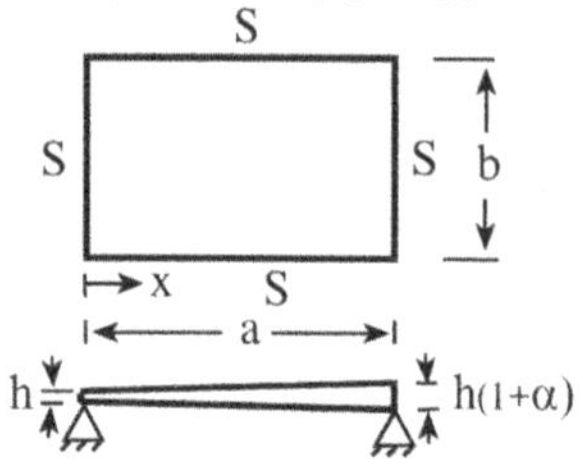

Fundamental mode λ^2 — Refs [75, 76]

a/b	α = 0.0	0.2	0.4	0.6	0.8	1.0
0.25	10.49	11.51	12.48	13.42	14.34	
0.5	12.33	13.55	14.73	15.89	16.99	
1.0	19.74	21.69	23.61	25.30	27.36	29.21
2.0	49.35	54.15	58.75	63.21	67.56	

Plates are linearly tapered on x axis. thickness = $h(1+\alpha x/a)$

---

33. Tapered, Clamped, Simply Supported, Clamped, SS

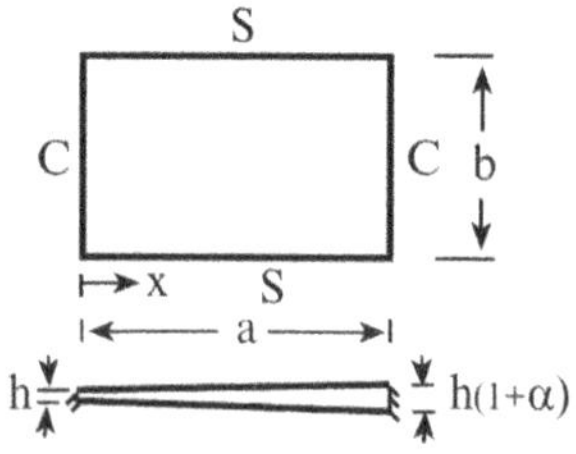

Fundamental mode α^2 — Ref. [77]

a/b	α = 0.0	0.2	0.4	0.6	0.8	1.0
0.5	23.82	26.15	28.40	30.58	32,73	34.81
2/3	25.04	27.50	29.86	32.17	34.41	36.62
1.0	28.95	31.80	34.53	37.19	39.80	42.36
2.0	54.75	60.11	65.24	70.21	75.05	79.79

Plates are linearly tapered on x axis. thickness=$h(1+\alpha x/a)$

---

34. Tapered, Clamped Edges

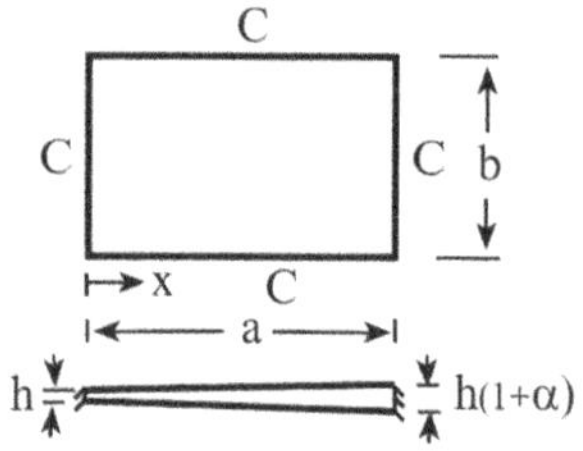

Fundamental mode λ^2 — Ref. [78]; also see Refs [79–80]

a/b	α = 0.0	0.2	0.4	0.6	0.8	1.0
0.5	24.59	27.00	29.32	31.58	33.90	35.96
2/3	27.02	29.67	32.22	34.71	37.13	39.52
1.0	36.00	39.52	42.93	46.24	49.47	52.64
2.0	98.33	107.8	116.6	124.9	132.9	140.5

Plates are linearly tapered on x axis. Thickness = $h(1+\alpha x/a)$.

---

35. Plate with In-Plane Loads

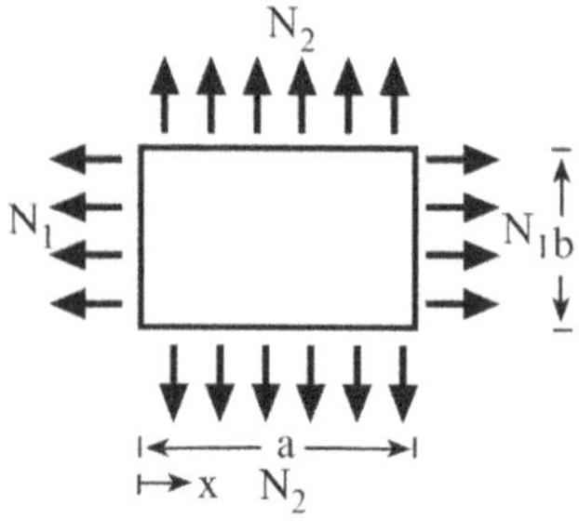

Natural frequency of a rectangular plate with in-plane loads.

$$f_{ij} = \left[f_{ij}^2 \Big|_{\substack{\text{no in-plane loads} \\ N_1 = N_2 = 0}} + \frac{N_1 J_i}{4\gamma a^2} + \frac{N_2 J_j}{4\gamma b^2} \right]^{1/2}, \text{ Hz}$$

J_i and J_j given in cases 1 through 21. They are dependent on the boundary conditions. Mode shape is independent of load. N_1 and N_2 are tensile loads per unit length of edge. They are negative for compression Refs [61, 81]. See Eqs. 5.8, 5.20.

Table 5.3 Natural frequency of rectangular plates, continued

$$\text{Natural Frequency, } f_{ij} = \frac{\lambda_{ij}^2}{2\pi a^2}\left[\frac{Eh^2}{12\rho(1-\nu^2)}\right]^{1/2} \text{Hz, } i=1,2,3..;\ j=1,2,3...$$

Boundary Conditions	Natural Frequency Parameter, λ^2

36. Simply Supported Square Torsion Springs on 4 Edges

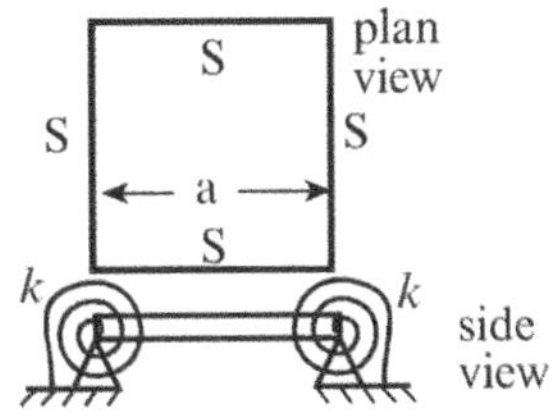

λ^2 Ref. [82]

mode	$ka/(Eh^3/[12(1-\nu^2)]$					
ij	0	1	10	100	1000	∞
11	19.74	21.50	28.50	34.67	35.84	35.99
21	49.35	51.87	60.21	70.78	73.10	73.41
12	49.35	51.87	60.21	70.78	73.10	73.41
22	78.96	80.82	90.81	104.5	107.8	108.3

37. Center Mass, Fund. Mode

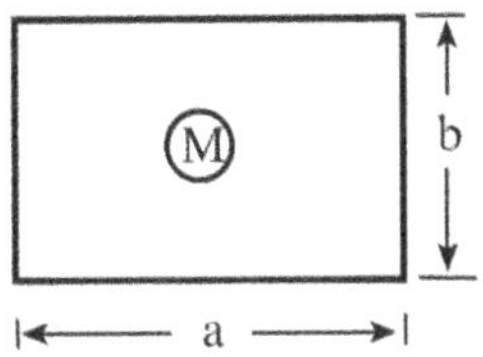

Fundamental mode λ^2 Appendix A

$$\lambda^2 = \frac{\lambda_{M=0}^2}{\sqrt{1+\alpha\dfrac{M}{\gamma A}}}, \quad \text{where } \alpha \approx \frac{A\tilde{w}_{11}^2(x_o,y_o)}{\int_A \tilde{w}_{11}^2(x,y)dA}, \ A = ab = \text{area plate}$$

Center mass, $x_o = a/2$, $y_o = b/2$ Also see Refs [83, 84]

Boundary	λ_{11} for M=0	λ
SSSS	see case 2	4
CCCC	see case 3	6.35
CSCS	see case 8	5.04

38. Regular polygonal Plate Center Mass, N =3,4,... sides

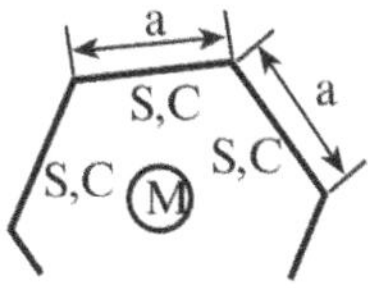

$$\lambda^2 = \frac{\lambda_{M=0}^2}{\sqrt{1+\alpha\dfrac{M}{\gamma A}}}, \quad \text{where } \alpha \approx \begin{cases} 4 \text{ simply supported} \\ 5.6 \text{ clamped} \end{cases}$$

$\lambda|M=0$ is in case 25 of Table 5.4 for N=3,4,5,...
Approximate solution, A = area of plate, Table 2.1, case 21.
Based in part on Refs [83, 84]

about two axes. Four material constants describe the stress–strain relationships of an orthotropic plate:

$$\sigma_x = \frac{(E_x\varepsilon_x + \nu_x E_y \varepsilon_y)}{(1-\nu_x\nu_y)}$$

$$\sigma_y = \frac{(E_y\varepsilon_y + \nu_y E_x \varepsilon_x)}{(1-\nu_x\nu_y)}, \quad \sigma_{xy} = G\varepsilon_{xy} \tag{5.21}$$

E_x and E_y are the elastic moduli along the perpendicular x- and y-axes. G is the shear modulus. It can be shown by a symmetric argument that $\nu_y E_x = \nu_x E_y$ ([2], p. 364; [99–102]).

Table 5.4 Natural frequencies of parallelogram, triangular, and other plates

Notation: a = length of plate; b = width of plate; h = thickness of plate; i = modal index; E = modulus of elasticity; f = natural frequency, Hz; γ = mass per unit area of plate. ρh where ρ is plate material density plus any non-structural mass per unit area, if any; ν = Poisson's ratio; λ = dimensionless natural frequency parameter; β angle, degrees. S = simply supported edge; C = clamped edge; F = free edge. Consistent sets of units are in Table 1.2. [85–98].

$$\text{Natural Frequency, } f_i = \frac{\lambda_i^2}{2\pi a^2}\left[\frac{Eh^3}{12\gamma(1-\nu^2)}\right]^{1/2} \quad \text{Hz}, \ i = 1,2,3..$$

Boundary Conditions	Natural Frequency Parameter, λ_i^2

1. Clamp-Free- Rhombus

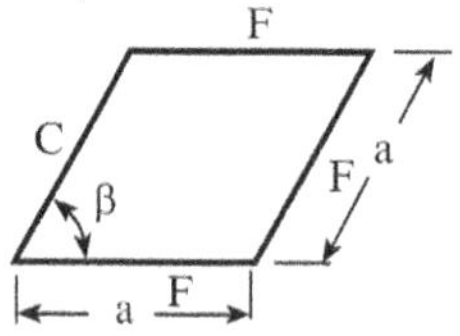

λ^2 ν=0.3   Refs [85, 86]

β	Mode 1	Mode 2
90	3.492	8.525
75	3.601	8.872
60	3.961	10.19
45	4.824	13.75

2. Simply Supported Rhombus

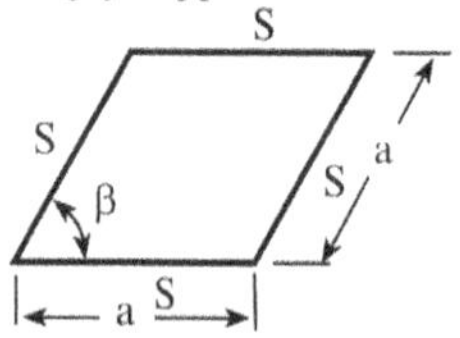

β	Mode		λ^2		Ref. [87]	
	1	2	3	4	5	6
90	19.74	49.35	49.35	78.96	98.70	98.70
70	21.82	49.04	60.07	79.94	108.5	116.9
60	24.91	52.67	71.79	83.92	122.0	123.1
45	34.79	66.36	100.5	107.3	141.0	168.3

3. Simply Supported Parallelogram

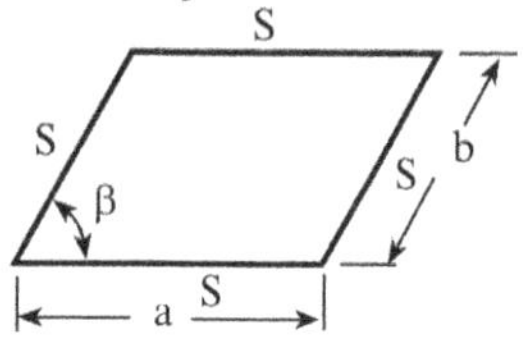

β	Mode		λ^2		Ref. [87]	
	a/b = 2/3		a/b = 1/2		a/b = 1/3	
	1	2	1	2	1	2
90	14.26	27.42	12.33	19.74	10.97	14.26
70	15.82	29.44	13.76	21.44	12.31	15.72
60	18.15	32.48	15.89	23.96	14.35	17.88
45	25.69	42.17	22.87	32.05	21.06	24.95

4. Clamped Rhombus

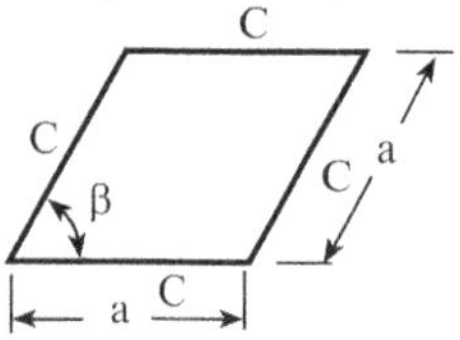

β	Mode		λ^2		Ref. [88]	
	1	2	3	4	5	6
90	35.99	73.41	73.41	108.3	131.6	132.2
70	40.05	74.65	88.32	111.6	143.5	155.5
60	46.14	81.20	105.5	119.5	165.7	165.5
45	65.93	106.6	149.0	158.9	199.4	231.9

5. Clamped Parallelogram

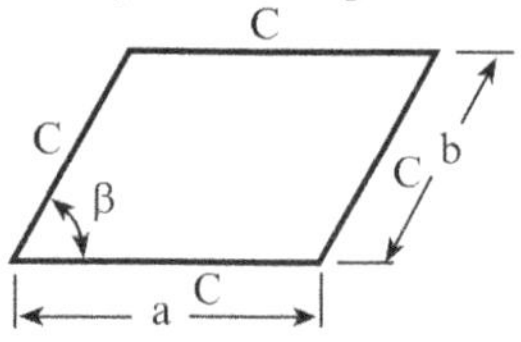

β	Mode		λ^2		Ref. [88]	
	a/b = 2/3		a/b = 1/2		a/b = 1/3	
	1	2	1	2	1	2
90	27.01	41.72	24.58	31.84	23.20	25.87
70	30.19	45.44	27.62	35.10	26.18	28.92
60	35.01	51.04	32.22	39.98	30.70	33.52
45	50.79	69.19	47.37	56.06	45.71	48.76

Table 5.4 Natural frequencies of parallelogram, triangular, and other plates, continued

$$\text{Natural Frequency, } f_i = \frac{\lambda_i^2}{2\pi a^2}\left[\frac{Eh^3}{12\gamma(1-\nu^2)}\right]^{1/2}, \text{ Hz, } i = 1,2,3..$$

Boundary Conditions | Natural Frequency Parameter, λ_{ij}^2

6. Clamped-Simple-Simple-Simple Rhombus

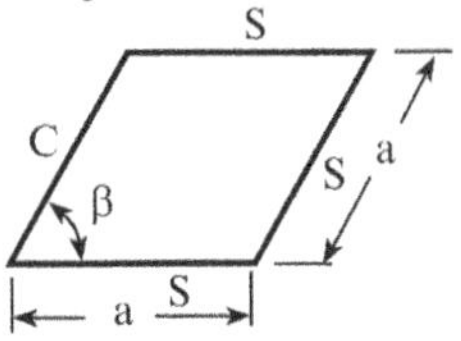

β	Mode 1	2	3	4	5	6
	λ^2					Ref. [89]
90	23.65	51.68	58.65	86.15	100.3	113.2
70	26.47	54.43	68.55	88.16	114.9	130.1
60	30.79	59.72	82.37	94.06	133.0	136.1
50	38.80	69.99	105.9	108.0	151.4	164.8
40	54.24	90.01	132.5	154.0	182.6	224.0

7. Clamped-Simple-Clamped Simple Rhombus

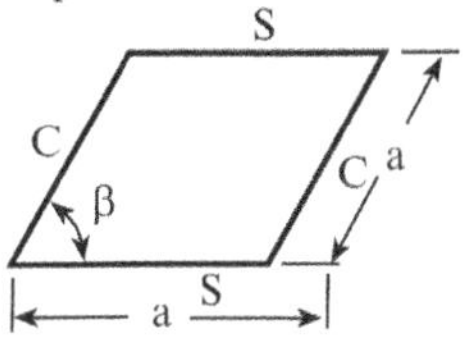

β	Mode 1	2	3	4	5	6
	λ^2					Ref. [89]
90	28.95	54.76	69.33	94.61	102.3	129.1
70	32.33	58.77	79.35	95.73	119.4	138.7
60	37.47	65.12	94.15	101.7	139.2	145.1
50	46.91	76.89	116.5	120.4	161.9	173.7
40	65.04	99.68	144.1	171.7	194.4	236.9

8. Clamped-Simple-Clamped-Simple Parallelogram

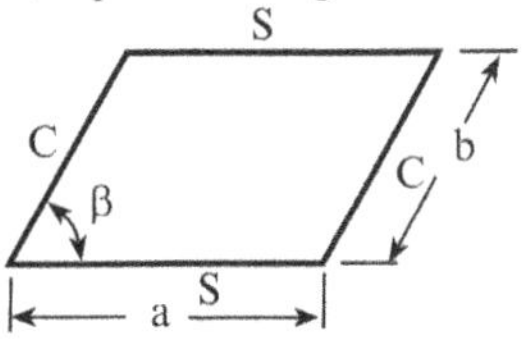

β	Mode a/b = 2/3			a/b = 1/2		
	λ^2					Ref. [89]
	1	2	3	1	2	3
90	25.04	35.11	54.76	23.82	28.96	39.10
75	26.71	36.94	56.65	25.45	30.66	40.91
60	32.77	43.59	63.81	31.41	36.91	47.53
45	48.18	60.26	82.06	46.59	52.73	64.22

9. Clamped-Simple-Simple-Clamped Rhombus

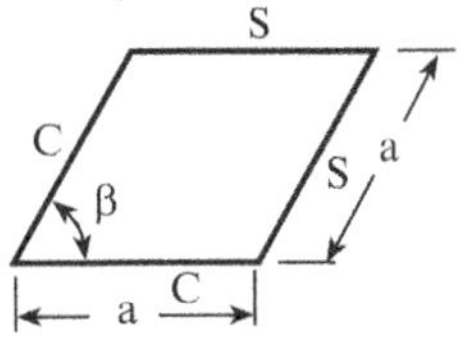

β	Mode 1	2	3	4	5	6
	λ^2					Ref. [89]
140	61.61	105.2	144.1	165.8	194.4	239.8
120	35.23	66.94	88.94	102.2	144.0	146.4
90	27.05	60.54	60.80	92.86	114.6	114.7
70	30.07	61.45	73.44	95.63	126.6	135.5
50	43.30	78.41	114.4	116.2	162.2	177.7
40	59.70	100.5	144.3	163.8	194.4	239.8

10. Clamped-Clamped-Simple-Clamped Rhombus

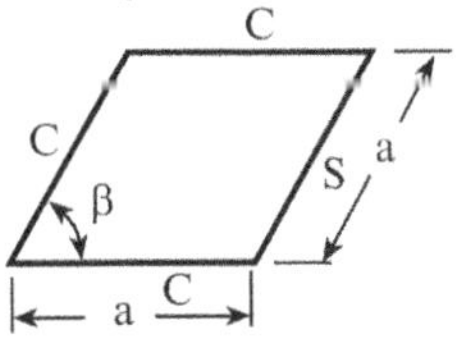

β	Mode 1	2	3	4	5	6
	λ^2					Ref. [88]
90	31.83	63.34	71.08	100.9	116.4	130.4
70	35.47	66.89	82.53	103.5	132.6	148.5
60	40.97	73.51	98.40	110.7	153.1	155.5
45	58.95	96.43	138.2	149.0	187.5	216.1
40	70.51	110.8	156.1	179.6	207.3	254.6

Table 5.4 Natural frequencies of parallelogram, triangular, and other plates, continued

Triangular Plate Natural Frequency, $f_i = \frac{\lambda_i^2}{2\pi a^2}\sqrt{\frac{Eh^3}{12\gamma(1-\nu^2)}}$, i =0,1,2,3... Hertz

Boundary — λ_i^2

11. Clamp-Free -Free Isosceles

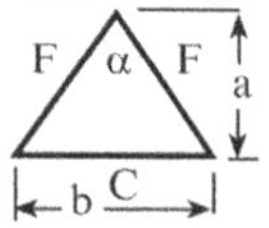

a/b	α, deg.	mode ν=0.3 1	2	3
0.5	90	6.322	17.97	26.47
0.866	60	6.692	26.32	28.86
1.866	30	6.879	29.80	46.84

12. Clamp-Free Free right Angle

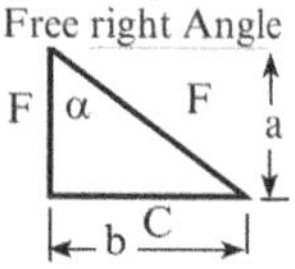

a/b	α, deg.	mode ν=0.3 1	2	3
0.5	63.4	5.502	15.06	28.42
1.0	45.0	6.169	23.46	32.67
2.0	26.6	6.632	28.42	49.43

13. SimpleSimple Simple Isosceles

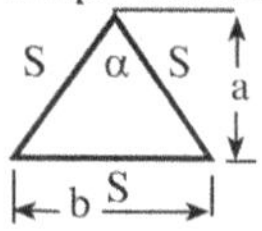

a/b	α, deg.	mode 1	2	3
0.5	90	24.68	49.34	64.15
0.866	60	39.48	92.13	92.13
1.866	30	97.93	183.1	273.9

14 Simple Sup. Right Triangle

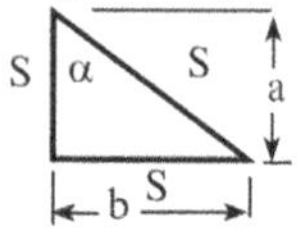

For a = b, $\lambda_{ij}^2 = \pi^2(i^2 + j^2)$

for i, j > 0 but i ≠ j.

i,j = (1,2), (2,1),(3,1), (1,3),..

$\lambda_{12}^2 = \lambda_{21}^2 = 5\pi^2, \lambda_{13}^2 = 10\pi^2$

For b = $(1/3^{1/2})$a, $\lambda_1^2 = 92.11$

15. Clamped Isosceles Triangle

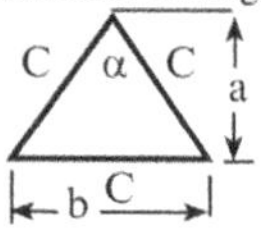

a/b	α, deg.	mode 1	2	3
0.5	90	46.89	78.89	97.41
0.866	60	74.27	141.7	141.7
1.866	30	187.3	298.4	412.1

16. Clamp Simple Simple Isosceles

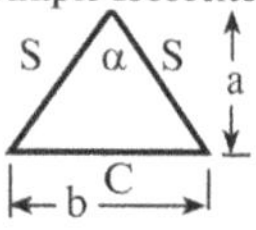

a/b	α, deg.	mode 1	2	3
0.5	90	32.89	60.54	77.22
0.866	60	49.62	107.1	107.6
1.866	30	113.3	204.5	298.9

17. Simple-Clamp -Clamp Isosceles

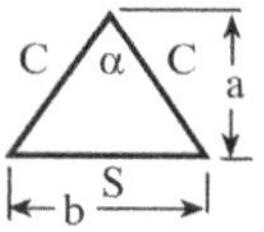

a/b	α, deg.	mode 1	2	3
0.5	90	36.69	65.78	82.50
0.866	60	61.21	123.7	124.0
1.866	30	165.8	271.9	381.3

18. Free-Simple Simple Isosceles

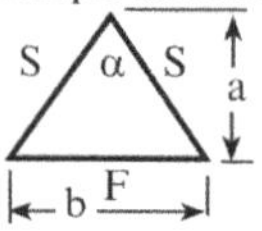

a/b	α, deg.	mode ν=0.3 1	2	3
0.5	90	4.90	17.31	29.28
0.866	60	12.06	43.21	51.22
1.866	30	47.20	122.1	172.6

19. Free-Clamped Clamped Isosceles

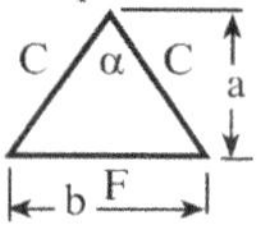

a/b	α, deg.	mode ν=0.3 1	2	3
0.5	90	14.45	31.78	44.92
0.866	60	30.01	71.86	76.34
1.866	30	105.5	196.8	270.6

20. Free-Clamp- Free Isosceles

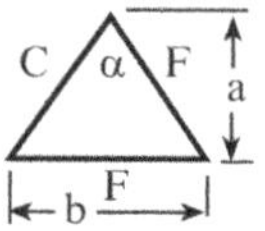

a/b	α,, deg.	mode ν=0.3 1	2	3
0.5	90	3.082	11.73	16.34
0.866	60	6.693	26.32	28.86
1.866	30	22.12	60.54	106.2

21. Free Isosceles Triangle

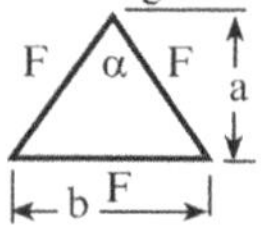

a/b	α, deg.	mode 1	2	3
0.5	90	12.88	20.34	31.80
0.866	60	24.58	55.01	56.78
1.866	30	55.05	109.4	176.4

Refs. [90–93]

These references also include effect of shear deformation.

Table 5.4 Natural frequencies of parallelogram, triangular, and other plates, continued

$$\text{Natural Frequency, } f_i = \frac{\lambda_i^2}{2\pi a^2}\left[\frac{Eh^3}{12\gamma(1-\nu^2)}\right]^{1/2} \text{ Hz, } i = 1,2,3..$$

Boundary Conditions	Natural Frequency Parameter, λ^2

22. Simply-Supported Asymmetric Triangle

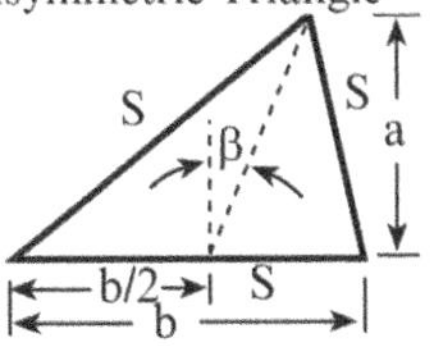

λ^2 Fundamental Mode Ref. [94], has higher modes

a/b	β (degree) 0	10	20	30	45
0.5	24.69	24.78	25.06	25.64	27.78
1.0	45.85	46.28	47.71	50.57	60.22
1.5	73.66	74.64	77.85	84.21	105.1

23. Simply-Supported Symmetric Trapezoid

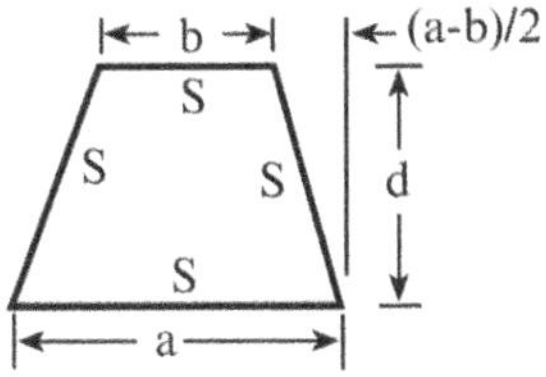

λ^2 Fundamental Mode Ref. [95]

d/a	b/a 0.0	0.2	0.4	0.6	0.8	1.0
0.5	98.78	76.50	63.18	55.97	51.85	49.35
2/3	69.70	55.09	44.70	38.38	34.54	32.08
1.0	45.85	37.75	30.79	25.64	22.13	19.74
1.5	32.74	28.04	23.64	19.72	16.58	14.26

24. Unsymmetric Simply Supported Trapezoid

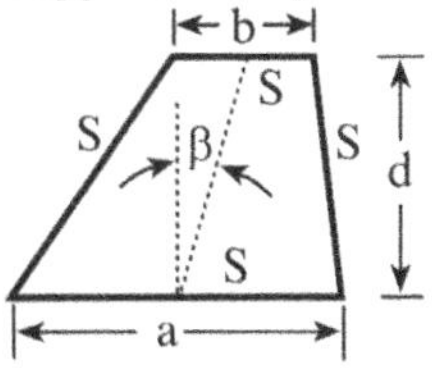

λ^2 Fundamental Mode Ref. [94]

β deg	d/a 0.5, b/a=.4	0.5, b/a=.8	1.0, b/a=.4	1.0, b/a=.8	1.5, b/a=.4	1.5, b/a=.8
10	63.42	52.00	31.20	22.37	24.05	16.88
20	64.26	52.47	32.50	23.18	25.39	17.86
30	66.02	53.44	35.14	24.87	28.02	19.85
45	72.81	56.94	44.06	30.97	36.59	26.81

25. Regular Polygon with n Clamped or Simply S. Sides

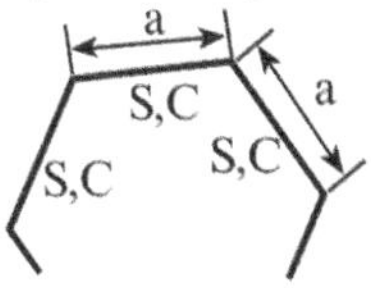

Refs [96, 97]

Number of sides n	3	4	5	6	7	8
S. Supported, λ^2	39.9	19.74	11.01	7.152	5.068	3.794
Clamped, λ^2	74.4	35.08	19.71	12.81	9.081	6.787

All sides have equal length and same support conditions. Fundamental Mode. N=4 is a square.

26. Arbitrary Plate with Clamp or Simply Supported Sides

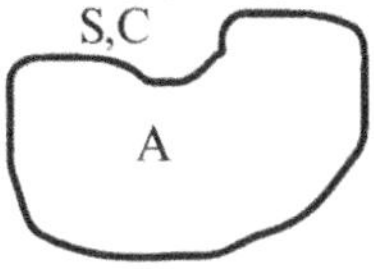

Lower bound for $\lambda_i^2 = 4.977i$; i=1,2,3, for simply supported edges. Lower bound for $\lambda_i^2 = 10.22i$; i=1,2,3, for clamped edges.

The characteristic length is defined as $a = A^{1/2}/\pi^{1/2}$
A is area of plate and i is the mode number. i=1 is fundamental mode. Ref. [98]

Table 5.5 Natural frequencies of grillages, stiffened, and orthotropic plates

Notation: a = length of plate, as subscript, fiber length; b = width of plate, as subscript, fiber length; C = torsion constant, Table 4.15; h = thickness of plate; G_i, H_i, J_i from cases 1 to 21 of Table 5.3; I = moment of inertia about neutral axis, Table 2.1; i, j = modal indices; E = modulus of elasticity; γ = mass per unit area of plate including non-structural mass, if any; ν = Poisson's ratio. Consistent sets of units are given in Table 1.2.

$$\text{Nat. Frequency, } f_{ij} = \frac{\pi}{2\gamma^{1/2}}\left[\frac{G_i^4 D_x}{a^4} + \frac{G_j^4 D_y}{b^4} + \frac{2[H_iH_jD_{xy} + 2D_k(J_iJ_j - H_iH_j)]}{a^2b^2}\right]^{1/2}, \text{Hz}, i, j = 1, 2, \ldots$$

Orthotropic Plate	Orthotropic Constants D_x, D_y, D_{xy}, D_k
1. Orthotropic Rectangular Plate 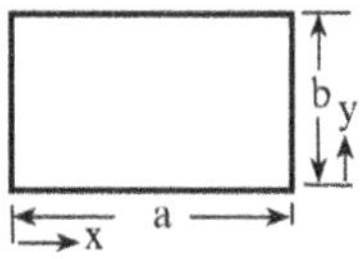	$D_x = \frac{E_x h^3}{12(1-\nu_x\nu_y)}$, $D_y = \frac{E_y h^3}{12(1-\nu_x\nu_y)}$ $D_k = G_{shear}h^3/12$, $D_{xy} = D_x\nu_y + 2D_k$, G_{shear} = shear modulus E_x and E_y are the elastic modulii for stress and strain in the x and y directions, respectively. ν_x and ν_y are the associated Poisson's ratio.
2. Grillage of interlocking Beams 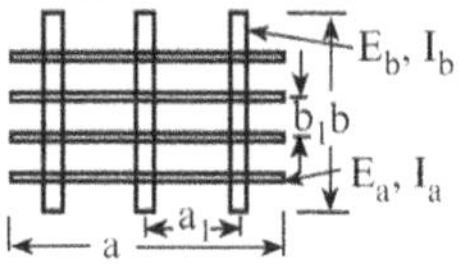	$D_x = E_aI_a/b_1$, $D_y = E_bI_b/a_1$ $D_{xy} = 2D_k = \frac{E_aC_a}{2b_1} + \frac{E_bC_b}{2a_1}$ = 0 to neglect beam torsion stiffness Subscripts a, b refer to horizontal and vert. stiffeners, respectively. For plate with stiffeners in two directions, add cases 1 and 2. Ref. [2]
3. Plate with Integral Stiffeners In the y direction 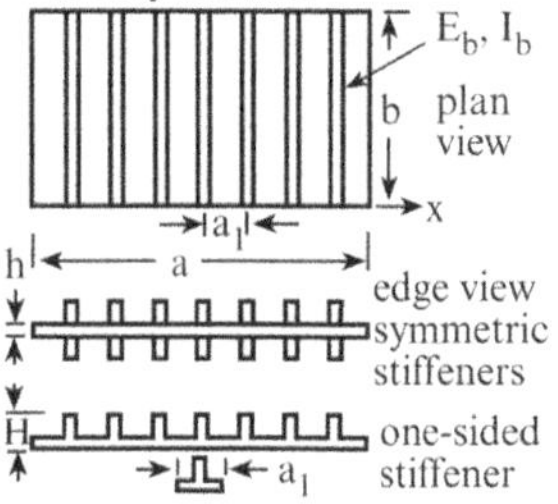	For symmetric (two-sided) stiffeners, Ref. [2] $D_x = \frac{Eh^3}{12(1-\nu^2)}$, $D_y = \frac{Eh^3}{12(1-\nu^2)} + \frac{E_bI_b}{a_1}$ $D_{xy} = 2D_k = \frac{Eh^3}{12(1-\nu^2)}$ For one-sided stiffeners (see figure for I_r), Ref. [2]. See Ref. [98] $D_x = \frac{Ea_1h^3}{12(a_1 - t + t(h/H)^3)}$, $D_y = \frac{EI_r}{a_1}$, $D_{xy} = \frac{Eh^3}{12(1+\nu)} + \frac{EC_r}{a_1}$. C_r = rib torsion constant, Table 4.15.
4. Corrugated Plate 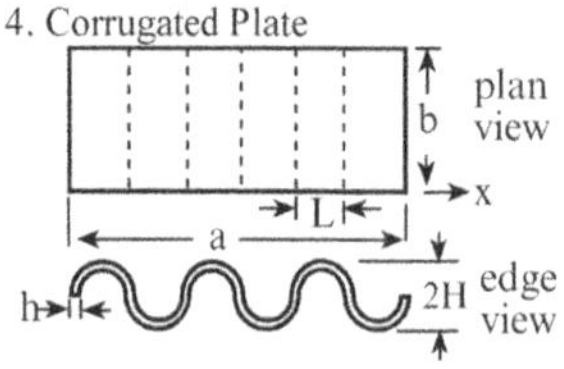	$D_x = \frac{Eh^3}{12(1-\nu^2)}\frac{a}{s}, D_{xy} \approx 2D_k = \frac{Eh^3}{12(1+\nu)}\frac{s}{a}, \frac{s}{a} \approx 1 + \frac{\pi^2H^2}{4L^2}$ $D_y \approx \frac{EhH^2}{2}\left[1 + \frac{\pi H^2}{L^2}\right]$, note D_y in Ref. [2] is too low Corrugations are sine waves, H sin $n\pi x/L$. Ref. [2]
5. Fiber Reinforced Plate 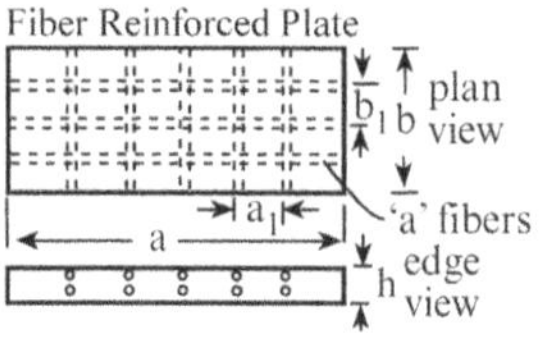	$D_x = \frac{E}{1-\nu^2}\left[\frac{h^3}{12} + \left(\frac{E_a}{E} - 1\right)\frac{I_a}{b_1}\right], D_{xy} = (D_xD_y)^{1/2}, E_a, E_b$ = fiber $D_y = \frac{E}{1-\nu^2}\left[\frac{h^3}{12} + \left(\frac{E_b}{E} - 1\right)\frac{I_b}{a_1}\right], D_k \approx \frac{D_{xy}}{2}, \nu = \nu_{plate}$ I_a, I_b = moment of inertia of fibers about mid surface. Ref. [2]

It is conventional to use the four orthotropic plate stiffnesses defined in Table 5.5 for the analysis of rectangular orthotropic plates:

$$D_x = \frac{E_x h^3}{12(1 - \nu_x \nu_y)}, \quad D_y = \frac{E_y h^3}{12(1 - \nu_x \nu_y)}$$

$$D_{xy} = D_x \nu_y + 2D_k, \quad D_k = \frac{Gh^3}{12} \tag{5.22}$$

D_{xy} is the sum of torsion and shear stiffnesses. In the limit case of an isotropic material, $E = E_x = E_y$, and $\nu_x = \nu_y$ and $D = D_x = D_y = D_{xy} = Eh^3/[12(1 - \nu^2)]$, $G = E/[2(1 + \nu)]$. h is the plate thickness.

Table 5.5 provides natural frequency and orthotropic properties D_x, D_y, and D_{xy} of stiffened and fiber-reinforced orthotropic rectangular plates and grillages. The natural frequency formula at the top of the table uses boundary conditions and constants in cases 1–21 of Table 5.3; Eq. 5.14. The exact solution for the natural frequencies of a simply supported orthotropic rectangular plate [1, 61, 103] is

$$f_{ij} = \frac{1}{2\pi}\left[\frac{\pi^4}{\rho a^4}\left(i^4 D_x + 2i^2 j^2 D_{xy}\frac{a^2}{b^2} + j^4 D_y \frac{a^4}{b^4}\right)\right]^{1/2} \text{Hz}, \quad i,j = 1, 2, 3, \ldots \tag{5.23}$$

Its mode shape is given by Equation 5.6. Since the natural frequencies are functions of orthotropic properties, it is possible to use measurements of plate natural frequency to determine the orthotropic properties [100]. The effect of in-plane loads on orthotropic plates and grillages is given in case 35 of Table 5.3. Additional solutions and reviews are provided in Refs [1, 2, 102, 103, 104].

Grillage is a rectangular network of mutually perpendicular, interlocking beams (case 2 of Table 5.5). Grillages can be analyzed by matrix analysis of discrete beams [104] or with smeared equivalent orthotropic properties (case 2 of Table 5.5). The moment of inertia of the beams is calculated about the neutral axis of the grillage, which is equivalent to the mid surface of the plate. The cross term D_{xy} accounts for beam twisting about its own axis, and setting $D_{xy} = 0$ neglects this usually small effect.

5.2.7 *Stiffened Plates*

Beams, called stiffeners, are attached to the plates to prevent large deformations under pressure loads [104, 105]. The result is called a stiffened plate. Stiffened plates with symmetric patterns of stiffeners are orthotropic. Equivalent orthotropic properties for stiffened plates are given in Table 5.5.

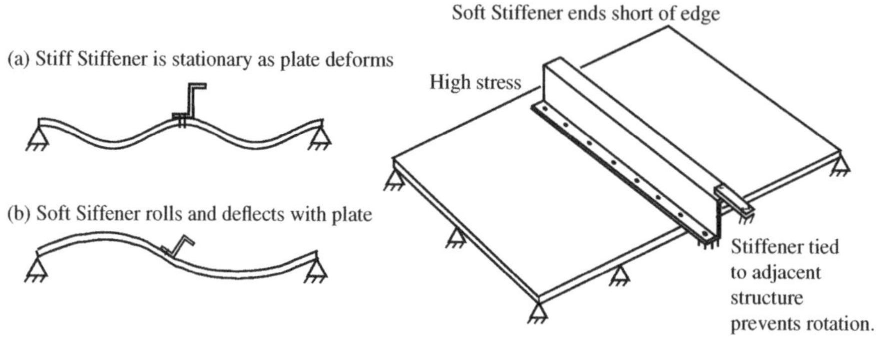

Figure 5.2 A rectangular plate with a riveted Z stiffener. Attaching the stiffener to the surrounding structure increases the stiffener's ability to resist plate deformation

The stiffener should reduce the stress and deformation of the unstiffened plate (Figure 5.2). An efficient stiffener breaks the plate into subpanels as shown in Figure 5.2a that deform independently. To do this, stiffener should be stiffer than the supported plate [105, 106]:

$$\frac{k_{\text{stiffener}}}{k_{\text{plate}}} >> 1 \text{ which implies } E_{\text{stiff.}} I_{\text{stiff.}} >> \frac{E_{\text{plate}} I_{\text{plate}}}{b}$$

$$\text{hence, } I_{\text{stiff}} >> \left(\frac{1}{12}\right) ah^3 \frac{E_{\text{plate}}}{E_{stiffener} b}$$

k is the lateral stiffness of the plate and the stiffener (cases 13 and 45 of Table 3.1). E is the modulus of elasticity, h is the plate thickness, a is the span of plate between stiffeners, b is the stiffener spacing, and I_{stiff} is the moment of inertia of the stiffener about an axis perpendicular to the axis of the stiffener and parallel to the plate. Similarly, torsion stiffness of the stiffener about its own axis should exceed that of the plate to restrain the plate edge from rotation (Figure 5.2b).

It is tempting to minimize fabrication costs by terminating stiffeners on the plate rather than tying the stiffener into the surrounding structure (see Figure 5.2). Rao and Nair [107] showed that stopping the stiffener short of the surrounding structure produces a significant stress concentration in the plate at the toe of stiffener.

5.2.8 *Perforated Plates*

Perforations are round holes with diameter D that are centered distance P apart and cover the surface of a uniform-thickness plate in a regular square or equilateral triangle pattern. Perforations reduce the natural bending frequencies of plates [108, 109]. The frequency reduction is primarily a function of the minimum ligament $h = P - D$ between holes and $\eta = h/P = P/D - 1$, which is called ligament efficiency. The ratio of natural frequency of a perforated plate f_{perf} to that of a similar unperforated solid plate f_{solid} in the same mode is a function of ligament efficiency. Typical values are as follows.

$\eta = P/D - 1$	0.05	0.1	0.2	0.3	0.5	> 0.7
$f_{\text{perf}}/f_{\text{solid}}$	0.78	0.82	0.87	0.90	0.94	1.0

The holes just touch each other at $P/D = 1, \eta = 0$. Myung and Jong [109] provide formulas for the square of $f_{\text{perf}}/f_{\text{solid}}$.

Example 5.1 Natural Frequency and stress of Round Plate

Consider a round clamped steel plate, 3 mm (0.118 in.) thick and 200 mm (7.87 in.) in diameter. Determine its natural frequency. What is the maximum stress in the plate when it vibrates 1 mm (0.040 in.) out of plane at the center in the first mode?

Solution: The modulus of elasticity of steel at room temperature is approximately $E =$ 190E9 Pa (27.5E6 psi), its density is 7.97 g/cc (0.288 lb/cubic inch), and Poisson's ratio is 0.29. The equation for the natural frequencies of the round clamped plate is given in case 3 of Table 5.2. For the fundamental mode, $i = j = 0$ and $\lambda^2 = 10.22$. Using SI units (case 1 of Table 1.3), the fundamental natural frequency is as follows:

$$f = \frac{\lambda^2}{2\pi a^2}\sqrt{\frac{Eh^3}{12\gamma(1-\nu^2)}} = \frac{10.22}{2\pi(0.1\,\text{m})^2}\sqrt{\frac{190\text{E}9\,\text{Pa}\,(0.003\,\text{m})^3}{12(7970\,\text{kg/m}^3 \times 0.003\,\text{m})(1-0.29^2)}}$$
$$= 718.7\ \text{Hz}$$

Note that the diameter is 200 mm but the 100 mm radius enters Equation 5.15. The mass per unit area of the plate is density times thickness, $\gamma = \rho h$. In US customary units of case 7 of Table 1.3, $h = 0.118$ in., $a = 3.937$ in., $E = 27.5\text{E}6$ psi, and $\rho\text{h} = 0.288\ \text{lb/in.}^3(0.118\ \text{in.})/386.1\ \text{in./s}^2 = 0.8802\text{E}-4\ \text{lb-s}^2/\text{in.}^3$. The same result emerges.

Plate stress is determined by the mode shape and the modal amplitude. The fundamental mode shape is given in case 3 of Table 5.2. This is a symmetric mode. As the plate deforms in its fundamental mode, the plate displacement is

$$W(r,t) = A\left[J_0\left(\frac{\lambda r}{a}\right) - \left\{\frac{J_0(\lambda)}{I_0(\lambda)}\right\} I_0\left(\frac{\lambda r}{a}\right)\right]\sin(2\pi f t)$$

$$W(0,t) = A\left[J_0(0) - \left\{\frac{J_0(\lambda)}{I_0(\lambda)}\right\} I_0(0)\right]\sin(2\pi f t + \phi) = 1.056A\sin(2\pi f t)$$

The second line is evaluated at the center of the plate, $r = 0$, $J_0(0) = I_0(0) = 1$, $J_0(3.197) = -0.319$, and $I_0(3.197) = 5.73$ [15]. If the displacement at the center of the plate is 1 mm, then $W = 1.056 \times A = 1$ mm and so $A = 0.947$ mm. The maximum stress in the plate is the radial bending stress at the surface of the clamped edge, $r = a$.

This bending stress is found using case 1 of Table 5.2 for a 10 mm displacement at the center of the plate:

$$\sigma_r = -E\frac{h}{2}\frac{\partial^2 W(r)}{\partial r^2}, \quad |W(r)| = A\left[J_0\left(\frac{\lambda r}{a}\right) + 0.0557 I_0\left(\frac{\lambda r}{a}\right)\right]$$

$$\begin{aligned} d^2 J_0 \frac{(\lambda r/a)}{dr^2} &= \left(\frac{\lambda}{a}\right)^2 \frac{\left(-J_0\left(\frac{\lambda r}{a}\right) + J_2\left(\frac{\lambda r}{a}\right)\right)}{2} \\ &= 0.401\left(\frac{\lambda}{a}\right)^2/2 \text{ at } r = a \\ d^2 I_0 \frac{(\lambda r/a)}{dr^2} &= \left(\frac{\lambda}{a}\right)^2 \frac{(I_0(\lambda r/a) + I_2(\lambda r/a))}{2} \\ &= 4.25\left(\frac{\lambda}{a}\right)^2 \text{ at } r = a, \lambda = 10.22^{1/2} \end{aligned}$$

$$\begin{aligned} \sigma_r &= 0.64AE\frac{h}{2}\frac{\lambda^2}{a^2} = (0.64)(0.000947\,\text{m})(190\text{E}9\,\text{Pa})\frac{0.003\,\text{m}}{2}\frac{10.22}{0.1\,\text{m}^2} \\ &= 176.\text{E}6\text{Pa}\ (25{,}600\,\text{psi}) \end{aligned}$$

This oscillating stress amplitude could fatigue a metallic plate.

5.3 Cylindrical Shells

A cylindrical shell is a thin sheet of elastic material formed into a cylinder as shown in Figure 5.3. The classical thin shell theory follows the assumptions first made by Love [110, p. 6; 111, 112]: the thickness of the shell is small compared with the shell radius and terms of order of thickness squared over radius squared are neglected.

Different shell theories retain different linear terms – see Refs [4, 6, 110–118]. For example, the circumferential strain computed by the Donnell and Flugge cylindrical shell theories differs by terms on the order of z/R [110, 112]:

$$\varepsilon_\theta|_{\text{Donnell}} = \frac{1}{R}\frac{\partial v}{\partial \theta} + \frac{w}{R} - \frac{z}{R^2}\frac{\partial^2 w}{\partial \theta^2}, \qquad \varepsilon_\theta|_{\text{Flugge}} = \frac{1}{R}\frac{\partial v}{\partial \theta} + \frac{w}{R+z} - \frac{z}{R(R+z)}\frac{\partial^2 w}{\partial \theta^2} \tag{5.24}$$

The two strain theories are identical if the through-the-thickness coordinate z is negligible in comparison to the radius R, $z << R$. The Donnell shell theory (Table 5.6) is the simplest shell theory. The Sanders [117] and Flugge [118] theories are the most accurate. All reduce to eight-order partial differential equations [110, 112].

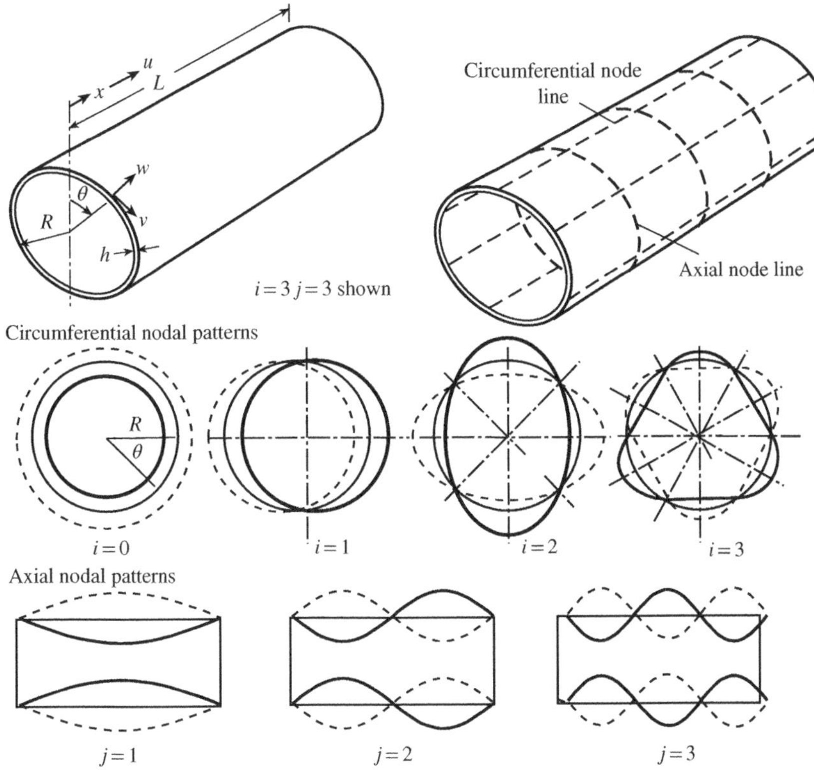

Figure 5.3 Coordinates for a circular cylinder, nodal patterns, and deformation of a simply supported cylinder without axial constraint. After Ref. [110] in part

5.3.1 Donnell Thin Shell Theory

The Donnell thin shell theory in Table 5.6 describes the dynamic motion of thin elastic cylindrical shells by three coupled equations of motion, one each for the axial deformation u, the circumferential deformation v, and the radial outward deformation w (Figure 5.3). The Donnell [116] shell equations of motion in matrix format are ([112, p. 297]; [110, p. 32])

$$\begin{bmatrix} \frac{\partial^2}{\partial x^2} + \frac{1-\nu}{2R^2}\frac{\partial^2}{\partial \theta^2} - \frac{\rho\left(1-\nu^2\right)}{E}\frac{\partial^2}{\partial t^2} & \frac{1+\nu}{2R}\frac{\partial^2}{\partial x \partial \theta} & \frac{\nu}{R}\frac{\partial}{\partial x} \\ \frac{1+\nu}{2R}\frac{\partial^2}{\partial x \partial \theta} & \frac{\partial^2}{R^2 \partial \theta^2} + \frac{1-\nu}{2}\frac{\partial^2}{\partial x^2} - \frac{\rho(1-\nu^2)}{E}\frac{\partial^2}{\partial t^2} & \frac{1}{R^2}\frac{\partial}{\partial \theta} \\ -\frac{\nu}{R}\frac{\partial}{\partial x} & -\frac{\partial}{R^2 \partial \theta} & -\frac{1}{R^2} - R^2 k \nabla^4 - \frac{\rho(1-\nu^2)}{E}\frac{\partial^2}{\partial t^2} \end{bmatrix} \begin{pmatrix} u \\ v \\ w \end{pmatrix} = \begin{pmatrix} -\frac{(1-\nu^2)}{Eh} q_x \\ -\frac{(1-\nu^2)}{Eh} q_y \\ -\frac{(1-\nu^2)}{Eh} p \end{pmatrix} \tag{5.25}$$

Table 5.6 Cylindrical shell stress and strain theory

Notation: h = plate thickness; θ = circumferential angle, radians; x = axial coordinate; z = out-of-plane coordinate; u = axial deformation; v = circumferential deformation; w = out-of-plane deformation, in z direction; E = modulus of elasticity; σ = stress, force/area; ε = strain, unitless; ν = Poisson's ratio; $G = E/[2(1+\nu)]$, shear modulus. Refs [110, 112, 116].

Description	Parameter in Donnell Shell Theory
1. Strain and Deformation 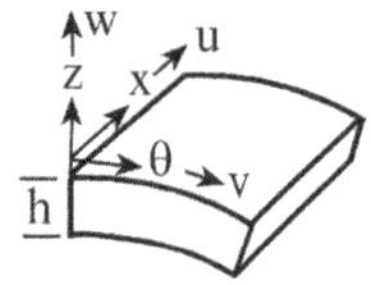	$\varepsilon_x = \frac{\partial u}{\partial x} - z\frac{\partial^2 w}{\partial x^2}$, $\varepsilon_\theta = \frac{1}{R}\frac{\partial v}{\partial \theta} + \frac{w}{R} - \frac{z}{R^2}\frac{\partial^2 w}{\partial \theta^2}$, $\varepsilon_{x\theta} = \frac{\partial v}{\partial x} + \frac{1}{R}\frac{\partial u}{\partial \theta} - \frac{2z}{R}\frac{\partial^2 w}{\partial x \partial \theta}$, $\varepsilon_{xz} = \varepsilon_{\theta z} = \varepsilon_{zz} = 0$
2. Stress and Strain 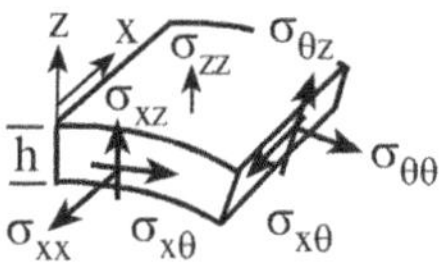	$\sigma_{xx} = \frac{E}{1-\nu^2}(\varepsilon_x + \nu\varepsilon_\theta)$, $\sigma_{yy} = \frac{E}{1-\nu^2}(\varepsilon_\theta + \nu\varepsilon_x)$, $\sigma_{x\theta} = \sigma_{x\theta} = G\varepsilon_{x\theta}$, $\sigma_{xz} = \sigma_{\theta z} = \sigma_{zz} = 0$
3. Moment Resultants 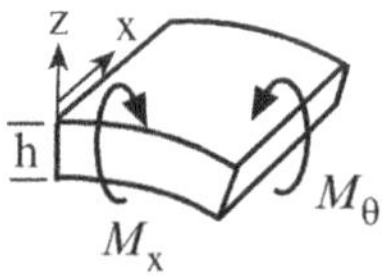 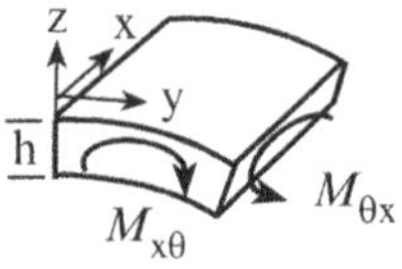	$M_x = -\int_{-h/2}^{h/2} \sigma_{xx}\left(1+\frac{z}{R}\right)dz = \frac{Eh^3}{12(1-\nu^2)}\left(\frac{\partial^2 w}{\partial x^2} + \frac{\nu}{R^2}\frac{\partial^2 w}{\partial \theta^2}\right)$ $M_\theta = -\int_{-h/2}^{h/2} \sigma_{\theta\theta} z\,dz = \frac{Eh^3}{12(1-\nu^2)}\left(\frac{1}{R^2}\frac{\partial^2 w}{\partial \theta^2} + \nu\frac{\partial^2 w}{\partial x^2}\right)$ $M_{x\theta} = -\int_{-h/2}^{h/2} \sigma_{xy}\left(1+\frac{z}{R}\right)dz = \frac{Eh^3}{12(1+\nu)}\frac{1}{R}\frac{\partial^2 w}{\partial x \partial \theta}$ $M_{\theta x} = -\int_{-h/2}^{h/2} \sigma_{\theta x}\, z\,dz = M_{x\theta}$
4. Shear and Normal Resultants 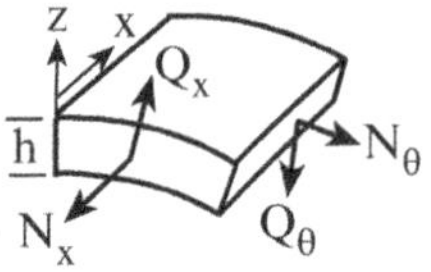 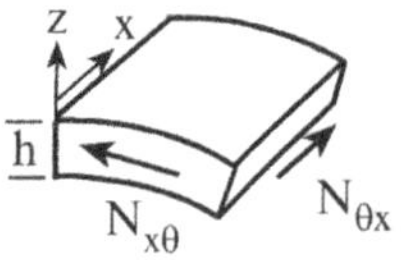	$Q_x = -\int_{-h/2}^{h/2} \sigma_{xz}\left(1+\frac{z}{R}\right)dz = \frac{Eh^3}{12(1-\nu^2)}\frac{\partial}{\partial x}\left(\frac{1}{R^2}\frac{\partial^2 w}{\partial \theta^2} + \frac{\partial^2 w}{\partial x^2}\right)$ $Q_\theta = -\int_{-h/2}^{h/2} \sigma_{\theta z}\,dz = \frac{Eh^3}{12(1-\nu^2)}\frac{1}{R}\frac{\partial}{\partial \theta}\left(\frac{\partial^2 w}{\partial x^2} + \frac{1}{R^2}\frac{\partial^2 w}{\partial \theta^2}\right)$ $N_x = \int_{-h/2}^{h/2} \sigma_{xx}\left(1+\frac{z}{R}\right)dz = \frac{Eh^3}{(1-\nu^2)}\left(\frac{\partial u}{\partial x} + \frac{\nu}{R}\frac{\partial v}{\partial \theta} + \nu\frac{w}{R}\right)$ $N_\theta = \int_{-h/2}^{h/2} \sigma_{\theta\theta}\,dz = \frac{Eh^3}{(1-\nu^2)}\left(\frac{1}{R}\frac{\partial v}{\partial \theta} + \frac{w}{R} + \nu\frac{\partial u}{\partial x}\right)$ $N_{x\theta} = \int_{-h/2}^{h/2} \sigma_{x\theta}\left(1+\frac{z}{R}\right)dz = \frac{Eh^3}{2(1+\nu)}\left(\frac{\partial v}{\partial x} + \frac{1}{R}\frac{\partial u}{\partial \theta}\right)$ $N_{\theta x} = \int_{-h/2}^{h/2} \sigma_{\theta x}\,dz = N_{x\theta}$

where E = modulus of elasticity; R = radius to mid surface; h = shell thickness; $k = h^2/(12R^2)$; u = axial deformation of mid surface; υ = circumferential deformation of mid surface; w = radial deformation of mid surface, positive in outward, z, direction; x = axial coordinate; t = time; ν = Poisson/s ratio of shell material; $\nabla^2 = \partial^2/\partial x^2 + \partial^2/R^2\partial\theta^2$. The loads per unit surface area of the shell, q_x, q_y, and p, are positive in the axial (x), circumferential (θ), and outward radial (w) directions, respectively.

The inertial terms associated with the axial, tangential, and radial deformation are proportional to ρ, the density of shell material, including any nonstructural material density. The shell thickness h appears, through the parameter k, only in the last line of the Donnell shell equations (Eq. 5.25). As a consequence, many modes of thin shells are independent of the shell thickness. These are called "membrane" or "extensional" modes because there is no bending, only stretching of the shell walls.

As shown in the remainder of this section, the natural frequencies of the shell can be reduced to the solution of a cubic characteristic polynomial, and the mode shapes, the relative deformation of u, v, and w, are found from free vibration solution to Equation 5.25.

5.3.2 *Natural Frequencies of Cylindrical Shells*

The natural frequencies of cylindrical shells are given in Table 5.7 in terms of a dimensionless natural frequency parameter λ_{ij} [6, 107, 110, 114–121]:

$$f_{ij} = \frac{\lambda_{ij}}{2\pi R}\sqrt{\frac{E}{\rho(1-\nu^2)}}, \quad \text{Hz}, \quad i = 0, 1, 2, \ldots, \quad j = 1, 2, \ldots \tag{5.26}$$

where E is the modulus of elasticity; R is the radius to the mid surface; ρ is the mass density of the material, including any nonstructural mass; and ν is Poisson's ratio. Consistent sets of units are given in Table 1.2. Shell natural frequencies increase with the modulus of elasticity and decrease with increasing density ρ and radius to the mid surface R. The nondimensional frequency parameter λ_{ij} for each i and j is a function of the boundary conditions on the ends of the shell, the aspect ratio L/R, the relative thickness h/R, and Poisson's ratio for simply supported and free-edged shells.

Equation 5.26 is similar in form to equations for natural frequency of beams (Tables 4.2, 4.12, 4.15 and 4.16). A long cylindrical shell has beam-like modes, but it also has modes unique to shells because curvature couples the extensional and flexural deformations of shells. As a result of curvature, the natural frequencies of the shell do not increase monotonically with mode number, as is the case with beams and plates. The lowest-frequency mode of a thin shell is not generally the $i = j = 1$ beam-like mode. Very often, it is a *shell mode* with higher order circumferential waves (see Figures 5.3 and 5.4, and Figure 5.5).

Shell modes are classified by the modal indices i and j; $i = 0, 1, 2, 3, \ldots$ is the number of circumferential half waves in the mode shape; $j = 0, 1, 2, \ldots$ is the number of axial (x-axis) half waves (see Figure 5.3).

Table 5.7 Natural frequencies of cylindrical shells

Notation: E = modulus of elasticity; i = 0, 1, 2, … = number of circumferential waves; j = 0, 1, 2, … number of axial half waves; h = shell thickness; L = length of shell; R = cylinder radius to mid surface; x = axial distance; u, v, w = axial, circumferential, and radial displacements; ρ = material density plus any nonstructural mass per unit volume; θ = circumferential angle; ν = Poisson's ratio. λ_{ij} = dimensionless shell natural frequency parameter; $\underline{\lambda}_j$ = dimensionless beam natural frequency parameter, Table 4.2. S = simply supported edge, C = clamped edge, F = free edge. Over bar (~) denotes mode shape. Consistent sets of units are in Table 1.2.

$$\text{Natural Frequency,}\quad f_{ij} = \frac{\lambda_{ij}}{2\pi R}\sqrt{\frac{E}{\rho(1-\nu^2)}},\quad i=0,1,2,3\ldots\ j=0,1,2,3\ldots\ \text{Hertz}$$

Boundary Conditions	λ	Mode Shape and Remarks
Cylindrical shell of infinite length, j=0, Modes that are independent of axial length 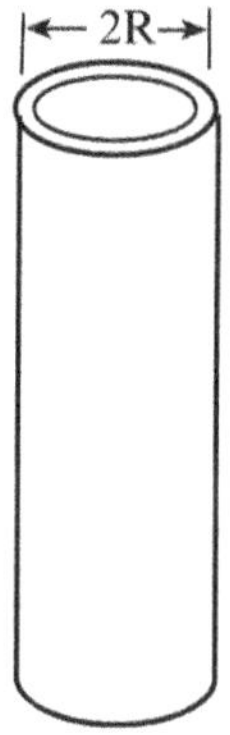	1. Axial Modes: $\lambda_{i,j=0} = \frac{i(1-\nu)^{1/2}}{2^{1/2}}$ $i = 1,2,3..\quad j=0$ Extensional modes	$\begin{Bmatrix}\tilde u\\ \tilde v\\ \tilde w\end{Bmatrix}_i = \begin{Bmatrix}\cos i\theta\\ 0\\ 0\end{Bmatrix}$ Ref. [110], p. 38, Ref. [115], p. 585
	2. Radial Modes: $\lambda_{i,j=0} = (1+i^2)^{1/2}$ $i = 0,1,2,\ldots, j=0$ Extensional modes	$\begin{Bmatrix}\tilde u\\ \tilde v\\ \tilde w\end{Bmatrix}_i = \begin{Bmatrix}0\\ i\sin i\theta\\ \cos i\theta\end{Bmatrix}$ Extensional modes i=0 is fundamental radial mode Ref. [110], p. 39 for h<<R. Ref. [115], p. 585.
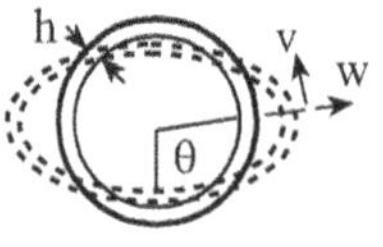	3. Ovaling Modes $\lambda_{i,j=0} = \frac{h}{12^{1/2}R}\frac{i(i^2-1)}{(1+i^2)^{1/2}}$ $i = 2,3,4,\ldots, j=0$ Bending modes	$\begin{Bmatrix}\tilde u\\ \tilde v\\ \tilde w\end{Bmatrix}_i = \begin{Bmatrix}0\\ -(1/i)\cos i\theta\\ \sin i\theta\end{Bmatrix}$ Inextensional modes. 16, case 3. Ref. [115], p. 585.
4. Love Modes of Free-Free Cylindrical Shell without Axial Constraint, j=0 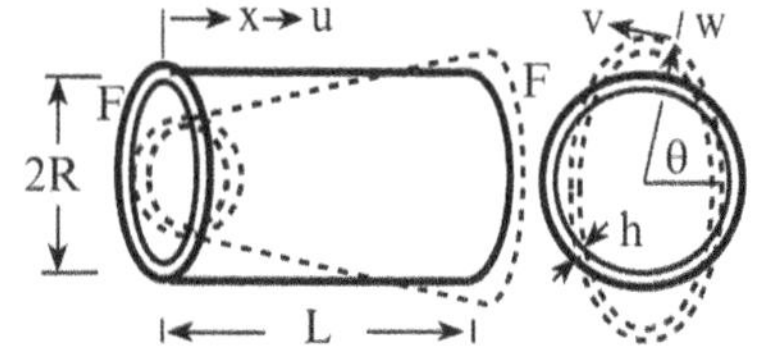	$\lambda_{ij} = \left[\frac{i^2(i^2-1)^2}{(i^2+1)}\frac{h^2}{12R^2}\frac{1+24(1-\nu)R^2/(i^2L^2)}{1+12R^2/[i^2(i^2+1)L^2]}\right]^{1/2}\quad i=2,3,..$ $\begin{Bmatrix}\tilde u\\ \tilde v\\ \tilde w\end{Bmatrix}_i = \begin{Bmatrix}(R/i^2)\cos i\theta\\ [(x-L/2)/i]\sin i\theta\\ (x-L/2)\cos i\theta\end{Bmatrix}$ bending modes also see Case 3 Refs [1, 111] Love modes toe in, Rayleigh modes, case 3, do not .	
5. Torsional Modes, i=0, Simply Supported Cylindrical Shell without Axial Constraint 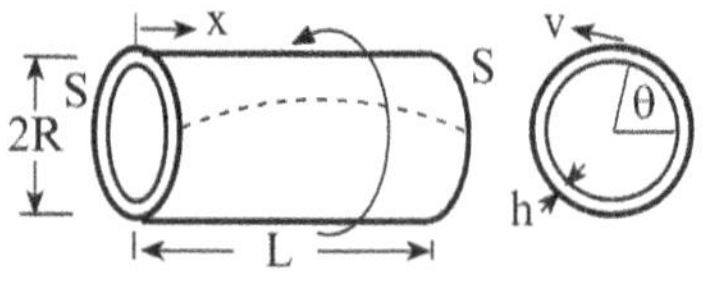	$\lambda_{i=0,j} = \frac{(1-\nu)^{1/2}}{2^{1/2}}\frac{j\pi R}{L}$ $i=0$ $j=1,2,3..$	$\begin{Bmatrix}\tilde u\\ \tilde v\\ \tilde w\end{Bmatrix}_j = \begin{Bmatrix}0\\ \sin j\pi x/L\\ 0\end{Bmatrix}$ Unlike cases 1, 2, 3, this is a finite shell of length L. Ref. [110], p. 67.

Table 5.7 Natural frequencies of cylindrical shells, continued

6. Axial, i=0, Modes of Simply Supported Cylindrical Shell without AxialConstraint 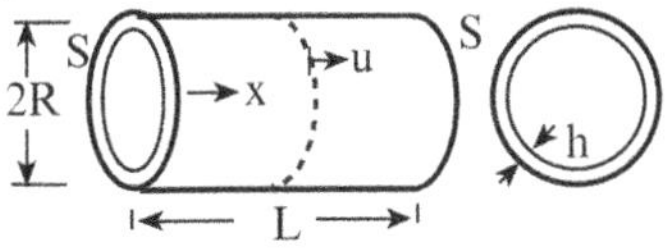 	$\lambda_{i=0,j} = j\pi(1-\nu^2)^{1/2}\dfrac{R}{L}$ $i = 0$ $j = 1,2,3..$ $\begin{Bmatrix}\tilde{u}\\ \tilde{v}\\ \tilde{w}\end{Bmatrix}_j = \begin{Bmatrix}\cos j\pi x/L\\ 0\\ 0\end{Bmatrix}$ Extension mode Ref. [115], p. 81. Table 4.13, case 1.
7. Radial, i=0, Mode Simply Supported deformation at x=L/(2j) 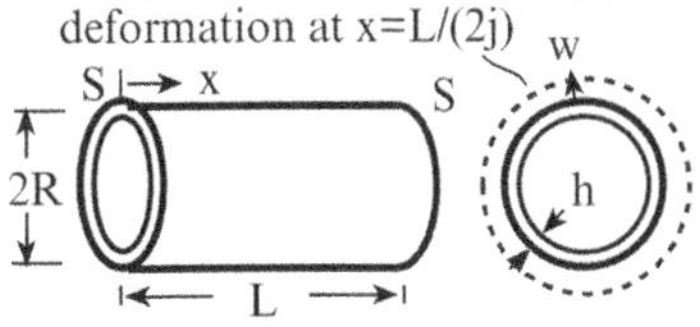 	$\lambda_{i=0,j} = 1, \quad i = 0,\ j = 1,2,3..$ λ is independent of j for long cylinders, L >> jR, Extensional mode. $\begin{Bmatrix}\tilde{u}\\ \tilde{v}\\ \tilde{w}\end{Bmatrix}_j = \begin{Bmatrix}0\\ 0\\ \sin j\pi x/L\end{Bmatrix}$
8. Beam Bending, i=1, Modes of Simply Supported Cylindrical Shell deformation at x=L/(2j) 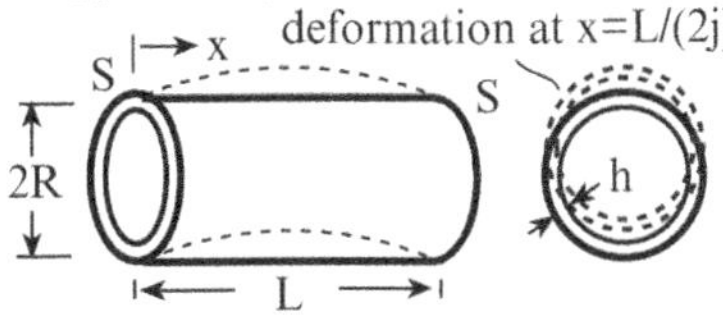 	$\lambda_{i=1,j} = \left(j\pi\dfrac{R}{L}\right)^2 \dfrac{(1-\nu^2)^{1/2}}{2^{1/2}}$ $i = 1$ $j = 1,2,3..$ $\begin{Bmatrix}\tilde{u}\\ \tilde{v}\\ \tilde{w}\end{Bmatrix}_i = \begin{Bmatrix}0\\ \cos\theta \sin j\pi x/L\\ \sin\theta \sin j\pi x/L\end{Bmatrix}$ Beam bending theory result. Case 4 of Table 4.2. Valid for long cylinders, L>8jR. Also see Ref. [115], p. 584.
9. Simply Supported Cylindrical Shell without Axial Constraint i =2,3,4.,j=1,2, 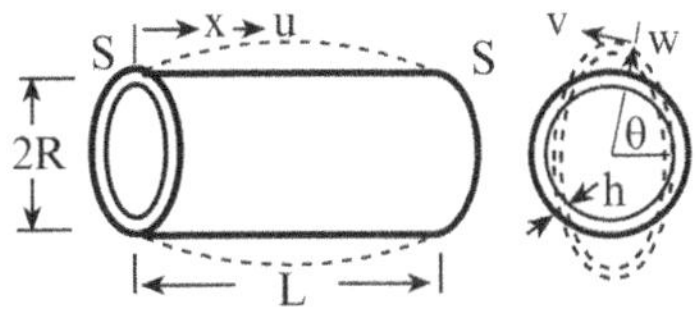 	$\lambda_{ij} = \dfrac{\left\{(1-\nu^2)\left(\dfrac{j\pi R}{L}\right)^4 + \dfrac{h^2}{12R^2}\left(i^2 + \left(\dfrac{j\pi R}{L}\right)^2\right)^4\right\}^{1/2}}{i^2 + (j\pi R/L)^2}$ $\begin{Bmatrix}\tilde{u}\\ \tilde{v}\\ \tilde{w}\end{Bmatrix}_i = \begin{Bmatrix}A\cos i\theta \cos j\pi x/L\\ B\sin i\theta \sin j\pi x/L\\ C\cos i\theta \sin j\pi x/L\end{Bmatrix}$ i = 2,3,4... Ref. [110], p. 74 j = 1,2,3... Ref. [115], p. 588 bending mode Ref. [6], p. 187
10.Simply Support-Free Cylindrical Shell Shell modes i=2,3,4.., j=1,2,3 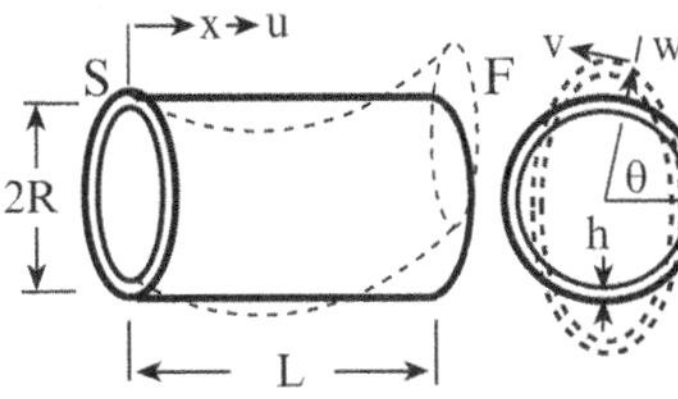 j = 1, i = 2 deformation shown Mode shape is given by Eq. 5.31. Beam modes given in case 4 of Table 4.2.	$\lambda_{ij} \approx \left\{\dfrac{(1-\nu^2)(\underline{\lambda}_j R/L)^4}{\left(i^2 + (\underline{\lambda}_j R/L)^2\right)^2} + \dfrac{h^2}{12R^2}\left(i^2 + \left(\underline{\lambda}_j \dfrac{R}{L}\right)^2\right)^2\right\}^{1/2}$ Approximate solution with beam bending mode shapes. $\underline{\lambda}_j$ is in case 1 of Table 4.2 and below. Ref. [6] p. 187 Exact solution is Eq. 5.39 with the following parameters. j 1 2 3 4 bending modes $\underline{\lambda}_j$ 3.926 7.0680 10.210 13.352 α_1 0.7467 0.8585 0.9021 0.9251 $\alpha_1=(\sigma_j/\underline{\lambda}_j)(\sigma_j \underline{\lambda}_j-1)$ α_2 1.7662 1.4244 1.2938 1.2247 $\alpha_2=(\sigma_j/\underline{\lambda}_j)(\sigma_j \underline{\lambda}_j+3)$

Table 5.7 Natural frequencies of cylindrical shells, continued

Natural Frequency, $f_{ij} = \dfrac{\lambda_{ij}}{2\pi R}\sqrt{\dfrac{E}{\rho(1-\nu^2)}}$, i =0,1,2,3... j =0,1,2,3... Hertz

Boundary Conditions	Natural Frequency Parameter, λ_{ij}

11. Free-Free Cylindrical Shell i=2,3,4, j=1,2

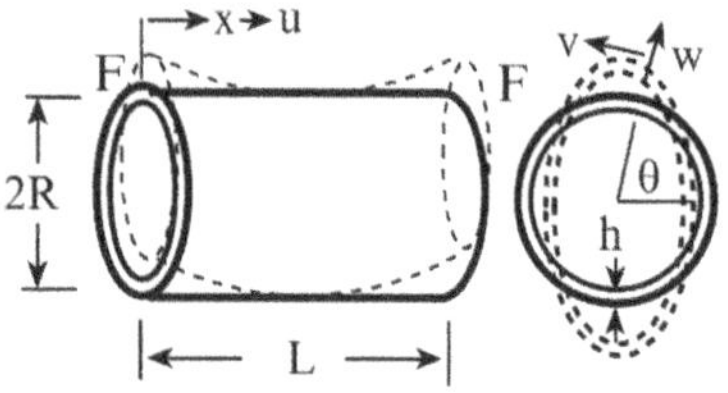

j = 1, i = 2 deformation shown

j=0 modes in cases 3 and 4.

Mode shape is given by Eqs. 5.31, 5.36.

$$\lambda_{ij} \approx \left\{ \frac{(1-\nu^2)(\underline{\lambda}_j R/L)^4}{\left(i^2+(\underline{\lambda}_j R/L)^2\right)^2} + \frac{h^2}{12R^2}\left(i^2+\left(\underline{\lambda}_j \frac{R}{L}\right)^2\right)^2 \right\}^{1/2}$$

Approximate solution uses free-free beam mode shapes. $\underline{\lambda}$ is in case 1 of Table 4.2 and below. Ref. [6] p. 187

Exact solution is Eq. 5.39 with parameters α_1, α_2.

j	1	2	3	4	bending modes
$\underline{\lambda}$	4.7304	7.853	10.995	13.352	
α_1	0.5499	0.7467	0.8180	0.8585	$\alpha_1=(\sigma_j/\underline{\lambda}_j)(\sigma_j\ \underline{\lambda}_j-2)$
α_2	2.2116	1.17662	1.5456	1.4244	$\alpha_2=(\sigma_j/\underline{\lambda}_j)(\sigma_j\ \underline{\lambda}_j+6)$

12. Clamp-Free Cylindrical Shell, i=2,3,4

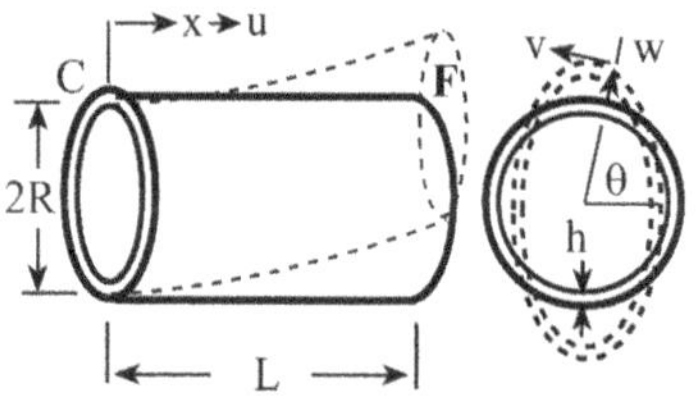

j = 1, i = 2 deformation shown

Mode shape is given by Eqs. 5.31, 5.36.

$$\lambda_{ij} \approx \left\{ \frac{(1-\nu^2)(\underline{\lambda}_j R/L)^4}{\left(i^2+(\underline{\lambda}_j R/L)^2\right)^2} + \frac{h^2}{12R^2}\left(i^2+\left(\underline{\lambda}_j \frac{R}{L}\right)^2\right)^2 \right\}^{1/2}$$

Approximate solution uses clamp free beam mode shapes. $\underline{\lambda}$ is in case 3 of Table 4.2 and below. Ref. [6] p. 187

Exact solution is Eq. 5.39 with the following parameters.

j	1	2	3	4	bending modes
$\underline{\lambda}$	1.8751	4.6964	7.8547	10.995	
α1	-0.2441	0.6033	0.7740	0.8182	$\alpha_1=(\sigma_j/\underline{\lambda}_j)(\sigma_j\ \underline{\lambda}_j-2)$
α2	1.3219	1.4712	1.2529	1.1820	$\alpha_2=(\sigma_j/\underline{\lambda}_j)(\sigma_j\ \underline{\lambda}_j+2)$

13. Clamp-Pinned Cylindrical Shell, i=2,3.

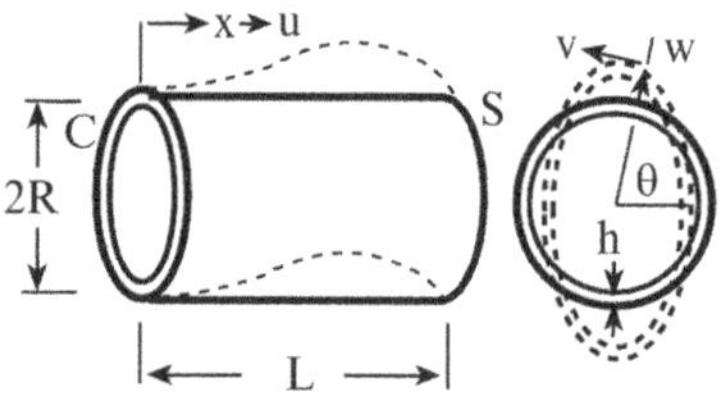

j = 1, i = 2 deformation shown

Mode shape is given by Eqs. 5.31, 5.36.

$$\lambda_{ij} \approx \left\{ \frac{(1-\nu^2)(\underline{\lambda}_j R/L)^4}{\left(i^2+(\underline{\lambda}_j R/L)^2\right)^2} + \frac{h^2}{12R^2}\left(i^2+\left(\underline{\lambda}_j \frac{R}{L}\right)^2\right)^2 \right\}^{1/2}$$

Approximate solution uses clamp-pin beam mode shapes $\underline{\lambda}$ in case 6 of Table 4.2 and below. Exact solution is Eq. 5.39 with parameters α_1, α_2. Ref. [6] p. 187

j	1	2	3	4	
$\underline{\lambda}$	3.9266	7.0686	10.210	13.352	bending modes
α1	0.7467	0.8585	0.9021	0.9251	$\alpha_1=(\sigma_j/\underline{\lambda}_j)(\sigma_j\ \underline{\lambda}_j-1)$
α2	0.7467	0.8585	0.9021	0.9251	$\alpha_2=(\sigma_j/\underline{\lambda}_j)(\sigma_j\ \underline{\lambda}_j-2)$

Table 5.7 Natural frequencies of cylindrical shells, continued

Natural Frequency, $f_{ij} = \dfrac{\lambda_{ij}}{2\pi R}\sqrt{\dfrac{E}{\rho(1-\nu^2)}}$, i =0,1,2,3... j =0,1,2,3... Hertz

Boundary Conditions	Natural Frequency Parameter, λ_{ij}
14. Clamped-Clamped Cylindrical Shell i=2,3,4,,,, j=1,2,3.. 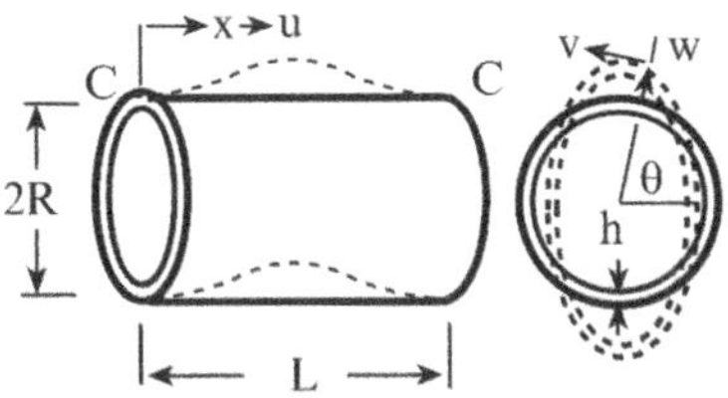 j = 1, i = 2 deformation shown Mode shape is given by Eqs. 5.31, 5.36.	$\lambda_{ij} \approx \left\{ \dfrac{(1-\nu^2)(\underline{\lambda}_j R/L)^4}{\left(i^2+(\underline{\lambda}_j R/L)^2\right)^2} + \dfrac{h^2}{12R^2}\left(i^2+\left(\underline{\lambda}_j \dfrac{R}{L}\right)^2\right)^2 \right\}^{1/2}$ Approximate solution uses clamp-clamp beam modes. $\underline{\lambda}$ is in case 7 of Table 4.2 and below. Ref. [6] p. 187 Exact solution is Eq. 5.35 with the following parameters. j 1 2 3 4 bending modes $\underline{\lambda}$ 4.7304 7.8553 10.995 14.137 α1 0.5499 0.7467 0.8180 0.85 $\alpha_i=(\sigma_j/\underline{\lambda}_j)(\sigma_j \underline{\lambda}_j-2)$ α2 0.5499 0.7467 0.8180 0.8585 $\alpha_i=(\sigma_j/\underline{\lambda}_j)(\sigma_j \underline{\lambda}_j-2)$
15. Cylindrically Curved Panel with Simply Supported Edges 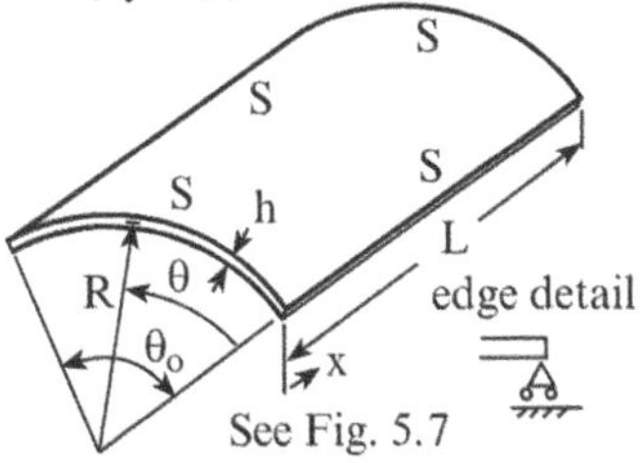 	$\lambda_{ij} = \left\{ \dfrac{\pi^4(hR)^2}{12}\left[\dfrac{i^2}{R^2\theta_o^2}+\dfrac{j^2}{L^2}\right]^2 + \dfrac{(j/L)^4(1-\nu^2)}{\left((j/L)^2+(i/R\theta_o)^2\right)^2} \right\}^{1/2}$ i= 2,3... j=1,2,3 Bending modes. The straight edges move freely in circumferential direction. The curved edges can rotate about circumferential axis. i,,j =1,2,3... .Refs [6, 120, 122] Equation 5.42 in text has the natural frequencies for other boundary conditions.
16. Pressurized and Preloaded Cylinder, Simply Supported without Axial Const. 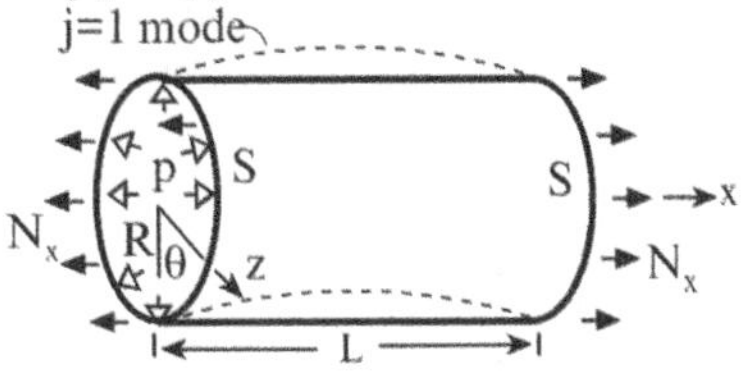 	$f_{ij}^2\Big\vert_{Nx,Ny\neq 0} = f_{ij}^2\Big\vert_{Nx=Ny=0} + \dfrac{1}{4\pi^2}\left[\dfrac{N_x}{\rho h}\left(\dfrac{j\pi}{L}\right)^2 + \dfrac{N_\theta}{\rho h}\left(\dfrac{i\pi}{R}\right)^2\right]$, Hz $i = 1,2,3..,\ j = 1,2,3..$ bending modes f_{ij} for unstressed shell given in case 8 through15 according to the boundary conditions. Based on pressurized membrane is given in case 16 of Table 3.7. Nx = axial load per unit length of edge. N_θ = circumferential load per unit length, often results from differential pressure p.

5.3.3 *Infinitely Long Cylindrical Shell Modes* (j = 0)

Infinitely long cylindrical shell modes ($j = 0$) axial modes are independent of shell length and the axial coordinate x, so they are called infinitely long cylindrical shells. The axial, circumferential, and radial deformations of the mid surfaces u, v, and w, respectively, oscillate in phase in time with amplitudes A, B, and C, respectively:

$$u = A\cos i\theta \cos\omega t, \quad v = B\sin i\theta\cos\omega t, \quad w = C\cos i\theta\cos\omega t, \quad i = 0, 1, 2, 3, \ldots \tag{5.27}$$

The integer index i is the number of circumferential waves (Figure 5.5). These mode shapes also can be obtained from Equation 5.29 with $j = 0$. Equation 5.27 is substituted into the

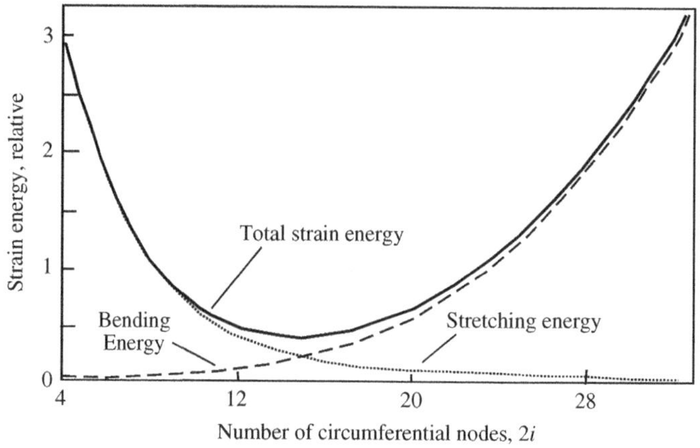

Figure 5.4 Variation in strain energy of a cylindrical shell with increasing number of waves [119]

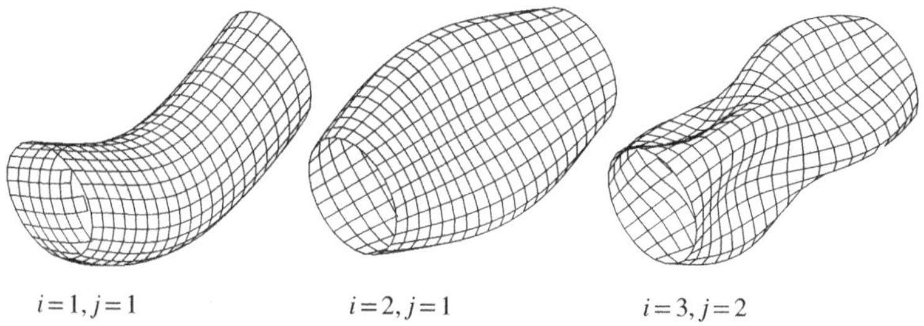

Figure 5.5 Deformed shapes for simply supported shell without axial constraint

equation of motion (Eq. 5.25), which is then multiplied through by $-R^2$. The resultant free vibrations ($q_x = q_y = p = 0$) are compactly described by a matrix equation for A, B, and C ([110], p. 38):

$$\begin{bmatrix} \left(\frac{1}{2}\right)(1-\nu)i^2-\lambda^2 & 0 & 0 \\ 0 & i^2-\lambda^2 & i \\ 0 & i & 1+ki^4-\lambda^2 \end{bmatrix} \begin{bmatrix} A \\ B \\ C \end{bmatrix} = \begin{bmatrix} 0 \\ 0 \\ 0 \end{bmatrix}$$

$$\lambda^2 = \left[\frac{\rho\left(1-\nu^2\right)R^2}{E}\right]\omega^2 \quad \text{and} \quad k = \frac{h^2}{12R^2} \tag{5.28}$$

Nontrivial solutions are found by setting the determinant of the coefficient matrix on the left-hand side to zero [123]. This gives three sets of $j = 0$ modes in cases 1–3 of Table 5.7 ([110], pp. 38–39) that are independent of the axial coordinate x:

1. *Axial shearing modes* (case 1 of Table 5.7). $B = C = 0$, $A = 1$. $(1/2)(1-\nu)i^2 - \lambda^2 = 0$. The solution is extension axial waves and no radial or circumferential motion.

2. *Radial modes* (case 2). $A = 0$ so $B/C = i/(\lambda^2 - i^2)$. The solution has no axial motion.
3. *Radial–circumferential bending modes* (case 3 of Table 5.7). $A = 0$. $B/C = i/(\lambda^2 - i^2)$.

The natural frequencies of these radial–circumferential bending modes are proportional to the shell thickness; they have the lowest natural frequencies of the three sets of modes; they are similar to the ring ovaling solution (case 3 of Table 4.16).

5.3.4 *Simply Supported Cylindrical Shells without Axial Constraint*

Consider a cylindrical shell whose edges are supported by thin planar diaphragms that are stiff in their own plane and so they restrict the circumferential and radial motion of the ends, $v = w = 0$ at $x = 0$ and $x = L$, but allow axial motion and rotation of the edges so the axial load and edge moments are zero, $M_x = N_x = 0$ at $x = 0, L$ (Table 5.6). These boundary conditions are also called simply supported without axial constraint. They are satisfied by the following sinusoidal deformations that oscillate in phase with circular frequency ω:

$$
\begin{aligned}
u &= A \cos\left(\frac{j\pi x}{L}\right) \cos i\theta \cos \omega t \\
v &= B \sin\left(\frac{j\pi x}{L}\right) \sin i\theta \cos \omega t \\
w &= C \sin\left(\frac{j\pi x}{L}\right) \cos i\theta \cos \omega t, \quad i = 0, 1, 2, 3, \ldots, \quad j = 1, 2, 3, \ldots
\end{aligned}
\tag{5.29}
$$

The integer index i is the number of circumferential waves, one half the number of circumferential nodes. The integer j is the number of axial half waves along the length of the cylinder. The interaction of axial and circumferential waves produces complex and beautiful modal patterns (Figure 5.5).

Equation 5.29 is substituted into Equation 5.25. The resultant free vibrations, $q_x = q_y = p = 0$, can be put into the form of a symmetric matrix equation for the amplitudes A, B, and C ([110, p. 37], with right-hand matrix multiplied through by -1):

$$
\begin{bmatrix} a_{11} - \alpha_2 \lambda^2 & a_{12} & a_{13} \\ a_{12} & a_{22} - \lambda^2 & a_{23} \\ a_{13} & a_{23} & a_{33} - \lambda^2 \end{bmatrix} \begin{bmatrix} A \\ B \\ C \end{bmatrix} = \begin{bmatrix} 0 \\ 0 \\ 0 \end{bmatrix} \tag{5.30}
$$

The mode shapes are then found by solving two rows given λ_{ij}:

$$
\begin{aligned}
\frac{A}{C} &= \frac{[a_{23}^2 - (a_{33} - \lambda^2)(a_{22} - \lambda^2)]}{(a_{13}(a_{22} - \lambda^2) - a_{12}a_{23})} \\
\frac{B}{C} &= \frac{[a_{12}(a_{33} - \lambda^2) - a_{13}a_{23}]}{(a_{13}(a_{22} - \lambda^2) - a_{12}a_{23})}
\end{aligned}
\tag{5.31}
$$

The matrix entries for a simply supported cylindrical shell with the Donnell shell theory are ([6], p. 95; [110], p. 37)

$$a_{11} = -\xi^2 - \frac{(1-\nu)}{2} i^2 + \lambda^2, \quad a_{21} = a_{12}, \quad a_{31} = -\nu\xi,$$

$$a_{12} = \frac{(1+\nu)}{2}\xi i, \quad a_{22} = -\frac{(1-\nu)}{2}\xi^2 - i^2 + \lambda^2, \quad a_{32} = i,$$

$$a_{13} = \nu\xi, \quad \alpha_{23} = -i, \quad a_{33} = 1 + k(\xi^2 + i^2)^2 - \lambda^2$$

$$\lambda^2 = \left[\frac{\rho\left(1-\nu^2\right)R^2}{E}\right]\omega^2, \ k = \frac{h^2}{(12R^2)}, \ \beta_j = \frac{j\pi R}{L} \tag{5.32}$$

The determinant of the matrix on the left-hand side of Equation 5.31 is set equal to zero to generate nontrivial solutions [123], and thickness (k) is assumed to be small for thin shells. The result is a cubic characteristic polynomial for the square of the natural frequency parameter λ [110], p. 44):

$$\lambda^6 + a_1\lambda^4 + a_2\lambda^2 + a_3 = 0 \tag{5.33}$$

where for a simply supported cylindrical shell,

$$a_1 = -1 - \left(\frac{1}{2}\right)(3-\nu)(i^2 + \beta_j^2) - k(i^2 + \beta_j^2)^2$$

$$a_2 = \left(\frac{1}{2}\right)(1-\nu)\left[(3+2\nu)\beta_j^2 + i^2 + (i^2+\beta_j^2)^2 + \left\{\frac{(3-\nu)}{(1-\nu}\right\}k(i^2+\beta_j^2)^3\right]$$

$$a_3|_{\text{Donnell}} = -\left(\frac{1}{2}\right)(1-\nu)[(1-\nu^2)\beta_j^4 + k(i^2+\beta_j^2)^4]$$

$$a_3|_{\text{Flugge}} = a_3|_{\text{Donnell}}$$

$$-\left(\frac{k}{2}\right)(1-\nu)[2(2-\nu)\beta_j^2 i^2 + i^4 - 2\nu\beta_j^6 - 6\beta_j^4 i^2 - 2(4-\nu)\beta_j^2 i^4 - 2i^6] \tag{5.34}$$

The parameters a_1 and a_2 do not change in the Flugge theory but a_3 gains additional terms proportional to k. (In Ref. [110] notation, $a_3 = -K_0$, $a_1 = -K_2$, $K_1 = a_2$, and the first two lines of Equation 5.31 are multiplied through by -1.) The corresponding entries a_1, a_2, $a_3, \ldots$ for other boundary conditions are provided in Equation 5.37.

The natural frequencies of the simply supported cylinder are found by solving the cubic Equation 5.33 exactly (Eq. 5.39) or approximately (Eq. 5.35) for three values of λ_{ij} for each i and j. The natural frequencies are classified by the modal circumferential wave index i and the axial index j.

Solutions for $i = 0$ and $j \neq 0$. The cross section stays round and centered for $i = 0$. There are two sets of modes:

1. *Torsional mode.* $i = 0, A = C = 0$, and $B = 1$. Thus, $a_{22} = 0$, and $\lambda^2 = (1/2)(1-\nu)\beta_j^2$. The torsion mode is given in case 5 of Table 5.7. It is independent of shell thickness.

2. Coupled axial–radial modes. $i = 0$, and $B = 0$. $A/C = \nu\beta/(\beta^2 - \lambda^2)$:

$$\lambda^2 = \left(\frac{1}{2}\right)\{(1 + \beta_j^2 + k\beta_j^4) \mp [(1 - \beta_j^2)^2 + 2\beta_j^2(2\nu^2 + k\beta_j^2 - k\beta_j^4)]^{1/2}\}$$

The coupled axial–radial modes decouple into an axial mode and a radial mode for long cylinders, $L/(jR) > 8$. These are given in cases 6 and 7 of Table 5.7.

Solutions for $i = 1$. For $i = 1$, the cross section remains round and translates, as shown in Figure 5.3, in a beam-like mode. The beam theory (case 5 of Table 4.3 with $i = 1$) is identical to the $i = 1$ Donnell shell theory for long cylinders, $L/(jR) > 10$ (case 7 of Table 5.7).

Solutions for $i = 2, 3, 4 \ldots$. Approximate solutions exist for the higher shell modes, $i = 2, 3, \ldots$ where the shell is dominated by the radial (w) deformation. If we retain only the inertia associated with radial deformation by dropping the terms in λ in the first two lines of Equation 5.31, then Equation 5.33 reduces to $a_2\lambda^2 + a_3 = 0$. The approximate solution is ([110, p. 74])

$$\lambda_{ij}^2 = -\frac{a_3}{a_2} \approx \frac{(1 - \nu^2)(j\pi R/L)^4 + k(i^2 + (j\pi R/L)^2)^4}{(i^2 + (j\pi R/L)^2)^2} \tag{5.35}$$

The spectrum of natural frequency parameters is given in Figure 5.6 for a simply supported cylindrical shell with $h/R = 0.002$. The lowest-frequency mode is not generally that

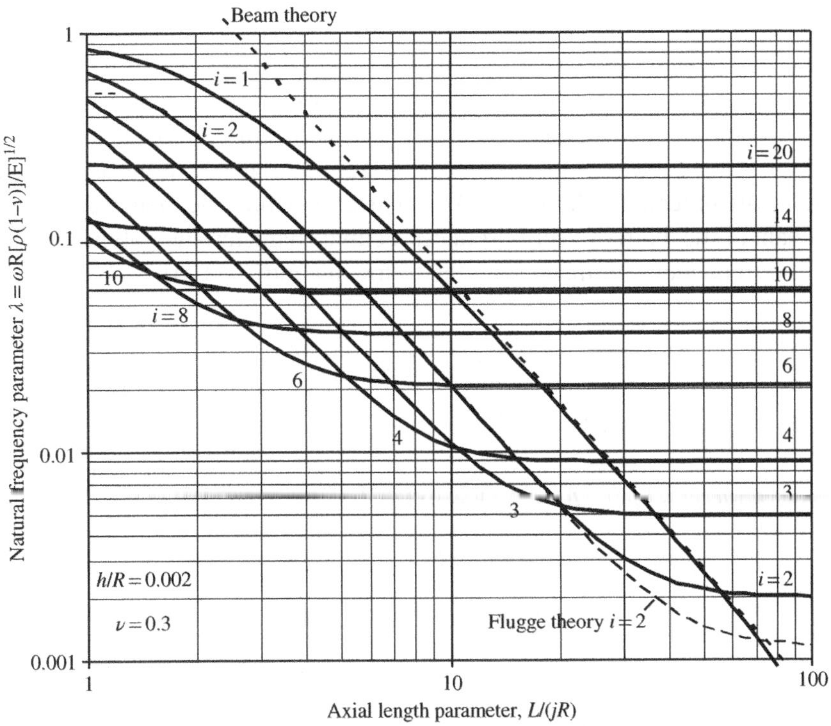

Figure 5.6 Natural frequencies of vibration of a cylindrical shell with ends simply supported without axial constraint using the Donnell shell theory and computed using Equation 5.39

with the lowest modal indices owing to the balance of stretching and bending energies (Figure 5.4). For long cylinders, the $i = 1$ solution approaches the simply supported beam which is given in case 8 of Table 5.7, and the $i = 2, 3, 4$ and $j = 0$ modes approach those of infinite cylinders given in case 1 of Table 5.7. Including the Flugge theory terms provides a lower and more accurate estimate than the Donnell shell theory [110]. The exact solution is provided in Eq. 5.39.

5.3.5 Cylindrical Shells with Other Boundary Conditions

Sharma and Johns [124] developed solutions for other boundary conditions by approximating the axial shape function $\widetilde{\varphi}_j(x/L)$ with single-span beam mode shapes (Table 4.2; [119, 123–126]; 110, p. 85):

$$u = A\,\widetilde{\varphi}_j'\left(\frac{\underline{\lambda}_j x}{L}\right)\cos i\theta\cos\omega t$$

$$v = B\,\widetilde{\varphi}_j\left(\frac{\underline{\lambda}_j x}{L}\right)\sin i\theta\cos\omega t$$

$$w = C\,\widetilde{\varphi}_j\left(\frac{\underline{\lambda}_j x}{L}\right)\cos i\theta\cos\omega t, \quad i = 0, 1, 2, 3, \ldots, \quad j = 1, 2, 3, \ldots \tag{5.36}$$

The index i is the number of circumferential waves in the mode shape (one half the number of circumferential nodes), and j is the number of longitudinal half waves. The primes ($'$) denote differentiation by $\underline{\lambda}_j x/L$ where $\underline{\lambda}_j x/L$ is the beam nondimensional natural frequency parameter in Table 4.2. Including the Flugge shell theory terms and then applying the Rayleigh–Ritz procedure (Appendix A) produces a symmetric matrix equation. Setting the determinant of the matrix to zero produces a cubic characteristic polynomial (Eq. 5.33) for the natural frequency parameter with the following entries [124]:

$$a_{11} = \beta_j^2 + \frac{1}{2}(1+k)(1-\nu)i^2\alpha_2, \qquad a_{12} = -\nu i\beta_j\alpha_1 - \frac{(1-\nu)i\beta_j\alpha_2}{2}$$

$$a_{22} = i^2 + \frac{1}{2}(1+3k)(1-\nu)\beta_j^2\alpha_2, \qquad a_{23} = i + ki\beta_j^2\left[\nu\alpha_1 + \frac{3}{2}(1-\nu)\,\alpha_2\right]$$

$$a_{13} = -\nu\beta_j\alpha_1 + k\beta_j\left[-\beta_j^2 + \frac{1}{2}(1-\nu)\,i^2\alpha_2\right]$$

$$a_{33} = 1 + k[\beta_j^4 + (i^2-1)^2 + 2\nu i^2\beta_j^2\alpha_1 + 2(1-\nu)i^2\beta_j^2\alpha_2]$$

$$a_1 = \frac{(-a_{11} - \alpha_2 a_{22} - \alpha_2 a_{33})}{\alpha_2}$$

$$a_2 = \frac{-a_{12}^2 - a_{13}^2 - \alpha_2 a_{23}^2 + \alpha_2 a_{22}a_{33} + a_{11}a_{22} + a_{11}a_{33}}{\alpha_2}$$

$$a_3 = \frac{-a_{11}a_{22}a_{33} - 2a_{12}a_{13}a_{23} + a_{12}^2 a_{33} + a_{11}a_{23}^2 + a_{13}^2 a_{22}}{\alpha_2}$$

$$\lambda = \left[\frac{\rho\left(1-\nu^2\right)R^2\omega^2}{E}\right]^{1/2}, \qquad k = \frac{h^2}{12R^2}, \qquad \beta_j = \frac{\underline{\lambda}_j R}{L} \tag{5.37}$$

$$\alpha_1 = -\frac{\int_0^L \widetilde{\varphi}_j''(\underline{\lambda}_j x/L)\widetilde{\varphi}_j(\underline{\lambda}_j x/L)\,dx}{\int_0^L \widetilde{\varphi}_j^2(\underline{\lambda}_j x/L)\,dx}$$

$$\alpha_2 = \frac{\int_0^L [\widetilde{\varphi}_j'(\underline{\lambda}_j x/L)]^2\,dx}{\int_0^L \widetilde{\varphi}_j^2(\underline{\lambda}_j x/L)\,dx} \tag{5.38}$$

The exact solution of Equation 5.33, with the entries of Equation 5.34 for simply supported cylinders or the entries of Equation 5.37 for other boundary conditions ([6], p. 98; [126]) is:

$$\lambda_{ij} = \left(-\frac{2}{3}\left(a_1^2 - 3a_2\right)^{1/2}\cos\left[\frac{\alpha+\eta}{3}\right] - \frac{a_1}{3}\right)^{1/2}, \qquad \eta = 0,\ 2\pi, 4\pi$$

$$\text{where } \alpha = ArcCos\left[\frac{\left(27a_3 + 2a_1^3 - 9a_1a_2\right)}{2(a_1^2 - 3a_2)^{3/2}}\right], \qquad 0 < \alpha < \pi \tag{5.39}$$

There are three positive real values of the natural frequency parameter λ for each i and j. The lowest of these is for $\eta = 0$ and it is dominated by out-of-plane deformation. $\eta = 2\pi$ and 4π are higher-frequency in-plane modes. Computation of arc cosine to six digits of accuracy is needed in some cases to resolve λ to 3 digits. Once λ is computed, the natural frequencies are given by Equation 5.26, and the corresponding mode shapes are found inserting λ into Equation 5.36. α_1 and α_2 (Eq. 5.38) are integrals over the beam mode shape (Appendix D and [125], amended), given in cases 9 through 14 of Table 5.7. The derivatives ($'$) are with respect to the parameter $\lambda_j x/L$.

There are approximate solutions for cylindrical shell natural frequencies. When the axial and tangential mass inertial terms are deleted from Equation 5.33, the approximate solution is $\lambda = (-a_3/a_2)^{1/2}$, where a_2 and a_3 are defined in Equation 5.37. This approximate solution is most applicable to the higher modes, $i = 2, 3, 4, \ldots$ of long shells $(L/jR) > 10$ that are dominated by radial deformation. Soedel's [126] approximate solution generalizes Equation 5.35 to other mode shapes:

$$\lambda_{ij} \approx \left\{\frac{\left(1-\nu^2\right)}{(i^2 + (\underline{\lambda}_i R/L)^2)^2}\left(\frac{\underline{\lambda}_j R}{L}\right)^4 + \left(\frac{h^2}{12R^2}\right)\left(i^2 + \left(\underline{\lambda}_i \frac{R}{L}\right)^2\right)^2\right\}^{1/2} \tag{5.40}$$

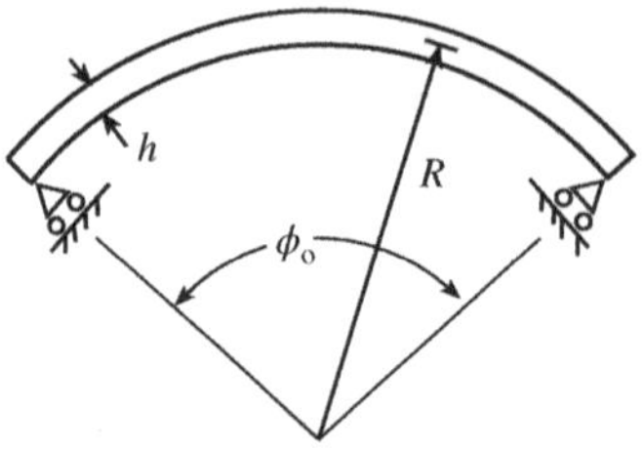

Figure 5.7 Section of a cylindrical panel. Simply supported edge conditions without circumferential constraint prevents radial displacement but does not restrict circumferential motion and rotation

The term $\underline{\lambda}_j$, which appears in Equations 5.39 and 5.37, is the beam dimensionless natural frequency parameter in Table 4.2 for beam mode shape (here denoted by $\underline{\lambda}_j$) that satisfies the boundary conditions on the ends of the cylinder. For simply supported ends without axial constraint, $\alpha_1 = \alpha_2 = 1$, $\underline{\lambda}_j = j\pi$, and Equation 5.40 reduces to Equation 5.35.

5.3.6 Free–Free Cylindrical Shell

There are no force, moment, or shear resultants on the ends of a completely free cylindrical shell (Table 5.7, cases 4 and 11). The boundary conditions at $x = 0, L$, are $M_x = N_x = Q_x = 0$ and $N_{x\theta} + M_{x\theta}/R = 0$, which is the *Kirchhoff free edge condition* ([2], pp. 83–84; [4], p. 108).

The mode shapes of a free–free cylinder can be grouped by their axial behavior. Higher modes, where the modal indices $i = 2, 3, 4, \ldots$ (i is the number of circumferential waves) and $j = 1, 2, 3$ (j is the number of axial half waves), are similar in shape to the modes of the simply supported shells without axial constraint (Figures 5.3 and 5.5), and their natural frequencies are fairly well predicted by Equations 5.39 and 5.40 as shown in Figure 5.7.

Free–free cylindrical shells also have two sets of low-frequency inextensional bending modes corresponding to $j = 0$. These are the *Rayleigh modes* (case 3 of Table 5.7) and *Love modes* (case 9 of Table 5.7; [110], pp. 543–547). The Rayleigh and Love modes have nearly the same natural frequency for long cylinders. The principal difference between the two sets of modes is that the ends of the cylinders *toe* in or out in the Love modes, while the cylinder flexes uniformly along its length in the Rayleigh modes. Both modes have been observed experimentally as shown and discussed by Warburton [127].

Example 5.2 Calculate the natural frequencies of a clamped–free cylindrical shell with the following properties: $L = 502$ mm (19.26 in.), $R = 63.5$ mm (2.5 in.), $h = 1.63$ mm (0.0641 in.), $E = 2.1 \times 10^{11}$ Pa (30.45×10^6 lb/in.^{2}), $\rho = 7800$ kg/m^3 (0.2818 lb/in.^{3}), and $\nu = 0.28$.

Solution: $R/L = 502\ \text{mm}/63.5\ \text{mm} = 7.905$, $h/R = 1.63\ \text{mm}/63.5\ \text{mm} = 0.025$. The natural frequency (Eq. 5.27) is proportional to the dimensionless parameter λ:

$$f_{ij} = \frac{\lambda_{ij}}{2\pi R}\sqrt{\frac{E}{\rho(1-\nu^2)}} = \frac{\lambda_{ij}}{2\pi \times 0.0635\,\text{m}}\sqrt{\frac{2.1\times 10^{11}\,\text{N/m}^2}{78{,}000\,\text{kg/m}^3\,(1-0.28^2)}} = 13547\,\lambda_{ij}\ \text{Hz}$$

$$= 3193\ \text{Hz},\ \text{for}\ i = 2,\ j = 1,\ \lambda = 0.02358$$

λ is given in case 10 of Table 5.7 or Eq. 5.39. For $i = 2$ and $j = 1$, Equation 5.39 gives $\lambda_{21} = 0.02358$ and the natural frequency for this mode is $f_{21} = 319.3$ Hz.

Natural frequencies are computed with these equations that are generally in good agreement with measurements by Sharma [128] on a clamped area shell in the following table, although the theory overpredicts the lower modes.

			Natural	Freq.	Hertz	
	i	$j = 1$	$j = 2$	$j = 3$	$j = 4$	$j = 5$
Experiment	2	293	827	1894	n/a	n/a
Case 10 of Table 5.7	2	320.9	1095	2922	5566	8922
Equation 5.39	2	319.3	1019	2400	3963	5471
Experiment	3	760	886	1371	1371	3208
Case 10 of Table 5.7	3	769.9	941.5	1638	2883	4596
Equation 5.39	3	769.9	930.3	1515	2428	3486
Experiment	4	1451	1503	1673	2045	2713
Case 10 of Table 5.7	4	1466	1526	1776	2284	3149
Equation 5.39	4	1466	1525	1730	2158	2783
Experiment	5	2336	2384	2480	2667	2970
Case 10 of Table 5.7	5	2367	2410	2519	2755	3169
Equation 5.39	5	2367	2409	2513	2723	3059
Experiment	6	3429	3476	3546	3667	3880
Case 10 of Table 5.7	6	3470	3509	3589	3734	3970
Equation 5.39	6	3470	3509	3587	3724	3936

5.3.7 *Cylindrically Curved Panels*

A cylindrically curved panel (Figure 5.7) subtends the circumferential angle θ between 0 and θ_o. The solutions for natural frequency and mode shape of complete cylindrical shells also apply to cylindrical panels provided the straight edges at $\theta = 0$, θ_o are supported by knife edges that allow circumferential motion:

$$w = N_\theta = M_\theta = 0 \text{ on the edges } \theta = 0, \theta_o \text{ from } x = 0 \text{ to } L$$
$$i = \frac{n\pi}{\theta_o}, \quad n = 1, 2, 3 \tag{5.41}$$

The circumferential wave index is allowed to take on noninteger values: $i = 0, 1, 2, 3, \ldots$ is replaced by $n\pi/\theta_o$. The mode shape is given by Equations 5.31 and 5.36.

Approximate solutions for the natural frequencies of shallow ($\theta_o < 1$ rad) cylindrically curved panels in higher modes ($j > 1$) have been developed by neglecting their tangential inertia [110, p. 158], approximating the mode shapes with beam mode shapes [120–123], and applying the Rayleigh–Ritz technique (Appendix A):

$$f_{ij} = \frac{1}{2\pi}\left[4\pi^2 f_{ij}^2|_{\substack{\text{flat plate,}\\ \text{Table 5.3}}} + \frac{\alpha_{ij}E}{R^2\rho\left(1-\nu^2\right)}\right]^{1/2}, \text{ Hz} \quad i = 1, 2, \ldots, \quad j = 1, 2, \ldots$$

$$\alpha_{ij} = \frac{\begin{array}{c}\{(G_1^4 - \nu^2 H_1^2)(G_2^4 - H_2^2) + 1/2(1-\nu)J_1J_2[G_1^4(R\theta_o/L)^2 - \nu^2 H_1^2(R\theta_o/L)^2 \\ + G_2^4(L/R\theta_o)^2 - H_2^2(L/R\theta_o)^2]\}\end{array}}{\begin{array}{c}\{G_1^4G_2^4 + 1/2(1-\nu)J_1J_2[G_1^4(R\theta_o/L)^2 + G_2^4(L/R\theta_o)^2] \\ -\nu^2H_1^2H_2^2 - \nu(1-\nu)J_1J_2H_1H_2\}\end{array}} \tag{5.42}$$

Parameters G_1, G_2, H_1 and H_2 are given in cases 1 through 19 of Table 5.3. This solution is equivalent to case 15 of Table 5.7 [120, 121] for simply supported and for clamped edges; it is less accurate for free edges because the beam modes do not accurately reproduce the mode shape of a cylindrically curved plate with free edges [64] (Eq. 5.21).

5.3.8 Effect of Mean Load on Natural Frequencies

The exact solutions for the natural frequency of a simply supported cylindrical shell without axial constraint but with mean load are given in case 16 of Table 5.7 ([110, p. 238]). Studies of tubes show that a uniform internal pressure has no effect on the beam bending natural frequencies ($i = 1, j = 1, 2, \ldots$) if the ends are capped so the cylinder supports the internal pressure p, $N_x = p\pi R^2/2\pi R = pR/2$ [129]. Additional discussion of the effect of mean load on the natural frequencies of cylindrical shells is provided in [129–132].

5.4 Spherical and Conical Shells

5.4.1 Spherical Shells

Table 5.8 has formulas for the natural frequencies and mode shapes of thin spherical shells. Unlike beams and plates, a complete spherical shell does not have inextensional bending modes ([133, p. 423; 114, p. 323]). Cases 1, 2, and 3 in Table 5.8 give the natural

Table 5.8 Natural frequencies of spherical shells

Notation: h = shell thickness; E = modulus of elasticity; i = 0, 1, 2, ... = modal index for circumferential waves; j = 0, 1, 2, ... modal index for polar waves; h = shell thickness, h << R; P_i, P_i^1 = Legendre polynomials; R = cylinder radius to mid surface; u,v,w = polar, circumferential, and radial displacements; ρ = material density plus any nonstructural mass per unit volume; θ = circumferential angle; ϕ = polar (inclination) angle; ν = Poisson's ratio. Formulas are valid for thin shells, ih << R. Over bar (~) denotes mode shape. Consistent sets of units are given in Table 1.2.

$$\text{Natural Frequency,}\quad f_i = \frac{\lambda_i}{2\pi R}\sqrt{\frac{E}{\rho(1-\nu^2)}},\qquad \text{Hertz}$$

Spherical Shell	Natural Frequency Parameter, λ_i	Mode Shape
Spherical Shell Axisymmetric modes Independent of θ	1. Fundamental Radial Extensional Mode i=0 $\left[\dfrac{2(1+\nu)}{1+h^2/(12R^2)}\right]^{1/2}$	$\begin{Bmatrix}\tilde\phi\\ \tilde\theta\\ \tilde r\end{Bmatrix}=\begin{Bmatrix}0\\ 0\\ 1\end{Bmatrix}$ Ref. [115], p. 616
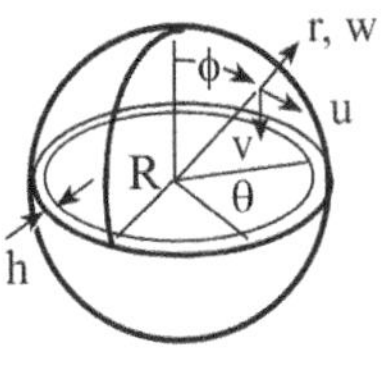	2. Radial-Tangential Modes, i = 1,2,3.. $\dfrac{1}{2^{1/2}}\left\{i^2+i+1+3\nu)\mp \left[(i^2+i+1+3\nu)^2-4(1-\nu^2)(i^2+i-2)\right]^{1/2}\right\}^{1/2}$ These modes are extensional. The modes with bending effects are computed from the cubic equation given in Ref. [135].	$\begin{Bmatrix}\tilde\phi\\ \tilde\theta\\ \tilde r\end{Bmatrix}_i=\begin{Bmatrix}\varepsilon\\ \varepsilon\\ P_i\cos\phi\end{Bmatrix}$ Ref. [115], p. 616
	3. Torsional Axisymmetric modes $\left[\dfrac{(1-\nu)(i^2+i-2)}{2+5h^2/(6R^2)}\right]^{1/2}$ i=1,2,3.., j=1,2,3.. Modes with bending effects included are given in Ref. [134].	$\begin{Bmatrix}\tilde\phi\\ \tilde\theta\\ \tilde r\end{Bmatrix}_i=\begin{Bmatrix}--\\ -j\csc\phi P_i^1(\cos\phi)\\ 0\end{Bmatrix}$ Ref. [115], p. 620
Deep Spherical Dome with Free Edge 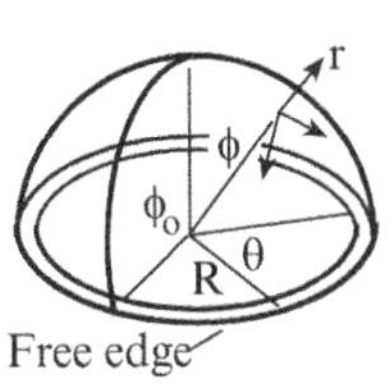 Free edge	4. Inextensional Bending Modes of open dome, i = 2,3,4,.., $\lambda_i=\left[\dfrac{(1-\nu^2)(i^2-1)i^2}{3(1+\nu)}\right]^{1/2}\dfrac{h}{R}\left(\dfrac{g_1(i,\phi_o)}{g_2(i,\phi_o)}\right)^{1/2}$	$\begin{Bmatrix}\tilde\theta\\ \tilde\phi\\ \tilde r\end{Bmatrix}_i=\begin{Bmatrix}-\sin\phi(\tan(\phi/2))^i\cos i\theta\\ -(\tan(\phi/2))^i\sin i\theta\\ R(i+\cos\phi)(\tan(\phi/2))^i\cos i\theta\end{Bmatrix}$

g_1/g_2

ϕ_o, deg.	10	30	60	90	120
i=2	34.32	1.569	0.3070	0.1907	0.3101
i=3	13.23	0.6382	0.1363	0.09412	0.1720
i=4	7.153	0.3535	0.07838	0.05626	0.1056
i=5	4.506	0.2257	0.0510	0.03724	0.07036

dome extends from ϕ=0 to ϕ_o

$$g_1=\frac{1}{8}\left[\frac{\{\tan(\phi_o/2)\}^{2i-2}}{i-1}+\frac{2\{\tan(\phi_o/2)\}^{2i}}{i}+\frac{\{\tan(\phi_o/2)\}^{2i+2}}{i+1}\right]=(i^3-i)(2i^2-1)\ \text{for}\ \phi_o=90\ \text{deg.}$$

$$g_2=\int_0^{\phi_o}\{\tan(\phi_o/2)\}^{2i}[(i+\cos\phi)^2+2(\sin\phi)^2]\sin\phi\, d\phi$$

Ref. [133], pp. 419-432.
Ref. [114], pp. 323-334

frequencies and mode shapes of complete, closed, thin spherical shells as they vibrate in axisymmetric modes that are either purely extensional (cases 1, 2, and 3) or coupled bending–extensional. Krauss [112] (p. 341) gives the following solution for natural frequencies of spherical shells that includes bending:

$$-\lambda_i^2 = 3(1+\nu) + i(i+1) + \left(\frac{1}{k}\right)(i(i+1)+3)(i(i+1)+1+\nu)$$
$$\pm\left\{\left[3(1+\nu) + i(i+1) + \left(\frac{1}{k}\right)(i(i+1)+3)(i(i+1)+1+\nu)\right]^2\right.$$
$$\left.-4i(i+1)\left[(1-\nu^2) + \left(\frac{1}{k}\right)(i^2(i+1)^2 + 2i(i+1) + 1 - \nu^2)\right]\}^{1/2}\right\} \qquad (5.43)$$

$k = h^2/12R^2$, $i = 0, 1, 2, 3, \ldots$ is the inclination wave index, and the natural frequency f_i (Hz) $= \lambda_i/(2\pi R)[E/\rho(1-\nu^2)]^{1/2}$; for $h = 0$, this reduces to cases 1 and 2 of Table 5.8. Wilkinson [134] gives a cubic characteristic equation for spherical shell modes that include bending and stiffness deformation.

5.4.2 *Open Shells and Church Bells*

Rayleigh developed the case 4 solution for inextensional bending modes of a deep open spherical shell with a free edge and applied it to the tones of church bells ([133], p. 391; [136]). These bending modes are also given by Niordson [114] (pp. 323–334), Kraus [112] (pp. 35–336), and Kalnis [137]. Perrin and Charnley [138] measured vibration modes of modern English church bells.

5.4.3 *Shallow Spherical Shells*

The shallow shell theory describes thin shells, such as shown in Figure 5.8, whose rise is less than about 1/8 of their typical lateral dimension. Soedel [6, 139] developed an analogy between transverse vibration of flat plates and shallow spherically curved shells with the

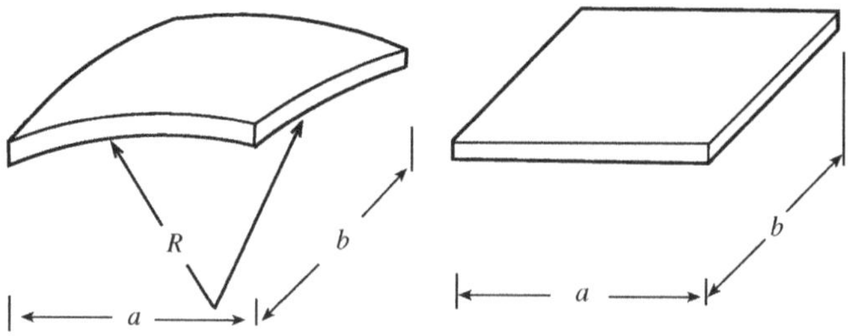

Figure 5.8 Shallow spherically curved shell and a flat rectangular plate with the same planform

Table 5.9 Natural frequencies of conical shells

Notation: h = shell thickness; E = modulus of elasticity; i = modal index for circumferential waves; j = 0, 1, 2, … modal index for circumferential waves; h = shell thickness; u, v, w = axial, circumferential, and radial displacements; ρ = material density plus any non-structural mass per unit volume; θ = circumferential angle; ν = Poisson's ratio. Consistent sets of units are given in Table 1.2. Formulas are valid for thin shells, h << R. Over bar (~) denotes mode shape [110, 142].

Natural Frequency, $f_{ij} = \frac{\lambda_{ij}}{2\pi R}\sqrt{\frac{E}{\rho}}$, Hertz

R_2 replaces R in case 4

$L(1-\nu)^{1/2}$ replaces R in case 5

Cone	Natural Frequency Parameter, λ_{ij}	Mode Shape

1. Free Base

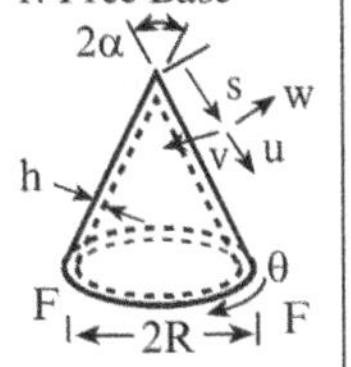

Natural Frequency Parameter: $0.875(h/R)^{7/8}(\sin\alpha)^{1/4}[i^4\nu^2 + (1/2)\underline{\lambda}_j^4 \sin^4\alpha \cdot (1-4\nu^2\sin^4\alpha) - 2\nu(i-2)(i-3)(2i-5)]^{1/2}$,

$\underline{\lambda}_j = 1.875, 4.694, (j-1/2)\pi,\ j>2.\ j=1,2,3,\ i=2,3,4..$

Radial-circumferential modes. Lowest frequency modes. ν=0.3.

Mode Shape: $\begin{Bmatrix}\tilde{u}\\ \tilde{v}\\ \tilde{w}\end{Bmatrix} = \begin{Bmatrix}0\\ \varepsilon \ll 1\\ F_j(s)\cos i\theta\end{Bmatrix}$

Tip of cone fixed, $F_j(0)=0$. Ref. [142]. Also see Ref. [110].

2. Cone Clamped Base

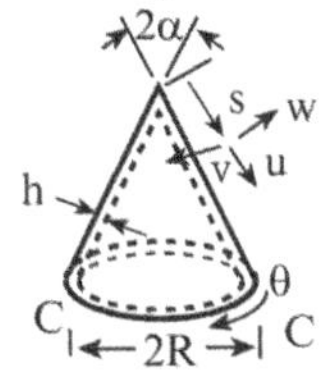

$\frac{12R^2(1-\nu^2)}{h^2\tan^4\alpha}$	j = 1	2	3	4
0.4	266.9	3964.	19860	62570
1.0	110.4	1591.	7954.	25040
4.0	32.07	404.8	1999.	6275.
10.	16.33	167.5	808.2	2521.
100.	6.096	24.69	93.37	269.0
1000.	3.574	7.648	18.58	40.76
100000.	1.802	2.431	3.449	4.604

Mode Shape: $\begin{Bmatrix}\tilde{u}\\ \tilde{v}\\ \tilde{w}\end{Bmatrix} = \begin{Bmatrix}G_j(s)\cos i\theta\\ 0\\ F_j(s)\cos i\theta\end{Bmatrix}$

ν = 0.3
i = 0, j = 1,2,3..

3 Free Edges

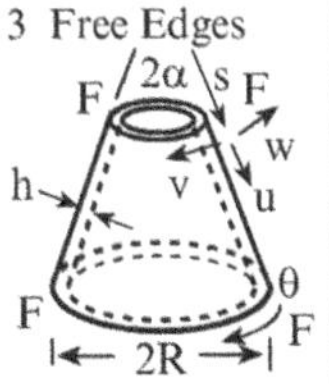

Natural Frequency Parameter: $\frac{h}{12^{1/2}R}\ \frac{i^2-i}{(i^2+1)^{1/2}}\ \frac{i+1-4\sin(3\alpha/2)}{(1-\nu^2)^{1/2}}$

Frustrum of a cone.
Radial-Circumferential inextensional modes
i = 2,3,4.., j = 0, α<60 deg.
$S_2 = R/\sin\alpha$, B = constant

Mode Shape: $\begin{Bmatrix}\tilde{u}\\ \tilde{v}\\ \tilde{w}\end{Bmatrix} = \begin{Bmatrix}\sin\alpha\cos\alpha\cos i\theta\\ 1+B\frac{s}{S_2}i\cos\alpha\sin i\theta\\ i^2-\sin^2\alpha+B\frac{si^2}{S_2}\cos i\theta\end{Bmatrix}$

4. Frustrum Clamp–Free

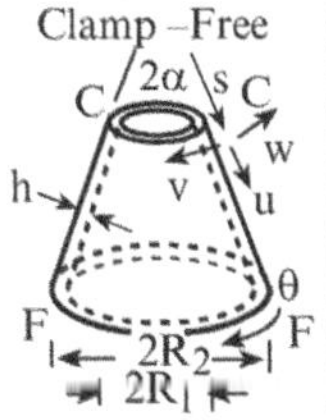

$\frac{12R^2(1-\nu^2)}{h^2\tan^4\alpha}$	R_1/R_2 = 0.8	0.7	0.5	0.1
0.1	260.0	113.	40.2	13.2
1.0	81.7	36.2	13.2	4.28
10.	26.8	12.1	4.45	1.61
100	8.74	4.00	1.71	0.96
1000	3.22	1.57	1.09	0.83

Mode Shape: $\begin{Bmatrix}\tilde{u}\\ \tilde{v}\\ \tilde{w}\end{Bmatrix} = \begin{Bmatrix}G(s)\\ 0\\ F(s)\end{Bmatrix}$

ν=0.3
Fundamental Axisymmetric mode, i = j =0.

5 Simple Support

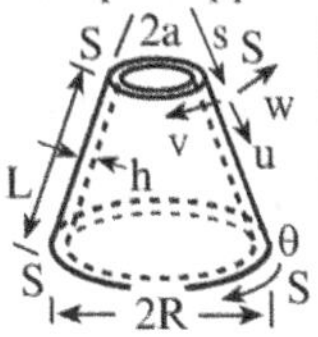

α degree	h/R = 0.03	0.01	0.005	0.001
10	0.891	0.553	0.432	0.229
30	0.776	0.479	0.350	0.172
50	0.652	0.386	0.282	0.138
70	0.479	0.287	0.199	0.0950
85	-	0.141	0.0967	0.0448

Mode Shape: $\begin{Bmatrix}\tilde{u}\\ \tilde{v}\\ \tilde{w}\end{Bmatrix} = \begin{Bmatrix}\cos(s-S_1)/L\cos i\theta\\ \sin(s-S_1)/L\sin i\theta\\ \sin(s-S_1)/L\cos i\theta\end{Bmatrix}$

$S_1 = R/\sin\alpha - L,\ j=1$

$f = [\lambda/(2\pi L)][E/\{\rho(1-\nu^2)\}]^{1/2}$, Hz

λ_i and i of lowest freq mode at left.

same thickness, material, and projected lateral dimensions (cast same shadow on a plane with a light source at infinity):

$$f_{ij}^2\Big|_{\substack{\text{shallow} \\ \text{spherical shell}}} = f_{ij}^2\Big|_{\substack{\text{flat} \\ \text{plate}}} + \frac{1}{4\pi^2}\frac{E}{\rho R^2}, \quad \text{Hz} \tag{5.44}$$

where E is the modulus of elasticity, ρ is the material density, and R is the radius of curvature of the shallow shell. The natural frequencies of flat plates are given in Table 5.3 through 5.8.

However, the analogy of Equation 5.44 assumes both systems have identical mode shapes and so it neglects inextensional shell modes that are unique to the shell. Thus, Equation 5.44 applies most exactly to shells with simply supported or clamped edges in their fundamental mode. It will not provide an accurate prediction of the modes of shells with free edges. Seide [115] provides simple analysis of the torsional modes of shallow spherical caps. Additional analysis of shallow shells is given by Young and Dickinson [140] and Leissa [141].

5.4.4 *Conical Shells*

The natural frequencies and mode shapes of several complete conical shells and frustrums of conical shells are provided in Table 5.9 [110, 142]. As the conical angle becomes increasingly shallow (cone semiangle β approaches zero), the natural frequencies of a frustrum of a conical shell approach the natural frequencies of a similar cylindrical shell. This suggests that the natural frequencies of shallow frustrums of conical shells may be calculated from the natural frequencies of a cylindrical shell using an average radius $R_{avg} = (R_1 + R_2)/2$, where R_1 and R_2 are the radii of the ends of the conical shell. This approximation is discussed in Refs 143, 144 and appears to generally yield a result within about 20% of the true natural frequency if the $R_1/R_2 < 1.5$, where R_2 is the larger end radius.

References

[1] Leissa, A. W., Vibration of Plates, Office of Technology Utilization, NASA, NASA SP-160, 1969.
[2] Timoshenko, S., and S. Woinowsky-Krieger, Theory of Plates and Shells, McGraw-Hill, New York, 1959.
[3] Meirovitch, L., Analytical Methods in Vibrations, MacMillian Co., N.Y., 1967.
[4] Reddy, J. N., Theory and Analysis of Elastic Plates and Shells, Taylor and Francis, 2007.
[5] Rao, J. S. Dynamics of Plates, Marcel Dekker, N.Y. 1998.
[6] Soedel, W., Vibrations of Shells and Plates, 3rd ed., Marcel Decker, N.Y., 2004.
[7] Wan, F. Y. M., Stress Boundary Conditions for Plate Bending, International Journal of Solids and Structures, vol. 40, pp. 4107–4123, 2003.
[8] Gorman, D. J., Free Vibration Analysis of Rectangular Plates, Elsevier, New York, 1982.
[9] Abramowitz, M., and I. A. Stegun, Handbook of Mathematical Functions, National Bureau of Standards, Applied Mathematics Series 55, Washington, D.C., 1964.
[10] Leissa, A. W., The Free Vibration of Rectangular Plates, Journal of Sound and Vibration, vol. 31, pp. 257–293, 1973.
[11] Gontkevich, V. S., Natural Vibrations of Plates and Shells, A. P. Filippov (ed.), Nauk Dumka (Kiev), 1964. Translated by Lockheed Missiles and Space Company. Also see Ref. [1].
[12] Stafford, J. W., Natural Frequencies of Beams and Plates on an Elastic Foundation with a Constant Modulus, Journal of the Franklin Institute, vol. 284, pp. 262–264, 1967.
[13] Mindlin, R. D., A. Schacknow, and H. Deresiewicz, Flexural Vibrations of Rectangular Plates, Journal of Applied Mechanics, vol. 23, pp. 430–436, 1956.
[14] Reissner, E., The Effect of Transverse Shear Deformation on the Bending of Elastic Plates, Journal of Applied Mechanics, vol. 12, pp. A68–A77, 1969.

[15] Martincek, G., The Determination of Poisson's Ratio and Modulus from the Natural Vibration in Thick Circular Plates, Journal of Sound and Vibration, vol. 2, 116–127, 1965.
[16] Wang, C. M., Lim, G. T. and Reddy, J. N., The Relationships between Bending Solution of Reissner and Mindlin Plate Theories, Engineering Structures, vol. 23, pp. 838–849, 2001.
[17] Liew, K. M., Research on Thick Plate Vibration: a Literature Survey, Journal of Sound and Vibration, vol. 180, pp. 163–176, 1995.
[18] Rao, S. S., and A. S. Prasad, Vibrations of Annular Plates Including the Effects of Rotary Inertia and Transverse Shear Deformation, Journal of Sound and Vibration, vol. 42, pp. 305–342, 1975.
[19] Weisensel, G. N., Natural Frequency Information for Circular and Annular Plates, Journal of Sound and Vibration, vol. 133, pp. 129–137, 1989.
[20] McGee, O. G., C. S. Huang and A. W. Lessia, Comprehensive Exact Solutions for Free Vibrations of Thick Annular Sectorial Plates with Simply Supported Radial Edges, International Journal of Mechanical Sciences, vol. 37, 537–566, 1995.
[21] Liew, K. M., J. B. Han and Z. M. Xiao, Vibration Analysis of Circular Mindlin Plates, Journal of Sound and Vibration, vol. 205, pp. 617–630, 1997.
[22] Huang, C. S., M. J. Chang and A. W. Leissa, Vibrations of Mindlin Sectorial Plates Using the Ritz Method Considering Stress Singularities, Journal of Vibration and Control, vol. 12, 635–667, 2006.
[23] Taissarini, L., C. C. Gentilini, Unified Formulation for Reissner-Mindlin Plates: A Comparison with Numerical Results, Proceedings of the International Association for Shell and Spatial Structures (IASS) Symposium 2009, Valencia.
[24] Colwell, R. C., and H. C. Hardy, The Frequencies and Nodal Systems of Circular Plates, Philosophical Magazine, vol. 24(7), pp. 1041–1055, 1937.
[25] Itao, K. and S. H. Crandall, Natural Modes and Natural Frequencies of Uniform Circular Free-Edge Plates, Journal of Applied Mechanics, vol. 48, pp. 447–453, 1979.
[26] Leissa, A. W. and Y. Narita, Natural Frequencies of Simply Supported Circular Plates, Journal of Sound and Vibration, vol. 70, pp. 221–229, 1980.
[27] Carrington, H., The Frequencies of Vibration of Flat Circular Plates Fixed at the Circumference, Philosophical Magazine, vol. 50(6), pp. 1261–1264, 1925.
[28] Bauer, H. F. and W. Eode, Determination of the Lower Natural Frequencies of Circular Plates with Mixed Boundary Conditions, Journal of Sound and Vibration, vol. 293, pp. 742–764, 2006.
[29] Noble, B., The Vibration and Buckling of a Circular Plate Clamped on Part of Its Boundary and Simply Supported on the Remainder, Proceedings 9th Midwest Conference on Solid and Fluid Mechanics, August 1965.
[30] Sakharov, I. E., Dynamic Stiffness in the Theory of Axisymmetric Vibration of Circular and Annular Plates, Izy. An SSSR, OTN. Mech. I Mashin, vol. 5, pp. 90–98, 1959. Also see Ref. [1].
[31] Bodine, R. Y., The Fundamental Frequencies of a Thin Flat Plate Simply Supported Along a Circle of Arbitrary Radius, Journal of Applied Mechanics, vol. 26, pp. 90–98, 1959.
[32] Sato, K., Free-Flexural Vibrations of an Elliptical Plate with a Free Edge, Journal of the Acoustical Society of America, vol. 54, pp. 547–550, 1973.
[33] Leissa, A. W., Vibration of a Simply Supported Elliptical Plate, Journal of Sound and Vibration, vol. 6, pp. 145–148, 1967.
[34] Sato, K., Free Flexural Vibration of an Elliptical Plate with a Simply Supported Edge, Journal of the Acoustical Society of America, vol. 52, pp. 919–922, 1972.
[35] McNitt, R. P., Free Vibration of a Clamped Elliptical Plate, Journal of Aerospace Sciences, vol. 29, pp. 1124–1125, 1962.
[36] Kantham, C. L., Bending and Vibration of Elastically Restrained Circular Plates, Journal of the Franklin Institute, vol. 265, pp. 483–491, 1959.
[37] Zagrai, A. and D. Donskoy, A Soft Table for the Natural Frequency of Circular Plates with Elastic Edge Supports, Journal of Sound and Vibration, vol. 287, pp. 343–351, 2005.
[38] Laura, P. A. A., J. C. Paloto, and R. D. Santos, A Note on the Vibration of a Circular Plate Elastically Restrained Against Rotation, Journal of Sound and Vibration, vol. 41, pp. 177–180, 1975.
[39] Gutierrez, R. H. and P. A. A. Laura, Note on Fundamental Frequency of Vibration of Plates Supporting Masses Distributed over a Finite Area, Applied Acoustics, vol. 10, pp. 303–313, 1977.
[40] Roberson, R. E., Transverse Vibrations of a Free Circular Plate Carrying Concentrated Mass, Journal of Applied Mechanics, vol. 18, pp. 280–282, 1951.
[41] Roberson, R. E., Vibrations of a Clamped Circular Plate Carrying Concentrated Mass, Journal of Applied Mechanics, vol. 18, pp. 349–352, 1951.
[42] Wah, T., Vibration of Circular Plates, Journal of the Acoustical Society of America, vol. 34, pp. 275–281, 1962.

[43] Rao, S. S. and A. S. Prasad, Vibration of Annular Plates including Rotatory Inertia and Transverse Shear Deformation, Journal of Sound and Vibration, vol. 42, pp. 305–324, 1975.
[44] Vogel, S. M., and D. W. Skinner, Natural Frequencies of Transversely Vibrating Uniform Annular Plates, Journal of Applied Mechanics, vol. 32, pp. 926–931.
[45] Ben-Amoz, M., Note on Deflections and Flexural Vibrations of Clamped Sectorial Plates, Journal of Applied Mechanics, vol. 26, 1959, pp. 136–137.
[46] Kennedy, W. and D. Gorman, Vibration Analysis of Variable Thickness Disks Subjected to Centrifugal and Thermal Stresses, Journal of Sound and Vibration, vol. 53, pp. 83–101, 1977.
[47] Conway, H. D., Some Special Solutions for the Flexural Vibrations of Discs of Varying Thickness, Ingr-Arch., vol. 26, pp. 408–410, 1958.
[48] Thurston, E. G., and Y. T. Tsui, On the Lowest Flexural Resonant Frequency of a Circular Disc of Linearly Varying Thickness Driven at Its Center, Journal of the Acoustical Society of America, vol. 27, pp. 926–929, 1955.
[49] Jain, R. K., Vibrations of Circular Plates of Variable Thickness under an InPlane Force, Journal of Sound and Vibration, vol. 23, pp. 407–414, 1972.
[50] Ramaiah, G. K., and K. Vijayakumar, Vibration of Annular Plates with Linear Thickness Profiles, Journal of Sound and Vibration, vol. 40, pp. 293–298, 1975.
[51] Grossi, R. O. and P. A. A. Laura, Transverse Vibrations of Circular Plates of Linearly Varying Thickness, Applied Acoustics, vol. 13, pp. 7–18, 1980.
[52] Krauter, A. T., and P. Z. Bulkeley, Effect of Central Clamping on Transverse Vibrations of Spinning Membrane Disks, Journal of Applied Mechanics, vol. 37, pp. 1037–1042, 1970.
[53] Eversman, W. and R. O. Dodson, Free Vibration of a Centrally Clamped Spinning Circular Disk, AIAA Journal, vol. 7, pp. 2010–2012, 1969.
[54] Parker, R. G. and P. J. Sathe, Exact Solutions for a Rotating Dick-Spindle System, Journal of Sound and Vibration, vol. 222, pp. 445–465, 1999.
[55] Wook Kang, et al., Approximate Closed Form Solution for Free Vibration of Polar Orthotropic Circular Plates, Applied Acoustics, vol. 66, pp. 1162–1179, 2005.
[56] Woo, H. H., P. G. Kirmser, and C. L. Huang, Vibration of an Orthotropic Plate with an Isotropic Cores, AIAA Journal, vol. 11, pp. 1421–1422, 1973.
[57] Rao, K. S., K. Ganapathi, and G. V. Rao, Vibration of Cylindrically Orthotropic Circular Plates, Journal of Sound and Vibration, vol. 36, pp. 433–434, 1973.
[58] Grossi, R. O. and P. A. A. Laura, Additional Results of Transverse Vibration of Polar Orthotropic Circular Plates Carrying Concentrated Masses, Applied Acoustics, vol. 21, 225–233, 1987.
[59] Narita, Y., Natural Frequencies of Completely Free Annular and Circular Plates having Polar Orthotropy, Journal of Sound and Vibration, vol. 92, pp. 33–38, 1984.
[60] Warburton, G. B., The Vibration of Rectangular Plates, Proceedings of the Institution of Mechanical Engineers, vol. 168(19), pp. 371–381, 1954.
[61] Dickinson, S. M., The Buckling and Frequency of Flexural Vibration of Rectangular, Isentropic and Orthotropic Plates Using Rayleigh's Method, Journal of Sound and Vibration, vol. 61, pp. 1–8, 1978.
[62] Li, K. M. and Z. Yu, A Simple Formula for Predicting Resonant Frequencies of a Rectangular Plate with Uniformly Restrained Edges, Journal of Sound and Vibration, 10, 1016, 2009.
[63] Gorman, D. J., Free Vibration Analysis of Completely Free Rectangular Plates by the Superposition-Galerkin Method, Journal of Sound and Vibration, vol. 237, 901–914, 2000.
[64] Waller, M. D., Vibration of Free Square Plates, Proceedings of the Physical Society (London), vol. 51, pp. 831–844, 1939.
[65] Reed, R. E., Jr., Comparison of Methods in Calculating Frequencies of Corner-Supported Rectangular Plates, NASA Report TN-D-3030, 1965.
[66] Srinivasan, R. S., and K. Munaswamy, Frequency Analysis of Skew Orthotropic Point Supported Plates, Journal of Sound and Vibration, vol. 39, 207–216, 1975.
[67] Rao, G. V., Fundamental Frequency of a Square Panel With Multiple Point Supports on Edges, Journal of Sound and Vibration, vol. 38, p. 271, 1975.
[68] Johns, D. J., and R. Nataroja, Vibration of a Square Plate Symmetrically Supported at Four Points, Journal of Sound and Vibration, vol. 39, pp. 207–216, 1975.
[69] Petyt, M., and W. H. Mirza, Vibration of Column-Supported Floor Slabs, Journal of Sound and Vibration, vol. 21, pp. 355–364, 1972.
[70] Nowacki, W., Dynamics of Elastic Systems, John Wiley, New York, p. 228, 1963.
[71] Cox, H. L., Vibration of Certain Square Plates Having Similar Adjacent Edges, Quarterly Journal of Mechanics and Applied Mathematics, vol. 8, pp. 454–456, 1955.

[72] Hegarty, R. F., and T. Ariman, Elasto-Dynamic Analysis of Rectangular Plates with Circular Holes, International Journal of Solids and Structures, vol. 11, pp. 895–906, 1975.
[73] Anderson, R. G., B. M. Irons, and O. C. Zienkiewicz, Vibration and Stability of Plates Using Finite Elements, International Journal of Solids and Structures, vol. 4, pp. 1031–1055, 1968.
[74] Paramasivam, P., Free Vibration of Square Plates with Square Openings, Journal of Sound and Vibration, vol. 30, pp. 173–178, 1973.
[75] Appl, F. C., and N. R. Byers, Fundamental Frequencies of Simply Supported Rectangular Plates with Linearly Varying Thickness, Journal of Applied Mechanics, vol. 32, pp. 163–167, 1965.
[76] Chopra, I., and S. Durasula, Natural Frequencies and Modes of Tapered Skew Plates, Journal of Sound and Vibration, vol. 13, pp. 935–944, 1971.
[77] Ashton, J. E., Natural Modes of Vibration of Tapered Plates, ASCE Journal of the Structural Division, vol. 95, pp. 787–790, 1969.
[78] Ashton, J. E., Free Vibration of Linearly Tapered Clamped Plates, ASCE Journal of the Structural Division, vol. 95, pp. 497–500, 1969.
[79] Soni, S. R., and K. S. Rao, Vibrations of Non-Uniform Rectangular Plates: A Spline Technique Method of Solution, Journal of Sound and Vibration, vol. 35, pp. 35–45, 1974.
[80] Cheung, Y. K. and D. Zhou, Vibrations of Tapered Mindlin Plates in Terms of Static Timoshenko Beam Functions, Journal of Sound and Vibration, vol. 260, pp. 693–708, 2003.
[81] Bassily, S. F., and S. M. Dickinson, Buckling and Lateral Vibration of Rectangular Plates Subject to Inplane Loads-A Ritz Approach, Journal of Sound and Vibration, vol. 24, pp. 219–239, 1972.
[82] Li, W. L. et al., An Exact Series Solution for vibration of Rectangular Plates with Elastic Boundary, Journal of Sound and Vibration, vol. 321, pp. 254–269, 2009.
[83] Laura, P. A. A., R. H. Gutierrez, and D. S. Steinberg, Vibration of Simply-Supported Plates Carrying Concentrated Masses, Journal of Sound and Vibration, vol. 55, pp. 49–53, 1977.
[84] Pombo, J. L. et al., Analytical and Experimental Investigation of Free Vibrations of Clamped Plates Carrying Concentrated Masses, Journal of Sound and Vibration, vol. 55, pp. 521–532, 1977.
[85] Barton, M. V., Vibration of Rectangular and Skew Cantilever Plates, Journal of the Acoustical Society of America, vol. 18, pp. 129–134, 1951.
[86] McGee, O. G. and A. W. Lessia, Natural Frequencies of Shear Deformable Rhombic Plates with Clamped and Simply Supported Edges, International Journal of Solids and Structures, vol. 36, 1133–1148, 1994.
[87] Durvasula, S., Natural Frequencies and Modes of Skew Membranes, Journal of the Acoustical Society of America, vol. 44, pp. 1636–1646, 1968.
[88] Durvasula, S., Natural Frequencies and Modes of Clamped Skew Plates, AIAA Journal, vol. 7, pp. 1164–1167, 1969.
[89] Nair, P. S., and S. Durvasula, Vibration of Skew Plates, Journal of Sound and Vibration, vol. 26, pp. 1–19, 1973.
[90] Kim, C. S., and S. M. Dickinson, The Free Flexural Vibration of Isotropic and Orthotropic General Triangular Shaped Plates, Journal of Sound and Vibration, vol. 152, 383–403, 1992.
[91] Kim, C. S., and S. M. Dickinson, The Free Flexural Vibration of Right Triangular Plates, Journal of Sound and Vibration, vol. 141, 291–311, 1990.
[92] Karunasna, W., S. Kitipornchai, and F. G. A. AL-Bermani, Free Vibration of Cantilevered Arbitrary Triangular Mindlin Plates, International Journal of Mechanical Sciences, vol. 38, 431–442, 1996.
[93] Cheung, Y. K., and D. Zhou, Three-Dimensional Vibration Analysis of Cantilevered and Free Isosceles Triangular Plates, Journal of Solids and Structures, vol. 39, 673–687, 2002.
[94] Chopra, I. and S. Durvasula, Vibration of Simply-Supported Trapezoidal Plates, I. Symmetric Trapezoids, Journal of Sound and Vibration, vol. 19, pp. 379–392, 1971.
[95] Chopra, I. and S. Durvasula, Vibration of Simply-Supported Trapezoidal Plates, II. Unsymmetric Trapezoids, Journal of Sound and Vibration, vol. 20, pp. 125–134, 1972.
[96] Shahady, P. A., R. Pasarelli, and P. A. A. Laura, Application of Complex Variable Theory to the Determination of the Fundamental Frequency of Vibrating Plates, Journal of the Acoustical Society of America, vol. 42, pp. 806–809, 1967.
[97] Laura, P. A. A., and R. Gutierrez, Fundamental Frequency of Vibration of Clamped Plates of Arbitrary Shape Subjected to Hydrostatic State of In-Plane Stress, Journal of Sound and Vibration, vol. 48, pp. 327–333, 1976. Also see Journal of Sound and Vibration, vol. 70, 1980, pp. 77–84.
[98] Pnueli, D., Lower Bounds to the Gravest and All Higher Frequencies of Homogeneous Vibrating Plates, Journal of Applied Mechanics, vol. 42, pp. 815–820, 1975.
[99] Sundara, K. T., and R. N. Iyengdar, Determination of Orthotropic Plate Parameters, Applied Scientific Research, vol. 17, pp. 422–438, 1967.
[100] Fallstrom, K.-E., and N.-E. Molin, A Nondestructive Method to Determine Material Properties in Orthotropic Plates, Polymer Composites, vol. 8, pp. 103–108, 1987.

[101] Vijayakumar, K., Natural Frequencies of Rectangular Orthotropic Plates with a Pair of Parallel Edges Simply Supported, Journal of Sound and Vibration, vol. 35, pp. 379–394, 1974.
[102] Magrab, E., Natural Frequencies of Elastically Supported Orthotropic Rectangular Plates, Journal of the Acoustical Society of America, vol. 67, pp. 79–83, 1977.
[103] Hearmon, R. F. S., The Frequency of Flexural Vibration of Rectangular Orthotropic Plates with Clamped or Supported Edges, Journal of Applied Mechanics, vol. 26, pp. 537–540, 1959.
[104] Ellington, J. P. and H. McCallion, Free Vibration of Grillages, Journal of Applied Mechanics, vol. 26, pp. 663–607, 1959.
[105] Niu, M. C. Y., Air Frame Structural Design, Conmilit Press, Hong Kong, 1999.
[106] Lin, Y. K., Free Vibration of Continuous Skin-Stringer Panels, Journal of Applied Mechanics, pp. 669–676, 1960.
[107] Rao, M. S. and P. S. Nair, Effect of Boundary Conditions on the Frequencies of Plates with Varying Stiffener Length, Journal of Sound and Vibration, vol. 179, pp. 900–913, 1995.
[108] Slot, T., and W. J. O'Donnell, Effective Elastic Constants for Thick Perforated Plates with Square and Triangular Penetration Patterns, Journal of Engineering for Industry, vol. 93, pp. 935–942, 1971.
[109] Jhung M. J., and J. C. Jo, Equivalent Material Properties of Perforated Plate with Triangular or Square Pattern for Dynamic Analysis, Nuclear Engineering and Technology, vol. 38, pp. 689–696, 2006.
[110] Leissa, A. W., Vibrations of Shells, NASA Report NASA-SP-288, Ohio State University, 1973.
[111] Love, A. E. H., A Treatise on the Mathematical Theory of Elasticity, 4th ed., Dover Publications, N.Y., 1944, pp. 491–549. Also see Love, A. E. H, The Small Free Vibrations of a Thin Elastic Shell, Phil. Trans. Roy. Soc. (London), Ser. A, vol. 179, 1888, pp. 491–549.
[112] Kraus, H., Thin Elastic Shells, John Wiley, N.Y., pp. 202,297, 1967.
[113] Forsberg, K., A Review of Analytical Methods Used to Determine the Modal Characteristics of Cylindrical Shells, NASA Report NASA CR-613, Lockheed Aircraft Company, CA, 1966.
[114] Niordson, F. I., Shell Theory, North-Holland, N.Y., 1985.
[115] Seide, P., Small Elastic Deformations of Thin Shells, Noordhoff International Publishing, Leyden, 1975.
[116] Donnell, L. H., Stability of Thin Walled Tubes under Torsion, NACA Report no. 479, 1933.
[117] Sanders, J. L., An Improved First Approximation Theory for Thin Shells, NASA Report NASA-TR-R24, 1959.
[118] Flugge, W., Stresses in Shells, 2nd ed., Springer-Verlag, N.Y., 1973.
[119] Arnold, R. N., and G. R. Warburton, Flexural Vibrations of Thin Cylindrical Shells having Freely Supported Ends, Proceedings of Royal Society of London, Series A197, pp. 238–256, 1949.
[120] Webster, J. J., Free Vibration of a Rectangular Curved Panel, International Journal Mechanical Science, vol. 10, pp. 571–582, 1968.
[121] Szechenyi, E., Approximate Formulas for the Determination of the Natural Frequencies of Stiffened and Curved Panels, Journal of Sound and vibration, vol. 14, pp. 401–418, 1971.
[122] Blevins, R. D., Natural Frequencies of Shallow Cylindrically Curved Panels, Journal of Sound and Vibration, vol. 75, pp. 145–149, 1981.
[123] Bellman, R. Introduction to Matrix Analysis, 2nd ed., McGraw-Hill, 1970.
[124] Sharma, C. B., and D. J. Johns, Vibration Characteristics of a Clamped-Free Circular Cylindrical Shell, Journal of Sound and Vibration, vol. 14, pp. 459–474, 1971.
[125] Sharma, C. B., Vibration Characteristics of Thin Circular Cylinders, Journal of Sound and Vibration, vol. 63, pp. 581–593, 1979.
[126] Soedel, W., A New Frequency Formula for Cylindrical Shells, Journal of Sound and Vibration, vol. 70, pp. 309–317, 1980.
[127] Warburton, G. B., Vibration of Thin Cylindrical Shells, Journal of Mechanical Engineering Science, vol. 7, pp. 399–407, 1965.
[128] Sharma, C. B., Frequencies of Clamped-Free Cylindrical Shells, Journal of Sound and Vibration, vol. 30, pp. 525–528, 1973.
[129] Blevins, R. D., Flow Induced Vibration, 2nd ed., Kreiger, p. 387, 1994.
[130] Bozick, W. F., The Vibration and Buckling Characteristics of Cylindrical Shells Under Axial Load and External Pressure, AFFDL-TR-67-28, 1967.
[131] Armenakas, A. E. and G. Herrmann, Vibrations of Infinitely Long Cylindrical Shells Under Initial Stress, AIAA Journal, vol. 1, pp. 100–106, 1963.
[132] Fung, Y. C., E. E. Sechler, and A. Kaplan, On the Vibration of Thin Cylindrical Shells Under Internal Pressure, Journal of Aeronautical Sciences, vol. 24, pp. 650–671, 1957.
[133] Rayleigh, J. W. S., Theory of Sound, vol. 1, 2nd ed., Macmillan, 1894, reprinted Dover, N.Y., 1945.
[134] Wilkinson, J. P., Natural Frequencies of Closed Spherical Shells, Journal of the Acoustical Society of America, vol. 38, pp. 367–368, 1965.

[135] Baker, W. E., Axisymmetric Modes of Vibration of Thin Spherical Shells, Journal Acoustical Society of America, vol. 33, pp. 1749–1758, 1961.
[136] Hwang, C., Some Experiment on the Vibration of a Hemispherical Shell, Journal of Applied Mechanics, vol. 33, pp. 817–824, 1966; also see discussion vol. 34, pp. 792–794, 1967.
[137] Kalnis, A. Effect of Bending on Vibration of Spherical Shells, Journal Acoustical Society of America, vol. 36, pp. 74–81, 1964.
[138] Perrin R. and T. Charnley, Normal Modes of the Modern English Church Bell, Journal of Sound and Vibration, vol. 90, pp. 29–49, 1983.
[139] Soedel, W., A Natural Frequency Analogy between Spherically Curved Panels and Flat Plates, Journal of Sound and Vibration, vol. 29, pp. 457–461, 1973.
[140] Young, P. G., and S. M. Dickinson, Vibration of a Class of Shallow Shells Bounded by Edges Described by Polynomials, Journal of Sound and Vibration, vol. 181, pp. 203–230, 1995.
[141] Leissa, A. W., Vibrations of Completely Free Shallow Shells of Rectangular Planform, Journal of Sound and Vibration, vol. 96, pp. 207–218, 1984.
[142] Jager, E. H., An Engineering Approach to Calculating the Lowest Natural Frequencies of Thin Conical Shells, Journal of Sound and Vibration, vol. 63, pp 259–264, 1979.
[143] Hartung, R. F. and W. A. Loden, Axisymmetric Vibration of Conical Shells, Journal of Spacecraft and Rockets, vol. 7, 1153–1159, 1970.
[144] Herrmann, G. and I. Mirsky, On Vibrations of Conical Shells, Journal Aerospace Sciences, vol. 25, 451–458, 1958.

6

Acoustics and Fluids

This chapter has formulas for acoustic natural frequencies and propagation of acoustic waves in compressible fluids. This chapter also has formulas for incompressible fluids: the natural periods of surface waves in basins, periods of ship motions, and added mass of structures oscillating in fluids. All of these formulas are solutions to the *wave equation*, which was first applied to acoustics by Euler in 1759 [1–3].

6.1 Sound Waves and Decibels

Linear acoustic theory is based on small oscillatory pressures (p) that propagate at the speed of sound (c) in compressible fluids. The general assumptions are that (1) the acoustic particle velocity (u) is much less than the speed of sound, (2) the acoustic pressure (p) is much less than the mean atmospheric pressure (p_o), and (3) the density oscillation $\delta\rho$ is much less than the mean density ρ:

$$\frac{|p|}{p_o} << 1, \quad \frac{|u|}{c} << 1, \quad \frac{|U|}{c} << 1, \quad \frac{|\delta\rho|}{\rho} << 1 \tag{6.1}$$

In addition, the fluid is assumed to be *inviscid*, zero viscosity.

6.1.1 Speed of Sound

Table 6.1 [5–7, 10–23] has formulas for the speed of sound propagation in gases, liquids, and solids. The speed of sound in air at room temperature air and pressure is 343 m/s (1128 ft/s) and 1483 m/s (4860 ft/s) in pure water. Figure 6.1 shows that the speed of sound in fluids generally increases with temperature. An exception is when liquid water boils into steam [4]. Additional data for speed of sound is in Chapter 8.

Rapid changes in acoustical pressure (p) during wave propagation produce isentropic, adiabatic changes in density (ρ). The *speed of sound* (c) in fluids is the square root of the

Formulas for Dynamics, Acoustics and Vibration, First Edition. Robert D. Blevins.
© 2016 John Wiley & Sons, Ltd. Published 2016 by John Wiley & Sons, Ltd.

Table 6.1 Speed of sound

Notation: B = bulk modulus, units of pressure; c = speed of sound; p = static pressure; R = universal gas constant; T = absolute temperature; W = molecular weight; ρ = fluid density; ν = Poisson's ratio; γ = ratio of specific heats. Refs [5–7, 10–13]. Chapter 8 has additional data. Consistent units are in Table 1.2.

Medium	Speed of Sound, c	Examples and Comments
1. Air Water	air 343.3 m/s (1128 ft/s) H_2O 1483 m/s (4860ft/s)	1 atmosphere pressure, 20 °C (68 °F) See Fig. 6.1 and cases 2 and 3. More data is provided in Chapter 8.
2. Perfect Gas	$\left(\frac{\gamma p}{\rho}\right)^{1/2}$ or $\left(\frac{\gamma RT}{W}\right)^{1/2}$ see Chapter 8.	Adiabatic gas law. c is nearly independent of static pressure. For dry air c is only dependent on temperature. See Ref. [4] moist air. $c_{air} = 20.05T^{1/2}$ m/s where T = °K = °C + 273.2, $= 49.02T^{1/2}$ ft/s where T = °R = °F + 459.69.
3. Liquids	$\left(\frac{B}{\rho}\right)^{1/2}$ Refs. 5, 6. see Chapter 8.	For pure water, where T = temperature in Centigrade, 0<T<95, $c_{pure\,water} = 1402.385 + 5.0388T - 0.057991T^2 + 0.00032872T^3 - 1.3988\times10^{-6}T^4 + 2.78786\times10^{-9}T^5$ m/s $c_{saltwater} = 1448.96 + 4.591T - 0.05304T^2 + 0.0002374T^3 + 1.340(S-35) + 0.0163D + 1.675\times10^{-7}D^2 - 0.01025T(S-35) - 7.139\times10^{-13}TD^3$, m/s For salt water, 2<T<30, S = salinity in parts per thousand (25<S<40), D = depth in meters, 0<D<8000, c in meters per sec.
4. Liquids with Gas Bubbles	$\left(\frac{B_e}{\rho_e}\right)^{1/2}$ (1% gas bubbles in water reduces sound speed by over 50%.)	$B_e = \frac{B_{liq}}{1+(V_g/V_t)(B_{liq}/B_g - 1)}, \rho_e = \rho_g \frac{V_g}{V_t} + \rho_l \frac{V_{liq}}{V_t}, V_g + V_{liq} = V_t$ $B_{liq} = \rho_{liq} c_{liq}^2$ = bulk modulus of liquid, ρ_{liq} = density of liquid, $B_g = \rho_g c_g^2$ = bulk modulus of gas, ρ_g = density of gas. Fig. 6.2. V_t = total volume, V_g = gas volume, V_{liq} = liquid volume, Ref. [7]
5. Fluid in Pipe 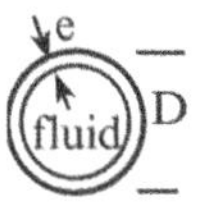	$c_o \sqrt{\frac{1}{1+\frac{B}{E}\frac{D}{e}\alpha}}$ Waves propagate longitudinally in pipe.	c_o = speed of sound in fluid, cases 1 to 4, above. Ref. [7] $B = \rho c_o^2$ = bulk modulus of fluid, ν = Poisson's ratio of pipe E = pipe modulus of elasticity, D = pipe diameter, e = pipe wall, $\alpha = 1-\nu/2$, pipe anchored at upstream end; $1-\nu^2$, pipe anchored to stop all axial movement; 1, pipe with free axial expansion.
6. General Fluid	$\sqrt{p/\rho}\,\Big\rvert_{adiabatic}$	Applies to both liquids and gasses for small amplitude isentropic perturbations. See Eq. 6.3 for shock waves.
7. Solids 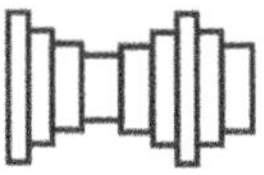  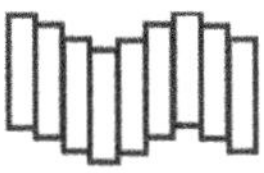	Slender Solid axial $\left(\frac{E}{\rho}\right)^{1/2}$ Bulk Solid axial $\left(\frac{E(1-\nu)}{\rho(1+\nu)(1-2\nu)}\right)^{1/2}$ Shear Wave $\left(\frac{E}{2\rho(1+\nu)}\right)^{1/2}$ Refs. 8, 9, 10.	Waves propagate longitudinally in slender solids such as rods. In bulk solids there is a volume change as dilatational waves. (See sound speed table below.)

Material at 20 C	Sound Speed c, m/s Slender Solid	Bulk Solid Dilatation	Bulk Solid Shear wave
Aluminum	5000	6240	3040
Brass	3480	4700	2110
Copper	3750	5010	2270
Iron	5120	5950	3240
Nickel	4900	6040	3000
Carbon Steel	5180	5940	3220
Glass	3720	3980	2380
Lucite	1840	2680	1100
Concrete	3400	5500	2250

Table 6.1 Speed of sound, continued

Additional rotation: λ = wave length.

Structure	Speed of Wave Propagation, c	Equation of Motion
8. Tensioned String 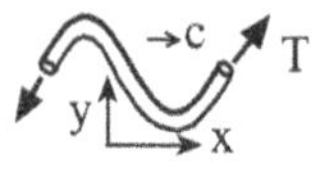	$\left(\frac{T}{m}\right)^{1/2}$ See Section 3.6	$\frac{\partial^2 y(x,t)}{\partial x^2} = \frac{1}{c^2}\frac{\partial^2 y(x,t)}{\partial t^2}$ T = tension in string; m = mass per unit length of string
9. Longitudinal waves in thin Rod 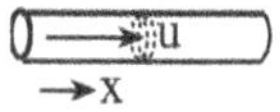	$\left(\frac{E}{\rho}\right)^{1/2}$ see case 7 and Section 4.6	$\frac{\partial^2 u(x,t)}{\partial x^2} = \frac{1}{c^2}\frac{\partial^2 u(x,t)}{\partial t^2}$ E = modulus of elasticity; ρ = mass density
10. Torsion waves In Beam or Rod 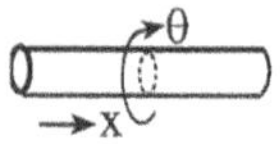	$\left(\frac{GC}{\rho I_p}\right)^{1/2}$ see Section 4.6.	$\frac{\partial^2 \theta(x,t)}{\partial x^2} = \frac{1}{c^2}\frac{\partial^2 \theta(x,t)}{\partial t^2}$, I_p = polar moment in. G = shear mod.; C = torsion constant, Table 4.14
11. Flexure Beam 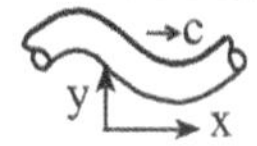	$\frac{2\pi}{\lambda}\left(\frac{EI}{m}\right)^{1/2} = (2\pi f)^{1/2}\left(\frac{EI}{m}\right)^{1/4}$ see Section 4.6	$\frac{\partial^4 y(x,t)}{\partial x^4} + \frac{m}{EI}\frac{\partial^2 y(x,t)}{\partial t^2} = 0$ λ = wavelength; I = area mom. inertia Table 1.5
12. Flexure Beam with Rotary inertia 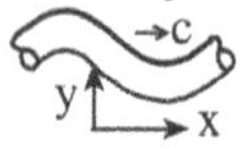	$\frac{2\pi\left(\frac{EI}{m}\right)^{1/2}}{\left(\lambda^2 + (2\pi)^2 I/A\right)^{1/2}}$	$EI\frac{\partial^4 y(x,t)}{\partial x^4} - \rho I\frac{\partial^4 y(x,t)}{\partial x^2 \partial y^2} + m\frac{\partial^2 y(x,t)}{\partial t^2} = 0$ A = cross section area, Table 1.5. See case 11. see Section 4.6
13. Shear Beam 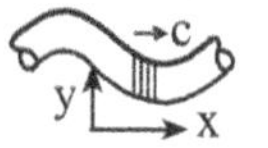	$\left(\frac{KAG}{m}\right)^{1/2}$ see Sections 4.4, 4.6	$\frac{\partial^2 y(x,t)}{\partial x^2} + c^2\frac{\partial^2 y(x,t)}{\partial t^2} = 0$ K = shear coefficient, dim'lss, Table 4.11
14. Membrane 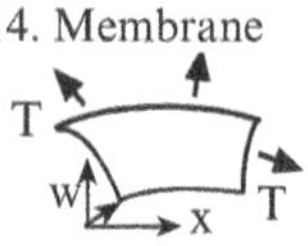	$\left(\frac{T}{\gamma}\right)^{1/2}$ see Section 3.6	$\nabla^2 w(x,y,t) = \frac{1}{c^2}\frac{\partial^2 w(x,y,t)}{\partial t^2}$ T = tension per unit edge; γ = mass per unit area
15. Thin Plate 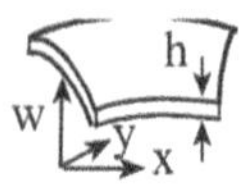	$\frac{2\pi}{\lambda}\left(\frac{Eh^2}{12\rho(1-\nu^2)}\right)^{1/2} = (2\pi f)^{1/2}\left(\frac{Eh^2}{12\rho(1-\nu^2)}\right)^{1/4}$	$\frac{Eh^3}{12(1-\nu^2)}\nabla^4 w(x,y,t) + \rho h\frac{\partial^2 w(x,y,t)}{\partial t^2} = 0$ $\nabla^2 = \frac{\partial^2}{\partial x^2} + \frac{\partial^2}{\partial y^2}$, see Section 5.1.
16. Surface Wave 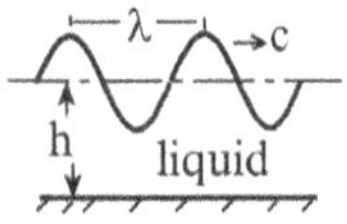	Wave speed is solution to transcendental equation. There are limit cases: $c = \frac{g}{\omega}\tanh\frac{\omega h}{c} = \begin{cases}(gh)^{1/2}, & \text{shallow liquid wave, } \omega h/c = 2\pi h/\lambda < \pi/10 \\ g/\omega, & \text{deep liquid wave, } \omega h/c = 2\pi h/\lambda > \pi\end{cases}$ h = mean depth of liquid, $\omega = 2\pi f$. see Section 6.5. g = acceleration of gravity.	

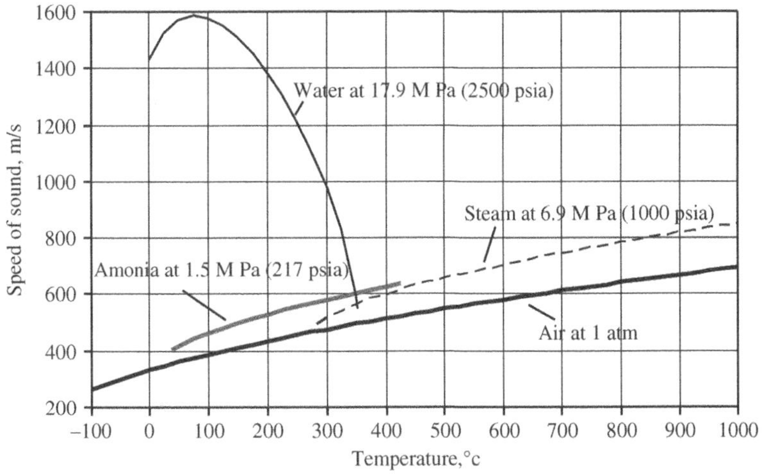

Figure 6.1 Speed of sound in air, water, steam, and ammonia as a function of temperature. 1 m/s = 3.28 ft/s [4]

incremental change in pressure divided by the incremental change in density (cases 3 and 6 of Table 6.1; [1, 2]):

$$c^2 = \left.\frac{\partial p}{\partial \rho}\right|_{\substack{\text{constant}\\ \text{entropy}}} = \begin{cases} \dfrac{B}{\rho}, \text{ general} \\ \dfrac{\gamma p}{\rho} = \gamma \dfrac{RT}{W}, \text{ perfect gas} \end{cases} \tag{6.2}$$

Additional data for speed of sound is in Chapter 8.

B is the *bulk modulus* of the fluid. It has units of pressure. Dimensionless γ is the *ratio of specific heats* of gas at constant pressure to that at constant volume. γ is between 1 and 5/3 [14]: $\gamma = 1.4$ for room temperature air, 5/3 for a monatomic gas such as helium, and 7/5 for a diatomic gas such as gas oxygen. R is the universal gas constant (Chapter 8).

The speed of sound is independent of the sound pressure if the sound pressure is much smaller than the mean fluid pressure. High-pressure sound waves, called *shock waves*, propagate faster than the speed of sound in proportion to their *shock strength* $P - P_o$, which is the difference in the elevated static pressure P behind the shock wave and ambient pressure P_o in front of the shock [3, 15, 16].

$$\frac{c_{\text{shock wave}}}{c} = \sqrt{1 + \frac{\gamma + 1}{2\gamma}\frac{P - P_o}{P_o}} \tag{6.3}$$

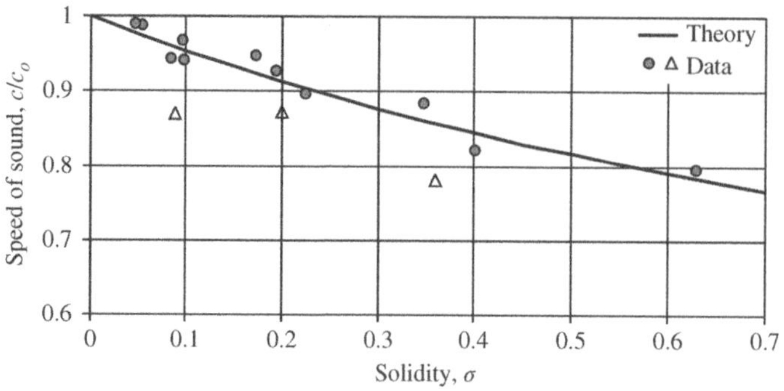

Figure 6.2 Reduction in speed of sound for propagation through an array of rigid rods [8, 9]

On the other hand, long wavelength sound propagates slower than the speed of sound (c) through a regular array of fixed-rigid cylinders in fluid [8, 9]:

$$c|_{\text{rigid array}} = \frac{c}{\sqrt{1+\sigma}}, \quad \sigma \geq 0 \tag{6.4}$$

where σ is the ratio of the cylinders' volume to the total volume. This equation has been verified for fill fractions up to 0.5 for sound propagation normal to the cylinders' axes (Figure 6.2). Gas bubbles in liquid and elastic duct walls also slow the speed of sound (cases 4 and 5 of Table 6.1).

6.1.2 Acoustic Wave Equation

Table 6.2 [1–3, 17] shows a partial differential equation that governs acoustic wave propagation. It is derived with the fluid element in Figure 6.3. A negative pressure gradient accelerates the fluid element along the positive x-axis (Eq. 2.4): $\{p - [p + (dp/dx)dx]\}dy = -dxdy(du/dt)$. Dividing by $dxdy$ gives the *fluid momentum equation* in the x-direction:

$$\frac{\partial p}{\partial x} = -\rho \frac{\partial u}{\partial t} \tag{6.5}$$

where u is acoustic velocity in the x-direction, ρ is the mean fluid density, and t is time.

Conservation of mass requires that mass entering the element, less mass leaving the element, equals the change in mass of the element: $\{\rho u - \rho[u + (du/dx)dx]\}\, dydt = \{(\rho + (d\rho/dt)dt - \rho\}dxdy$; inserting Equation 6.2 and dividing by $dxdydt$ gives the fluid *equation of conservation of mass*:

$$\rho \frac{\partial u}{\partial x} = -\frac{\partial \rho}{\partial t} = -\frac{\partial \rho}{\partial p}\frac{\partial p}{\partial t} = -\frac{1}{c^2}\frac{\partial p}{\partial t} \tag{6.6}$$

Table 6.2 Acoustic wave equation

Notation: c = speed of sound (Table 6.1); M = U/c Mach number; p = acoustic pressure; u,v, w = acoustic velocity in x,y,z directions, respectively; r = radial coordinate; **u**, **U** = vector acoustic an mean velocity; u_r = radial velocity component; t = time; U = mean flow velocity; a constant, |U| >> |u|; x, y, z = Cartesian coordinates; ρ = mean fluid density; ϕ = polar angle, θ = circumferential angle; Φ = potential function; ∇ = vector gradient operator. Velocity potential, acoustic velocity and instantaneous density also satisfy the wave equation. Consistent sets of units are given in Table 1.2. Refs [1–3, 17].

Wave Equation	Momentum Equation	Velocity Potential, Φ
1. One-Dimensional Cartesian $\frac{\partial^2 p}{\partial x^2}=\frac{1}{c^2}\frac{\partial^2 p}{\partial t^2}$	$\frac{\partial u}{\partial t}=-\frac{1}{\rho}\frac{\partial p}{\partial x}$ mass $\rho\frac{\partial u}{\partial x}=-\frac{1}{c^2}\frac{\partial p}{\partial t}$ momentum	$u=\frac{\partial \Phi}{\partial x},\ p=-\rho\frac{\partial \Phi}{\partial t}$
2. Vector form in 3 Dimensions $\nabla^2 p=\frac{1}{c^2}\frac{\partial^2 p}{\partial t^2}$	$\rho\frac{\partial \mathbf{u}}{\partial t}=-\nabla p$ momentum $\rho\nabla\cdot\mathbf{u}=-\frac{1}{c^2}\frac{\partial p}{\partial t}$ mass	$\mathbf{u}=\nabla\Phi$ $p=-\rho\frac{\partial \Phi}{\partial t}$
3. Cartesian Coordinates $\frac{\partial^2 p}{\partial x^2}+\frac{\partial^2 p}{\partial y^2}+\frac{\partial^2 p}{\partial z^2}=\frac{1}{c^2}\frac{\partial^2 p}{\partial t^2}$	$\frac{\partial u}{\partial t}=-\frac{1}{\rho}\frac{\partial p}{\partial x},\ \frac{\partial v}{\partial t}=-\frac{1}{\rho}\frac{\partial p}{\partial y}$ $\frac{\partial w}{\partial t}=-\frac{1}{\rho}\frac{\partial p}{\partial z}$	$u=\frac{\partial \Phi}{\partial x},\ v=\frac{\partial \Phi}{\partial y}$ $w=\frac{\partial \Phi}{\partial z},\ p=-\rho\frac{\partial \Phi}{\partial t}$
4. Cylindrical Coordinates $\frac{1}{r}\frac{\partial}{\partial r}(r\frac{\partial p}{\partial r})+\frac{1}{r^2}\frac{\partial^2 p}{\partial \theta^2}+\frac{\partial^2 p}{\partial z^2}=\frac{1}{c^2}\frac{\partial^2 p}{\partial t^2}$	$\frac{\partial u_r}{\partial t}=-\frac{1}{\rho}\frac{\partial p}{\partial r},\ \frac{\partial u_\theta}{\partial t}=-\frac{1}{\rho r}\frac{\partial p}{\partial \theta}$ $\frac{\partial u_z}{\partial t}=-\frac{1}{\rho}\frac{\partial p}{\partial z}$	$u_r=\frac{\partial \Phi}{\partial r},\ u_\theta=\frac{\partial \Phi}{r\partial \theta}$ $u_z=\frac{\partial \Phi}{\partial z},\ p=-\rho\frac{\partial \Phi}{\partial t}$
5. Spherical Coordinates $\frac{\partial^2 p}{\partial r^2}+\frac{2}{r}\frac{\partial p}{\partial r}+\frac{1}{r^2\sin\theta}\frac{\partial}{\partial \theta}(\sin\theta\frac{\partial p}{\partial \theta})+$ $\frac{1}{r^2\sin^2\theta}\frac{\partial^2 p}{\partial \phi^2}=\frac{1}{c^2}\frac{\partial^2 p}{\partial t^2}$	$\frac{\partial u_r}{\partial t}=-\frac{1}{\rho}\frac{\partial p}{\partial r},\ \frac{\partial u_\theta}{\partial t}=-\frac{1}{\rho r}\frac{\partial p}{\partial \theta}$ $\frac{\partial u_\phi}{\partial t}=-\frac{1}{\rho r\sin\theta}\frac{\partial p}{\partial \phi}$	$u_r=\frac{\partial \Phi}{\partial r},\ u_\theta=\frac{\partial \Phi}{r\partial \theta}$ $u_\phi=\frac{1}{r\sin\theta}\frac{\partial \Phi}{\partial \phi},\ p=-\rho\frac{\partial \Phi}{\partial t}$
6. Vector form, 3D, Mean Flow $\nabla^2 p=\frac{1}{c^2}\frac{D^2 p}{Dt^2}$	$\rho\frac{D\mathbf{u}}{Dt}=-\nabla p$ momentum $\rho\nabla\cdot\mathbf{u}=-\frac{1}{c^2}\frac{Dp}{Dt}$ mass	U = vector mean velocity Substantive derivative $\frac{D}{Dt}=\frac{\partial}{\partial t}+\mathbf{U}\cdot\nabla$ Ref. [3], p. 701
7. One-Dimension, Mean Flow $\frac{\partial^2 p}{\partial x^2}=\frac{1}{c^2}\left(\frac{\partial^2 p}{\partial t^2}+2U\frac{\partial^2 p}{\partial x\partial t}+U^2\frac{\partial^2 p}{\partial x^2}\right)$ U = mean flow in +x direction	$\frac{\partial u}{\partial t}+U\frac{\partial u}{\partial x}=-\frac{1}{\rho}\frac{\partial p}{\partial x}$ moment. $\rho\frac{\partial u}{\partial x}=-\frac{1}{c^2}\left(\frac{\partial p}{\partial t}+U\frac{\partial p}{\partial x}\right)$ mass	$\frac{D}{Dt}=\frac{\partial}{\partial t}+U\frac{\partial}{\partial x}$ $\frac{D^2}{Dt^2}=\frac{\partial^2}{\partial t^2}+2U\frac{\partial^2}{\partial x\partial t}+U\frac{\partial^2}{\partial x^2}$
8. One-Dimensional Mean Flow and Duct Friction $\frac{\partial^2 p}{\partial x^2}=\frac{1}{c^2}\left(\frac{\partial^2 p}{\partial t^2}+2U\frac{\partial^2 p}{\partial x\partial t}+U^2\frac{\partial^2 p}{\partial x^2}\right)$ $+\frac{1}{c^2}\frac{Uf}{d}\left(\frac{\partial p}{\partial t}+U\frac{\partial p}{\partial x}\right)$	$\frac{Du}{Dt}=-\frac{1}{\rho}\frac{\partial p}{\partial x}-Uu\frac{f}{d}$ moment. $\rho\frac{\partial u}{\partial x}=-\frac{1}{c^2}\frac{Dp}{Dt}$ mass	*f* = dimensionless friction factor . Ref. [17]. U = mean velocity *d* = duct diameter Mean pressures drop: $d\bar{p}/dx=-\frac{1}{2}\rho U^2(f/d)$

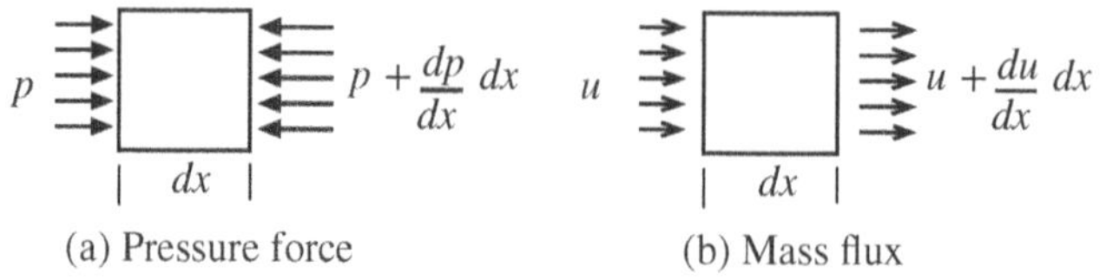

Figure 6.3 A fluid element for derivation of the one-dimensional wave equation

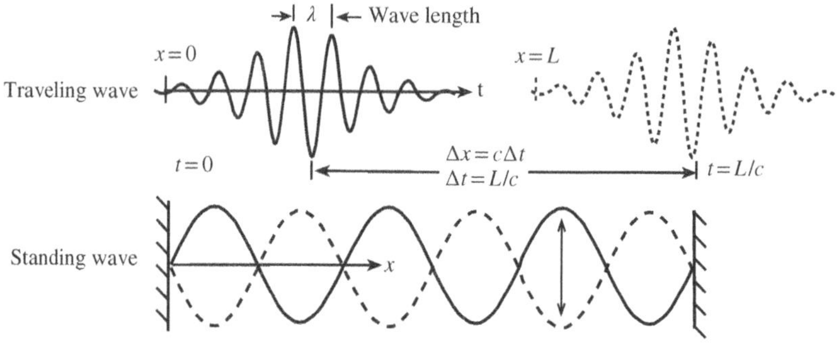

Figure 6.4 Wavelength, traveling, and standing waves

The *one-dimensional wave equation* for acoustic pressure is generated by subtracting the derivative of the momentum equation with respect to x from the derivative of mass conservation equation with respect to time to eliminate velocity ([1, 2]; case 1 of Table 6.2):

$$c^2\frac{\partial^2 p(x,t)}{\partial x^2} - \frac{\partial^2 p(x,t)}{\partial t^2} = 0 \tag{6.7}$$

Single and *multidimensional wave* and *momentum equations* are shown in Table 6.2 [3].

The wave equation has both *standing wave* and *propagating wave* solutions (Figure 6.4). A continuous function of the time–space argument $x - ct$, where x = distance and t = time, is a *one-dimensional traveling wave* solution to the wave equation (case 1 of Table 6.3; Eq. 6.7):

$$p(x,t) = P(x - ct) = \text{forward traveling wave} \tag{6.8}$$

Differentiation with respect to the argument $\xi = (x - ct)$ is done with the chain rule for differentiation:

$$\begin{aligned}
&\frac{\partial P(\xi)}{\partial t} = \frac{(\partial P(\xi)/\partial \xi)\partial \xi}{\partial t} = -\frac{cP'(x-ct), \partial^2 P(\xi)}{\partial t^2} = \partial P(\xi) = c^2 P''(x-ct) \\
&\frac{\partial P(\xi)}{\partial x} = P'(x-ct), \frac{\partial^2 P(\xi)}{\partial x^2} = P''(x-ct)
\end{aligned} \tag{6.9}$$

Substituting the second derivatives into Equation 6.7 shows that the Equation 6.8 the traveling wave satisfies the wave equation, $c^2P''(x - ct) - c^2P''(x - ct) = 0$. The forward-traveling wave shape $P(x - ct)$ at $x = 0$ and $t = 0$ duplicates itself downstream

Table 6.3 Acoustic wave propagation

Notation: c = speed of sound (Table 6.3); Exp(x) = e^x, e = 2.781; $e^{-i\omega t} = \cos\omega t - i\sin\omega t$; $f = \omega/2\pi$ = frequency, Hz; I = acoustic intensity; $i = (-1)^{1/2}$ = imaginary constant; $k = \omega/c = 2\pi/\lambda$ = wave number; m = mass per unit area; **n** = unit vector in direction of propagation; p = acoustic pressure; P_o = pressure amplitude; Q = volume velocity, volume/time; r = radius; **r** = radius vector; R = radius of cylinder or sphere; S = propagation area; t = time; u = acoustic velocity in x or r direction; U_o = velocity amplitude; W = IS = acoustic power; x,y = Cartesian coordinates; α_m = molecular absorption, units of 1/meter; α = reflection-absorption, dim'lss; ρ = mean density; θ = angle; $\omega = kc = 2\pi f$ = circular frequency, rad./sec; $\lambda = c/f$ = wavelength. Subscripts: i = incident; o = amplitude; r = reflected or radial; t = transmitted. Table 1.2 has consistent units. Refs [1–3, 13, 17, 19–28].

Acoustic Wave	Acoustic Pressure and Velocity	Intensity and Power
1. One Dimensional Wave propagates out + x axis 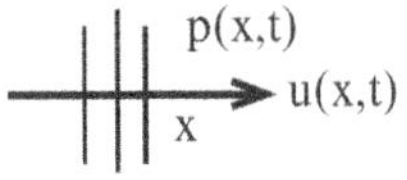	$p(x,t) = P_o e^{i(kx-\omega t)} = \rho c\, u(x,t)$ $= P_o \cos(kx - \omega t)$, real $u(x,t) = \frac{P_o}{\rho c}\cos(kx-\omega t)$, real	$I = \frac{P_o^2}{2\rho c}$, $W = IS = \frac{P_o^2}{2\rho c}S$ S = area perpendicular to x. Pressure and velocity in phase.
2. Plan Wave in direction of Unit vector **n**. vector form 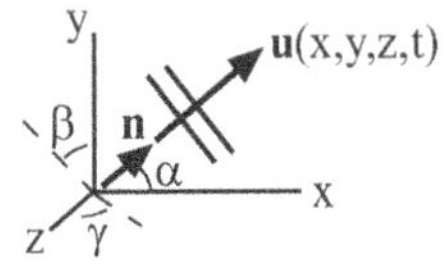	$p(x,y,z,t) = P_o \mathrm{Exp}[i(k\mathbf{n}\bullet\mathbf{r} - \omega t)]$ $\mathbf{u}_r(r,t) = \frac{P_o}{\rho c}\mathrm{Exp}[i(k\mathbf{n}\bullet\mathbf{r}-\omega t)]\mathbf{n}$ $\mathbf{r} = x\mathbf{i} + y\mathbf{j} + z\mathbf{k}$ $\mathbf{n} = \mathbf{i}\cos\alpha + \mathbf{j}\cos\beta + \mathbf{k}\cos\gamma$ • = vector dot product, Eq. 1.25. α,β,γ = direction cosine angles	$I = \frac{P_o^2}{2\rho c}$, $W = IS = \frac{P_o^2}{2\rho c}\pi r_o$ W = power per unit length Propagation in direction **n**. S = plane area normal to **n**. Pressure and velocity in phase.
3. Cylindrical Wave (2D) 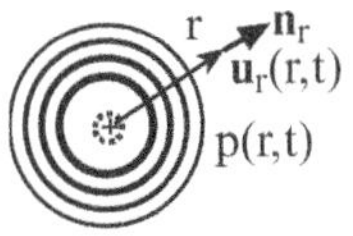	$p(r,t) = P_o \frac{r_o^{1/2}}{r^{1/2}} e^{-i(kr-\omega t)}$, $u_r(r,t) = U_o \frac{r_o^{1/2}}{r^{1/2}} e^{-i(kr-\omega t)}$	$W = \frac{P_o^2}{\rho c}\pi r_o$, Ref. [3], p. 357 far field, case 11 has near field $U_o = P_o/\rho c$. P_o = pressure at $r=r_o$.
4. Spherical Wave (3D) 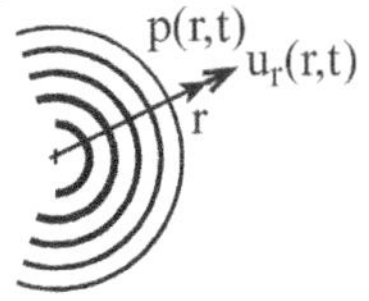	$p(r,t) = P_o \frac{r_o}{r} e^{i(kr-\omega t)}$ $= P_o(r_o/r)\cos(kr-\omega t)$, real $u_r(r,t) = \frac{P_o}{\rho c}\frac{r_o}{r}e^{i(kr-\omega t)}\left(1 + i\frac{1}{kr}\right)$ P_o = pressure amplitude at $r=r_o$.	$I = \frac{P_o^2 r_o^2}{2\rho c r^2}$, $W = IS = \frac{2\pi P_o^2 r_o^2}{\rho c}$ Far field. Case 13 has near field. Refs [1], p. 47; [3], p. 311 pressure and velocity are not completely in phase.
5. One-Dimensional Propagation with Absorption 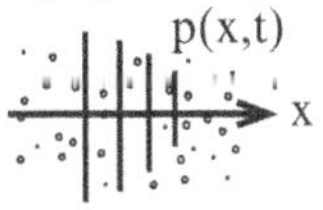	$p_{rms}(x) = P_{orms} e^{-\alpha_m (x-x_o)}$, α_m = fluid absorption /unit length $W = P_{orms}^2/\rho c$ = source power per unit area at $x=x_o$ where $p_{rms} = P_{orms}$. See case 1.	$SPL(x_o) - SPL(x) = \alpha_{dB}(x - x_o)$ $\alpha_{m,dB} = (20\log_{10} e)\,\alpha_m = 8.6859\,\alpha_m$ Fig 6.6, Eq. 6.17 for α_{dB} in dB/km. r in kilometers. Ref. [19].
6. Cylindrical Spreading with Absorption 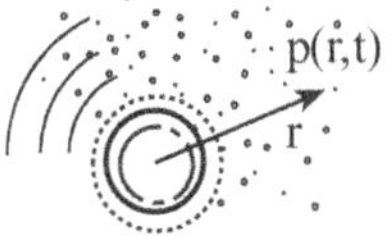	$p_{rms}(r) = P_{orms}\frac{r_o^{1/2}}{r^{1/2}} e^{-\alpha_m (r-r_o)}$ α_m = fluid absorption /unit length $W = \pi r_o P_{orms}^2/\rho c$ = source power per unit length at $r=r_o$. Far field. See case 11 for near field	$SPL(r_o) - SPL(r) = 10\log_{10} r/r_o + \alpha_{dB}(r - r_o)$ $\alpha_{dB} = (20\log_{10} e)\,\alpha_m = 8.6859\,\alpha_m$ See Fig 6.6, Eq. 6.17 for α_{dB} in db /km. r in kilometers, Ref. [19]

Table 6.3 Acoustic wave propagation, continued

Acoustic Wave	Acoustic Pressure and Velocity	Intensity and Power
7. Spherical Spreading with Absorption 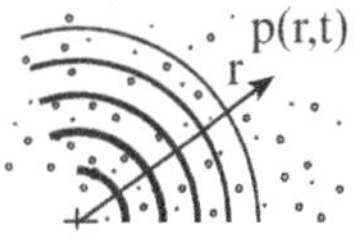	$p_{rms}(r) = P_{orms}\frac{r_o}{r}e^{-\alpha_m(r-r_o)}$ α_m = fluid absorption /unit length $W = 2\pi P_{orms}^2 r_o^2/\rho c$ source energy at $r=r_o$ where $p_{rms}=P_{orms}$. Refs. [3, 29]	$SPL(r_o) - SPL(r) = 20\log_{10} r/r_o + \alpha_{dB}(r-r_o), r \ge r_o$ $\alpha_{dB} = (20\log_{10}e)\alpha_m = 8.6859\,\alpha_m$ See Fig 6.6, Eq. 6.17 for α_{dB} in dB /km. r in kilometer. far field.
8. Plane wave passing cylinder 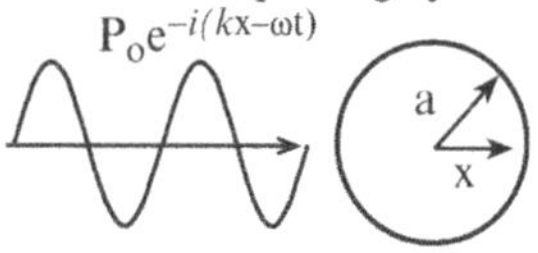	Force on cylinder/unit length $F = -\frac{P_o c}{\omega}\frac{e^{-i\omega t}}{dH_m^1(ka)/dka}$ $= i\,2\pi P_o k a^2 e^{-i\omega t},\ \lambda >> 2\pi a$	Ref. [3], pp. 401-405, with correction. wavelength $\lambda=\omega/(2\pi c)$ $\omega=2\pi f$, f = frequency, Hz. H_m^1 = Hankel function.
9. Oscillating Circular Piston in Rigid Planar Baffle 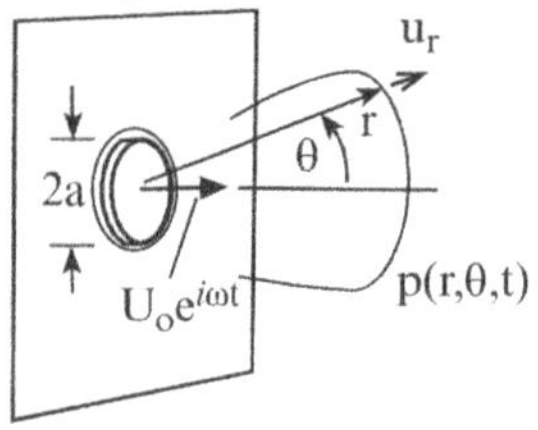	$p(r,\theta,t) = i\rho c U_o ka\frac{a}{r}\left[\frac{2J_1(ka\sin\theta)}{ka\sin\theta}\right]Exp[-i(kr-\omega t)]$ $u_r(r,\theta,t) \approx p(r,\theta,t)/(\rho c)$, Ref. [3], p. 381 $I = \frac{\rho c}{8}U_o^2(ka)^2\left(\frac{a}{r}\right)^2\left[\frac{2J_1(ka\sin\theta)}{ka\sin\theta}\right]^2,\ W = \frac{\rho c}{2}U_o^2\pi a^2\left[1-\frac{2J_1(2ka)}{ka}\right]$ $= \frac{\rho c}{2}U_o^2\pi a^2 \times \begin{cases} 1-\frac{\cos(2ka-3\pi/4)}{\pi^{1/2}(ka)^{3/2}}, & ka >> 1, \text{ high frequency} \\ (ka)^2/2, & ka << 1, \text{ low frequency} \end{cases}$	
10. Exponential Horn Forced by Oscillating Piston 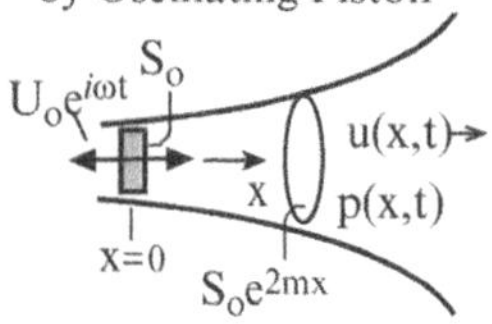 m = duct exp. coefficient	$p(x,t) = \rho c U_o \frac{k'+im}{k}e^{-mx} \times e^{-i(k'x-\omega t)}$ $u(x,t) = U_o e^{-mx}e^{-i(k'x-\omega t)}$ $S(x) = S_o e^{2mx}$ = area of horn m = 0 uniform duct, $k' = \sqrt{k^2-m^2}$ m has units of 1/length.	$I = \begin{cases} \frac{1}{2}\rho c U_o^2 \frac{k'}{k}e^{-2mx} & k > m \\ 0 & k < m \end{cases}$ $W = \begin{cases} \frac{1}{2}\rho c U_o^2 \frac{k'}{k}S_o & k > m \\ 0 & k < m \end{cases}$ Ref. [17]., p. 67
11. Expanding - Contracting Cylinder 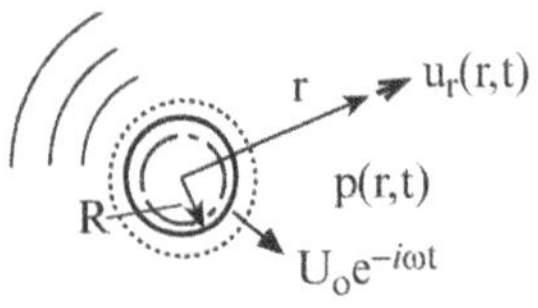 U_o = amplitude of surface velocity at R This is a 2D monopole, also a line of 3D monopoles.	$p(r,t) = i\rho c U_o \frac{H_0(kr)}{H_0'(kr_o)}e^{-i\omega t}$ $\quad H_0'(kr_o) = H_1(kr_o)$ $= P_o \frac{r_o^{1/2}}{r^{1/2}}e^{i(kr-\omega t-\pi/4)}$, far field $r >> \lambda >> r_o$, $P_o = \pi\rho c U_o\sqrt{\frac{kr_o}{2\pi}}$ $u_r(r,t) = U_o\frac{H_0'(kr)}{H_0'(kR)}e^{-i\omega t} = U_o e^{-i\omega t}$ at $r=r_o$. P_o= press. amp at r_o. $= p(r,t)/(\rho c)$ far field $r >> \lambda >> r_o, kR << 1$ Per unit length Ref. [3], p. 358. In far field $r >> \lambda >> r_o, kR << 1$ $I = \frac{\pi}{2r}\rho c k r_o^2 U_o^2,\ W = \pi^2 \rho c k r_o^2 U_o^2,\ Q = Q_o e^{-i\omega t},\ Q_o = 2\pi r_o U_o$ H_0 = Hankel function has real and imaginary components	

Table 6.3 Acoustic wave propagation, continued

Acoustic Wave	Acoustic Pressure, Velocity, Intensity, and Power
12. Vibrating Cylinder (2D) 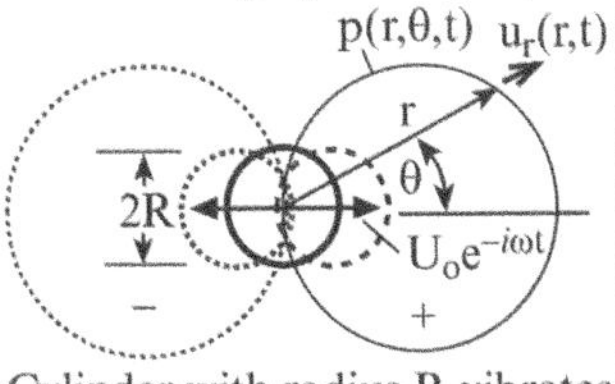 Cylinder with radius R vibrates along θ = 0, normal to its axis, with velocity amplitude U_o and circular frequency ω.	$p(r,t) = i\rho c U_o \cos\phi \frac{H_1(kr)}{H_1'(kR)} e^{-i\omega t}$, H_1= Hankel fn. Ref. [3], pp. 358–359 $= \sqrt{\frac{\pi}{2}}\rho c U_o \frac{(kR)^2}{(kr)^{1/2}} \cos\phi\, e^{-i(kr-\omega t-3\pi/4)}$, far field r>>l>>R, kR<<1 $u_r(r,t) = U_o \cos\phi \frac{H_1'(kr)}{H_1'(kR)} \text{Exp}[-i\phi t] = p(r,t)/(\rho c)$ $I = \frac{\pi}{2}\rho c U_o^2 (kR)^3 \frac{R}{r}\cos^2\phi$, $W = \frac{\rho c}{2} U_o^2 (kR)^3 \pi^2 R$ Force on cylinder/unit length $F = -\rho\pi R^2 dU/dt$
13. Expanding and Contracting Sphere - 3D Monopole 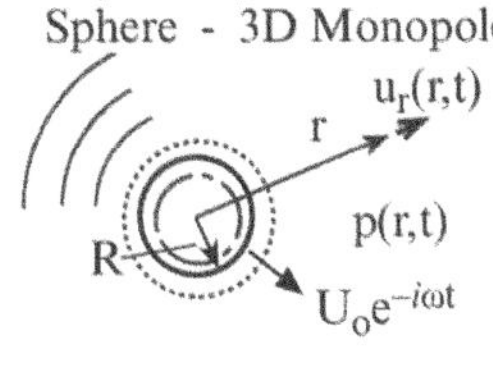 R = means radius of sphere $U(t)=U_o \text{Exp}[-i\omega t]$ =surface velocity Q_o = amplitude of volume velocity	$p(r,t) = \frac{kR^2}{r}\frac{kR-i}{1+(kR)^2}\rho c U_o e^{i\{k(r-R)-\omega t\}} = -i\frac{R}{r}P_o e^{i(kr-\omega t)}, r >> \lambda >> R$ $u_r(r,t) = \left[\left(kR+\frac{1}{kr}\right)+i(\frac{R}{r}-1)\right]\frac{(kR^2/r)}{1+(kR)^2}U_o e^{i\{k(r-R)-\omega t\}} = U_o e^{-i\omega t}$ at r = R $= \frac{p(r,t)}{\rho c}\left(1+\frac{i}{kr}\right)$, far field $r >> \lambda >> R$, $P_o = \rho U_o \omega R$ The pressure and velocity are not exactly in phase. Ref. [1], p. 154. $I = \frac{1}{2}\rho c U_o^2 (kR)^2 \frac{R^2}{r^2}$, $W = 2\pi\rho c U_o^2 (kR)^2 R^2$, $Q_o = 4\pi R^2 U_o$ Pressure on sphere $p(R,t) = -\frac{4\rho c\pi R^2 (kR)^2}{1-(kR)^2}U(t) - \frac{4\rho\pi R^3}{1-(kR)^2}\frac{dU(t)}{dt}$
14. Doublet 3D Dipole - Two 3D Monopoles Opposite Sign 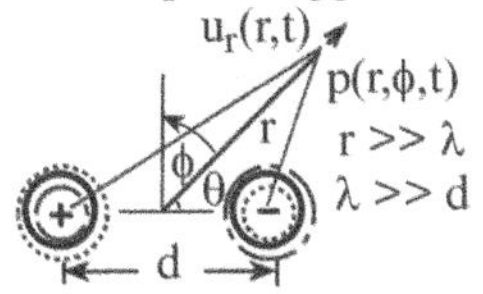 	$p(r,t) = \frac{\rho kd}{4\pi r}\omega Q_o \sin\phi\, e^{i(kr-\omega t)}$, Far field: $r >> \lambda >> d, kd << 1$ $u_r(r,t) = p(r,t)/(\rho c)$, Force on Fluid $= i\rho k\omega Q_o e^{-i\omega t}$ Q_o = amplitude of volume velocity of one monopole (volume/time) Equivalent to case 15 in far field. Ref. [3], p. 312 $I = \frac{1}{32\pi^2 c r^2}\rho (kd)^2 \omega^2 Q_o^2 \cos^2\theta$, $W = \frac{1}{24\pi c}\rho (kd)^2 \omega^2 Q_o^2$
15. Vibrating Sphere, 3D Dipole 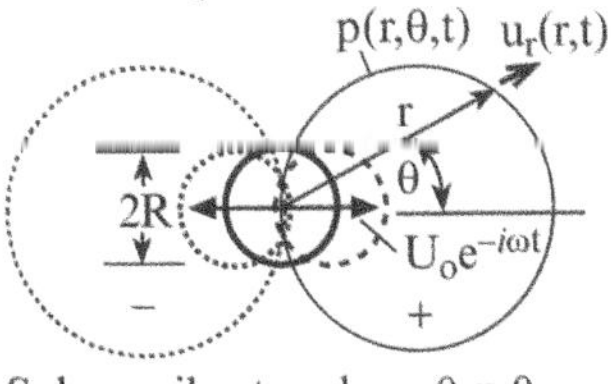 Sphere vibrates along θ = 0. U_o = Velocity amplitude R = Radius of sphere	$p(r,t) = -\frac{\rho U_o (kR)^2 \cos\theta}{[2-(kR)^2-2ikR]}\frac{R}{r}\left(1-\frac{R}{ikr}\right)e^{i(k(r-R)-\omega t)}, r \ge R$ $= -\rho U_o \frac{R}{2r}\cos\theta\, e^{i(kr-\omega t)}$, far field: $r >> \lambda >> R$, $kR << 1$ $u_r(r,t) = U_o \frac{R^3}{r^3}\frac{2-2ikr-(kr)^2}{2-2ikR-(kR)^2}\cos\theta\, e^{i(k(r-R)-\omega t)}, r \ge R$ $I = \frac{1}{4}\rho c U_o^2 (kR)^2 \frac{R^2}{r^2}\cos^2\theta$, $W = \frac{\pi}{6}\rho c U_o^2 (kR)^4 R^2$ Ref. [1], pp. 156–157 Fluid force on sphere for $\lambda >> R$, $F = -\rho\frac{2\pi R^3}{3}\frac{dU}{dt}$

Table 6.3 Acoustic wave propagation, continued

Acoustic Wave	Acoustic Pressure, Velocity, Intensity, and Power
16. Refraction Velocity Grad. 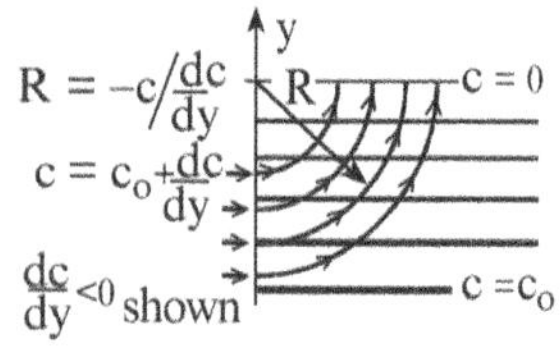	sound speed, radius of sound refraction dc/dy = constant $c=c_o+\frac{dc}{dy}y$ $R=-c(x,y)\frac{dy}{dc}$ An initially horizontally propagating wave in vertical sound speed gradient bends in a circular arc towards the lower speed of sound. The center of the arc is at the point where the sound speed projects to zero. Ref. [2], p. 402.
17. Reflection Fluid Interface 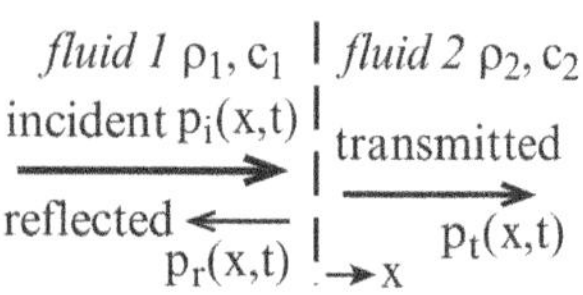	$p_i(x,t)=P_i\,\mathrm{Exp}[i(\omega t-k_1x)]$ $p_r(x,t)=P_r\,\mathrm{Exp}[i(\omega t+k_1x)]$ $p_t(x,t)=P_t\,\mathrm{Exp}[i(\omega t-k_2x)]$ $\frac{P_r}{P_i}=\frac{\rho_2c_2-\rho_1c_1}{\rho_1c_1+\rho_2c_2},\ \frac{P_t}{P_i}=\frac{2\rho_2c_2}{\rho_1c_1+\rho_2c_2}$ $p_i/u_i=-p_r/u_r=\rho_1c_1,$ $\frac{I_r}{I_i}=\left[\frac{\rho_1c_1-\rho_2c_2}{\rho_1c_1+\rho_2c_2}\right]^2$ Ref. [2], p.126 $\frac{I_t}{I_i}=\frac{4\rho_1c_1\rho_2c_2}{(\rho_1c_1+\rho_2c_2)^2}$ $k_1=\omega/c_1, k_2=\omega/c_2,$ $p_t/u_t=\rho_2c_2$
18. Oblique Incidence and Reflection at Fluid Interface 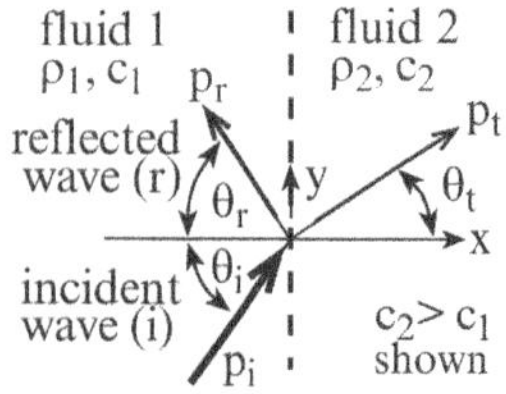 Wave bends towards slower c.	$p_i(x,t)=P_ie^{i(\omega t-k_1x\cos\theta_i-k_1y\sin\theta_i)}$, $p_r(x,t)=P_re^{i(\omega t+k_1x\cos\theta_i-k_1y\sin\theta_i)}$ $p_t(x,t)=P_te^{i(\omega t-k_2x\cos\theta_t-k_2y\sin\theta_t)}$, Ref. [2], pp. 132–133 Snell's law: $\frac{\sin\theta_{incident}}{\sin\theta_{transmit}}=\frac{c_1}{c_2},\quad \theta_r=\theta_i, k_1=\frac{\omega}{c_1}, k_2=\frac{\omega}{c_2}$ $\frac{P_r}{P_i}=\frac{\rho_2c_2\cos\theta_i-\rho_1c_1\cos\theta_t}{\rho_2c_2\cos\theta_i+\rho_1c_1\cos\theta_t},\ \frac{P_t}{P_i}=\frac{2\rho_2c_2\cos\theta_i}{\rho_2c_2\cos\theta_i+\rho_1c_1\cos\theta_t}$ $\frac{I_r}{I_i}=\left(\frac{P_r}{P_i}\right)^2,\ \frac{I_t}{I_i}=\left(\frac{P_t}{P_i}\right)^2\frac{\rho_1c_1\cos\theta_t}{\rho_2c_2\cos\theta_i}$, if $1-\frac{c_2^2}{c_1^2}\sin^2\theta_i>0$, else 0
19. Propagation Wind-Shadow 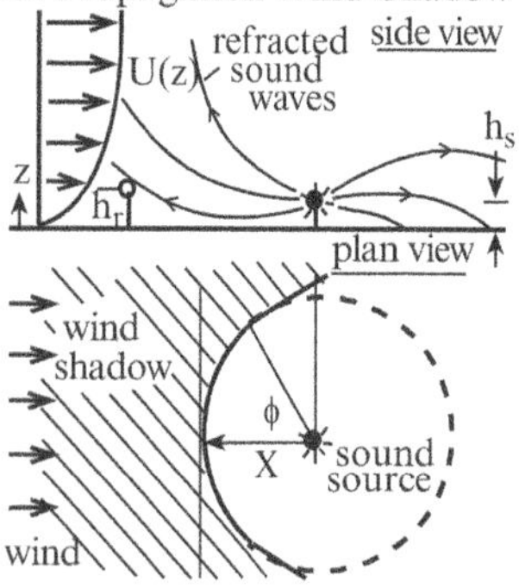	Radius R from source to wind shadow and critical angle ϕ. Ref. [20] $X=\left(\frac{2c_o}{\frac{dU}{dz}-\frac{dc}{dz}}\right)^{1/2}\left(h_s^2+h_r^2\right)^{1/2},\ \phi=\cos^{-1}\frac{-\frac{dc}{dz}}{\frac{dU}{dz}}$ =crit angle (max) Wind velocity U(z) and temperature T(z) have vertical gradients. $U(z)=(z/z_o)^\beta U_o,\quad z_o=10\text{m}(33\text{ ft}),\ U_o=$ wind at 10 m $dU/dz=(\alpha/z)(z/z_o)^\beta U_o,\quad 0.1<\beta<0.4,\ \beta=0.28$ typical $dc/dz=[c_o/(2T_o)](dT/dz)$, $c_o=$ sound speed at abs. temp T_o $dT/dz=-6.5°\text{C}/1000\text{ m}(-3.57°\text{F}/1000\text{ ft})$ standard lapse rate Sound cannot be heard radius X upwind from source.

Table 6.3 Acoustic wave propagation, continued

20. Reflection from Surface

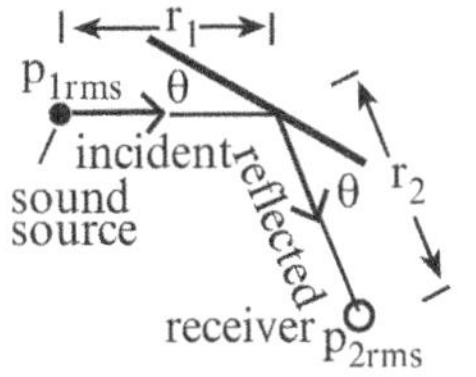

$$\frac{p_{2rms}}{p_{1rms}} = (1-\alpha)^{1/2}$$

α = surface reflection absorption coefficient, below, 0<α<1, dim'lss
Plane waves, no divergence.
Sound wave front reflects off surface that absorbs part of the incident wave energy. Refs [21, 22]

20. continued Acoustic Configuration					Reflection coefficient α, Ref. [21]					
	Thickness		Weight		Frequency, Hz					
Reflecting Surface	inch	mm	lb/ft²	kg/m²	125	250	500	1000	2000	4000
1 Ceiling Tile, textured	0.5	13	0.7	3.42	0.69	0.76	0.65	0.86		0.75
3 Ceiling Tile, mineral wool	0.625	16	0.65	3.17	0.35	0.27	0.47	0.65	0.74	0.82
4 Ceiling tile, cloth faced	1.5	38	0.7	3.42	.74	.97	.85	1.03	1.05	1.00
5 Duct Liner, fiberglass	1	25	0.12	0.59	.34	.42	.47	.84	.8	.79
6 Duct liner, fiber board	1	25	0.25	1.22	.03	.22	.60	.84	.98	.97
7 Duct liner, fiberglass	1.5	38	.18	0.88	.29	.55	.72	.85	.89	.89
8 Duct liner, fiber board	1.5	38	.37	1.81	.16	.39	.91	1.01	1.01	1.01
9 Duct Liner, fiberglass	2	51	.24	1.17	32	.71	.83	.91	.95	1.01
10 Duct Liner, fiber-board	2	51	.5	2.44	.24	.79	1.13	1.13	1.04	1.05
11 Foam panel, urethane	1	25	.17	0.83	.20	.81	.61	.73	.71	.69
12 Foam Panel, polyimide	2	51	.23	1.12	.23	.51	.96	1.04	.93	.96
13 Wedge type foam panel	2	51	.17	0.83	.08	.25	.61	.92	.95	.92
14 Wedge-type foam panel	3	76	.51	2.49	.14	.43	.98	1.03	1.00	1.00
15 Wedge-type foam panel	4	102	.67	3.27	.2	.7	.98	1.06	1.01	100
16 Blanket, unfaced fiberglass	1	25	.19	0.93	.06	.2	.29	.4	.5	.54
17 Blanket, unfaced fiberglass	2	51	.11	0.54	.25	.48	.81	.90	.97	.94
18 Blanket, unfaced fiberglass	3	76	.16	0.78	.36	.76	1.04	.94	.98	1.00
19 Blanket, unfaced fiberglass	6	152	.33	1.61	1.18	1.36	1.02	1.02	1.12	1.07
20 Perforate sheet-fiberglss blanket	2	51	1.5	7.32	.44	.62	.94	.99	.79	.59
21 Perforated sheet-fiberglss blan't	4.26	108	5	24.4	.86	1.09	1.22	1.06	1.05	1.04
22 Sintered metal panel	0.1	3	0.9	4.39	.44	.81	.87	.52	.69	.8
23 Sintered metal-fiberglass blanket	2.1	53	1.9	9.27	.77	1.03	.93	1.07	1.03	1.07
24 Brick, unglazed					.03	.03	.03	.04	.00	.07
25 Brick, painted					.01	.01	.02	.02	.02	.03
26 Concrete block, unpainted					.36	.44	.31	.29	.39	.25
27 Concrete block, painted					.10	.05	.06	.07	.09	.08
28 Marble or glazed tile					.01	.01	.01	.01	.02	.02
29 Concrete Floor					.01	.01	.015	.02	.02	.02
30 Wood Floor					.15	.11	.10	.07	.06	.07
31 Linoleum or cork floor					.02	.02	.03	.03	.03	.02
32 Heavy Glass panels					.18	.06	.04	.03	.02	.02
33 Window Glass, ordinary					.35	.25	.18	.12	.07	.04
34 Gypsum Board over 2x4 studs 16in centers	0.5	13			.29	.1	.05	.04	.07	.09
35 Open window					1	1	1	1	1	1
36 Plaster on blocks					.013	.015	.02	.03	.04	.05
37 Plaster on wood lath					.14	.10	.06	.05	.04	.03
38 Water surface of pool					.008	.008	.013	.015	.0	.025
39 Drapes, light fabric, flat			10(a)	338(b)	.03	.04	.11	.17	.24	.35
40 Drapes, medium fabric, hung to half area			14(a)	475(b)	.07	.31	.49	.75	.70	.60
41 Drapes, heavy fabric, hung to half area			18(a)	610(b)	.13	.35	.55	.72	.70	.65
42 Persons seated on wood chairs	2(c)				0.24	.4	.75	.96	.96	.87

(a) Ounces per square yard, (b) gram per square meter, (c) 2 persons per square meter
Reflection coefficient by reverberation room method, ASTM C423-02A

Table 6.3 Acoustic wave propagation, continued

Additional notation: IL = Insertion loss = $20 \log_{10} p_{s\text{-}rms}/p_{r\text{-}rms}$, additional attenuation between source and receiver due to insertion of buildings, foliage, or ground plane between source and receiver, in decibels; a, d, z = distances, meter; f = frequency, Hz; α = loss coefficient, dim'less. Subscripts: r = receiver, s = source.

Acoustic Wave	Insertion Loss, $20 \log_{10} p_{s\text{-}rms}/p_{r\text{-}rms}$, dB
21. Multiple Reflections 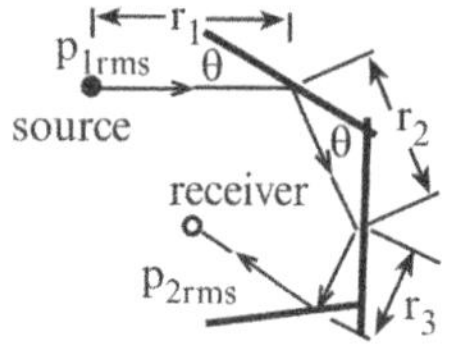	$IL = 20\log_{10}[(1-\alpha_1)^{1/2}(1-\alpha_2)^{1/2}(1-\alpha_3)^{1/2}..]$, dB Plane waves, no divergence. α_i = surface reflection-absorption coefficient of ith surface, i=1,2,3, 1<α<, from case 20 Three reflections shown.
22. Sound Propagates over Long Barrier 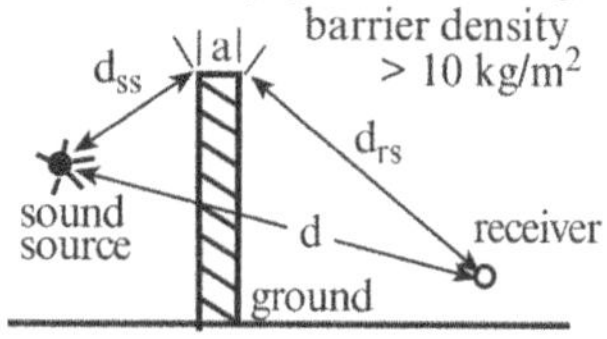 Barrier width, perpendicular to plane, is greater than wavelength.	$IL_{barrier} = 10 \log_{10} [3 + 20(z/\lambda)K]$, dB Distances d, d_{ss}, d_{rs}, λ, z in meters. $\lambda = c/f$ = wavelength. $z = [(d_{ss}+d_{rs})^2 + a^2]^{1/2} - d$ If $z \le 0$, then K=1 If $z > 0$, then $K = Exp\left[-\frac{1}{2000m}\sqrt{\frac{d_{ss}d_{rs}d}{2z}}\right]$, Ref. [23]
23. Propagation through Trees and Foliage and Between Buildings 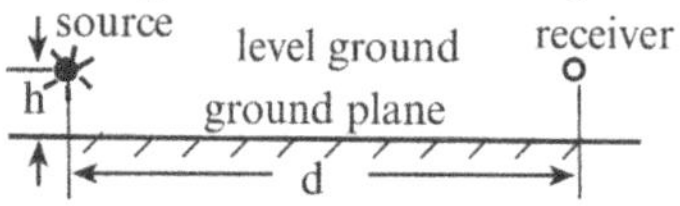 foliage, if 10 m< d< 20 m, d = 20 m 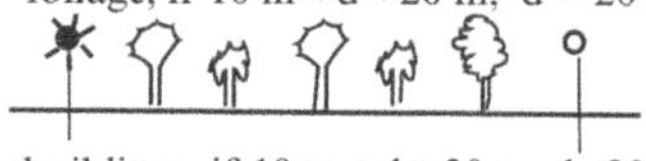 buildings, if 10 m< d< 20 m, d=20 m 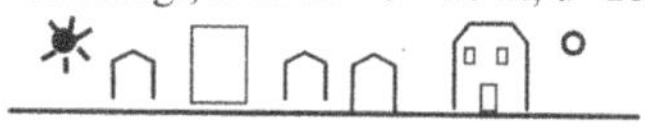	Above ground, parallel to earth $IL_{ground} = 4.8 - (2h/d)(17 + 300/d)$, dB $IL_{foliage} = \alpha_{foliage} d$, dB, $\alpha_{foliage}$ dB per km below $IL_{houses} = 0.1Bd$, dB, houses in a general random pattern freq, Hz | 63 | 125 | 250 | 500 | 1000 | 2000 | 4000 $\alpha_{foliage}$ dB/km | .03 | 0.03 | 0.04 | 0.05 | 0.06 | 0.08 | .09 h = mean height above ground in meters, d = distance between source and receiver in meters., d≥10 m B = ground area covered by houses / total area, B<1 Ref. ISO9613-2, Ref. [23]
24. Reflection and Absorption from an Impedance Wall 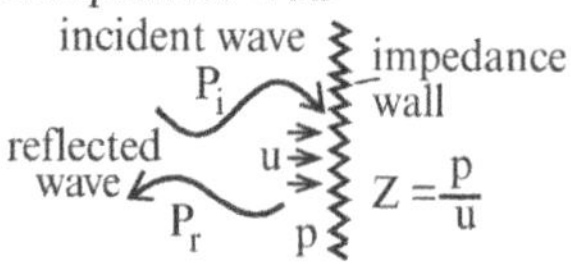	$\frac{P_i}{P_r} = \frac{Z/\rho c + 1}{Z/\rho c - 1}$, $\alpha_s = 1 - \left\lvert\frac{Z/\rho c - 1}{Z/\rho c + 1}\right\rvert^2$ $\frac{\text{pressure p on wall}}{\text{incident pressure}} = \frac{P_i + P_r}{P_i} = \frac{2Z}{Z + \rho c} = \begin{cases} 2, & Z=\infty \text{ rigid wall} \\ 1, & Z=\rho c \text{ absorbing wall} \end{cases}$ Z = p/u is the impedance of the face of the wall.

Table 6.3 Acoustic wave propagation, continued

Notation: f = frequency, Hz; p(t) = pressure; P = amplitude of pressure; i = imaginary constant; x = distance; k = stiffness per unit area; m = mass per unit area; $\alpha = 1 - |Pr/Pi|^2$ = absorption coefficient; ρ = density; $\omega = 2\pi f$ = circular frequency, rad./sec. Subscripts : i = incident wave, approaching wall, r = reflected wave; t = transmitted wave. $P_i - P_r = P_t$, Table 1.2 has consistent units.

Acoustic Wave	Step Impedance Z_s, Absorption Coefficient α, and TL

25. Spring-Supported, Damped Panel with mass and stiffness

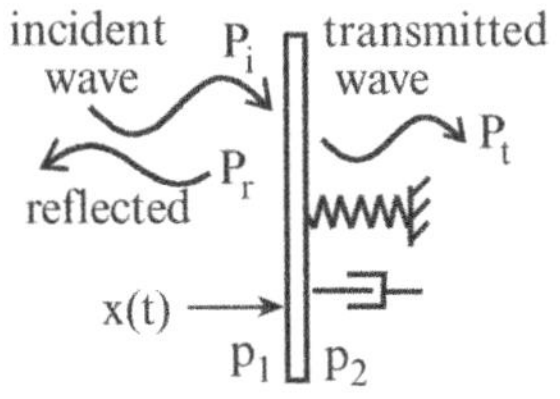

Refs [13, 24, 30] + authors result
Panel is rigid plate on springs.
m = panel mass per unit area
k = stiffness per unit area
ζ = damping factor
$\omega_n = (k/m)^{1/2}$ panel nat fequency on spring support, rad/s.

step impedance across panel,

$$Z_s = \frac{p_1 - p_2}{dx/dt} = 2m\zeta\omega_n - i\frac{(k - m\omega^2)}{\omega}$$

impedance ahead of panel,

$$Z = \frac{p_1}{dx/dt} = \rho c + 2\zeta m\omega_n - i\frac{\omega_n^2}{\omega}\left(1 - \frac{\omega^2}{\omega_n^2}\right)$$

$$TL = 10\log_{10}\left[\frac{m\omega_N^2}{2\rho c}\right]^2\left\{\left(1 - \frac{\omega^2}{\omega_n^2}\right)^2 + \left(2\zeta\frac{\omega}{\omega_n} + \frac{2\rho c\,\omega}{m\omega_n^2}\right)^2\right\}$$

$$= \begin{cases} 10\log_{10}[1 + k/(2\rho c\omega)]^2, \lim m \to 0 \\ 10\log_{10}[1 + (m\omega/2\rho c)^2], \lim k \to 0, \text{mass law} \\ 10\log_{10}[1 + \zeta m\omega/(\rho c)]^2, \omega = \omega_n, \zeta < 1 \end{cases}$$

$$\alpha = \frac{4}{MF}\frac{\rho c\omega}{m\omega_n^2}\left(2\zeta\frac{\omega}{\omega_n} + \frac{\rho c\omega}{m\omega_n^2}\right) / \left\{\left(1 - \frac{\omega^2}{\omega_n^2}\right)^2 + \left(2\zeta\frac{\omega}{\omega_n} + \frac{2\rho c\,\omega}{m\omega_n^2}\right)^2\right\}$$

26. Porous Resistive Wall.

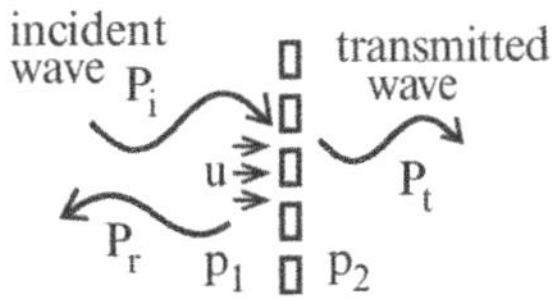

See cases 27,28

X_w=0, fine porosity.

step impedance across wall

$$Z_s = \frac{p_1 - p_2}{u} = \rho c R_w$$

impedance for incident wave

$$Z = p_1/u = \rho c - \rho c R_w$$

Case	P_r/P_i	P_t/P_i	α	Z_s
Porous Wall Resistive	$\frac{R_w}{2 + R_w}$	$\frac{2}{2 + R_w}$	$\frac{4(1 + R_w)}{(2 + R_w)^2}$	$\rho c R_w$ $(X_w = 0)$

$$\text{Transmission Loss} = 10\log_{10}\frac{P_i^2}{P_t^2} = 10\log_{10}\left|1 + \frac{Z_s}{2\rho c}\right|^2$$

Resistance of various materials in air (Refs [13], p. 231).

Material	wire dia, in. wires/inch	Mass, lb/ft²	mass g/m²	$\rho c R_w$ Ns/m²
Wire Cloth	.0087 in, 50/in	0.25	1200	5.9
"	.0045 in 100/in	0.13	630	9.0
"	.00225 200/in	0.63	480	24.6
Fiberglass cloth		1.87oz/yd	57	10
"		9.60	293	13
"		24.6	750	200
Sintered metal	.04 in thick	0.79	3.9	100

Table 6.3 Acoustic wave propagation, continued

Additional notation: d = hole diameter; e = plate thickness; $i = (-1)^{1/2}$, imaginary constant; $k = \omega/c$ = wave number; p = acoustic pressure; P = mean pressure; R = resistance, component of $Z/\rho c$ in phase with velocity; u(t) = orifice acoustic velocity; $u_1(t)$ = approach acoustic velocity; t = time; X = reactance, component of $Z/\rho c$ in phase with acceleration; Z = impedance; $\alpha = 1 - |P_r/P_i|^2$ = absorption coefficient; σ = open area/total area; ρ = density; ω = circular frequency, rad./sec; ν = kinematic viscosity; λ = wavelength, c/f. Table 1.2 contains consistent units.

Orifice Plate Configuration	Impedance Step across Plate, $Z_s=(p_1-p_2)/u = \rho c\,(R+iX)$
27. Perforated and Rigid Walls 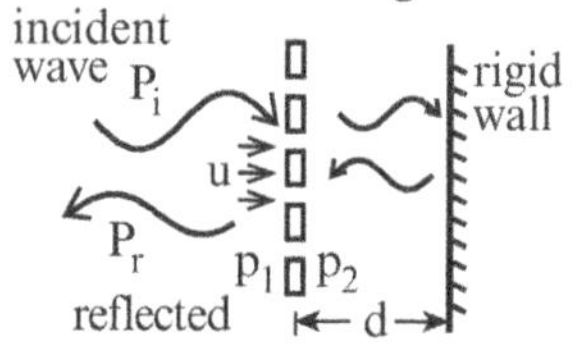 $Z_s = \frac{p_1-p_2}{u} = \rho c(R_w + iX_w)$ $Z = p_1/u = \rho c - Z_s$ Ref. [31], p. 32, in part. See cases 26 and 28 for R_w and Case 28 for X_w	P_r/P_i: $\frac{Z_s}{2\rho c + Z_s}$; P_t/P_i: $\frac{2\rho c}{2\rho c + Z_s}$; α: $1-\left\|\frac{Z_s}{2\rho c + Z_s}\right\|^2$; Z_s: Z_s $Z_s = \frac{p_1-p_2}{u} = \rho c R_w + i\rho c X_w$ = step impedance across plate $Z = \frac{p_1}{u} = \rho c R_w + i\rho c(X_w - \cot kd)$ = impedance in front of plate $\alpha = 1 - \left\|\frac{P_r}{P_i}\right\|^2 = \frac{4R_w}{(\cot kd - X_w)^2 + (1+R_w)^2}$ = absorption coefficient Tranmission Losss, dB $= 10\log_{10} P_i^2/P_t^2 = 10\log_{10}\|1 + Z_s/2\rho c\|^2$
28. Circular Orifice and Perforated Plates 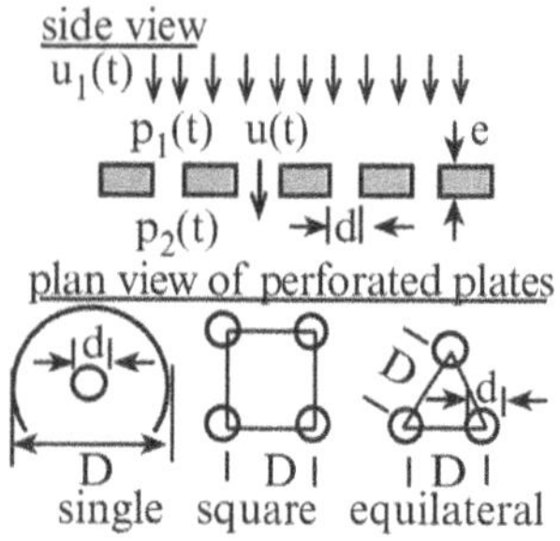 	$R = \frac{\sqrt{8\nu\omega}}{\sigma c}\left(1+\frac{e}{d}\right) + \frac{\pi^2}{2\sigma}\left(\frac{d}{\lambda}\right)^2$, σ = porosity, d = orifice diameter $X = \frac{\omega}{\sigma c}(e+\delta)$, where $\delta = \frac{8d}{3\pi}\left(1-0.7\sigma^{1/2}\right) + \sqrt{\frac{8\nu}{\omega}}\left(1+\frac{e}{d}\right)$ $\sigma = \begin{cases} (d/D)^2 \text{ single hole with diameter d, in circular plate dia D} \\ (\pi/(2\sqrt{3}))(d/D)^2 \text{ square pattern perforations, spacing D} \\ (\pi/(4)(d/D)^2 \text{ equalateral triangle perforations, spacing D} \end{cases}$ $u_1 = \sigma u$ = approach acoustic velocity, u = orifice acoustic velocity. Refs [25, 26]
29. Perforate or Orifice with Steady Normal Mean flow U_o 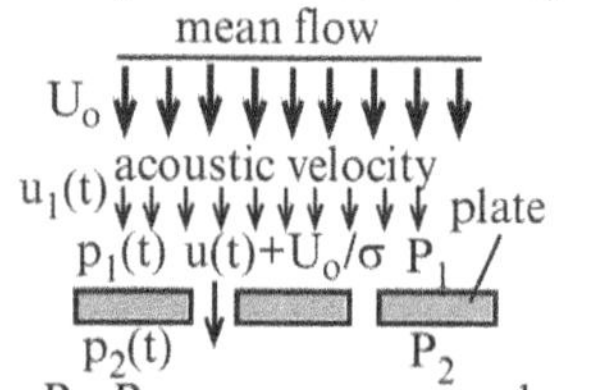 P_1-P_2 = mean pressure drop	$R = \frac{2\overline{\Delta P}}{\rho c U_o} + \frac{\pi^2}{2\sigma}\left(\frac{d}{\lambda}\right)^2$, includes resistance due to pressure drop X = inertance, see case 29 above. $u_1(t) = \sigma u(t)$ $\sigma = \frac{\text{open area}}{\text{total area}} < 1$ $\overline{\Delta P}$ = Mean pressure drop across plate ~ ½ $\rho(U_o/\sigma)^2$ Total approach velocity = $U_o + u_1(t)$
30. Perforated plate or Orifice with Grazing Flow	Grazing flow increases resistance (R) and decreases inertance (X). These effects are a function of mean flow Mach number. See Figure 6.12 See Refs [27, 28]

point x_1 at the later time $t_1 = x_1/c$, as it propagates energy out the x-axis (Figure 6.4a). Similarly, $P(x + ct)$ is a backward-traveling wave that propagates in the $-x$-direction.

Specific acoustic impedance, ρc, is the ratio of acoustic pressure to acoustic particle velocity in a one-dimensional sound wave:

$$\frac{p}{u} = \begin{cases} \rho c & \text{forward traveling wave } (+\,x\text{-direction}) \\ -\rho c & \text{backward traveling wave } (-x\text{-direction}) \end{cases} \tag{6.10}$$

This is shown by substituting first-order derivatives in Equation 6.9 into Equation 6.5. The metric unit of specific impedance, $\text{kg/m}^2\text{-s} = \text{N-s/m}^3 = \text{Pa-s/m}$, is called 1 mks rayl to honor Lord Rayleigh [32].

At atmospheric pressure and 20 °C, the specific impedances of air and freshwater are shown below (Table 6.1; also see Chapter 8): Specific acoustic impedance has units of mass/length^2-time. See Table 1.2.

Fluid	Density (ρ)	Speed of sound (c)	Specific impedance (ρc)
Air	1.204 kg/m^3	343.4 m/s	413.4 kg/m^2-s
	0.07516 lb/ft^3	1127. ft/s	2.633 lb-s/ft^3
Water	998.2 kg/m^3	1482. m/s	1.479×10^6 kg/m^2-s
	62.32 lb/ft^3	4863. ft/s	9420 lb-s/ft^3

The difference between acoustic impedance of water and air $\rho c|_{\text{water}}/\rho c|_{\text{air}} = 3775$ results in a 29 dB loss during air–water transmission (Eq. 6.15c; case 15 of Table 6.2). This is similar to decibel difference between reference pressure of 20 μPa in air and 1 μPa in water, $20 \log_{10}(20/1) = 26\,\text{dB}$ [33].

Wavelength λ is the distance between the propagating crests (Figure 6.4). *Wave number* k is 2π over wavelength. $\omega = 2\pi f$ is the circular frequency in radian per second; frequency f is in Hertz:

$$\text{Wavelength}: \ \lambda = \frac{c}{f} \qquad \text{Wave number}: \ k = \frac{\omega}{c} = \frac{2\pi f}{c} = \frac{2\pi}{\lambda} \tag{6.11}$$

Wavelength decreases as frequency increases. Wavelengths in room temperature air (above) are as follows:

Frequency (Hz)	1	10	100	1000	10,000
Wavelength (ft)	1128	112.8	11.28	1.128	0.1128 (1.3 in.)
Wavelength (m)	343	34.3	3.43	0.343	0.0343 (34.3 mm)

Wavelength scales acoustic interaction with physical objects. Sound reflects off objects larger than a wavelength and diffracts around objects much smaller than a wavelength. Human hearing is most acute at 1000 Hz where the acoustic wavelength is about twice the distance between our ears, which helps in direction finding. See decibel A-weightings in Appendix C.

6.1.3 Decibels and Sound Power Level

Sound pressure level (SPL) uses the *decibel* logarithmic unit of sound pressure, abbreviated dB and named after Alexander Graham Bell, to express the wide range of sound pressures perceptible to humans in two or three digits [1, 2, 33–36] (Figure 6.5):

$$\text{SPL} = 20\log_{10}\frac{p_{\text{rms}}}{p_{\text{ref}}} = 10\log_{10}\frac{p_{\text{rms}}^2}{p_{\text{ref}}^2}\text{dB, inversely}, p_{\text{rms}} = p_{\text{ref}}10^{\text{SPL}/20} \quad (6.12)$$

SPL in decibels is 20 times the logarithm to base 10 $(\log_{10})$ of the rms (square root of the mean square pressure acoustic pressure) relative to a reference pressure, which is a pressure of 0 dB. The reference pressure in air is 20 μPa rms (2.9E−9 psi), [1, 2], p. 113; [34, 35]; the threshold of human hearing in air at 1000 Hz. The reference pressure for underwater sound is 1 μPa [36].

The difference between two SPL in decibels is twenty times the logarithm of the ratio of their rms pressures:

$$\text{SPL}_2 - \text{SPL}_1 = 20\log_{10}\frac{p_{2\text{rms}}}{p_{1\text{rms}}},\ \text{dB, inversely,}\ \frac{p_{2\text{rms}}}{p_{1\text{rms}}} = 10^{(\text{SPL}_2-\text{SPL}_1)/20} \quad (6.13)$$

$\text{SPL}_2 - \text{SPL}_1$, dB	1/10	1	2	3	6	10	20	30	100
$p_{2\text{rms}}/p_{1\text{rms}}$	1.011	1.122	1.256	1.412	1.995	3.162	10	31.6	40

SPL changes 6 dB for a factor of two change in rms sound pressure, which mimics the pressure sensitivity of the human ear [2] (p. 115). See Appendix C. As a rule of thumb,

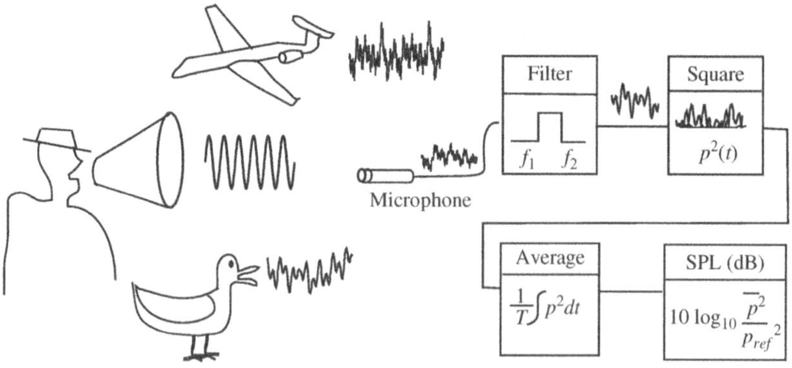

Figure 6.5 Measurement of sound pressure level

3 dB is the threshold of difference in sound pressure detectable by the human ear; a 6 dB difference is readily detectable.

Example 6.1 Decibels, Pascal and Psi

Convert 65 dB into Pascals and then into pounds per square inch.

Solution: Equation 6.12 is applied. The metric reference rms pressure in air is 20 μPa (Table 6.1):

$$p_{\text{rms}} = p_{\text{ref}} 10^{\text{SPL}/20} = 0.00002 \quad 10^{65/20} \text{ Pa} = 0.03557 \text{ Pa } (0.00000516 \text{ psi}) \text{ rms}$$

The rms pressure is the square root of the mean (average) square pressure, averaged over many cycles. *Acoustic intensity* **I** is the rate at which energy is transported by sound waves through a unit area at speed of sound. *Acoustic power* **W** is acoustic intensity times the normal area S, through which the sound passes:

$$\begin{aligned}
&\text{Mean square pressure } p_{\text{rms}}^2 = \frac{1}{T}\int_0^T p^2(t)\, dt \\
&\text{Intensity} \quad \mathbf{I} = \frac{1}{T}\int_0^T p\mathbf{u}dt = \frac{\text{acoustic energy}}{\textit{time} \times \text{area}} = \frac{p_{\text{rms}}^2}{\rho c}\mathbf{n} \\
&\text{Power } \mathbf{W} = \mathbf{I}S = \frac{S}{T}\int_0^T p\mathbf{u}dt = \frac{\text{acoustic energy}}{\textit{time}} = \frac{p_{\text{rms}}^2}{\rho c} S\mathbf{n} \\
&\text{Energy density} \quad E = \frac{|I|}{c} = \frac{\text{acoustic energy}}{\text{volume}} = \frac{p_{\text{rms}}^2}{\rho c^2}
\end{aligned} \tag{6.14}$$

where ρ is the mean fluid density. Acoustic particle velocity vector is $\mathbf{u}(t)$. T is a time period that contains many cycles. The formulas on the far right-hand side apply to one-dimensional propagating sound waves. **n** is a unit vector in the direction of wave propagation, normal to the area S.

Sound pressure, intensity, power, and energy density have interrelated decibel scales. A uniaxial sound wave in standard air with the reference sound pressure of 20 μPa rms has an intensity of 10^{-12} watts rms per meter square (the reference intensity); it delivers a sound power of $W_{\text{ref}} = 10^{-12}$ watts rms (the reference acoustic power) through a 1 m^2 window perpendicular to direction of propagation. The reference energy density is, $p_{\text{rms}}^2/\rho c^2 = |I|/c = 2.9 \times 10^{-15}$ joule/m^3 [1], p. 39.

6.1.4 Standards for Measurement

Sound pressure measurements are often casually made with handheld sound level meters, whereas measurements taken to demonstrate noise compliance require calibration and documentation [34–35]. The following organizations publish standards for sound measurement, equipment, and procedures:

International Organization for Standardization	www.ISO.org
American National Standards Institute	www.ANSI.org
American Society of Testing and Materials	www.ASTM.org
SAE International	www.SAE.org
British Standards Institution	www.bsi-global.com
International Electrotechnical Commission	www.IEC.com

Acoustical standards exist for measurement of sound pressure [18, 39, 40], sound intensity [41, 42], sound power [31, 43–45], transmission loss (TL) [46–48], and absorption [19, 48]. Reference [49] has a compilation of standards. Appendix C has standard 1/3-octaves and A-weighting for noise assessment.

6.1.5 *Attenuation and Transmission Loss (TL)*

Sound changes magnitude, but not frequency, as it propagates. Three decibel scales for the change magnitude of sound pressure across a component are *attenuation*, *insertion loss*, and *Transmission loss*:

$$\text{Attenuation} = 10\log_{10}\left(\frac{p_{1\text{rms}}^2}{p_{2\text{rms}}^2}\right), \text{dB} \tag{6.15a}$$

$$\text{Insertion loss} = 10\log_{10}\frac{p_{2\text{rms}}^2|_{\text{without component}}}{p_{2\text{rms}}^2|_{\text{with component}}}, \text{ dB} \tag{6.15b}$$

$$\text{Transmission loss} = 10\log_{10}\frac{I_{1\rightarrow}S_1}{I_{2\rightarrow}S_2} = 10\log_{10}\left.\frac{p_{1\rightarrow}^2}{p_{2\rightarrow}^2}\right|_{S1=S2}, \text{ dB} \tag{6.15c}$$

$$\text{Absorption coefficient} \quad \alpha_s = 1 - \frac{p_{r-\text{rms}}^2}{p_{i-\text{rms}}^2}, \text{ dimensionless} \tag{6.15d}$$

where p_1 is source sound pressure and p_2 is receiver sound pressure. *Attenuation* is the difference, in decibels, between two sound pressures. *Insertion loss* is the change in receiver SPL, in decibels, when the component is inserted in the transmission path with an anechoic termination [2, 42].

Transmission Loss (Eq. 6.15c) is the change in sound power (intensity I times area S, Eq. 6.14) during propagation through a component. It is a true measure of loss because it is independent of the acoustic source and termination. However, measurement of sound intensity of forward-traveling waves (→) in a complex acoustic environment requires simultaneous measurement with two closely spaced microphones (case 6 of Table 6.4; [37, 38, 41, 44, 55, 56]).

Table 6.4 Duct and room acoustics

Notation: c = speed of sound (Table 6.1); d = diameter; f = frequency Hz = $\omega/2\pi$; e = 2.718; $e^{-i\omega t} = \cos\omega t - i\sin\omega t$; f = $\omega/2\pi$, frequency, Hertz; I = acoustic intensity Equation 6.15c; i = imaginary constant $(-1)^{1/2}$; $k = \omega/c = 1/\lambda$, wave number; L = length; M = U_o/c, Mach number, case 9; P = pressure amplitude; p = acoustic pressure; u = acoustic particle velocity; x = direction of wave propagation; u = acoustic particle velocity; U = amplitude of acoustic particle velocity; S = cross sectional area; t = time, transmitted wave; TL = transmission loss (dB); V = volume; Z = impedance; $\alpha_S = 1 - |p_r/p_i|^2$ absorption coefficient, Equation 6.15d; ρ = fluid density; ω = circular natural frequency, radians per second. Subscripts: rms = root mean square, Equation 6.14; i = incident (+x); g = gas bag; r = reflected (−x); t = transmitted. See Table 1.2 for consistent units. Refs [1–3, 17, 18, 22, 24, 29–31, 37, 38, 46, 50–53].

Acoustical Duct	Acoustic Pressure p and velocity u, Acoustic Intensity I
1. Long Duct Forced by Oscillating Piston at x=0 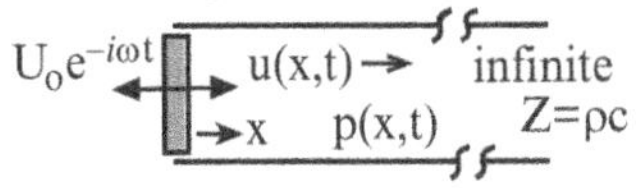	$p(x,t) = \rho c U_o e^{i(kx-\omega t)}$ $\text{Real}\{p(x,t)\} = \rho c U_o \cos(kx-\omega t)$ $u(x,t) = U_o e^{i(kx-\omega t)}$ $I = \dfrac{P_o^2}{2\rho c} = \dfrac{p_{rms}^2}{\rho c} = \dfrac{1}{2}\rho c U_o^2$ $W = IS = (1/2)\rho c U_o^2 S$ S = duct cross section
2. Duct with Closed End at x=L Forced by Oscillating Piston 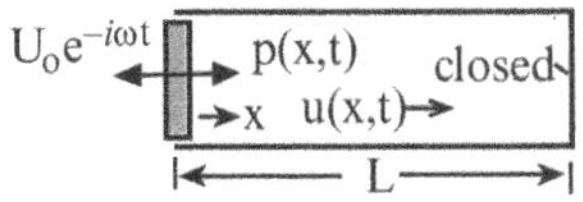	$p(x,t) = i\rho c U_o \dfrac{\cos k(L-x)}{\sin kL} e^{-i\omega t}$ $\text{Real}\{p(x,t)\} = \rho c U_o (\cos k(L-x)/\sin kL)\sin\omega t$ $u(x,t) = U \dfrac{\sin k(L-x)}{\sin kL} e^{-i\omega t}$ u = 0 at x=L. (same as case 4 with $Z = \infty$)
3. Duct with Open End at x=L Forced by Oscillating Piston 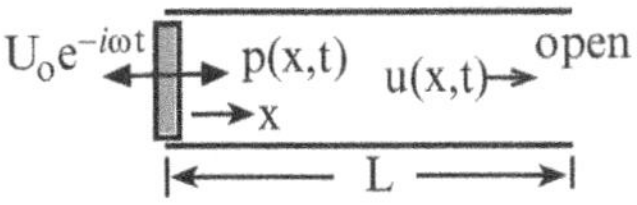	$p(x,t) = -i\rho c U_o \dfrac{\sin k(L-x)}{\cos kL} e^{-i\omega t}$ $u(x,t) = U_o \dfrac{\cos k(L-x)}{\cos kL} e^{-i\omega t}$ p = 0 at x = L (same as case 4 with Z = 0)
4. Duct with Impedance Z at x=L Forced by Oscillating Piston 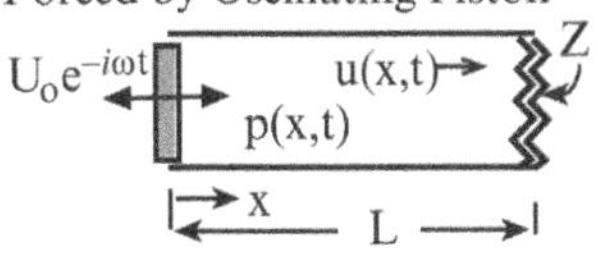	$p(x,t) = \rho c U_o \left\{ \dfrac{i(Z/\rho c)\cos k(L-x) + \sin k(L-x)}{i\cos kL + (Z/\rho c)\sin kL} \right\} e^{-\omega t}$ $u(x,t) = U_o \left\{ \dfrac{i\cos k(L-x) + (Z/\rho c)\sin k(L-x)}{i\cos kL + (Z/\rho c)\sin kL} \right\} e^{-\omega t}$ Ref. [3], p. 468 has same result in different form.
5. One Microphone Measurement of Sound Intensity 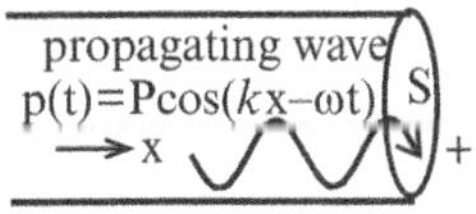	$I = \dfrac{p_{rms}^2}{\rho c}$, $W = IS = \dfrac{p_{rms}^2 S}{\rho c}$, $p_{rms}^2 = \dfrac{1}{T}\int_0^T p^2(t)\,dt$, Forward traveling wave propagates through surface S, perpendicular to wave propagation. See case 1. Refs [1, 2], p. 110; [37], p. 173, [38]. $p_{rms}^2 = P^2/2$ for sine amplitude P
6. Two Microphone Measurement of Sound Intensity in +x direction 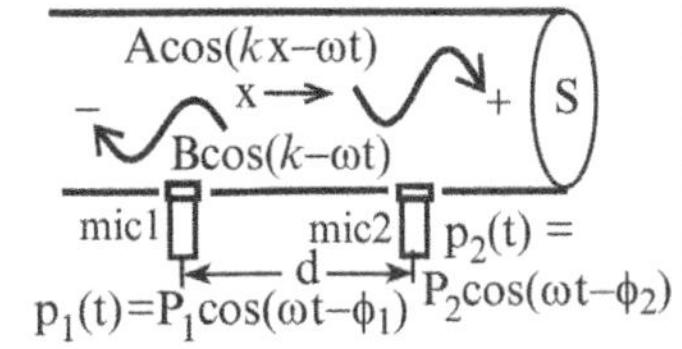	$I = \begin{cases} A^2/2\rho c - B^2/2\rho c, & \text{Refs [18, 37, 38]} \\ p_{1rms}p_{2rms}(\phi_1-\phi_2)/(2\pi\rho f d), & fd < 43 \\ \text{Im}[S_{12}(f)]/(2\pi\rho f d), & fd < 43 \\ (1/2)\text{Real}\{PU^*\}, & U^* = \text{complex conjugate velocity ampl.} \end{cases}$ +x Intensity of fwd wave A plus backward wave B, by mag., or by relative phase of measured pressures $p_1(t)$ and $p_2(t)$, or $\text{Im}[S_{12}(f)]$ = imaginary part of cross spectral density p_1 and p_2.

Table 6.4 Duct acoustics, continued

Duct Acoustical Element	Transfer Matrix	Transmission Loss, dB
7. Acoustic Circuit Forced by Piston at Point 0 with Boundary at point N+1 Transfer Matrix Solutions 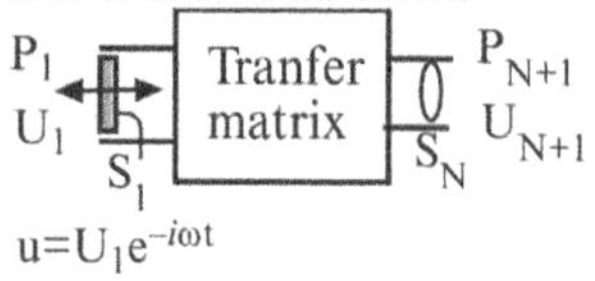 $u=U_1e^{-i\omega t}$ Transmission loss, Eq. 6.29	$\begin{Bmatrix} P_1 \\ U_1 \end{Bmatrix} = \begin{bmatrix} T_{11} & T_{12} \\ T_{21} & T_{22} \end{bmatrix} \begin{Bmatrix} P_{N+1} \\ U_{N+1} \end{Bmatrix}$ Transform matrices [T] are given in cases below. For multiple components in series the transfer matrix is $[T]=[T_1]..[T_n][T_{n+1}]..[T_N]$ For $P_{N+1}=0$ (open) then $P_1=(T_{12}/T_{22})U_1$ and $U_{N+1}=U_1/T_{22}$ For $U_{N+1}=0$ (wall) then $P_1=(T_{12}/T_{21})U_1$ and $P_{N+1}=U_1/T_{21}$ For $P_{N+1}/U_{N+1}=Z$ (imped.) then $P_1=(ZT_{11}/+T_{12})U_1/(ZT_{11}/+T_{12})$ $P_{N+1}=ZU_1$. $Z=\rho c$ For no reflection from exit.	
8. One-Dimensional Duct 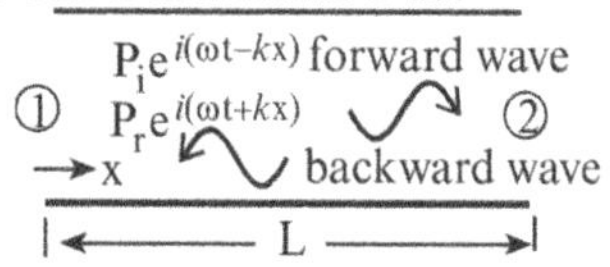	$\begin{Bmatrix} P_1 \\ U_1 \end{Bmatrix} = \begin{bmatrix} \cos kL & -i\rho c \sin kL \\ -\frac{i}{\rho c}\sin kL & \cos kL \end{bmatrix} \begin{Bmatrix} P_2 \\ U_2 \end{Bmatrix}$ $k=2\pi f/c$	TL = 0, Ref. [17], p. 77 $\alpha_S=0$ See Fig.6.4 and Case 1.
9. Duct with Mean Flow Velocity, U_o 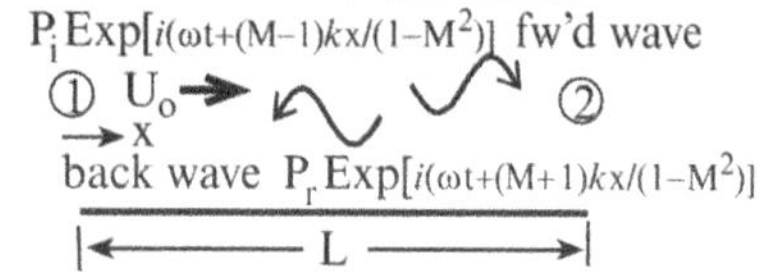	$\begin{Bmatrix} P_1 \\ U_1 \end{Bmatrix} = \text{Exp}\left[-\frac{iMkL}{1-M^2}\right] \begin{bmatrix} \cos\frac{kL}{1-M^2} & i\rho c \sin\frac{kL}{1-M^2} \\ \frac{i}{\rho c}\sin\frac{kL}{1-M^2} & \cos\frac{kL}{1-M^2} \end{bmatrix} \begin{Bmatrix} P_2 \\ U_2 \end{Bmatrix}$ Mach number, $M=U_o/c<1$ TL = 0 Ref. [17], p.123	
10. Duct with Mean Flow Velocity, U_o, and Flow Friction Factor f 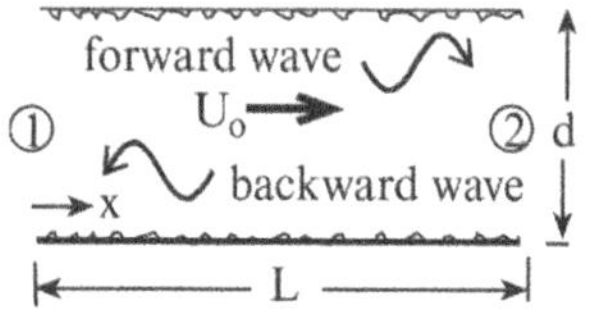 mean pressure loss $\Delta P=\frac{1}{2}\rho U_o^2 f L/d$	$\begin{Bmatrix} P_1 \\ U_1 \end{Bmatrix} = \text{Exp}\left(-\frac{iMkL}{1-M^2}\right)\text{Exp}(-M\xi) \begin{bmatrix} A & i\rho c^{*}B \\ \frac{i}{\rho c^{*}}B & A \end{bmatrix} \begin{Bmatrix} P_2 \\ U_2 \end{Bmatrix}$ $A=\cos\frac{kL}{1-M^2}+i\xi\sin\frac{kL}{1-M^2},\quad B=\xi\cos\frac{kL}{1-M^2}+i\sin\frac{kL}{1-M^2}$ $c^{*}=c[1-iM(1+M)\frac{Lf}{2d}],\quad TL=\frac{10}{\ln(10)}\frac{M}{1+M}\frac{Lf}{2d}$ $\xi=\frac{kL}{1-M^2}\frac{Uf}{2\omega d}=\frac{1}{2}\frac{M}{1-M^2}\frac{L}{d}f<<1,\ \ln(10)=2.3025$ *f = dimensionless friction factor*, $f<<1$. Refs [17], p.123; [50].	
11. Transmission Perforated Plate 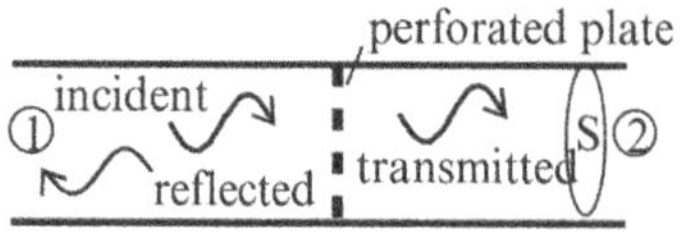	$\begin{Bmatrix} P_1 \\ U_1 \end{Bmatrix} = \begin{bmatrix} 1 & Z_s \\ 0 & 1 \end{bmatrix} \begin{Bmatrix} P_2 \\ U_2 \end{Bmatrix}$ Z_s = step impedance, Table 6.3. Also applies to orifice.	$TL=10\log_{10}\left\|1+\frac{Z_s}{2\rho c}\right\|^2$ $\alpha_s=1-\|Z_s/(2\rho c+Z_s)\|$
12. Transmission between Two Fluids 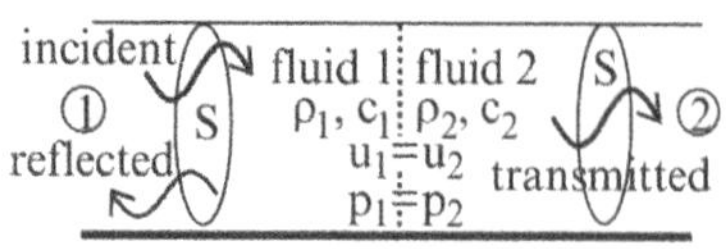 Note: transmission loss is associated with change of speed of sound.	$\begin{Bmatrix} P_1 \\ U_1 \end{Bmatrix} = \begin{bmatrix} 1 & 0 \\ 0 & 1 \end{bmatrix} \begin{Bmatrix} P_2 \\ U_2 \end{Bmatrix}, TL=20\log_{10}\frac{1}{2}\left[\left(\frac{\rho_1c_1}{\rho_2c_2}\right)^{1/2}+\left(\frac{\rho_2c_{12}}{\rho_1c_1}\right)^{1/2}\right]$ $\frac{p_t}{p_i}=\frac{2\rho_2c_2}{\rho_1c_1+\rho_2c_2}=\frac{u_t}{u_i}\frac{\rho_2c_2}{\rho_1c_1},\quad \alpha_S=\frac{2[(\rho_1c_1)^2+(\rho_2c_2)^2]}{(\rho_1c_1+\rho_2c_2)^2}$ $\frac{p_r}{p_i}=\frac{\rho_2c_2-\rho_1c_1}{\rho_1c_1+\rho\rho_2c_2}=-\frac{u_r}{u_i}$ Also see case 17 Table 6.3.	

Table 6.4 Duct acoustics, continued

Duct Acoustical Element	Transfer Matrix	Transmission Loss, dB
13. Abrupt Expansion or Contraction 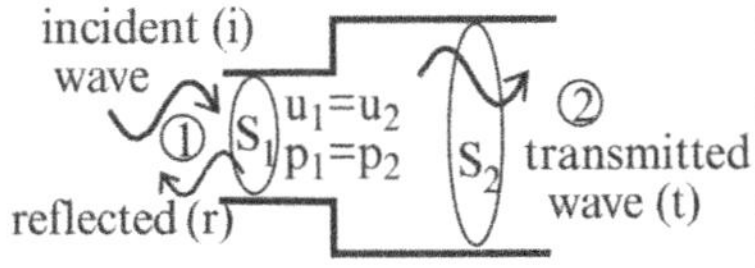	$\begin{Bmatrix} P_1 \\ U_1 \end{Bmatrix} = \begin{bmatrix} 1 & 0 \\ 0 & S_2/S_1 \end{bmatrix} \begin{Bmatrix} P_2 \\ U_2 \end{Bmatrix}$ $\frac{p_r}{p_i} = -\frac{u_r}{u_i} = \frac{1 - S_2/S_1}{1 + S_2/S_1}$ $\frac{p_t}{p_i} = \frac{u_t}{u_i} = \frac{2}{1 + S_2/S_1}$	$TL = 10\log_{10}\left[\frac{(S_1+S_2)^2}{4S_1S_2}\right]$ $\alpha_S = 4S_2/S_1[1+(S_2/S_1)^2]$ Applies to both contractions $S_1 \ge S_2$ and expansions $S_1 < S_2$.
14. Reactive Muffler 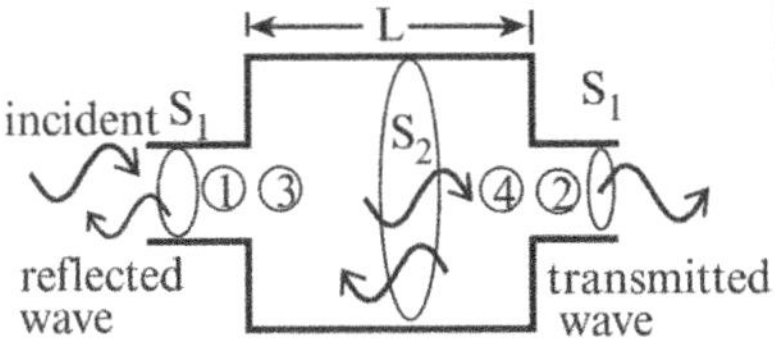 $u_1S_1 = u_3S_2$ $u_4S_2 = u_2S_1$ $p_1 = p_3$ $p_4 = p_2$ See text for unequal inlet and outlet areas.	$\begin{Bmatrix} P_1 \\ U_1 \end{Bmatrix} = \begin{bmatrix} \cos\frac{2\pi fL}{c} & i\rho c\frac{S_1}{S_2}\sin\frac{2\pi fL}{c} \\ i\frac{S_2}{S_1}\frac{\sin\frac{2\pi fL}{c}}{\rho c} & \cos\frac{2\pi fL}{c} \end{bmatrix} \begin{Bmatrix} P_2 \\ U_2 \end{Bmatrix}$, Ref. [22], p. 418 $TL = 10\log_{10}\left[1 + \frac{1}{4}(S_2/S_1 - S_1/S_2)^2 \sin^2\frac{2\pi fL}{c}\right]$ $p_i/p_t = \cos 2\pi fL/c + (i/2)(S_1/S_2 - S_2/S_1)\sin 2\omega fL/c$ $p_r/p_i = (S_1/S_2 - S_2/S_1)/(1 - 2i/\tan(2\pi fL/c))$	
15. Side Branch with Impedance Z 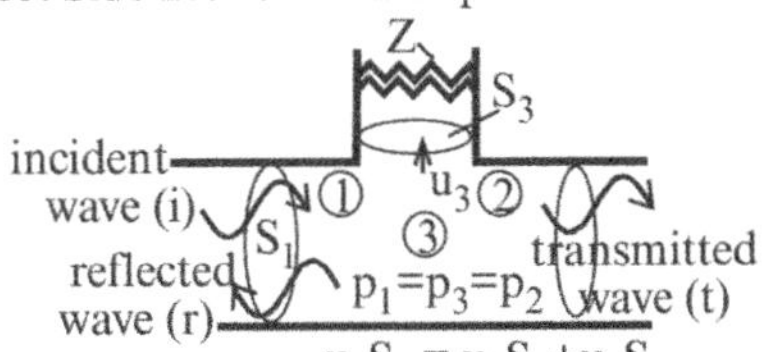 $u_1S_1 = u_3S_3 + u_2S_2$	$\begin{Bmatrix} P_1 \\ U_1 \end{Bmatrix} = \begin{bmatrix} 1 & 0 \\ \frac{S_3}{ZS_1} & 1 \end{bmatrix} \begin{Bmatrix} P_2 \\ U_2 \end{Bmatrix}$ $p_r/p_i = -S_3\rho c/(2S_1Z + S_3\rho c)$ $p_t/p_i = 2S_1Z/(2S_1Z + S_3\rho c)$ $\frac{u_r}{u_i} = -\frac{p_r}{p_i}, \frac{u_t}{u_i} = \frac{p_t}{p_i}, Z = \frac{p_3}{u_3}$	$TL = 10\log_{10}\left\|1 + \frac{S_3}{2S_1}\frac{\rho c}{Z}\right\|^2$ $\alpha_S = 1 - \left\|\frac{S_3\rho c}{2S_1Z + S_3\rho c}\right\|^2$ For Z, see Cases 17 to 22. |..| denotes magnitude
16. n Multiple Identical Side Branches 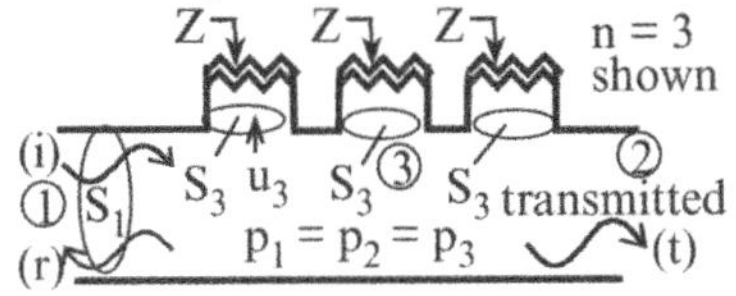	$\begin{Bmatrix} P_1 \\ U_1 \end{Bmatrix} = \begin{bmatrix} 1 & 0 \\ \frac{nS_3}{ZS_1} & 1 \end{bmatrix} \begin{Bmatrix} P_2 \\ U_2 \end{Bmatrix}$ For Z, see Cases 17 to 22, below. |..| = magnitude. n = number of side branches	$TL = 10\log_{10}\left\|1 + \frac{nS_3}{2S_1}\frac{\rho c}{Z}\right\|^2$ $Z = \frac{p_3}{u_3}$ = branch impedance Case 15 transfer matrix.
17. Side Branch to Infinity 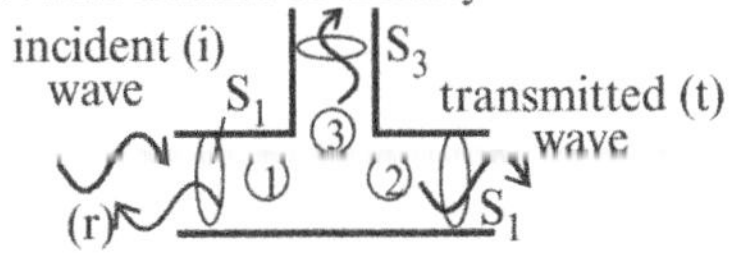	$Z = \rho c$ $u_1S_1 = u_2S_1 + u_3S_3$ $p_1 = p_2 = p_3$ case 15 has transfer matrix.	$TL = 10\log_{10}\left(1 + \frac{S_3}{2S_1}\right)^2$ $\alpha_S = 8S_1S_3/(2S_1 + S_3)^2$
18. Side Branch to Gas Bag 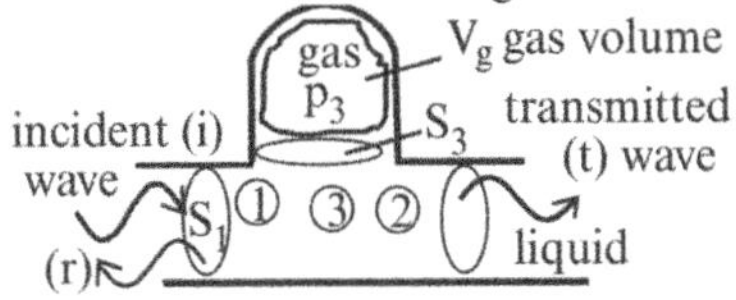	$Z = \frac{p_3}{u_3} = \frac{i\rho_g c_g^2 S_3}{2\pi f V_g}$, See Case 15 above, for transfer matrix. $u_1S_1 = u_2S_3 + u_2S_2$ c_g, ρ_g = speed of sound and density of gas in bag. $p_1 = p_2 = p_3$ Mean pressure in gas bag equals mean pressure in duct.	$TL = 10\log_{10}\left(1 + \left(\pi\frac{\rho}{\rho_g}\frac{c}{c_g^2}\frac{fV_g}{S_1}\right)^2\right)$

Table 6.4 Duct acoustics, continued

Duct Side Branch Element	Branch Impedance Z and Transmission Loss, TL(dB)
19. Helmholtz Resonator Side Branch 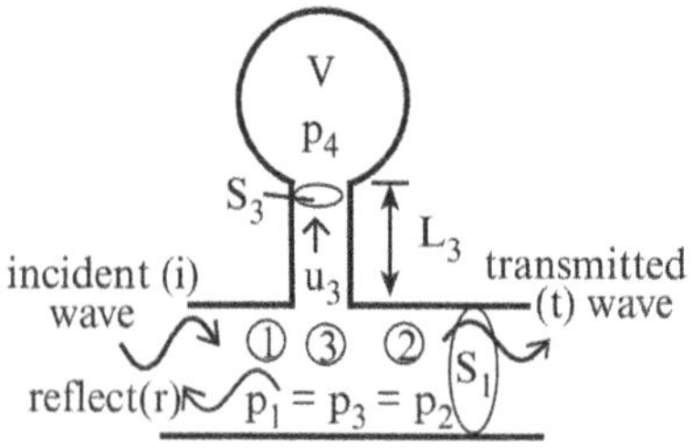 	$Z = \rho c\left(\frac{S_3 L_{3e}}{V}\right)^{1/2}\left(i\left(\frac{f}{f_n}-\frac{f_n}{f}\right)+2\zeta\right),\ f_n = \frac{c}{2\pi}\sqrt{\frac{S_3}{L_{3e}V}}$ $TL = 10\log_{10}\left[1+\left(\frac{S_3}{2S_1}\right)^2\frac{V}{S_3L_{3e}}\left(\frac{f}{f_n}-\frac{f_n}{f}\right)^{-2}\right]$, for $\zeta = 0$ $L_{3e} = L + 0.61\,(S_3)^{1/2},\ p_1/p_4 = 1/(1-f^2/f_n^2)$ ζ = dim'lss damping factor, $\zeta \approx 0.02$. Text has TL with ζ. Case 15 has transfer matrix and ratios. $u_1S_1=u_2S_3+u_2S_2$ f_n is resonator fundamental natural frequency in Hertz.
20. Side Branch Quarter Wave Resonator 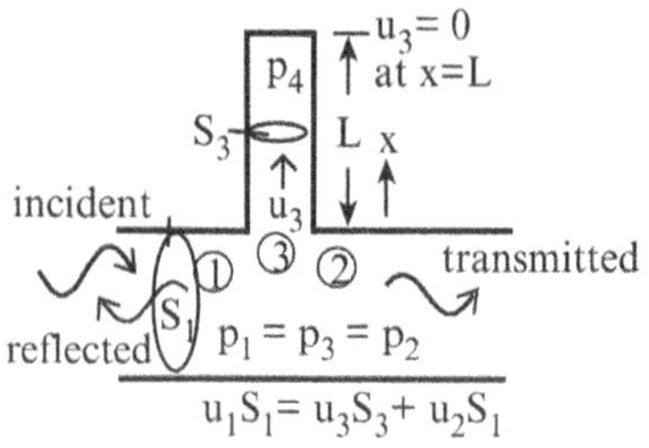 	$Z = -i\rho c/\tan(2\pi L_e/c)$ $TL = 10\log_{10}\left[1+\frac{S_3^2}{4S_1^2}\tan^2\frac{2\pi f L_e}{c}\right]$ $L_e = L + 0.61\,(S_3)^{1/2},\ p_1/p_4 = \cos 2\pi f L_e/c$ $\frac{p_i}{p_t} = 1 + i\frac{S_3}{2S_1}\tan\frac{2\pi f L_e}{c},\ \frac{p_i}{p_r} = -\frac{S_2}{2S_1}\frac{i\tan 2\pi f L_e/c}{1+i\frac{S_3}{2S_1}\tan\frac{2\pi f L_e}{c}}$ See Case 15 above for transfer matrix and ratios.
21. Prismatic Helmholtz Side Branch 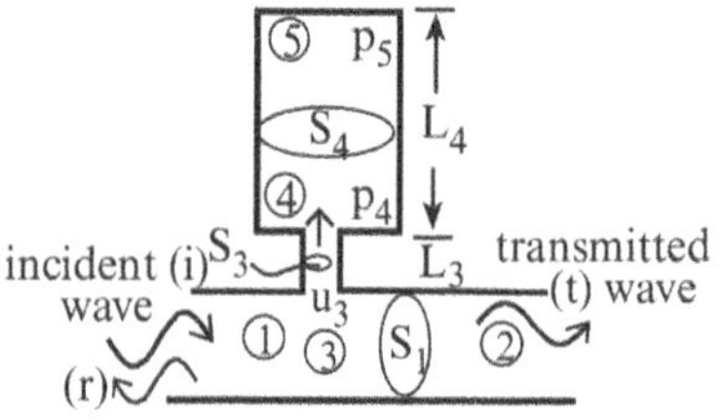 $u_1S_1 = u_2S_1 + u_3S_3,\ p_1 = p_2 = p_3$ $p_4 - p_3 = \rho L_4 S_3 du_3/dt$	$Z = i\rho c\left[\frac{L_{3e}\omega S_4}{cS_3} - \frac{1}{\tan kL_4}\right]$ $TL = 10\log_{10}\left[1+\frac{\frac{1}{4}\left(\frac{S_3^2}{S_1S_4}\frac{c}{\omega L_3}\right)^2}{\left[1-\frac{c}{\omega L_3}\frac{S_3}{S_4}\cot kL_4\right]^2}\right]$ $\frac{p_i}{p_t} = 1 + i\frac{S_4}{2S_1}\frac{1}{\cot kL_4 - (\omega L_3/c)(S_4/S_3)}$ $L_{3e} = L_3 + 0.61\,(S_3)^{1/2},\ \omega = 2\pi f,\ k = \omega/c$ Case 15 transfer matrix and ratios and αs.
22. Acoustically Lined Duct 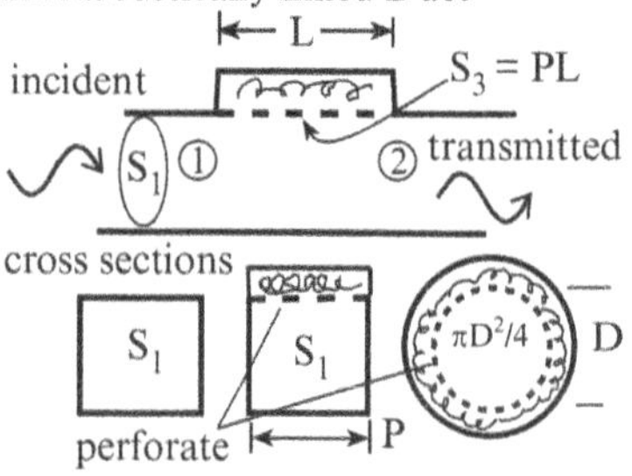 	$TL \approx 1.5\frac{S_3}{S_1}\alpha_s = 1.5\frac{PL}{S_1}\alpha_s$ $= 6\frac{L}{D}\alpha_s$, 360 degree liner in circular duct diameter D P = lined perimeter of section, S_1 = flow area of duct L = axial length of liner, S_3 = PL = acoustically lined surface. α_s = surface absorption coefficient, case 20 of Table 6.3. Approximate formulas. Case 15 has exact solution. Also see text for extrapolating liners. Ref. [17], p. 242.

Table 6.4 Duct and room acoustics, continued

Additional notation: $A_\alpha = \Sigma\alpha_i A_i$; = sum surface areas surface area that is acoustically absorptive times its surface absorption coefficient; p_{rms} = root mean square acoustic pressure averaged over room, diffuse; $S = \Sigma S_i$; = window or virtual surface area; V = room volume; W = source sound power (watts); $W_{dB} = 10 \log_{10} W/W_{ref}$; α = surface absorption coefficient, dim'lss, case 20 Table 6.3. Room acoustic solutions on this page assume a diffuse sound pressure field with wave lengths smaller than room typical dimensions.

Room Configuration	Average rms Room Pressure	Room SPL, dB
23. Room Acoustic Pressure 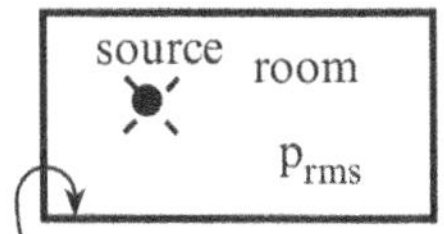 surface absorption area A_α	$p_{rms}^2 = \frac{4\rho c W}{A_\alpha}$ Sabine' law $A_\alpha = \Sigma\alpha_i A_i$ = room surface absorption area in meter^2 α_i = abs'tn coef, case 20 Table 6.3 W = sound power of source	$SPL = W_{dB} + 10\log_{10}\left[\frac{4}{A_\alpha}\right]$ $W_{dB} = 10\log_{10} W/W_{ref}$, $W_{ref} = 10^{-12}$ watts Refs [1], p. 253; [2], p316; [24, 30]
24. Room, Near Field Source 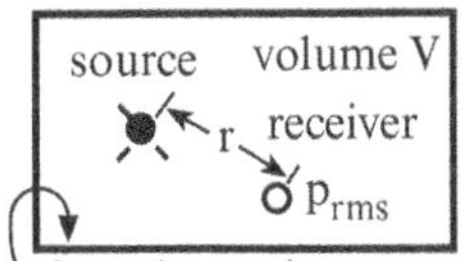 surface absorption area A_α	$p_{rms}^2 = \rho c W\left(\frac{1}{4\pi r^2} + \frac{4}{A_\alpha}\right)$ W = sound energy of source $A_\alpha = \Sigma\alpha_i A_i$ = room surface absorption area in meter^2 α_i= abs'tn coef, case 20 Table 6.3. r in meter, A_α in meter^2, V in meter^3	$SPL = W_{dB}$ $+10\log_{10}\left[\frac{1}{4\pi r^2} + \frac{4}{A_\alpha}\right]$, Also see empirical correlations in Refs [29], p. 4.20 and [30, 51]. Ref. [2], p.325
25.Transmission to Room from External Sound 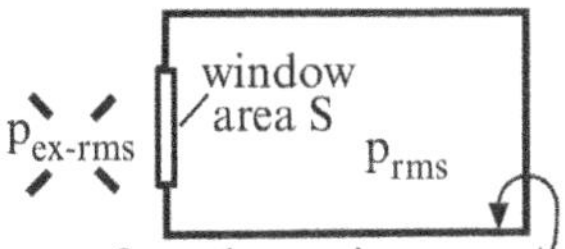 surface absorption area A_α	$p_{rms}^2 = 10^{-TL/10}\frac{S}{A_\alpha}p_{ex-rms}^2$ $p_{ex\text{-}rms}$ = external rms sound. Diffuse pressure; SPL_{ex} = Sound Pressure Level just outside of window. Transmission loss TL (dB), case 28. α_i = reflection absorption, Table 6. 3, case 20.	$SPL = SPL_{ex}$ Ref. [51] $+10\log_{10}\left[\frac{S}{A_\alpha}\right] - TL(db)$ $A_\alpha = \Sigma\alpha_i A_i$ = absorption area. S and A_α have units of area.
26. Room Reverberation Time and Absorption 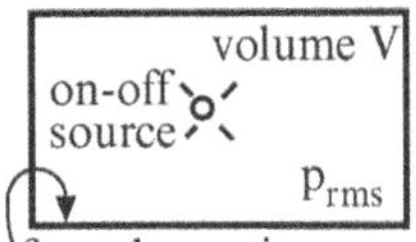 surface absorption area A_α	T_{60dB} = reverberation time for 60 dB decay in sound after sound is turned off. d = sound decay rate in dB per second. $T_{60dB} = 55.2\frac{V}{cA_\alpha}$, $A_\alpha = 55.2\frac{V}{cT_{60dB}} = 0.921\frac{Vd}{c}$ V = room volume This technique computes absorption area from measured reverberation time. Refs [2], p. 318; [52]	
27. Sound Power Measurement (moving microphone method) virtual surface box 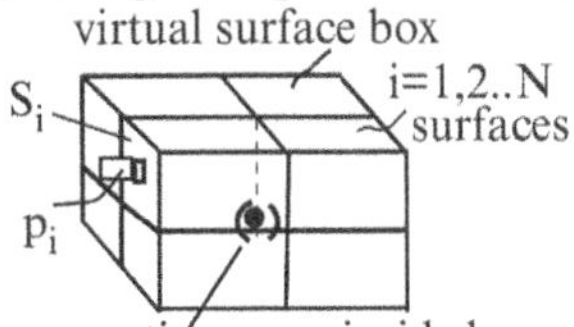 acoustic source inside box	$I_i = \frac{p_{irms}^2}{\rho c}$ for each i surface, $I_{av} = \sum_{i=1}^{N} I_i \frac{S_i}{S}$, $S = \sum_i S_i$ $W_i = I_i S_i$ $W_{av} = I_{av} S$ Applied in anechoic chamber or open air. Virtual surface S is has i=1 to N surface elements, area S_i. A sound measurement is made at each surface element. See Refs [31, 43, 53].	

Table 6.4 Duct and room acoustics, continued

Additional notation: p_{rms} = average root mean square acoustic pressure in room; TL = transmission loss between diffuse pressure field in room 1 and diffuse pressure field in room 2 with compensation for absorption in room 2 (dB).

Diffuse Transmission	rms Sound Pressures	Transmission Loss, dB
28. Sound Transmission through Panel between Two Rooms 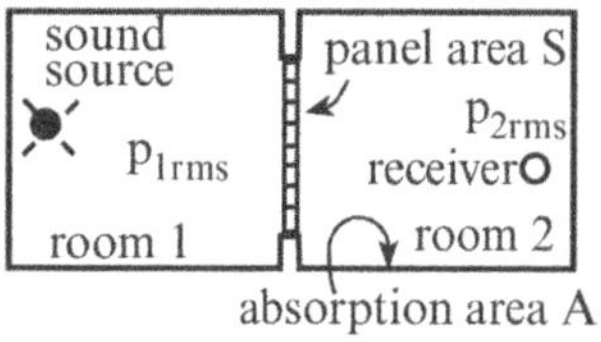	$p_{2rms}^2 = p_{1rms}^2 \frac{S}{A_2} 10^{-TL/10}$ S = area of panel $A_2 = \Sigma\alpha_i A_i$ = surface absorption area of room 2. α = surface absorption coef., case 20, Table 6.3 Diffuse sound fields.	Transmission Loss, dB = $SPL_1 - SPL_2 + 10\log_{10} S/A$ Transmission Loss between 2 rooms with compensation for absorption of wall in receiver room. Ref. [47]. See case 25.

29. Transmission Loss through Panels — Transmission Loss, dB, Ref. [29]

	Panel Description	Thickness		STC	Frequency, Hz						
		inch	mm	dB	63	125	250	500	1000	2000	4000
1	Gypsum board	0.5	10	28		15	20	25	29	32	27
2	Gypsum board, 2 layers	1	25	31		19	26	30	32	29	37
3	Gypsum boards (2@0.5 inch) nailed to 2x4 in studs 16 inch ctr	4	100	33		12	23	32	41	44	39
4	Gypsum boards (2 at.5 in) each side nailed to 2x4 in studs on 16 inch ctr. fiberglass blanket in between	4.5	112.5	52		28	37	53	65	68	71
5	Gypsum boards (2@ 0.5in each side) nailed to staggered 2x4 studs at 8inch	5	125	39		20	27	39	43	42	55
6	Gypsum board (2 at 0.5 inch), each side steel studs at 24 in fiberglass blanket	5	125	45		21	35	48	55	56	.43
7	Concrete block30 lb/ft^2,147kg/m^2	4	90	37		29	30	30	37	35	38
8	Concreteblock14 lb/ft^2,202 kg/m^2	6	140	45		30	34	41	48	56	55
9	Concrete block-solid, 62 lb/ft^2	6	140	50	38	38	37	46	55	62	69
10	Concrete block, 5/8 in gypsum board	8	190	50	36	35	35	46	55	58	62
11	Concrete block63lg/ft^2, 306 kg/m^2	12	290	49		31	40	44	51	57	61
12	Concrete panel, hollow core 45lb/f^{t2}	6	150	48		33	37	43	51	57	60
13	Concrete Panel, hollow core 57 lb/ft^2	8	200								
14	Concrete panel, solid, 54 lb/ft^2	4	100	49		48	42	45	56	57	67
15	Concrete panel, solid, 75lb/ft^2	6	150	55		44	48	55	58	63	67
16	Concrete slab, 95 lb/ft2, 460 kg/m^2	8	200	58		44	48	55	58	63	67
17	Wood Joist Floor, 5/8 inch plywoodon	0.625	15.6	24	12	10	16	22	25	24	27
	2x10 inch joints with 3-5/8 in) fiberglass										
	blanketwith carpet and 2 gypsum			33	19	13	33	35	43	47	48
	boards			42	18	24	30	38	50	60	68
18	Glass pane	1/8	3	29		21	21	28	31	34	25
19	Glass pane	1/4	6	31		25	28	31	34	30	37
20	Glass pane	1/2	13	36		30	33	36	32	40	50
21	Two ¼ in glass panes laminated	1/2	13	38		29	33	36	37	41	51
22	Two 1/8 in panes ¼ in. air gap	1/2	13	28		24	24	24	34	39	32
23	Two 1/8 in. panes ½ in. air gap	3/4	18	31		24	19	29	40	45	35
24	Two 1/4 in. panes ½ in. air gap	1	25	36		29	27	36	43	36	42
25	Door, wood solid core 4.9 lb/ft^2 no seal			22	16	19	22	26	24	23	20
26	Door, wood solid core, foam seal			26	18	22	25	29	25	26	28
27	Door, steel, hollow core, 4.9lb/ft^2	18ga.		17	12	13	15	16	17	18	20

Surface absorption coefficient (Eq. 6.15d) is the fraction of incident sound energy absorbed by a reflecting surface [29, 30]. The incident (i, or →) sound power equals the sum of the reflected (r, or ←) and absorbed (a) sound power. Since energy is conserved, amplitude of incident wave (P_i), reflected wave (P_r), and absorbed waves (P_a) are related by $P_i^2 = P_a^2 + P_r^2$. Dividing by the incident intensity gives the absorption coefficient $\alpha_S = P_a^2/P_i^2 = 1 - P_r^2/P_i^2$. $\alpha_S = 0$ for a perfect reflection.

6.2 Sound Propagation in Large Spaces

6.2.1 Acoustic Wave Propagation

Table 6.3 [1–3, 13, 17, 19–28] has solutions for acoustic wave propagation. Acoustic waves propagate outward into space from acoustic sources. The most common acoustic source models are:

Baffled piston: A one-dimensional source of oscillating velocity and pressure
Monopole or simple source: Two-dimensional or three-dimensional oscillating volume
Dipole: Two adjacent equal monopoles that oscillate out-of-phase model a vibrating surface or oscillating fluid force; a directional source of sound
Quadrapole: Two adjacent dipoles in various geometric combinations; used to model sound from turbulence

A three-dimensional monopole is a sphere with an oscillating radius. Sound radiates equally in all directions. On the surface of the sphere, the surface radial velocity equals the acoustic particle velocity (case 13 of Table 6.3). The monopoles' acoustic pressure decreases as inversely with distance r from the source, but its particle velocity terms decrease as $1/r^2$ and $1/r$. The $1/r^2$ terms dominate the *near field* ($r << \lambda$, case 13 of Table 6.3), and the $1/r$ terms dominate the *far field*, $r >> \lambda$ (case 4). Monopoles are created by pumping fluid in and out of a port smaller than a wavelength or by popping a balloon.

Sound pressure p_{rms} decreases inversely with increasing radius r from a three-dimensional source monopole (case 4 of Table 6.3):

$$p_{\text{rms}} = p_{o\text{rms}} \frac{r_o}{r} \tag{6.16}$$

Wave spreading is called *divergence*. The reference rms source pressure is $p_{o\text{rms}}$ at the reference radius $r = r_o$.

Divergence is one of the five factors that decrease sound pressure with increasing distance from the source:

1. *Divergence*, geometric spreading with distance from a compact source
2. *Attenuation* by molecular absorption α_m
3. *Reflection–absorption* as sound reflects off an absorbing surface, α_s
4. *Refraction* by speed of sound gradients
5. *Ground effects* by propagation through small obstacles such as trees and grass

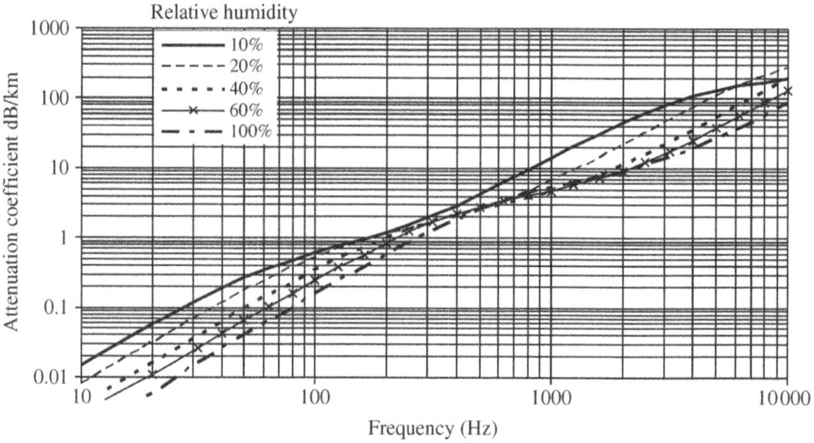

Figure 6.6 Sound absorption in atmospheric air at 1 atm and 20 °C [19]

Divergence and refraction (cases 1–4 and 8–19 of Table 6.3) do not change sound energy. Molecular absorption, surface absorption, and ground obstacles (cases 5, 6, and 7 and 20–31 of Table 6.3) absorb energy as the wave propagates.

Molecular absorption of sound in air is a function of frequency and humidity [19] as shown in Figure 6.6. SPL drops by 4.7 dB/km at 1000 Hz in 50% air humidity. Approximate formulas for absorption in air and freshwater at 20 °C (68 F) indicate that sound absorption increases with frequency [57]:

$$\alpha_m \begin{pmatrix} \text{in units of} \\ \text{dB/km} \end{pmatrix} \approx \begin{cases} 4.7 \times 10^{-6} f^2, & \text{air, 50\% humidity} \\ 2.3 \times 10^{-4} f^2, & \text{freshwater} \end{cases} \tag{6.17}$$

where f = frequency in Hz, d = depth in meters, T = temperature in degree centigrade, and km = kilometer. Absorption in seawater is a function of salinity [57]. To convert α_m in dB/km (Eq. 6.17) to1/km, divide by $20 \log_{10} e = 8.6859$.

Molecular absorption can be introduced into acoustic wave analysis by replacing the wave number with a complex speed wave number k^* (Eq. 6.13) or, equivalently, by using a complex speed of sound c^* such that $k^*/k = c/c^* = 1 + i\alpha_m/k$ [2] (p. 146) where α_m is the molecular absorption, a real number with units of 1/length. Pressure of a plane traveling wave with absorption is $p(x, t) = P_o e^{i(kx-\omega t)} e^{-\alpha m \cdot x}$ (case 4 of Table 6.3).

Ray tracing acoustic solutions are shown in cases 1–7 and 20 and 23 of Table 6.3. Sound in large spaces can be most easily modeled by *sound rays* that propagate in straight lines perpendicular to the wave front from simple monopole sources. In Figure 6.7a, the measured reference monopole source pressure is p_o at distance r_o (say, 1 m) from the source. After traveling distance r_1 from r_o, the sound ray encounters a large (greater than 1 wavelength in width) planar surface that absorbs part of the wave energy and reflects the remainder. The reflected wave then travels distance r_2 from the reflecting surface to the receiver. The mean square pressure at the receiver is the source mean square reference pressure at $r = r_o$ times

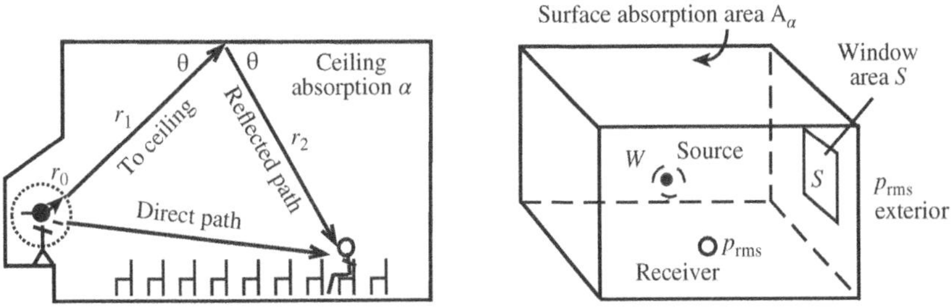

(a) Acoustic rays in auditorium (b) Room acoustics with exterior noise

Figure 6.7 Ray and room acoustics

factors for (1) divergence (cases 2 and 4 of Table 6.3), (2) absorption–reflection (α_s, case 20), and (3) atmospheric molecular absorption (α_m, case 7):

$$p_{r-\mathrm{rms}}^2 = p_{o\mathrm{rms}}^2(1-\alpha_s)\frac{r_o^2}{r^2}e^{-2\alpha_m(r-r_o)}, \text{ spherical spreading, } r > r_o \tag{6.18a}$$

$$p_{r-\mathrm{rms}}^2 = p_{o\mathrm{rms}}^2(1-\alpha_s)e^{-2\alpha_m(r-r_o)}, \text{ plane wave, no spreading} \tag{6.18b}$$

The sound field from a source can be built up with sound rays at various propagation angles.

Ray acoustic calculations are simplest when made in decibels. Dividing by square of reference pressure and taking the base 10 logarithm transform Equation 6.18 to decibels:

$$10\log_{10}\frac{p_{r-\mathrm{rms}}^2}{p_{\mathrm{ref}}^2} = 10\log_{10}\frac{p_{0\mathrm{rms}}^2}{p_{\mathrm{ref}}^2} + 10\log_{10}\left[\frac{r_o^2}{r^2}\right] + 10\log_{10}(1-\alpha_s) - 2\alpha_m(10\log_{10}e)(r-r_o) \tag{6.19}$$

Loss factors in decibels subtract from the source SPL in dB for $r > r_o$ as shown in following example.

Example 6.2 Ray acoustics of a siren

Sirens for outdoor warning systems force pressurized air through a rotating slotted disk to create an intense tone that varies with the speed of the motor [58]. A large warning siren makes 512 Hz and 124 dB tone 100 ft (30 m) from the siren. What is SPL at ground level 2000 ft (610 m) from the siren?

Solution: Equation 6.19 shows that the received SPL is the source SPL less the propagation path losses of spherical divergence, air absorption, and ground attenuation in decibels:

$$\begin{matrix}\text{Receiver sound} \\ \text{SPL, dB}\end{matrix} = \begin{matrix}\text{source} \\ \text{SPL, dB}\end{matrix} - \begin{matrix}\text{divergence} \\ \text{SPL, dB}\end{matrix} - \begin{matrix}\text{air atten.} \\ \text{SPL, dB}\end{matrix} - \begin{matrix}\text{ground atten.} \\ \text{SPL, dB}\end{matrix} \tag{6.20}$$

Air absorption is about 3 dB/1000 m at 512 Hz (Figure 6.6). The divergence reduction is 20 $\log_{10}(r/r_o)$ dB, where r_o = 100 ft (30) m (case 4 of Table 6.3). Ground level attenuation by buildings and foliage is estimated at 2 dB per 1000 m (case 24 of Table 6.3).

Distance r from source (ft)	100	500	1000	1500	2000	3000	5000	10,000
Distance r from source (m)	30	152	305	457	610	915	1524	3049
Source SPL (dB)	124	124	124	124	124	124	124	124
Distance effect (diverge) (dB)	0	14.0	20.0	23.5	26.0	29.5	34.0	40.0
Air absorption (no wind) (dB)	0	0.4	0.8	1.3	1.7	2.7	4.5	9.1
Ground absorption (dB)	0	2.4	5.5	8.5	11.6	17.7	29.9	60.4
Total attenuation (dB)	0	16.8	26.3	33.3	39.3	49.9	68.3	109.4
Remaining SPL level (dB)	124	107.2	97.7	90.7	84.7	74.1	55.7	14.6

To be clearly heard outdoors, the siren sound must rise about typical urban background levels of 50–75 dB (Chapter 8); this limits useful range of a single siren to 3000–5000 ft (600–1500 m). Multiple sirens are used to warn larger areas.

Refraction by speed of sound gradients is shown in cases 14 and 17 of Table 6.3. Sound propagating into speed of sound gradients turns toward the slower speed of sound. This is called *refraction*. It is described by *Snell's law of refraction*:

$$\frac{\sin\theta}{c} = \text{constant} \tag{6.21}$$

where c is the speed of sound in the fluid and θ is the acute angle between the speed of sound gradient and the direction of propagation; it is also the angle of the reflected wave. A vertical (z) speed of sound gradient dc/dz due to temperature turns a longitudinally propagating sound vertically about the following radius [1] (p. 387):

$$R = \frac{c_o}{(dc/dz)} \tag{6.22}$$

The speed of sound relative to the ground is the sum wind speed U and speed of sound c. A positive wind vertical gradient, $dU/dz > 0$, turns sound propagating with the wind downward toward lower $U + c$ which increases ground level sound. Sound propagating along the ground into the wind turns upward toward the lower values of $c - U$, and this results in a zone of silence on the upwind ground. See case 19 of Table 6.3.

6.2.2 *Sound Pressure on Rigid Walls*

Generally (case 24 of Table 6.3), the wall pressure produced by a normal approaching wave is a function of the wall surface impedance Z. The acoustic pressure on a rigid wall, $Z = \infty$, by a the normally impinging sound wave is twice the incident wave sound pressure owing to wave reflection.

6.2.3 *Mass Law for Sound Transmission*

Sound impinging on a flexible plate causes the plate to vibrate and retransmit sound from their back surface. The Transmission loss (TL) is given in case 26 of Table 6.3. The TL of an approaching acoustic wave through limp (low stiffness, low natural frequency) massive plate increases with plate mass:

$$\mathrm{TL, dB} = 10\log_{10}\left[1+\left(\frac{m\omega}{2\rho c}\right)^2\right] \approx 20\log_{10}\frac{m\omega}{2\rho c}, \quad \text{for } \frac{m\omega}{2\rho c} = \frac{\pi m}{2\rho\lambda} >> 1$$

Doubling the mass of a limp massive plate produces 6 dB increase attenuation in the sound transmission through the plate. This is called the *mass law*. Specifically, the attenuation across the massive plate is high if the sound frequency is higher than the plate natural frequency (Chapter 5) and if the ratio of the plate mass per unit area m is much greater than the fluid density times the wavelength of the sound wave $\lambda = c/f = 2\pi c/\omega$. For example, the TL of 100 Hz sound in air through a 3 mm (0.12 in.) thick steel plate [$\lambda = c/f = 3.4$ m, $m\omega/2\rho c = 8000\ \mathrm{kg/m^3} \times 0.003\ \mathrm{m} \times 2 \times \pi \times 100/(2 \times 1.2\ \mathrm{kg/m^3} \times 343\ \mathrm{m/s}) = 18.3$], is 25 dB.

However, if the frequency of the impinging acoustic wave is at the natural frequency of a thin flexible plate, such as a pane of glass, then there can be substantial retransmission of sound by the vibrating plate (also see [13], p. 283; [59, 60]).

6.3 Acoustic Waves in Ducts and Rooms

Whereas the previous section and Table 6.3 apply to wave propagation in free space to infinity, this section and Table 6.4 apply sound waves that are confined to propagate in ducts and finite volumes.

6.3.1 *Acoustic Waves in Ducts*

6.3.1.1 Duct Acoustics

Cases 1–22 of Table 6.4 [1–3, 17, 18, 22, 24, 29–31, 37, 38, 50–53] show formulas for propagation of low-frequency long wavelength (wavelength much greater than duct diameter) sound in ducts. The acoustic pressure and velocity in a duct with circular frequency ω are the sum of forward-traveling and backward-traveling waves (Figure 6.8a):

$$p(x,t) = Ae^{i(kx-\omega t)} + Be^{i(-kx-\omega t)}, \quad u(x,t) = \left(\frac{A}{\rho c}\right)e^{i(kx-\omega t)} - \left(\frac{B}{\rho c}\right)e^{i(-kx-\omega t)} \qquad (6.23)$$

where A and B are the amplitudes of the forward-traveling wave and aft-traveling long wavelength waves. The pressure amplitude $P(x)$ and velocity $U(x)$ amplitude are

$$\begin{aligned} p(x,t) &= [(A+B)\cos kx + i(A-B)\sin kx]e^{-i\omega t} = P(x)e^{-i\omega t} \\ \rho c u(x,t) &= \rho c[(A-B)\cos kx + i(A+B)\sin kx]e^{-i\omega t} = \rho c U(x)e^{-i\omega t} \end{aligned} \qquad (6.24)$$

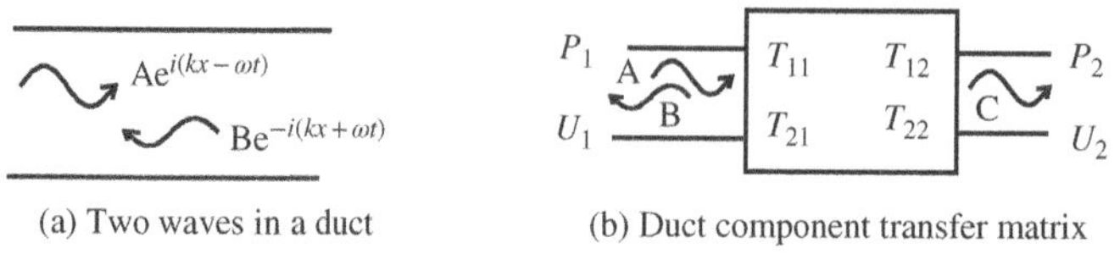

Figure 6.8 Waves in ducts and transfer matrix model for a duct acoustical component

where ρ = fluid density, c = speed of sound, and $k = \omega/c = 1/\lambda$ is wave number. x is the distance along the duct axis. Complex notation simplifies steady-state acoustic analysis. $i = (-1)^{1/2}$ is the *imaginary constant*. The unit-magnitude oscillatory term $\exp(-i\omega t) = e^{-i\omega t} = \cos\ \omega t - i\ \sin\omega t$ contains in-phase (real) and out-of-phase (imaginary) components relative to the source.

Wave amplitudes A and B in Equation 6.24 can be fixed with boundary conditions. For example, setting the piston velocity amplitude to $U_o = U(0)$ at $x = 0$ and the closed end velocity at $x = L$ to zero $U(L) = 0$, in Equation 6.24, gives two equations that are solved for A and B to produce the solution in case 2 of Table 6.4:

$$p(x,t) = i\,\rho c U_o \frac{\cos k(L-x)}{\sin kL} e^{-i\omega t} \tag{6.25}$$

At $\sin kL = 0$, the frequencies are the duct natural frequencies, $\omega L/c = \pi/2$, $3\pi/2$, and $5\pi/2$ (case 3 of Table 6.5 for open–closed duct), and the pressure is predicted to become large as shown in Figure 6.9. This is called a *duct resonance*. Duct resonance pressure is limited by damping.

Duct component transfer matrix expresses the relationships between pressure amplitude P_1 and velocity amplitude U_1 at a point x_1 and pressure amplitude P_2 and velocity amplitude U_2 at x_2. The transfer matrix for a straight duct (Eq. 6.24) with $x_2 - x_1 = L$ is

$$\begin{pmatrix} P_1 \\ U_1 \end{pmatrix} = \begin{bmatrix} \cos kL & -i\rho c \sin kL \\ -(i/\rho c)\sin kL & \cos kL \end{bmatrix} \begin{pmatrix} P_2 \\ U_2 \end{pmatrix} \tag{6.26}$$

The transfer matrix between the entrance and exit of a component has four entries called *pole parameters*, T_{11}, T_{12}, T_{21}, and T_{22} (Figure 6.8b):

$$\begin{pmatrix} P_1 \\ U_1 \end{pmatrix} = [T] \begin{pmatrix} P_2 \\ U_2 \end{pmatrix} \quad \text{where } [T] = \begin{bmatrix} T_{11} & T_{12} \\ T_{21} & T_{22} \end{bmatrix}, \text{ or inversely,}$$

$$\begin{pmatrix} P_2 \\ U_2 \end{pmatrix} = [T]^{-1} \begin{pmatrix} P_1 \\ U_1 \end{pmatrix} \quad \text{where } [T]^{-1} = \frac{1}{T_{11}T_{22} - T_{12}T_{21}} \begin{bmatrix} T_{22} & -T_{12} \\ -T_{21} & T_{11} \end{bmatrix} \tag{6.27}$$

T_{11} and T_{22} are dimensionless. T_{12} and $1/T_{21}$ have the dimensions of specific impedance (Eq. 6.10).

Table 6.5 Acoustic natural frequencies

Notation: A = area; a = radius; c = speed of sound (Table 6.3); D = diameter; L = length; L_x, L_y, L_z = lengths along x, y, and z coordinates, respectively; $J_j()$ = Bessel function of 1st kind and jth order; p = pressure; R, r = radius; $Y_i()$ = Bessel function of second kind and ith order; x, y, z = orthogonal coordinates; u = acoustic particle velocity; S = area of duct; α, θ = angles, radians; λ = nat. freq. parameter. Overbar (~) denotes shape function. See Table 1.2 for consistent sets of units. Refs [1, 60–66].

Acoustic Fluid Volume	Natural Freq. f_i, Hz	Mode Shape, Potential $\tilde{\Phi}$
1. Slender Duct with Both Ends Open 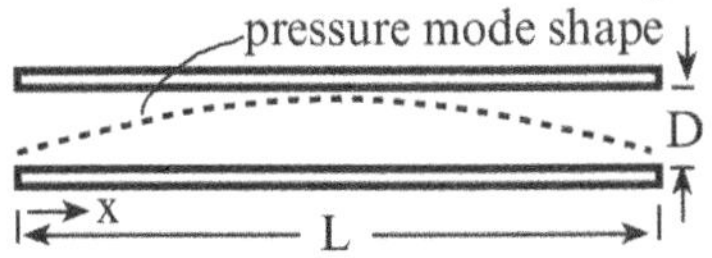	$\frac{ic}{2L}$, i = 1,2,3,.. L>>D. See case 18 for end corrections.	$\tilde{\Phi}_i = \sin(i\pi x/L)$ $\tilde{p} = \tilde{\Phi}_i = \sin(i\pi x/L)$ $\tilde{u} = (P_o/\rho c)\cos(i\pi x/L)$
2. Slender Duct Closed-Open Ends 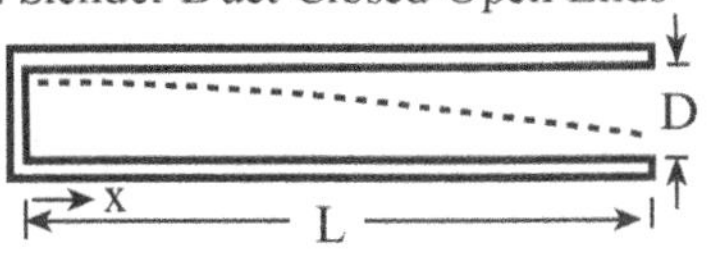	$\frac{ic}{4L}$, i = 1,3,5,.. (odd) L>>D, case 18 has end corrections. Quarter wave resonator.	$\tilde{\Phi}_i = \cos(i\pi x/2L)$ $\tilde{p} = P_o\cos(i\pi x/2L)$ $\tilde{u} = -(P_o/\rho c)\sin(i\pi x/2L)$
3. Slender Duct with Both Ends Closed 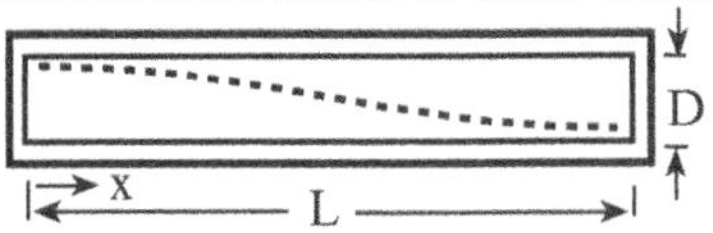	$\frac{ic}{2L}$, i = 1,2,3,.. L>>D	$\tilde{\Phi}_i = \cos(i\pi x/L)$ $\tilde{p} = P_o\cos(i\pi x/L)$ $\tilde{u} = -(P_o/\rho c)\sin(i\pi x/L)$
4. Rectangular Volume with Open Ends 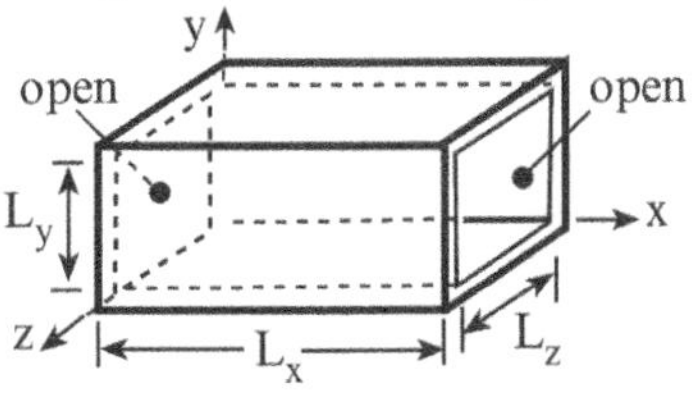	$\frac{c}{2}\left(\frac{i^2}{L_x^2}+\frac{j^2}{L_y^2}+\frac{k^2}{L_z^2}\right)^{1/2}$ i = 0,1,2,3.. j = 0,1,2,3.. k = 0,1,2,3..	$\tilde{\Phi}_{ijk} = \sin\frac{i\pi x}{L_x}\cos\frac{j\pi y}{L_y}\cos\frac{k\pi z}{L_z}$
5. Rectangular Volume One End Open 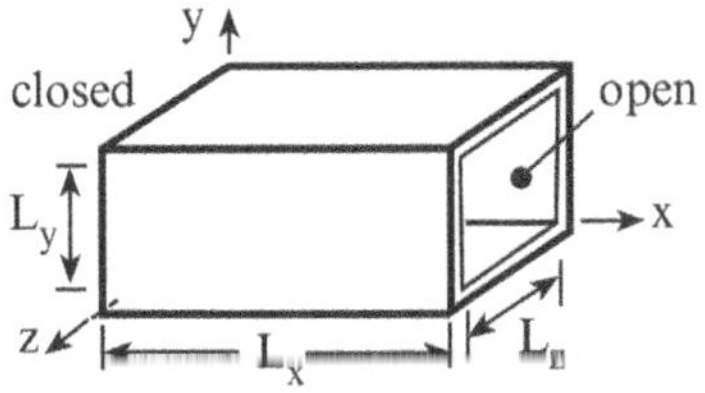	$\frac{c}{2}\left(\frac{i^2}{4L_x^2}+\frac{j^2}{L_y^2}+\frac{k^2}{L_z^2}\right)^{1/2}$ i = 0,1,3,5.. (odd) j = 0,1,2,3.. k = 0,1,2,3..	$\tilde{\Phi}_{ijk} = \cos\frac{i\pi x}{2L_x}\cos\frac{j\pi y}{L_y}\cos\frac{k\pi z}{L_z}$
6. Closed Rectangular Volume 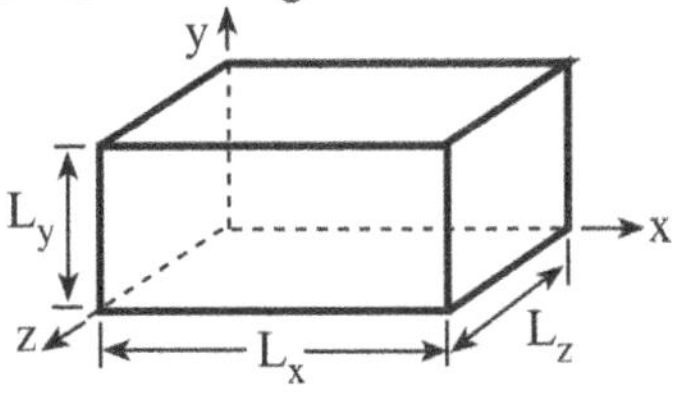	$\frac{c}{2}\left(\frac{i^2}{L_x^2}+\frac{j^2}{L_y^2}+\frac{k^2}{L_z^2}\right)^{1/2}$ i = 0,1,2,3.. j = 0,1,2,3.. k = 0,1,2,3..	$\tilde{\Phi}_{ijk} = \cos\frac{i\pi x}{L_x}\cos\frac{j\pi y}{L_y}\cos\frac{k\pi z}{L_z}$ Ref. [32], p. 70; Ref. [61], pp. 258–259

Table 6.5 Acoustic natural frequencies, continued

Acoustic Fluid Volume	Natural Freq. f_i, Hz	Velocity Potential $\tilde{\Phi}$
7. Right Triangular Wedge 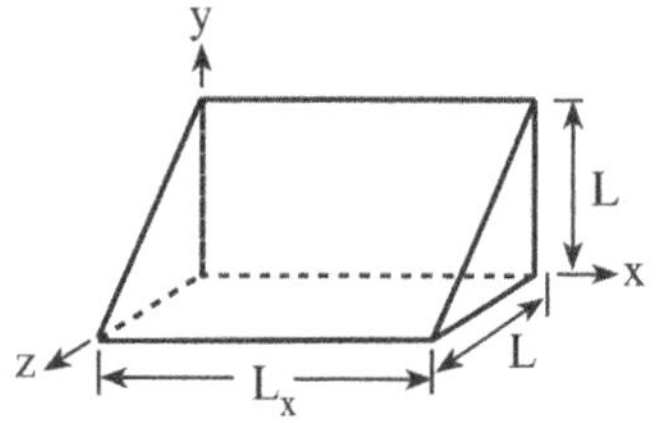	7a Ends at x=0, L_x open $\frac{c}{2}\left(\frac{i^2}{L_x^2}+\frac{j^2}{L^2}+\frac{k^2}{L^2}\right)^{1/2}$ i = 0,1,2,3..	$\tilde{\Phi}_{ijk} = \sin\frac{i\pi x}{L_x}$ $\left(\cos\frac{j\pi y}{L}\cos\frac{k\pi z}{L} \pm \cos\frac{j\pi z}{L}\cos\frac{k\pi y}{L}\right)$
j,k = (0,0), (1,2),(2,1)(3,1)(1,3), (2,3),,j ≠ k unless j = k =0	7b x=0 Closed, L_x Open $\frac{c}{2}\left(\frac{i^2}{4L_x^2}+\frac{j^2}{L^2}+\frac{k^2}{L^2}\right)^{1/2}$ i = 0,1,3,5.. (odd)	$\tilde{\Phi}_{ijk} = \cos\frac{i\pi x}{2L_x}$ $\left(\cos\frac{j\pi y}{L}\cos\frac{k\pi z}{L} \pm \cos\frac{j\pi z}{L}\cos\frac{k\pi y}{L}\right)$
Velocity potential formulas use + if j + k = even – if j – k = odd Lateral sides are closed	7c Ends x=0,L_x Closed $\frac{c}{2}\left(\frac{i^2}{L_x^2}+\frac{j^2}{L^2}+\frac{k^2}{L^2}\right)^{1/2}$ i = 0,1,2,3..	$\tilde{\Phi}_{ijk} = \cos\frac{i\pi x}{L_x}$ $\left(\cos\frac{j\pi y}{L}\cos\frac{k\pi z}{L} \pm \cos\frac{j\pi z}{L}\cos\frac{k\pi y}{L}\right)$
8. Open Cylindrical Volume 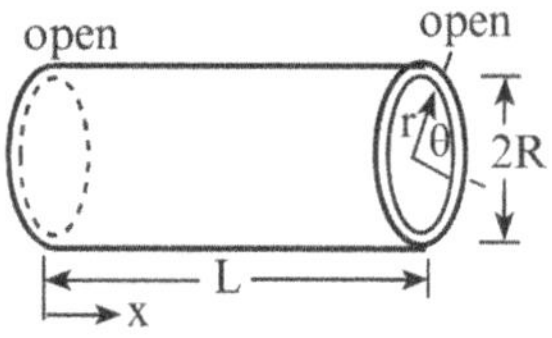	$\frac{c}{2\pi}\left(\frac{\lambda_{jk}^2}{R^2}+\frac{i^2\pi^2}{L^2}\right)^{1/2}$ i = 1,2,3.. j = 0,1,2,3.. k = 0,1,2,3.. λ_{jk} given in case 10b.	$\tilde{\Phi}_{ijk} = J_j(\lambda_{jk}\frac{r}{R})\sin\frac{i\pi x}{L}\left\{\begin{matrix}\sin j\theta\\ \text{or}\\ \cos j\theta\end{matrix}\right\}$ i = number of axial nodes j = number of nodal diameters k = number of nodal circles λ_{01} = 1.842 is fund. transverse mode.
9. Open-Closed Cylinder Volume 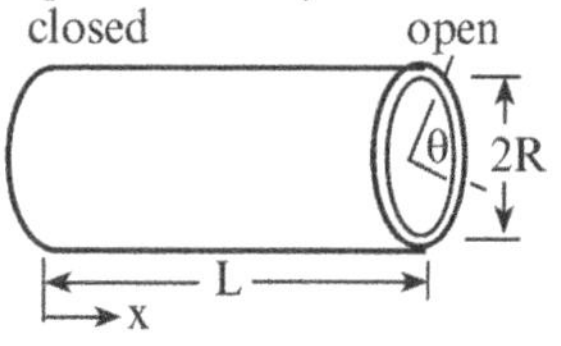	$\frac{c}{2\pi}\left(\frac{\lambda_{jk}^2}{R^2}+\frac{i^2\pi^2}{4L^2}\right)^{1/2}$ i = 1,3,5.. (odd) j = 0,1,2,3.. k = 0,1,2,3.. λ_{jk} given n case 10b.	$\tilde{\Phi}_{ijk} = J_j(\lambda_{jk}\frac{r}{R})\cos\frac{i\pi x}{2L}\left\{\begin{matrix}\sin j\theta\\ \text{or}\\ \cos j\theta\end{matrix}\right\}$ i = number of axial nodes j = number of nodal diameters k = number of nodal circles λ_{01} = 1.842 is fund. transverse mode.
10. Closed Cylinder Volume 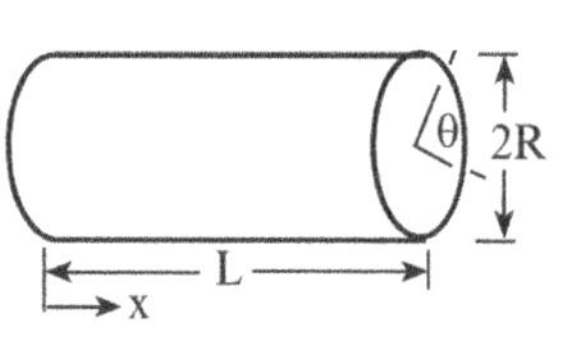	$\frac{c}{2\pi}\left(\frac{\lambda_{jk}^2}{R^2}+\frac{i^2\pi^2}{L^2}\right)^{1/2}$ i = 0,1,2,3,.. j = 0,1,2,3.. k = 0,1,2,3.. λ_{jk} given in case 10 b	$\tilde{\Phi}_{ijk} = J_j(\lambda_{jk}\frac{r}{R})\cos\frac{i\pi x}{L}\left\{\begin{matrix}\sin j\theta\\ \text{or}\\ \cos j\theta\end{matrix}\right\}$ i = number of axial nodes j = number of nodal diameters k = number of nodal circles λ_{01} = 1.842 is fund. transverse mode.

10b Transcendental Equation for Circular Cylinders

$J_i'(\lambda_{ik}) = 0$

Roots are tabulated at right.

$\lambda_{j=0,k} \approx \pi(k+1/4)$ for $k \geq 3$

Refs [32], pp. 300–301; [61], p. 263–262, p. 411,468

k	$\lambda_{jk,}$				j = nodal diameters		
	0	1	2	3	4	5	6
0	0	1.8412	3.0542	4.2012	5.3176	6.4156	7.5013
1	3.8317	5.3314	6.7061	8.0152	9.2824	10.5199	11.735
2	7.0156	8.5363	9.9695	11.3459	12.6819	13.9872	15.268
3	10.173	11.706	13.1704	14.5859	15.9641	17.3128	18.637

Table 6.5 Acoustic natural frequencies, continued

Acoustic Fluid Volume	Natural Freq. f_i Hz	Velocity Potential $\tilde{\Phi}$
11. Segment of Right Circular Cylinder or Annulus 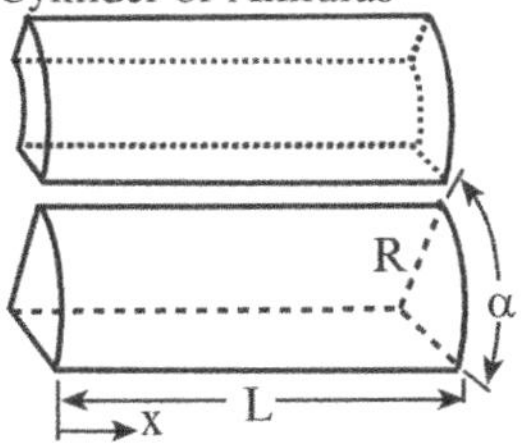	Natural frequencies given in cases 8 to 10, or 12 to 14 if annulus, with the following choices of index j. j=180 n/α, n=0,1,2.. α in degrees. Index j integer if α is sub multiple of 180 n deg.	Mode shapes can be adapted from those of cases 8 through 10, or 12 through 14 . For non integer j, the shapes will involve Bessel functions of fractional order. Ref. [62], pp. 299-300. Note α=360 deg gives circle with a baffle.
12. Annular Volume Open at Ends 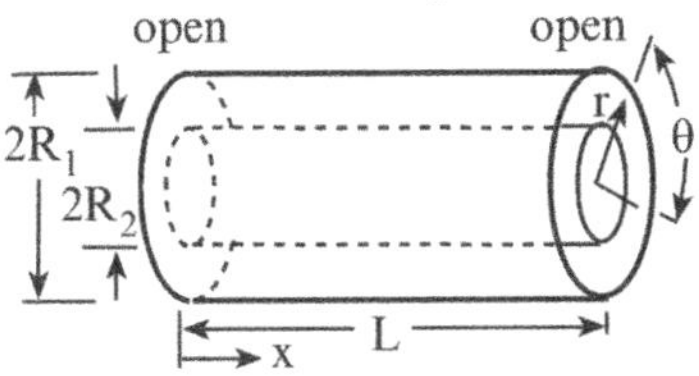	$\frac{c}{2\pi}\left(\frac{\lambda_{jk}^2}{R_1^2}+\frac{i^2\pi^2}{L^2}\right)^{1/2}$ i = 1,2,3.. j = 0,1,2,3.. k = 0,1,2,3.. λ_{jk} given in case 14b.	$\tilde{\Phi}_{ijk}=G_{jk}(\lambda_{jk}\frac{r}{R_1})\sin\frac{i\pi x}{L}\begin{Bmatrix}\sin j\theta\\ \text{or}\\ \cos j\theta\end{Bmatrix}$ j i = number of nodal diameters J = number of nodal diameters k = number of nodal circles G_{jk} from case 14b
13. Annular Volume Open One End 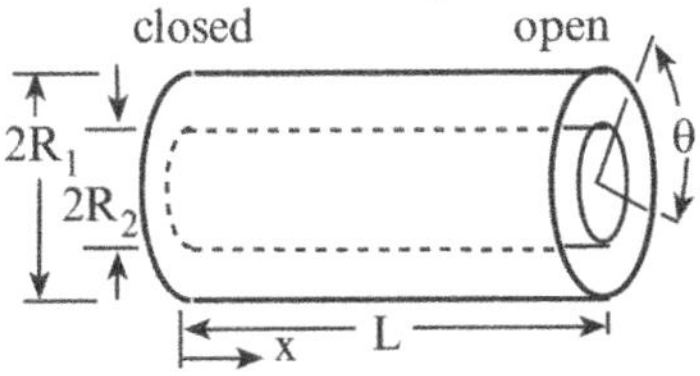	$\frac{c}{2\pi}\left(\frac{\lambda_{jk}^2}{R_1^2}+\frac{i^2\pi^2}{4L^2}\right)^{1/2}$ i = 1,3,5.. (odd) j = 0,1,2,3.. k = 0,1,2,3.. λ_{jk} given in case 14b	$\tilde{\Phi}_{ijk}=G_{jk}(\lambda_{jk}\frac{r}{R})\cos\frac{i\pi x}{2L}\begin{Bmatrix}\sin j\theta\\ \text{or}\\ \cos j\theta\end{Bmatrix}$ i i = number of axial nodes j = number of nodal diameters k = number of nodal circles G_{ik} from case 14b.
14. Closed Annular Volume 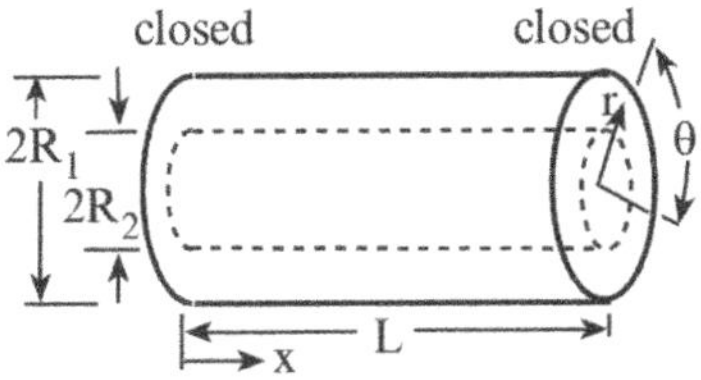	$\frac{c}{2\pi}\left(\frac{\lambda_{jk}^2}{R_1^2}+\frac{i^2\pi^2}{L^2}\right)^{1/2}$ i = 0,1,2,3.. j = 0,1,2,3.. k = 0,1,2,3.. λ_{jk} given in case 14b	$\tilde{\Phi}_{ijk}=G_{jk}(\lambda_{jk}\frac{r}{R_1})\cos\frac{i\pi x}{L}\begin{Bmatrix}\sin j\theta\\ \text{or}\\ \cos j\theta\end{Bmatrix}$ i = number of axial nodes j = number of nodal diameters k = number of nodal circles G_{ik} from case 14b.

14b Transcendental Equation for Parameter λ for Annulus

$$J_j'(\lambda_{jk})Y_j'(\lambda_{jk}\frac{R_2}{R_1}) - J_j'(\lambda_{jk}\frac{R_2}{R_1})Y_j'(\lambda_{jk}) = 0$$

Roots are tabulated at right.
Refs [62], p. 374; [63, 64]
For zero pressure on inner radius,

$$J_j'(\lambda_{jk}\frac{R_2}{R_1})Y_j(\lambda_{jk}) - J_j(\lambda_{jk})Y_j'(\lambda_{jk}\frac{R_2}{R_1}) = 0$$

R_2/R_1	k	λ_{jk} j = modal diameters					
		0	1	2	3	4	5
0.3	0	0	1.5821	2.9685	4.1801	5.3130	6.4147
	1	4.7058	5.1374	6.2738	7.7213	9.1526	10.475
0.5	0	0	1.3547	2.6812	3.9577	5.1752	6.3389
	1	6.3932	6.5649	7.0626	7.8401	8.8364	9.9858

$$\lambda_{jk}^2 \approx \left(\frac{k\pi R_1}{R_1-R_2}\right)^2+\left(\frac{2jR_1}{R_1+R_2}\right)^2, \quad \text{for } R_2/R_1 \ge 0.5$$

$$G_{jk}=Y_j'(\lambda_{jk})J_j(\lambda_{jk}r/R_1)-J_j'(\lambda_{jk})Y_j(\lambda_{jk}r/R_1)$$

$$G_{jk}\approx\cos\frac{k\pi(r-R_2)}{R_1-R_2}, \quad \text{for } R_2/R_1 \ge 0.8$$

Table 6.5 Acoustic natural frequencies, continued

<table>
<tr><th>Acoustic Fluid Volume</th><th>Natural Freq. f_i Hz</th><th>Velocity Potential $\tilde{\Phi}$</th></tr>
<tr><td rowspan="3">15. Narrow Annulus
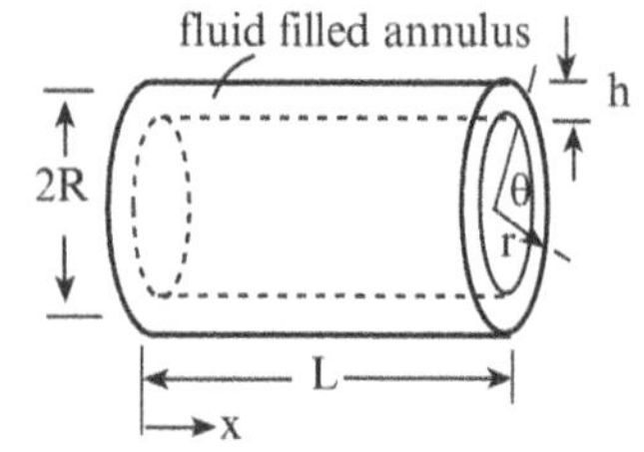
$h/R \ll R$
This result is obtained from previous case for k=0 and $h \ll R$.
$G_{jk}=1$.
R = average radius of annulus.</td>
<td>15a Ends at x=0, L Open
$\frac{c}{2\pi}\left(\frac{\pi^2 i^2}{L^2}+\frac{j^2}{R^2}\right)^{1/2}$
i = 1,2,3.., j = 1,2,3,..</td>
<td>$\tilde{\Phi}_{ijk} = \sin\frac{i\pi x}{L}\begin{Bmatrix}\sin j\theta \\ \text{or} \\ \cos j\theta\end{Bmatrix}$</td></tr>
<tr><td>15b End at x=0 Closed
End at x=L is Open
$\frac{c}{2\pi}\left(\frac{\pi^2 i^2}{4L^2}+\frac{j^2}{R^2}\right)^{1/2}$
i = 1,3,5.. (odd)</td>
<td>$\tilde{\Phi}_{ijk} = \cos\frac{i\pi x}{2L}\begin{Bmatrix}\sin j\theta \\ \text{or} \\ \cos j\theta\end{Bmatrix}$
j = 0,1,2,3..</td></tr>
<tr><td>15c Ends at x=0,L Closed
$\frac{c}{2\pi}\left(\frac{\pi^2 i^2}{L^2}+\frac{j^2}{R^2}\right)^{1/2}$
i = 0,1,2,3.., j=0, 1, 2, 3..</td>
<td>$\tilde{\Phi}_{ijk} = \cos\frac{i\pi x}{L}\begin{Bmatrix}\sin j\theta \\ \text{or} \\ \cos j\theta\end{Bmatrix}$</td></tr>
<tr><td>16. Spherical Volume
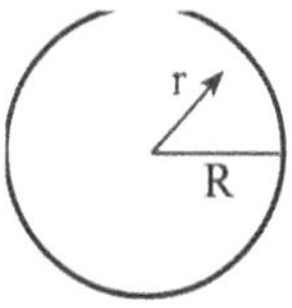</td>
<td>$\frac{c\,\lambda_i}{2\pi\,R}$ i =1, 2, 3..
Modes symmetric about r=0
<table><tr><td>i</td><td>1</td><td>2</td><td>3</td><td>4</td></tr><tr><td>λ_i</td><td>4.4934</td><td>7.725</td><td>10.904</td><td>14.</td></tr></table>tan λ=λ. $\lambda_i \approx \pi(i+1/2)$ for i>4. Transverse modes
<table><tr><td>i</td><td>1</td><td>2</td><td>3</td><td>4</td></tr><tr><td>λ_i</td><td>2.081</td><td>5.940</td><td>9.205</td><td>12.</td></tr></table></td>
<td>Modes symmetric about r=0
$\tilde{\Phi}_i = [R/(r\lambda_i)]/\sin(\lambda_i r/R)$
Transverse modes
$\tilde{\Phi}_i = \left\{\cos\frac{\lambda_i r}{R} - \frac{R}{\lambda_i r}\sin\frac{\lambda_i r}{R}\right\}\cos\theta$
See tables at left for λ.
Ref. [32], pp. 264-268
Ref. [62], pp. 259-263.</td></tr>
<tr><td>17. Elliptical Cross Section
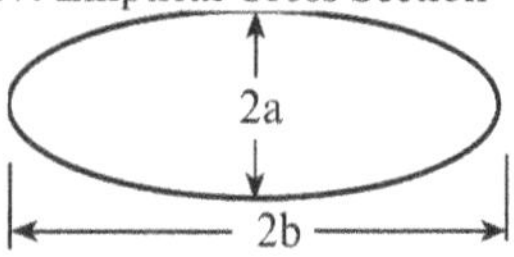</td>
<td colspan="2">$\frac{c\,\lambda}{2\pi\,b}$
<table><tr><td>a/b</td><td>1</td><td>.979</td><td>.916</td><td>.8</td><td>0.6</td></tr><tr><td>λ ↕</td><td>1.841</td><td>1.843</td><td>1.850</td><td>1.857</td><td>1.862</td></tr><tr><td>λ ↔</td><td>1.841</td><td>1.848</td><td>2.001</td><td>2.277</td><td>2.984</td></tr></table>Fundamental longitudinal and transverse in-plane modes. Ref. [64]</td></tr>
<tr><td>18. Helmholtz Resonator
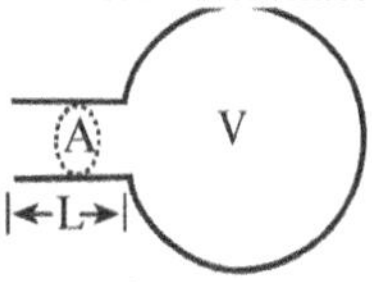
A = area of neck, V = volume,</td>
<td>$\frac{c}{2\pi}\sqrt{\frac{A}{V(L+l_c)}}$
Fundamental mode.
Dimension of volume is less than wavelength/4.</td>
<td>$L_c = l_{c1}+l_{c2}$ is sum of neck length corrections each end of the neck
For round neck shown at left
$L_c = 0.61a + 0.82a = 1.42a$, $A = \pi a^2$
For non-circular necks $a = (A/\pi)^{1/2}$.
$L_c=0$ is upper bound on frequency.</td></tr>
<tr><td colspan="2">18b. Neck Length Corrections, l_c for Helmholtz resonators
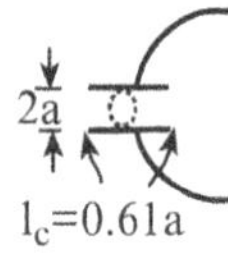
Protruding neck
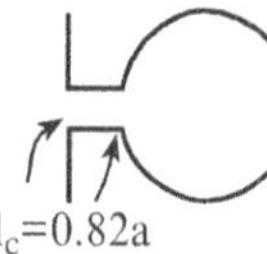
Baffled neck
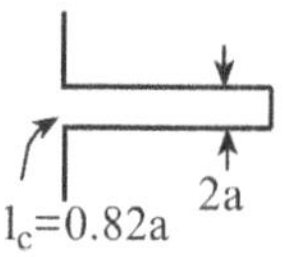
Baffled neck</td>
<td>The inertance (entrained fluid mass) of the neck ends can be increase in effective neck length. For long wave lengths this is 0.61a for a protruding neck and 0.82a for a baffled neck. Effective neck radius is a. Ref. [1], pp.349–350</td></tr>
</table>

Table 6.5 Acoustic natural frequencies, continued

Acoustic Fluid Volume	Natural Freq., f_i, Hz	Mode Shape, Remarks
19. Coupled Resonators 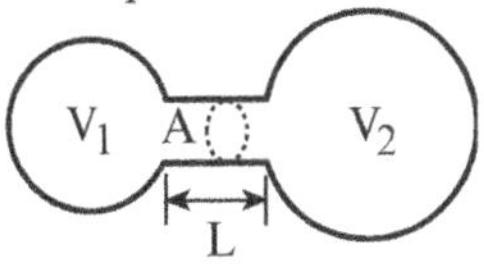	$\frac{c}{2\pi}\left(\frac{A}{L}\left[\frac{1}{V_1}+\frac{1}{V_2}\right]\right)^{1/2}$	See case 18b for end corrections.
20. Resonator with Multiple Vents 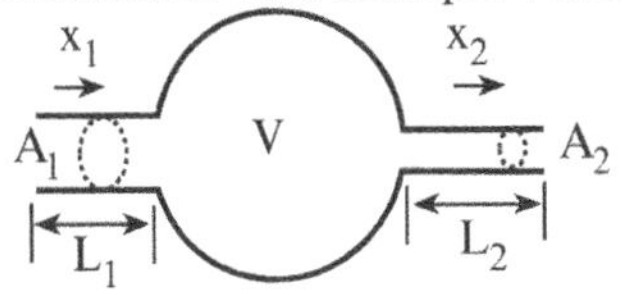	$0,\ \frac{c}{2\pi}\sqrt{\frac{A_1}{VL_1}+\frac{A_2}{VL_2}}$ See case 18b for neck end corrections.	$\begin{bmatrix}\tilde{x}_1\\ \tilde{x}_2\end{bmatrix}=\begin{bmatrix}1\\ -A_1/A_2\end{bmatrix},\begin{bmatrix}1\\ -L_1/L_2\end{bmatrix}$ For multiple necks, $f=\frac{c}{2\pi}\sqrt{\sum_i \frac{A_i}{VL_i}}$
21. Coupled Vented Resonator 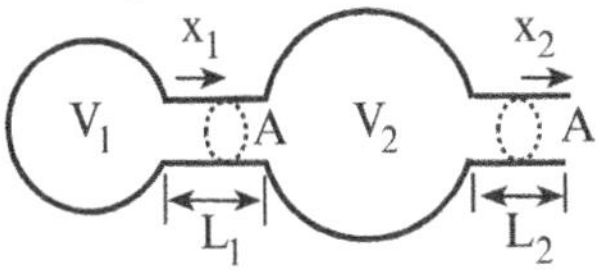	$f_i=\frac{cA^{1/2}}{2^{3/2}\pi}\left\{B\mp\left[B^2-\frac{4}{V_1L_1V_2L_2}\right]^{1/2}\right\}^{1/2}$ $\begin{bmatrix}\tilde{x}_1\\ \tilde{x}_2\end{bmatrix}_i=\begin{bmatrix}1\\ 1+\frac{V_2}{V_1}-\frac{L_1V_2}{Ac^2}(2\pi f_i)^2\end{bmatrix},\ i=1,2$	$B=1/(V_1L_1)$ $+1/(V_2L_2)$ $+1/(V_2L_1)$
22. Double Vent Coupled Resonator 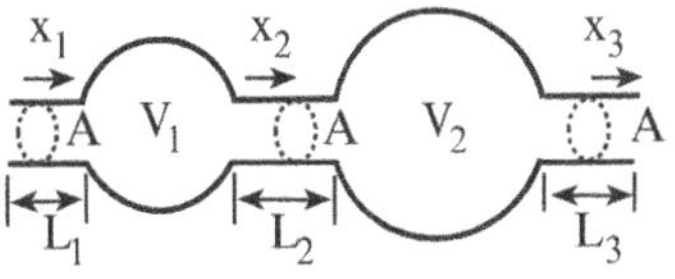	See case 15 of Table 3.2 with substitutions, $k_1=\frac{c^2A^2}{V_1},\quad k_1=\frac{c^2A^2}{V_2},$ $M_1=AL_1,\quad M_2=AL_2$	Various other resonators can be solved as spring mass systems using cases 1 through 18 of Table 3.2. See case 18b for end corrections.
23. Gas Bubble in a Liquid 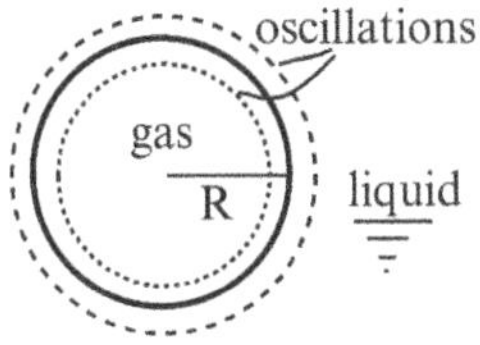For air bubbles in 20 °C water. R, mm≥10 1 0.1 0.01 $R<0.001$ κ 1.4 1.39 1.29 1.06 1.0	$\frac{1}{2\pi R}\left(\frac{1}{\rho}\left[3\kappa p-\frac{2\sigma}{R}\right]\right)^{1/2}$ for large bubbles, $\frac{1}{2\pi R}\left(\frac{3\gamma p}{\rho}\right)^{1/2}$ Fig. 6.14. Refs [65, 66]	p = mean pressure inside bubble p −2σ/R = mean pressure in liquid R = mean radius of bubble σ = surface tension, 0.0727 N/m water/air at 20 °C. (Ref. [65]) ρ = liquid density κ = polytropic gas constant. $pV^{\kappa}=C$. γ = ratio of specific heats, 1.4 for air.
24. Free Piston in a Cylinder 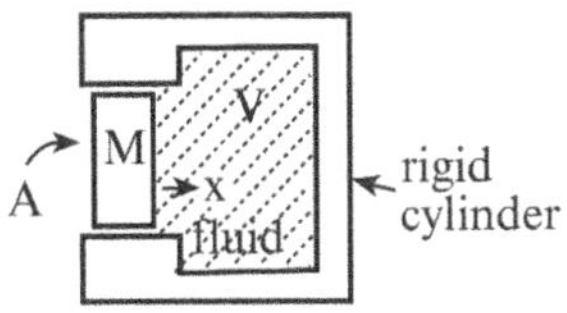	$\frac{c}{2\pi}\left(\frac{\rho A^2}{MV}\right)^{1/2}$ No heat transfer into or out of fluid during oscillation.	V = mean volume of cylinder M = mass of piston ρ = mean density of fluid A = area of piston face c = speed of sound in fluid Fluid compression is spring.

Table 6.5 Acoustic natural frequencies, continued

Natural Frequency of Slender Ducts, $f_i = \lambda c/(2\pi L)$, Hz ($L=L_1+L_2$, cases 27,28,29)

Boundary Conditions	Transcendental Eqn for λ	Mode Potential $\tilde{\Phi}$
25. Closed and Piston with spring 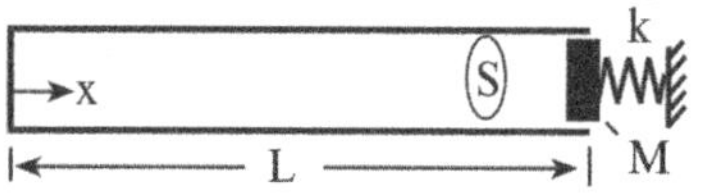	$\tan\lambda = \left(\frac{M\lambda}{\rho S L} - \frac{kL}{\rho c^2 S\lambda}\right)^{-1}$ ($\lambda_1 = \pi/2$ for k = M = 0) k = spring constant S = cross section area of duct	$\tilde{\Phi} = \sin(\lambda x/L)$ $\tilde{p} = P_o \sin(\lambda x/L)$ $\tilde{u} = (P_o/\rho c)\cos(\lambda x/L)$ k = 0 for no spring. .
26. Open and Piston with Spring 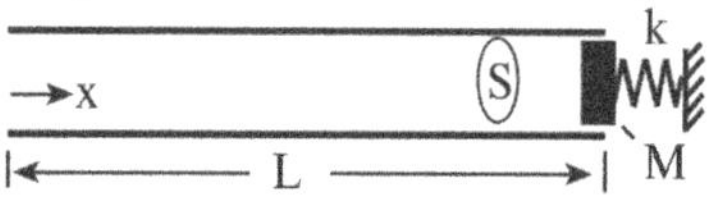	$\tan\lambda = \frac{M\lambda}{\rho S L} - \frac{k}{S}\frac{L}{\rho c^2 \lambda}$ ($\lambda_1 = \pi$ for k = M = 0) k = spring constant M = mass of piston	$\tilde{\Phi} = \cos(\lambda x/L)$ $\tilde{p} = P_o \cos(\lambda x/L)$ $\tilde{u} = (P_o/\rho c)\sin(\lambda x/L)$ M = 0 for no mass.
27. closed Two Joined Ducts 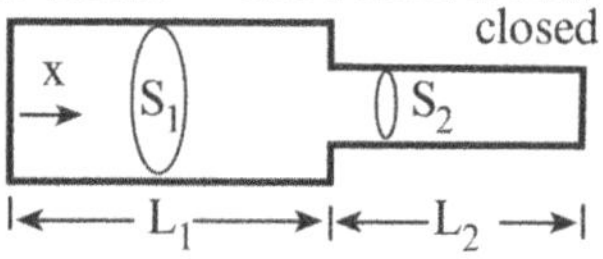	$\tilde{\Phi} = \begin{cases} \cos\lambda x/L_1, & 0 \le x < L_1 \\ \frac{\cos\lambda}{\cos(\lambda L_2/L_1)}\cos\frac{\lambda(L_1+L_2-x)}{L_1}, & L_1 \le x < L_2 \end{cases}$ Equation for λ: $\tan\lambda + (S_2/S_1)\tan(\lambda L_2/L_1) = 0$	
28. closed Two Joined Ducts 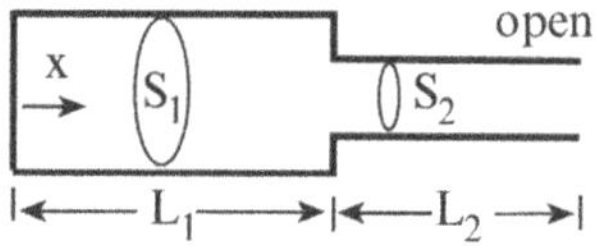	$\tilde{\Phi} = \begin{cases} \cos\lambda x/L_1, & 0 \le x < L_1 \\ \frac{\cos\lambda}{\sin(\lambda L_2/L_1)}\sin\frac{\lambda(L_1+L_2-x)}{L_1}, & L_1 \le x < L_2 \end{cases}$ Equation for λ: $\tan\lambda + \tan(\lambda L_2/L_1) = S_2/S_1$	
29. open Two Joined Ducts 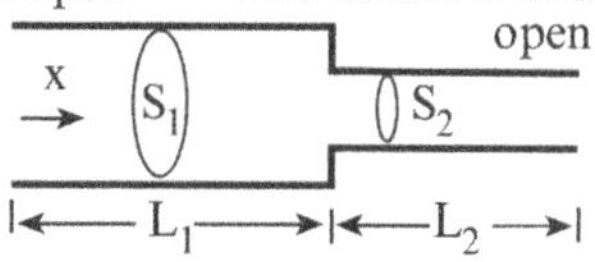	$\tilde{\Phi} = \begin{cases} \sin\lambda x/L_1, & 0 \le x < L_1 \\ \frac{\sin\lambda}{\sin(\lambda L_2/L_1)}\sin\frac{\lambda(L_1+L_2-x)}{L_1}, & L_1 \le x < L_2 \end{cases}$ Equation for λ: $\tan\lambda + (S_1/S_2)\tan(\lambda L_2/L_1) = 0$	
30. Three Joined Ducts Closed Ends 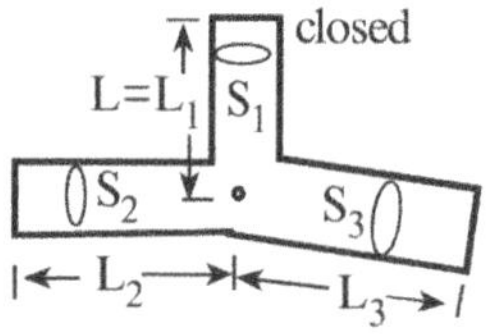	$S_1\tan\lambda + S_2\tan\left(\lambda\frac{L_2}{L_1}\right)$ $+ S_3\tan\left(\lambda\frac{L_3}{L_1}\right) = 0$	This can be extended to N closed jointed ducts by added additional trailing terms to the above equation.
31. Three Joined Ducts, Two Closed Ends and One Open End 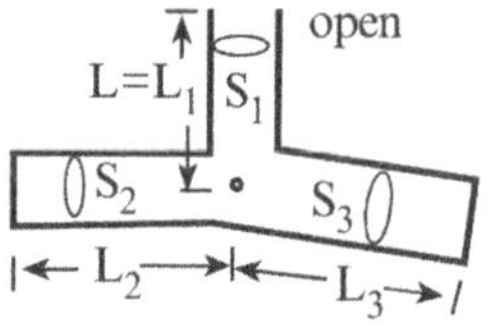	$S_1 - S_2\tan\lambda\tan\left(\lambda\frac{L_2}{L_1}\right)$ $- S_3\tan\lambda\tan\left(\lambda\frac{L_3}{L_1}\right) = 0$	This can be extended to N closed jointed ducts by added additional terms to the trailing terms to the above equation. Authors results this page

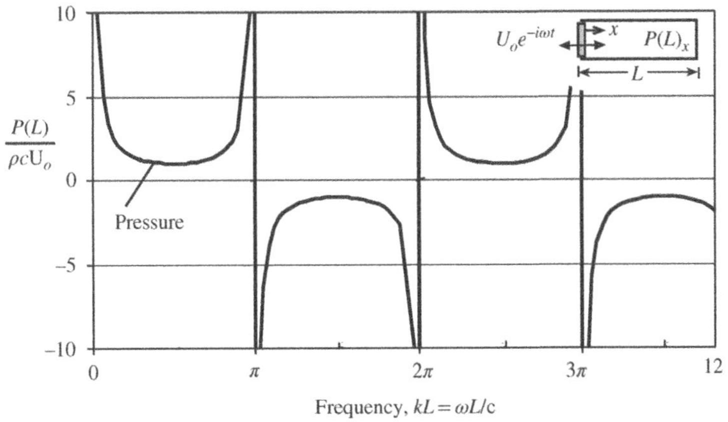

Figure 6.9 Acoustic pressure on the closed end of a duct as a function of piston velocity amplitude U_o and circular frequency (case 2 of Table 6.4)

The transfer matrix of an *acoustic circuit* of joined components is the product of their component transfer matrix:

$$\text{one component}: \begin{pmatrix} P_1 \\ U_1 \end{pmatrix} = [T_1] \begin{pmatrix} P_2 \\ U_2 \end{pmatrix}, \text{two components}: \begin{pmatrix} P_1 \\ U_1 \end{pmatrix} = [T_1]\,[T_2] \begin{pmatrix} P_3 \\ U_3 \end{pmatrix}$$

$$n \text{ components}: \begin{pmatrix} P_1 \\ U_1 \end{pmatrix} = [T_1][T_2]\cdots[T_n] \begin{pmatrix} P_{n+1} \\ U_{n+1} \end{pmatrix}, \text{or}, \tag{6.28}$$

$$\begin{pmatrix} P_{n+1} \\ U_{n+1} \end{pmatrix} = [T_n]^{-1}[T_{n-1}]^{-1}\cdots[T_1]^{-1} \begin{pmatrix} P_1 \\ U_1 \end{pmatrix}$$

P_1 and U_1 are the amplitude acoustic pressure and velocity at the entrance component. P_{n+1} and U_{n+1} are at the exit of the nth component. Transfer matrices for the ducts, expansion, and contraction and side branch components in cases 11 through 22 of Table 6.4 were developed by applying continuity of mass and pressure continuity of interfaces and boundaries. For a side branch, the in and out pressures are equal, $P_1 = P_2$, and independent of velocity so $T_{11} = 1$ and $T_{12} = 0$. The transfer matrix of the expansion chamber in case 14 is found by multiplying the transfer matrices of an expansion (case 13), the duct (case 8), and a contraction (case 13 with $S_2 < S_1$).

Duct Transmission Loss (TL) (Eq. 6.15c) is the ratio of intensity of forward-traveling wave (amplitude A, Eq. 6.23) that enters a component through area S_1 to the intensity of forward-traveling wave (amplitude C) that exits the component through area S_2. The transmission loss (Eq. 6.15d) is $10\log_{10}([S_1/\rho_1 c_1]A^2]/[S_2/\rho_2 c_2]C^2)$, which includes density and speed of sound changes across the component.

Equation 6.27 is solved for the TL (Eq. 6.15c) and absorption coefficient α (Eq. 6.15d) in terms of the transfer matrix pole parameters [17] (p. 82):

Transmission Loss, TL

$$= 20\log_{10}\left[\frac{1}{2}\left(\frac{S_1}{S_2}\right)^{1/2}\left(\frac{\rho_2 c_2}{\rho_1 c_1}\right)^{1/2}\left|T_{11} + \frac{T_{12}}{\rho_2 c_2} + \rho_1 c_1 T_{21} + \frac{\rho_1 c_1}{\rho_2 c_2}T_{22}\right|\right], \text{dB}$$

$$\text{Absorption Coefficient } \alpha = 1 - \frac{\left|T_{11} + \frac{T_{12}}{\rho_2 c_2} - \rho_1 c_1 T_{21} - \frac{\rho_1 c_1}{\rho_2 c_2}T_{22}\right|^2}{\left|T_{11} + \frac{T_{12}}{\rho_2 c_2} + \rho_1 c_1 T_{21} + \frac{\rho_1 c_1}{\rho_2 c_2}T_{22}\right|^2} \le 1, \text{dimensionless} \tag{6.29}$$

$\rho_1 c_1$ and $\rho_2 c_2$ are the specific acoustic impedances at the entrance and exit. *Absorption coefficient* α is the fraction of incident sound energy absorbed. Note that a change in density alone generates TL. α is zero, and TL is infinite for perfect reflection such as for large impedance or area changes.

Frictional pressure drop in a duct with *mean flow velocity* U and acoustic velocity u damps propagating sound waves (cases 9 and 10 of Table 6.4). The frictional mean static pressure loss over span L, $\Delta p = (1/2)\rho(U+u)^2 f\ L/D \approx \Delta p_o + 2\Delta p_o u/U$, has a damping term proportional to acoustic velocity u times mean drop Δp_o. The friction coefficient is f. Case 10 of Table 6.4 shows the TL:

$$\text{Transmission loss} \approx 10\log_{10}\left[1 + \frac{\Delta p_o}{\rho c U}\right]^2$$
$$\approx 10\log_{10}\left[1 + f\frac{ML}{2D}\right]^2 \tag{6.30}$$

Frictional damping can be incorporated into wave propagation analysis for low Mach $M = U/c$, numbers ($M < 1$) with a complex speed of sound $c * /c \approx 1 + iU\ f/(2d\omega)$. Resonant frequency shift by flow is discussed in Section 6.4 (also see [50, 56]).

Radiused bend is not included in Table 6.4. If the wavelength of the sound is much greater than the bend radius, the bend radius has no effect other than to lengthen the sound path [32] (p. 62). However, sharp miter bends reflect sound waves and produce TL [54, 67–69]. Additional acoustical duct circuit components are given by Munjal [17] and Mechel [69] (see Section 6.3.2).

6.3.2 *Mufflers and Resonators*

Mufflers, duct liners, and resonators are installed in ducts to reduce noise and vibration. Large mufflers are more effective at low frequencies and have lower flow pressure drop

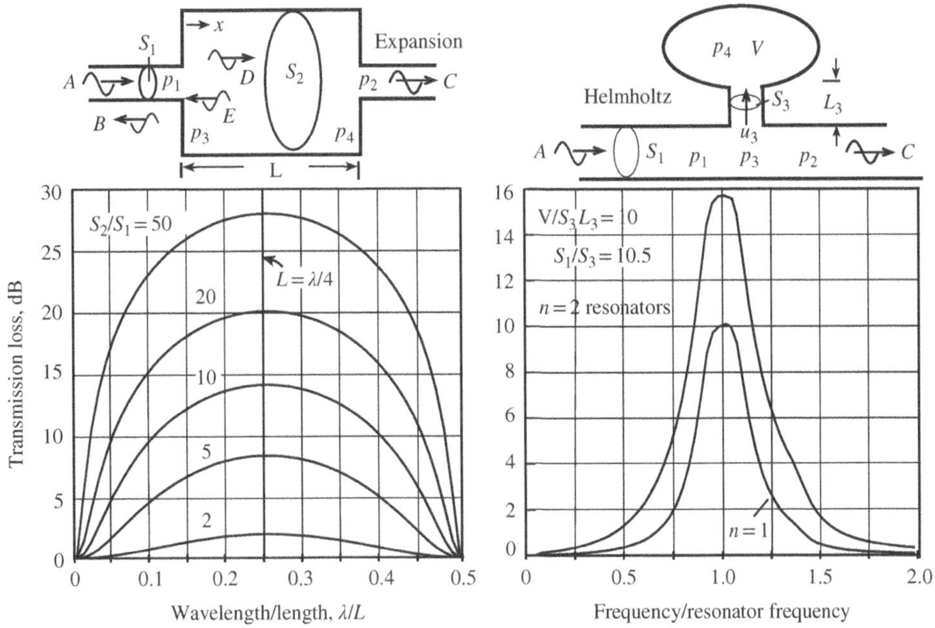

Figure 6.10 Expansion muffler and Helmholtz resonator transmission loss. S_2 = cross-sectional area of chamber; S_1 = cross-sectional area of inlet and exit pipes. Curves between $\lambda/L = 0$ and 0.5 repeat between $\lambda/L = 0.5$ and 1 and its multiples

than small mufflers. Reactive mufflers reflect sound back into the source. Tuned resonators absorb sound at discrete frequencies. Fibrous liners absorb higher frequencies. Minimum sizes for tuned mufflers are about one quarter to one eight of a wavelength at the frequency of interest. Modern automobile mufflers can attenuate 40 dB above 100 Hz with about one dynamic head flow pressure loss.

Expansion chamber reactive muffler (case 14 of Table 6.4 and Figure 6.10a) is a chamber that is modeled with five propagating pressure waves: forward-traveling waves A, C, and D and backward-traveling waves B and E. (See notation of Eq. 6.22: $P_1 = A + B$, $P_3 = D + E$.) The outlet area, S_3, and the inlet area, S_1, differ from the expansion chamber cross-sectional area, S_2. Pressure and fluid mass are conserved across the entrance expansion at $x = 0$ and the exit contraction at $x = L$:

$$\text{at } x = 0 : p_1 = p_3,\ u_1 S_1 = u_3 S_2, \quad \text{at } x = L : p_4 = p_2,\ u_4 S_2 = u_2 S_3 \tag{6.31}$$

With these boundary conditions and Equation 6.24, the TL, $10 \log_{10}|A/C|^2$ (Eq. 6.29; [22]), is found:

$$\text{TL} = 10 \log_{10}\left(\frac{1}{4}\frac{S_1}{S_3}\left[\left(1 + \frac{S_3}{S_1}\right)^2 \cos^2\frac{2\pi f L}{c} + \left(\frac{S_2}{S_1} + \frac{S_3}{S_2}\right)^2 \sin^2\frac{2\pi f L}{c}\right]\right), \text{dB} \tag{6.32}$$

TL is maximized at the frequency of interest if expansion chamber length L is an odd multiple of a quarter of wavelength, $L = i\pi/4$, $i = 1, 3, 5$, (odd) so the wave (E) reflected back from the end of the chamber is exactly out of phase with the entering wave (A):

$$\text{TL}_{\substack{S_1=S_3,\\ L=i\lambda/4;\ i=1,3,5}} = 10\log_{10}\left[1+\frac{1}{4}\left(\frac{S_2}{S_1}-\frac{S_1}{S_2}\right)^2\right], \text{dB} \tag{6.33}$$

Area ratio, S_2/S_1	1	2	4	6	10	20	40	80	100
Diameter ratio, d_2/d_1	1	1.41	2	2.45	3.16	4.47	6.32	8.94	10
Transmission loss (dB)	0	1.74	6.54	9.78	14.1	20.0	26.0	32.0	34.0

A practically effective TL of 6 dB requires a chamber diameter that is at least twice the duct diameter.

Example 6.3 Sizing a muffler

An engine exhausts 200 °C (392 °F) air (speed of sound 435 m/s, 1426 ft/s) through a 50 mm (2 in.) diameter pipe. What is the length and diameter of an expansion chamber muffler that will have a 20 dB TL at 250 Hz.

Solution: The quarter wave chamber length is $\pi/4 = c/(4f) = 435\ \text{m/s}/(4 \times 250\ \text{Hz}) = 0.435\ \text{m}\ \ (1.43\ ft)$. To achieve 20 dB attenuation, the previous table shows that the muffler diameter must be 4.47 times the pipe diameter, 4.47(50 mm) = 224 mm (8.8 in.).

A **side branch Helmholtz resonator** (Figure 6.10b and case 19 of Table 6.4) dissipates low-frequency acoustic energy at the resonator natural frequency (case 18 of Table 6.5) by viscous resistance, which is modeled as a neck impedance step, $Z_n = (p_3 - p_4)/u_3$ (cases 26–28 of Table 6.3). The fluid mass in the neck accelerates in response to the pressure differential across the neck.

$$p_3 - p_4 = \rho S_3 L_3 \frac{du_3}{dt} + R\rho c u_3, \text{ where } p_4 = \left(\frac{\gamma p_o}{V_o}\right) S_3 \xi_3,\ u_3 = \frac{d\xi_3}{dt}$$
$$S_1 u_1 = S_1 u_2 + S_3 u_3, \qquad p_1 = p_2 = p_3, \tag{6.34}$$

where S_1 is the duct area; S_3 is the resonator neck area; L_3 is the resonator neck length, including the end correction, and R is a dimensionless viscous neck resistance (cases 28 and 29 of Table 6.3). The acoustic duct pressures are p_1 upstream of the neck, p_3 at the neck, and p_2 downstream of the neck, and their acoustic velocities are u_1, u_3, and u_2, respectively. p_4 is the chamber pressure. The TL and resonator natural frequency f_n in Hertz are as follows.

Damping factor is typically $\zeta \approx 0.02$:

$$\mathrm{TL} = 10\ \log_{10}\left[1 + \left\{\left(\frac{nS_3}{2S_1}\right)^2\left(\frac{V}{S_3L_3}\right) + 2\zeta\left(\frac{nS_3}{S_1}\right)\left(\frac{V}{S_3L_3}\right)^{1/2}\right\} \times \left[\left(\frac{f}{f_n} - \frac{f_n}{f}\right)^2 + (2\zeta)^2\right]^{-1}\right]$$

$$f_n = \left\{\frac{c}{(2\pi)}\right\}\left[\left[\frac{S_3}{(L_3V_o)}\right]\right]^{1/2},\ \mathrm{Hz} \qquad \zeta = \left(\frac{R}{2}\right)\left[\frac{V}{(L_3S_3)}\right]^{1/2} \tag{6.35}$$

Resonator loss is maximum at the resonator natural frequency, and it increases with resonator size, but to hold the resonator natural frequency f_n constant while increasing resonator size creates a difficult constraint between S_3, L_3, and V, so multiple ($n = 1, 2, 3, \ldots$) resonators are often installed.

Acoustical duct liners, shown in Figures 6.11 and 6.12, consist of fibrous bulk material or an array of small Helmholtz resonators attached to the inside of the duct with a perforate facing the flow. The exact liner TL (case 15 of Table 6.4) is a function of liner impedance. There are two approximate solutions that do not require the liner impedance.

A known TL of lined duct A can be extrapolated to another duct B with same liner but different area and diameters with the ratio of their lined areas S_3 to the duct cross sections S_1:

$$\begin{aligned} \mathrm{TL}_B &\approx \mathrm{TL}_A + 20\ \log_{10}\frac{(S_3/S_1)_B}{(S_3/S_1)_A} && \mathrm{TL}_A > 10\,\mathrm{dB} \\ &\approx (\mathrm{TL})_A\frac{(S_3/S_1)_B}{(S_3/S_1)_A} && \mathrm{TL}_A < 1\ \mathrm{dB} \end{aligned} \tag{6.36}$$

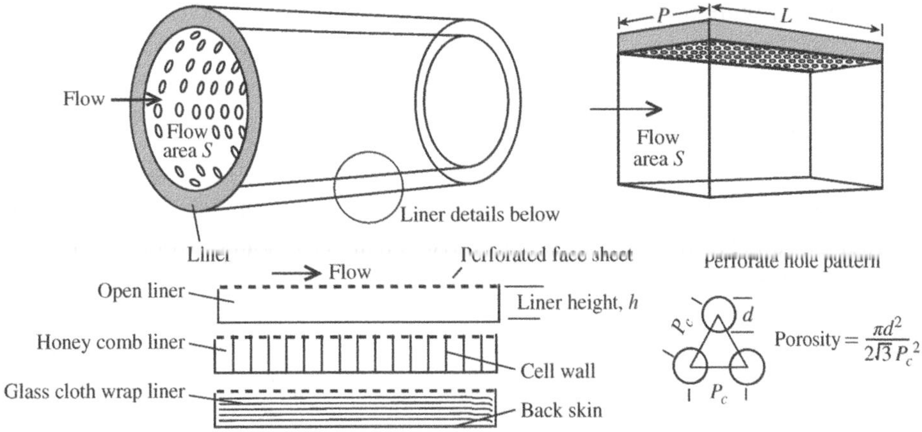

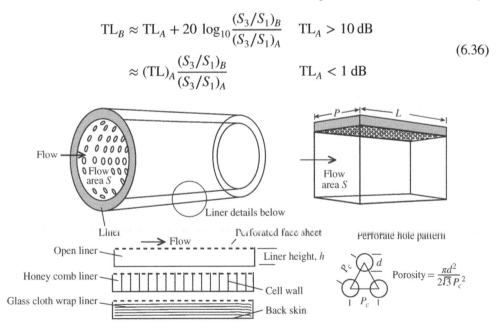

Figure 6.11 Lined ducts. The circular duct has a 360° liner, whereas the rectangular duct shown has only one side lined

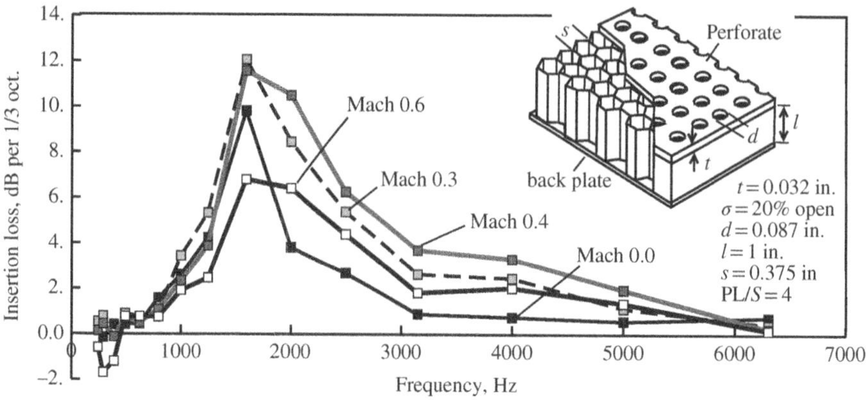

Figure 6.12 Insertion loss of a lined duct in air as a function of the mean airflow velocity. The liner duct has 20% opening and PL/S = 4 and is tuned to 1600 Hz

Piening's approximate formula for TL of lined ducts is (case 21 of Table 6.4; [17], p. 242)

$$\text{TL} \approx 1.5\frac{S_{\text{lined}}}{S_1}\alpha_s,\ \text{dB} \tag{6.37}$$

S_{lined} is the internal surface area of the duct liner. S_1 is the duct flow area and α_s is the liner surface absorption coefficient for normal impingement given in case 20 of Table 6.3. $\alpha_s = 0.5$ is a typical value. The formula implies that lining the inner surface of a round duct with diameter D ($S_1 = \pi D^2/4$) over one diameter liner ($S_{\text{lined}} = \pi D^2$) achieves a 3 dB loss, which is about the minimum differential that can be detected by the human ear.

Figure 6.12 shows the measured TL in a lined duct with various duct flow velocities, $S_3/S_1 = 4$. Increasing the flow Mach number beyond about 0.4 decreases the maximum liner loss but increases the bandwidth of frequencies over which there is some loss. The liner thickness in Figure 6.12 is one-eighth of the wavelength at 1600 Hz.

6.3.3 *Room Acoustics*

Diffuse sound analysis (also called *room acoustics*) in cases 23–28 of Table 6.4 assumes that wavelengths are much shorter than room dimensions, which is often the case at frequencies on the order of 1000 Hz where humans are most sensitive to sound. Room acoustic solutions apply to diffuse sound waves that propagate in all directions with high modal density (Table 6.6; see Example 6.6).

Sabine's law of room acoustics ([13, 24, 51–53, 71]; case 23 of Table 6.4) states that the *average steady-state volumetric acoustic energy density* (Eq. 6.14) *of random directional*

Table 6.6 Cumulative modal density

Notation: A = two dimensional area; c = speed of sound (Table 6.3); E = elastic modulus; f = frequency Hz; f_n = fundamental frequency Hz; h = plate thickness; Int(x) = integer part of x, for example, Int(π) = 3; I = area moment of inertia; L = length; m = mass per unit length; N = cumulative number of vibration modes below frequency f; P = perimeter; S = surface area;. γ = mass/area. ν = Poisson's ratio. Asymptotic = approximate. Modal density = dN/df. Ref. [81].

Element	Cumulative Modal Density, N
1. Acoustic Rectangular Volume rigid closed sides 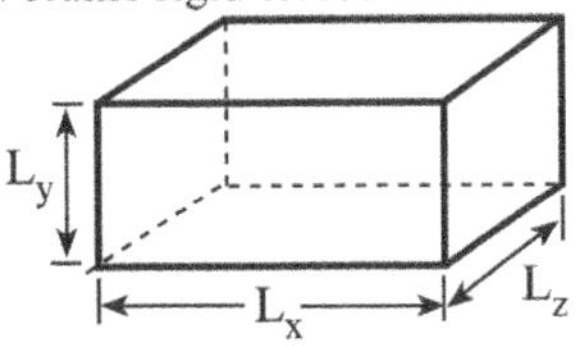 	$N_{asymptotic} \approx 4\pi V f^3/3c^3 + \pi S f^2/4c^2 + Lf/8c$ $N_{exact} = \sum_{k=0}^{K}\sum_{j=0}^{Int[J(k)]}\left\{\left[Int\left(1-\frac{j^2}{j_{max}^2}-\frac{k^2}{k_{max}^2}\right)^{1/2}\right]+1\right\}, f>f_n$ where $i_{max}=2fL_x/c$, $j_{max}=2fL_y/c$, $k_{max}=2fL_z/c$, $K=Int(k_{max})$ $J(k)=j_{max}(1-k^2/k_{max}^2)^{1/2}$, $f_n = c/2L_n$, L_n = max of L_x, L_y, L_z $L=4(L_x+L_y+L_z)$, $S = 2(L_xL_y+L_xL_z+L_yL_z)$, $V = L_xL_yL_z$.
2. Acoustic Two Dimensional Volume between Rigid Walls 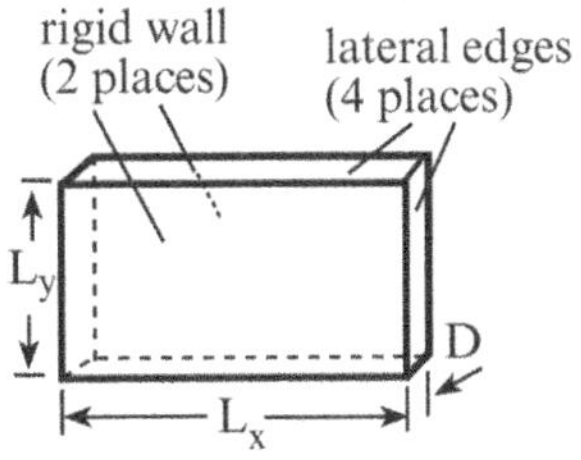 $L_y, L_x >> D$, $f< c/(2D)$	Lateral edges closed: $N_{asymptotic} \approx \pi A f^2/c^2 + Pf/2c$, $f_n=c/2L_n$, L_n = max of L_x, L_y $N_{exact} = \sum_{i=0}^{Int[2L_xf/c]} Int\left[\frac{2L_yf}{c}\left(1-\frac{i^2}{(2L_xf/c)^2}\right)^{1/2}+1\right]$, $f>f_n$, else 0 Lateral edges open: $N_{asymptotic} \approx \pi A f^2/c^2 - Pf/2c$, $f_n = (c/2)(1/L_x^2+1/L_y^2)^{1/2}$ $N_{exact} = \sum_{i=1}^{Int[2L_xf/c]} Int\left[\frac{2L_yf}{c}\left(1-\frac{i^2}{(2L_xf/c)^2}\right)^{1/2}\right]$ $A = L_xL_y$, $P = 2(L_x+L_y)$, Int(x) = integer part of x. $f< c/(2D)$
3. One-Dimension DuctL>>D 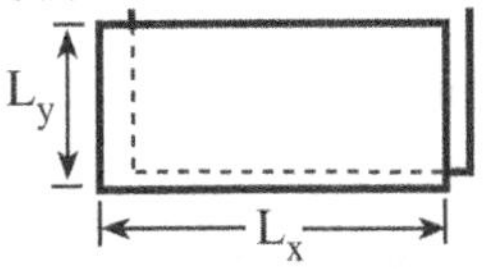 	$N = \frac{2Lf}{c}$, $f< c/(2D)$, $f_n=c/(2L)$, Int[N] is exact solution. Formula applied to both ends open, or, both closed. Also applies to tensioned string of length L with $c = (T/m)^{1/2}$, T = tension, m = mass/ length of string.
4. Simply Supported Flat Plate 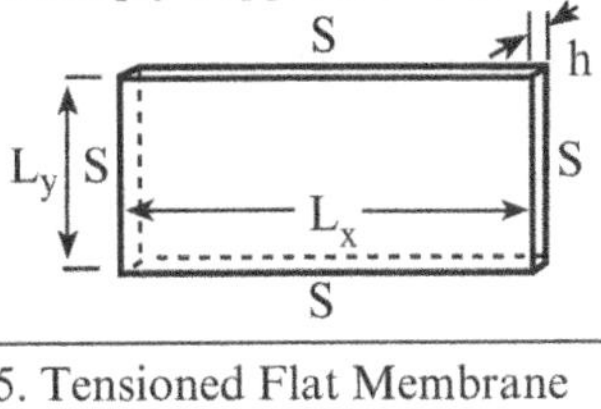 	$N_{asymptotic} \approx \pi A f/c_e^2 - Pf^{1/2}/2c_e$ $N_{exact} = \sum_{i=1}^{Int[2f^{1/2}L_x/c_e]} Int\left[\frac{2L_yf^{1/2}}{c_e}\left(1-\frac{i^2}{(2L_xf^{1/2}/c_e)^2}\right)^{1/2}\right]$ $A = L_xL_y$, $P = 2(L_x+L_y)$, $c_e = [2\pi\{(Eh^3/(12\gamma(1-\nu^2))\}^{1/2}]^{1/2}$, Table 5.3.
5. Tensioned Flat Membrane with Stationary Edgoo 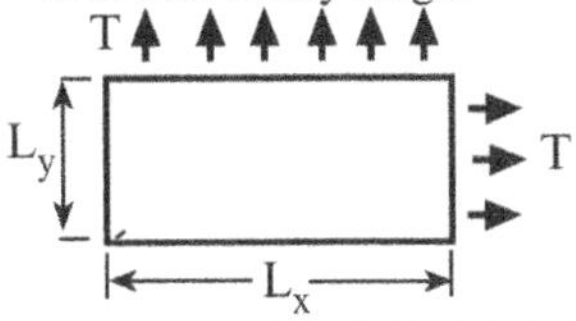 	$N_{asymptotic} \approx \pi A f^2/c_m^2 - Pf/2c_m$ $N_{exact} = \sum_{i=0}^{Int[2L_xf/c_m]} Int\left[\frac{2L_yf}{c_m}\left(1-\frac{i^2}{(2L_xf/c_m)^2}\right)^{1/2}\right]$ $A = L_xL_y$, $P=2(L_x+L_y)$, $c_m = (T/\gamma A)^{1/2}$, $f_n=(c_e/2)(L_y/L_x+L_x/L_y)^{1/2}$ T = tension per unit length of edge, γ = mass per unit area
6. Simply Supported Beam 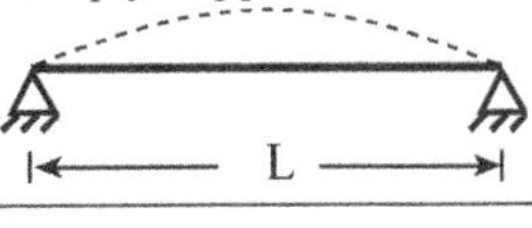 	$N = \left(\frac{f}{f_1}\right)^{1/2}$, $f_1 = \frac{\pi^2}{2\pi L^2}\left(\frac{EI}{m}\right)^{1/2}$, Int[N] is exact solution I = beam moment of inertia.. See Table 4.2.

traveling waves in a room is equal to the sound power input to the room (W) *over the absorption area of sound* of the room (A_α):

$$\begin{aligned}
\text{Energy density} &= \frac{p_{rms}^2}{\rho c^2} \\
\text{rms pressure}, p_{\text{rms}}^2 &= \frac{4\rho c W}{A_\alpha} \\
\text{Sound power}, W &= \frac{p_{rms}^2 A_\alpha}{4\rho c} \\
\text{Absorption area}, A_\alpha &= \sum_{i=1}^{N} \alpha_{si} A_i, \quad i = 1, 2, 3, \ldots \text{ surfaces}
\end{aligned} \tag{6.38}$$

Room sound absorption is primary by reflection–absorption from walls, furniture, people, and windows. The room *surface absorption* area A_α is the sum of room surfaces times their dimensionless *surface absorption coefficient* α_s (case 20 of Table 6.3). The units of absorption area are Sabines-ft^2 or Sabines-m^2. W is the sound power of the source (cases 1–13 of Table 6.3). See case 27 of Table 6.4 and Refs [31, 53, 71, 72].

The time for diffuse sound to decrease by 60 dB in a room after the sound source stops is the room's *reverberation time* (case 21 of Table 6.3), which is a function of room absorption area and volume. In standard air ([2], p. 317; [27]; case 26 of Table 6.4; Figure 6.7b),

$$T_{60\,\text{dB}} = 55.2\frac{V}{A_\alpha c} = \begin{cases} \dfrac{0.161V}{A_\alpha}, & \text{air metric SI}, A_\alpha, V \text{ in meter units} \\ \dfrac{0.049V}{A_\alpha}, & \text{air US units}, A_\alpha, V \text{ in feet units} \end{cases} \tag{6.39}$$

Reverberation time can be used to calculate room absorption area and absorption coefficient, $\alpha_{\text{avg}} = A_\alpha/\Sigma A_i$ [52, 73]. Absorption is increased and reverberation time is lowered by increasing the absorption of the ceiling, walls, and floor with rugs, drapery, and acoustic tiles. Reverberation times for large music concert halls are typically between 1.6 and 2.5 s at 512 Hz [74] to favor the melody. Reverberation times of 0.5–1.2 s are desirable in lecture rooms to maximize *speech intelligibility* [24] (p. 194). Architecture acoustics is discussed in Refs [74–76].

The sound energy flow in a circuit of $i = 1$ to N rooms joined by windows is described by N-coupled ordinary differential equations [24], p. 121):

$$V_i\frac{dE_i}{dt} + \frac{c_i A_{\alpha i}}{4}E_i + \frac{c_i S_i}{4}E_i = W_i + \sum_{j\neq i}\frac{c_i S_{ij}}{4}E_j, \quad i = 1, 2, \ldots, N \tag{6.40}$$

where V^i = volume of the ith room, $E_i = p_{i\text{rms}}^2/\rho c^2$ = average acoustic energy density (Eq. 6.14; [24], p. 11) in the ith room, and $V_i dE_i/dt$ = the rate of change in acoustical energy in the ith room; it is set to zero for steady state. W_i = power (watts in metric units) of sources in the ith room; S_{ij} = window area between the ith and jth rooms; and S_i = the total window area in the ith room. The damping term $c_i A_{\alpha i} E_i/4$ is the power absorbed by surface reflection in the ith room (Eq. 6.38). The summation term is energy entering the room through windows in the ith room with an area S_{ij} that connects to the adjoining j room where acoustic energy density is E_j. Solutions for the $N = 1$ room are shown in cases 23–27 of Table 6.4. Case 28 of Table 6.4 shows TL through walls and windows.

Example 6.4 Sound enters a room through a single-pane window that overlooks a busy street where the noise level is 75 dB(A) at 1000 Hz. Determine the change in room sound level by changing the window to double-pane glass.

Solution: Case 25 of Table 6.4 shows the room acoustic solution. Case 28 shows the window TL. The interior sound level in decibels is the exterior sound level less room absorption and the window TL (case 28 of Table 6.4):

$$\text{SPL}_{\text{int}},\ \text{dB} = \text{SPL}_{\text{ext}} + 10\log_{10}\left(\frac{S}{A}\right) - \text{TL}$$

The TL of a single-pane window at 1000 Hz is 31 dB. For constant S/A, a double-pane window with ½ in. (13 mm) spacing increases TL by 9 to 40 dB and lowers the interior sound by 9 dB, which is readily detected by the human ear.

6.4 Acoustic Natural Frequencies and Mode Shapes

Inviscid acoustic particle motions are *irrotational*, $\nabla \times p = 0$, where $\times$ is the cross product (Eq. 1.24), p is fluid pressures, and ∇ is the vector gradient operator (Table 6.5). Acoustic velocity is the gradient of a scalar function called a *velocity potential* $\Phi(x, y, z, t)$ that is a separation-of-variables solution to the homogeneous wave equation in a volume (Table 6.2, [2]):

$$\begin{aligned} &\Phi(x, y, z, t) = \tilde{\varphi}_{ijk}(x, y, z)P(t), \qquad && c^2\nabla^2\Phi + \frac{\partial^2\Phi}{\partial t^2} = 0 \\ &\mathbf{u} = \nabla\Phi, && p = -\frac{\rho\partial\Phi}{\partial t} \end{aligned} \tag{6.41}$$

The speed of sound (Table 6.1) is c and $\mathbf{u}$ is the acoustic particle velocity vector. There are reflective *boundary conditions*:

$$\text{Open boundary}: \ p = 0, \quad \text{closed boundary}: \ \mathbf{n} \bullet \mathbf{u} = 0, \ \text{so} \ \frac{dp}{d\mathbf{n}} = 0 \tag{6.42}$$

A closed boundary is a rigid wall where the normal component of velocity ($\mathbf{u} \bullet \mathbf{n}$) is zero. The unit outward normal is $\mathbf{n}$. Open boundaries relieve pressure to zero. Acoustic energy does not transmit across either of these ideal boundaries. The acoustic mode shapes in Table 6.5 are *orthogonal* over the volume V [1] (p. 286):

$$\int_V \tilde{\varphi}_{ijk}\tilde{\varphi}_{mno}dV = 0 \quad \text{unless } i = m,\ j = n,\ k = o \tag{6.43}$$

Acoustic pressure in a mode is proportional to the *acoustic mode shape*, $\tilde{\varphi}(x, y, z)$, a function of space, but not of time.

Table 6.5 [1, 32, 61–66, 68, 76] has **Acoustic natural frequencies and mode shapes.** That was found by applying boundary conditions to solutions (Eq. 6.42) of the wave equation (Table 6.2). Consider a duct with length L with a mode shape of the form $\sin \omega x/c$. Zero pressure boundary condition $x = L$ requires $\sin \omega L/c = 0$. Hence, $\omega L/c$ must be an integer multiple of π. $\omega_i = i\pi c/L$, $i = 0, 1, 2, 3, \ldots$ are the circular natural acoustic frequencies of the open–open duct (case 1 of Table 6.5):

$$f_i = \frac{\omega_i}{2\pi} = \frac{ic}{2L}, \quad i = 1, 2, 3, \ldots \text{ Hz} \tag{6.44}$$

Acoustic natural frequencies are proportional to the speed of sound c. The fundamental acoustic period in a duct of length L, $T = 1/f_1 = 2L/c$, is the time for sound to make a round trip of the duct.

6.4.1 Structure-Acoustic Analogy

The one-dimensional wave equation (Eq. 6.7; case 1 of Table 6.2) describes both acoustic wave propagation in ducts and longitudinal vibration of elastic beams (Chapter 4; Eq. 4.36). Beam displacement is analogous to acoustic pressure. The beam modulus of elasticity (E) and density (ρ) is chosen to replicate the acoustic speed of sound $c^2_{\text{acoustic}} = E/\rho$ in the duct. The open acoustic boundary, $p = 0$, is zero beam displacement ($X = 0$), and the acoustic wall, $dp/dx = 0$, is an unconstrained (free) structural boundary ($dX/dx = 0$). For equal cross-sectional areas, the beam longitudinal natural frequencies are also duct acoustic natural frequencies. This analogy allows structural finite element codes to solve acoustic problems, but it only applies to structural solutions without shear stress as inviscid fluids do not support shear. See Cory [77] and MSC/NASTRAN [78].

Example 6.5 Room acoustic natural frequencies

Calculate the six natural frequencies of air in a 13.1 by 16.4 by 8 ft high (4 by 5 by 2.43 m) room with rigid walls.

Solution: The acoustic natural frequencies are shown in case 6 of Table 6.5: $f_{ijk} = (c/2)(i^2/L_x^2 + j^2/L_y^2 + k^2/L_z^2)^{1/2}$, where $L_x = 13.1$ ft (4 m), $L_y = 16.4$ ft (5 m), $L_z = 8$ ft (2.43 m), and $c = 1125$ ft/s (343 m/s). The natural frequencies are a function of the three spatial indices i, j, and k. There are three classes of acoustic modes: one-dimensional (axial) modes with two indices zero; two-dimensional modes with one of i, j, and k zero; and three-dimensional modes with all i, j, and k greater than zero:

i	j	k	f_{ijk} (Hz)	Mode shape
0	0	0	0	Static mode
1	0	0	42.87	x-Axis 1D mode
0	1	0	34.30	y-Axis 1D mode
0	0	1	70.57	z-Axis 1D mode
1	1	0	54.91	xy 2D mode
1	1	1	89.42	First xyz 3D mode

Example 6.6 computes the cumulative modal density of this room.

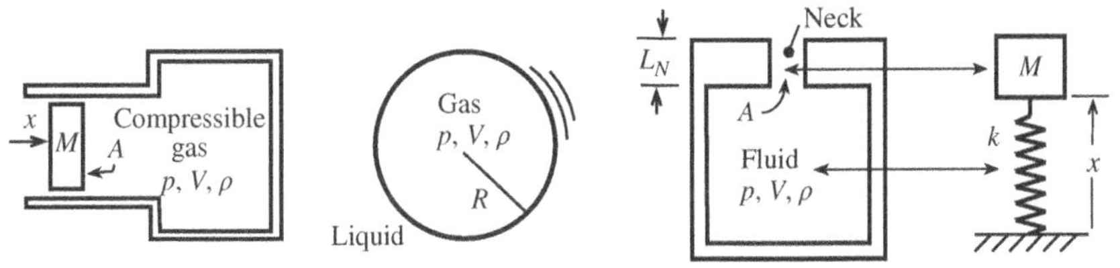

(a) Piston and gas cylinder (b) Gas bubble in liquid (c) Helmholtz resonator and Spring-mass

Figure 6.13 Systems with compressible fluid in variable volume cavities

Bubbles and air spring natural frequencies are shown in cases 23 and 24 of Table 6.5. The gas trapped within the variable volume cavity in Figure 6.13 obeys the *polytropic relationship* between gas absolute static pressure p and cavity volume V:

$$pV^{\kappa} = \text{constant} \quad 1\ (\text{isothermal}) \leq \kappa \leq \gamma(\text{adiabatic}) \tag{6.45}$$

The dimensionless *polytropic exponent* κ depends on heat transfer into the gas [14]. Without heat transfer κ is γ, the ratio of specific heat at constant pressure to that at constant volume; $\kappa \approx 1.4$ for air [12, 14]. The derivative of Equation 6.45 shows that static pressure decreases as the gas volume increases:

$$dp = -\left(\frac{\kappa p}{V}\right) dV \tag{6.46}$$

An incremental change dx in piston position results in an incremental pressure dp and force dF on the piston (area S) that can be expressed as the gas-spring constant:

$$dF = dpS = \left(\frac{\kappa p}{V}\right) S^2 dx$$

$$k = \frac{dF}{dx} = \left(\frac{\gamma p}{V}\right) S^2 = c^2 S^2 \left(\frac{\rho}{V}\right), \quad \text{for adiabatic process} \tag{6.47}$$

The speed of sound c in gases is given by Equation 6.2, case 2 of Table 6.1, and Figure 6.1. Gieck [79] discusses applications of gas springs to vehicle suspensions.

The free piston's natural frequency on the air spring (Figure 6.13a) is found from case 1 of Table 3.1 with Equation 6.47:

$$f = \frac{1}{2\pi}\sqrt{\frac{k}{M}} = \frac{c}{2\pi}\sqrt{\frac{\rho S^2}{VM}}, \quad \text{Hz} \tag{6.48}$$

where c = speed of sound (Table 6.1), S = area of piston face, ρ = mean density of gas in the cylinder, V = mean volume of the cavity, and M = the piston mass. Consistent units are shown in Table 1.2.

Gas-spring analysis also applies to spherical gas bubbles that oscillate axisymmetrically in liquid (Figure 6.8b):

$$R(t) = R_o + \delta R \sin \omega t \tag{6.49}$$

where R_o is the mean radius of the bubble and δR is the amplitude of radial oscillation, which is assumed to be much smaller than R_o. The oscillating bubble surface reacts against the

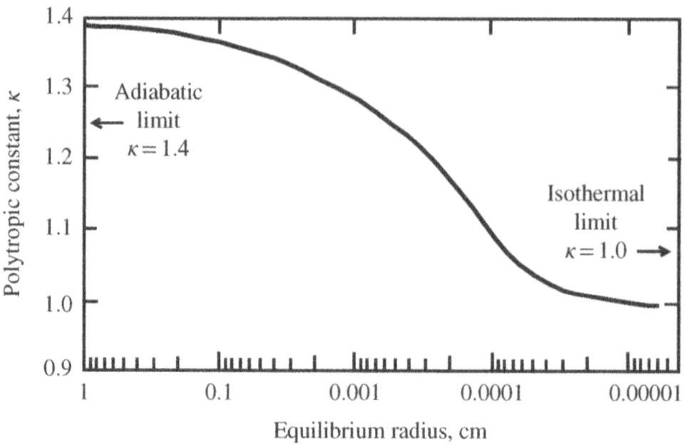

Figure 6.14 Polytropic constant as a function of the size of an air bubble in water at 20°C [65]

entrained mass ($4\rho\pi R^3$) of liquid surrounding the bubble. The resultant natural frequency of bubble oscillation is [66].

$$f = \frac{(3\kappa p_o/\rho)^{1/2}}{2\pi R_o} \quad \text{Hz} \tag{6.50}$$

The *polytropic exponent* κ depends on the heat transfer into the bubble from the surrounding fluid during oscillation; it is a function of the size of the bubble (Figure 6.14; [65]). Very small air bubbles in water, $R_o < 10^{-2}$ mm, oscillate isothermally, $\kappa \approx 1$. Large bubbles oscillate adiabatically, $\kappa \approx 1.4$.

The equilibrium bubble radius is proportional to the surrounding fluid's hydrostatic static pressure (p_o), which increases with increasing depth d below the surface. As bubbles rise to the surface, they grow in size, $R_o \sim 1/d^{1/3}$, and their natural frequency (Eq. 6.50) decreases, $f \sim d^{5/6}$. Surface tension further decreases the natural frequency of very small bubbles (case 23 of Table 6.5).

Helmholtz resonators (Figure 6.13c) are named to honor Herman Helmholtz's experiments on acoustic cavities with one or more small openings (necks) to the atmosphere ([80]; cases 18 through 22 of Table 6.5). Blowing across the open mouth of a bottle excites the low tones of the bottle resonator. Helmholtz resonators passively dissipate low-frequency acoustic energy by viscous fluid shear in the neck.

The resonator cavity is the gas spring (Eq. 6.48), and the mass is the mass of fluid in the neck, $M = \rho A L_N$ (Figure 6.13c). The Helmholtz resonator's acoustic natural frequency in Hertz is proportional to the speed of sound:

$$f_n = \frac{c}{2\pi}\sqrt{\frac{A}{VL_N}}, \quad \text{Hz} \tag{6.51}$$

where c = speed of sound (Table 6.1), A = area of the neck, V = mean volume of the cavity, and L_N = length of the neck. Oscillations in the neck entrain fluid mass just outside the neck. The effective neck length of a baffle-ended neck (case 18b of Table 6.5) is the length of the neck L_N plus 0.82 times the radius of the neck r_N ([1], p. 348), $L_{Ne} = L_N + 0.82r_N$. With the neck length correction, the natural frequency of a closed–open pipe with a baffled neck is

$$f_{\text{quarter wave}} = \frac{c}{4(L_N + 0.82r_n)}, \quad \text{Hz} \tag{6.52}$$

$r_N \approx (A/\pi)^{1/2}$ is the neck length correction for noncircular necks with area A. The length correction for a protruding (unbaffled) neck is $0.61r$ (case 18 of Table 6.5).

Alster [81] measured the natural acoustic frequency of a 2.5 cm radius closed–open pipe in air. The open end is baffled.

Length of pipe, L (cm)	50	40	30	20	10	5
Length/diameter	20	16	12	8	4	2
f (experiment) (Hz)	169	211	279	411	778	1380
f (Eq. 6.52) (Hz)	170.6	213.0	283.4	423.3	836.1	1631
f (Eq. 6.52 with $r_N = 0$) (Hz)	171.5	214.4	285.8	428.8	857.5	1715

The end correction improves the agreement of theory with his data by lowering the predicted natural frequency.

As suggested in Figure 6.13, the frequencies of systems of resonators (cases 18–22 of Table 6.5) can be found from analogous spring–mass systems (Table 3.2). Sound absorption by Helmholtz resonators is discussed in Section 6.3.3.

Cumulative modal density, N, in Table 6.6 [70] is the total number of acoustic modes with natural frequencies less than frequency f (Hz). Table 6.6 has exact and asymptotic approximate formulas, valid at high frequency, for cumulative modal density. dN/df is modal density. For a closed rectangular volume (case 1 of Table 6.6), the asymptotic expressions are

$$N \approx \frac{4\pi V f^3}{3c^3} + \frac{\pi S f^2}{4c^2} + \frac{Lf}{8c} \tag{6.53a}$$

$$\frac{dN}{df} \approx \left(\frac{4\pi V}{c^3}\right) f^2 + \left(\frac{\pi S}{2c^2}\right) f + \left(\frac{L}{8c}\right) \tag{6.53b}$$

There are three terms on the right-hand sides: The first term is three-dimensional acoustic modes, the second is the two-dimensional acoustic modes, and the third is the one-dimensional (axial) acoustic modes. The modal density of the three-dimensional modes (those with nonzero indices i, j, and k in case 6 of Table 6.5) increases with the square of frequency, whereas one-dimensional modes (case 3 of Table 6.5) are evenly spaced in frequency as shown in Figure 6.9. Equation 6.53b is solved for the minimum for frequency for a modal density of three-dimensional acoustic modes:

$$f = \left(\frac{c^3}{4\pi V} \frac{dN}{df}\right)^{1/2} \tag{6.54}$$

The minimum frequency for 1 Hz modal density ($dNdf = 1$) is $f = [c^3/(4\pi V\{1\text{ Hz}\})]^{1/2}$; it is the number of acoustic modes in a 1 Hz band. This frequency is 194 Hz for the room in Example 6.6. That is, above about 194 Hz, there is more than one acoustic mode in each 1 Hz band and so diffuse room theory applies.

Example 6.6 Room modes

Calculate the cumulative number (N) of acoustic modes of a 4 m by 5 m by 2.43 m high (13.1 by 16.4 by 8 ft high) room with rigid walls and filled with air at frequencies of 20, 50, 100, 500, 1000, and 2000 Hz. The speed of sound c is 343 m/s (1125 ft/s).

Solution: Case 1 of Table 6.6 is applied. $L_x = 4\,\text{m}$, $L_y = 5\,\text{m}$, $L_z = 2.43\,\text{m}$. For the asymptotic solution, the total length of the twelve edges is $L = 4(L_x + L_y + L_z) =$ 54.88 m (180 ft), the total surface area is $S = 120.8\ \text{m}^2$ (1300 ft^2), and the volume is $L_xL_yL_z = 85\ \text{m}^3$ (3002 ft^3). The results with the asymptotic and exact formulas are

Frequency (Hz)	20	50	100	200	500	1000	2000
N exact	1	4	19	107	1306	9644	73,857
N asym.	0.79	4.1	19.9	106.9	1314	9652	73,873
dN/df asym.	0.062	0.167	0.446	1.401	7.44	28.1	109.1

In the normal hearing range near 1000 Hz, there are approximately 28 modes per hertz. Room acoustic methods apply at this high frequency.

6.5 Free Surface Waves and Liquid Sloshing

Once disturbed, the free surface of liquid in a tank flows back and forth in a series of standing waves at discrete natural frequencies, as shown in Table 6.7 and Figure 6.15. These are called *free surface waves*, *gravity waves*, and *sloshing waving*. The most familiar form of sloshing is the oscillations of coffee or tea in a cup while walking [95]. Sloshing waves are also important in design of fuel tanks in vehicles, aircraft, and ships and in earthquake

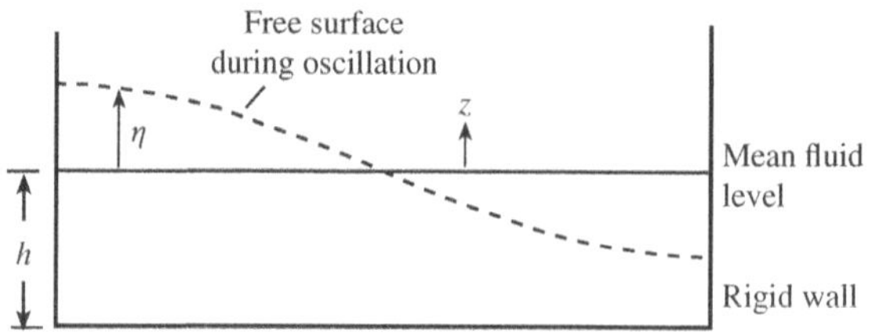

Figure 6.15 Fundamental sloshing mode in a rectangular basin (case 1 of Table 6.7). The x–y plane is parallel to the mean fluid level and perpendicular to the paper

Table 6.7 Natural frequencies of surface waves in basins

Notation: g = acceleration of gravity (Table 1.3); h = depth of liquid in basin; f_{ij} = natural frequency of i,j modes in Hertz, = 1/T where T is natural period; $J_j()$ = Bessel function of 1st kind and jth order; p = pressure; R, r = radius; $Y_j()$ = Bessel function of second kind and ith order; α, θ = angles, degrees; θ = angles, radians. Overbar (~) denotes shape function. See Table 1.2 for consistent sets of units. Pressure, velocity, and fluid displacement are expressed in terms of the mode shape using the formulas in the text. .... = mean fluid level; - - - - - = fluid during oscillation. Refs [32, 62, 82–93].

Basin	Natural Frequency f_{ij}, Hz	Potential $\tilde{\phi}(x,y)$, Eq. 6.57
1. Rectangular Basin with Constant Depth 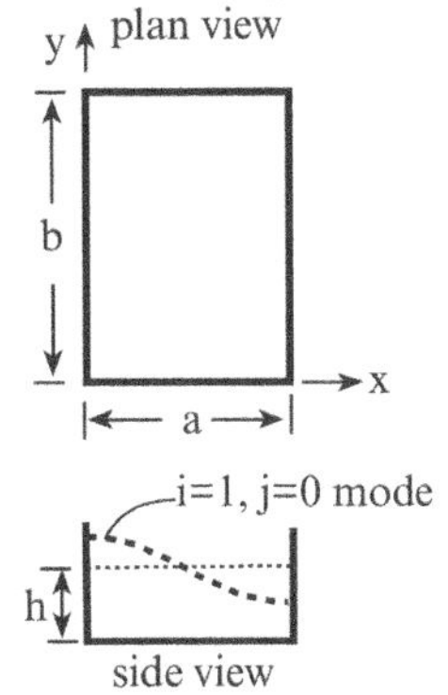	General case, $\frac{1}{2}\left[\frac{g}{\pi}\left(\frac{i^2}{a^2}+\frac{j^2}{b^2}\right)^{1/2}\tanh \pi h\left(\frac{i^2}{a^2}+\frac{j^2}{b^2}\right)^{1/2}\right]^{1/2}$ $i=0,1,2,..,\ j=0,1,2...$ Shallow liquid, $h\left((i/a)^2+(j/b)^2\right)^{1/2}<1/10$ $\frac{(gh)^{1/2}}{2}\left(\frac{i^2}{a^2}+\frac{j^2}{b^2}\right)^{1/2}$, $i=0,1,2..,\ j=0,1,2..$ Deep liquid, $h\left((i/a)^2+(j/b)^2\right)^{1/2}>1$, $\frac{g^{1/2}}{2\pi^{1/2}}\left(\frac{i^2}{a^2}+\frac{j^2}{b^2}\right)^{1/4}$, $i=0,1,2..,\ j=0,1,2..$	$\cos\frac{i\pi x}{a}\cos\frac{j\pi y}{b}$ i = number of half waves along x axis j = number of half waves along y axis Period $T_{ij}=1/f_{ij}$ all cases $\Phi=\frac{Ag}{\omega}\frac{\cosh\omega(h+z)}{\cosh\omega h/c}\cdot\tilde{\varphi}(x,y)\sin\omega t$ Ref. [82], p. 440
2. Isosceles Right Triangle with Constant Depth 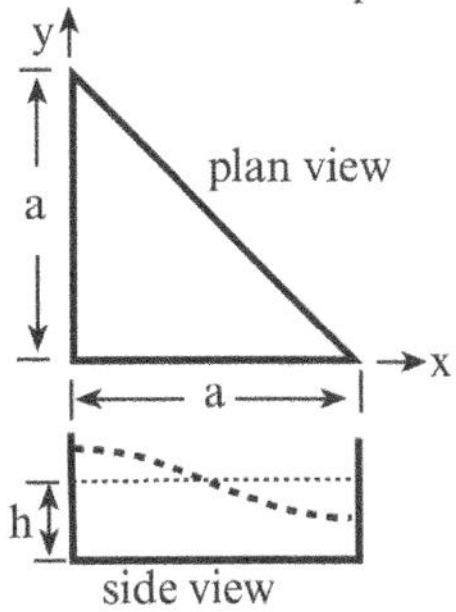	General case, $\frac{1}{2}\left[\frac{g}{\pi a}\left(i^2+j^2\right)^{1/2}\tanh\frac{\pi h}{a}\left(i^2+j^2\right)^{1/2}\right]^{1/2}$ Shallow liquid, $(h/a)\left(i^2+j^2\right)^{1/2}<1/10$, $\frac{(gh)^{1/2}}{2a}\left(i^2+j^2\right)^{1/2}$ Deep liquid, $(h/a)\left(i^2+j^2\right)^{1/2}>1$, $\frac{g^{1/2}}{2\pi^{1/2}a^{1/2}}\left(i^2+j^2\right)^{1/4}$,	$\cos\frac{i\pi x}{a}\cos\frac{j\pi y}{b}$ $\pm\cos\frac{j\pi x}{a}\cos\frac{i\pi y}{b}$ – if i+j = even + if i+j = odd (i,j) = (1,2). (2,1), (2,3), (3,2).. i.j >0 but i≠j.
3. Circular Const. Depth 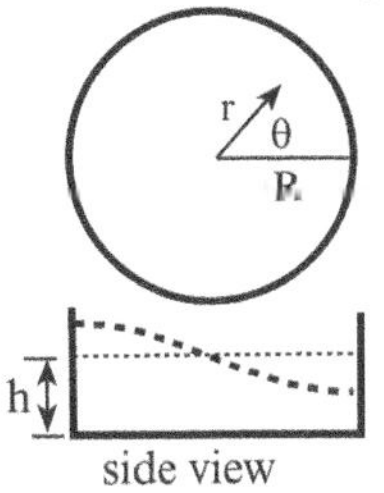 Additional values of λ_{ij} are tabulated in Refs [62], pp. 411,468; [82], pp. 284-285.	General case, $\frac{1}{2\pi}\left(\frac{\lambda_{ij}g}{R}\tanh\frac{\lambda_{ij}h}{R}\right)^{1/2}$ $i=0,1,2,3,..\ j=0,1,2,3,\ \lambda_{ij}$ given at right Shallow liquid, $\lambda_{ij}h/R<\pi/10$, $\frac{\lambda_{ij}(gh)^{1/2}}{2\pi R}$, $i=0,1,2,3..,\ j=0,1,2,3..$ Deep liquid, $\lambda_{ij}h/R>\pi$, $\frac{1}{2\pi}\left(\frac{\lambda_{ij}g}{R}\right)^{1/2}$, $i=0,1,2,3..,\ j=0,1,2,3..$	$J_i(\lambda_{ij}r/R)\cos i\theta$, $[J_i'(\lambda_{ij})=0]$ i = 0,1,2,.. nodal diameters j = 0,1,2,.. nodal circles λ (see table below) $\lambda_{i=0,j}\approx\pi(j+0.25),\ j\geq 3$

λ

i	j=0	j=1	j=2
0	0	3.8317	7.0156
1	1.8412	5.3316	8.5363
2	3.0542	6.7061	9.970
3	4.2012	8.0152	11.346
4	5.3175	9.2824	12.682
5	6.4156	10.520	13.987

$\lambda_{i=0,j}\approx\pi(j+0.25),\ j\geq 3$

Table 6.7 Natural frequencies of surface waves, continued

Basin Plan Form	Natural Frequency f_{ij}, Hz	Mode Shape Potential $\tilde{\Phi}$
4. Annular Constant Depth 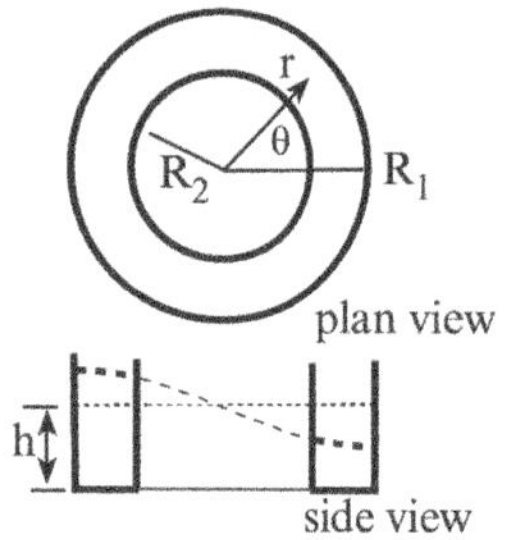 i = 0,1,2,.. nodal diameters j = 0,1,2,.. nodal circles Additional values of λ_{ij} are tabulated in Ref. [62] Eqn for λ: $d\Phi/dr\|_{r=R2}=0$	General case, $\frac{1}{2\pi}\left(\frac{g\lambda_{ij}}{R_1}\tanh\frac{\lambda_{ij}h}{R_1}\right)^{1/2}$ $i=0,1,2,3,..\ j=0,1,2,3,\ \lambda_{ij}$ at right - - - - - - - - - - Shallow liquid, $\lambda_{ij}h/R_1<\pi/10$, $\frac{\lambda_{ij}(gh)^{1/2}}{2\pi R_1},\quad i=0,1,2,3..,\ j=0,1,2,3..$ - - - - - - - - - - Deep liquid, $\lambda_{ij}h/R_1>\pi$, $\frac{1}{2\pi}\left(\frac{\lambda_{ij}g}{R_1}\right)^{1/2},\quad i=0,1,2,3..,\ j=0,1,2..$	$R_1\le r\le R_2$ $\left[Y_i'(\lambda_{ij})J_i(\lambda_{ij}r/R_1)\right.$ $\left.-J_i'(\lambda_{ij})Y_i(\lambda_{ij}r/R_1)\right]\cos i\theta$ $\approx\cos\left[\frac{j\pi(r-R_2)}{R_1-R_2}\right]\cos i\theta,\ \frac{R_1}{R_2}>0.8$ (table of λ below) $\lambda_{ij}^2\approx\left(\frac{j\pi}{R_1}\right)^2+\left(\frac{2iR_1}{R_1+R_2}\right)^2,\ \frac{R_1}{R_2}>0.5$

R_1/R_2	j	λ, i = 0	1	2	3	4
0.3	0	0	1.582	2.969	4.180	5.310
	1	4.706	5.137	6.274	7.721	9.153
0.5	0	0	1.355	2.681	3.958	5.175
	1	6.393	6.565	7.063	7.840	8.836

5. Annular, Circular Sector 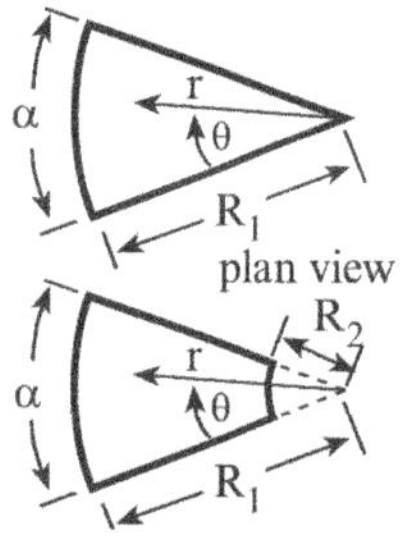 	The natural frequencies can be adapted from those in cases 4 and 5 with the following choices of the index i, $i=\frac{180n}{\alpha},\quad n=0,1,2,3..$ 0° <=α<360°, α in degrees. j and given in cases 4 and 5. i will be a submultiple of 180n. For non integer i, λ_{ij} can be interpolated from the tables in these cases.	See cases 3 and 4. Sectors are constant depth. Ref. [32]
6. Conical Basin 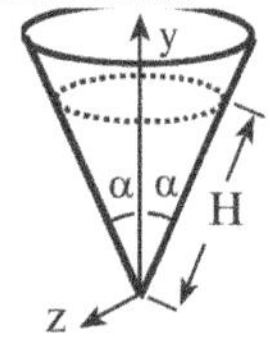 	$f=\frac{\lambda^{1/2}}{2\pi}\left(\frac{g}{H}\right)^{1/2}$ Fundamental mode α, deg. 0 20 3 45 50 60 80 90 λ sin α 1.8 1.55 1.18 1.05 .95 0.77 0.21 0	Ref. [83]
7. Inclined Right Circular Cylinder 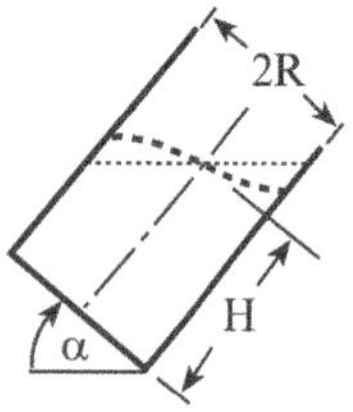 	$f=\frac{\lambda^{1/2}}{2\pi}\left(\frac{g}{R}\right)^{1/2}$ fundamental mode Ref. [84]	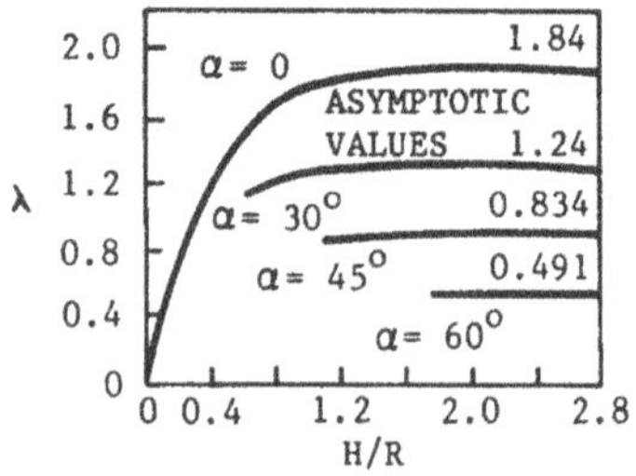

Table 6.7 Natural frequencies of surface waves, continued

Basin Description	Natural Frequency f_i Hz

8. Partially Filled Sphere of Radius R

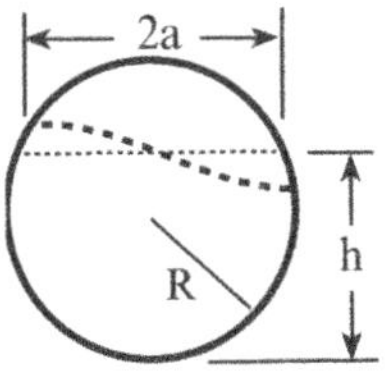

i = 1, lowest frequency mode, shown.
$a^2 = 2hR - h^2$
Refs [82]; [85], p. 292

$$f_i = \frac{\lambda_i^{1/2}}{2\pi}\left(\frac{g}{R}\right)^{1/2}, \quad i = 1,2,3,.\ i\text{-}1 = \text{num. nodal diameters}$$

h/R	λ_0	λ_1	λ_2	λ_3
0	4.000	1.000	-	-
0.2	3.861	1.073	2.108	3.129
0.4	3.7080	1.158	2.234	3.282
0.6	3.650	1.263	2.388	3.466
0.8	3.658	1.392	2.577	3.696
1.0	3.7452	1.560	2.820	3.994
1.2	3.938	1.7881	3.149	4.403
1.4	4.3010	2.123	3.6334	5.012
1.6	5.0073	2.686	4.451	6.055
1.8	6.7642	3.959	6.315	8.463
2.0	∞.	∞	∞	∞

9. Round or Rectangular Hole in a Sheet Covering a Reservoir

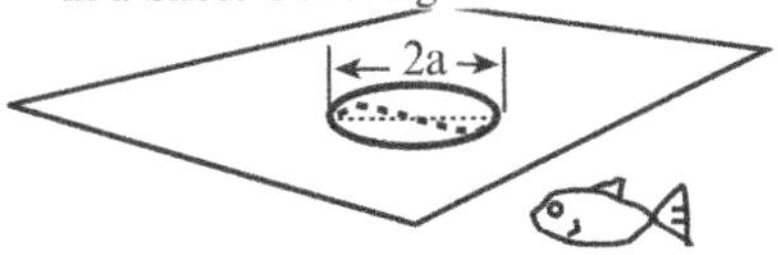

$$f_i = \frac{\lambda_i^{1/2}}{2\pi}\left(\frac{g}{a}\right)^{1/2}, \quad i = 1,2,3,.$$ Ref. [86]

λ_1=2.75 for round opening shown
λ_1=2.006 for square opening, side 2a.
Fundamental mode for limit h→2R from case 8 and 14

10. Angular Canal, transverse modes

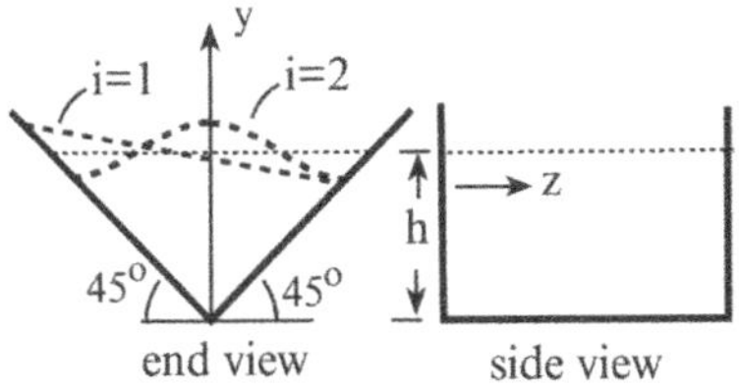

$$f_i = \frac{\lambda_i^{1/2}}{2\pi}\left(\frac{g}{h}\right)^{1/2}, \quad i = 1,2,3,..,$$ Ref. [82], pp. 442–444.

$\lambda_1 = 1.0$, $\lambda_2 = 2.324$. $\lambda_3 = 3.924$, $\lambda_4 = 5.498$
$\lambda_i = \alpha \tanh \alpha$, where $\cos(2\alpha)\cosh(2\alpha) = 1$, $i > 1$
Solutions are given in Table 4.3 (free-free beam).
Modes with i odd (1,3,..) are anti-symmetric about y-z plane Modes with i=even (2,4,..) are symmetric and apply also to a channel between vertical wall and a 45 degree wall. For angles other than 45 degrees see Ref. [87].

11. Trapezoidal Channel Transverse modes

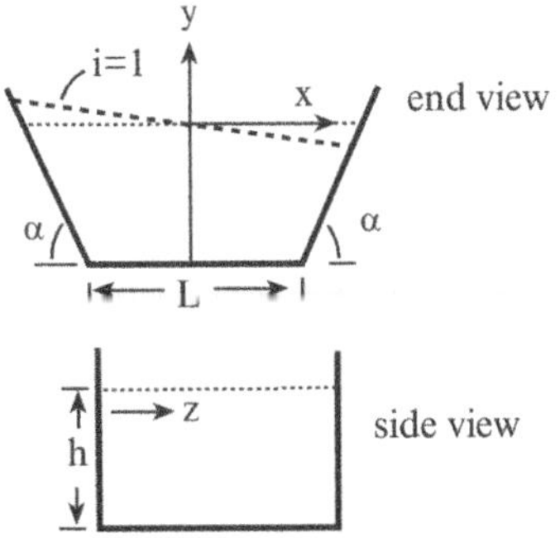

See Ref. [88] for longitudinal modes.

$$f = \frac{\lambda^{1/2}}{2\pi}\left(\frac{g}{h}\right)^{1/2}$$

anti-symmetric modes independent of z

Ref. [89] and case 1

$\frac{L}{h}$	λ	α, degree 30°	45°	60°	90°
0.5	λ_1	0.477	0.882	1.538	6.283
	λ_2	2.116	3.426	5.369	18.84
	λ_3	3.887	5.816	8.832	31.41
1.0	λ_1	0.413	0.721	1.162	3.130
	λ_2	1.861	2.844	4.120	6.283
	λ_3	3.411	4.834	6.783	9.425
1.5	λ_1	0.364	0.600	0.910	2.032
	λ_2	1.674	2.444	3.349	6.283
	λ_3	3.034	4.145	5.529	15.71
2.0	λ_1	0.303	0.490	0.717	1.441
	λ_2	1.48	2.094	2.795	4.712
	λ_3	2.667	3.572	4.626	7.854

Table 6.7 Natural frequencies of surface waves, continued

Additional notation: $J_0()$ = Bessel function of first kind and zero order.

Basin Description	Natural Frequency f_i Hz

12. Right Circular Canal

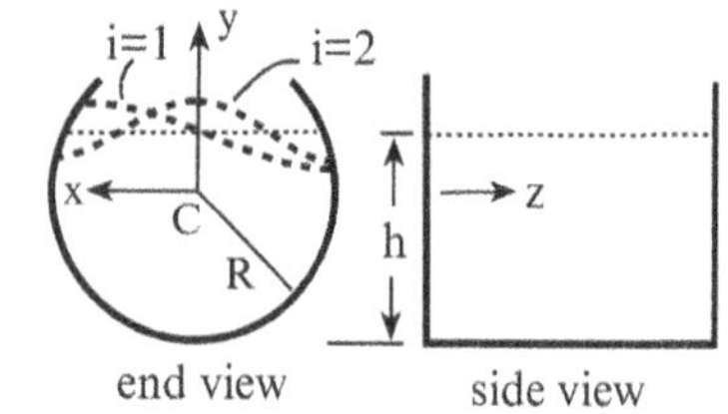

R = radius of the canal
These modes are independent of z.
Modes λ_1 and λ_3 are anti-symmetric and modes λ_2 and λ_4 are symmetric about y-z plane. $\lambda_i=(i/2)(i+1)$ for $h/R \to 0$. Ref. [70]. See case 14 for $h/R \to 2$.

$$f_i = \frac{\lambda_i^{1/2}}{2\pi}\left(\frac{g}{R}\right)^{1/2}, \quad i = 1,2,3,.., \qquad \text{Ref. [85]}$$

h/R	λ_1	λ_2	λ_3	λ_4
0	1.0000	3.000	6.000	10.000
0.2	1.0439	2.929	5.354	8.030
0.4	1.0970	2.890	4.937	6.991
0.6	1.1627	2.889	4.700	6.461
0.8	1.2461	2.932	4.607	6.236
1.0	1.3557	3.033	4.651	6.239
1.2	1.5075	3.216	4.851	6.468
1.4	1.7434	3.538	5.277	7.000
1.6	2.1237	4.143	6.139	8.103
1.8	3.0214	5.627	8.313	10.906
2.0	∞	∞	∞	∞

13. Circular Basin with Parabolic Bottom

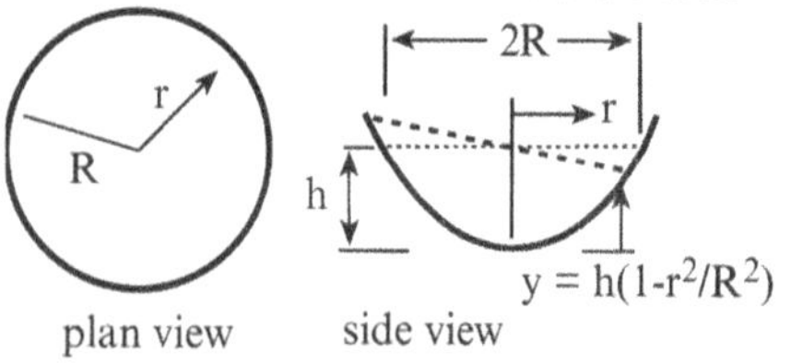

$$f_{ij} = \frac{\lambda_{ij}^{1/2}}{2\pi}\frac{(gh)^{1/2}}{R}$$

$\lambda_{ij}= i(4j-2)+4j(j-1)$
i = 0,1,2, = number of nodal diameters
j = 1,2,3,, = number of nodal circles
fundamental mode, i=1, j=1: $\lambda_{11}=2$,
Refs [82], pp. 291–293; [90].

14. Variable depth Channel, transversemode

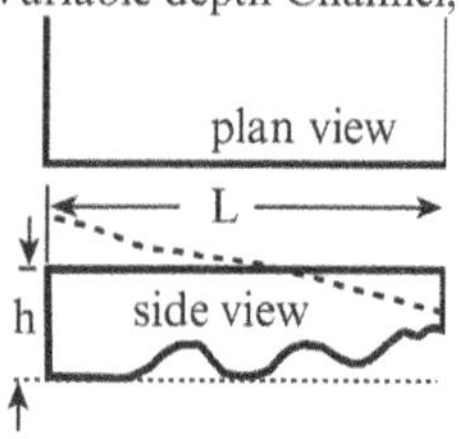

$$f \le \begin{cases} \dfrac{(gh)^{1/2}}{2L}, & \dfrac{h}{L} < \dfrac{1}{10} \\ \dfrac{1}{2}\left(\dfrac{g}{\pi L}\right)^{1/2}, & \dfrac{h}{L} > 1 \end{cases}$$

Approximate formula for fundamental anti-symmetric transverse mode. Refs [91, 92]

15. Shallow Elliptical Const. Depth

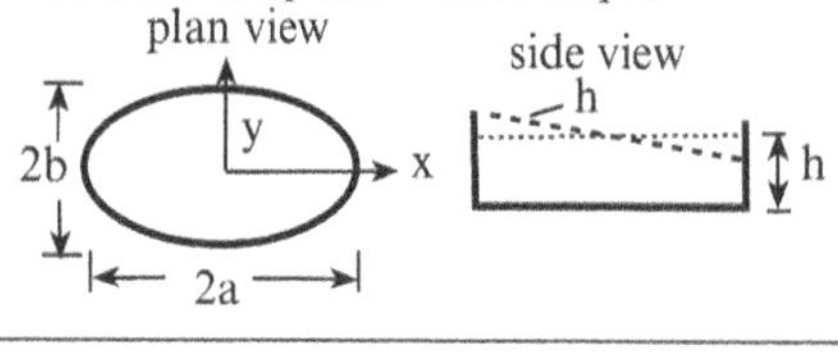

Fundamental mode for shallow liquid, h<<a, b/a<1

$$\frac{(gh)^{1/2}}{2\pi a}\left[\frac{18+6(b/a)^2}{5+2(b/a)^2}\right]^{1/2}, \qquad \frac{x^2}{a^2}+\frac{y^2}{b^2}=1$$

Shallow basin (long wave length).
Surface elevation, $\eta \sim x$. Ref. [82], p. 290.

16. Shallow Rectangular Tapered depth

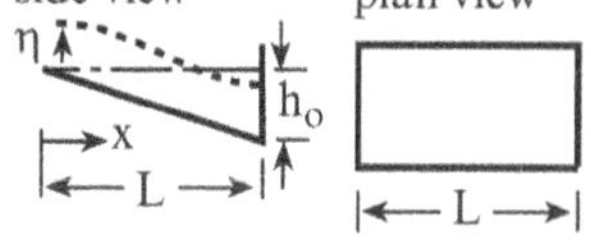

$$\frac{\lambda_i (gh_o)^{1/2}}{4\pi L}, \; i = 1,2,3.., \qquad \tilde{\eta}_i(x) = J_o\left(2\lambda_i\left(\frac{x}{L}\right)^{1/2}\right)$$

$\lambda_i = 3.8317, 7.0156, 10.1735,... \quad J_1(\lambda_i) = 0$

Shallow basin (long wave length), h_o<<L
Ref. [82], p. 276 in part

Table 6.7 Natural frequencies of surface waves, continued

Shallow Basin Description, h/L<0.1	Natural Frequency f_i Hz; Period $T_i=1/f_i$
17 Shallow Rectangular basin, Open end plan view side view 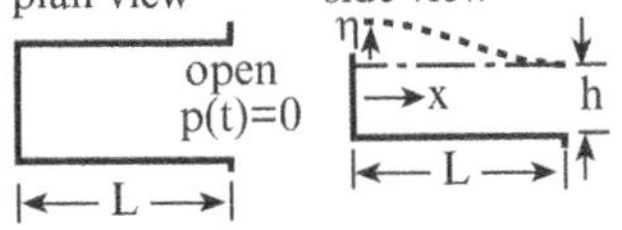	$\frac{(gh)^{1/2} i}{4L}$, $i=1,3,5..(\text{odd})$ Mode shape $\tilde{\eta}_i(x)=\cos\frac{i\pi x}{2L}$ Shallow basin (long wave length), h<<L Open at x=L, η=0 at x=L. Adapted from case 1
18. Shallow Rectangular with Open End and Sloping depth plan view side view 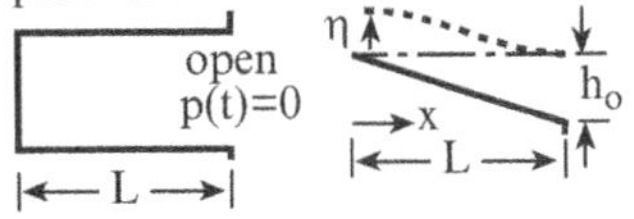	$\frac{\lambda_i (gh_o)^{1/2}}{4\pi L}$, $i=1,2,3..$ Mode shape $\tilde{\eta}_i(x)=J_o\left(2\lambda_i\left(\frac{x}{L}\right)^{1/2}\right)$ $\lambda_i=2.408, 5.520, 8.654,...$ $J_o(\lambda_i)=0$ Shallow basin (long wave length), h_o<<L. η=0 at x=L. Ref. [82], p. 276; [93]
19. Shallow Open Angular Basin plan view side view 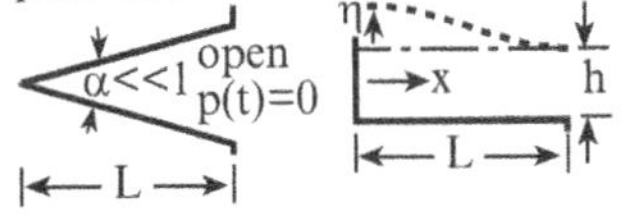	$\frac{\lambda_i (gh)^{1/2}}{\pi L}$, $i=1,2,3..$ Mode shape $\tilde{\eta}_i(x)=J_o\left(\lambda_i\frac{x}{L}\right)$ $\lambda_i=2.408, 5.52007, 8.653...$ $J_o(\lambda_i)=0$ Shallow basin (long wave length), h<<L, α<<1 radian. Ref. [82], p. 259
20. Shallow Open Semi-Circular Basin 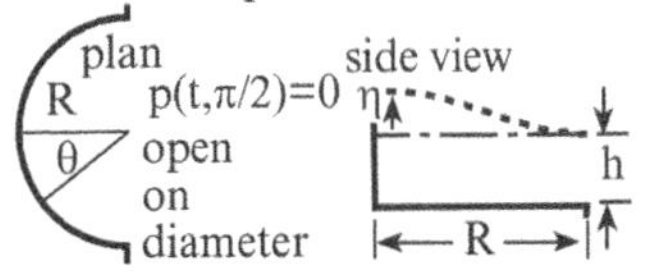	$\frac{\lambda_i (gh)^{1/2}}{\pi R}$, $i=1,2,3,..$ Mode shape $\tilde{\eta}_i(x)=J_1\left(\lambda_i\frac{r}{R}\right)\cos\theta$ $\lambda_i=1.8412, 5.3316, 8.53...$ $(J_1'(\lambda_i)=0)$ Shallow basin (long wave length), h<<R. η=0 at θ=π/2. Adapted from case 3.
21. Shallow Rectangular Double Angle plan view side view 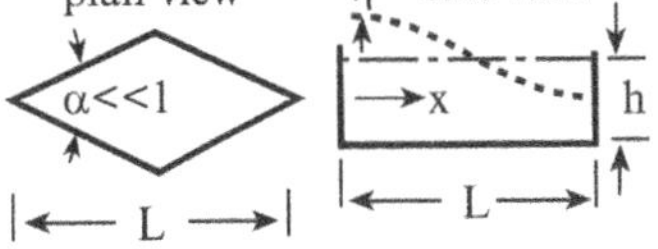	$\frac{\lambda_i (gh_o)^{1/2}}{\pi L}$, $i=1,2,3$ Mode shape $\tilde{\eta}_i(x)=J_o\left(-\lambda_i\left(\frac{x}{L}\right)^{1/2}\right)$ anti-symmetric modes $\lambda_i=3.8317, 7.0156, 10.172...$ $(J_1(\lambda_i)=0)$ symmetric modes $\lambda_i=2.408, 5.5201, 8.653...$ $(J_o(\lambda_i)=0)$ Shallow basin (*long* wave length), h<<L. Ref. [82], p. 276
22 Rectangular basin, Sloping beach side view plan view 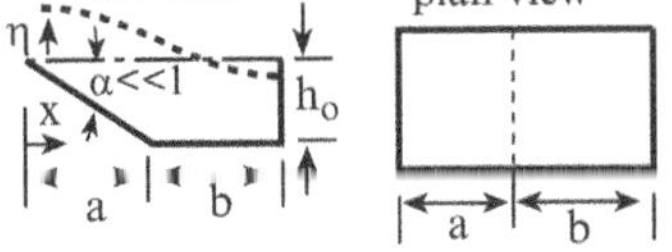	$\frac{\lambda_a (gh_o)^{1/2}}{2\pi a}$ or, $\frac{\lambda_b (gh_o)^{1/2}}{2\pi b}$ first mode a/b 0 0.1 0.2 0.5 1 2 5 10 ∞ λ_a - 0.285 .5197 .9839 1.321 1.560 1.749 1.827 1.91 λ_b π 2.852 2.598 1.967 1.321 .7797 .350 .1827 - $J_1(2\lambda_a)\cos((b/a)\lambda_a)+J_0(2\lambda_a)\sin((b/a)\lambda_a)=0$ Authors's result requested by Charles Coughran

Table 6.7 Natural frequencies of waves in U-tubes, continued

Additional notation: g = acceleration of gravity (Table 1.3); h = height of fluid in tank; u, v = displacement of fluid from level; A = cross-sectional area; L = mean length of duct along duct centerline. Overbar (~) denotes the mode shape. Ducts need not have circular cross section.

U-Tube Description	Natural Frequency f_i, Hz	Mode Shape
23. Uniform U-Tube 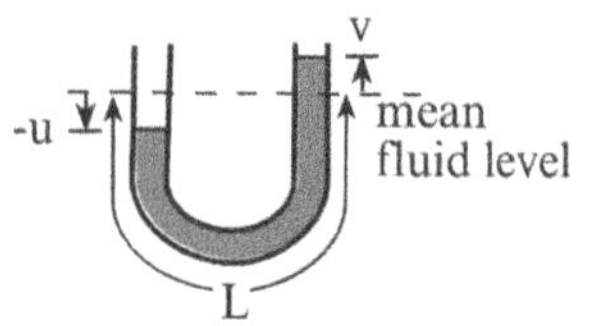	$\frac{1}{2\pi}\sqrt{\frac{2g}{L}}$	$\begin{bmatrix}\tilde{u}\\ \tilde{v}\end{bmatrix}=\begin{bmatrix}1\\ -1\end{bmatrix}$ L = mean fluid-filled length of pipe along centerline.
24. Tank and Duct 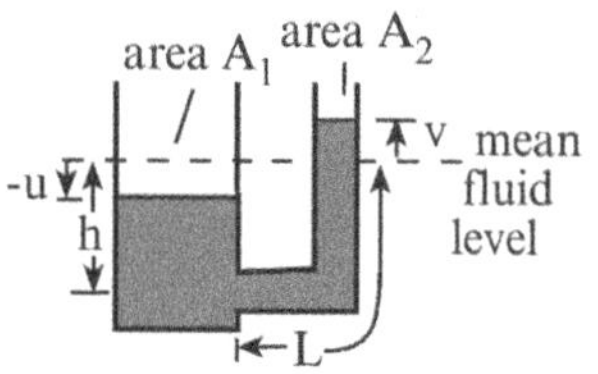	$\frac{1}{2\pi}\left[\frac{g\left(1+\frac{A_2}{A_1}\right)}{h\frac{A_2}{A_1}+L}\right]^{1/2}$	$\cdot\ \begin{bmatrix}\tilde{u}\\ \tilde{v}\end{bmatrix}=\begin{bmatrix}1\\ -\frac{A_1}{A_2}\end{bmatrix}$ L = mean fluid-filled length of pipe along centerline
25. Two Tanks Connected Same Level 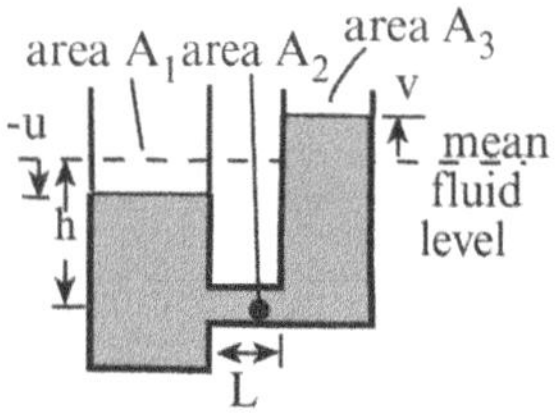	$\frac{1}{2\pi}\left[\frac{g\left(1+\frac{A_2}{A_1}\right)}{h\left(1+\frac{A_2}{A_1}\right)+\frac{A_2}{A_3}L}\right]^{1/2}$	$\cdot\ \begin{bmatrix}\tilde{u}\\ \tilde{v}\end{bmatrix}=\begin{bmatrix}1\\ -\frac{A_1}{A_2}\end{bmatrix}$ Ref. [93] L = mean fluid-filled length of pipe along centerline
26. Two Tanks Connected on Different Levels 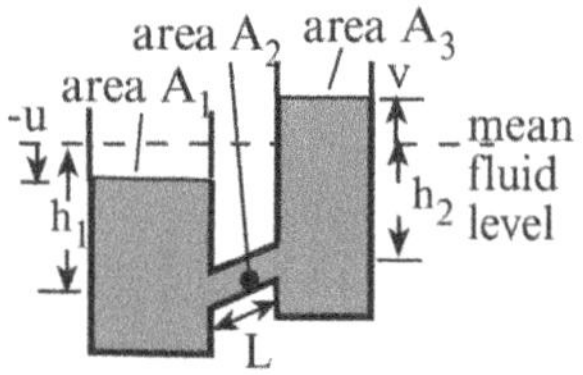	$\frac{1}{2\pi}\left[\frac{g\left(1+\frac{A_2}{A_1}\right)}{h_2+h_1\frac{A_2}{A_1}+\frac{A_2}{A_3}L}\right]^{1/2}$	$\cdot\ \begin{bmatrix}\tilde{u}\\ \tilde{v}\end{bmatrix}=\begin{bmatrix}1\\ -\frac{A_1}{A_2}\end{bmatrix}$ L = mean fluid-filled length of pipe along centerline
27. Pipe into ocean 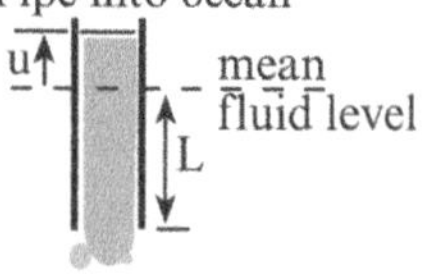	$\frac{1}{2\pi}\sqrt{\frac{g}{L}}$	Open ended pipe extends distance L into reservoir.
28 Pipe in Ocean with Restriction 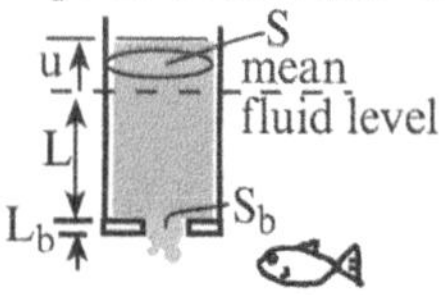	$\frac{1}{2\pi}\sqrt{\frac{gS_b}{L_bS+S_bL}}$	S_a = cross section area of main tank. S_b = area of bottom vent.

excitation of liquid storage tanks [96–99]. For example, fuel sloshing in the liquid oxygen tanks of the Saturn I missile induced unexpected roll oscillations during the late stages of powered flight, and landslides in the lake behind the dam in Longarone, Italy, on October 9, 1963, induced such violent sloshing that the lake overspilled the dam, causing a catastrophic flood [96].

The general assumptions for sloshing wave analysis are that (1) liquid is homogeneous, (2) inviscid (zero viscosity), and (3) incompressible; (4) the boundaries are rigid; (5) the wave amplitudes are small so linear analysis applies; (6) the density of the surrounding atmosphere is negligible; and (7) the surface tension is negligible. The inviscid model usually is a good approximation if the Reynolds number based on maximum velocity is order of 1000 or greater. Surface tension increases wave speed by the factor $(1 + (2\pi)^2\sigma/\rho g\lambda^2))^{1/2}$, which is significant for ripples in water with wavelengths on the order of 1.7 cm or smaller at 20 °C [99, 100]. σ is the surface tension, ρ is liquid density, and g is acceleration due to gravity (Table 1.2). Surface tension has little effect on long wavelength water waves.

Surface wave propagation speed is a function of circular frequency ω (rad/s), depth h, and gravity acceleration g [98, 99, 101]:

$$c = \frac{g}{\omega}\tanh\frac{\omega h}{c} = \begin{cases} (gh)^{1/2}, & \text{shallow liquid wave, } \dfrac{\omega h}{c} = \dfrac{2\pi h}{\lambda} < \dfrac{\pi}{10} \\ \dfrac{g}{\omega}, & \text{deep liquid wave, } \dfrac{\omega h}{c} = \dfrac{2\pi h}{\lambda} > \pi \end{cases} \tag{6.55}$$

Wave speed increases with fluid depth until the deep water limit is reached. In shallow liquid waves, the wave oscillations extend to the bottom of the basin, whereas deep liquid waves decay exponentially with depth. These limits are associated with mathematical limits of hyperbolic functions: $\cosh 2\pi h/\lambda = \sinh 2\pi h/\lambda x = (1/2)\exp(2\pi h/\lambda)$ for $2\pi h/\lambda >> 1$, $\cosh 2\pi h/\lambda = 1$, and $\sinh 2\pi h/\lambda = 2\pi h/\lambda$ for $2\pi h/\lambda << 1$. Wavelength is $\lambda = 2\pi c/\omega$.

Table 6.7 has Sloshing natural frequencies and mode shapes (Table 6.7) [32, 62, 82–94] that apply to liquid in free surface basins. The surface wave displacement, liquid pressure, and liquid velocity are expressed by a velocity potential function Φ:

$$\begin{aligned} &\text{Wave height}: \ \eta = \frac{1}{g}\frac{\partial \Phi}{\partial t} \ \text{at free suface } z = 0 \\ &\text{Pressure}: \ p = \rho\frac{\partial \Phi}{\partial t} - \rho g z \\ &\text{Velocity}: \ u = \nabla\Phi \\ &\text{Rigid boundary}: \ \mathbf{u}\cdot\mathbf{n} = 0, \text{ or, } \nabla\Phi\cdot\mathbf{n} = 0 \end{aligned} \tag{6.56}$$

The natural frequencies and mode shapes of surface waves in basins are found by solving the Laplace equation $\nabla^2\Phi = 0$, which is the wave equation in incompressible fluid (look ahead to Section 6.7). The waves propagate in x–y plane. The potential function [98, 99, 101] in constant depth basins is

$$\Phi = \frac{Ag}{\omega}\frac{\cosh\omega(h+z)}{\cosh\omega h/c}\tilde{\varphi}(x,y)\sin\omega t \tag{6.57}$$

where A = amplitude of the wave, c = wave speed, g = acceleration due to gravity (Table 1.2), h = mean depth of liquid in basin, $\tilde{\varphi}(x, y)$ = dimensionless mode shape, t = time, and z = depth, positive upward from bottom.

In shallow basins, the wavelength is much longer than the depth, the vertical acceleration and vertical velocity are neglected, and the wave speed is $(gh)^{1/2}$. The equations of motion and pressure for one-dimensional shallow water waves are [82] (pp. 255–274)

$$\frac{\partial^2 \eta}{\partial t^2} = \frac{g}{b}\frac{\partial}{\partial x}\left(S(x)\frac{\partial \eta}{\partial x}\right), \quad \frac{\partial p}{\partial x} = \rho g \frac{\partial \eta}{\partial x}, \quad \frac{\partial u}{\partial t} = -g\frac{\partial \eta}{\partial x}$$

$S(x) = bh$ is the basin cross-sectional area, perpendicular to the direction of propagation (x), where b is the breadth and h is mean depth. Solutions are shown in cases 15 through 22 of Table 6.7. Additional solutions are shown in Refs [83, 98, 99].

The fundamental sloshing natural frequency can be estimated:

$$f(\text{Hz}) = \begin{cases} \dfrac{(gh)^{1/2}}{(2L)}, & \dfrac{h}{L} < \dfrac{1}{10} \\[2ex] \dfrac{(g/(\pi L))^{1/2}}{2}, & \dfrac{h}{L} > 1 \end{cases} \tag{6.58}$$

L is the maximum lateral dimension of the basin, h is the mean depth, and g is the acceleration due to gravity. For example, consider a circular basin with radius R whose maximum depth equals the radius and four variations in that depth. The fundamental natural frequency is

$$f\ (\text{Hz}) = \frac{\beta}{2\pi}\left(\frac{g}{R}\right)^{1/2} \quad \text{where } \beta = \begin{cases} 1.32 & \text{flat bottom} \\ 1.23 & \text{conical bottom} \\ 1.41 & \text{parabolic bottom} \\ 1.25 & \text{spherical bottom} \end{cases} \tag{6.59}$$

The difference between the highest and lowest of these is only 14%. Thus, constant depth basin solutions are often good estimates of the fundamental natural frequencies of the surface of variable depth basins. See case 10 of Table 6.7. If the maximum lateral dimension of the basin is on the order of kilometers, such as for a lake, the fundamental natural period ($1/f$) is on the order of minutes or even hours.

Epstein [97], Ibrahim [98], and Faltinsen [99] discuss sloshing as a result of forced basin motion and the effect of basin wall elasticity.

Harbors and channels open to the sea (cases 17–20 of Table 6.7) are differentiated from the tanks and basins by an opening to the sea. Exact solutions for shallow waves set the pressure at the open boundary to the hydrostatic pressure in the sea, which implies that the wave height at the opening is zero. Numerical solutions for circular and rectangular harbors in Refs [98, 99, 101–104] and Figure 6.16 suggest that the open channel results in

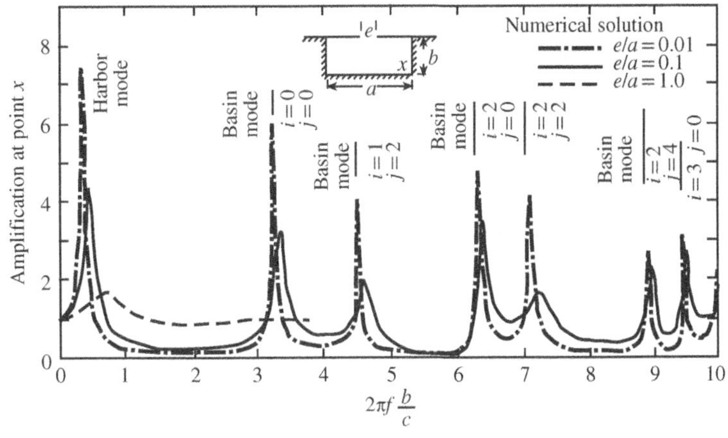

Figure 6.16 Response of a rectangular harbor with various size openings to sea waves at frequency f for $a/b = 2$. The response is measured at x [101]

at least *harbor mode* or *pumping* modes that is lower in frequency than the modes of the basin formed by closing the opening to the sea.

U-tubes (cases 23–28 of Table 6.7) are filled with incompressible liquid to the mean level. Once disturbed, liquid flows back and forth at the U-tube natural frequency. The natural frequencies and mode shapes were derived from principles of conservation of mass and energy [94]. The velocity is assumed to be uniform over the cross section and viscosity is neglected. Tanks connected with U-tubes have been used to damp roll of ships ([98], p. 677; [99]).

6.6 Ships and Floating Systems

6.6.1 Ship Natural Frequencies (1/Period)

Table 6.8 [105, 106] has the natural frequencies of ships. A ship has six rigid body degrees of freedom. There are three displacements, surge (η_1), sway (η_2), and heave (η_3), and three rotations, roll (η_4), pitch (η_5), and yaw (η_6), as shown in Table 6.8. Heave, pitch, and roll possess hydrostatic stiffness and natural frequencies. For example, at surge equilibrium, the weight of water displaced by the hull equals the dry weight of the ship. A small heave displacement ($-\eta_3$) into the water results an incremental increase in buoyancy on the ship equal to the weight of the incremental displaced water, $F_3 = \rho g S \eta_3$, where ρ is the liquid density in mass units, g is the acceleration due to gravity, and S is the area enclosed by the ship's waterline. This buoyancy stiffness gives rise to a surge natural frequency (case 1 of Table 6.8).

Buoyancy force is applied at the center of buoyancy, which is the centroid of hull volume below the waterline. Roll (η_4) and pitch (η_5) shift the center of buoyancy laterally relative to the ship's center of mass (Figure 6.17) and generate a hydrostatic roll moment on the

Table 6.8 Ship motions

Notation: b = beam of ship at water line; b_m = maximum beam of ship at waterline; B = centroid of volume of the liquid displaced by hull; d = depth of hull below waterline; f = natural frequency in Hertz, = 1/T where T is natural period; g = acceleration of gravity (Table 1.2); x = fore-and-aft coordinate; y = lateral coordinate; z = vertical coordinate; ρ = density of liquid (water); A_z = added mass of ship in heave; G = center of gravity of dry ship; I_{yy} = moment of inertial of water plane area about the roll axis; J_{xx} = mass moment of inertial of ship about x axis; $M = \rho\nabla$ = mass of ship; R = radius of hull plan form; S = water plane area; θ = roll angle, radian; ∇ = hull displacement volume. See Table 1.2 for consistent sets of units.

Motion	Natural Frequency f, Hz	Approximate Natural Frequency for Slender Ships, Hz
1. Heave –vertical motion (z) axis	$\frac{1}{2\pi}\left[\frac{\rho g S}{M+A_z}\right]^{1/2}$	$0.12\left(\frac{g}{d}\right)^{1/2}$ Ref. [105]
2. Pitch – rotation about transverse (y) axis	$\frac{1}{2\pi}\left[\frac{\rho g I_{yy}}{J_{yy}+A_{yy}}\right]^{1/2}$	$0.13\left(\frac{g}{d}\right)^{1/2}$ Ref. [106]
3. Roll – rotation about longitudinal (x) axis	$\frac{1}{2\pi}\left[\frac{Mg\overline{GM}}{J_{xx}+A_{xx}}\right]^{1/2}$	$0.35\left(\frac{g\overline{GM}}{b_m^2}\right)^{1/2}$

Roll Stability	Pitch Moment M_x and Metacentric Height GM
4. Ship Rolls 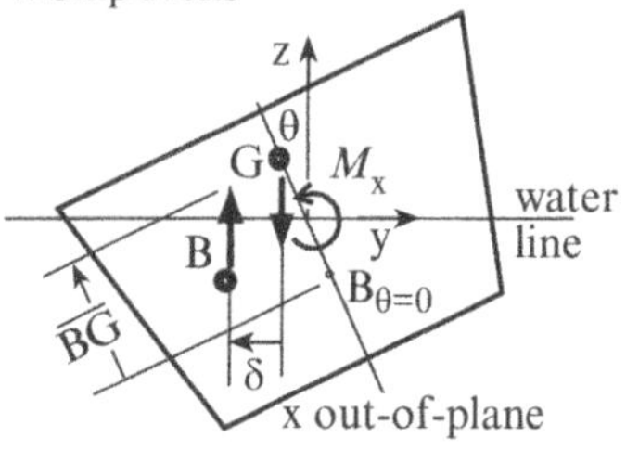	Roll moment is result of buoyancy and gravity forces. $M_x = -(\rho g\nabla)\delta$ $\approx -(\rho g\nabla)\left.\frac{d\delta}{d\theta}\right\|_{\theta=0}\theta = -(\rho g\nabla)\overline{GM}\theta$ for small angles $\overline{GM} = \left.\frac{d\delta}{d\theta}\right\|_{\theta=0} > 0$ for positive roll stability $\overline{GM} = -\overline{BG} + \frac{I_{yy}}{\nabla}$, for ship with flat vertical sides I_{yy} = area moment of inertia of waterline area at $\theta = 0$ δ is the horizontal distance from the ship center of gravity G to the center of buoyancy B, positive in direction of roll.

5. Ship Motion Nomenclature

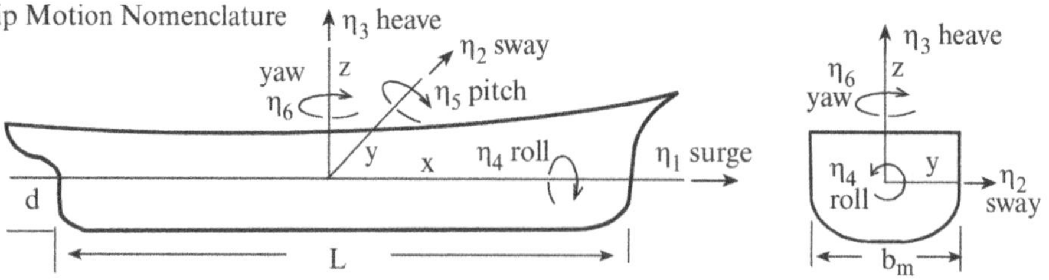

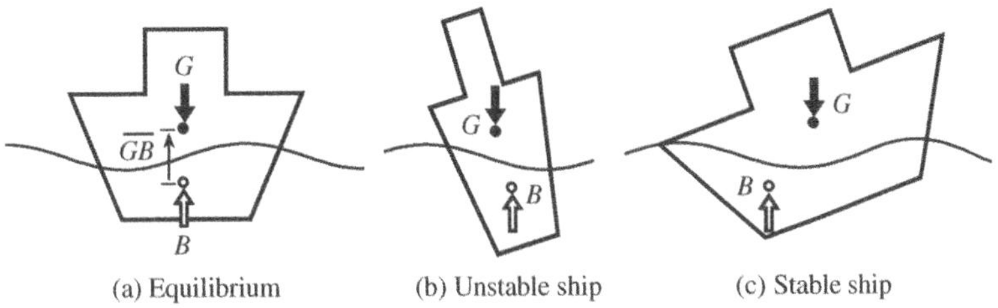

Figure 6.17 Stability of a ship in roll. The moment increases roll of an unstable ship

hull. If roll moves the center of buoyancy toward the leeward side of the center of gravity then the roll moment is stabilizing since it counters roll:

$$M_4 = -\overline{\mathrm{GM}}\ \rho g \nabla \eta_4 \tag{6.60}$$

where η_4 is the small roll angle from vertical equilibrium in radians. Here, ∇ is the displaced water volume of the hull; thus, $\rho g \nabla$ is the mass of the ship. The length GM is *metacentric height*. If GM is negative, the ship is unstable and it will turn turtle, a fate which befell the warship Vasa on her maiden voyage from Stockholm in 1628.

Metacentric height GM has units of length. For small angles of symmetric ships with flat vertical sides (at the waterline), the metacentric height is the sum of the equilibrium distance BG, which is positive upward from the center of buoyancy upward to the center of gravity, plus the tilting waterline effect:

$$\overline{\mathrm{GM}} \approx -\overline{\mathrm{BG}} + \frac{I_{yy}}{\nabla} \tag{6.61}$$

BG is positive if the center of gravity is above the center of buoyancy, which is generally the case for ships. I_{yy} is the moment of inertia of the water plane about the roll axis. GM is positive for a stable ship. Stability is enhanced by having the ship's center of gravity as low as possible and the ship's beam as wide as possible to increase I_{yy}. Submarines, spar buoys, and floating platforms achieve stability not through the tilting water plane but by adding mass to the keel, via ballast or chains, so that the center of gravity is below the center of buoyancy so BG is always negative.

6.7 Added Mass of Structure in Fluids

Tables 6.9 and 6.10 have the added mass of two dimensional sections and three dimensional bodies respectively. The natural frequency of an elastic structure that is submerged in water is lower than its natural frequency in air or owing to the entrained of fluid mass, called *added mass* or *virtual mass*.

Table 6.9 Added mass of sections

Notation: m_a = added mass per unit length; J_a = added mass moment of inertia per unit length about centroid; ρ = fluid mass density. t = theory; e = experimental result.

<table>
<tr><th>Section</th><th>Added Mass per Unit Length, m_a</th><th>Added Mass Moment Inertia J_a</th></tr>
<tr><td>1. Circle with Radius R
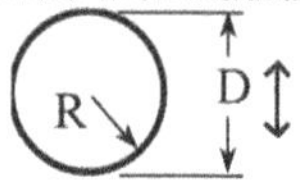
</td><td>$\rho\pi R^2 = \rho\frac{\pi}{4}D^2$
↕ direction of acceleration</td><td>0 t Ref. [107]</td></tr>
<tr><td>2. Ellipse
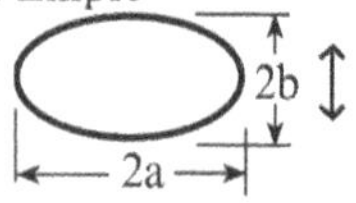
</td><td>$\rho\pi a^2$
$0 < a/b < \infty$</td><td>$\rho\frac{\pi}{8}(a^2 - b^2)^2$</td></tr>
<tr><td>3. Square
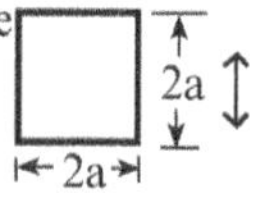
</td><td>$1.51\rho\pi a^2$</td><td>$0.234\rho\pi a^4$ t Ref. [107]</td></tr>
<tr><td>4. Rectangle
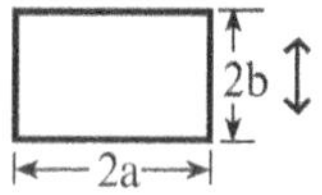
</td><td colspan="2">$m_a = \alpha\rho\pi a^2$, $J_a = \beta\rho\pi a^4$ t Ref. [107]
<table>
<tr><td>a/b</td><td>0.1</td><td>0.2</td><td>0.5</td><td>1.0</td><td>2.0</td><td>5.0</td><td>10.0</td><td>∞</td></tr>
<tr><td>α</td><td>2.23</td><td>1.98</td><td>1.70</td><td>1.51</td><td>1.36</td><td>1.21</td><td>1.14</td><td>1.0</td></tr>
<tr><td>β</td><td>-</td><td>-</td><td>-</td><td>0.234</td><td>0.15</td><td>0.15</td><td>0.14</td><td>1/8</td></tr>
</table></td></tr>
<tr><td>5. Diamond
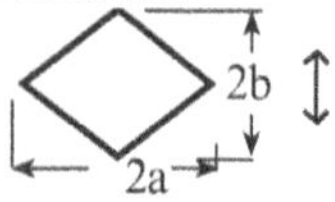
</td><td>$\alpha\rho\pi a^2$
<table>
<tr><td>a/b</td><td>0.2</td><td>0.5</td><td>1.0</td><td>2.0</td></tr>
<tr><td>α</td><td>0.61</td><td>0.67</td><td>0.76</td><td>0.85</td></tr>
</table></td><td>$0.057\rho\pi a^4$
t Ref. [107],</td></tr>
<tr><td>6. Regular Polygon
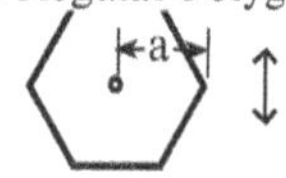
</td><td colspan="2">$\alpha\rho\pi a^2$, $J = 0.055\rho\pi a^4$, n= number sides, hexagon n=6 shown t Refs [107, 108]
<table>
<tr><td>n</td><td>3</td><td>4</td><td>5</td><td>6</td><td>∞</td></tr>
<tr><td>α</td><td>0.654</td><td>0.787</td><td>0.823</td><td>0.867</td><td>1.0</td></tr>
</table></td></tr>
<tr><td>7. Lens from two Arcs
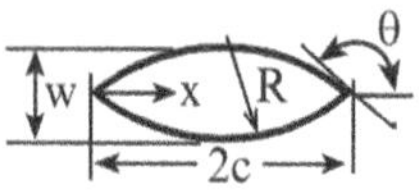
</td><td colspan="2">$m_{ax} = \rho[(2\pi/3)\{(\pi/\theta)^2 - 1\}c^2 - A]$ ↔ $w = 2R(1 - \cos\theta)$; t Ref. [110], p. 216
$m_{ay} = \rho[(2\pi/3)\{(\pi/\theta)^2/2 + 1\}c^2 - A]$ ↕ $c = R\sin\theta$
$\pi/2 \le \theta \le \pi$, θ = angle (radian), $A = \text{area} = (c/\sin\theta)^2 R^2[2(\pi - \theta) + \sin 2\theta]$,</td></tr>
<tr><td>8. I beam
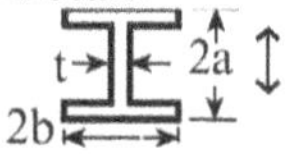
</td><td colspan="2">$2.11\rho\pi a^2$ e Ref. [109]
for a/t =2.6, b/t = 3.6</td></tr>
<tr><td>9. Plate
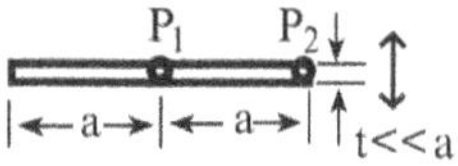
</td><td>$\rho\pi a^2$</td><td>$\rho\frac{\pi}{8}a^4$ about P_1,
$\rho\frac{9\pi}{8}a^4$ about P_2
t Refs [107, 110]</td></tr>
</table>

Note: m_a and J_a in this table are per unit length for two dimensional sections.

Table 6.9 Added mass of sections, continued

<table>
<tr><th>Section</th><th>Added Mass per Unit Length, m_a</th></tr>
<tr><td>10. Multiple Equally Spaced Fins. n=4 shown
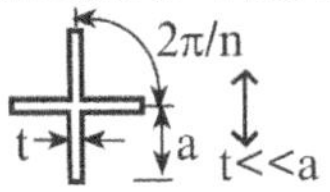
</td><td>$\frac{2\pi\rho a^2}{2^{4/n}}$ $n \geq 3$ $J_a = \beta\rho\pi a^4$ t Refs [108, 111]
n = number of fins, $n \geq 3$, $t \ll a$
$\beta = 0.533$ if n=3; $2/\pi$ if n= 4 see case 9 for n=2 fins</td></tr>
<tr><td>11. Circle with Two symmetric Fins
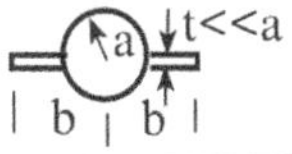
</td><td>$\rho\pi b^2(1-(a/b)^2+(a/b)^4)$ ↕ t Ref. [112]
$\rho\pi a^2$ ↔</td></tr>
<tr><td>12. Circle with n Fins
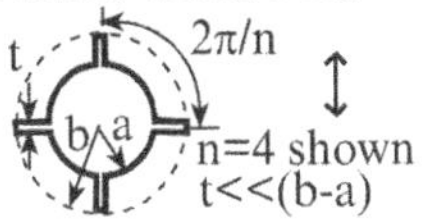
</td><td>$2\rho\pi b^2\left\{\left[\frac{1+(a/b)^n}{2}\right]^{4/n} - \frac{1}{2}\left(\frac{a}{b}\right)^2\right\}$, n 3 t Refs [108, 111]</td></tr>
<tr><td>13. Finned Square
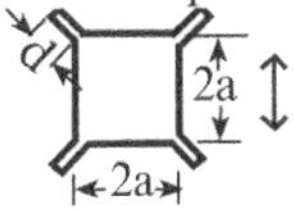
</td><td>$m_a = \alpha\rho\pi a^2$, $J_a = \beta\rho\pi a^4$ t Ref. [113]
<table><tr><td>d/a</td><td>0</td><td>0.05</td><td>0.1</td><td>0.25</td></tr><tr><td>α</td><td>1.51</td><td>1.61</td><td>1.72</td><td>2.19</td></tr><tr><td>β</td><td>0.25</td><td>0.31</td><td>0.40</td><td>0.69</td></tr></table></td></tr>
<tr><td>14. Rotated Section
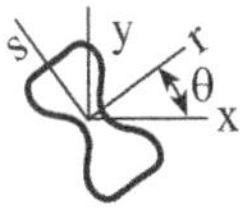
</td><td>$A_y = A_{rr}\sin^2\theta + A_{ss}\cos^2\theta$ = added mass y direction
$A_{xy} = \frac{1}{2}(A_{ss} - A_{rr})\sin 2\theta$
Ref. [116], p. 24</td></tr>
<tr><td>15. Cylinder in Cylinder
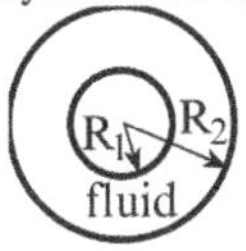
</td><td>$\rho\pi R_1^2\left[\frac{1+R_1^2/R_2^2}{1-R_1^2/R_2^2}\right]$ inner cylinder, outer fixed
t Ref. [115]
$\rho\pi R_2^2\left[\frac{1+R_1^2/R_2^2}{1-R_1^2/R_2^2}\right]$ outer cylinder, inner fixed</td></tr>
<tr><td>16. Cylinder Surrounded by Fixed Cylinders
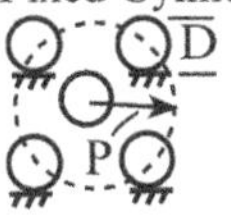
</td><td>$\frac{\rho\pi D^2}{4}\left[\frac{(D_e/D)^2+1}{(D_e/D)^2-1}\right]$, where $\frac{D_e}{D} = \left(1+\frac{1}{2}\frac{P}{D}\right)\frac{P}{D}$ Refs [105, 116]
approximate solution, t</td></tr>
<tr><td>17. Cylinder Next to Fixed Cylinder
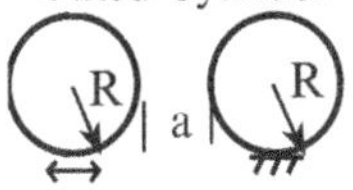
</td><td>$\alpha\rho\pi R^2$ t Ref. [117]
<table><tr><td>117a/R</td><td>0.1</td><td>0.2</td><td>0.4</td><td>1.2</td><td>∞</td></tr><tr><td>α</td><td>1.22</td><td>1.16</td><td>1.10</td><td>1.02</td><td>1.0</td></tr></table></td></tr>
</table>

Note m_a and J_a in this table are per unit length for two dimensional sections.

Table 6.9 Added mass of sections, continued

Notation: J_a = added mass moment of inertia about centroid; ρ = fluid mass density.

<table>
<tr><th>Section</th><th>Added Mass, m_a</th></tr>
<tr><td>18. Double Cylinder, or, Cylinder Parallel to Wall
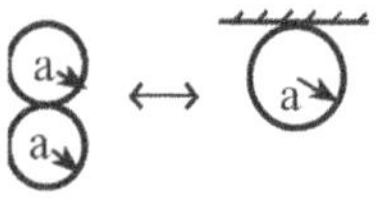</td><td>Double cylinder
$2\alpha\rho\pi a^2$
↕ $\alpha = \pi^2/6-1=0.645$
↔ $\alpha = \pi^2/3-1=2.290$

Single cylinder moving parallel to wall. t Ref. [107]
$\alpha\rho\pi a^2$ α =2.290</td></tr>
<tr><td>19. Cylinder near a Wall
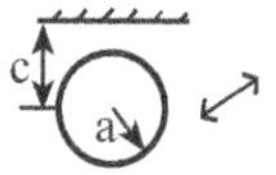</td><td>$\rho\pi a^2\left(1+\frac{a^2}{2c^2}\right)$ any direction , a/c<<1 t Ref. [110]</td></tr>
<tr><td>20. Symmetric Floating Section or Body
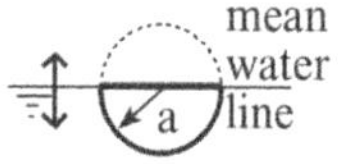</td><td>The added masse ½ that of the fully submerged body. Free surface remains plane.</td></tr>
<tr><td>21. Circle in Channel
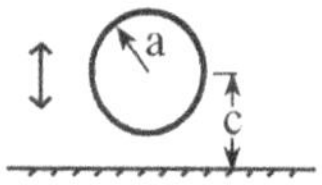</td><td>$\alpha\rho\pi a^2$
α given in case 22.
For rectangle or plate in channel use twice the values in cases 23 and 24.</td></tr>
<tr><td>22. Floating Cylinder
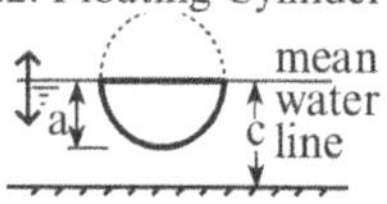</td><td>$\frac{\alpha}{2}\rho\pi a^2$ circular cylinder t Ref. [118]
<table>
<tr><td>c/a</td><td>1.2</td><td>1.5</td><td>2.0</td><td>3.0</td><td>>5</td></tr>
<tr><td>α</td><td>1.83</td><td>1.45</td><td>1.22</td><td>1.09</td><td>1.0</td></tr>
</table></td></tr>
<tr><td>23. Floating Rectangle
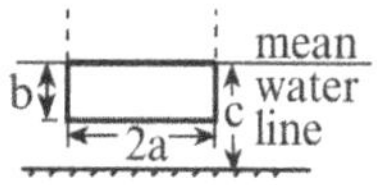</td><td>↕ $2\alpha\pi a^2$, t Ref. [107]
<table>
<tr><td>c/b</td><td>1.1</td><td>1.2</td><td>1.5</td><td>3.0</td><td>8.0</td><td>∞</td></tr>
<tr><td>α</td><td>5.52</td><td>3.49</td><td>2.11</td><td>1.35</td><td>1.21</td><td>1.19</td></tr>
</table>
Free surface remains plane during oscillations.

↔ $2\beta\rho ab$, , t Ref. [118, 119]
<table>
<tr><td rowspan="2">b/c</td><td colspan="2">β</td><td colspan="2">a/b</td></tr>
<tr><td>0.2</td><td>0.5</td><td>0.1.0</td><td>2.0</td></tr>
<tr><td>0.2</td><td>4.92</td><td>2.23</td><td>1.29</td><td>9,783</td></tr>
<tr><td>0.4</td><td>5.43</td><td>2.63</td><td>1.66</td><td>1.16</td></tr>
<tr><td>0.6</td><td>6.63</td><td>3.56</td><td>2.53</td><td>2.02</td></tr>
<tr><td>0.8</td><td>10.5</td><td>6.45</td><td>5.23</td><td>4.62</td></tr>
</table></td></tr>
<tr><td>24. Floating Thin Plate In Circular Channel
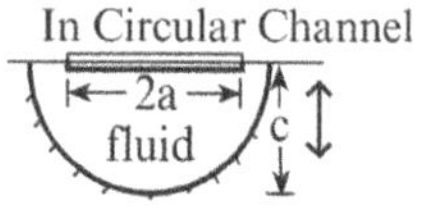</td><td>$\alpha\rho\pi a^2/2$ t Ref. [114]
<table>
<tr><td>c/a</td><td>1.25</td><td>1.67</td><td>2.5</td><td>5.0</td><td>∞</td></tr>
<tr><td>α</td><td>1.60</td><td>1.22</td><td>1.10</td><td>1.04</td><td>1.0</td></tr>
</table></td></tr>
<tr><td>25. Two Plates in Tandem
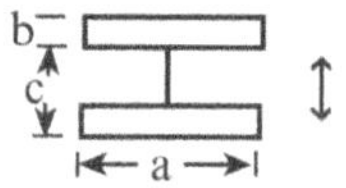</td><td>$2\alpha\rho ab$ values of α e Ref. [120]
<table>
<tr><td rowspan="2">c/a</td><td colspan="4">b/a</td></tr>
<tr><td>0.1</td><td>0.2</td><td>0.4</td><td>1.0</td></tr>
<tr><td>0.5</td><td>4.7</td><td>2.6</td><td>1.3</td><td>-</td></tr>
<tr><td>1.0</td><td>5.2</td><td>3.2</td><td>1.7</td><td>0.6</td></tr>
<tr><td>2.0</td><td>6.4</td><td>4.0</td><td>2.0</td><td>0.9</td></tr>
<tr><td>4.0</td><td>-</td><td>4.8</td><td>-</td><td>-</td></tr>
</table></td></tr>
</table>

Table 6.10 Added mass of bodies

Notation: J_a = added mass moment of inertia; ρ = fluid mass density.

Body | **Added Mass, M_a**

1. Thin Circular Disk

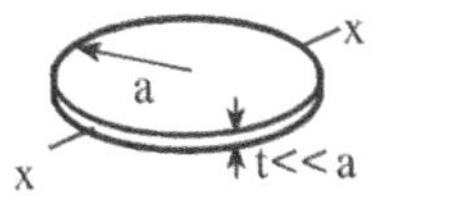

$\updownarrow \ \frac{8}{3}\rho a^3$ $\quad \leftrightarrow 0$ t Ref. [107]

Added mass moment of inertial for rotation about x-x axis $J_a = 0.37\ \rho a^5$.

2. Thin Elliptical Disk

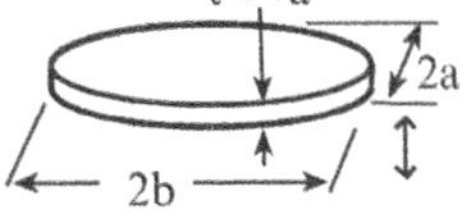

$\alpha\rho\pi(4/3)a^2b$ t Ref. [121]

b/a	1	1.5	2.0	3.0	4.0	6.0	10.	14.3	∞
α	0.637	0.758	0.826	0.90	0.933	0.964	0.984	0.991	1.0

3. Thin Rectangle Plate

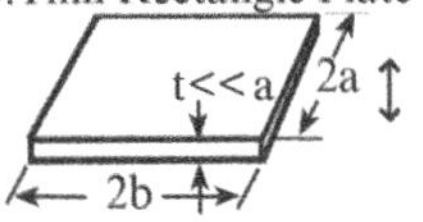

$\alpha\rho(\pi/4)a^2b$ t Ref. [122]

b/a	1.0	1.25	2.0	2.5	3.17	5.0	10.	∞
α	0.5790	0.6419	0.7568	0.8008	0.8404	0.8965	0.9469	1.0

4. Isosceles Triangular

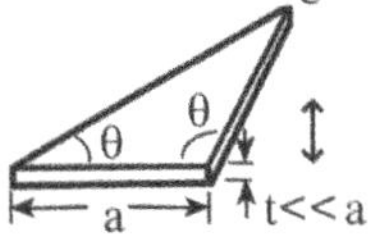

$\frac{\rho a^3}{3\pi}(\tan\theta)^{3/2}$ t Ref. [108]

5. Cube

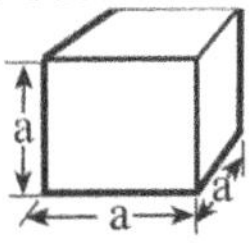

$\alpha\rho a^3$

$$\alpha = \begin{cases} 0.67 & \text{e Ref. [123]} \\ 0.7 & \text{e Ref. [120]} \\ 0.64 & \text{t Ref. [124]} \end{cases}$$

6. Rectangular Solid

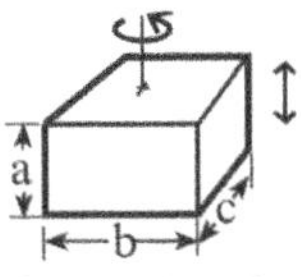

Ma for acceleration in vertical axis, J_a for rotation about vertical axis through centroid.

$\alpha\rho abc \ \updownarrow$ values of $\alpha \approx 0.64\sqrt{bc}/a$ t Ref. [124]

c/a	α							b/a
	b/a=0.5	0.6	0.8	1.0	1.2	1.6	2.0	2.8	3.6
0.5	0.34	0.37	0.41	0.44	0.47	0.51	0.53	0.57	0.59
0.6	0.37	0.40	0.45	0.49	0.52	0.68	0.60	0.64	0.67
0.8	0.41	0.45	0.52	0.57	0.62	0.78	0.73	0.79	0.83
1.0	0.44	0.49	0.57	0.64	0.69	0.86	0.84	0.92	0.97
1.2	0.47	0.52	0.62	0.69	0.76	0.98	0.93	1.03	1.10
1.6	0.51	0.57	0.68	0.78	0.86	1.08	1.08	1.23	1.32
2.0	0.53	0.60	0.73	0.84	0.93	1.23	1.21	1.39	1.51
3.6	0.59	0.67	0.83	0.97	1.10	1.32	1.51	1.82	2.06

$J = J_a - \beta\rho abc(a^2 + b^2)/12$, values of β t Ref. [124]

c/a	β							b/a
	b/a=0.5	0.6	0.8	1.0	1.2	1.6	2.0	2.8	3.6
0.5	0.26	0.27	0.36	0.46	0.57	0.76	0.91	1.13	1.29
0.6	0.27	0.25	0.29	0.36	0.45	0.61	0.74	0.94	1.08
0.8	0.38	0.29	0.24	0.26	0.31	0.42	0.52	0.68	0.80
1.0	0.46	0.36	0.26	0.24	0.25	0.31	0.39	0.53	0.63
1.2	0.57	0.45	0.31	0.25	0.23	0.26	0.31	0.42	0.51
1.6	0.75	0.60	0.42	0.31	026	0.22	0.23	0.29	0.36
2.0	0.91	0.74	0.52	0.39	0.31	0.23	0.20	0.23	0.27
3.6	1.29	1.08	0.80	0.63	0.51	0.36	0.27	0.19	0.17

Table 6.10 Added mass of bodies, continued

Notation: J_a = added mass moment of inertia; ρ = fluid mass density.

Body	Added Mass, M_a

7. Right Circular Cylinder Finite Length

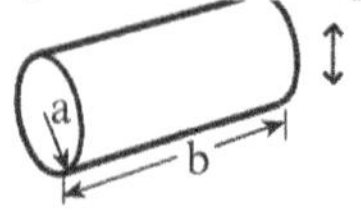

$\alpha\rho\pi a^2 b$ e Ref. [107]

b/2a	0	1.2	2.5	5	9	>10
α	0	0.62	0.78	0.90	0.96	1.0

8. Sphere

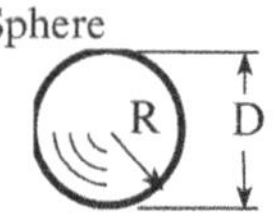

$$\frac{2}{3}\rho\pi R^3 = \frac{\pi}{12}\rho D^3$$

9. Sphere in a Sphere

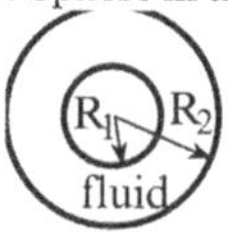

$$\frac{2}{3}\rho\pi R_1^2\left[\frac{1+2R_1^3/R_2^3}{1-R_1^3/R_2^3}\right]$$

inner sphere, outer fixed

$$\frac{2}{3}\alpha\rho\pi R_2^2\left[\frac{1+R_1^3/R_2^3}{1-R_1^3/R_2^3}\right]$$

outer sphere, inner cylinder fixed

t Ref. [125]

10. Sphere, Surface

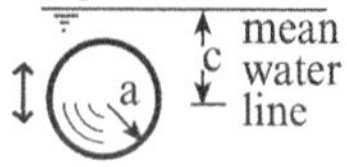

$(2/3)\alpha\rho\pi a^3$ e Refs [109, 126]

c/2a	0	0.5	0.75	1	1.5	2	3	4	>4
α	0.5	0.7	0.88	0.94	1.03	1.08	1.16	1.05	1.0

11. Ellipsoid of Revolution

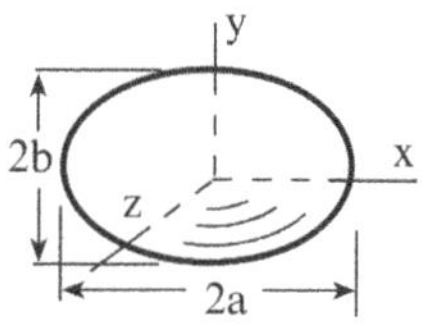

x-axis is axis of revolution.

$$J_{zz} = \frac{4}{15}\beta\rho\pi b^5$$

t Ref. [82], pp. 155–156, 700–701; Ref. [121]

$M_{xx} = \frac{4}{3}\alpha\rho\pi ab^2 = \frac{4}{3}\beta\rho\pi b^3$

acceleration in x direction

a/b	α	β
0.01	-	0.6348
0.1	6.148	0.6148
0.2	3.008	0.6016
0.4	1.428	0.5712
0.6	0.9078	0.5447
0.8	0.6514	0.5211
1.0	0.5	0.5
1.5	0.3038	0.4557
2.	0.2100	0.4200
3.	0.1220	0.3660
5.	0.0591	0.2956
10.	.02071	0.2071

$\frac{4}{3}\alpha\rho\pi ab^2$ or $\frac{4}{3}\beta\rho\pi b^3$

acceleration in y

a/b	α	β
0.01	.00781	-
0.1	0.0748	.00748
0.2	0.1425	0.0285
0.4	0.2593	0.1037
0.6	0.3552	0.2131
0.8	0.4343	0.3474
1.0	0.5	0.5
1.5	0.6221	0.9331
2.	0.7042	1.048
3.	0.8039	2.412
5.	0.8943	4.472
10.	0.9602	9.602

$J_{zz} = \frac{4}{15}\alpha\rho\pi ab^2(a^2+b^2)$

rotation about z or y axis

a/b	α	β
0.01	42.33	0.2117
0.1	4.022	.02031
0.2	1.793	0.1865
0.4	0.5862	0.1360
0.6	0.1843	0.0752
0.8	.03455	.02266
1.0	0	0
1.5	0.0951	0.2318
2.	0.2394	1.197
3.	0.4657	6.990
5.	0.6999	45.49
10.	0.8835	-

12. Two, or more (n) square plates in Tandem

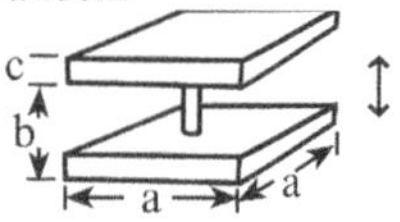

$2\alpha\rho a^3$, α in table below, eRef. [120]

c/a	b/a n=2 0.1	0.2	0.4	1.0
0.5	3.2	1.6	0.75	-
1.0	3.8	2.0	1.1	0.35
2.0	4.4	2.7	1.5	0.55
4.0	5.6	3.0	1.6	0.7

$n\alpha\rho a^3$, α in table below, e Ref. [127]

n thin plates	c/a (n = number of plates) 0.25	0.5	1.0	1.5
2	0.435	0.52	0.53	0.55
3	0.37	0.46	0.52	0.54
4	0.33	0.45	0.515	0.526
6	0.30	0.42	0.51	0.52

Table 6.10 Added mass of bodies, continued

Plate or Shell	Added Mass
13. Baffled Piston 	$\frac{8}{3}\rho a^3$ fluid on **one** side of baffled piston t Ref. [3] p. 380–384 Piston is rigid and oscillates perpendicular to baffle. The added mass per unit area is $\rho(8/(3\pi))a$.
14. Baffled Rectangular 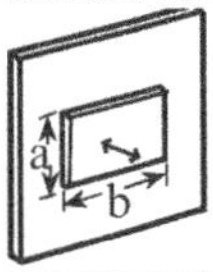	$\frac{\alpha\rho}{2\pi}(ab)^{3/2}$ $\alpha = 3.9162$ for a=b, fluid on **one** side t Ref. [128]; [1], p. 248 $\alpha = 2\left(\frac{a}{b}\right)^{1/2}\sinh^{-1}\frac{b}{a} + 2\left(\frac{b}{a}\right)^{1/2}\sinh^{-1}\frac{a}{b} + \frac{2}{3}\left[\left(\frac{a}{b}\right)^{3/2} + \left(\frac{b}{a}\right)^{3/2} - \left(\frac{a}{b}+\frac{b}{a}\right)^{1/2}\right]$

15. Elastic Circular Plate Fluid on One Side and Baffled edges

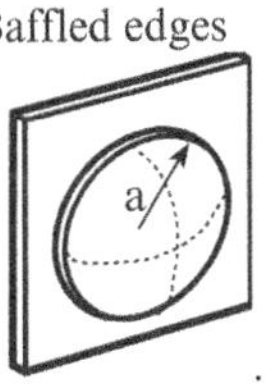

Plate Boundary	Mode	Added mass
Free edge	1 nodal diameter	$(16/15\pi)\rho\pi a^3$ Ref. [129]
Free edge	1 nodal circle	$(24/35\pi)\rho\pi a^3$ Ref. [129]
Free edge with center point support	axi-symmetric mode	$(120/63\pi)\rho\pi a^3$ Ref. [129]
Simply supportedv=0.3	fundamental mode	$0.7554\ \rho\pi a^3$ Ref. [130]
Clamped edges	fundamental mode	$0.6689\ \rho\pi a^3$ Ref. [131]
Clamped edge	1 nodal diameter	$0.3087\ \rho\pi a^3$ Ref. [131]

16. Cantilever Rectangular Plate Fluid on **Both** Sides

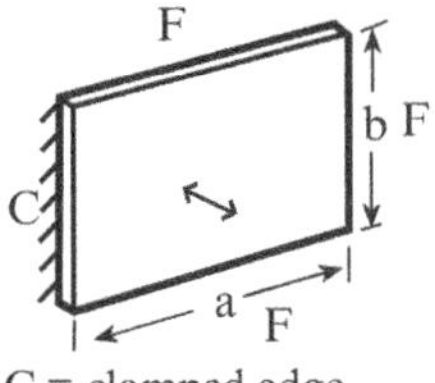

C = clamped edge
F = free edge

Mode	i, j	Added mass
1st cantilever	i=1, j=1	$\frac{1}{4}\pi\rho ab^2$
1st torsion	i=1, j=2	$\frac{3}{32}\pi\rho ab^2$
2nd cantilever	i=1, j=2	$\frac{1}{4}\pi\rho ab^2$
2nd transverse	i=1, j=3	$0.0803\,\pi\rho ab^2$
2-2 mixed mode	i=2, j=2	$\frac{1}{4}\pi\rho ab^2$
3rd cantilever	i=2, j=2	$\frac{1}{4}\pi\rho ab^2$

e Ref. [132] Strip theory result. Also see Ref. [133]. Agreement with data is improved by multiplying the added mass by the factor 2(a/b)/(1+2a/b) for $5 \geq a/b \geq 0.5$.

17. Sinusoidal PlateStrip with fluid on **both** sides

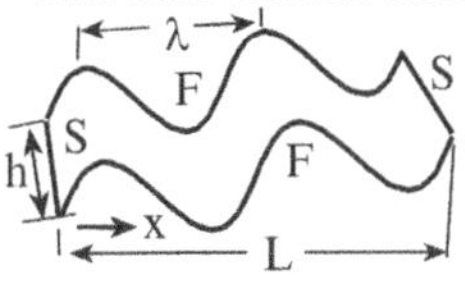

$\rho\frac{\pi}{4}Lh^2$, L >> h

$\rho\lambda hL/\pi = \rho(2/i\pi)hL^2$, h >> L

Plate with two ends Simply Supported (S) and two sides Free (F).
Mode shape $w(x) = \sin(i\pi x/L)$; i=1,2,3.. . Wavelength $\lambda = 2L/i$.
Fluid on both sides. Divide by Lh for added mass per surface. t Ref. [134]

18. Added Mass for Cylindrical Shell modes

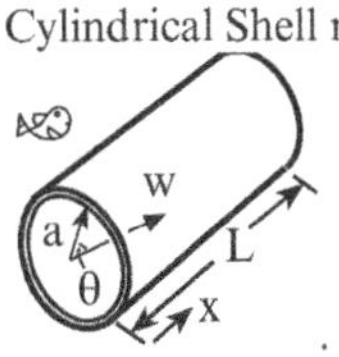

cylinder mode shape $\tilde{w}(x,\theta) = \sin(n\theta)\sin i\pi x/L$, n = 1,2,3, i = 1,2,3... Ref. [142]

add mass fluid outside / unit area: $m_a = \frac{-\rho a K_n(i\pi a/L)}{(i\pi a/L)K'_n(i\pi a/L)} = \rho a/n,\ i\pi a \ll L$

add mass fluid inside / unit area: $m_a = \frac{\rho a I_n(i\pi a/L)}{(i\pi a/L)I'_n(i\pi a/L)} = \rho a/n,\ i\pi a \ll L$

$K_n(z)$ = Modified Bessel function order n 2nd kind $K'_n(z) = dK_n(z)/dz$
$I_n(z)$ = Modified Bessel function order n 1st kind $I'_n(z) = dI_n(z)/dz$

6.7.1 Added Mass Potential Flow Theory

The incompressible fluid equation of conservation of mass, $\nabla \bullet \mathbf{u} = 0$, is satisfied if the vector fluid velocity $\mathbf{u}$ is gradient of a *velocity potential* Φ that satisfies the wave equation ([82] and Table 6.2):

$$\nabla^2\Phi = 0, \quad \mathbf{u} = \nabla\Phi(x, y, z, t) \tag{6.62}$$

The boundary condition on a rigid body with velocity $\mathbf{U}$ in an inviscid fluid is that the body surface's normal outward velocity equals the adjacent fluid velocity, $\mathbf{U} \bullet \mathbf{n} = \partial\Phi/\partial\mathbf{n}$. $\mathbf{n}$ is the unit outward normal to the surface.

Fluid conservation of momentum (case 2 of Table 6.2) implies that the fluid static pressure is proportional to the time derivative of the velocity potential:

$$p = -\frac{\rho\partial\Phi}{\partial t} \tag{6.63}$$

The velocity potential in each direction is proportional to the body's velocity U_i in each direction [112]:

$$\Phi_i = \phi_i(x, y, z)U_i(t), \quad i = 1, 2, 3 \tag{6.64}$$

The vector fluid force $\mathbf{F}$ on a rigid body is the resultant of fluid pressure on its surface S:

$$\mathbf{F} = \int_S p\mathbf{n}\,dS = -\rho\frac{\partial}{\partial t}\int_S \Phi n\,dS = -[M]\frac{d\mathbf{U}}{dt} \tag{6.65}$$

where dS is the surface element area with outward normal $\mathbf{n}$. The components of the outward normal are $n_i = \partial\phi_i/\partial n$ as a consequence of $\mathbf{U} \bullet \mathbf{n} = \partial\Phi/\partial\mathbf{n}$. $[M]$ is the added mass matrix.

The added mass M_i and added mass force F_i on the accelerating body in each direction are scalar versions of Equation 6.65 ([112], p. 139; [108], p. 379):

$$F_i = -M_i\frac{dU_i}{dt}, \quad M_i = \rho\int_S \phi_i n_i\,dS = \rho\int_S \phi_i\frac{\partial\phi_i}{\partial n}\,dS, \quad i = 1, 2, 3 \tag{6.66}$$

Added mass force is proportional to fluid density times body acceleration. The negative sign denotes that it opposes the direction of acceleration. Added mass force accounts for the energy imparted to fluid by an accelerating body. This is shown by multiplying Equation 6.66 by an increment of body displacement ($dx_i = U_i dt$) and integrating in time:

$$\frac{1}{2}\rho\int_V u_i u_j dV = \frac{1}{2}M_i M_i U_i U_j \tag{6.67}$$

where U_i is the velocity of the body and u_i is fluid velocity, both in the ith direction. The increase in fluid kinetic energy during acceleration in i and j directions is proportional to added mass [112].

The equation of motion for free vibration of a single-degree-of-freedom spring-supported rigid body in an otherwise still fluid is the same as if the structural mass M is increased by the added mass M_a:

$$M\ddot{x} + kx = F = -M_a\ddot{x}, \quad or, \quad (M + M_a])\ddot{x} + kx = 0 \tag{6.68}$$

Added mass increases the effective mass of the body and decreases natural frequencies of elastic structures in dense fluids. The natural frequencies in fluid are lower than those in a vacuum by the factor

$$\frac{f_{Ma}}{f_{Ma=0}} = \frac{M^{1/2}}{(M_a + M)^{1/2}} = \frac{1}{\sqrt{1 + M_a/M}} \tag{6.69}$$

where M_a is the added mass, M is the structural mass, and $f|_{Ma=0}$ is the natural frequencies in vacuum.

Fluid viscosity increases the added mass. The theoretical added masses of a circular cylinder, length L and diameter D, and a sphere, diameter D, at frequency f in Hertz are expressed as series of the dimensionless parameter fD^2/υ [82, p. 644; 135]:

$$m_a = \begin{cases} \frac{1}{4}\rho\pi D^2 L \left[1 + 4\left(\frac{\nu}{\pi f D^2}\right)^{1/2} + \cdots + O\left(\frac{\nu}{\pi f D^2}\right)\right] & \text{cylinder} \\ \frac{1}{12}\rho\pi D^3 \left[1 + 9\left(\frac{\nu}{\pi f D^2}\right)^{1/2} + \cdots + O\left(\frac{\nu}{\pi f D^2}\right)\right] & \text{sphere} \end{cases} \tag{6.70}$$

These suggest that the inviscid fluid ($\nu = 0$) added masses in Table 6.9 are a good approximations if fD^2/υ is greater than a few hundred, which is the case for larger structures and higher frequencies. Shed vortices and separation complicate the added mass in viscous fluids for large amplitude oscillations [113].

6.7.2 *Added Mass*

Tables 6.9 and 6.10 show the theoretical and experimental solutions for added mass of accelerating rigid section and bodies, respectively, in an otherwise still, incompressible, inviscid fluid. Additional solutions are shown in Refs [107–110, 136]. In general, the added mass is a function of shape and the direction of acceleration. A flat plate has the same added mass as the circular cylinder of the same maximum width, but a thin plate has zero added mass for in-plane acceleration. An added mass coefficient can be defined that is the added mass divided by the displaced fluid mass. This has not been done here because of the difficulty of defining displaced mass of plates and fins.

Added mass of rigid bodies is given in Table 6.10. *Strip theory* is an approximate method used to predict the added mass of slender bodies by sectioning them into two-dimensional strips. Two-dimensional theory (Table 6.9) is applied to each strip and the results are summed. For example, the strip theory result is compared with the exact result for the finite length circular cylinder:

Cylinder length, diameters	1	2.5	5	9	> 10
Added/displaced mass exact	0.62	0.78	0.90	0.96	1.0
Added mass/displaced mass strip a	1.0	1.0	1.0	1.0	
Added mass/displaced mass strip b	0.62	0.82	0.94	0.96	1.0

Strip theory (a) uses the infinitely long cylinder solution (case 1 of Table 6.9). Strip theory (b) is improved by the use of a half sphere (case 27 of Table 6.9) for the free ends.

Added mass is a resultant of the acoustic pressure on oscillating bodies in cases 11, 12, 13, and 15 of Table 6.3. The real part of the acoustic solution is in phase with velocity and it represents radiated energy damping; the imaginary part is in phase with damping and it is added mass in the low-frequency, incompressible flow limit. For example, case 13 of Table 6.3 demonstrates that the added mass pressure on an expanding sphere is $\rho R\, dU/dt$ for long wavelengths where R is the sphere radius and U is its surface velocity. See discussion of bubbles in Section 6.4.

6.7.3 *Added Mass of Plates and Shells*

The added mass fluid pressure varies over the surface of the plate or shell (cases 13–18 of Table 6.10; Refs [128–143]). Since the mode shapes are not much effected by fluid added mass [128], an effective modal added mass for plates and shell can be developed with an assumed mode shape by setting the computed kinetic energy of fluid equal to the kinetic energy of added mass of the body (Eq. 6.67; [137]).

Fluid may be on one or both sides of a plate. The added mass with two sides exposed to fluid is twice the added mass of a single side. But the added mass of fluid inside a cylinder differs from the added mass of external fluid (case 41 of Table 6.9) except in the limit of large radius and high frequencies. Coupled fluid-and-structural finite element methods have been successfully applied to added mass calculation [98, 99, 138]. Reviews of vibration of fluid-coupled shells are shown in Refs [139, 140]. Submerged solid and perforated plates have been used lower frequency and damp oscillation of offshore platforms [127, 141, 143].

Added mass matrix $[M_a]$ of a six-degree-of-freedom body is a 6×6 symmetric matrix. The entries m_{ij} are the added mass in each (ith) coordinate (three translations plus three rotations) due to acceleration in the jth coordinate [108]:

$$m_{ij} = m_{ji} = \rho \int_S \phi_j \frac{\partial \phi_i}{\partial n} dS \quad \text{for } i,j = 1,2,3 \tag{6.71}$$

The symmetric added mass matrix has 21 independent terms: 6 diagonal and 15 off-diagonal (see Chapter 2, Eq. 2.73):

$$\begin{bmatrix} F_x \\ F_y \\ F_z \\ M_x \\ M_y \\ M_z \end{bmatrix} = - \begin{bmatrix} A_{11} & A_{12} & A_{13} & A_{14} & A_{15} & A_{16} \\ A_{12} & A_{22} & A_{23} & A_{24} & A_{25} & A_{26} \\ A_{13} & A_{23} & A_{33} & A_{34} & A_{35} & A_{36} \\ A_{14} & A_{24} & A_{34} & A_{44} & A_{45} & A_{46} \\ A_{15} & A_{25} & A_{35} & A_{46} & A_{55} & A_{56} \\ A_{16} & A_{26} & A_{36} & A_{46} & A_{56} & A_{66} \end{bmatrix} \begin{bmatrix} \ddot{X} \\ \ddot{Y} \\ \ddot{Z} \\ \dot{\omega}_x \\ \dot{\omega}_y \\ \dot{\omega}_z \end{bmatrix} \tag{6.72}$$

Off-diagonal terms couple degrees of freedom. For example, if a body is not symmetric about the x-axis, then acceleration in the x direction generally induces added mass forces in both the x and y directions and moments. If the body has symmetry, some or all off-diagonal coupling terms are zero. If the body is symmetric about the x–z-axis, then 18 of the off-diagonal cross-coupling terms must be zero [131, 132]. If the body is symmetric about two axes, x–y and y–z, then 26 of the off-diagonal terms are zero:

$$\begin{bmatrix} A_{11} & 0 & 0 & 0 & 0 & 0 \\ 0 & A_{22} & 0 & 0 & 0 & A_{26} \\ 0 & 0 & A_{33} & 0 & A_{35} & 0 \\ 0 & 0 & 0 & A_{44} & 0 & 0 \\ 0 & 0 & A_{35} & 0 & A_{55} & 0 \\ 0 & A_{26} & 0 & 0 & 0 & A_{66} \end{bmatrix}, \begin{bmatrix} A_{11} & 0 & 0 & 0 & 0 & 0 \\ 0 & A_{22} & 0 & 0 & 0 & 0 \\ 0 & 0 & A_{33} & 0 & 0 & 0 \\ 0 & 0 & 0 & A_{44} & 0 & 0 \\ 0 & 0 & 0 & 0 & A_{55} & 0 \\ 0 & 0 & 0 & 0 & 0 & A_{66} \end{bmatrix} \quad (6.73)$$

If the body is symmetric about x–y, x–z, and y–z planes, an ellipsoid, for example, then cross coupling is eliminated and only the diagonal terms remain.

References

[1] Pierce, A. D., Acoustics, McGraw-Hill, N.Y., 1981.
[2] Kinsler, L. E., A. Frey, A. B. Coppens, and J. V. Sanders, Fundamentals of Acoustics, 3rd ed., Wiley, 1992.
[3] Morse, P. M. and K. U. Ingard, Theoretical Acoustics, McGraw-Hill, N.Y., pp. 603–605, 879, 1968.
[4] National Institute of Standards and Technology, NIST, REFPROP Code, version 7.1, USA.
[5] Cramer, O., The Variation of the Specific Heat Ratio and the Speed of Sound in Air with Temperature, Pressure, Humidity, and CO_2, Journal of the Acoustical Society of America, vol. 93, pp. 2510–2516, 1993.
[6] W. Marczak, Water as a Standard in the Measurements of Speed of Sound in Liquids, Journal of Acoustic Society America, vol. 102(5), pp. 2776–2779, 1997.
[7] K. V. Mackenzie, Nine-term Equation for the Sound Speed in the Oceans, Journal of Acoustical Society America, vol. 70(3), pp. 807–812, 1981.
[8] Cervera, F. *et al.*, Refractive Acoustic Devices for Airborne Sound, Physical Review Letters, vol. 88(2), 023902, 2002.
[9] Blevins R. D., Acoustic Modes of Heat Exchanger Tube Bundles, Journal of Sound and Vibration, vol. 109, pp. 19–31, 1986.
[10] Wylie, B. E., and V. L. Streeter, Fluid Transients in Systems, Prentice Hall, New Jersey, p. 9, 10, 1993.
[11] Graf, K. F., Wave Motion in Elastic Solids, Ohio State University Press, p. 77, 276, 277, 1975.
[12] Chemical Rubber Company, CRC Handbook of Chemistry and Physics, 78th ed., pp. 14–36, 1998.
[13] Beranek, L., and I. Ver, Noise and Vibration Control Engineering, John Wiley, N.Y., pp. 230–231, 1992.
[14] Jones, J. B., and G. A. Hawkins, Engineering Thermodynamics, John Wiley, N.Y., p. 169, 1960.
[15] Shapiro, A. H., Compressible Fluid Flow, John Wiley, vol. 1, p. 48, 1953.
[16] Liepman, H., and A. Roshko, Elements of Gas Dynamics, John Wiley, N.Y., pp. 64–65, 1957.
[17] Munjal, M. L., Acoustics of Ducts and Mufflers, John Wiley, N.Y., 1987.
[18] ASTM standard E 2459 – 05, Standard Guide for Measurements of In-Duct Sound Pressure Levels from Large Industrial Gas Turbines and Fans, 2005.
[19] ANSI S1.26 Method for Calculation of the Absorption of Sound by the Atmosphere. Also ISO 9613.
[20] Weiner, F. M., and D. N. Keast, Experimental Study of Propagation of Sound Over Ground, Journal of Acoustical Society of America, vol. 31, pp. 724–733, 1959.
[21] Harris, C. R., Noise Control in Buildings, The Institute of Noise Control Engineering. Reprint of 1994 McGraw-Hill edition. ISBN 0-9022072-1-7, 1997.
[22] Beranek, L. L., Noise Reduction, Reprint by Robert E. Krieger, 1980.
[23] International Standard ISO 9613-2, Acoustics – Attenuation of Sound Propagation Outdoors, Part 2.
[24] Kuttruff, H., Room Acoustics, 2nd ed., Applied Science Publishers, London. 1979.

[25] Guess, A. W., Calculation of Perforated Plate Liner Parameters, Journal of Sound and Vibration, vol. 40, pp. 119–137, 1973.

[26] Melling, T. H., The Acoustic Impedance of Perforates, Journal of Sound and Vibration, vol. 29, pp. 1–65, 1973.

[27] Malmary, C. *et al.*, Acoustic Impedance Measuring with Grazing Flow, AIAA Paper AIAA-2001-2193, 2001.

[28] Dickey, N. S. and A Selaet, An Experimental Study of the Impedance of Perforated Plates with Grazing Flow, Journal of the Acoustical Society of America, vol. 110, pp. 2230–2370, 2001.

[29] Harris, C. R., Handbook of Acoustical Measuring and Noise Control, McGraw-Hill, N.Y., 1991, Chapter 4.

[30] Jones, R. E., Intercomparisons of Laboratory Determinations of Airborne Sound transmission Loss, Journal of Acoustical Society of America, vol. 66, pp. 148–164, 1979.

[31] ASTM E 1124 Standard Test Method for Field Measurement of Sound Power by the Two Surface Method.

[32] Rayleigh, J. W. S., The Theory of Sound, vol. 2, 2nd ed., Dover, N.Y., 1945.

[33] American National Standard ANSI S1.4-1983, Sound Level Meters, N.Y. Also European Standard EN 60651, Sound Level Meters and IEC 651, January 1979 and British Standard BS EN 60651, Specification for Sound Level Meters, 1994.

[34] American National Standard ANSI S1.1-1994, Acoustic Terminology, N.Y., Reaffirmed 1999. Also American National Standard C634-02, Standard Terminology Relating to Environmental Acoustics, 2002.

[35] American National Standard ANSI S1.8-1998, Reference Quantities for Acoustical Levels, N.Y., Reaffirmed 2001. Also, Internationals Standards Organization, ISO 1683:1983, Acoustics – Preferred Reference Quantities for Acoustic Levels.

[36] Carey, W. M., Standard Definitions for Sound Levels in the Ocean, IEEE Journal of Ocean Engineering, vol. 20, pp. 109–113, 1995.

[37] Fahy, F. J., Sound Intensity, 2nd ed., Chapman & Hall, London, 1995.

[38] Seybert, A. F., Two-Sensor Methods for the Measurement of Sound Intensity and Acoustic Properties in Ducts, Journal of the Acoustical Society of America, vol. 83, pp. 2233–2239, 1988.

[39] ASTM Standard E1780 Standard Guide for Measuring Outdoor Sound Received from a Nearby Fixed Source.

[40] Society of Automotive Engineers, Practical Methods to Obtain Free Field Sound Pressure Levels from Acoustical Measurements over Ground Surfaces, Aerospace Information Report AIR, 1672B, 1983.

[41] American National Standard ANSI S1.9-1996, Instruments for Measurement of Sound Intensity. Also see IEC Standard 1043.

[42] International Standard ISO 5136 Acoustics – Determination of Sound Power Radiated into a Duct by Fans and Other Air-Moving Devices – In-Duct Method.

[43] ISO 3740 Acoustics – Determination of Sound Power Levels of Noise Sources – Guidelines for the Use of Basic Standards. Also ISO 3740 to 3747.

[44] ISO 9614-1,-2, Determination of Sound Power Levels of Noise Sources by Sound Intensity Measurement. Also see ANSI 12.12.

[45] American National Standard ANSI A12.50-2002,Acoustics – Determination of Sound Power Levels of Noise Sources – Guidelines for the use of Basic Standards, Also S12.12, S12.35, S12.50-S12.57, ASTM F1334.

[46] International Standard ISO 15186, Measurement of Sound Insulation in Buildings and of Building Elements.

[47] ASTM Test Standard E90-04, Laboratory Measurement of Airborne Sound Transmission Loss of Building Partitions and Elements, 2004.

[48] ASTM Standard C423-07a, Standard Test Method for Sound Absorption and Sound Absorption Coefficients by the Reverberation Room Method. Also ISO 3741, 2007.

[49] *Technology for a Quieter America*, 2010, National Academies Press, Washington D.C., http://www.nap.edu/catalog.php?record_id=12928

[50] Ville, Jean-Michel and F. Foucart, Experiment set up for Measurement of Acoustic Power with Air Mean Flow, Journal of the Acoustical Society of America, vol. 114, pp. 1742–1748, 2003.

[51] Schultz, T. J., Relationship Between Sound Power and Sound Pressure Level in Dwellings and Offices, ASHRAE Transactions, vol. 91, pp. 124–153, 1985.

[52] ASTM E 2235–04, Determination of Decay Rates for Use in Sound Insulation Test Methods, 2004.

[53] American National Standard ANSI S12.12 – Engineering Method for Determination of Sound Power Levels of Sound Sources using Sound Intensity.

[54] Miles, J. W., The Diffraction of Sound due to Right-angled Joints in Rectangular Tubes, Journal of the Acoustical Society of America, vol. 19, pp. 572–579, 1947.

[55] Tao, Z. and A. F. Seybert, A Review of Current Techniques for Measuring Muffler Transmission Loss, SAE Transactions Journal of Passenger Cars-Mechanical Systems, vol. 112, pp. 2096–2100, 2004.

[56] Neise, W. and F. Arnold, On Sound Power Determination in Flow Ducts, Journal of Sound and Vibration, vol. 244, 481–503, 2001.

[57] Ainslie, M. A., and J. G. McColm, A Simplified Formula for Viscous and Chemical Absorption in Sea Water, Journal of the Acoustical Society of America, vol. 103, pp. 1671–1673, 1998.
[58] Allen, C. H., and B. G. Watters, Siren Design for Producing Controlled Wave Form, Journal of the Acoustical Society of America, vol. 31, pp. 463–469, 1959. Also see ASTM Standard S12.14-Sirens.
[59] Junger, M. C., and D. Feit, Sound Structures and Their Interaction, MIT Press, 1986.
[60] Richards, E. J., and D. J. Meade, Noise and Acoustic Fatigue in Aeronautics, John Wiley, London, pp. 230–231, 1968.
[61] Lamb, H., The Dynamical Theory of Sound, 2nd ed., Dover, N.Y., 1960. Reprint of 1925 edition.
[62] Abramowitz, M., and I. A. Stegun, Handbook of Mathematical Functions, Dover, New York, p. 411, 468, 1970.
[63] Tyler, J., and T. Sofrin, Axial Flow Compressor Noise Studies, SAE Transactions, vol. 70, pp. 309–332, 1962.
[64] Hong, K. and J. Kim, Natural Mode Analysis of Elliptic Cylindrical Cavities, Journal of Sound and Vibration, vol. 183, pp. 327–351, 1995.
[65] Chapman, R. B., and M. S. Plesset Thermal Effects in the Free Oscillation of Gas Bubbles, Journal of Basic Engineering, vol. 93, pp. 373–376, 1971.
[66] Plesset, M. S. and A. Prosperetti, Bubble Dynamics and Cavitation, Annual Reviews of Fluid Mechanics, M. Van Dyke (ed.), Palo Alto, CA, Annual Review, Inc., 1977.
[67] Lippert, W. K. R., Wave Transmission around Bends of Different Angles in Rectangular ducts, Acustica, vol. 5, pp. 274–278, 1955.
[68] Morse, P. M. and H. Feshback, Methods of Theoretical Physics, McGraw-Hill, N.Y., part II, pp. 1474–1452, 1953.
[69] Mechel, F. P. (ed.), Formulas for Acoustics, Springer-Verlag, Berlin, 2001.
[70] Blevins, R. D., Modal Density of Rectangular Volumes, Areas, and Lines, Journal of the Acoustical Society of America, vol. 119, pp. 788–791, 2006.
[71] International Standard IS0-3744, Determination of Sound Power Levels of Noise Sources. ISO 3740 to 3747. ISO 5136, ISO 9614.
[72] Vinokur, R., Vibroacoustic Measurements without Transducers, Sound and Vibration, vol. 41, p. 5, 2007.
[73] Dowell, E. H., Reverberation Time, Absorption, and Impedance, Journal of the Acoustical Society of America, vol. 64, pp. 181–191, 1978.
[74] Beranek, L., Concert Halls and Opera Houses, Music, Acoustics, and Architecture, 2nd ed., Springer, N.Y., 2004.
[75] Long, M., Architectural Acoustics, Elsevier, Academic Press, Amsterdam, p. 250, 796, 2006.
[76] Egan, M. D., Architectural Acoustics, J. Ross Publishing, Fort Lauderdale, Florida, 2007, Reprint of 1988 McGraw-Hill edition.
[77] Cory, J. F. and F. J. Hatfield, Force-Flow Analogy for Pulsation in Piping, in Finite Element Applications in Acoustics, ASME, N.Y., pp. 121–126, 1981.
[78] MscSoftware, 2004, NASTRAN Reference Manual, Section 13.5, Los Angeles California, MacNeal Schwindler.
[79] Gieck, J., Riding on Air, A History of Air Suspension, SAE International, Warrendale, PA, USA, 1999.
[80] Helmholtz, H., On the Sensations of Tone, Dover, N.Y., 1954, revised edition of 1877.
[81] Alster, M., Improved Calculation of Resonant Frequencies of Helmholtz Resonators, Journal of Sound and Vibration, vol. 24, pp. 64–85, 1972.
[82] Lamb, H., Hydrodynamics, 6th ed., Dover, New York, 1945, reprint of 1932 edition.
[83] Moiseev, N. N., and A. A. Petrov, The Calculation of Free Oscillations of a Liquid in a Motionless Container, in Advances in Applied Mechanics, G. Kuerti (ed.), Academic Press, New York, vol. 9, pp. 91–155, 1966.
[84] McNeill, W. A., and J. P. Lamb, Fundamental Sloshing Frequency for an Inclined, Fluid-Filled Right Circular Cylinder, Journal of Spacecraft and Rockets, vol. 7, pp. 1001–1002, 1970.
[85] McIver, P. Sloshing Frequencies for Cylindrical and Spherical Containers Filled to an Arbitrary Depth, Journal of Fluid Mechanics, vol. 201, pp. 243–257, 1989.
[86] Kozlov, V., and N. Kuznetsov, The 'Ice Fishing Problem' the Fundamental Sloshing Frequency Versus Geometry of Holes, Mathematical Methods in the Applied Sciences, vol. 27, pp. 289–312, 2004.
[87] Packham, B. A., Small-Amplitude Waves in a Straight Channel of Uniform Triangular Cross Section, Quarterly Journal of Mechanics and Applied Mathematics, vol. 33, pp. 179–187, 1980.
[88] McIver, P and M. McIver, Sloshing Frequencies of Longitudinal Modes for a Liquid Contained in a Trough, Journal of Fluid Mechanics, vol. 252, pp. 525–541, 1993.
[89] Shi, J. S. and C.-S. Yih, Waves in Open Channels, Journal of Engineering Mechanics, vol. 110, pp. 847–870, 1984.
[90] Goldsbrough, G. R., The Tidal Oscillations in an Elliptic Basin of Variable Depth, Proceedings of the Royal Society of London, Series A, vol. 130, pp. 157–167, 1931.
[91] Moiseev, N. N., Introduction to the Theory of Oscillations of Liquid-Containing Bodies, Advances in Applied Mechanics, vol. 8, pp. 233–289, 1964.
[92] Yih, Chia-Shun, Comparison Theorems for Gravity Waves in Basins of Variable Depth, Quarterly Journal of Applied Mechanics, 33, pp. 387–394, 1976.

[93] Housner, G. W., and D. E. Hudson, Applied Mechanics: Dynamics, 2nd ed., D. Van Nostrand Co., Princeton, N.J., p. 128, 1959.
[94] Wilson, B. W., Seiches, Advances in Hydroscience, vol. 8, pp. 1–94, 1972.
[95] Mayer, H. C., Walking with Coffee: Why Does it Spill?, Physical Review, E, vol. 85, 046177, 2012.
[96] Abramson, H. N. (ed.), Dynamic Behavior of Liquids in Moving Containers, NASA Report NASA SP-106, 1966.
[97] Epstein, H. I., Seismic Design of Liquid Storage Tanks, ASME Journal of the Structural Division, vol. 102, pp. 1659–1672, 1976.
[98] Ibrahim, R. A., Liquid Sloshing Dynamics, Cambridge University Press, p. 677, 2005.
[99] Faltinsen, O. M., Sloshing, Cambridge University Press, 2009.
[100] Sabersky, R. H., A. J. Acosta, and E. G. Hauptman, Fluid Flow: A First course in Fluid Mechanics, 2nd ed., Macmillan, pp. 388–389, 1971.
[101] Raichlen, F., Harbor Resonance, in Estuary and Coastline Hydrodynamics, A. T. Ippen (ed.), McGraw-Hill, pp. 281–340, 1966.
[102] Xing, X. Y., J. J. Lee, and F. Raichlen. Comparison of Computed Basin Response at San Pedro Bay with Long Period Wave Records, Proceedings of ICCE 2008, vol. 2, 1223–1235, 2008.
[103] Lee, J-.J. and F. Raichlen, 1972, Oscillation in Harbors with Connected Basins, Journal of Waterways, Harbors and Coastal Engineering, ASCE, vol. 91, 311–331.
[104] Miles, J. W., Harbor Seiching, in Annual Review of Fluid Mechanics, M. Van Dyke *et al.* (eds), pp. 17–36, 1974.
[105] Blevins, R. D., Flow-Induced Vibration, Van Nostrand Reinhold, reprinted by Krieger Publishing, 1990.
[106] Blasoveschensky, S. N., Theory of Ship Motions, Dover, N.Y., 1962.
[107] Wendel, K., Hydrodynamische Massen und hydrodynamische Massentragheitmoment, Jahrbuch Schiffbautechnisches Gesellschaft 44, 207 (1950). Translation in English "Hydrodynamic Masses and Hydrodynamic Moments of Inertia." David Taylor Model Translation No 260, 1950.
[108] Nielsen, J. N., Missile Aerodynamics, McGraw-Hill, N.Y., pp. 371–379, 1960.
[109] Patton, K. T., Tables of Hydrodynamic Mass Factors for Translation Motion, ASME Paper 65-WA/UNT-2, 1965.
[110] Kennard, E. H., Irrotational Flow of Frictionless Fluids, Mostly of Invariable Density, David Taylor Model Basin, DTMB Report 2299, Washington, D.C., p. 380, 1967.
[111] Bryson, A. E., Evaluation of the Inertia Coefficients of the Cross Section of a Slender Body, Journal of Aeronautical Sciences, vol. 21, 424–427, 1954.
[112] Newman, J. N., Marine Hydrodynamics, MIT Press, pp. 132–145, 1977.
[113] Lee, T. and R. Budwig, The Onset and Development of Circular Cylinder Vortex Wakes in Uniformly Accelerating Flows, Journal of Fluid Mechanics, vol. 232, 611–627, 1991.
[114] Sedov, L. I., Two-Dimensional Problems in Hydrodynamics and Aerodynamics, John Wiley, Interscience, 1965.
[115] Chen, S. S. M. W. Wambsganss, and J. A. Jendrzejczyk, Added Mass and Damping of a Vibrating Rod in Confined Viscous Fluid, Journal of Applied Mechanics, vol. 98, pp. 325–329, 1976.
[116] Moretti, P. M. and R. Lowery, Hydrodynamic Inertia Coefficient for a Tube Surrounding by Rigid Tube, Journal of Pressure Vessel Technology, vol. 97, 1975.
[117] Chen, S. S., Dynamic Response of Two Parallel Circular Cylinders in a Liquid, Journal of Pressure Vessel Technology, vol. 97, pp. 78–83, 1975.
[118] Bai, J. K., The Added-Mass of Two Dimensional Cylinders Heaving Water of Finite Depth, Journal of Fluid Mechanics, vol. 81, 85–105, 1977.
[119] Flagg, C. N. and J. N. Newman, Sway Added-Mass Coefficients for Rectangular Profiles in Shallow Water, Journal of Ship Research, vol. 15, pp. 257–265, 1971.
[120] Sarpkaya, T., Added Mass of Lenses and Parallel Plates, ASCE Journal of Engineering Mechanics Division, pp. 141–151, 1960.
[121] Munk, M. M. Aerodynamic Theory, in Fluid Dynamics Part II, W. F. Durand (ed.), Springer, Berlin, vol. I, p. 302; Aerodynamics of Airships, vol. IV, pp. 32–36, 1934.
[122] Meyerhoff, W. K. Added Mass of Thin Rectangular Plates Calculated from Potential Theory, Journal of Ship Research, vol. 14, pp. 100–111, 1970.
[123] Stelson, T. T. and F. T. Mavis, Virtual Mass and Acceleration in Fluids, Proceedings of American Society of Civil Engineers, vol. 81, pp. 670-1–670-9, 1955.
[124] Fernandes, A. C., F. P. S. Mineiro, Assessment of Hydrodynamic Properties of Bodies with Complex Shapes, Applied Ocean Research, vol. 29, 155–156; vol. 30, 341–344, 2007&2008.
[125] Ackermann, N. L., and A. Arbhabhirama, Viscous and Boundary Effects on Virtual Mass, Journal of Engineering Mechanics Division, Society of Civil Engineers, vol. 90, pp. 123–130, 1964.
[126] Waugh, J. G. and A. T. Ellis, Fluid Free-Surface Proximity Effect on a Sphere Vertically from Rest, Journal of Hydronautics, vol. 3, 175–179, 1969.

[127] Prislin, I, R. D. Blevins, and J. E. Halkyard, Viscous Damping and Added Mass of a Square Plate, 17th International Conference on Offshore Mechanics and Arctic Technology Conference, Lisbon, 1998, paper OMAE-98-316.
[128] Greenspon, J. E., Vibrations of Cross-Stiffened and Sandwich Plates with Application to Underwater Sound Radiators, Journal of the Acoustical Society of America, vol. 33, pp. 1485–1497, 1961.
[129] McLachlan, N. W., The Accession to Inertia of Flexible Discs Vibrating in a Fluid, Proceedings of Physical Society (London), vol. 44, pp. 546–555, 1932.
[130] Amabili, M and M. K. Kwak, Free Vibrations of Circular Plates Coupled with Liquids: Revising the Lamb Problem, Journal of Fluids and Structures, vol. 10, 743–761, 1996.
[131] Lamb, H. On the Vibrations of an Elastic Plate in Contact with Water, 1921.
[132] Lindholm, U. S., D. D. Kana and H. N. Abransom, Elastic Vibration of Cantilever Plates in Water, Journal of Ship Research, vol. 9, 11–22, 1965.
[133] Liang, C.-C., *et al.*, The Free Vibration Analysis of Submerged Cantilever, Proceedings of Royal Society (London), Series A 98, 205–216. Plates, Ocean Engineering, vol. 28, 1225–1245, 2001.
[134] Yadykin, Y., V. Tenetov, and D. Levin, 2003, The Added Mass of a Flexible Plate Oscillating in a Fluid, Journal of Fluid and Structures, vol.17, pp. 115–123.
[135] Bearman, P. W. *et al.*, Forces on Cylinder in Viscous Oscillatory Flow at Low Keulegan-Carpenter Numbers, Journal of Fluid Mechanics, vol. 154, pp 337–356, 1985.
[136] Brennen, C. E., A Review of Mass and Fluid Inertial Forces, Naval Civil Engineering Laboratory Report CR 82.010, Port Hueneme, California, 1982.
[137] Peake, W. H., and E. G. Thurston, The Lowest Resonant Frequency of a Water-Loaded Circular Plate, Journal of Acoustical Society of America, vol. 26, pp. 166–168, 1954.
[138] Marcus, M. S., 1978 Finite Element Method Applied to the Vibration of Submerged Plates, Journal of Ship Research, vol. 22, 94–99.
[139] Brown, S. J. 1982, A Survey of Studies into Hydrodynamic Response of Fluid Coupled Circular Cylinders, Journal of Pressure Vessel Technology, vol. 104, pp. 2–19.
[140] Amabili, E. P and M. P. Paidoussis, Nonlinear Vibrations of Simply Supported Circular Cylindrical Shells, Coupled to Quiescent Fluid, Journal of Fluids and Structures, vol. 12, 883–918, 1998.
[141] Lake, M. *et al.*, Hydrodynamic Coefficient Estimation for TLP and Spar Structures, Journal of Offshore Mechanics and Arctic Engineering, vol. 122, pp.118–124, 2000.
[142] Au-Yang, M. K., 1986, Dynamics of Coupled Fluid-Shells, Journal of Vibration, Acoustics, Stress, and Reliability in Design, vol. 108, pp. 339–347.
[143] De Santo, D. F., 1981, Added Mass and Hydrodynamic Damping of Perforated Plates Vibrating in Water, Journal of Pressure Vessel Technology, vol. 103, pp. 175182.

Further Reading

Salvesen, N. E., O. Tuck, and O. Faltinsen, 1970, Ship Motions and Sea Loads, Transactions Society of Naval Architects and Marine Engineers, vol. 78, 250–287.
Seldov, L. I. A Course in Continuum Mechanics, vol. III, Wolters-Noordhoff Publishing, Groningen, The Netherlands, pp. 203–707 (translated from Russian), 1972.

7

Forced Vibration

This chapter provides the dynamic response to forced vibration. Steady-state response to sinusoidal excitation, generated by motors, pumps, and rotating machinery, is discussed in Section 7.1. Section 7.2 shows formulas for shock and transient response to time history loading. Vibration isolation is discussed in Section 7.3. Section 7.4 has random response to spectral loads. Section 7.5 has simplified methods for quick estimation of dynamic response. The natural frequencies and mode shapes of elastic structures that support these formulas are in Chapters 3, 4, and 5.

7.1 Steady-State Forced Vibration

7.1.1 Single-Degree-of-Freedom Spring–Mass Response

Many practical systems consist of a massive component on an elastic support as shown in case 1 of Table 7.1 and Figure 7.1. The displacement of mass M at time t is $x(t)$. Newton's second law (Eq. 2.4) applied to the spring–mass systems gives the equation of motion:

$$M\ddot{x} + 2M\zeta\omega_n\dot{x} + kx = F(t) = F_o \cos \omega t \tag{7.1}$$

F_o $\cos(\omega t)$ = external harmonic force with amplitude F_o. $\omega = 2\pi f$ = frequency in radians per second where f = forcing frequency f in Hertz, k = the spring stiffness (Table 3.2), and $\omega_n = 2\pi f_n = (k/M)^{1/2}$ = the circular natural frequency in radians per second (case 1 of Table 3.3). Overdot denotes derivative with respect to time. The middle term $2M\zeta\omega_n\, dx/dt$ on the left-hand side of Equation 7.1 is a *linear viscous damping force* that is proportional to velocity times the dimensionless *damping factor* ζ.

Steady-state response is at the forcing frequency. Displacement $x(t)$ has amplitude X_o and phase φ, relative to the harmonic force:

$$x(t) = X_o \cos(\omega t - \varphi) \tag{7.2}$$

Formulas for Dynamics, Acoustics and Vibration, First Edition. Robert D. Blevins.
© 2016 John Wiley & Sons, Ltd. Published 2016 by John Wiley & Sons, Ltd.

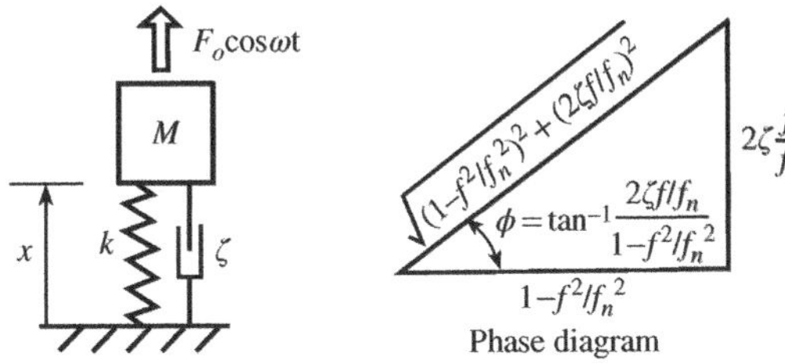

Figure 7.1 Forced vibration of a spring–mass damped system and phase of response relative to the force

Equation 7.2 is substituted into Equation 7.1 and factored by the harmonic terms with the identity $\cos(\omega t-\varphi)=\cos\varphi\sin\omega t+\sin\varphi\cos\omega t$:

$$(-M\omega^2X_o\cos\varphi+2M\zeta\omega_n\omega X_o\sin\varphi+kX_o\cos\varphi-F_o)\cos\omega t$$
$$+(-M\omega^2X_o\sin\varphi-2M\zeta\omega_n\omega X_o\cos\varphi+kX_o\sin\varphi)\sin\omega t=0 \qquad (7.3)$$

The factors of sin ωt and cos ωt must be zero for a solution. Setting the sine factor to zero produces the phase triangle in Figure 7.1 that is solved for phase angle φ. This phase is substituted into the cosine factor, which is solved for amplitude X_o.

The assembled solution to Equation 7.1 is the displacement in time as a function of forcing frequency and damping:

$$x(t)=\frac{F_o}{k}\frac{1}{\sqrt{(1-f^2/f_n^2)^2+(2\zeta f/f_n)^2}}\cos(\omega t-\varphi),\quad \tan\varphi=\frac{2\zeta(f/f_n)}{1-f^2/f_n^2} \qquad (7.4)$$

The magnitude of $x(t)$ is amplitude X_o:

$$X_o=\frac{F_o}{k}\frac{1}{\sqrt{(1-f^2/f_n^2)^2+(2\zeta f/f_n)^2}} \qquad (7.5)$$

X_o is equal to the static displacement F_o/k at zero frequency times the dynamic amplification factor. The *dynamic amplification factor* is amplitude X_o divided by the static displacement F_o/k:

$$\frac{X_ok}{F_o}=\frac{1}{\sqrt{(1-f^2/f_n^2)^2+(2\zeta f/f_n)^2}}=\begin{cases}1, & f\ll f_n,\quad \varphi=0\\ 1/2\zeta, & f=f_n,\quad \varphi=\pi/2\\ f_n^2/f^2, & f\gg f_n,\quad \varphi=\pi\end{cases} \qquad (7.6)$$

There are three limit cases associated with amplitude and dynamic amplification factor:

1. At *low frequency* ($f\ll f_n$). At zero frequency ($f=0$), Equation 7.4 reduces to the static solution

$$\lim_{f\to 0}X_o=\frac{F_o}{k}$$

$x = F_o/k$. At low frequency the mass moves without amplification. Dynamic amplification factor is unity and phase is zero.

2. At *high frequency* ($f \gg f_n$), mass acceleration dominates damping and stiffness. Amplitude becomes inversely proportional to frequency squared:

$$\lim_{f \gg f_n} X_o = \frac{F_o}{k}\frac{f_n^2}{f^2}$$

This is Newton's law (Eq. 2.4) applied to an unsupported mass. The dynamic amplification factor is less than unity, which is called isolation (see Section 7.3).

3. At *resonance*, ($f \approx f_n$), the forcing frequency equals, or approximately equals, the natural frequency f_n. The mass inertial and stiffness terms cancel each other. The phase is 90 degrees, so the damping force ($2\zeta\omega_n dx/dt$) equals the forcing. Maximum amplitude increases greatly. Maximum resonant response is limited by damping:

$$X_o|_{f=f_n} = \frac{F_o}{k}\frac{1}{2\zeta}, \quad \zeta \ll 1 \tag{7.7}$$

Dynamic amplification factor at resonance is called $Q = 1/2\zeta$. Light damping implies large amplitude resonant vibration. For example, for $\zeta = 0.02$, 2% damping, the amplification factor $Q = 1/2\zeta = 25$ and the resonant response is 25 times larger than the static response.

The vibrating mass exerts force on its base through the spring and the damper:

$$F_{\text{base}} = 2M\zeta\omega_n\dot{x} + kx = kX_o\sqrt{1 + \left(\frac{2\zeta f}{f_n}\right)^2}\cos(\omega t - \psi)$$
$$\tan\psi = \frac{2\zeta(f/f_n)^3}{[1 - (f/f_n)^2 + (2\zeta f/f_n)^2]} \tag{7.8}$$

The force on the bases includes both stiffness and damping terms.

7.1.1.1 Damping Factor

ζ is called the *damping factor*, *damping ratio*, and *fraction of critical damping*.

Built-up structures, assembled with fasteners from multiple pieces, typically have damping between $\zeta = 0.002$ and 0.05; $\zeta = 0.02$ is typical of built-up structures vibrating elastically at low amplitude in air. The ASME Boiler and Pressure Vessel Code, section III appendix N, provides damping factors from 0.002 for welded-in heat exchange tubes for low amplitude flow-included vibration to 0.05 for piping system during strong-motion earthquakes.

The dimensionless *damping factor* ζ for lightly damped systems, $\zeta < 1$, is proportional to the *bandwidth* Δf across the resonant response at an amplitude equal to $1/2^{1/2}$, which

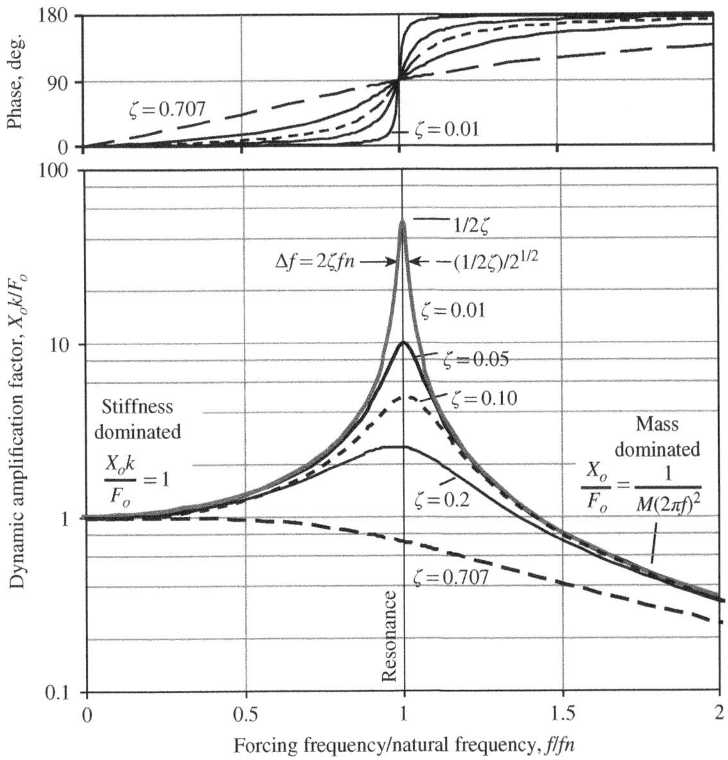

Figure 7.2 Dynamic amplification factor and phase

are called half-power points, times the resonant amplitude (Fig. 7.2; [1]):

$$\Delta f = 2\zeta f_n, \text{ equivalently, } \zeta = \frac{\Delta f}{2f_n} \tag{7.9}$$

Bandwidth is also the band of excitation frequencies about resonance where large responses will occur. Damping can be determined from measurements of bandwidth during forced excitation sine sweeps. Damping factor can also be measured from free decay after impact with an abrupt load (Eqs. 7.36 and 7.37).

7.1.1.2 Base Excitation

An important class of problems is due to vibration of the base [3]. Examples include earthquake excitation of buildings and equipment mounted on unbalanced rotating machinery. The mass on the vibrating base in Figure 7.3a displaces $x(t)$ with respect to the moving base. The base displaces $w(t)$ with respect to inertial (fixed) coordinates. The position of mass with respect to inertial coordinates is $y(t)$, the sum of $x(t)$ and $w(t)$:

$$y(t) = x(t) + w(t) \tag{7.10}$$

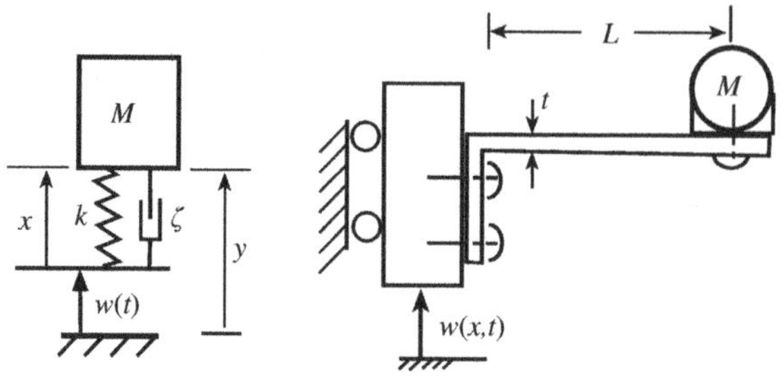

Figure 7.3 (a) and (b) Spring–mass systems on vibrating base

Application of Newton's second law requires acceleration in inertial coordinates. d^2y/dt^2 is substituted for d^2x/dt^2 in Equation 7.1 to produce the equation of motion for base excitation:

$$M(\ddot{x} + \ddot{w}) + 2M\zeta\omega_n\dot{x} + kx = 0 \tag{7.11a}$$

$$M\ddot{x} + 2M\zeta\omega_n\dot{x} + kx = -M\ddot{w}(t) \tag{7.11b}$$

This equation has the same form as Equation 7.1 with external force exchanged for base inertial load. Solutions for relative displacement $x(t)$ and absolute displacement $y(t)$ are shown in cases 3 and 4 of Table 7.1.

There are three limit cases for base excitation. As frequency f approaches zero ($f \ll f_n$), then the mass moves in lockstep with the foundation without amplification, $x = 0$. At resonance, $f \approx f_n$, the mass displacement is maximum for small damping. During high-frequency base excitation ($f \gg f_n$), Eq. 7.11a reduces to $\ddot{y} = \ddot{x} + \ddot{w} = 0$. At high frequency, the mass is nearly still as the base vibrates under it, which is called isolation (look ahead to Section 7.3 and Fig. 7.13).

Example 7.1 A 2.8 kg (6.17 lb) valve is attached to a vibrating machine with a flexible bracket (Fig 7.3b). The machine vibrates with vibration velocity amplitude $U_o = 3$ cm/s at 600–3000 revolutions per minute (10–50 Hz). The bracket stiffness is $k = 40$ N/mm (228.4 lb/inch). What is the maximum valve vibration amplitude over the excited frequency range? Use damping of 2%, that is, $\zeta = 0.02$.

Solution: The natural frequency of the valve on the bracket is computed from case 1 of Table 3.2 with units of case 1 and case 7 of Table 1.2:

$$f_n = \frac{1}{2\pi}\sqrt{\frac{k}{M}} = \frac{1}{2\pi}\sqrt{\frac{40000\,\text{N/m}}{2.8\,\text{kg}}} = \frac{1}{2\pi}\sqrt{\frac{228.4\,\text{lb/in}}{6.17\,\text{lb}/386.1\,\text{in/s}^2}} = 19.02\,\text{Hz}$$

Turbine case amplitude is computed from harmonic motion (Table 3.1) at this frequency: $Y_o = U_o/(2\pi f) = 3\,\text{cm/s}/(2\pi\ 19.02\,\text{Hz}) = 0.0425$ cm (0.167 in)

Table 7.1 Harmonically forced vibration of spring-mass systems

Notation: $c = 2M\zeta\omega_n$ = linear viscous damping coefficient, force/ velocity; e = radius of rotating unbalance; F_o = amplitude of force; f = excitation frequency; f_n, f_i = natural frequency, Hz, ith mode; k = spring stiffness, force/displacement, Table 3.2; M = mass of body; m = mass; t = time; $x = X_o \cos(\omega t - \varphi)$, displacement, or axial coordinate, X_o = single amplitude; $w = W_o \cos \omega t$, base displacement; y = transverse coordinate; ζ = damping factor, dimensionless; φ = phase of displacement relative to forcing $0 \le \varphi \le \pi$; $\omega = 2\pi f$ = circular frequency, radians/s. Overdot $(\cdot)$ denotes derivative with respect to time; subscript n = natural frequency. Consistent units are in Table 1.2. [1, 2] and author's results.

Harmonically Forced System	Equation of Motion and Response x(t)	Amplitude of Response, X_o
1. Spring Mass Damper Oscillating Force 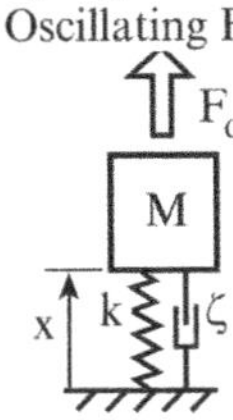 Force wall = $2M\zeta\omega_n\dot{x}+kx$ Ref. [1], p. 52	$M\ddot{x} + 2M\zeta\omega_n\dot{x} + kx = F_o\cos 2\pi ft$ $x(t) = \frac{F_o}{k}\frac{\cos(2\pi ft-\varphi)}{\sqrt{\left(1-\frac{f^2}{f_n^2}\right)^2+\left(2\zeta\frac{f}{f_n}\right)^2}}$ $\tan\varphi = \frac{2\zeta f/f_n}{1-(f/f_n)^2},\ 0 \le \varphi \le \pi$ $f_n = \frac{\omega_n}{2\pi} = \frac{1}{2\pi}\sqrt{\frac{k}{M}}$ Hz, $k = M\omega_n^2$	$X_o = \frac{F_o}{k}\frac{1}{\sqrt{\left(1-\frac{f^2}{f_n^2}\right)^2+\left(2\zeta\frac{f}{f_n}\right)^2}}$ $= \begin{cases} \frac{F_o}{k}, \text{at } f=0, \\ \frac{F_o}{k}\frac{1}{2\zeta(1-\zeta^2)^{1/2}}, \text{ for } \zeta < \frac{1}{2^{1/2}} \\ \frac{F_o}{k}\frac{1}{2\zeta}, \text{ at } f = f_n \end{cases}$
2. Spring Mass Damper Rotating Imbalance m 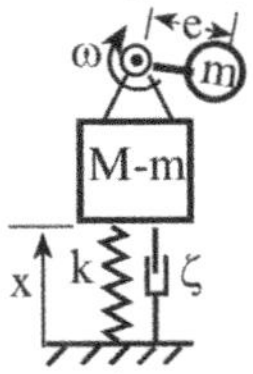 M = total mass non-rotating mass + unbalance mass m that rotates with frequency f at radius e.	$M\ddot{x} + 2M\zeta\omega_n\dot{x} + kx = em\omega^2\cos 2\pi ft$ $x(t) = \frac{m}{M}\frac{f^2}{f_n^2}\frac{e\cos(2\pi ft-\varphi)}{\sqrt{\left(1-\frac{f^2}{f_n^2}\right)^2+\left(2\zeta\frac{f}{f_n}\right)^2}}$ $\tan\varphi = \frac{2\zeta f/f_n}{1-(f/f_n)^2},\ 0 \le \varphi \le \pi$ $f_n = \omega_n/(2\pi) = (1/2\pi)\sqrt{k/M}$ Same as case 1 with $F_o = em\omega^2$. Ref. [1], p. 56	$X_o = \frac{e\frac{m}{M}\left(\frac{f}{f_n}\right)^2}{\sqrt{\left(1-\frac{f^2}{f_n^2}\right)^2+\left(2\zeta\frac{f}{f_n}\right)^2}}$ $= \begin{cases} 0, \text{at } f=0, \\ e\frac{m}{M}\frac{1}{2\zeta}, \text{ at } \frac{f}{f_n}=1 \\ \frac{m}{M}\frac{e}{2\zeta(1-\zeta^2)^{1/2}}, \text{ for } \zeta < \frac{1}{2^{1/2}} \end{cases}$
3. Spring Supported Mass with Base Displacement 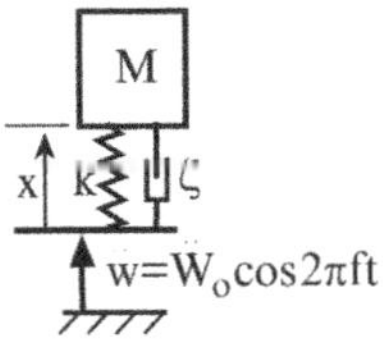 Note that x is relative to accelerating base. Force base $F_{base} = 2M\zeta\omega_n\dot{x}+kx$	$M\ddot{x} + 2M\zeta\omega_n\dot{x} + kx$ Ref. [1] $= -Md^2w/dt^2 = -M\ddot{W}_o\cos 2\pi ft$ $x(t) = \frac{\ddot{W}_o}{\omega_n^2}\frac{\cos(2\pi ft-\varphi)}{\sqrt{\left(1-\frac{f^2}{f_n^2}\right)^2+\left(2\zeta\frac{f}{f_n}\right)^2}}$ $\tan\varphi = \frac{2\zeta f/f_n}{1-(f/f_n)^2},\ 0 \le \varphi \le \pi$ $f_n = \omega_n/2\pi = (1/2\pi)\sqrt{k/M}$, Hz $\ddot{W}_o = (2\pi f)^2 W_o$ = amp base accl.	$X_o = \frac{\ddot{W}_o}{(2\pi f_n)^2}\frac{1}{\sqrt{\left(1-\frac{f^2}{f_n^2}\right)^2+\left(\frac{2\zeta f}{f_n}\right)^2}}$ $= \begin{cases} 0, \text{at } f=0 \\ W_o\frac{1}{2\zeta}, \text{ at } f=f_n \\ \frac{W_o}{2\zeta(1-\zeta^2)^{1/2}}, \text{ for } \zeta < \frac{1}{2^{1/2}} \end{cases}$ Same as case 1 with $F_o = M(2\pi f)^2$.

Table 7.1 Harmonically forced vibration of spring-mass systems, continued Plates

Harmonically Forced	Equations of Motion and Response
4. Spring Mass Damper With Base Displacement 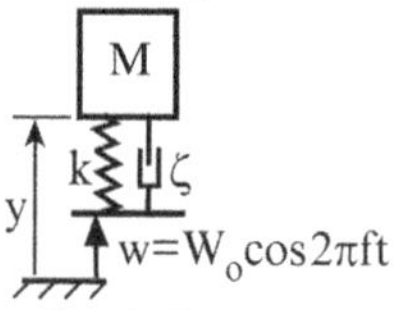 Note y is relative to ground	$M\ddot{y} + 2M\zeta\omega_n\dot{y} + ky = 2M\zeta\omega_n\dot{w} + kw$ $w(t) = W_o\cos 2\pi ft,\ f_n = \omega_n/2\pi = (1/2\pi)\sqrt{k/M}$, Hz $y(t) = \dfrac{W_o\sqrt{1+\left(2\zeta\dfrac{f}{f_n}\right)^2}\cos(2\pi ft-\varphi)}{\sqrt{\left(1-\dfrac{f^2}{f_n^2}\right)^2+\left(\dfrac{2\zeta f}{f_n}\right)^2}},\quad \tan\varphi = \dfrac{2\zeta\dfrac{f^3}{f_n^3}}{\left(1-\dfrac{f^2}{f_n^2}\right)^2+\left(\dfrac{2\zeta f}{f_n}\right)^2}$
5. Two Spring, Two Mass, Damped System with Base Displacement 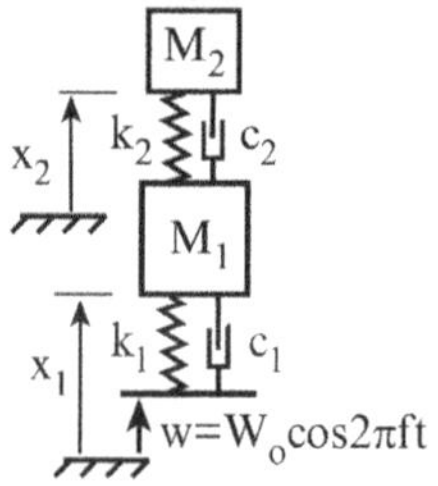 Absolute displacements Force on moving base $=c_1(\dot{x}_1-\dot{w})+k_1(x_1-w)$ $\omega = 2\pi f$ $f_1 = \dfrac{\omega_1}{2\pi} = \dfrac{1}{2\pi}\sqrt{\dfrac{k_1}{M_1}}$ $f_2 = \dfrac{\omega_2}{2\pi} = \dfrac{1}{2\pi}\sqrt{\dfrac{k_2}{M_2}}$ $c_1 = 2M_1\zeta_1\omega_1$ $c_2 = 2M_2\zeta_2\omega_2$ natural frequencies f_{n1}, f_{n1} case 3 of Table 3.2 $2f_n^2 = \alpha \mp \sqrt{\alpha^2 - 4f_1^2 f_2^2}$ $\alpha = f_1^2 + f_2^2(1+M_2/M_1)$ f_1 is nat freq with M_2 removed. f_2 is nat. freq. with M_1 stationary.author's result. Reduces to case 4 for $f_2 >> f$, f_1 and $M = M_1 + M_2$ Reduce to case 6 for $k_1 = 0$.	$M_1\ddot{x}_1 + (c_1+c_2)\dot{x}_1 + (k_1+k_2)x_1 - c_2\dot{x}_2 - k_2x_2 = c_1\dot{w} + k_1w$ $M_2\ddot{x}_2 + c_2\dot{x}_2 + k_2x_2 = c_2\dot{x}_1 + k_2x_1,\quad w(t) = W_o\cos 2\pi ft$ $x_1(t) = \dfrac{W_o}{\sqrt{G1}}\sqrt{\left[\left(1-\dfrac{f^2}{f_2^2}\right)^2+\left(2\zeta_2\dfrac{f}{f_2}\right)^2\right]\left[1+\left(2\zeta_1\dfrac{f}{f_1}\right)^2\right]}\cos(2\pi ft-\varphi_1),$ $x_2(t) = \dfrac{W_o}{\sqrt{G1}}\sqrt{\left[1+\left(2\zeta_1\dfrac{f}{f_1}\right)^2\right]\left[1+\left(2\zeta_2\dfrac{f}{f_2}\right)^2\right]}\cos(2\pi ft-\varphi_2)$ $G1 = \left[\left(1-\dfrac{f^2}{f_1^2}\right)\left(1-\dfrac{f^2}{f_2^2}\right)-\dfrac{M_2}{M_1}\dfrac{f^2}{f_1^2}\right]^2 + 8\zeta_1\zeta_2\dfrac{M_2}{M_1}\left(\dfrac{f^2}{f_1\,f_2}\right)^3$ $+\left(2\zeta_1\dfrac{f}{f_1}\right)^2\left\{\left(1-\dfrac{f^2}{f_2^2}\right)^2+\left(2\zeta_2\dfrac{f}{f_2}\right)^2\right\}+\left(2\zeta_2\dfrac{f}{f_2}\right)^2\left(1-\left(1+\dfrac{M_2}{M_1}\right)\dfrac{f^2}{f_1^2}\right)^2$ $=(1-f^2/f_{n1}^2)^2(1-f^2/f_{n2}^2)^2,\quad \text{for } \zeta_1=\zeta_2=0$ $\tan\varphi_1 = \dfrac{2\zeta_1\dfrac{f^3}{f_1^3}\left(1-\dfrac{f^2}{f_2^2}\right)\left(1+\dfrac{M_2}{M_1}-\dfrac{f^2}{f_2^2}\right)+8\zeta_1\zeta_2^2\dfrac{M_2}{M_1}\dfrac{f^5}{f_1^2f_2^3}+2\zeta_2\dfrac{M_2}{M_1}\dfrac{f^5}{f_1^2f_2^3}}{G2}$ $G2 = \left(1-\dfrac{f^2}{f_2^2}\right)\left(1-(1+\dfrac{M_2}{M_1})\dfrac{f^2}{f_1^2}\right)+4\zeta_1\zeta_2\dfrac{M_2}{M_1}\dfrac{f^6}{f_1^3f_2^3}+\left(2\zeta_1\dfrac{f}{f_1}\right)^2\left(1-\dfrac{f^2}{f_2^2}\right)^2$ $+\left(2\zeta_2\dfrac{f}{f_2}\right)^2\left(1-\left(1+\dfrac{M_2}{M_1}\right)\dfrac{f^2}{f_1^2}+\left(2\zeta_1\dfrac{f}{f_1}\right)^2\right)$ $\tan\varphi_2 = \dfrac{G3}{G4}\qquad G3 = 2\zeta_2\dfrac{f^3}{f_2^3}\left(1-\dfrac{f^2}{f_1^2}\right)+2\zeta_1\dfrac{f^3}{f_1^3}\left(1+\dfrac{M_2}{M_1}-\dfrac{f^2}{f_2^2}\right)$ $+8\zeta_1\zeta_2\dfrac{f^5}{f_1^2f_2^3}\left(\zeta_1+\zeta_2\dfrac{f_2}{f_1}\left(1+\dfrac{M_2}{M_1}\right)\right)$ $G4 = \left(1-\dfrac{f^2}{f_1^2}\right)\left(1-\dfrac{f^2}{f_2^2}\right)-\dfrac{M_2}{M_1}\dfrac{f^2}{f_1^2}+\left(2\zeta_2\dfrac{f}{f_2}\right)^2\left(1-\left(1+\dfrac{M_2}{M_1}\right)\dfrac{f^2}{f_1^2}\right)$ $-4\zeta_1\zeta_2\dfrac{f^6}{f_1^3f_2^3}+\left(2\zeta_1\dfrac{f}{f_1}\right)^2\left[1-\dfrac{f^2}{f_2^2}+\left(2\zeta_2\dfrac{f}{f_2}\right)^2\right]$

Table 7.1 Harmonically forced vibration of spring-mass systems, continued

Harmonically Forced	Equation of Motion and Response
6. Two Free Masses, Oscillating Force on M_1 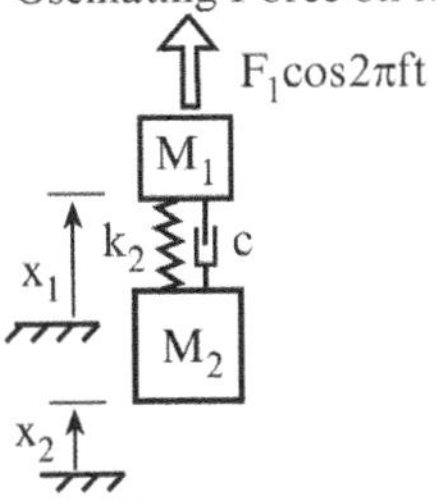 Natural frequency, f_n $f_n = \frac{1}{2\pi}\sqrt{\frac{(M_1+M_2)k_2}{M_1 M_2}}$ $f_2 = \frac{1}{2\pi}\sqrt{\frac{k_2}{M_2}}$	$M_1\ddot{x}_1 + c\dot{x}_1 + k_2 x_1 - c\dot{x}_2 - k_2 x_2 = F_1\cos 2\pi ft, \qquad c = 2M_2(2\pi f_2)\zeta.$ $M_2\ddot{x}_2 + c\dot{x}_2 + k_2 x_2 = c\dot{x}_1 + k_2 x_1$ $x_1(t) = \frac{F_1}{(2\pi f)^2(M_1+M_2)} \frac{\sqrt{\left(1-f^2/f_2^2\right)^2 + (2\zeta f/f_2)^2}}{\sqrt{\left(1-f^2/f_n^2\right)^2 + (2\zeta f/f_2)^2}} \cos(2\pi ft + \varphi_1)$ $x_2(t) = \frac{F_1}{(2\pi f)^2(M_1+M_2)} \frac{\sqrt{1+(2\zeta f/f_2)^2}}{\sqrt{\left(1-f^2/f_n^2\right)^2 + (2\zeta f/f_2)^2}} \cos(2\pi ft + \varphi_2)$ $\tan\varphi_1 = \frac{2\zeta f^3/f_2^3}{(1-f^2/f_2^2)\left[1-f^2/f_2^2 + M_1/M_2\right] + (1+M_1/M_2)(2\zeta f/f_2)^2}$ $\tan\varphi_2 = \frac{-2\zeta(M_1/M_2)^2 f^3/f_2^3}{(1+M_1/M_2)\left[1+(2\zeta f/f_2)^2\right] - (M_1/M_2) f^2/f_2^2}$
7. Two Spring, Two Mass, Oscillating Force on M_1 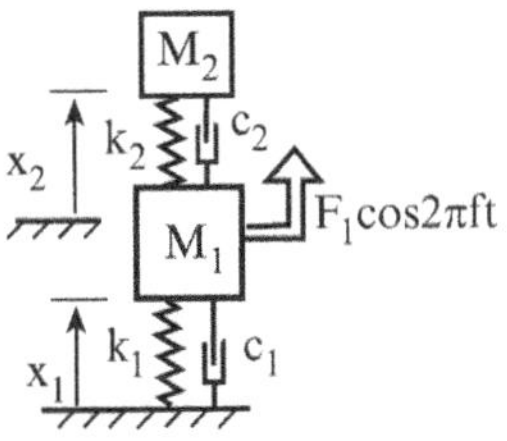 $\omega_1 = 2\pi f_1 = (k_1/M_1)^{1/2}$ $\omega_2 = 2\pi f_2 = (k_2/M_2)^{1/2}$ $c_1 = 2M_1\zeta_1\omega_1$ $c_2 = 2M_2\zeta_2\omega_2$ Reduces to case 1, $f_2 >> f, f_1$. For $\zeta_1=0$ reduces to Ref. [2], Eq. 3.24. Natural freqs. in case 5 Table 3.2. $x_2 - x_1$ = gap. author's result.	$M_1\ddot{x}_1 + (c_1+c_2)\dot{x}_1 + (k_1+k_2)x_1 - c_2\dot{x}_2 - k_2 x_2 = F_1\cos 2\pi ft$ $M_2\ddot{x}_2 + c_2\dot{x}_2 + k_2 x_2 = c_2\dot{x}_1 + k_2 x_1$ $x_1(t) = \frac{F_1}{k_1\sqrt{G1}}\sqrt{\left[\left(1-\frac{f^2}{f_2^2}\right)^2 + \left(2\zeta_2\frac{f}{f_2}\right)^2\right]}\cos(2\pi ft - \varphi_1)$ $x_2(t) = \frac{F_1}{k_1\sqrt{G1}}\sqrt{\left[1+\left(2\zeta_2\frac{f}{f_2}\right)^2\right]}\cos(2\pi ft - \varphi_2)$, G1 in case 5 $\tan\varphi_1 = \frac{\zeta_2\frac{M_2}{M_1}\frac{f^5}{f_1^2 f_2^3} + 2\zeta_1\frac{f}{f_1}\left[\left(1-\frac{f^2}{f_2^2}\right)^2 + \left(2\zeta_2\frac{f}{f_2}\right)^2\right]}{\left(1-\frac{f^2}{f_2^2}\right)\left(1-\frac{f^2}{f_1^2}\right)\left[\left(1-\frac{f^2}{f_2^2}\right) - \frac{M_2}{M_1}\frac{f^2}{f_1^2}\right] + \left(2\zeta_2\frac{f}{f_2}\right)^2\left[1-\left(1+\frac{M_2}{M_1}\right)\frac{f^2}{f_1^2}\right]}$ $\tan\varphi_2 = \frac{2\zeta_2\frac{f^3}{f_2^3}\left(1-\frac{f^2}{f_1^2}\right) + 2\zeta_1\frac{f}{f_1}\left(1-\frac{f^2}{f_2^2} + \left(2\zeta_2\frac{f}{f_2}\right)^2\right)}{\left(1-\frac{f^2}{f_1^2}\right)\left(1-\frac{f^2}{f_2^2}\right) - \frac{M_2}{M_1}\frac{f^2}{f_1^2} + 2\zeta_2\frac{f^2}{f_2^2}\left(2\zeta_2\left(1-\frac{f^2}{f_1^2} - \frac{M_2}{M_1}\frac{f^2}{f_1^2}\right) - 2\zeta_1\frac{f^2}{f_1 f_2}\right)^2}$

The response amplitude of the valve is shown in case 2 of Table 7.1. The maximum response occurs at resonance when the rotation frequency equals the 19.02 Hz natural frequency:

$$X_o = Y_o\frac{f^2}{f_n}\left[\left(1-\frac{f^2}{f_n^2}\right)^2+\left(\frac{2\zeta f}{f_n}\right)^2\right]^{-1/2} = \frac{Y_o}{2\zeta} = \frac{0.0425\ \text{cm}}{2(0.02)}$$

$$= 1.06\ \text{cm}\ (0.41\ \text{in.})\ \text{at}\ 19.02\ \text{Hz}$$

The resonant bracket displacement is 25 times the turbine case displacement. Its large motion is likely to alarm the operators and fatigue the bracket. Increasing the bracket stiffness by a factor of 10, such by increasing its thickness by a factor of 2.15, raises the valve natural frequency to 60 Hz, above the 50 Hz forcing frequency, and decreases response by a factor of ten.

7.1.2 *Multiple-Degree-of-Freedom Spring–Mass System Response*

The equations of motion of multi-degree-of-freedom discrete (spring–mass) elastic systems are placed in *matrix form* for simultaneous solution. Each matrix row is the ordinary differential equation of motion (Eq. 7.1) of a mass degree of freedom.

$$[M]\{\ddot{x}\} + [C]\{\dot{x}\} + [K]\{x\} = \{F(t)\} \tag{7.12}$$

The *mass matrix* $[M]$ and the *stiffness matrix* $[K]$ are square symmetric N by N matrices for $i = 1, 2, \ldots N$ mass degrees of freedom [3, 4]. Their entries are m_{ij} and k_{ij}, $i, j = 1, \ldots N$. (A mass element can have 1–6 degrees of freedom.) The N by 1 vectors $\{F\}$ and $\{x\}$ contain the $i = 1, \ldots N$ loads and displacements respectively, of the N nodes (masses), respectively.

Steady-state loads and response oscillate harmonically in time at the circular forcing frequency ω. Their in- and out-of-phase components can be expressed compactly as the real and imaginary components of a complex number or, equivalently, with real components with a phase angle φ:

$$\begin{aligned}\text{Complex}: x_i(t) &= X_i e^{j\omega t} = X_i \cos\omega t + jX_i \sin\omega t,\\ \text{Quadrature}: x_i(t) &= X_i\cos(\omega t - \varphi_i) = X_i\cos\varphi_i\cos\omega t + X_i\sin\varphi_i\sin\omega t\end{aligned} \tag{7.13}$$

In complex number form, the amplitude, $\{X\} = \{X_R\} + j\{X_I\}$, has a real (cosine) component X_R and an imaginary (sine) component X_I. j is the imaginary constant $(-1)^{1/2}$. Quadrature form conveys the same information with phase angle φ. Oscillating force is usually represented by its real (cosine) component, so φ is the phase of response relative to the real component of excitation. The matrix equation of steady-state motion is found by substituting Equation 7.13 into Equation 7.12:

$$(-\omega^2[M] + [C]\omega + [K])\{X\} = \{F\} \tag{7.14}$$

Off-diagonal terms ($i \neq j$) in the matrices couple the degrees of freedom. This system of equations must be solved simultaneously by direct matrix solution or by modal series solution as discussed in the following sections.

7.1.2.1 Direct Matrix Modal Solution

Direct matrix solution solves Equation 7.14 by *harmonic balance* (Eqs. 7.3 and 7.4; Table 7.1). The equations of motion are sorted into real and imaginary, or equivalently, in-phase and out-of-phase, components. Their factors are set to zero to produce two N by N linear algebraic equations in $2N$ unknown real and imaginary amplitudes. The equations are solved frequency by frequency. Harmonic balance does not require modal analysis; it can solve nonsymmetric matrices, some nonlinear equations, and systems with discreet viscous dampers. But exact solution of matrices is complex for larger systems (e.g., case 5 of Table 7.1). Numerical Gauss–Jordan elimination [4] is often used to solve the matrices of larger systems.

7.1.2.2 Modal Transformation Solution

Unlike direct matrix solution, modal solution allows each mode to be viewed and solved separately. It is preceded by a *free vibration eigenvalue analysis* of Equation 7.14, $(-\omega^2[M]+[K])\{x\}=0$, for N *eigenvalues* (natural frequencies) ω_i^2, $i=1, \ldots, N$ and their associated *unique mode shape* vectors $\{\tilde{x}\}_i$. These are discussed in Chapters 3–6 for spring–mass systems, membranes, beams, plates, shells, and acoustic cavities.

Modal transformation uncouples the Eq. 7.14 mode-by-mode. The dynamic displacement is expanded in a *modal series*. Each term is a mode shape multiplied by its time-dependent *scalar modal coordinates* $X_j(t)$:

$$x_j(t)=\sum_{i=1}^{N}\tilde{x}_{ji}X_i(t), \quad j=1, \ldots, N \tag{7.15}$$

Each mode shape is placed in an N by 1 column vector $\{x\}_i$. The nonsymmetric N by N mode shape matrix $[\tilde{x}]=[\{\tilde{x}\}_1, \{\tilde{x}\}_2, .\{\tilde{x}\}_j.\{\tilde{x}\}_N]$ is the collection of the modal column vectors. The mode shapes are orthogonal with respect to the mass and stiffness matrices [5, 6]:

$$\begin{aligned}\{\tilde{x}\}_i^T[M]\{\tilde{x}\}_j &= \left[\sum_{r=1}^{N}\sum_{s=1}^{N}\tilde{x}_{ri}m_{rs}\tilde{x}_{sj}\right] = M_i, \quad \text{if } i=j, \text{else } 0, \\ \{\tilde{x}\}_i^T[K]\{\tilde{x}\}_j &= \left[\sum_{r=1}^{N}\sum_{s=1}^{N}\tilde{x}_{ri}k_{rs}\tilde{x}_{sj}\right] = K_i, \quad \text{if } i=j, \text{else } 0\end{aligned} \tag{7.16}$$

The modal series (Eq. 7.15) is inserted into Equation 7.14, which is premultiplied by the transform (T means exchange rows and columns) of the eigenvector matrix whose entries are mode shapes $\tilde{x}_{ij}$ where i is degree of freedom and j the mode. Postmultiplying by the eigenvector matrix and using orthogonality (Eq. 7.16) eliminates off-diagonal terms (terms with $i \neq j$ are zero), which uncouples the N equations:

$$[\tilde{x}]^T[M][\tilde{x}]\{\ddot{X}\}+[\tilde{x}]^T[C][\tilde{x}]\{\dot{X}\}+[\tilde{x}]^T[M][\tilde{x}]\{X\}=[\tilde{x}]^T\{F\} \tag{7.17a}$$

diagonal ($i=j$) entries are one degree of freedom oscillators,

$$M_i\ddot{x}_i + 2\zeta_i M_i \omega_i \dot{x}_i + M_i \omega_i^2 x_i = \{\tilde{x}\}_i^T \{F\} = \sum_{j=1}^{N} \tilde{x}_{ji} F_j, \quad i = 1, 2, \ldots, N \tag{7.17b}$$

The diagonal terms in the transformed mass and stiffness matrices are the modal masses M_i and modal stiffness $K_i = M_i \omega_i^2$, where ω_i are the natural frequencies in rad/s. Each uncoupled modal equation (Eq. 7.17ab) of motion is a single-degree-of-freedom oscillator equation of motion that can be solved independently, which is the advantage of the modal superposition.

The damping terms in Eq. 7.17a uncouple only if the damping matrix is proportional to a linear combination of mass and stiffness matrices [6, 7]: $[C] = a[M] + b[K]$, which is not a constant modal damping. Choosing damping at two frequencies determines the proportionality constant and the proportional damping. If we pick the same damping factor $\zeta_1 = \zeta_2 = \zeta_{12}$ at two frequencies ω_1 and ω_2, then the proportional damping factor ζ_i is less than ζ_{12} between ω_1 and ω_2 and otherwise above ζ_{12}.

$$\zeta_i = \frac{\omega_1 \omega_2 + \omega_i^2}{\omega_i(\omega_1 + \omega_2)} \zeta_{12} \tag{7.18}$$

However, for light damping, damping modal coupling is small and one is more or less free to independently choose a modal damping for each mode [6, p. 432].

Static solutions to Equation 7.17ab (Eq. 7.21 for base excitation) are found by neglecting the time derivatives as the forcing frequencies approach zero:

$$x_i^s = \frac{\{\tilde{x}\}_i^T \{F^s\}}{K_i} = \begin{cases} \left(\frac{1}{M_i \omega_i^2}\right) \sum_{j=1}^{N} \tilde{x}_{ji} F_j^s & \text{forced vibration} \\ -\left(\frac{1}{M_i \omega_i^2}\right) \sum_{j=1}^{N} \tilde{x}_{ji} m_{ij} \ddot{w}_j^s & \text{base excitation} \end{cases} \tag{7.19}$$

These quasistatic displacements are important because they scale the modal dynamic response just as the displacement F_o/k scales the dynamic response of single-degree-of-freedom spring–mass system (Eq. 7.4).

7.1.2.3 Base Excitation with Multiple Degrees of Freedom Systems

Consider masses supported on an accelerating base that has inertial displacement $w(t)$. The inertial mass acceleration vector $\{\ddot{y}\} = \{1\}\ddot{w} + \{\ddot{x}\}$ replaces $\{\ddot{x}\}$ for application of Newton's second law. This is inserted in Equation 7.12 to produce an equation for displacement $\{x\}$ relative to base displacement $w(t)$:

$$[M]\{\ddot{x}\} + [K]\{x\} = -[M]\{1\}\ddot{w} \tag{7.20}$$

Modal transformation (Eq. 7.16) diagonalizes (uncouples) the modal equations of motion for base excitation:

$$M_i\ddot{x}_i + M_i\omega_i^2 x_i = \sum_{r=1}^{N}\sum_{s=1}^{N} \tilde{x}_{ri} m_{rs}\ddot{w}(t) = -M_i\Gamma_i\ddot{w}(t), \quad i = 1, 2, \ldots N$$

$$\Gamma_i = \frac{\{\tilde{x}_i\}^T[M]\{1\}}{M_i} = \frac{\sum_{r=1}^{N}\sum_{s=1}^{N}\tilde{x}_{ri}m_{rs}}{\sum_{r=1}^{N}\sum_{s=1}^{N}\tilde{x}_{ri}m_{rs}\tilde{x}_{si}} \tag{7.21}$$

Modal participation factors Γ have dimensions of one over the dimensions of mode shape. The effective modal masses, $m_{e,i} = M_i\Gamma_i^2$, have units of mass, and they are proportional to the modal response. The sum of the effective masses is equal to the sum of the entries in the mass matrix [8, 9]. $M_T = \sum_r^N \sum_{s=1}^N m_{rs} = \sum_{i=1}^N m_{e,i}$. Building codes require that sufficient number of modes be included in an earthquake response analysis so that the sum of the effective modal masses is 90% of the total mass [10].

7.1.3 Forced Harmonic Vibration of Continuous Systems

7.1.3.1 Beams Response

Table 7.2 [3, 5, 11–13] has modal solutions for the harmonic (sinusoidal) forced vibration of membranes (Chapter 3), beams (Chapter 4), and plates and shells (Chapter 5). Equation 4.6 in Chapter 4 shows a partial differential equation of bending of a uniform beam under an external force $F(x, t)$ in Figure 7.4a:

$$EI\frac{\partial^4 y(x,t)}{\partial x^4} + m\frac{\partial^2 y(x,t)}{\partial t^2} = F(x,t) = F(x)\cos\omega t \tag{7.22}$$

where $y(x,t)$ is the transverse displacements, m is the mass per unit length, E is the modulus of elasticity, and I (Table 1.5) is the area moment of inertia about the neutral axis.

The solution is sought with a modal series. Each term in the series is the product of a spatial mode shape, $\tilde{y}_i(x)$ (Tables 4.2 and 4.4), which is a function of the dimensionless parameter $\lambda x/L$ and its time-dependent mode coordinate, $Y_i(t)$:

$$y(x,t) = \sum_{i=1}^{N} \tilde{y}_i(x) Y_i(t) \tag{7.23}$$

Equation 7.23 is substituted into the partial differential equation of motion (Eq. 7.22).

$$\sum_{i=1}^{N}\left[EI\left(\frac{\lambda_i^4}{L^4}\right)\tilde{y}_i(x)\tilde{y}_j(x)y_i(t) + \tilde{y}_i(x)\tilde{y}_j(x)\ddot{y}_i(t)\right] = \tilde{y}_j(x)F(x)\cos\omega t$$

The result is multiplied by the mode shape of the jth mode and integrated over the span of the beam.

Table 7.2 Harmonic forced vibration of continuous systems

Notation: c = distance in cross section from neutral axis to stress point on edge of section; E = modulus of elasticity (Chapter 8); F_o = force amplitude; f = frequency of excitation, Hz; f_i = natural frequency, Hz, ith mode; i = integer mode index; I = area moment of inertia about neutral axis. Table 1.5; P, C, F, G = pinned, clamped, free, or guided boundary condition, Tables 4.2, 4.4; L = length of beam; N = maximum mode index; y = transverse displacement; m = mass of beam mass per unit length; t = time; W_o = amplitude of displacement of base; x = axial coordinate; y = transverse displacement; $\tilde{y}_i(x)$ = mode shape, Table 4.2; ζ_i =damping factor of ith mode; φ_i = phase of displacement relative to forcing, $\tan\varphi_i = 2\zeta_i(f/f_i)/(1-(f/f_i)^2)$, 0 to π radian; $\omega_i = 2\pi f_i$ = natural frequency, radian/s, ith mode; σ = axial bending stress in beam at point c. Table 1.2 has consistent units. Refs [6, 7, 11, 13], and author's results.

Total Displacement, $y(x,t)=\sum_{i=1}^{N} y_i(x,t)$

Harmonic Point Force F_o	Displacement in ith Mode, $y_i(x,t)$, i=1,2,3..
1. Clamp-Clamp Beam, i=1, Oscillating Force F_o at Point x_o 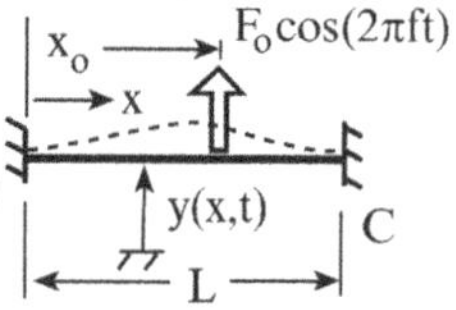 	$\frac{F_o}{m\omega_1^2 L}\frac{2}{3}\frac{(1-\cos 2\pi x_o/L)}{\sqrt{(1-f^2/f_1^2)^2+(2\zeta_1 f/f_1)^2}}(1-\cos\frac{2\pi x}{L})\cos(2\pi f t-\varphi_1)$ first mode response, nat. freq. $f_1=\frac{\omega_1}{2\pi}=\frac{4.73^2}{2\pi L^2}\sqrt{\frac{EI}{m}}$, Hz, i=1. Approx. mode shape $\tilde{y}_1(x)\approx(2/3)^{1/2}(1-\cos 2\pi x/L)$. case 7 Table 4.2. $\sigma=Ec\frac{\partial^2 y}{\partial x^2}=\frac{F_o Ec}{m\omega_1^2 L}\frac{4\pi^2}{L^2}\frac{2}{3}\frac{1.34(1-\cos 2\pi x_o/L)}{\sqrt{(1-f^2/f_1^2)^2+(2\zeta_1 f/f_1)^2}}\cos\frac{2\pi x}{L}\cos(\omega t-\varphi)$
2. Pinned-Pinned Beam Oscillating Force F_o at x_o 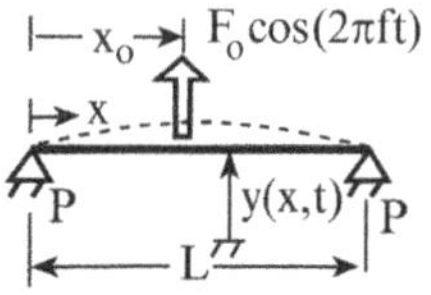 	$\frac{2F_o}{m\omega_i^2 L}\frac{\sin(i\pi x_o/L)}{\sqrt{(1-f^2/f_i^2)^2+(2\zeta_i f/f_i)^2}}\sin\frac{i\pi x}{L}\cos(2\pi f t-\varphi_i)$, i = 1,2,3.. first mode response, nat. freq. , $f_1=\frac{\omega_1}{2\pi}=\frac{(i\pi)^2}{2\pi L^2}\sqrt{\frac{EI}{m}}$, Hz, $\lambda_i = i\pi$ modeshape $\tilde{y}_1(x)=\sin(i\pi x/L)$, case 5 Table 4-2 $\sigma=Ec\frac{\partial^2 y}{\partial x^2}=\frac{2F_o Ec}{m\omega_i^2 L}\frac{i^2\pi^2}{L^2}\frac{\sin(i\pi x_o/L)}{\sqrt{(1-f^2/f_i^2)^2+(2\zeta_i f/f_i)^2}}\sin\frac{i\pi x}{L}\cos(\omega t-\varphi)$
3. Clamped-Free Beam Oscillating Force F_o at x_o 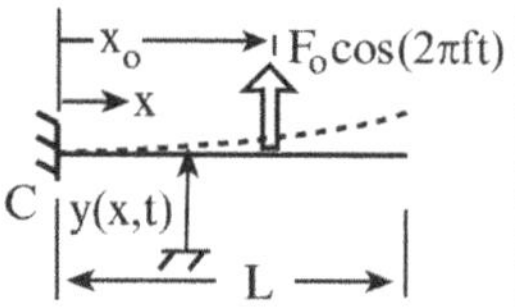 	$\frac{F_o}{m\omega_1^2 L}\frac{2.1^2(1-\cos\pi x_o/2L)}{\sqrt{(1-f^2/f_1^2)^2+(2\zeta_1 f/f_1)^2}}(1-\cos\frac{\pi x}{2L})\cos(2\pi f t-\varphi_1)$ first mode response, nat. freq. $f_1=\frac{\omega_1}{2\pi}=\frac{1.875^2}{2\pi L^2}\sqrt{\frac{EI}{m}}$, Hz, i=1 Approx. mode shape $\tilde{y}_1(x)\approx 2.1(1-\cos\pi x/2L)$, case 3 Table 4-2. $\sigma=Ec\frac{\partial^2 y}{\partial x^2}=\frac{F_o Ec}{m\omega_1^2 L}\frac{\pi^2}{4L^2}\frac{2.1^2(1.78)(1-\cos\pi x_o/2L)}{\sqrt{(1-f^2/f_1^2)^2+(2\zeta_1 f/f_1)^2}}\cos\frac{\pi x}{2L}\cos(\omega t-\varphi)$
4. Single Span Beam Oscil. Point Force F_o at point x_o 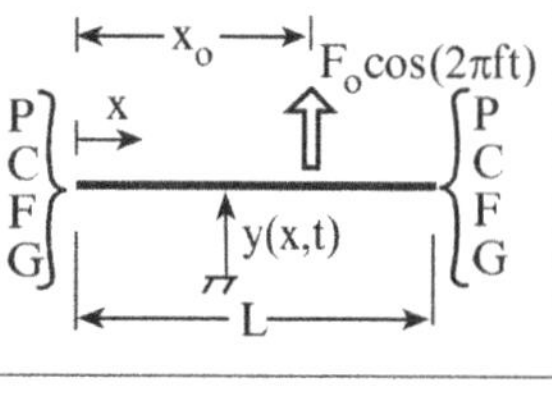 	$\frac{F_o}{m\omega_i^2}\frac{\tilde{y}_i(x_o)}{\int_0^L \tilde{y}_i^2(x)dx}\frac{1}{\sqrt{(1-f^2/f_i^2)^2+(2\zeta_i f/f_i)^2}}\tilde{y}_i(x)\cos(2\pi f t-\varphi_i)$, i = 1,2,. nat. freq. $f_i=\frac{\omega_i}{2\pi}=\frac{\lambda_i^2}{2\pi L^2}\sqrt{\frac{EI}{m}}$, Hz λ_i and $\tilde{y}_i(x)$ in Tables 4.2 and 4.4 according to boundary conditions. P,C,F,G = pinned, clamped, free and guided boundary conditions. Stress along span Eq. 4.2. $\sigma = Ec\partial^2 y/\partial x^2$.

Table 7.2 Harmonic forced vibration, continued, beams

Additional notation, F, F(x) = force per unit length.

Total Displacement, $y(x,t)=\sum_{i=1}^{N} y_i(x,t)$

Harmonic Force/Length	Displacement in ith Mode, $y_i(x,t)$, i=1,2,3..
5. Clamp-Clamp Beam, i=1, Uniform Oscillating Load ↦x $F\cos(2\pi ft)$ 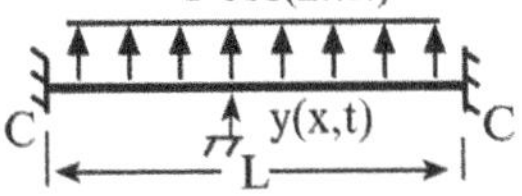 F = force per unit length stress correction factor=1.34	$\frac{F}{m(2\pi f_1)^2}\frac{2}{3}\frac{1}{\sqrt{(1-f^2/f_1^2)^2+(2\zeta_1 f/f_1)^2}}(1-\cos\frac{2\pi x}{L})\cos(2\pi f t-\varphi_1)$ first mode response, nat. freq. $f_1=\frac{\omega_1}{2\pi}=\frac{4.73^2}{2\pi L^2}\sqrt{\frac{EI}{m}}$,Hz, i=1 Approx mode shape $\tilde{y}_1(x)\approx(2/3)^{1/2}(1-\cos 2\pi x/L)$, case 7 Table 4.2. $\sigma=Ec\frac{\partial^2 y}{\partial x^2}=\frac{FEc}{m\omega_1^2}\frac{4\pi^2}{L^2}\frac{2}{3}\frac{1.34}{\sqrt{(1-f^2/f_1^2)^2+(2\zeta_1 f/f_1)^2}}\cos\frac{2\pi x}{L}\cos(\omega t\text{-}\varphi)$
6. Pinned-Pined Beam with Uniform Oscillating Load ↦x $F\cos(2\pi ft)$ 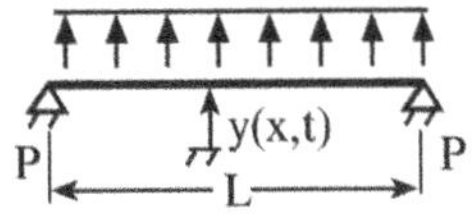 	$\frac{F}{m\omega_i^2}\frac{2}{i\pi}\frac{(1-(-1)^i)}{\sqrt{(1-f^2/f_i^2)^2+(2\zeta_i f/f_i)^2}}\sin\frac{i\pi x}{L}\cos(2\pi f t-\varphi_i)$, i=1,2,.. nat.freq. case 5 Table 4.2, $f_i=\frac{\omega_i}{2\pi}=\frac{(i\pi)^2}{2\pi L^2}\sqrt{\frac{EI}{m}}$, Hz. $\sigma=Ec\frac{\partial^2 y}{\partial x^2}=\frac{FEc}{m\omega_i^2}\frac{i^2\pi^2}{L^2}\frac{2}{i\pi}\frac{(1-(-1)^i)}{\sqrt{(1-f^2/f_i^2)^2+(2\zeta_i f/f_i)^2}}\sin\frac{i\pi x}{L}\cos(\omega t-\varphi_i)$
7. Clamped-Free Beam with Uniform Oscillating Load ↦x $F\cos(2\pi ft)$ 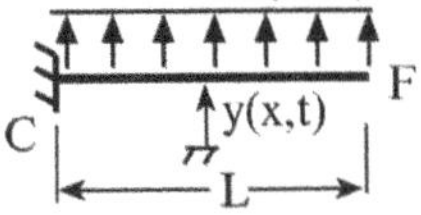 stress correction factor =1.78	$\frac{F}{m\omega_1^2}\frac{1.60}{\sqrt{(1-f^2/f_1^2)^2+(2\zeta_1 f/f_1)^2}}(1-\cos\frac{\pi x}{2L})\cos(2\pi f t-\varphi_1)$ first mode response, nat. freq. $f_1=\frac{\omega_1}{2\pi}=\frac{1.875^2}{2\pi L^2}\sqrt{\frac{EI}{m}},\frac{rad}{s}$, i=1 Approx. mode shape $\tilde{y}_1(x)\approx 2.1(1-\cos\pi x/2L)$, case 3 of Table 4.2. $\sigma=Ec\frac{\partial^2 y}{\partial x^2}=\frac{FEc}{m\omega_1^2}\frac{\pi^2}{4L^2}\frac{1.6(1.78)}{\sqrt{(1-f^2/f_i^2)^2+(2\zeta_i f/f_i)^2}}\cos\frac{\pi x}{2L}\cos(\omega t-\varphi)$
8. Beam with Uniform Oscillating Load F 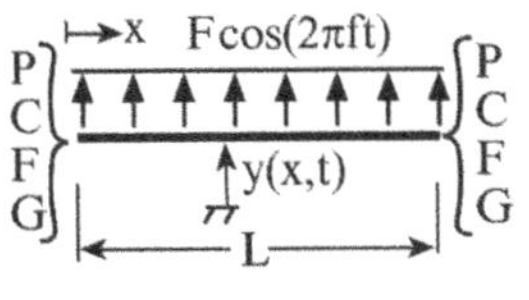 	$\frac{F\int_0^L \tilde{y}_i(x)dx}{m\omega_i^2\int_0^L \tilde{y}_i^2(x)dx}\frac{1}{\sqrt{(1-f^2/f_i^2)^2+(2\zeta_i f/f_i)^2}}\tilde{y}_i(x)\cos(2\pi f t-\varphi_i)$,i=1,2. λ_i and $\tilde{y}_i(x)$ in Tables 4.2, 4.4 according to boundary conditions. nat.freq, ,$f_i=\frac{f_i}{2\pi}=\frac{\lambda_i^2}{2\pi L^2}\sqrt{\frac{EI}{m}}$, Hz. Force F per unit length. Stress along span Eq. 4.2. $\sigma=Ec\partial^2 y/\partial x^2$.
9. Distributed Oscil. Force $F(x)\cos(2\pi ft)$ 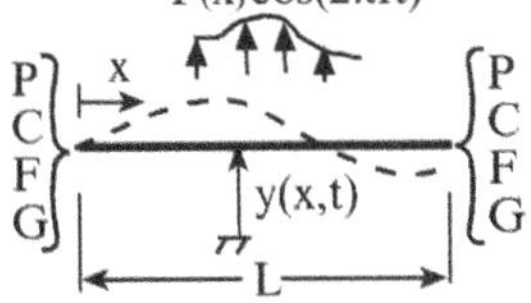 	$\frac{1}{m\omega_i^2}\frac{\int_0^L F(x)\tilde{y}_i(x)dx}{\int_0^L \tilde{y}_i^2(x)dx}\frac{1}{\sqrt{(1-f^2/f_i^2)^2+(2\zeta_i f/f_i)^2}}\tilde{y}_i(x)\cos(2\pi f t-\varphi_i)$ $\lambda_i,\tilde{y}_i(x)$ in Tables 4.2,4.4, nat.freq. $\omega_i=2\pi f_i=\frac{\lambda_i^2}{L^2}\sqrt{\frac{EI}{m}}$, rad / s. i = 1,2,. P,C,F,G = pinned, clamped, free and guided boundary conditions. Stresses in beam from Eq. 4.2.

Table 7.2 Harmonic forced vibration, continued, beams

Additional notation. W_o = amplitude of displacement of base; $\ddot{W}_o = (2\pi f)^2 W_o$ = amplitude of base acceleration; phase, $\tan\varphi_i = 2\zeta_i(f/f_i)/(1-(f/f_i)^2)$, m = beam mass per unit length. y(*t*) = displacement of beam relative to base

Displacement, $y(x,t)=\sum_{i=1}^{N} y_i(x,t)$	
Harmonic Base Vibration	Displacement in ith Mode, $y_i(x,t)$, i=1,2,3..
10. Clamped Beam, 1st Mode, Base Acceleration 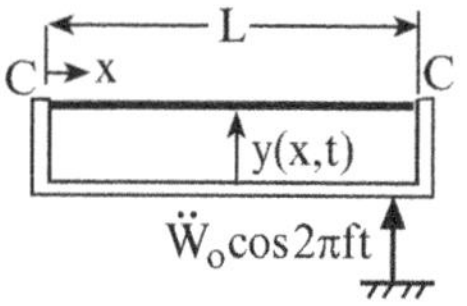 	$\frac{\ddot{W}_o}{(2\pi f_1)^2}\frac{2}{3}\frac{1}{\sqrt{(1-f^2/f_1^2)^2+(2\zeta_1 f/f_1)^2}}\left(1-\cos\frac{2\pi x}{L}\right)\cos(2\pi f t-\varphi_1)$ $\tilde{y}_1(x)\approx\sqrt{\frac{2}{3}}[1-\cos\frac{2\pi x}{L}],\ f_1=\frac{4.73^2}{2\pi L^2}\sqrt{\frac{EI}{m}}$, Hz. Table 4.2, case 7, i = 1 $\sigma=Ec\frac{\partial^2 y}{\partial x^2}=\frac{\ddot{W}_o Ec}{\omega_1^2}\frac{4\pi^2}{L^2}\frac{2}{3}\frac{1.34}{\sqrt{(1-f^2/f_1^2)^2+(2\zeta_1 f/f_1)^2}}\cos\frac{2\pi x}{L}\cos(\omega t-\varphi)$
11. Pinned-Pinned Beam Base Acceleration 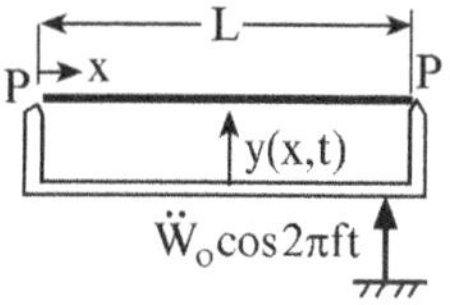 	$\frac{\ddot{W}_o}{\omega_i^2}\frac{2}{i\pi}\frac{(1-(-1)^i)}{\sqrt{(1-f^2/f_i^2)^2+(2\zeta_i f/f_i)^2}}\sin\frac{i\pi x}{L}\cos(2\pi f t-\varphi_i),\ i=1,2..$ $\tilde{y}_i(x)=\sin\frac{i\pi x}{L},\ f_i=\frac{\omega_i}{2\pi}=\frac{(i\pi)^2}{2\pi L^2}\sqrt{\frac{EI}{m}}$, Hz, case 5 of Table 4.2 $\sigma=Ec\frac{\partial^2 y}{\partial x^2}=\frac{\ddot{W}_o Ec}{\omega_i^2}\frac{i^2\pi^2}{L^2}\frac{2}{i\pi}\frac{(1-(-1)^i)}{\sqrt{(1-f^2/f_i^2)^2+(2\zeta_i f/f_i)^2}}\sin\frac{i\pi x}{L}\cos(\omega t-\varphi_i)$
12 Clamped Free Base Acceleration 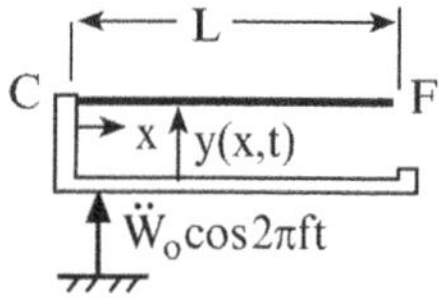 	$\frac{\ddot{W}_o}{\omega_1^2}\frac{1.6}{\sqrt{(1-f^2/f_1^2)^2+(2\zeta_1 f/f_1)^2}}(1-\cos\frac{\pi x}{2L})\cos(2\pi f t-\varphi_1)$ first mode response, i=1. nat. freq. $f_1=\frac{\omega_1}{2\pi}=\frac{1.875}{2\pi L^2}\sqrt{\frac{EI}{m}}$, Hz Approx. mode shape $\tilde{y}_1(x)\approx 2.1(1-\cos\pi x/2L)$, case 7 of Table 4.2. $\sigma=Ec\frac{\partial^2 y}{\partial x^2}=\frac{\ddot{W}_o Ec}{\omega_1^2}\frac{\pi^2}{4L^2}\frac{1.6(1.78)}{\sqrt{(1-f^2/f_1^2)^2+(2\zeta_1 f/f_1)^2}}\cos\frac{\pi x}{2L}\cos(\omega t-\varphi)$
13. Pinned-Pinned Beam One Moving Support 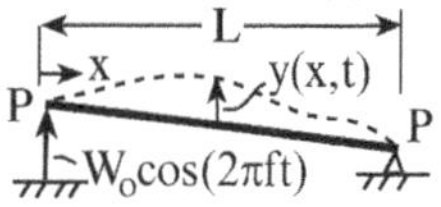 	$\frac{(2\pi f)^2 W_o}{\omega_i^2}\frac{2}{i\pi}\frac{(-1)^i}{\sqrt{(1-f^2/f_i^2)^2+(2\zeta_i f/f_i)^2}}\sin\frac{i\pi x}{L}\cos(2\pi f t-\varphi_i),\ i=1,2..$ $\tilde{y}_i(x)=\sin\frac{i\pi x}{L},\ f_i=\frac{\omega_i}{2\pi}=\frac{(i\pi)^2}{2\pi L^2}\sqrt{\frac{EI}{m}}$, Hz, case 5 of Table 4.2 $y(t)$ = beam displacement relative to $W_o(x/L)\cos(2pft)$.
14. Uniform Beam Base Acceleration 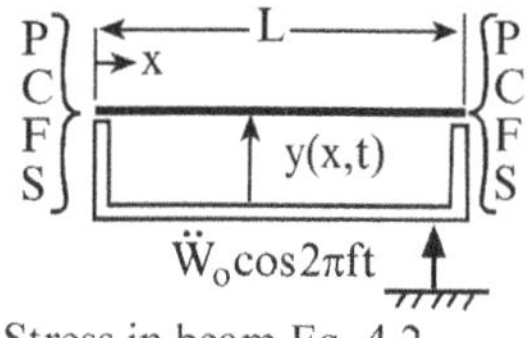 Stress in beam Eq. 4.2	$\frac{\ddot{W}_o}{(2\pi f_i)^2}\cdot\frac{\int_0^L\tilde{y}_i(x)dx}{\int_0^L\tilde{y}_i^2(x)dx}\cdot\frac{1}{\sqrt{(1-f^2/f_i^2)^2+(2\zeta_i f/f_i)^2}}\tilde{y}_i(x)\cos(2\pi f t-\varphi_i),$ $\approx\frac{\ddot{W}_o}{\omega_i^2\tilde{y}_{imax}(x)}\frac{1}{\sqrt{(1-f^2/f_i^2)^2+(2\zeta_i f/f_i)^2}}\tilde{y}_i(x)\cos(2\pi f t-\varphi_i), i=1,2,.$ $\tilde{y}_{imax}$ = max of $\tilde{y}_i(x)$, Eq. 4.18. See Section 7.1.3. nat freq $f_i=\omega_i/2\pi=\lambda_i^2(EI/m)^{1/2}/2\pi L^2$,Hz), $\lambda_i,\tilde{y}_i(x)$,Tables 4.2, 4.4 P,C,F,G = pinned, clamped, free and guided boundary conditions

Table 7.2 Harmonic forced vibration, continued

Additional Notation: A = area; a,b = length and width; S = simply supported; Eq. 5.2; C = clamped, Eq. 5.2; E = modulus of elasticity; F_o = force amplitude; h = plate thickness; i,j = integer modal indices; w = out of place displacement; F_o = amplitude of oscillating force; $\gamma = \rho h$ = mass per unit area of plate, ρ = density; ν = Poisson's ratio; $\omega_{ij} = 2\pi f_{ij}$ natural frequency, radian/s; phase, $\tan\varphi_{ij} = 2\zeta_{ij}(f/f_{ij})/(1-(f/f_{ij})^2)$.

Total Displacement $w(x,t)=\sum_i^N\sum_j^N w_{ij}(x,y,t)$

Harmonic Forcing	Displacement in i-j Mode, $w_{ij}(x,y,t)$
15. Clamp-Free Beam Axial Force F_o on Free End 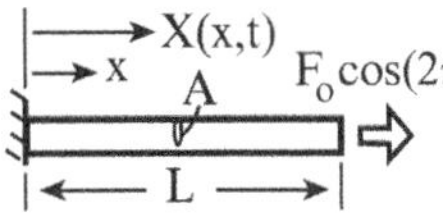	$X_i(x,t)=\frac{2F_oL}{EA}\left(\frac{2}{(2i-1)\pi}\right)^2\frac{(-1)^{i+1}}{\sqrt{(1-f^2/f_i^2)^2+(2\zeta_i f/f_i)^2}}\sin\frac{(2i-1)\pi x}{2L}$ $\times\cos(2\pi ft-\varphi_i), i=1,2,3..,$ Axial motion. Ref. [11] nat.freq. $f_i=\frac{1}{2\pi L}\left(\frac{(2i-1)\pi}{2}\right)\sqrt{\frac{EA}{m}}$, Hz. case 2 Table 4.13. A = area
16. Tension String, Pt. Force 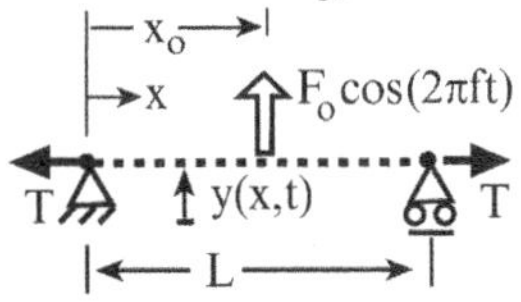	$\frac{F_oL}{T}\frac{2}{(i\pi)^2}\frac{\sin(i\pi x_o/L)}{\sqrt{(1-f^2/f_i^2)^2+(2\zeta_i f/f_i)^2}}\sin\frac{i\pi x}{L}\cos(2\pi ft-\varphi_i), i=1,2,3.$ mode shape and nat.freq. $\tilde{y}_i(x)=\sin\frac{i\pi x}{L}, f_i=\frac{i\pi}{2\pi L}\sqrt{\frac{P}{m}}$, Hz, T = string tension. Case 1 Table 3.6. Force = $F_o\cos(2\pi ft)$ at $x = x_o$.
17. Tensioned string Uniform oscillating load 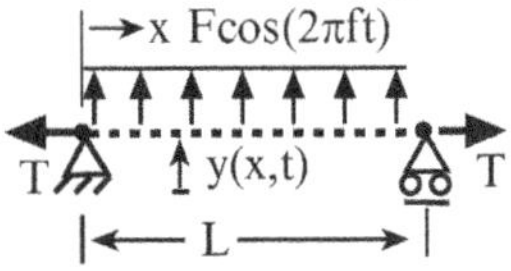	$\frac{FL^2}{T}\frac{2}{(i\pi)^3}\frac{1-(-1)^i}{\sqrt{(1-f^2/f_i^2)^2+(2\zeta_i f/f_i)^2}}\sin\frac{i\pi x}{L}\cos(2\pi ft-\varphi_i), i=1,2,3.$ mode shape and nat.freq. $\tilde{y}_i(x)=\sin\frac{i\pi x}{L}, f_i=\frac{i\pi}{2\pi L}\sqrt{\frac{P}{m}}$, Hz, T = string tension. Case 1 Table 3.6.
18. Circular Membrane with Oscillating Point Force F_o at $r=r_o$ and $\theta=0$. Tension T per unit length 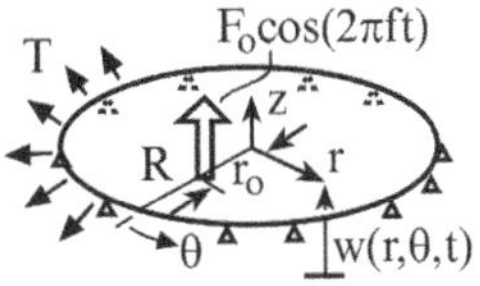	$\frac{F_o}{T}\frac{1}{\pi^2\lambda_{ij}^2}\frac{(1/\alpha_i)}{J_{i+1}^2(\pi^{1/2}\lambda_{0j})}\frac{J_i(\pi^{1/2}\lambda_{0j}r_o/R)\cos i\theta}{\sqrt{(1-\frac{f^2}{f_{ij}^2})^2+(\frac{2\zeta_{ij}f}{f_{ij}^2})^2}}J_i(\pi^{1/2}\lambda_{ij}\frac{r}{R})\cos(2\pi ft-\varphi_i)$ mode shape $J_i(\pi^{1/2}\lambda_{ij}r/R)\cos i\theta$, $f_{ij}=\frac{\lambda_{ij}}{2}\sqrt{\frac{T}{\gamma A}}$, Hz, $i=0,1,2,.,j=1,2,.$ λ_{ij} in case 8 of Table 3.7. $\lambda_{01}=1.357$. γ = mass/area. $\alpha_0=1$, $\alpha_i=2$, $i>0$. $J_i(x)$ = Bessel function of first kind. $J_0(0)=1$. $J_1(\pi^{1/2}1.357)=0.519$
19. Circular Membrane Oscillating pressure 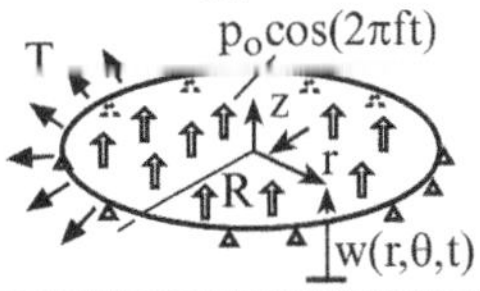	$\frac{2p_o}{\gamma\omega_{0j}^2}\frac{1}{\pi^{1/2}\lambda_{0j}}\frac{1}{\sqrt{(1-f^2/f_{0j}^2)^2+(2\zeta_{0j}f/f_{0j})^2}}J_0(\pi^{1/2}\lambda_{0j}\frac{r}{R})\cos(2\pi ft-\varphi_i)$ mode shape $J_0(\pi^{1/2}\lambda_{ij}r/R)$, $\omega_{0j}=2\pi f_{0j}=\pi\lambda_{0j}\sqrt{\frac{1}{\gamma A}}, \frac{rad}{s}$, Ref. [6], p. 296 λ_{0j} in case 8 of Table 3.7. γ = mass/area. $j=1,2.,,i=0$. see case 18.

Table 7.2 Harmonically forced vibration, continued, plates

Additional Notation: A = area of plate; a,b = length and width; S = simply supported; Eq. 5.2; C = clamped, Eq. 5.2; E = modulus of elasticity; F_o = force amplitude; h = plate thickness; i,j = integer modal indices; w = out of place displacement; F_o = amplitude of oscillating force; $\gamma = \rho h$ = mass per unit area of plate, ρ = density; ν = Poisson's ratio; $\omega_{ij} = 2\pi f_{ij}$ natural frequency, radian/s; phase, $\tan\varphi_{ij} = 2\zeta_{ij}(f/f_{ij})/(1-(f/f_{ij})^2)$.

Total Displacement $w(x,t) = \sum_i^N \sum_j^N w_{ij}(x,y,t)$

Harmonic Forced Plate	Displacement in i-j Mode, $w_{ij}(x,y,t)$
20. Clamped Circular Plate, Oscillating Point Force at r_o, $F_o\cos(2\pi ft)$ 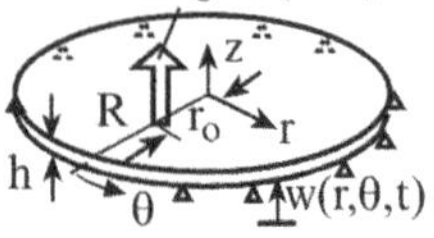	$\frac{1.562 F_o \left[J_0(\lambda_{00} r_o/R) - J_0(\lambda_{00}) I_0(\lambda_{00} r_o)/I_0(\lambda_{00})\right]}{\gamma R^2 \omega_{00}^2 \sqrt{(1-f^2/f_{00}^2)^2 + (2\zeta_{00} f/f_{00})^2}} \cos(2\pi ft - \varphi_{00})$ $\times\left[J_0(\lambda_{00} r/R) - J_0(\lambda_{00}) I_0(\lambda_{00} r)/I_0(\lambda_{00})\right]$, i=j=0, case 3 of Table 5.2 fund.nat. freq, $\omega_{00} = 2\pi f_{00} = \frac{\lambda_{00}^2}{R^2}\sqrt{\frac{Eh^3}{12\gamma(1-\nu^2)}}$, rad/s, $\lambda_{00} = 3.197$
21. Circular Plate, 1st mode, Simply Supported Edges Point Force at r_o, $F_o\cos(2\pi ft)$ 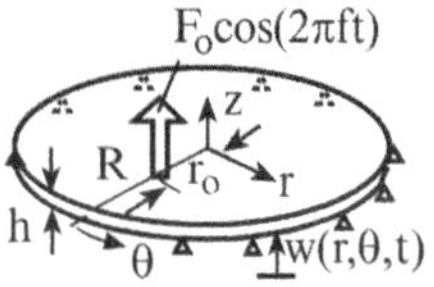	$\frac{1.201 F_o}{\gamma R^2 \omega_{00}^2} \frac{\tilde{w}_{ij}(r_o)}{\sqrt{(1-f^2/f_{00}^2)^2 + (2\zeta_{00} f/f_{00})^2}} \left[J_0\left(\frac{\lambda_{00} r_o}{R}\right) - \frac{J_0(\lambda_{00})}{I_0(\lambda_{00})} I_0\left(\frac{\lambda_{00} r_o}{R}\right)\right]$ $\times\left[J_0\left(\frac{\lambda_{00} r}{R}\right) - \frac{J_0(\lambda_{00})}{I_0(\lambda_{00})} I_0\left(\frac{\lambda_{00} r}{R}\right)\right] \cos(2\pi ft - \varphi_{00})$, case 2 of Table 5.2 fund. nat. freq. $\omega_{00} = 2\pi f_{00} = \frac{\lambda_{00}^2}{R^2}\sqrt{\frac{Eh^3}{12\gamma(1-\nu^2)}}$, rad/s, $\lambda_{00} = 2.2309$
22. Rectangular Clamp Plate Oscillating F_o at x_o, y_o 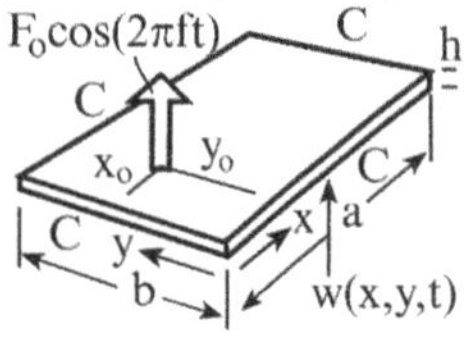	$\frac{F_o}{\gamma ab\omega_{11}^2} \frac{(1-\cos(2\pi x_o/a))(1-\cos(2\pi y_o/b))}{\sqrt{(1-f^2/f_{11}^2)^2 + (2\zeta_{11} f/f_{11})^2}} \left(1-\cos 2\pi\frac{x}{a}\right)\left(1-\cos\frac{2\pi y}{b}\right)$ $\times\cos(2\pi ft - \varphi_i)$, i = 1 approximate first mode shape, case 3 of Tabe 5.3. aprox nat. freq. $f_{11} = \frac{\omega_{11}}{2\pi} \approx \frac{2\pi}{3ab}\sqrt{3\frac{a^2}{b^2} + 2 + 3\frac{b^2}{a^2}}\sqrt{\frac{Eh^3}{12(1-\nu^2)}}$, Hz
23. Rectangular Plate Simply Supported Edges $F_o\cos(2\pi ft)$ 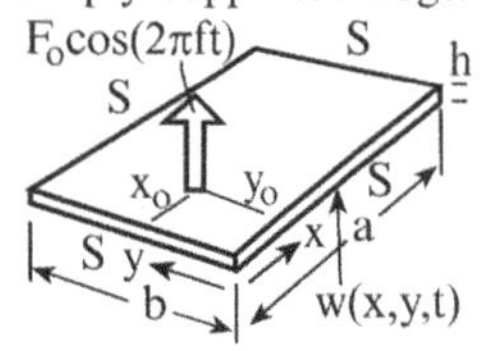	$\frac{4F_o}{\gamma ab\omega_{ij}^2} \frac{\sin(i\pi x_o/a)\sin(j\pi y_o/b)}{\sqrt{(1-f^2/f_{ij}^2)^2 + (2\zeta_{ij} f/f_{ij})^2}} \sin\frac{i\pi x}{a}\sin\frac{j\pi y}{b}\cos(2\pi ft - \varphi_i)$ natural frequency from case 2 of Table 5.3, i, j = 1,2,3.., Ref. [13], p. 253 $\omega_{ij} = 2\pi f_{ij} = \left(\left(\frac{i\pi}{a}\right)^2 + \left(\frac{j\pi}{b}\right)^2\right)^{1/2}\sqrt{\frac{Eh^3}{12\gamma(1-\nu^2)}}$, rad/s
24. Plate Oscillating Pt Force $F_o\cos(2\pi ft)$ 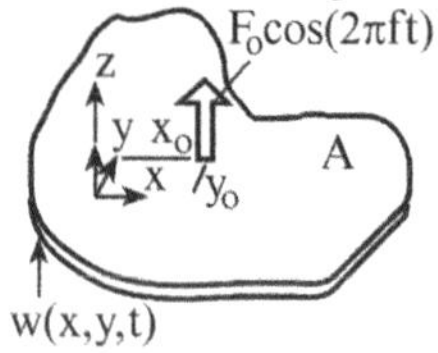	$\frac{F_o \tilde{w}_{ij}(x_o, y_o)}{\omega_{ij}^2 \int_A \gamma(x,y)\tilde{w}_{ij}^2(x,y)dxdy} \frac{\tilde{w}_{ij}(x,y)}{\sqrt{(1-f^2/f_{ij}^2)^2 + (2\zeta_{ij} f/f_{ij})^2}} \cos(2\pi ft - \varphi_{ij})$. nat. freq. $\omega_{ij}^2 = (2\pi f_{ij})^2 = \frac{Eh^3}{12(1-\nu^2)} \frac{\int_A (\nabla^4 \tilde{w}_{ij}(x,y))\tilde{w}_{ij}(x,y)dxdy}{\int_A \gamma(x,y)\tilde{w}_{ij}^2(x,y)dxdy}$, rad/s ω_{ij} and mode shapes $\tilde{w}_{ij}(x,y)$ Chapter 5. i, j = 1,2,3. Exact. See case32.

Table 7.2 Harmonic forced vibration, continued, plates

Additional Notation: A = area of plate; S = simply supported; Eq. 5.2; C = clamped, Eq. 5.2; E = modulus of elasticity; F_o = amplitude of oscillating force; h = plate thickness; p_o = pressure amplitude; w = out of plane displacement; $\gamma = \rho h$ = mass per unit area of plate, ρ = density; ν = Poisson's ratio; $\omega_{ij} = 2\pi f_{ij}$ natural frequency, radian/s.

Total Displacement $w(x,t) = \sum_i^N \sum_j^N w_{ij}(x,y,t)$

Harmonic Forced Plate Oscillating Pressure p_o	Displacement in i,j Mode, $w_{ij}(x,y,t)$
25. Circular Plate, 1st mode Clamped Edges Oscillating Pressure p_o, $p_o\cos(2\pi ft)$ 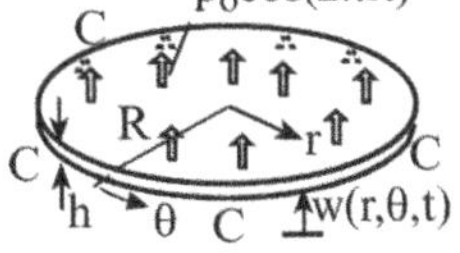	$W_{00} = \frac{1.614 p_o}{\gamma \omega_{00}^2} \frac{\tilde{w}_{ij}(r_o)}{\sqrt{(1-f^2/f_{00}^2)^2 + (2\zeta_{00} f/f_{00})^2}}$, i = j = 0, fund. mode $\tilde{w}_{00}(r) = \left[J_0\left(\frac{\lambda_{00} r}{R}\right) - \frac{J_0(\lambda_{00})}{I_0(\lambda_{00})} I_0\left(\frac{\lambda_{00} r}{R}\right)\right]$, case 3 of Table 5.2 $\omega_{00} = 2\pi f_{00} = \frac{\lambda_{00}^2}{R^2}\sqrt{\frac{Eh^3}{12\gamma(1-\nu^2)}}$, rad/s, $\lambda_{00} = 3.197$
26. Circular Plate, 1st mode Simply Supported Edges $p_o\cos(2\pi ft)$ 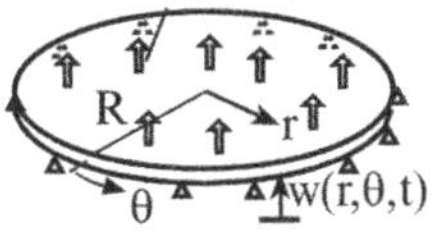	$\frac{1.635 p_o}{\gamma \omega_{00}^2} \frac{\tilde{w}_{00}(r_o)}{\sqrt{(1-f^2/f_{00}^2)^2 + (2\zeta_{00} f/f_{00})^2}} \left[J_0\left(\frac{\lambda_{00} r}{R}\right) - \frac{J_0(\lambda_{00})}{I_0(\lambda_{00})} I_0\left(\frac{\lambda_{00} r}{R}\right)\right]$ $\times\cos(2\pi ft - \varphi)$, case 2 of Table 5.3, fundamental mode, i = j = 0. $\omega_{00} = 2\pi f_{00} = \frac{\lambda_{00}^2}{R^2}\sqrt{\frac{Eh^3}{12\gamma(1-\nu^2)}}$, rad/s, $\lambda_{00} = 2.2309$
27. Rectangular Clamp Plate Oscillating Pressure p_o 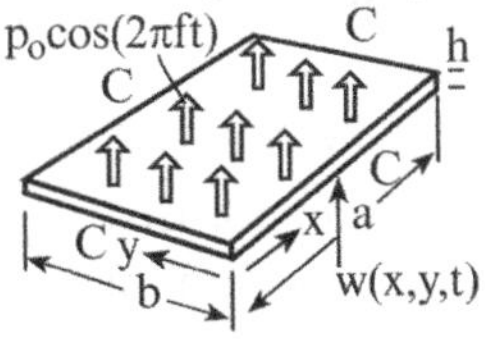	$\frac{p_o}{\gamma \omega_{11}^2} \frac{4}{9} \frac{(1-\cos 2\pi x/a)(1-\cos 2\pi y/b)}{\sqrt{(1-f^2/f_{11}^2)^2 + (2\zeta_{11} f/f_{11})^2}} \cos(2\pi f t - \varphi_{11})$ first mode, i = j = 1, approximate mode shape, case 3 of Table 5.3. nat. freq. $\omega_{11} = 2\pi f_{11} \approx \frac{4.73^2}{a^2}\left\{1 + (\frac{a}{b})^4 + 0.6(\frac{a}{b})^2]\right\}^{1/2} \sqrt{\frac{Eh^3}{12(1-\nu^2)}}$, rad./s
28. Rectangular Plate Simply Supported Edges 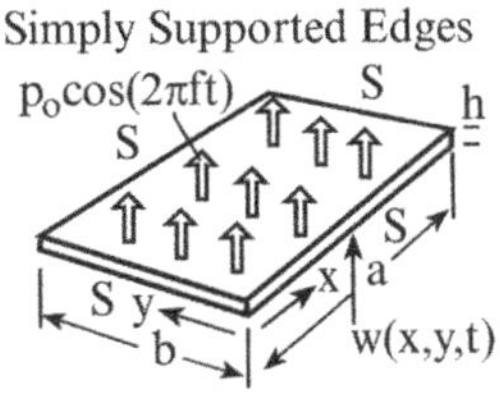	$\frac{4p_o}{\gamma \omega_{ij}^2} \frac{(1-(-1)^i)(1-(-1)^j)}{(i\pi)(j\pi)} \frac{\sin(i\pi x/a)\sin(j\pi y/b)}{\sqrt{(1-f^2/f_{ij}^2)^2 + (2\zeta_{ij} f/f_{ij})^2}} \cos(2\pi f t - \varphi_{11})$ i, j = 1,2,3.., case 2 of Table 5.3 nat. freq. $\omega_{ij} = 2\pi f_{ij} = \left(\left(\frac{i\pi}{a}\right)^2 + \left(\frac{j\pi}{b}\right)^2\right)^{1/2} \sqrt{\frac{Eh^3}{12\gamma(1-\nu^2)}}$, rad/s
29. Plate with uniform Oscillating Pressure 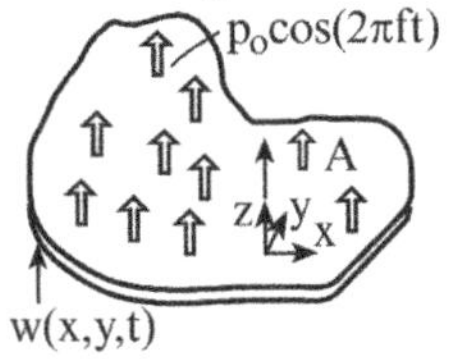	$\frac{p_o \int_A \tilde{w}_{ij}(x,y)dxdy}{\omega_{ij}^2 \int_A \gamma(x,y)\tilde{w}_{ij}^2(x,y)dxdy} \frac{\tilde{w}_{ij}(x,y)}{\sqrt{(1-f^2/f_{ij}^2)^2 + (2\zeta_{ij} f/f_{ij})^2}} \cos(2\pi f t - \varphi_{ij})$ pressure on plate = $p_o\cos(2\pi f t)$, i, j = 1,2,3... Exact. $\omega_{ij}^2 = \left(2\pi f_{ij}\right)^2 = \frac{Eh^3}{12(1-\nu^2)} \frac{\int_A (\nabla^4 \tilde{w}_{ij}(x,y))\tilde{w}_{ij}(x,y)dxdy}{\int_A \gamma(x,y)\tilde{w}_{ij}^2(x,y)dxdy}$, rad./s mode shapes $\tilde{w}_{ij}(x,y)$ from Tables in Chapter 5, according to shape.

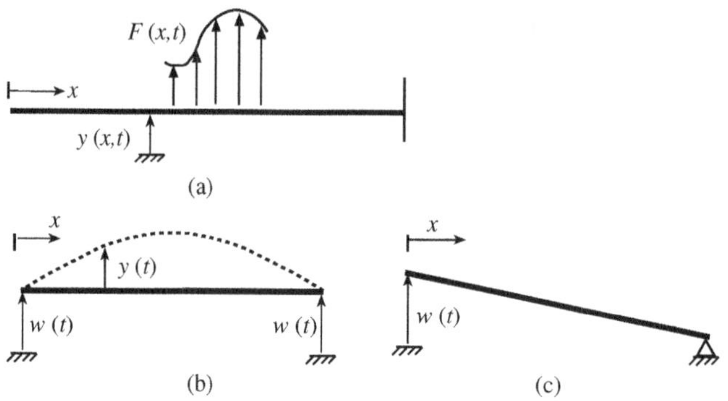

Figure 7.4 (a–c) Three cases of transverse excitation of a beam

Orthogonality of beam mode shapes (Eq. 4.17) is applied. Modal damping is added to the left-hand side. The result is a set of N uncoupled second-order linear ordinary differential equations for the modal coordinates:

$$m\frac{d^2Y_i(t)}{dt^2} + 2m\zeta_i\omega_i\frac{dY_i(t)}{dt} + m\omega_i^2Y_i(t) = \frac{\int_0^L F(x)\tilde{y}_i(x)dx}{\int_0^L \tilde{y}_i^2(x)dx}\cos\omega t, \quad i = 1, \ldots N \tag{7.24}$$

The solutions are modal coordinates $Y_i(t)$ that are dependent on mode normalization. However, the product of modal displacement and mode shape, $\tilde{y}_i(x)Y_i(t)$, is physical modal beam displacement, which is independent of mode normalization.

The exact solutions for the modal coordinate $Y_i(t)$ are shown in cases 2–4 and 6–9 of Table 7.2 for sinusoidal forcing. For example, beam modal displacement for distributed sinusoidal forcing is shown in case 9 of Table 7.2. Total displacement is the sum of modal displacements:

$$y(x,t) = \sum_{i=1}^{N}\left[\frac{\int_0^L F(x)\,\tilde{y}_i(x)dx}{m\omega_i^2\int_0^L \tilde{y}_i^2(x)dx}\right]\frac{1}{\sqrt{(1-f^2/f_i^2)^2+(2\zeta f/f_i)^2}}\tilde{y}_i(x)\cos\,(\omega t-\varphi_i) \tag{7.25}$$

Dynamic displacement in each mode is the product of static displacement, dynamic amplification factor, mode shape $\tilde{y}_i(x)$, and time history. A single mode often dominates the response, as discussed in Example 7.2.

7.1.3.2 Beam Base Excitation

A beam excited by vibrating supports is analyzed by modal superposition displacement $y(x,t)$ is defined as the displacement of the beam relative to the moving base (Fig. 7.4b).

The displacement of base with respect to around is $w(x,t)$. The modal coordinate equation of motion is

$$m\frac{d^2Y_i(t)}{dt^2} + 2m\zeta_i\omega_i\frac{dY_i(t)}{dt} + m\omega_i^2 Y_i(t) = -\frac{\int_0^L m\ddot{w}(x,t)\tilde{y}_i(x)dx}{\int_0^L \tilde{y}_i^2(x)dx} \tag{7.26}$$

If all supports move in unison, $w(x,t) = w(t)$. The first mode solution is shown in case 11 of Table 7.2 for sinusoidal base excitation of uniform beams with various boundary conditions. Also see the following example.

It is also possible to have only a one support moving. The pinned beam in Figure 7.4c left-hand support moves vertically from horizontal and the right-hand support is fixed. Thus, support motion induces a rigid body beam rotation: $w(x,t) = w(t)\,(x/L)$. The beam displacement $y(x,t)$ is relative to the quasistatic position of the beam $w(t)\ x/L$ where L is the span of the beam. Substituting $w(x,t) = w(t)\,(x/L)$ into Equation 7.26 and using sinusoidal mode shape ($\sin\ i\pi x/L$) for a pinned–pinned beam and the integrals from Appendix D result in the modal equation of motion with single-ended beam motion:

$$m\frac{d^2Y_i(t)}{dt^2} + 2m\zeta_i\omega_i\frac{dY_i(t)}{dt} + m\omega_i^2 Y_i(t) = -m\ddot{w}(t)\frac{\int_0^L (x/L)\sin(i\pi x/L)dx}{\int_0^L \sin^2(i\pi x/L)dx} = -m\ddot{w}(t)\frac{2(-1)^i}{i\pi}$$

The solutions are shown in case 13 of Table 7.2. Comparing with case 11, we see that the excitation in the first mode with one support moving is one half of that with two moving supports (base excitation). However, single support motion excites the second beam mode ($i = 2$) which is not excited at all when both supports move in unison.

Example 7.2 The ends of a $L = 0.5$ m (19.27 in.) long thin-walled straight steel tube with diameter $D = 1$ cm (0.4 inch) are pinned to a vibrating machine (case 11 of Table 7.2 and Fig. 7.4b). The machine vibrates sinsuoidally with acceleration amplitude of 20 g, twenty times the acceleration of gravity, $20(9.81\ \text{m/s}^2) = 196.1\ \text{m/s}^2$ between 50 and 5000 Hz. What are the stress and displacement of the tube? Assume 2% damping.

Solution: First, the natural frequency of the thin-walled pipe is evaluated with Equation 4.15 with an average diameter of $D = 0.01$ m, units of cases 1 and 7 of Table 1.2, and the pinned–pinned natural frequency parameter of case 5 of Table 4.2, $\lambda_i = i\pi$. E is the tube modulus of elastic, 200 MPa (29×10^6 psi), ρ is the tube material density

(8 gm/cc, 0.29 lb/in^3), and x is distance along the tube:

$$f_i = \frac{(i\pi)^2 D}{8\pi L^2}\sqrt{\frac{2E}{\rho}} = \begin{cases} \dfrac{(i\pi)^2 0.01\,m}{8\pi(0.5\,m)^2}\sqrt{\dfrac{2(200E9\ Pa)}{8000\ \ kg/m^3}} \\ \dfrac{(i\pi)^2\,0.4\,in.}{8\pi\,(19.7in.)^2}\sqrt{\dfrac{2\,(29E6\,lb/in.^3)}{0.29\,lb/in.^3/386.1\,in./s^2}} \end{cases} = 112.i^2\ Hz,\ \ i = 1,2,3..$$

The amplitude of base displacement at 112 Hz is computed with formulas in Table 3.1 for 20 g acceleration:

$$W_o = \frac{196.1\ \ \text{m/s}^2}{(2\pi \times 112\ \ \text{Hz})^2} = 0.000396\ \text{m}$$

The response in the first mode is calculated from cases 11 and 14 of Table 7.2. The result is worked out only for first mode resonance:

$$y_{1,\max} = \frac{\ddot{W}_o}{(2\pi f_i)^2}\cdot\frac{\int_0^L \tilde{y}_1(x)dx}{\int_0^L \tilde{y}_1^2(x)dx}\cdot\frac{1}{\sqrt{(1 - f^2/f_1^2)^2 + (2\zeta_1 f/f_1)^2}}$$

$$y_{1,\max}|_{f=fi} = W_o\cdot\frac{4}{\pi}\cdot\frac{1}{2\zeta_1} = 0.396\ \text{mm}\,\frac{4}{\pi}\frac{1}{2(0.02)}1 = 12.6\ \text{mm}$$

The integral factors are given in case 11 of Table 7.2 for a pinned–pinned beam.

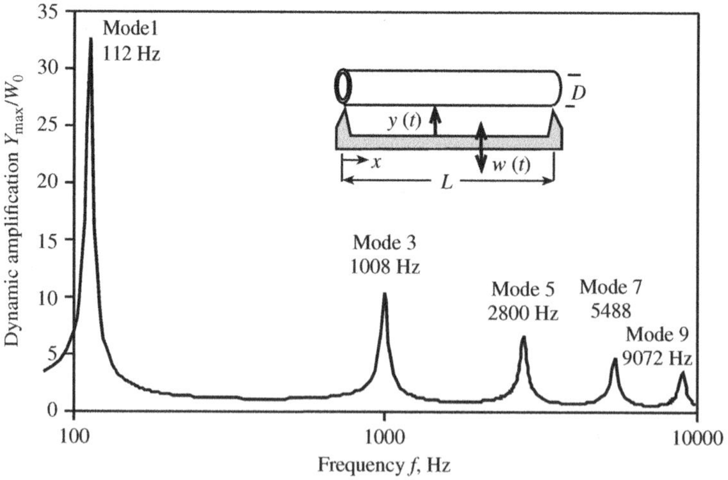

Figure 7.5 Dynamic response of a pinned–pinned tube on a vibrating support as a function of support vibration frequency f

Displacement is plotted as a function of forcing frequency in Figure 7.5. Only odd-numbered sinusoidal modes respond to uniform lateral acceleration because there is no net inertial force on symmetric modes. The fundamental mode, $i = 1$, at 112 Hz dominates the response. These results are independent of tube wall for thin tubes.

Maximum bending stress is also shown in Chapter 4, Eq. 4.2. Maximum is on the outer fiber at point of greatest curvature:

$$\sigma_{1,\max} = Ec \left. \frac{\partial^2 Y(x,t)}{\partial x^2} \right|_{\max} = Ec \cdot y_{1,\max} \cdot \frac{d^2 \tilde{y}_{1,\max}}{dx^2} = \frac{ED}{2} Y_1 \left(\frac{\pi}{L} \right)^2$$

$$= (200 \times 10^9 \text{ Pa})(0.01 \text{ m})(0.0126 \text{ m}) \frac{(1\pi)^2}{2(0.5 \text{ m})^2} = 500 \text{ MPa (72 ksi)}$$

One million vibration cycles at this stress amplitude can cause fatigue failure in steel. Adding an intermediate support to the tube halves the span, reduces the stress by 4, and eliminates the fatigue problem.

7.1.4 *General System Response*

The beam modal solution in the previous section can be generalized to an arbitrary elastic structure. A generic linear equation of motion based on Newton's second law includes stiffness, mass inertia, and forcing terms [5, 6, p. 287]:

$$L[w(P,t)] + M(P) \frac{\partial^2 w(P,t)}{\partial t^2} = F(P,t) = \overline{F}(P)F(t) \tag{7.27}$$

where P denotes the triplet of orthogonal spatial coordinates x, y, z, t is times. M is mass, and $L[]$ is a linear stiffness operator that consists of spatial partial derivatives of displacement $w(P,t)$ (see, e.g., Eq. 7.23 for beam bending). The force is a function of space $\overline{F}(P)$ times a separate function of time.

Modal analysis is applied to Equation 7.27. Solutions of the associated homogeneous eigenvalue equation $L[\tilde{w}(P)] = \omega^2 M(P)$ are a positive semi-infinite set of positive natural frequencies ω_i, and associated orthogonal mode shapes $\tilde{w}_i(p)$, $i = 1, 2 \ldots$ provided that (1) the mass M is positive and (2) the linear operator $L[]$ is positive definite and self-adjoint over the domain D for functions $u(P)$ and $v(P)$, which is the usual case [5, 6]:

$$\int_D uL[u]dD \geq 0, \int_D vL[u]dD = \int_D uL[v]dD, \int_D uL[u]dD > 0 \tag{7.28}$$

The mode shapes are orthogonal with respect to mass and stiffness over the domain:

$$\int_D \tilde{w}_r(P)L[\tilde{w}_s]dD = \int_D M(p)\tilde{w}_r(P)\tilde{w}_s dD = 0, \quad \text{unless} \quad r = s \tag{7.29}$$

No special normalization has been applied to the mode shapes in this book.

Dynamic solutions to Equation 7.27 are expanded in a *modal series* of mode shapes times time-dependent modal coordinates, $W_i(t)$:

$$w(P,t) = \sum_{j=i} \tilde{w}_i(P) W_i(t) \tag{7.30}$$

Modal transformation is as follows. Equation 7.31 is substituted into Equation 7.28, which is premultiplied by a mode shape of a unique mode, integrated over the domain D, and mode orthogonality (Eq. 7.29) is applied. The result is a series of uncoupled second-order ordinary differential equation of motion for the modal coordinate $W_i(t)$:

$$\frac{d^2 W_i(t)}{dt^2} + 2\zeta\omega_i \frac{dW_i(t)}{dt} + \omega_i^2 W_i(t) = \left[\frac{\int_D \tilde{w}_i(P)\overline{F}(P)dD}{\omega_i^2 \int_D M(P)\tilde{w}_i^2(P)dD} \right] \omega_i^2 F(t)$$

$$\text{natural frequency,} \quad \omega_i^2 = \frac{\int_D \tilde{w}_i(P,t) L[\tilde{w}_i(P)] dD}{\int_D M(P)\tilde{w}_i^2(P)dD}, \quad i = 1, 2, 3 \ldots \tag{7.31}$$

The exact modal solution for a harmonic oscillation, $F(t) = \cos\omega t$, is adapted from Equation 7.4:

$$W_i(t) = \left[\frac{\int_D \tilde{w}_i(P)\overline{F}(P)dD}{\omega_i^2 \int_D M(P)\tilde{w}_i^2(P)dD} \right] \left[\left(1 - \frac{f^2}{f_i^2}\right)^2 + \left(\frac{2\zeta f}{f_i}\right)^2 \right]^{-1/2} (P)\cos(\omega t - \varphi_i) \tag{7.32}$$

In the limit that the forcing frequencies approach zero ($f \ll f_i$), the quasistatic solution is

$$W_{i\text{-static}} = \left(\frac{1}{\omega_i^2}\right) \frac{\int_D \tilde{w}_i(P)\overline{F}(P)dD}{\int_D M(P)\tilde{w}_i^2(P)dD} \tag{7.33}$$

The static modal solution is important as it scales the dynamic solution. The two integrals over the domain of the structure can be difficult to evaluate. Approximations have been developed as discussed in section 7.6.

$$W_i(t) = W_{\text{i-static}} \left[\left(1 - \frac{f^2}{f_i^2}\right)^2 + \left(\frac{2\zeta f}{f_i}\right)^2 \right]^{-1/2} \cos(\omega t - \varphi_i) \tag{7.34}$$

In the limit that the forcing frequencies approach zero ($f \ll f_i$), the solution amplitude is quasistatic solution.

For harmonic base excitation $F(P,t) = -M(P)\ddot{W}(t) = M(P)\ddot{W}_o \cos(2\pi f t)$, the modal displacement is

$$W_i(t) = \left[\frac{\int_D \widetilde{w}_i(P) M(P) dD}{\omega_i^2 \int_D M(P) \widetilde{w}_i^2(P) dD} \right] \left[\left(1 - \frac{f^2}{f_i^2}\right)^2 + \left(\frac{2\zeta f}{f_i}\right)^2 \right]^{-1/2} \ddot{W}_o \cos(\omega t - \varphi_i) \quad (7.35)$$

This equation reduces to Equation 7.26 for beam base excitation and to case 3 of Table 7.2 for single degree of freedom.

7.2 Transient Vibration

7.2.1 Transient Vibration Theory

Table 7.3 [1] illustrates transient vibration of one-degree-of-freedom systems. Earthquakes, impact, and machinery start-up are the examples of transient loads. Unlike steady-state harmonic response, transient response starts from an initial condition and evolves over time. *Shock* is usually defined as a short duration transient load.

Consider a spring-supported mass M that is released from rest at an initial displacement $x(0)$. The equation of motion is Equation 7.1 with $F(t) = 0$. The transient solution decays exponentially in time (case 1 of Table 7.3):

$$x(t) = x(0)e^{-\zeta\omega_n t}(\cos[\omega_n(1-\zeta^2)^{1/2}t] + \zeta \sin[\omega_n(1-\zeta^2)^{1/2}t]), \quad \zeta \le 1 \quad (7.36)$$

The rate decay increases with the dimensionless *damping factor* ζ also called the *damping ratio* and *fraction of critical damping*. Zero damping ($\zeta = 0$) response is an upper bound. Overdamped, $\zeta > 1$, free vibration decays exponentially without oscillation, but it occurs only with high damping devices such as shock absorbers. Undamped, ($\zeta = 0$) underdamped, ($\zeta < 1$), and overdamped ($\zeta > 1$) solutions are shown in Table 7.3.

Damping factor can be measured from the amplitudes of successive peaks in the underdamped transient decay by taking the natural logarithm of the term $x(0)Exp(-\zeta\omega_n t)$ in the previous equation:

$$\zeta = \frac{1}{2\pi N} \log_e \frac{A_i}{A_{i+N}} \quad (7.37)$$

A_{i+N} is the amplitude after $i+N$ cycles. Negative damping ($\zeta < 0$) inputs rather than extracts energy, and it can occur in cases of flow-induced instability, such as flutter.

Transient solutions to linear equations of motion can be *superimposed* to create new solutions, as shown in Figure 7.7. The force impulse, $F\Delta t = M\Delta V$, produces velocity increment ΔV during the small time interval Δt (Chapter 2, Eq. 2.34):

$$\Delta V = \frac{\Delta x}{\Delta t} = \frac{F\Delta t}{M} \quad (7.38)$$

Table 7.3 Transient vibration of a spring-mass system

Notation: F = force; F_o = amplitude of force; f = excitation frequency, Hz; f_n = natural frequency, Hz; g(t) = dimensionless function of time; k = spring constant, force/displacement, Table 3.2; M = mass; t = time; τ = time delay; w(t) = displacement of base; x(t) displacement of mass M; ζ = damping factor, dim'lss; $\omega = 2\pi f$= circular frequency, radian/s; $\omega_n = 2\pi f_n (k/M)^{1/2}$, circular natural frequency, rad/s. Overdot $(\cdot)$ derivative with respect to time. Consistent units in Table 1.2.

Transient System	Equation of Motion and Transient Response x(t)
1. Free Vibration from Initial Displacement X_o 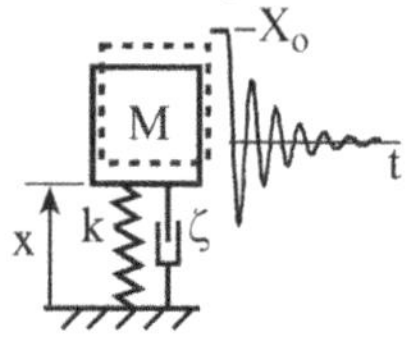	$M\ddot{x}+2M\zeta\omega_n\dot{x}+kx=0,\ x(0)=X_o,\ \dot{x}(0)=0,\ \omega_n=(k/M)^{1/2}$ Ref. [1], p. 31 undamped response, $\zeta=0$ $x(t)=X_o\cos\omega_n t$, $x_{max}=X_o$, all ζ under damped response, $\zeta\le 1$ $x(t)=X_o e^{-\zeta\omega_n t}\left(\cos[\omega_n(1-\zeta^2)^{1/2}t]+\{\zeta/(1-\zeta^2)^{1/2}\}\sin[\omega_n(1-\zeta^2)^{1/2}t]\right)$, over damped response, $\zeta\ge 1$ $x(t)=\frac{(\zeta+(\zeta^2-1)^{1/2})X_o}{2(\zeta^2-1)^{1/2}}e^{(-\zeta+(\zeta^2-1)^{1/2})\omega_n t}+\frac{(-\zeta+(\zeta^2-1)^{1/2})X_o}{2(\zeta^2-1)^{1/2}}e^{(-\zeta-(\zeta^2-1)^{1/2})\omega_n t}$
2. Free Vibration with Initial Velocity V_o 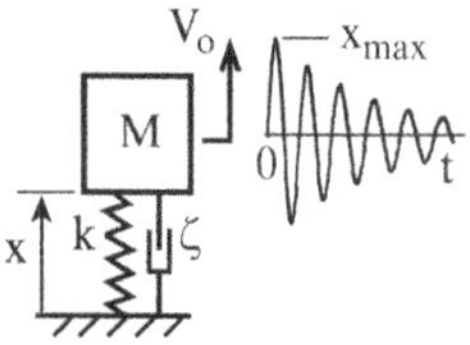	$M\ddot{x}+2M\zeta\omega_n\dot{x}+kx=0,\ x(0)=0,\ \dot{x}(0)=V_o$ Ref. [1], p. 31 undamped response, $\zeta=0$, $x(t)=(V_o/\omega_n)\sin\omega_n t$, $x_{max}=V_o/\omega_n$ under damped response, $\zeta\le 1$ $x(t)=\frac{V_o}{\omega_n}\frac{e^{-\zeta\omega_n t}}{(1-\zeta^2)^{1/2}}\sin\omega_n(1-\zeta^2)^{1/2}t$, $x_{max}=(V_o/\omega_n)\mathrm{Exp}\{-[\zeta/(1-\zeta^2)^{1/2})]\tan^{-1}[(1-\zeta^2)^{1/2}/\zeta]\}$. over damped response, $\zeta\ge 1$, $x(t)=\frac{V_o}{2\omega_n(\zeta^2-1)^{1/2}}\left[e^{(-\zeta+(\zeta^2-1)^{1/2})\omega_n t}-e^{(-\zeta-(\zeta^2-1)^{1/2})\omega_n t}\right]$
3. Free Vibration from Initial Displacement X_o and Initial Velocity V_o 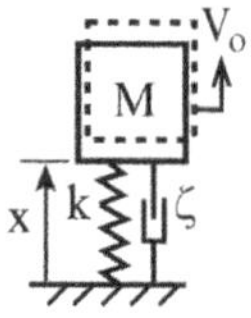	$M\ddot{x}+2M\zeta\omega_n\dot{x}+kx=0,\ x(0)=X_o,\ \dot{x}(0)=V_o$ Ref. [1], p. 32 undamped response, $\zeta=0$ $x(t)=X_o\cos\omega_n t+(V_o/\omega_n)\sin\omega_n t$, $x_{max}=V_o/\omega_n$ underdamped response, $\zeta\le 1$ $x(t)=e^{-\zeta\omega_n t}\left[X_o\cos\omega_n(1-\zeta^2)^{1/2}t+\frac{V_o+\zeta\omega_n X_o}{\omega_n(1-\zeta^2)^{1/2}}\sin\omega_n(1-\zeta^2)^{1/2}t\right],\ \zeta\le 1$ $x_{max}=e^{-\zeta\omega_n t_{max}}\left[V_o^2+(\omega_n X_o)^2+2\zeta\omega_n X_o V_o\right]^{1/2}/\omega_n$ at $t_{max}=\{1/[\omega_n(1-\zeta^2)^{1/2}]\}\tan^{-1}[V_o(1-\zeta^2)^{1/2}/(\omega_n X_o+\zeta V_o)]$
4. Force Impulse 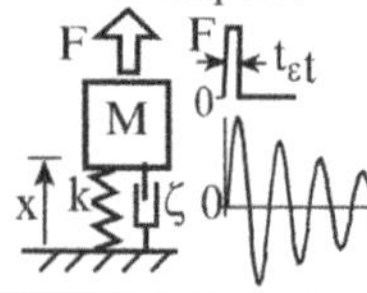	$M\ddot{x}+2M\zeta\omega_n\dot{x}+kx=\begin{cases}Ip, & 0<t<t_\varepsilon,\ x(0)=0,\ \dot{x}(0)=0\\ 0, & t>t_\varepsilon\end{cases}$ $Ip=\int_0^{t\varepsilon}F(t)dt$ = Force Impulse, $t_\varepsilon\omega_n<<1, t_e$ = time of impulse x(t) given in case 2 with $Ip=MV_o$, $V_o=Ip/M$
5. Step Force F_o from rest 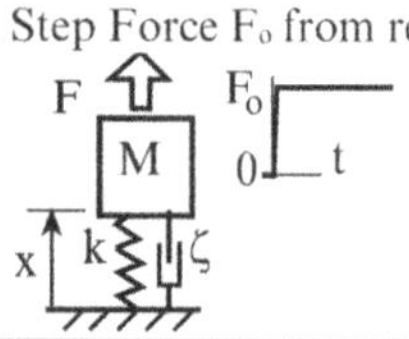	$M\ddot{x}+2M\zeta\omega_n\dot{x}+kx=\begin{cases}0, & t=0,\ x(0)=\dot{x}(0)=0.\\ F_o, & t>0\end{cases}$ Ref. [1], p. 91 undamped response, $\zeta=0$ $x(t)=\frac{F_o}{k}(1-\cos\omega_n t)$, $x_{max}=2\frac{F_o}{k}$ continued on next page

Table 7.3 Transient vibration of a spring-mass system, continued

Transient System	Equation of Motion and Transient Response x(t)
5. Step force F_o Continued 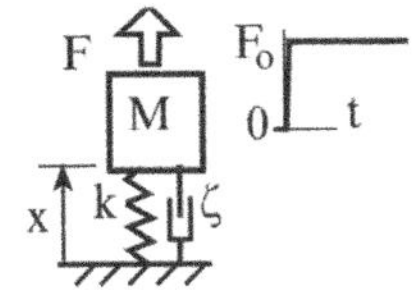 	under damped response, $\zeta \le 1$, $x(t)=\frac{F_o}{k}\left[1-e^{-\zeta\omega_n t}\left\{\cos\left(\omega_n(1-\zeta^2)^{1/2}t\right)+\frac{\zeta}{(1-\zeta^2)^{1/2}}\sin(\omega_n(1-\zeta^2)^{1/2}t)\right\}\right]$ $x_{max}=(F_o/k)(1+e^{-\pi\zeta/(1-\zeta^2)^{1/2}})$ over damped response, $\zeta \ge 1$ $x(t)=$ $\frac{F_o}{k}\left[1-\frac{e^{-\zeta\omega_n t}}{2(\zeta^2-1)^{1/2}}\left\{(\zeta+(\zeta^2-1)^{1/2})e^{(\zeta^2-1)^{1/2}\omega_n t}+(-\zeta+(\zeta^2-1)^{1/2})e^{-(\zeta^2-1)^{1/2}\omega_n t}\right\}\right]$
6. Sine Force from Zero 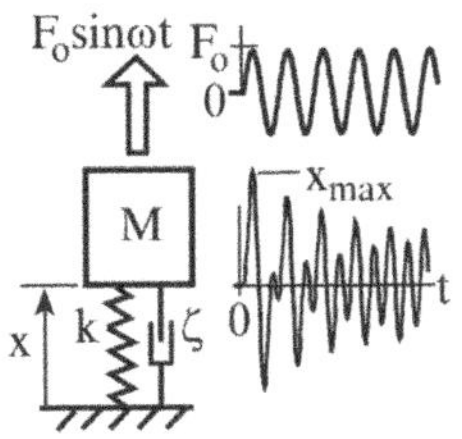 	$M\ddot{x}+2M\zeta\omega_n\dot{x}+kx=F_o\sin\omega t,\ x(0)=0,\ \dot{x}(0)=0.$ $x(t)=\frac{F_o}{k}\frac{1}{(1-\omega^2/\omega_n^2)}\left(\sin\omega t-\frac{\omega}{\omega_n}\sin\omega_n t\right), x_{max}=\frac{F_o}{k}\frac{1}{(1-\omega^2/\omega_n^2)},\ \zeta=0,$ $x(t)=\frac{F_o}{k}\frac{1}{(1-\omega^2/\omega_n^2)^2+(2\zeta\omega/\omega_n)^2}\left[\left(1-\frac{\omega^2}{\omega_n^2}\right)\sin\omega t-2\zeta\frac{\omega}{\omega_n}\cos\omega t+\right.$ $\left.e^{-\zeta\omega_n t}\left\{\frac{2\zeta\omega}{\omega_n}\cos\left(\omega_n(1-\zeta^2)^{1/2}t\right)-\frac{\omega(1-\omega^2/\omega_n^2-2\zeta^2)}{\omega_n(1-\zeta^2)^{1/2}}\sin\omega_n(1-\zeta^2)^{1/2}t\right\}\right], \zeta\le 1$
7. Time Limited Force Impulse From Rest 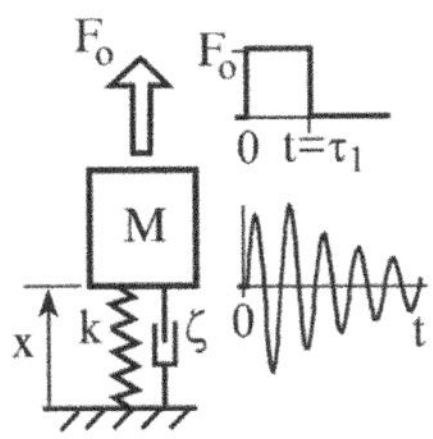 t_1 = time of impulse	$M\ddot{x}+2M\zeta\omega_n\dot{x}+kx=\begin{cases}F_o,\ 0<t\le t_1, & x(0)=\dot{x}(0)=0.\\ 0,\ t>t_1\end{cases}$ $x(t)=\frac{F_o}{k}\begin{cases}1-\cos\omega_n t, & t\le t_1\\ -\cos\omega_n t+\cos\omega_n(t-t_1), & t\ge t_1\end{cases}$ undamped response, $\zeta=0$. $x(t)=\frac{F_o}{k}\begin{cases}g(t), & 0\le t\le t_1\\ g(t)-g(t-t_1), & t\ge t_1\end{cases}$ damped response $g(t)=\begin{cases}1-e^{-\zeta\omega_n t}\left\{\cos\left(\omega_n(1-\zeta^2)^{1/2}t\right)+\frac{\zeta}{(1-\zeta^2)^{1/2}}\sin(\omega_n(1-\zeta^2)^{1/2}t)\right\}, & \zeta<1\\ 1-\frac{e^{-\zeta\omega_n t}}{2(\zeta^2-1)^{1/2}}\left\{(\zeta+(\zeta^2-1)^{1/2})e^{(\zeta^2-1)^{1/2}\omega_n t}+(-\zeta+(\zeta^2-1)^{1/2})e^{-(\zeta^2-1)^{1/2}\omega_n t}\right\}, & \zeta>1\end{cases}$
8. Half Sine Wave Pulse On Spring Mass 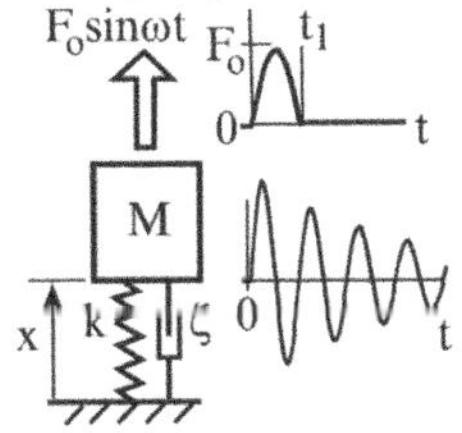 $t_1=1/(2f)$, $\omega=2\pi f=\pi/t_1$ for half sine. Solution for full sine wave pulse uses $t_1=1/f$ and $g(t)$-$g(t-t_1)$ for $t>t_1$. Solution is superposition of case 6.	$M\ddot{x}+2M\zeta\omega_n\dot{x}+kx=\begin{cases}0, & t=0 \text{ and } t>t_1,\ x(0)=0,\ \dot{x}(0)=0\\ F_o\sin\omega t, & 0\le t\le t_1\end{cases}$ undamped response, $\zeta=0$, $x(t)=\frac{F_o}{k}\begin{cases}(1-\omega^2/\omega_n^2)^{-1}\left(-(\omega/\omega_n)\sin\omega_n t+\sin\omega t\right), & t\le t_1,\\ \left[(1-\omega^2/\omega_n^2)^{-1}\left(-(\omega/\omega_n)\sin\omega_n t+\sin\omega t\right)+\right. & \\ \left.(1-\omega^2/\omega_n^2)^{-1}\left(-(\omega/\omega_n)\sin\omega_n(t-t_1)+\sin\omega(t-t_1)\right)\right], & t>t_1.\end{cases}$ under damped response, $\zeta\le 1$ by superposition of case 6 $x(t)=\frac{F_o}{k}\begin{cases}g(t), & 0\le t\le t_1\\ g(t)+g(t-t_1), & t\ge t_1,\end{cases}$ $t_1=\pi/\omega=1/(2f)$ $g(t)=\frac{1}{(1-\omega^2/\omega_n^2)^2+(2\zeta\omega/\omega_n)^2}\left[\left(1-\frac{\omega^2}{\omega_n^2}\right)\sin\omega t-2\zeta\frac{\omega}{\omega_n}\cos\omega t\right.$ $\left.+e^{-\zeta\omega_n t}\left\{\frac{2\zeta\omega}{\omega_n}\cos\left(\omega_n(1-\zeta^2)^{1/2}t\right)-\frac{\omega}{\omega_n}\frac{(1-\omega^2/\omega_n^2-2\zeta^2)}{(1-\zeta^2)^{1/2}}\sin\omega_n(1-\zeta^2)^{1/2}t\right\}\right]$

Table 7.3 Transient vibration of a spring-mass system, continued

Transient System	Equation of Motion, Transient Response x(t)
9. Force Ramp 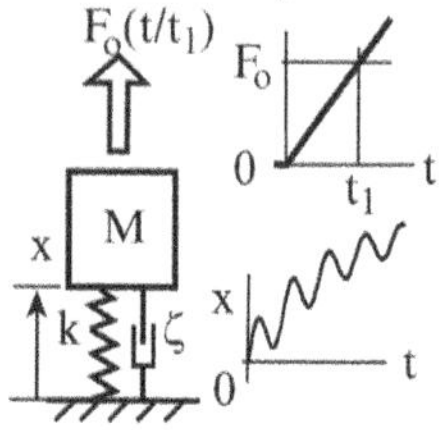	$M\ddot{x}+2M\zeta\omega_n\dot{x}+kx=(t/t_1)F_o, \quad x(0)=0,\ \dot{x}(0)=0$ undamped response, $\zeta=0$ $x(t)=(F_o/k)[t/t_1-(1/(\omega_n t_1))\sin\omega_n t]$, under damped response, $\zeta<1$ $\qquad \omega_d=\omega_n(1-\zeta^2)^{1/2}$ $x(t)=\frac{F_o}{k\omega_n t_1}\left[(\omega_n t\text{-}2\zeta)(1\text{-}\zeta^2)^{1/2}+e^{-\zeta\omega_n t}\left\{2\zeta(1\text{-}\zeta^2)^{1/2}\cos\omega_d t-(1\text{-}2\zeta^2)\sin\omega_d t\right\}\right]$
10. Ramp to const. Force 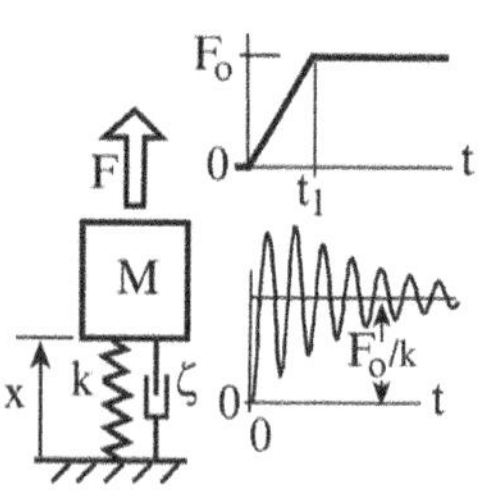 F_o = maxmum force	$M\ddot{x}+2M\zeta\omega_n\dot{x}+kx=\begin{cases}(t/t_1)F_o, & 0<t\le t_1, \quad x(0)=0,\ \dot{x}(0)=0\\ F_0,\ t>t_1\end{cases}$ undamped response, $\zeta=0$ $x(t)=\frac{F_o}{k}\begin{cases}[t/t_1-(1/(\omega_n t_1))\sin\omega_n t], & 0<t\le t_1\\ (1/(\omega_n t_1))[\omega_n t_1\cos\omega_n(t-t_1)+\sin\omega_n(t-t_1)-\sin\omega_n t], & t>t_1\end{cases}$ under damped response, $\zeta<1$ $\qquad \omega_d=\omega_n(1-\zeta^2)^{1/2}$ $x(t)=g_1(t),\ 0\le t\le t_1;\ g_2(t),\ t>t_1$, $g_1=\frac{F_o}{k\omega_n t_1}\left[\omega_d t-2\zeta(1-\zeta^2)^{1/2}+e^{-\zeta\omega_n t}\left\{2\zeta(1-\zeta^2)^{1/2}\cos\omega_d t-(1-2\zeta^2)\sin\omega_d t\right\}\right]$ $g_2=\frac{F_o}{k\omega_n t_1}\left[\omega_d t_1+2\zeta(1-\zeta^2)\left(e^{-\zeta\omega_n t}\cos\omega_d t-e^{-\zeta\omega_n(t-t_1)}\cos\omega_d(t-t_1)\right)\right.$ $\left.+(1-2\zeta^2)\left(-e^{-\zeta\omega_n t}\sin\omega_d t+e^{-\zeta\omega_n(t-t_1)}\sin\omega_d(t-t_1)\right)\right]$
11. Triangular Force 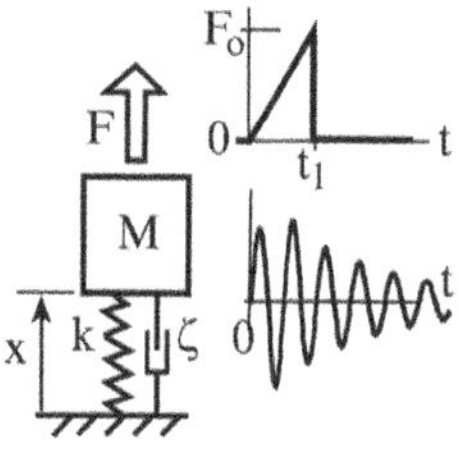	$M\ddot{x}+2M\zeta\omega_n\dot{x}+kx=\begin{cases}(t/t_1)F_o, 0\le t\le t_1\\ 0, t>t_1\end{cases}, \quad x(0)=0,\ \dot{x}(0)=0,$ under damped response, $\zeta\le 1$, $\omega_d=\omega_n(1-\zeta^2)^{1/2}$ $x(t)=\frac{F_o}{k}\frac{1}{\omega_n t_1}\{(\omega_d t-2\zeta)(1-\zeta^2)^{1/2}$ $+e^{-\zeta\omega_n t}[2\zeta(1-\zeta^2)^{1/2}\cos\omega_d t-(1-2\zeta^2)\sin\omega_d t])\},\ 0\le t\le t_1$ $x(t)=\frac{F_o}{k}\frac{1}{\omega_n t_1}\{e^{-\zeta\omega_n t}[2\zeta(1-\zeta^2)^{1/2}\cos\omega_d t-(1-2\zeta^2)\sin\omega_d t]$ $+e^{-\zeta\omega_n(t-t_1)}[(\omega_n t_1-2\zeta)(1-\zeta^2)^{1/2}\cos\omega_d(t-t_1)$ $+(1+\zeta\omega_n t_1-2\zeta^2)\sin\omega_d(t-t_1))],\ t>t_1$
12. General Forcing Spring-Mass-Damper 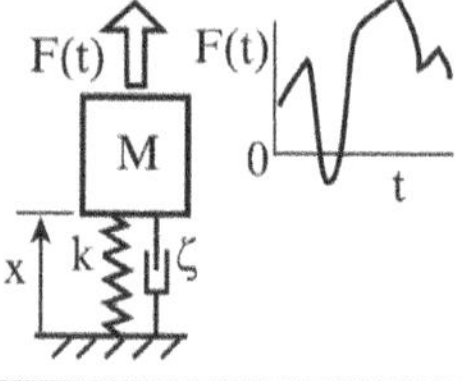	$M\ddot{x}+2M\zeta\omega_n\dot{x}+kx=F(t), \quad x(0)=0,\ \dot{x}(0)=0$ $\qquad$ Ref. [1], p. 90 $x(t)=\begin{cases}\frac{1}{k}\int_0^{\omega_n t}\frac{F(\tau)e^{-\zeta\omega_n(t-\tau)}}{(1-\zeta^2)^{1/2}}\sin\left(\omega_n(1-\zeta^2)^{1/2}(t-\tau)\right)d\omega_n\tau, & \zeta\le 1\\ \frac{1}{2k}\int_0^{\omega_n t}\frac{F(\tau)}{(\zeta^2-1)^{1/2}}[e^{(-\zeta+(\zeta^2-1)^{1/2})\omega_n(t-\tau)}-e^{(-\zeta-(\zeta^2-1)^{1/2})\omega_n(t-\tau)}]d\omega_n\tau, & \zeta>1\end{cases}$

Table 7.3 Transient vibration of a spring-mass system, continued, base excitation

Additional notation: $x(t)$ = position of mass M from base; $\omega_d = \omega_n(1-\zeta^2)^{1/2}$; V_o = vertical velocity of wheel ascending ramp; $w(t)$ displacement of base relative to ground.

Base Excitation	Equation of Motion, Transient Response x(t)
13. Abrupt base Step Over Short step 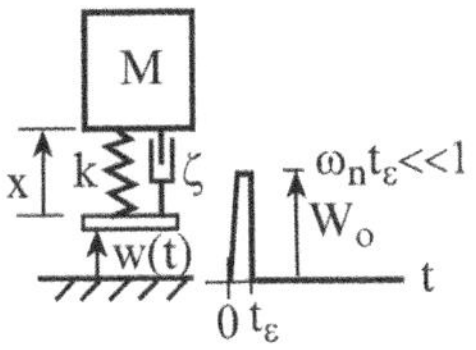 	$M\ddot{x} + 2M\zeta\omega_n\dot{x} + kx = -M\ddot{w},\ x(0) = \dot{x}(0) = 0,\ w(t) = W_o,\ 0 < t \le t_e$, else 0 For $t > t_e$ and $t_\varepsilon\omega_n << 1$, $x(t) = W_o\omega_n t_\varepsilon \sin\omega_n t$, undamped response, $\zeta = 0$. $x_{max} = W_o\omega_n t_\varepsilon$ under damped response, $\zeta < 1$, $x(t) = \frac{W_o\omega_n t_\varepsilon}{(1-\zeta^2)^{1/2}} e^{-\zeta\omega_n t}\sin(1-\zeta^2)^{1/2}\omega_n t$ $x_{max} = \left[W_o\omega_n t_\varepsilon/(1-\zeta^2)^{1/2}\right]\mathrm{Exp}\{-[\zeta/(1-\zeta^2)^{1/2})]\tan^{-1}[(1-\zeta^2)^{1/2}/\zeta]\}$.
14. Base Step to Plateau 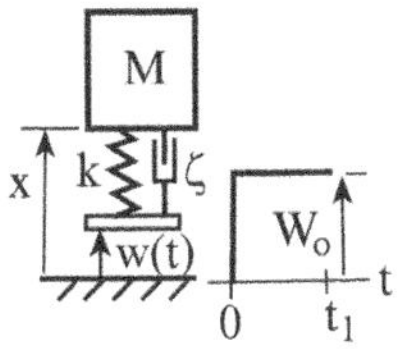 	$M\ddot{x} + 2M\zeta\omega_n\dot{x} + kx = -M\ddot{w},\ x(0) = \dot{x}(0) = 0,\ w(t) = W_o,\ t > 0$, else 0 undamped $\zeta = 0$: $x(t) = W_o(1-\cos\omega_n t)$ $\quad x_{max} = 2W_o$ under damped response, $\zeta < 1$: $\quad \omega_d = \omega_n(1-\zeta^2)^{1/2}$ $x(t) = W_o\left[1 - e^{-\zeta\omega_n t}(\cos\omega_d t + (\zeta/(1-\zeta^2)^{1/2})\sin\omega_d t)\right]$ $x_{max} = W_o(1 + e^{-\zeta\pi/(1-\zeta^2)^{1/2}})$
15.Triangular Base Shock Acceleration 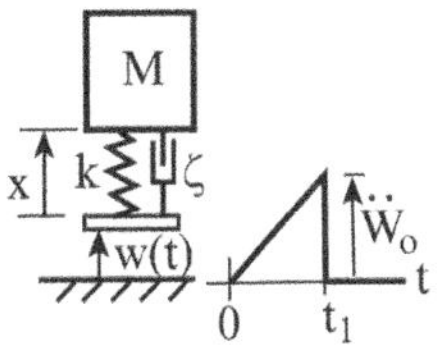 Note x is displacement Relative to the base	$M\ddot{x} + 2M\zeta\omega_n\dot{x} + kx = -\begin{cases} M(t/t_1)\ddot{W}_o, 0 \le t \le t_1 \\ 0, t > t_1 \end{cases},\quad x(0) = 0,\ \dot{x}(0) = 0,$ undamped response, $\zeta = 0$ $x(t) = -\frac{\ddot{W}_o}{\omega_n^2}\frac{1}{\omega_n t_1}\{\omega_n t - \sin\omega_n t\},\ 0 \le t \le t_1$ $x(t) = -\frac{\ddot{W}_o}{\omega_n^2}\frac{1}{\omega_n t_1}\{-\sin\omega_n t + \omega_n t_1\cos\omega_n(t-t_1) + \sin\omega_n(t-t_1)\},\ t > t_1$ under damped response, $\zeta \le 1$, $\omega_d = \omega_n(1-\zeta^2)^{1/2}$ $x(t) = -\frac{\ddot{W}_o}{\omega_n^2}\frac{1}{\omega_n t_1}\{(\omega_n t - 2\zeta)(1-\zeta^2)^{1/2}$ $+ e^{-\zeta\omega_n t}[2\zeta(1-\zeta^2)^{1/2}\cos\omega_d t - (1-2\zeta^2)\sin\omega_d t]\},\ 0 \le t \le t_1$ $x(t) = -\frac{\ddot{W}_o}{\omega_n^2}\frac{1}{\omega_n t_1}\{e^{-\zeta\omega_n t}[2\zeta(1-\zeta^2)^{1/2}\cos\omega_d t - (1-2\zeta^2)\sin\omega_d t]$ $+ e^{-\zeta\omega_n(t-t_1)}[(\omega_d t_1 - 2\zeta)(1-\zeta^2)^{1/2}\cos\omega_d(t-t_1)$ $+ (1+\zeta\omega_n t_1 - 2\zeta^2)\sin\omega_d(t-t_1)]\},\ t > t_1$
16. Start Sine Base 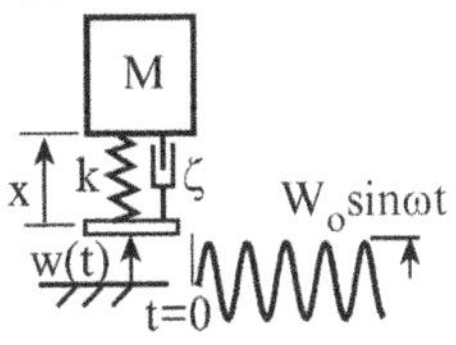 	$M\ddot{x} + 2M\zeta\omega_n\dot{x} + kx = \begin{cases} 0,\ t = 0,\ \ x(0) = 0,\ \dot{x}(0) = 0 \\ MW_o\omega^2\sin\omega t,\ t > 0 \end{cases}$ Response for $\zeta<1$ given in case 6 with substitution of $MW_o\omega^2$ for F_o. x = displacement of M relative to accelerating base. $x+W_o\sin\omega t$= absolute displacement relative to ground.

Table 7.3 Transient vibration of a spring-mass system, continued

Additional notation: $y(t)$ = position of mass M from ground; $\omega_d = \omega_n(1-\zeta^2)^{1/2}$; V_o = vertical velocity.

Vehicle Transient	Equation of Motion, Transient Response y(t)
17. Base Excitation of Spring-Mass Damper 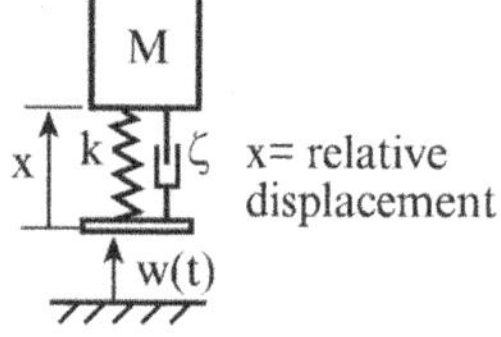	$M\ddot{x}+2M\zeta\omega_n\dot{x}+kx=-M\ddot{w}(t),\ t>0, x(0)=\dot{x}(0)=0.$ Ref. [1], p. 92 Integral expression for response to base acceleration time history, $\ddot{w}(t)$ $x(t)=\frac{1}{\omega_n^2}\int_0^{\omega_n t}\frac{\ddot{w}(\tau)}{(1-\zeta^2)^{1/2}}e^{-\zeta\omega_n(t-\tau)}\sin\omega_n(1-\zeta^2)^{1/2}(t-\tau)d\omega_n\tau,\ \zeta<1$ $\underset{\zeta>1}{x(t)}=\int_0^{\omega_n t}\frac{\ddot{w}(\tau)}{2\omega_n^2(\zeta^2-1)^{1/2}}\left[e^{(-\zeta+(\zeta^2-1)^{1/2})\omega_n(t-\tau)}-e^{(-\zeta-(\zeta^2-1)^{1/2})\omega_n(t-\tau)}\right]d\omega_n\tau$
18. Vehicle Ascends Ramp at velocity Vo to Plateau W_o 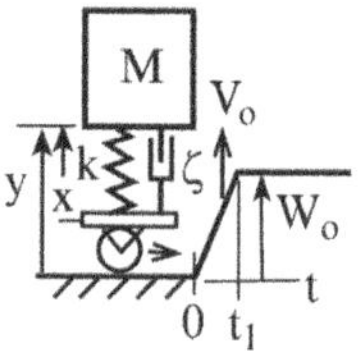	$M\ddot{x}+2M\zeta\omega_n\dot{x}+kx=-M\ddot{w},\ 0<t\le t_1,\ x(0)=0,\ \dot{x}(0)=0,$ undamped response, $\zeta=0$: Note: $V_o t_1=W_o,\ y(t)=x(t)+w(t)$ $y(t)=\begin{cases}(V_o/\omega_n)(\omega_n t-\sin\omega_n t),\ t<t_1\\ W_o+(V_o/\omega_n)(\sin\omega_n(t-t_1)-\sin\omega_n t),\ t>t_1\end{cases}$ under damped response, $\zeta\le 1$: $y(t)=\begin{cases}(V_o/\omega_n)(\omega_n t-e^{-\zeta\omega_n t}\sin\omega_n(1-\zeta^2)^{1/2}t),\ t<t_1\\ W_o+(V_o/\omega_n)e^{-\zeta\omega_n t}(\sin\omega_n(1-\zeta^2)^{1/2}(t-t_1)-\sin\omega_n(1-\zeta^2)^{1/2}t),t>t_1\end{cases}$
19. Abrupt Broad Step 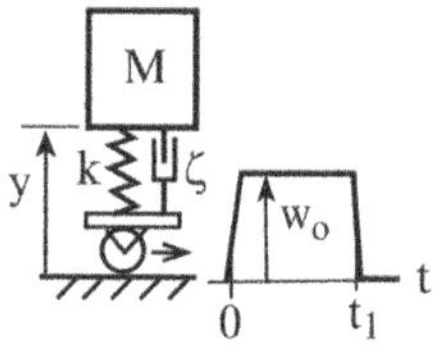 Wheel ascents and descends broad step	$M\ddot{x}+2M\zeta\omega_n\dot{x}+kx=-M\ddot{w},\ 0<t\le t_1,\ x(0)=0,\ \dot{x}(0)=0,$ undampedresponse, $\zeta=0\quad y(t)=x(t)+w(t)$ $y(t)=\begin{cases}W_o(1-\cos\omega_n t),\ t<t_1,\\ W_o[\sin\omega_n t\sin\omega_n t_1-\cos\omega_n t(1-\cos\omega_n t_1)],\ t>t_1\end{cases}$ under damped response, $\zeta<1$ $y(t)=W_o\begin{cases}g(t),\quad 0\le t\le t_1\\ g(t)-g(t-t_1), t\ge t_1\end{cases}$ $g(t)$ from case 7
20. Short Step 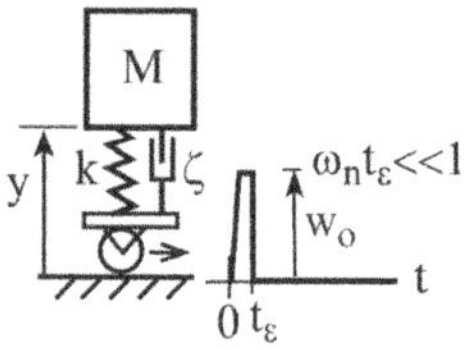	$M\ddot{x}+2M\zeta\omega_n\dot{x}+kx=-M\ddot{w},\ 0<t\le t_\varepsilon,\ x(0)=0,\ \dot{x}(0)=0,$ undamped response, $\zeta=0, t>t_\varepsilon$, $\quad t_\varepsilon\omega_n<<1$ $y(t)=w_o\,\omega_n\,t_\varepsilon\sin\omega_n t$ under damped response, $\zeta<1, t>t_\varepsilon$, $y(t)=\frac{W_o\omega_n t_\varepsilon}{(1-\zeta^2)^{1/2}}e^{-\zeta\omega_n t}\sin(1-\zeta^2)^{1/2}\omega_n t$
21. Vehicle Climbs Ramp 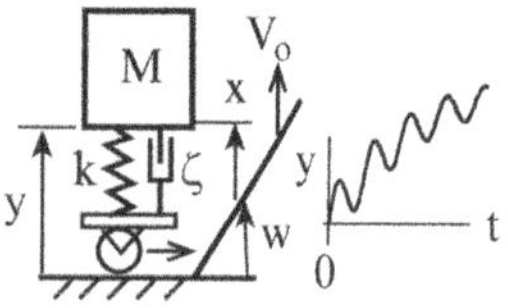 Wheel travels up ramp with vertical velocity V_o.	$M\ddot{x}+2M\zeta\omega_n\dot{x}+kx=-M\ddot{w}, 0<t\le t_1, x(0)=0,\ \dot{x}(0)=0$ undamped response, $\zeta=0 \qquad y(t)=x(t)+w(t)$ $y(t)=\frac{V_o}{\omega_n}[\omega_n t-\sin\omega_n t],$ under damped response, $\zeta<1$ $y(t)=\frac{V_o}{\omega_n}\left[\omega_n t-e^{-\zeta\omega_n t}\sin\omega_n(1-\zeta^2)^{1/2}t\right]$

The response to the instantaneous velocity at $t = 0$ is shown in case 2 of Table 7.3:

$$x(t) = \frac{\Delta V(0)}{\omega_n} \frac{e^{-\zeta\omega_n t}}{(1-\zeta^2)^{1/2}} \sin \omega_n (1-\zeta^2)^{1/2} t, \quad \zeta < 1 \tag{7.39}$$

Integrating (summing) this equation over a continuous set of timed impulses $\Delta V(\tau) = F(\tau)d\tau/M$ produces a *convolution integral* called *Duhamel's integral* [1–3] for the single degree of freedom system transient response to time-varying force:

$$x(t) = \begin{cases} \dfrac{1}{\omega_n^2} \displaystyle\int_0^{\omega_n t} \frac{F(\tau)}{M} \sin \omega_n (t-\tau) d(\omega_n \tau), & \zeta = 0 \\ \dfrac{1}{\omega_n^2} \displaystyle\int_0^{\omega_n t} \frac{F(\tau)}{M} \frac{e^{-\zeta\omega_n (t-\tau)}}{(1-\zeta^2)^{1/2}} \sin \omega_n (1-\zeta^2)^{1/2} (t-\tau) d(\omega_n \tau), & \zeta \le 1 \end{cases} \tag{7.40}$$

This exact general integral solution will evolve over time to steady state from zero initial displacement and velocity. Solutions for displacement due to step, ramp, rectangular, and sinusoid transient loads are shown in Table 7.3. Many undamped transient solutions are given by Ayres [15].

Force $F(t)$ is replaced by $-Md^2w/dt^2$ in Equation 7.40 to produce Duhamel's integral for base excitation:

$$x(t) = -\frac{1}{\omega_n^2} \int_0^{\omega_n t} \frac{d^2 w(\tau)}{d\tau^2} \frac{e^{-\zeta\omega_n (t-\tau)}}{(1-\zeta^2)^{1/2}} \sin \omega_n (1-\zeta^2)^{1/2} (t-\tau) \; d\omega_n t, \quad \zeta < 1 \tag{7.41}$$

where $w(t)$ is base displacement relative to ground and $x(t)$ is relative to the base. Transient numerical solution is shown in Appendix B.

Consider that force F_o is abruptly applied at $t = 0$ to a spring-supported mass and held indefinitely as shown in Figure 7.6 (case 5 of Table 7.3). The initial maximum transient displacement exceeds the static displacement F_o/k to the same load, which is called *overshoot*:

$$\frac{x_{\max} k}{F_o} = \begin{cases} 2, & \text{for } \zeta = 0, \\ 1 + e^{-\pi\zeta/(1-\zeta^2)^{1/2}}, & \text{for } \zeta < 1 \end{cases} \tag{7.42}$$

Figure 7.6 shows that this loading produces an initial overshoot above the statically computed displacement F_o/k. The maximum overshoot of an abruptly applied load is two times its static displacement to the same load.

7.2.2 Continuous Systems and Initial Conditions

Table 7.4 has Transient responses of continuous systems. These solutions were obtained with modal superposition and Duhamel's integral. The general solution for under damped response of a forced elastic system from stationary initial conditions is:

$$W_i(t) = \frac{\displaystyle\int_D \tilde{w}_i(P)\overline{F}(P) dD}{\omega_i^2 \displaystyle\int_D M(P)\tilde{w}_i^2(P) dD} \int_0^{\omega_n t} F(\tau) \frac{e^{-\zeta\omega_n (t-\tau)}}{(1-\zeta^2)^{1/2}} \sin \omega_i (1-\zeta^2)^{1/2} (t-\tau) d\omega_n \tau, \quad \zeta < 1 \tag{7.43}$$

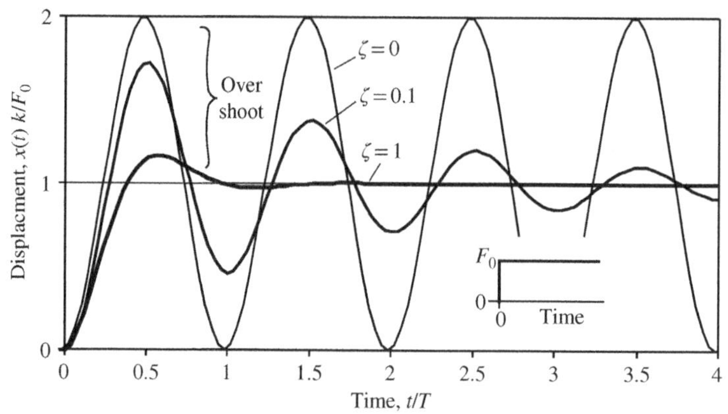

Figure 7.6 Response to a suddenly applied and held force (Fig. 7.3). T = natural period

Transient base excitation uses the inertial forcing. The solution is

$$W_i(t) = -\frac{\int_D \tilde{w}_i(P)dD}{\omega_i^2 \int_D M(P)\tilde{w}_i^2(P)dD} \int_0^{\omega_n t} \ddot{w}(\tau)\frac{e^{-\zeta\omega_n(t-\tau)}}{(1-\zeta^2)^{1/2}} \sin\omega_i(1-\zeta^2)^{1/2}(t-\tau)d\omega_n\tau, \quad \zeta < 1 \tag{7.44}$$

Approximate evaluation of the spatial integrals is presented in Section 7.2.4. Solutions of the time integral are shown in Table 7.3. The physical deflections in the ith mode are equal to the modal coordinate solution times the mode shape, $W_i(t)\tilde{w}_i(P)$, where P represents the triplet of x, y, z coordinates over the domain D of the structure.

Non zero initial conditions can be applied in transient modal analysis. Consider a plucked string (case 9 of Table 7.4). A quill raises the harpsichord string and then releases it. The initial position of the harpsichord string, just before release, is triangularly shaped displacement with the peak displacement $y(l/2, 0) = h$ at the quill. This displacement is expressed in a Fourier series of the sinusoidal mode shapes:

$$y(x, t=0) = \begin{cases} hx/L, & 0 \le x \le L/2 \\ 2h(L-x)/L, & L/2 \le x \le L \end{cases}$$

$$= h\sum_i \left(\frac{(-1)^{(i-1)/2}}{(i\pi)^2}\right) \sin i\pi/L, \quad i = 1, 3, 5 \ldots \tag{7.45}$$

This implies that the initial value of modal amplitudes in Equation 7.36 is $W_i(0) = \frac{h(-1)^{(2i-1)/2}}{(i\pi)^2}$. There is no force on the string after release at $t = 0$. The free decay solution is shown in case 9 of Table 7.4.

Piano strings are struck rather than plucked as in a harpsichord. Case 10 of Table 7.4 shows the response of a piano string which is given an initial velocity when a padded hammer strikes it. The piano hammer rebounds, and the string continues to vibrate at their

Table 7.4 Transient vibration of continuous structures

Notation: E = modulus of elasticity; F_o = force; $f_i = \omega_i/2\pi$ natural frequency, ith mode, Hz; I = moment of inertia of beam cross section (Table 1.5); L = span of beam; N = number of modes; m = beam mass per unit length (Table 1.5); λ = natural frequency parameter, Tables 4.2 and 4.4. $\tilde{y}_i(x)$ = mode shape of ith beam mode; x = distance along span; y = lateral displacement; P, C, F, G = Pinned, Clamped, Free and Guided beam boundary conditions; ζ_i = damping factor of ith mode, dim'less, $\zeta_i <= 1$. Consistent units in Table 1.2.

$$\text{Displacement, } y(x,t)=\sum_{i=1}^{N} y_i(x,t)$$

Transient Beam Load	Displacement in ith Mode, $y_i(x,t)$, i=1,2,3..
1. Pinned-Pinned Beam Step Off-Ctr Point Force 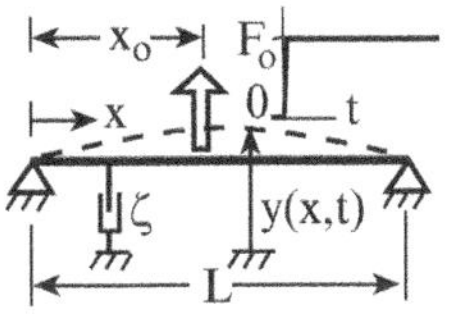	$\frac{2F_o}{m\omega_i^2 L}\sin(i\pi\frac{x_o}{L})\sin(i\pi\frac{x}{L})\times$ $\left[1-e^{-\zeta_i\omega_i t}\{\cos(\omega_i(1-\zeta_i^2)^{1/2}t)+\frac{\zeta_i}{(1-\zeta_i^2)^{1/2}}\sin(\omega_i(1-\zeta_i^2)^{1/2}t)\}\right]$ $m\omega_i^2=\frac{(i\pi)^4 EI}{L^4},\ f_i=\frac{\omega_i}{2\pi}=\frac{(i\pi)^2}{2\pi L^2}\sqrt{\frac{EI}{m}}$, Hz. i=1,2,3,4...case 5 of Table 4.2. F(t)=0 at t=0, F(t)=F_o,$t>0$, $\zeta_i<1, i=1,2,3..$
2. Clamped Beam Step Off-Ctr Point Force 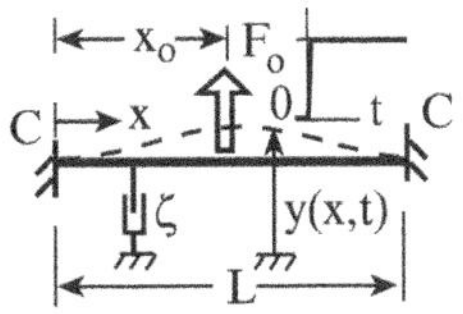	$\frac{2F_o}{m\omega_1^2 L}\left(\frac{2}{3}\right)\left(1-\cos 2\pi\frac{x_o}{L}\right)\left(1-\cos 2\pi\frac{x}{L}\right)\times$ $\left[e^{-\zeta_1\omega_1 t}\{\cos(\omega_1(1-\zeta_1^2)^{1/2}t)+\frac{\zeta_1}{(1-\zeta_1^2)^{1/2}}\sin(\omega_i(1-\zeta_i^2)^{1/2}t)\}\right]$ $m\omega_1^2=\frac{4.73^4 EI}{L^4},\ f_1=\frac{\omega_1}{2\pi}=\frac{4.73^2}{2\pi L^2}\sqrt{\frac{EI}{m}}$, Hz. i=1, case 7 of Table 4.2. Approximate first mode. F(t)=0 at t=0, F(t)=F_o,$t>0$, $\zeta_i<1, i=1,2,3..$
3. General Uniform Beam Off Ctr Point Step Force 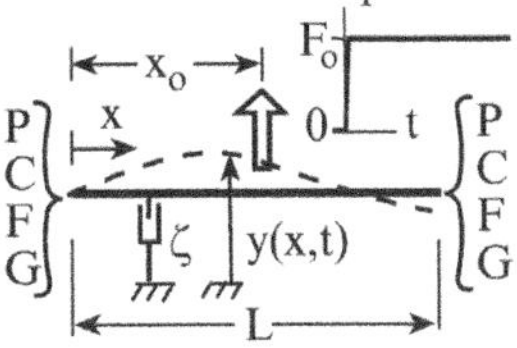	$y(x,t)=\sum_{i=1}^{N}\frac{F_o}{m\omega_i^2}\frac{\tilde{y}_i(x_o)}{\int_0^L \tilde{y}_i^2(x)dx}\tilde{y}_i(x)\times$ $\left[1-e^{-\zeta_i\omega_i t}\left\{\cos\left(\omega_i(1-\zeta_i^2)^{1/2}t\right)+\frac{\zeta_i}{(1-\zeta_i^2)^{1/2}}\sin\left(\omega_i(1-\zeta_i^2)^{1/2}t\right)\right\}\right]$ $m\omega_i^2=\frac{\lambda_i^4 EI}{L^4},\ f_i=\frac{\omega_i}{2\pi}=\frac{\lambda_i^2}{2\pi L^2}\sqrt{\frac{EI}{m}}$, Hz. λ_i, $\tilde{y}_i(x)$ = beam mode shape, P,C,F,G =beam boundary conditions.Tables 4.2, 4.4., $\zeta_i<1, i=1,2,3..$
4. Sinusoidal Point Force On Uniform Beam 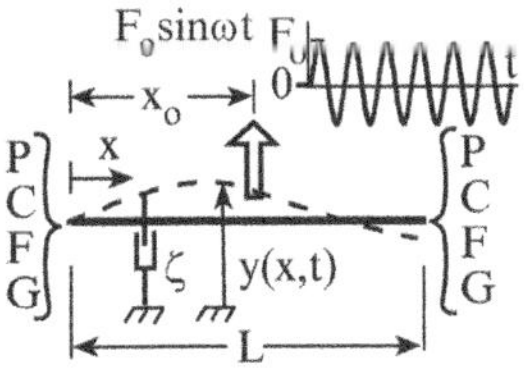	$\frac{F_o}{m\omega_i^2}\frac{\tilde{y}_i(x_o)}{\int_0^L \tilde{y}_i^2(x)dx}\frac{\tilde{y}_i(x)}{(1-\omega^2/\omega_i^2)^2+(2\zeta\omega/\omega_i)^2}\left[\left(1-\frac{\omega^2}{\omega_i^2}\right)\sin\omega t-2\zeta\frac{\omega}{\omega_i}\cos\omega t\right.$ $\left.+e^{-\zeta\omega_i t}\left\{\frac{2\zeta\omega}{\omega_i}\cos\left(\omega_i(1-\zeta_i^2)^{1/2}t\right)-\frac{\omega}{\omega_i}\frac{(1-\omega^2/\omega_i^2+2\zeta^2)}{(1-\zeta_i^2)^{1/2}}\sin\omega_i(1-\zeta_i^2)^{1/2}t\right\}\right]$ $m\omega_i^2=\frac{\lambda_i^4 EI}{L^4},\ f_i=\frac{\omega_i}{2\pi}=\frac{\lambda_i^2}{2\pi L^2}\sqrt{\frac{EI}{m}}$, Hz. λ_i, $\tilde{y}_i(x)$ = beam mode shape, P,C,F,G =beam boundary conditions.Tables 4.2, 4.4, $\zeta_i<1, i=1,2,3.$

Table 7.4 Transient vibration of beams, continued

Notation for beams: F = force per unit length; $f_i = \omega_i/2\pi$ natural frequency, ith mode, Hz; g(t) = dimensionless function of time. Consistent units in Table 1.2. See Appendix D and Eq. 4.17 for beam integrals.

Displacement, $y(x,t)=\sum_{i=1}^{N} y_i(x,t)$	
Transient Beam Load	Displacement in ith Mode, $y_i(x,t)$, i=1,2,3..
5. Time Limited Point Force on Uniform Beam 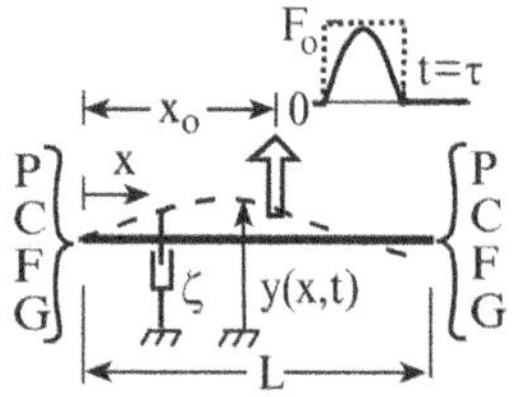	$\frac{F_o}{m\omega_i^2}\frac{\tilde{y}_i(x_o)}{\int_0^L \tilde{y}_i^2(x)dx}\begin{cases} g_i(t), 0<t\le t_1, \\ g_i(t)\pm g_i(t-t_1), t>t_1, \end{cases}$ t_1 = time period of impulse For half sine pulse use + and g(t) from case 8 Table 7.3 For rectangular pulses use – and g(t) from case 7 Table 7.3, $\omega_n=\omega_i$ $m\omega_i^2=\frac{\lambda_i^4 EI}{L^4},\ f_i=\frac{\omega_i}{2\pi}=\frac{\lambda_i^2}{2\pi L^2}\sqrt{\frac{EI}{m}}$, Hz. λ_i, $\tilde{y}_i(x)$ = beam mode shape, P,C,F,G =beam boundary conditions. Tables 4.2, 4.4
6. Pinned-Pinned Beam Uniform Step Force F 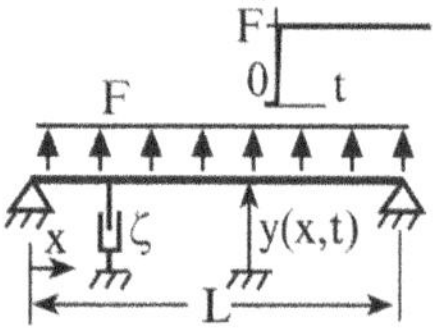	$\frac{F}{m\omega_i^2}\frac{2}{i\pi}(1-(-1)^i)\sin(i\pi\frac{x}{L})\times$ $\left[1-e^{-\zeta_i\omega_i t}\left\{\cos\left(\omega_i(1-\zeta_i^2)^{1/2}t\right)+\frac{\zeta_i}{(1-\zeta_i^2)^{1/2}}\sin\left(\omega_i(1-\zeta_i^2)^{1/2}t\right)\right\}\right]$ $m\omega_i^2=\frac{(i\pi)^4 EI}{L^4},\ f_i=\frac{\omega_i}{2\pi}=\frac{(i\pi)^2}{2\pi L^2}\sqrt{\frac{EI}{m}}$, Hz. i=1,2,3,..case 5 of Table 4.2. F(t)=0 at t=0, F(t)=F, $t>0$, $\zeta_i<1$, $i=1,2,3..$
7. Clamped Beam Uniform step Force/length F 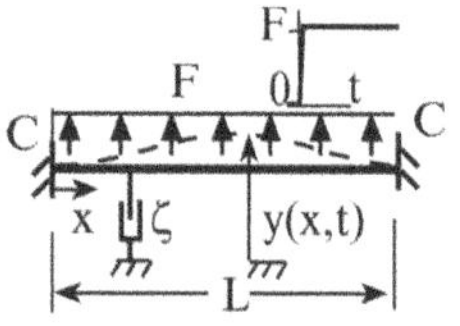	$\frac{F}{m\omega_1^2}\left(\frac{2}{3}\right)^{1/2}\left(1-\cos 2\pi\frac{x}{L}\right)\times$ (first mode, i=1) $\left[e^{-\zeta_1\omega_1 t}\{\cos(\omega_1(1-\zeta_1^2)^{1/2}t)+\frac{\zeta_1}{(1-\zeta_1^2)^{1/2}}\sin(\omega_1(1-\zeta_1^2)^{1/2}t)\}\right]$ $m\omega_1^2=\frac{4.73^4 EI}{L^4},\ f_1=\frac{\omega_1}{2\pi}=\frac{4.73^2}{2\pi L^2}\sqrt{\frac{EI}{m}}$, Hz. i=1, case 7 of Table 4.2. Approximate first mode. F(t)=0 at t=0, F(t)=F, $t>0$, $\zeta_i<1$, $i=1,2,3$.
8. Genera Beam uniform Step Force/length F 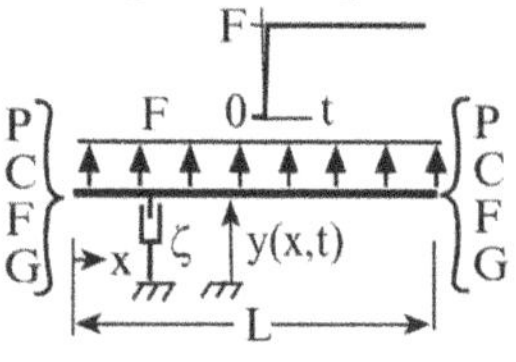 F(t)=0, t=0, F(t)=F, t>0	$\frac{F}{m\omega_i^2}\frac{\int_0^L \tilde{y}_i(x)dxy}{\int_0^L \tilde{y}_i^2(x)dx}\tilde{y}_i(x)\times$ $\left[1-e^{-\zeta_i\omega_i t}\left\{\cos\left(\omega_i(1-\zeta_i^2)^{1/2}t\right)+\frac{\zeta_i}{(1-\zeta_i^2)^{1/2}}\sin\left(\omega_i(1-\zeta_i^2)^{1/2}t\right)\right\}\right]$ $m\omega_i^2=\frac{\lambda_i^4 EI}{L^4},\ f_i=\frac{\omega_i}{2\pi}=\frac{\lambda_i^2}{2\pi L^2}\sqrt{\frac{EI}{m}}$, Hz. λ_i, $\tilde{y}_i(x)$ = beam mode shape, P,C,F,G = beam boundary conditions. Appendix D has beam integrals.

Table 7.4 Transient vibration of strings and plates, continued

	Displacement, $y(x,t)=\sum_{i=1}^{N} y_i(x,t)$
Transient Beam Load	Displacement in ith Mode, $y_i(x,t)$, i=1,2,3..
9. Sinusoidal Force on Uniform Beam 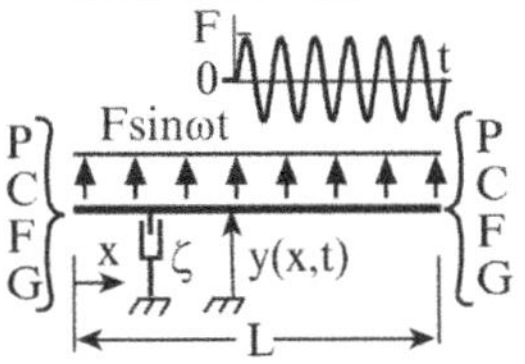 	$\frac{F}{m\omega_i^2}\frac{\int_0^L \tilde{y}_i(x)dx}{\int_0^L \tilde{y}_i^2(x)dx}\frac{\tilde{y}_i(x)}{(1-\omega^2/\omega_i^2)^2+(2\zeta\omega/\omega_i)^2}\left[\left(1-\frac{\omega^2}{\omega_i^2}\right)\sin\omega t-2\zeta_i\frac{\omega}{\omega_i}\cos\omega t\right.$ $\left.+e^{-\zeta_i\omega_i t}\left\{\frac{2\zeta_i\omega}{\omega_i}\cos\left(\omega_i(1-\zeta_i^2)^{1/2}t\right)-\frac{\omega}{\omega_i}\frac{(1-\omega^2/\omega_i^2)+2\zeta_i^2}{(1-\zeta_i^2)^{1/2}}\sin\omega_i(1-\zeta_i^2)^{1/2}t\right\}\right]$ $m\omega_i^2=\frac{\lambda_i^4 EI}{L^4}$, $f_i=\frac{\omega_i}{2\pi}=\frac{\lambda_i^2}{2\pi L^2}\sqrt{\frac{EI}{m}}$, Hz. λ_i, $\tilde{y}_i(x)$ = beam mode shape, P,C,F,G =beam boundary conditions.Tables 4.2, 4.4. $\zeta_i<1$
10. Time Limited Uniform Force on Beam 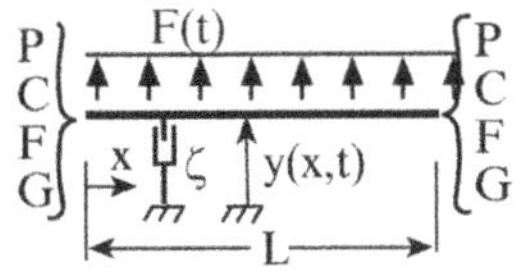 P, C, F, G = beam boundary conditions	$\frac{F}{m\omega_i^2}\frac{\int_0^L \tilde{y}_i(x)dx}{\int_0^L \tilde{y}_i^2(x)dx}\begin{cases} g_i(t), 0<t\le t_1, \\ g_i(t)\pm g_i(t-\tau), t>t_1, \end{cases}$ t_1 = time period of impulse For half sine pulse use + and g(t) from case 8 Table 7.3 For other pulses use – Table 7.3, $\omega_n=\omega_i, \zeta_n=\zeta_i$ $m\omega_i^2=\frac{\lambda_i^4 EI}{L^4}$, $f_i=\frac{\omega_i}{2\pi}=\frac{\lambda_i^2}{2\pi L^2}\sqrt{\frac{EI}{m}}$, Hz. λ_i, $\tilde{y}_i(x)$ = beam mode shape, P,C,F,G =beam boundary conditions.Tables 4.2, 4.4
11. Plucked Tension String 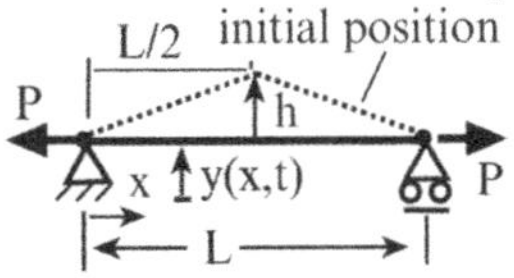 Initial Position, y(x,0)= hx/L, $0\le x\le L/2$; $2h(L-x)/L, L/2\le x\le L$	Initial position $y(x,0)=h\sum_i\left((-1)^{(i-1)/2}/(i\pi)^2\right)\sin i\pi/L$, $i=1,3,5...$ $y(x,t)=h\sum_{i=1,3,5.}\frac{(-1)^{(i-1)/2}}{(i\pi)^2}\sin\frac{i\pi x}{L}\times$ $\left[1-e^{-\zeta_i\omega_i t}\left\{\cos(\omega_i(1-\zeta_i^2)^{1/2}t)+\frac{\zeta_i}{(1-\zeta_i^2)^{1/2}}\sin(\omega_i(1-\zeta_i^2)^{1/2}t)\right\}\right]$ $\omega_i=\frac{i\pi}{L}\sqrt{\frac{P}{m}}$, rad / s. case 1 of Table 3.6, P = mean tension, $\zeta<1$
12. Initial Velocity Tensioned String 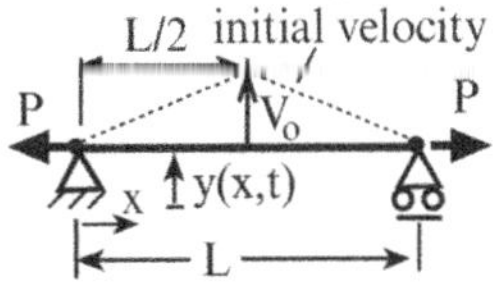 Initial velocity $2V_o x/L$, $0\le x\le L/2$; $2V_o(L-x)/L$, $L/2\le x\le L$	Initial velocity dy(x,0)/dt= $V_o\sum_i\left((-1)^{(i-1)/2}/(i\pi)^2\right)\sin i\pi/L$, $i=1,3,5...$ $y(x,t)=\frac{V_o}{\omega_i}\sum_{i=1,3,..}\frac{(-1)^{(i-1)/2}}{(i\pi)^4}\sin\frac{i\pi x}{L}$ $(\zeta_1<1)$, $\times\left[1-e^{-\zeta_i\omega_i t}\left\{\cos(\omega_i(1-\zeta_i^2)^{1/2}t)+\frac{\zeta_i}{(1-\zeta_i^2)^{1/2}}\sin(\omega_i(1-\zeta_i^2)^{1/2}t)\right\}\right]$ $\omega_i=\frac{i\pi}{L}\sqrt{\frac{P}{m}}$, rad / s. case 1 of Table 3.6, P = mean tension

Table 7.4 Transient vibration of strings and plates, continued

Transient Load	Response y(x,t)
13. Clamped Beam Acceleration shock 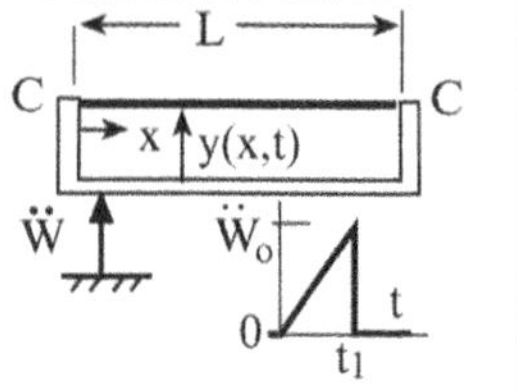 $\ddot{W}(t) = (t/t_1)\ddot{W}_o, 0 \le t \le t_1,$ $= 0, \quad t > t_1$	first mode, i=1, approx mode shape and $\zeta < 1$; for $0 \le t \le t_1$, $y(x,t) = -\frac{\ddot{W}_o}{\omega_n^2}\frac{1}{\omega_n t_1}\frac{2}{3}\left(1-\cos 2\pi\frac{x}{L}\right)\{(\omega_n t - 2\zeta)(1-\zeta^2)^{1/2}$ $+ e^{-\zeta\omega_n t}[2\zeta(1-\zeta^2)^{1/2}\cos\omega_d t - 2(1-2\zeta^2)\sin\omega_d t]\},$ $y(x,t) = -\frac{\ddot{W}_o}{\omega_n^2}\frac{1}{\omega_n t_1}\frac{2}{3}\left(1-\cos 2\pi\frac{x}{L}\right)\{e^{-\zeta\omega_n t}[2\zeta(1-\zeta^2)^{1/2}\cos\omega_d t$ $-(1-2\zeta^2)\sin\omega_d t]) + e^{-\zeta\omega_n(t-t_1)}[(\omega_n t_1 - 2\zeta(1-\zeta^2)^{1/2}\cos\omega_d(t-t_1)$ $+(1+\zeta\omega_n t_1 - 2\zeta^2)\sin\omega_d(t-t_1)])\}, \ t > t_1,$ $\omega_1 = 2\pi f_1 = \frac{4.73}{L^2}\sqrt{\frac{EI}{m}}$, rad / s. Table 4.2.
14. . Abrupt Point Force Simply Supported Plate 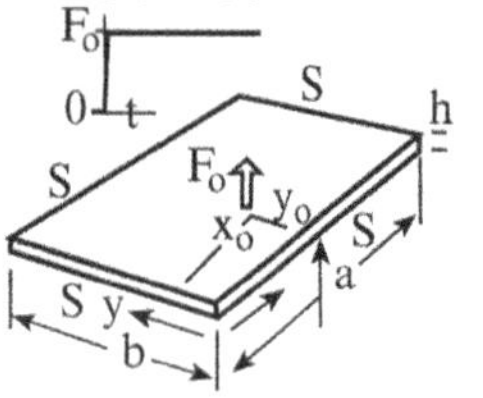 	$w(x,y,t) = \frac{4F_o}{\gamma ab\omega_{ij}^2}\sum_{i=1,2,..}\sum_{j=1,2,..}\sin\frac{i\pi x_o}{a}\sin\frac{j\pi y_o}{b}\sin\frac{i\pi x}{a}\sin\frac{j\pi y}{b}$ $\times\left[1 - e^{-\zeta_{ij}\omega_i t}\left\{\cos\left(\omega_{ij}(1-\zeta_{ij}^2)^{1/2}t\right) + \frac{\zeta_{ij}}{(1-\zeta_{ij}^2)^{1/2}}\sin\left(\omega_{ij}(1-\zeta_{ij}^2)^{1/2}t\right)\right\}\right], \zeta_{ij} < 1$ $\omega_{ij} = 2\pi f_{ij} = \left(\left(\frac{i\pi}{a}\right)^2 + \left(\frac{j\pi}{b}\right)^2\right)\sqrt{\frac{Eh^3}{12\gamma(1-\nu^2)}}$, rad/ s,case 2 of Table 5.3
15. Abruptly Applied Uniform Pressure on Simply Supported Plate 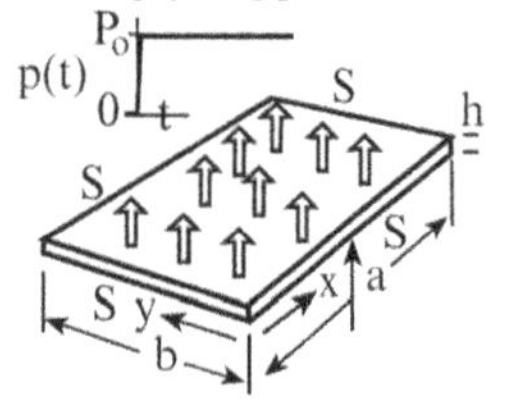 	$w(x,y,t) = \frac{4p_o}{\gamma\omega_{ij}^2}\sum_{i=1,2,..}\sum_{j=1,2,..}\sin\frac{i\pi x}{a}\sin\frac{j\pi y}{b}$ $\times\left[1 - e^{-\zeta_{ij}\omega_i t}\left\{\cos\left(\omega_{ij}(1-\zeta_{ij}^2)^{1/2}t\right) + \frac{\zeta_{ij}}{(1-\zeta_{ij}^2)^{1/2}}\sin\left(\omega_{ij}(1-\zeta_{ij}^2)^{1/2}t\right)\right\}\right], \zeta_{ij} < 1$ $\omega_{ij} = 2\pi f_{ij} = \left(\left(\frac{i\pi}{a}\right)^2 + \left(\frac{j\pi}{b}\right)^2\right)\sqrt{\frac{Eh^3}{12\gamma(1-\nu^2)}}$, rad/ s,case 2 of Table 5.3
16. Uniform Pressure Pulse On Plate 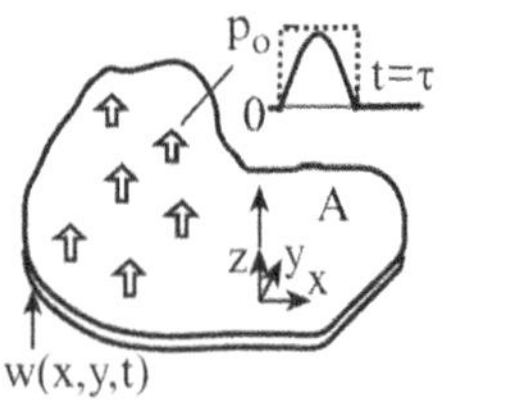 	$w(x,t) = \sum_{i=1}^{N}\frac{p_o}{\gamma\omega_{ij}^2}\frac{\int_0^L \tilde{w}_{ij}(x,y)dxdy}{\int_0^L \tilde{w}_{ij}^2(x,y)dxdy}\tilde{w}_{ij}(x,y)\begin{cases} g_i(t), 0 < t \le t_1, \\ g_i(t) \pm g_i(t-\tau), t > t_1, \end{cases}$ t_1 = time period of impulse For half sine pulse use + and g(t) from case 8 Table 7.3 For rectangular pulse use – and g(t) from case 7 Table 7.3, $\omega_n = \omega_i$, Plate natural frequencies, mode shapes $\tilde{w}_{ij}(x,y)$ and boundary conditions from Tables 5.2, 5.3.

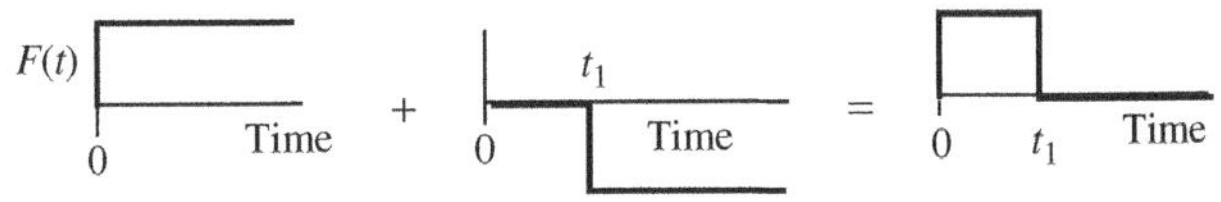

Figure 7.7 Superposition of loads to create a finite rectangular pulse

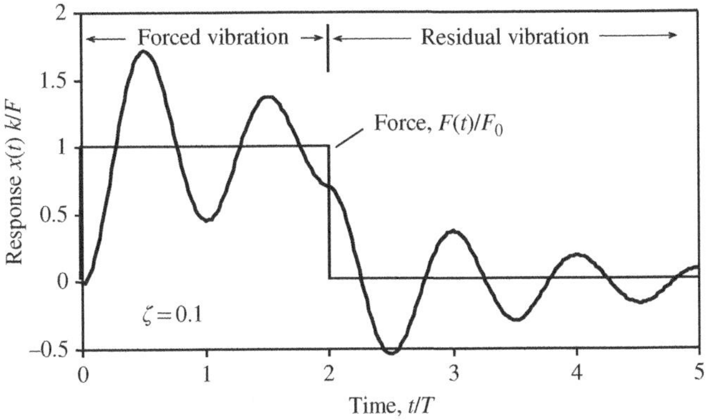

Figure 7.8 Transient response to a rectangular pulse lasting two natural periods

resonant frequencies. Both techniques excite the fundamental mode of the string. But higher string modes are excited more strongly with the harpsichord, as $1/i^2$, where i is the mode number. This gives the harpsichord its sharper harmonic sound, whereas in the piano, the higher mode excitation is proportional to $1/i^4$.

7.2.3 Maximum Transient Response and Response Spectra

Maximum response is important for design with transient loads. The maximum transient response can occur during while loading is applied, as is the case in Figures 7.6 and 7.8, or it can occur after the loading, during the *residual free vibration.* The maximum response during residual vibration can be determined exactly from cases 2 and 3 of Table 7.3 given the velocity and displacement at the start of vibration. The maximum of the absolute value of response of both forced and residual vibration is called the *maximax response.* Reviewing Figures 7.6–7.9, we see that the maximax response tends to occur about ½ period after loading begins.

Response Spectra are the maximum of the transient response to a given transient load of a family of single-degree-of-freedom oscillators plotted against their natural frequency. Response spectra in Figure 7.9, and Figure 7.10 were generated and made by calculating transient response at each point in time and then retaining only the maximum. Response spectra are used to simplify analysis of multiple systems that respond to the same excitation, such as earthquake excitation.

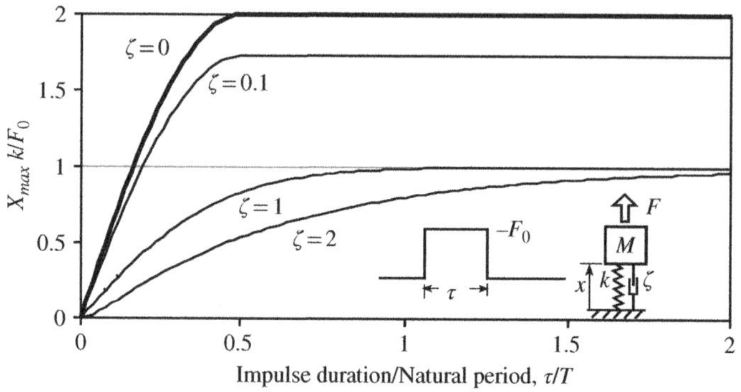

Figure 7.9 Response spectra (Maximum response) of a damped single-degree-of-freedom system to a duration limited step force

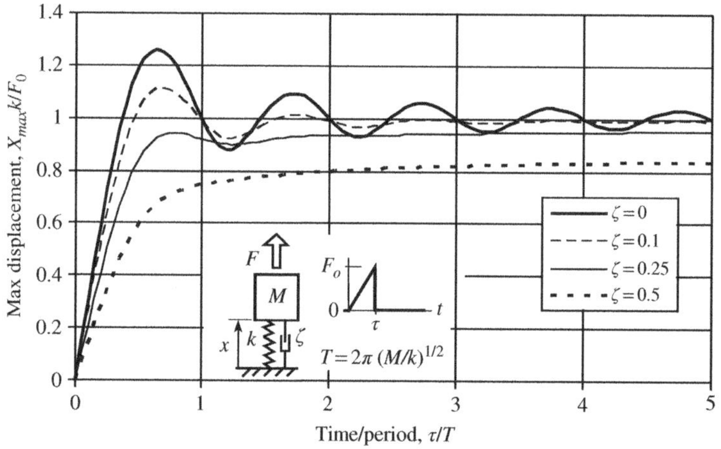

Figure 7.10 Maximum displacement of a damped single-degree-of-freedom oscillator to sawtooth force

Response spectra have three ranges according to natural frequency of the excited oscillator:

1. *Low natural frequency, infinite natural period.* The mass is stationary and relative displacement is just the upward motion of the base, and this is called *vibration isolation* (Section 7.3).
2. *High natural frequency and stiffness, zero period.* As the natural frequency f_n and stiffness k become large, displacement is just static response, $x = -M(d^2w/dt^2)/k$. A base-excited system with a natural frequency that is significantly higher than the frequencies of base excitation will move rigidly with the base and will not amplify the base displacement.
3. *Resonance.* When the natural frequency is at or near frequencies of the transient load amplification will occur that is limited by the time spent at near resonance and damping.

A unique part of response spectra is that they are single degree of freedom. The total responses are the sum of modal response. But because time and phase are lost in response spectrum analysis, the only rigorously conservative way to combine modes from response spectra is to sum their absolute values. A less conservative alternative is SRSS, which is endorsed by the ASME Boiler and Pressure Vessel Code [16]:

$$|x_j|_{\mathrm{SRSS}} = \left(\sum_{i=1}^{N} x_{Rs,ij}^2 \right)^{1/2} \tag{7.46}$$

SRSS is the square root of the sum of squares of modal responses.

Approximate velocity and accelerations are based on harmonic motion (Table 3.1):

Pseudo velocity magnitude $= 2\pi f_n$ (magnitude displacement).
Pseudo acceleration magnitude $= (2\pi f_n)^2$ (magnitude displacement).

They allow use of the three axis plotting, shown in Figure 3.2, for response spectra.

Example 7.3 The wooden platform, shown in Figure 7.11, supports a 2200 lb (1000 kg) load on two pinewood beams. The beams are 8 ft (2.44 m) long and 3.5 inch (8.89 cm) square in cross section. The upper ends of the beams are nailed to the platform. The lower end is embedded clamped to a foundation. An earthquake produces a 2 g sawtooth shock of 0.1 second duration. Will the platform fail?

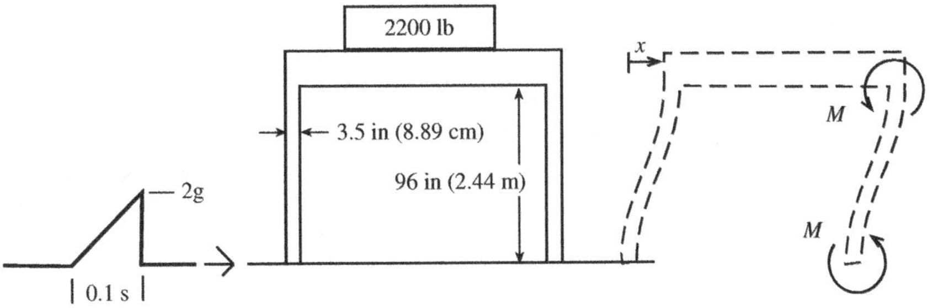

Figure 7.11 A portal frame with pine legs is loaded by a ground shock wave

Solution: First, the natural frequency is calculated. The legs provide lateral stiffness. The modulus of pine is approximately $E = 1.2 \times 10^6$ psi $(8.3 \times 10^9$ Pa). Case 4 of Table 1.5 illustrates the moment of inertia of the leg cross section $I = \frac{a^4}{12} = (3.5\,\text{in.})^{4/12} = 12.5\,\text{in}^4 (520.4\,\text{cm}^4)$. Case 18 of Table 3.2 shows the leg bending stiffness $k = 2(12EI/L^3) = 24(1.2 \times 10^6 \text{psi}\ 12.5\,\text{in.}^4)/(96\,\text{in.})^3 = 406.9\,\text{lb/in.}\ (712.6\,\text{N/cm})$. The fundamental natural frequency of the platform is shown in case 1 of Table 3.2:

$$f_n = \frac{1}{2\pi}\sqrt{\frac{k}{M}} = \frac{1}{2\pi}\sqrt{\frac{406.9\,\text{lb/in.}}{2200\,\text{lb}/386.1\,\text{in./s}^2}} = \frac{1}{2\pi}\sqrt{\frac{71260\,\text{N/m}}{1000\,\text{kg}}} = 1.3\,\text{Hz}$$

The natural period is 1/1.3 Hz. The ratio of shock duration to platform natural period is 0.1 s / 0.8 s = 0.13. The response is read from Fig 7.10. The nondimensional response spectra are about 0.41:

$$\text{Max displacement} = \frac{0.41\,\ddot{W}_o}{(2\pi f_n^2)} = 0.41 \times 2 \times 386.1\ \text{in./s}^2/(2\pi\,1.3\ \text{Hz})^2 = 4.75\ \text{in.}$$

$$\text{Max mass acceleration} = 0.41\,\ddot{W}_o = 0.41 \times 2\,\text{g} = 0.82\,\text{g} = 316\ \text{in./s}^2\ (8.04\ \text{m/s}^2)$$

The shear load in each leg is equal to its stiffness times the displacement. This is 1945 lb (8700 N) or 4350 lb (4350 N) per beam. The bending moment on each beam thus is 972.8 lb × 96 in. = 93400 in.-lb. Beam bending dominates the stresses in the legs. The bending stress in the beam is given in Chapter 4, Table 4.1 and Equation 4.3:

$$\sigma = \frac{Mc}{I} = \frac{93.4 \times 10^3\ \text{in.-lb}}{12.5\ \text{in.}^4}\frac{3.5\ \text{in.}}{2} = 13000\ \text{psi}\ (88.5\ \text{MPa})$$

This stress approximately equals the tensile strength of pinewood loaded parallel to grain. Thus, a bending stress failure is possible in the leg joint with the platform.

7.2.4 Shock Standards and Shock Test Machines

Shot and drop tests are performed to qualify equipment to withstand shocks and handling. There are industry standard shock tests. Shock waveforms which are specified by *shape*, such as half-sine, sawtooth, triangular, and rectangular formations; *amplitude*, which is expressed in g's (the force of gravity is 1 g); and *duration*, which includes pulse width and pulse number. Figure 7.12 shows the standard shock pulse MIL-STD-810. Standard shock test specifications include MIL-STD-202, RTCA/DO160, MIL-STD-883, IEC 68 2–27, GR-63-CORE, ANSI S2.14-1973, ANSI S2.15-1973, IEC 60068-2-27, MIL-STD-810D section 5.16.4, MIL-STD-167B, and ASTM F1931-98(2004). ASA S2.62 2009 provides shock isolation standards for gymnasium floor mats. Drop test specifications include MIL-STD-810D section 516.4 procedure IV, GR-63-CORE para 5.3.1, and ASTM D5276.

7.3 Vibration Isolation

7.3.1 Single-Degree-of-Freedom Vibration Isolation

Vibration isolation reduces vibration *transmissibility*. Figure 12.13a shows oscillating force on a spring-supported mass. Force is transmitted through the spring and the damper to the vibration sensitive base. *Force transmissibility* is the amplitude of force on the base (the vibration sensitive component) over the amplitude of force F_o on the vibrating mass. In Figure 7.12b, vibration sensitive component (output) is on a vibrating machine. The *displacement transmissibility* is the component amplitude X_o over the amplitude of the base displacement W_o. These force and displacement transmissibilities (cases 1 and 3 of

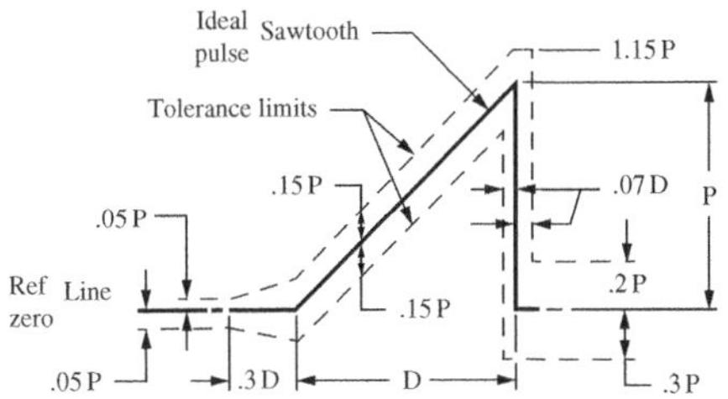

Procedure	Test	Peak value (P) g's		Nominal duration (D) ms	
		Flight vehicle equipment a	Ground equipment b	Flight vehicle equipment c	Ground equipment d
I	Basic design	20	40 2/	11	11
III	Crash safety	40	75	11	6
IV	High intensity	100	100	6	11

Figure 7.12 Sawtooth shock from Mil-Std M-810C

Table 7.1) are mathematically identical:

$$\text{Transmissibility} = \frac{F_{\text{out}}}{F_o} = \frac{X_o}{W_o} = \frac{\sqrt{1 + (2\zeta f/f_n)^2}}{\sqrt{(1 - f^2/f_n^2)^2 + (2\zeta f/f_n)^2}}$$
$$\approx \frac{f_n^2}{f^2} \text{ for } f \gg f_n \tag{7.47}$$

Transmissibility is plotted in Figure 7.14 as a function of ratio of the forcing frequency f to the natural frequency f_n, f/f_n, and damping factor ζ.

Good isolation at an excitation frequency f requires that the suspension system natural frequency f_n must be much lower than the has an excitation frequency, $f \gg f_n$. This is accomplished with a soft suspension. A soft suspension uses flexibility to lower the suspended mass natural frequency below the forcing frequency:

$$\frac{F_{\text{base}}}{F_o} \approx \frac{f_n^2}{f^2} = \frac{1}{(2\pi)^2} \frac{g}{\delta_s f^2} \tag{7.48}$$

A soft suspension deflects downward, $\delta_s = Mg/k$, under the weight of the component. g is the acceleration due to gravity (Table 1.2). The natural frequency in terms of the static

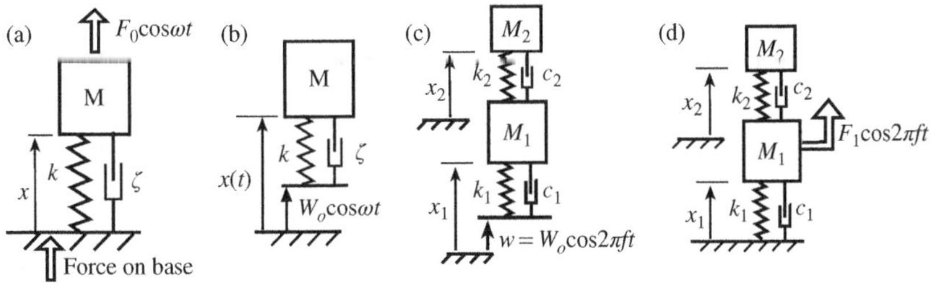

Figure 7.13 Four cases of isolation. Cases a and d minimize force on the base from motions of a vibrating mass, Cases b and c isolate mass from vibrating base

deflection is $f_n = (1/2\pi)(g/\delta_s)^{1/2}$ (Appendix A, Eq. A.1), so in general a soft suspension has a large static deflection under the weight of the supported component. The suspension must be capable of this deflection without bottoming out against stops. Further, a transient case, the start-up of a rotor, for example, will pass through the suspension natural frequency creating a briefly high response. The system must have sufficient damping to withstand this transient resonance. For this reason, rubber elements and dampers are often incorporated in isolation suspensions.

Example 7.4 Consider a 100 kg (220 lb) roof-mounted air-conditioning unit. Its unbalanced shaft rotates at 50 Hz. Design a soft support with transmissibility of 0.05 to isolate the vibrating air conditioner from the conference room directly below it.

Solution: Equation 7.48 is solved for the suspension natural frequency that has a transmissibility of 0.05 at 50 Hz:

$$\text{Transmissibility} \approx \frac{f_n^2}{f^2} = 0.05,$$

$$f_n = f\sqrt{Tr} = 50\,\text{Hz}\sqrt{0.05} = 11.18\,\text{Hz}$$

The suspension stiffness and the static deformation under the weight of the air conditioner are calculated:

$$k = M(2\pi f_n)^2 = 100\,\text{kg}(2\pi\ 11.18\,\text{Hz})^2 = 4.93 \times 10^5\,\text{N/m}\ (2811\,\text{lb/in.})$$

$$\delta_s = \frac{Mg}{k} = \frac{100\,\text{kg}\,(9.81\,\text{m/s})}{4.93 \times 10^5\,\text{N/m}} = 2\,\text{mm}\ (0.078\,\text{in.})$$

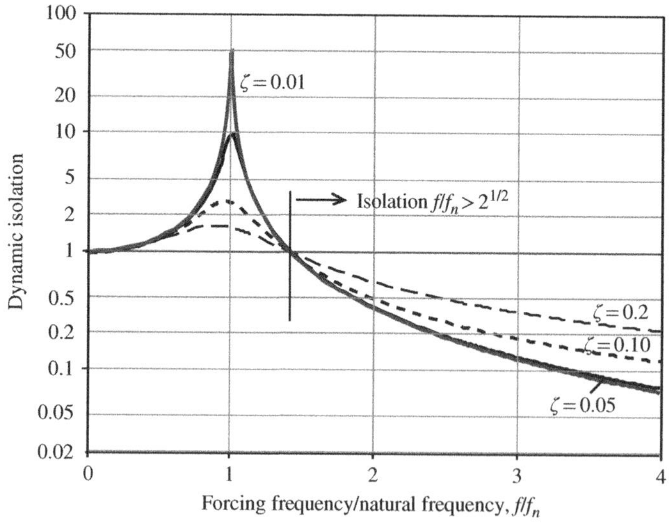

Figure 7.14 Transmissibility and isolation as a function of frequency and damping

The suspension geometry must accommodate the static deflection, vibration amplitude, plus a margin for design tolerances. In addition, most suspensions must be capable of high loads while passing through transient resonance and bear extraordinary loads during accidents and earthquakes, without releasing the component.

Figure 7.15 shows suspension designs that isolate vibrating components in three dimensions. Suspensions in the top row hang from overhead. In the bottom row, they are supported

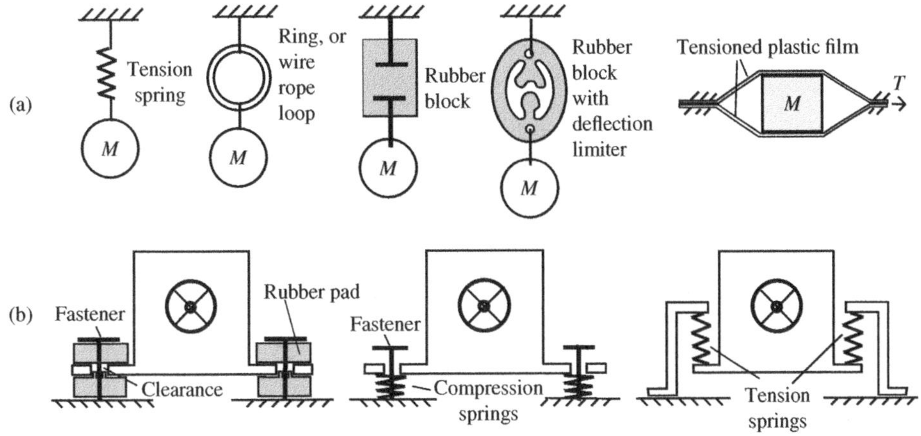

Figure 7.15 Vibration isolation suspensions. Upper row suspensions isolate vibration sensitive component (M) from support vibration. Bottom row isolates the floor from machinery vibration

from a floor, and these suspensions have stops for extreme loads. For example, brackets and fasteners restrain the component if the load capability of rubber isolation pads or springs is exceeded.

7.3.2 Two-Degree-of-Freedom Vibration Isolation

7.3.2.1 Vibration Isolation from Base Motion

Inertial mass isolation and the tuned mass dampener are two additional strategies for isolation and vibration reduction with two-degree-of-freedom systems. Consider the two-degree-of-freedom system in Figure 7.13c that isolates masses M_1 and M_2 from a vibrating base. Transmissibility of displacement from the base to each of the masses is shown in case 5 of Table 7.1. For zero damping, these simplify to

$$\left.\frac{X_1}{W_o}\right|_{\zeta 1=\zeta 2=0} = \left(1-\frac{f^2}{f_2^2}\right)\left[\left(1-\frac{f^2}{f_1^2}\right)\left(1-\frac{f^2}{f_2^2}\right)-\frac{M_2 f^2}{M_1 f_1^2}\right]^{-1} \approx \frac{f_1^2}{f^2} = \frac{k_1}{M_1}\frac{1}{f^2} \quad \text{if } f \gg f_1, f_2$$

$$\left.\frac{X_2}{W_o}\right|_{\zeta 1=\zeta 2=0} = \left[\left(1-\frac{f^2}{f_1^2}\right)\left(1-\frac{f^2}{f_2^2}\right)-\frac{M_2 f^2}{M_1 f_1^2}\right]^{-1} \approx \frac{f_1^2 f_2^2}{f^4} = \frac{k_1}{M_1}\frac{f_2^2}{f^4}, \quad \text{if } f \gg f_1, f_2 \tag{7.49}$$

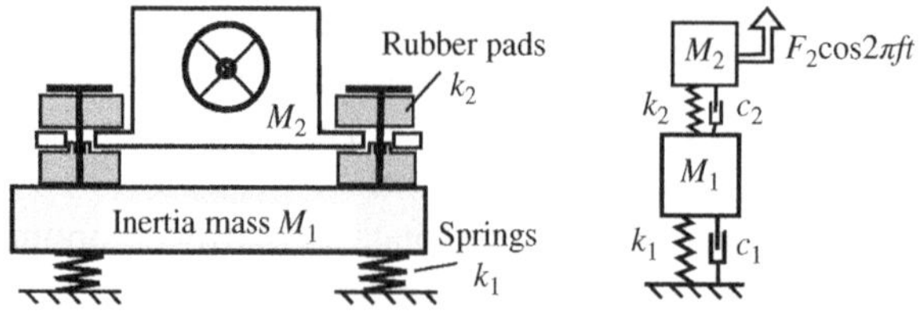

Figure 7.16 Two-stage vibration isolation of a rotating machine and its model

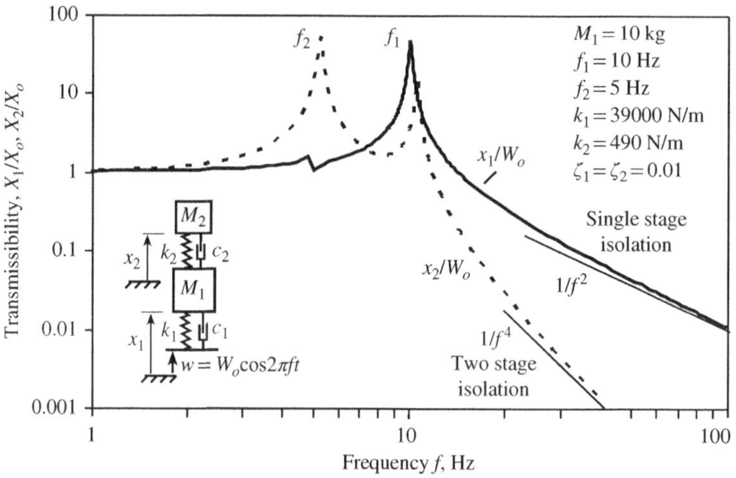

Figure 7.17 Two-stage vibration isolation. The upper mass has substantially higher isolation at high frequencies than the lower mass for forcing frequency above the natural frequencies Solution taken from case 5 of Table 7.1

$f_1 = (1/2\pi)(k_1/M_1)^{1/2}$ (Hz) is the natural frequency with M_2 removed. $f_2 = (1/2\pi)(k_2/M_2)^{1/2}$ is the natural frequency of M_2 with M_1 held stationary. The coupled natural frequencies are shown in case 3 of Table 3.3.

If the suspension natural frequencies are much less than the forcing frequency f, then the transmissibility of the mass, M_1, varies as $1/f^2$, which is single-degree-of-freedom isolation. But if both natural frequencies of the two-degree-of-freedom system are lower than the forcing frequency, then the isolation of the upper mass, M_2, varies as $1/f^4$, which is more isolation. This effect is shown in Figure 7.17.

Two-stage vibration isolation is further optimized with an inertia (heavy) foundation $M_1 > M_2$, because the vibration transmitted between M_2 and the base decreases as $1/M_1$. As shown in Figure 7.16, a vibrating machine is isolated from the floor by flexibly mounting the machine on a heavy inertial mass, which is in turn mounted with a soft suspension to the floor.

7.3.2.2 Tuned Mass Damper

A small tuned mass damper can significantly reduce the resonant vibration of a lightly damped large mass [2, 17–22]. The advantage of a tuned mass damper is that it is entirely supported off the vibrating structure at the point of maximum displacement. Its disadvantage is that it must be tuned to the larger mass and maintained. Also, the weight of effective tuned mass damper is about 3–5% of the mass of the damped structure, which is not insignificant for tall buildings.

Consider the cases shown in Figure 7.13. The large, lightly damped mass M_1 is excited. It is possible to tune upper mass, M_2, so that its natural frequency is approximately equal to the natural frequency of M_1 to reduce its vibrations.

Without the tuned mass damper ($M_2 = 0$), the resonant response of X_1 (Eq. 7.6) is inversely proportional to its small damping factor ζ_1, and this results in a large resonant. There are three unequal peaks of the two-mass system for light damping. Two are at natural frequencies. A third response peak arises at because the numerator of X_1 increases with increasing ζ_2, and denominator of X_1 has a minimum near $f_3 = (1/2\pi)[k_1/(M_1 + M_2)]^{1/2}$ Hz owing to the term proportional to $(1 - f^2/f_3{}^2)$.

Den Hartog's [2] strategy for optimal tuned mass dampers is to make the three X_1 response peaks near f_1, f_2, and f_3 equal by choosing optimal values of the mass damper tuning f_2/f_1 and damping ζ_2. He gives an approximate expression for tuning for equal peaks:

$$\frac{f_1}{f_2} \approx \frac{1}{1 + M_2/M_1} \tag{7.50}$$

Here, $f_1 = (1/(2\pi))(k_1/M_1)^{1/2}$ and $f_2 = (1/(2\pi))(k_2/M_2)^{1/2}$ are the natural frequencies of each mass by itself. With this tuning and small damping of the larger mass, $\zeta_1 << 1$, and larger damping of the small mass, $\zeta_2 > \zeta_1$, an approximate expression for the resonant amplitude of the large mass M_1 is

$$\left.\frac{X_1 k_1}{F_1}\right|_{\text{resonant}} \approx \begin{cases} \dfrac{1}{2\zeta_1}, & M_2 = 0, \text{no tuned damper} \\ \left[\dfrac{M_1}{M_2} + \dfrac{1}{(2\zeta_2)^2}\right]^{1/2}, & \text{with tuned damper} \end{cases} \tag{7.51}$$

These expressions suggest that the optimal tuned mass damper has (1) a mass of at least 3–5% of the damped mass; (2) a damping that is approximately $\zeta_2 = (M_2/M_1)^{1/2}/2$, which is $\zeta = 0.08$ to 0.010; and (3) a tuning that is given by Equation 7.51b. The effect of a tuned mass damper that meets these criteria is shown in Figure 7.18, which plots the formulas of case 7 of Table 7.2 and the same parameters as Figure 7.16, except that Figure 7.18 demonstrates the optimal tuning and damping.

7.4 Random Vibration Response to Spectral Loads

Random vibration describes response statisticaly interms of its as its overall root-mean-square (rms) displacement. Random vibration analysis is most useful when there are many

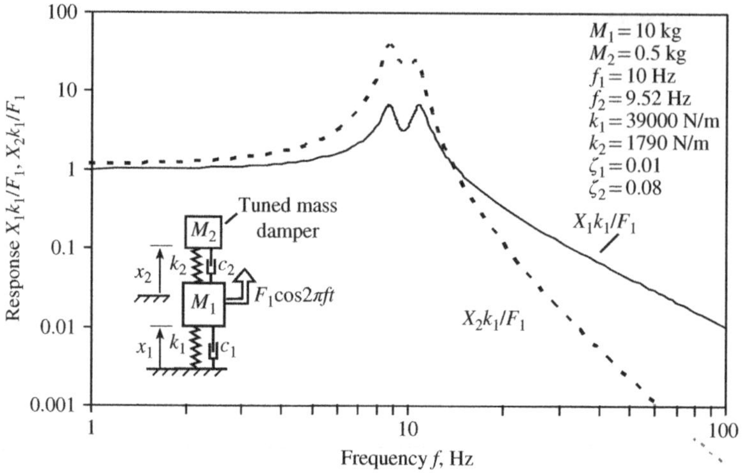

Figure 7.18 Tuned mass damper M_2 is tuned to the natural frequency of M_1. The tuned mass M_2 has 8% damping and 5% of the mass of the large mass M_1. It reduces M_1 resonant response by a factor of 8

random frequency components and the process is stationary and ergodic. That is, the process is statistically stationary and the sample is long enough to contain many cycles of vibration.

7.4.1 Power Spectral Density and Fourier Series

A finite Fourier series is used to describe a stationary, multifrequency process $y(t)$ over the time period $t = 0$ to $t = T$:

$$y(t) = a_o + \sum_{k=1}^{N} \left[a_k \cos \frac{2\pi k}{T} t + b_k \sin \frac{2\pi k}{T} t \right] \tag{7.52}$$

where the coefficients $a(k)$ and $b(k)$ are

$$a_o = \frac{1}{T} \int_0^T y(t) dt$$

$$a_k = \frac{2}{T} \int_0^T y(t) \cos \frac{2\pi kt}{T} dt$$

$$b_k = \frac{2}{T} \int_0^T y(t) \sin \frac{2\pi kt}{T} dt$$

There are N discrete frequencies, $f_k = k/T$, $k = 1, 2, 3$, in the finite Fourier series, not including the zero frequency average value a_0. Fourier frequencies are spaced at $1/T$. The lowest non-zero frequency represented is $1/T$, one over the sampling period, and the highest frequency is N/T. The rms of each individual component frequency is $[(a_k{}^2 + b_k{}^2)/2)]^{1/2}$. The overall rms is the sum of the root of the mean squares of the component:

$$y_{\text{rms}}^2 = \frac{1}{T}\int_0^T y^2(t)dt = a_0^2 + \frac{1}{2}\sum_{k=1}^{N}(a_k^2 + b_k^2) \tag{7.53}$$

The maximum possible value of a multifrequency random process is the sum of the magnitudes, $\sum (a_k{}^2 + b_k{}^2)^{1/2}$, which is unlikely but possible if all the maximums of each frequency component line up at the same time.

The ratio of maximum peak to overall rms (peak-to-rms ratio) is a measure of the randomness of the process. It is $2^{1/2}$ for a single sinusoidal time signal, and $[2N]^{1/2}$ for the sum of N equal amplitude sinusoids. Typically, real stationary random processes are often well represented by 6–20 random sine components, and their peak-to-rms ratios range from about 2.5 to 4 [23].

Power spectral density of a time history process is the rms of components contained in a frequency band divided by the width of the frequency band. For the Fourier series, this is

$$\text{Power spectral density, } S_p(f_k) = (a_k^2 + b_k^2)\frac{T}{2}, \quad f_k = \frac{k}{T}, \quad k = 1, 2, 3 \ldots N \tag{7.54}$$

Power spectral density has the units of the process squared divided by frequency. An equivalent definition is the mean square in a small frequency band divided by the width of the frequency band:

$$S_p(f_k) = \frac{\overline{y(t)^2}}{\delta f} \tag{7.55}$$

Here, $y(t)$ is the components of time history signal y that only includes components in the frequency band between $f - \delta f/2$ and $f + \delta f/2$. One consequence of these definitions is *Parseval's theorem* [24]:

$$\overline{y(t)^2} = \frac{1}{T}\int_0^T y(t)^2 dt = \int_0^\infty S_p(f)df \tag{7.56}$$

Parseval's theorem states that engineering power spectral density is defined such that the overall rms value of a single-sided engineering spectral density equals the overall rms of the time history signal, which implies the previous equations.

The spectral density $S_p(f)$ defined here is the single-sided autospectra in terms of Hertz. It is also called *engineering autospectral density*, since it can be obtained directly from physical measurements by engineers using analog or digital equipment. Double-sided spectra include the negative frequency range, and they have one half the amplitude of engineering spectra. Spectra can also be defined with frequency in radians per second.

7.4.2 Complex Fourier Transform and Random Response

Table 7.6 [13, 25] has the overall mean square response of systems to a power spectral density loading. These were obtained by solving the equation of motion for single-frequency excitation and then integrating its magnitudes over all frequencies using power spectral density of forcing. These random responses assume that components of the forcing spectral density near the resonant frequency provide the greatest contribution to the overall response.

The formulas in Table 7.5 were obtained by finite Fourier transform of the equation of motion of a spring–mass system (Eq. 7.1, Fig. 7.1):

$$M\ddot{x}(t) + 2M\zeta\omega_n\dot{x}(t) + kx(t) = F(t) \tag{7.57}$$

Finite Fourier transform is made to convert this equation from the time domain to the complex frequency domain by multiplying by the complex term $e^{-i\omega t}$ and integrating over the finite time interval 0 to T (note here that T is not the natural period but rather a times such that many cycles of oscillations are included between 0 and T). The quantity $i = (-1)^{1/2}$ is the complex constant, which permits the phase to be represented by real and imaginary terms.

The complex finite Fourier transform of $x(t)$ is denoted by $G_x(f)$ where $\omega = 2\pi f$ is the frequency in radians per second and f is the frequency in Hz. Like the real-number fourier transform, Eq. 7.52, the complex fourier transforms time-dependent functions into frequency-dependent functions. It has some unique properties for derivatives that can be shown by integrating the transform of dx/dt by parts:

$$G_x(f) = 2\int_0^T x(t)e^{-i\omega t}dt$$

$$\begin{aligned} G_{\dot{x}}(f) &= 2\int_0^T (dx/dt)e^{-i\omega t}dt = 2x(t)e^{-i\omega t}|_0^T + 2i\omega\int_0^T x(t)e^{-i\omega t}dt \\ &= i\omega G_x(f), \quad \text{if } x(0) = x(T) = 0 \end{aligned} \tag{7.58}$$

Fourier transform form of a derivative dx/dt with respect to time is just $i\omega$ time the transform of $x(t)$, provided that the $x(t)$ is zero at both ends of the sample. This condition is often enforced by setting the ends of a sample to zero, which is called windowing, and it reduces low frequency aliasing of low- and high-frequency components [24]. Fourier transforms over theoretically infinite time spans do not suffer this limitation. The finite Fourier transform used here is consistent with digital data sampling which by necessity is over a finite time period.

The finite Fourier transform of Equation 7.57 is taken by multiplying Equation 7.57 by $e^{-i\omega t}$ and integrating over time from 0 to T. Repeating this process $e^{+i\omega t}$ produces the conjugate transform denoted by *. The two are multiplied together to produce an equation in which the imaginary constant does not appear, and thus, phase information is lost:

$$[-\omega^2 + i\omega(2\zeta\omega_n) + \omega_n^2]G_x(f) = \frac{1}{M}G_F(f) \tag{7.59a}$$

$$[-\omega^2 - i\omega(2\zeta\omega_n) + \omega_n^2]G_x^*(f) = \frac{1}{M}G_F^*(f) \tag{7.59b}$$

$$[(\omega^2 - \omega_n^2)^2 + (2\zeta\omega\omega_n)^2]G_x(f)G_x^*(f) = \frac{1}{M^2}G_F(f)G_F^*(f) \tag{7.59c}$$

Table 7.5 Response to random broad band excitation

Notation: $S_F(f)$ = power spectral density of force with units of $(\text{force})^2/\text{Hz}$ in cases 1 and 2 units of $(\text{force/length})^2/\text{Hz}$ in cases 3 and 4; E = modulus of elasticity, Chapter 8; c = distance from neutral axis; h = plate thickness; k = spring constant, Table 3.2; M = mass; f = frequency, Hz; f_n = natural frequency, Hz; F(t) = random force or random force per unit length; I = area moment of inertia, Table 1.5; x_{rms}, y_{rms} = root-mean square displacement; ζ = damping factor $\ll 1$, dimensionless; ν = Poisson's ratio. Consistent units are in Table 1.2. Refs. [13, 25], Table 7.2.

System	RMS Displacement x_{rms}, y_{rms}
1. Spring-Mass External Force 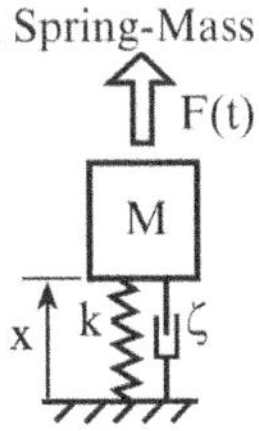	$\frac{1}{k}\sqrt{\frac{\pi f_n S_F(f_n)}{4\zeta}}$ $S_F(f_i)$ = power spectral density of force F(t) natural frequency, $f_n = \frac{1}{2\pi}\sqrt{\frac{k}{M}}$, Hz Ref. [25]
2. Spring-Mass, Base Exited 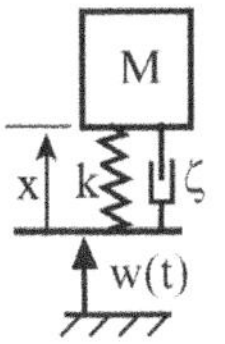	$\frac{1}{(2\pi f_n)^2}\sqrt{\frac{\pi f_n S_{\ddot{W}}(f_n)}{4\zeta}}$ natural frequency, $f_n = \frac{1}{2\pi}\sqrt{\frac{k}{M}}$, Hz $S_{\ddot{W}}(f)$=power spectra density of base acceleration
3. Pinned-Pined Beam Uniform Random Load 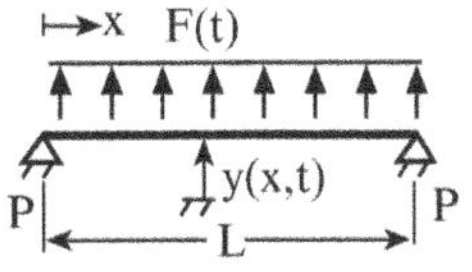 F = force per unit length	$y_{i-rms}(x) = \frac{1}{m\omega_i^2}\frac{2(1-(-1)^i)}{i\pi}\sin\frac{i\pi x}{L}\sqrt{\frac{\pi f_i S_F(f_i)}{4\zeta}}$ i = 1,2,.. $S_F(f_i)$ = power spectral density of force per unit length nat.freq. f_i in case 5 Table 4.2, $f_i = \frac{\omega_i}{2\pi} = \frac{(i\pi)^2}{2\pi L^2}\sqrt{\frac{EI}{m}}$, Hz. stress along span, $\sigma_{i-rms} = Ec(i\pi / L)^2 y_{i-rms}(x)$, Eqs. 4.2, 4.18. w_{i-rms} from above equation.
4. Clamp-Clamp Beam, Uniform Random Load First Mode 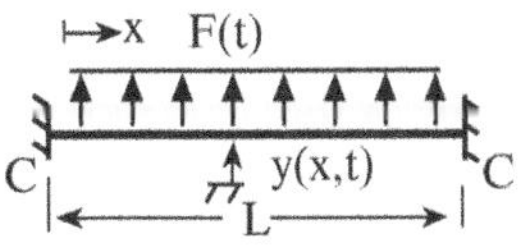 F = force per unit length	$y_{i-rms}(x) = \frac{1}{m(2\pi f_1)^2}\frac{2}{3}(1-\cos\frac{2\pi x}{L})\sqrt{\frac{\pi f_1 S_F(f_1)}{4\zeta}}$ $S_F(f_i)$ = power spectral density of force per unit length first mode response, i=1. nat. freq. $f_1 = \frac{\omega_1}{2\pi} = \frac{\lambda_1^2}{2\pi L^2}\sqrt{\frac{EI}{m}}$, Hz, $\lambda_1 = 4.73$ Approximate mode shape $\tilde{y}_1(x) \approx 1 - \cos 2\pi x / L$, case 7 of Table 4.2. Maximum bending stress along span is at clamped joints, $\sigma_{x=0,L} = 2\,Ec(\lambda_1 / L)^2 y_{x=L/2}$, Eqs. 4.2, 4.18. $y_{x=L/2}$ from above Eq.

Table 7.5 Response to random broad band excitation, continued

Notation: $S_p(f)$ = power spectral density of pressure with units of (pressure)^2/Hz; E = modulus of elasticity, Chapter 8, h = plate thickness. f = frequency, Hz; f_{ij} = natural frequency of I, j mode, Hz; w_{rms} = root mean square displacement; ζ = damping factor $\ll 1$, dimensionless; ν = Poisson's ratio. Consistent units are in Table 1.2. Ref. [25].

Plate or Shell	RMS Displacement w_{rms}
5. Pressure Excitation Simply Supported (pinned) Rectangular Plate 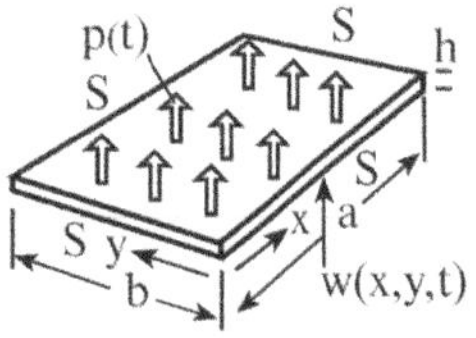	$w_{rms}(x,y) = \frac{192(1-\nu^2)}{\pi^6 Eh^3}\frac{a^4b^4}{(a^2+b^2)^2}\sqrt{\frac{\pi f_1 S_p(f_1)}{4\zeta}}\sin\frac{\pi x}{a}\sin\frac{\pi y}{b}$ First mode. nat. freq., $f_1 = \frac{\pi(a^2+b^2)}{2a^2b^2}\sqrt{\frac{Eh^3}{12\gamma(1-\nu^2)}}$, Hz. Table 4.2 $S_p(f)$=power spectra density of pressure Maximum rms stress at center of plate, x = a/2, = b/2 $\sigma_x = \frac{Eh}{2(1-\nu^2)}\left[\left(\frac{\pi}{a}\right)^2 + \nu\left(\frac{\pi}{b}\right)^2\right] w_{rms}(x = a/2, y = b/2)$ Ref. 13 $\sigma_y = \frac{Eh}{2(1-\nu^2)}\left[\left(\frac{\pi}{b}\right)^2 + \nu\left(\frac{\pi}{a}\right)^2\right] w_{rms}(x = a/2, y = b/2)$
6. Pressure Excitation Clamped Rectangular Plate, First Mode 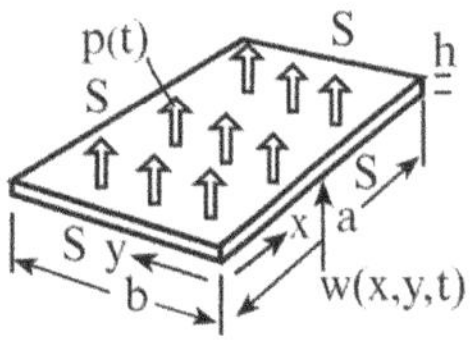	$w_{rms}(x = \frac{a}{2}, y = \frac{b}{2}) = \frac{12(1-\nu^2)}{Eh^3}\frac{0.6903}{500/a^4 + 302/a^2b^2 + 500/b^4}\sqrt{\frac{\pi f_1 S_p(f_1)}{4\zeta}}$ nat. freq. first mode, $f_1 = \frac{1}{2\pi ab}\left(500\frac{a^2}{b^2} + 302 + 500\frac{b^2}{a^2}\right)\sqrt{\frac{Eh^3}{12\gamma(1-\nu^2)}}$, Hz $S_p(f)$ = power spectra density of pressure Maximum rms stress at center of edges $\sigma_x = \frac{Eh}{2(1-\nu^2)}\frac{71.06}{a^2} w_{rms}(x = \frac{a}{2}, y = \frac{b}{2})$ $\sigma_y = \frac{Eh}{2(1-\nu^2)}\frac{71.06}{b^2} w_{rms}(x = \frac{a}{2}, y = \frac{b}{2})$
7. Pressure excitation of General Plate or shell 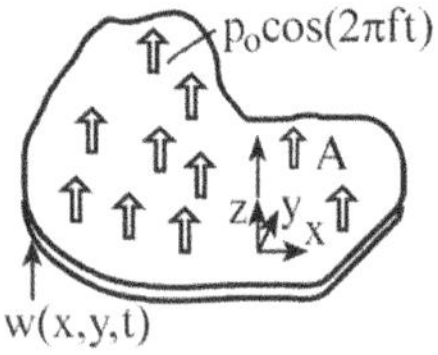	$w_{ij-rms}(x,y) = \frac{\tilde{w}_{ij}(x,y)\int_A \tilde{w}_{ij}(x,y)dxdy}{(2\pi f_{ij})^2\int_A \gamma(x,y)\tilde{w}_{ij}^2(x,y)dxdy}\sqrt{\frac{\pi f_{ij} S_p(f_{ij})}{4\zeta}}$ $S_p(f_{ij})$ = power spectra density of pressure at natural frequencies f_{ij}, Hz. Natural frequencies f_{ij} and mode shapes $\tilde{w}_{ij}(x,y)$ from Tables in Chapter 5, accoring to shape and boundary conditions. Ref. [13]

The power spectra, previously defined by the Fourier series in Equation 7.55, are now equivalently defined by the complex finite Fourier transform [24] as the real products of Fourier Transform and its conjugate:

$$S_x(x) = \frac{1}{2T} G_x(f) G_x^*(f), \quad S_F(x) = \frac{1}{2T} G_F(f) G_F^*(f) \tag{7.60}$$

These definitions are applied to Equation 7.59ac. The relationship between the *response power spectral density* $S_x(f)$ and the force power spectral density $S_F(f)$ is a transfer function in the frequency domain:

$$S_x(f) = \frac{1}{(M\omega_n^2)^2} \frac{1}{(1 - f^2/f_n^2)^2 + (2\zeta f/f_n)^2} S_F(f) \tag{7.61}$$

This equation is very similar to sinusoidal response (Eq. 7.4), except here that the frequency f varies continuously. Note that the term $M\omega_n{}^2$ is equal to k, the stiffness of the oscillator.

The displacement power spectral density, $S_x(f)$ in Equation 7.61, has units of displacement squared per Hertz. The force power spectral density, $S_F(f)$, has units of force squared per Hertz. Parseval's equation gives the overall mean square response displacement [25, p. 78]:

$$\begin{aligned} \overline{x(t)^2} &= \frac{1}{T} \int_0^T x(t)^2 dt = \int_0^\infty S_x(f) df = \frac{1}{(M\omega_n^2)^2} \int_0^\infty \frac{1}{(1 - f^2/f_n^2)^2 + (2\zeta f/f_n)^2} S_F(f) df \\ &\approx \frac{1}{(M\omega_n^2)^2} \frac{\pi f_n S_F(f_n)}{4\zeta}, \quad \text{for } S_F(f) \text{ approximately constant and } \zeta \ll 1 \end{aligned} \tag{7.62}$$

This last formula is called Miles Equation [14]. It is widely used for lightly damped systems dominated by broadband response near resonance. (1) These results are applied to base displacement excitation (Eq. 7.11a) by replacing the force power spectral density $S_F(f)$ with mass squared times the acceleration power spectral density $M^2 S_{\ddot{w}}(f)$, where the power spectral density of base acceleration has units of acceleration squared (length/second^2)^2 per hertz.

Comparing Eq 7.62 with Eq 7.4 we see that random rms response can be obtained from sinusoidol responses in Table 7.2 by replacing the sinusoidol force and dynamic amplication terms with the Miles equation term $[\pi f_n S_F(f_n)/(4\zeta)]^{1/2}$.

7.5 Approximate Response Solution

Sizing a structure for dynamics and vibration loads during design is difficult. Neither the structure nor the loads may be fully defined. Time for a multimode numerical analysis may be available. Yet, something must be done! Approximate vibration analysis methods are often used. Two of these are equivalent static loads and load scaling of mode shapes. The first requires only a static analysis, while second requires only a modal analysis. These approximate methods are given in Table 7.6.

Table 7.6 Equivalent static load and approximate response

Notation: e = radius of rotating unbalance mass m; F_o = amplitude of force; f = excitation frequency, Hz; f_n = natural frequency, Hz, ith mode; k = spring stiffness, force/displacement, Table 3.2; M = mass of body; m = rotating mass; t = time; x = displacement; $W_o \cos \omega t$ = base displacement; y = absolute displacement; ζ = damping factor, dimensionless, $\zeta <= 1$; φ = phase of displacement relative to forcing $0 \le \varphi \le \pi$; $\omega = 2\pi f$ = circular frequency, radians/s. Overdot $(\cdot)$ denotes derivative with respect to time; subscript i = natural frequency. Consistent units are in Table 1.2.

Forced Spring-Mass System	Equation of Motion and Exact Response x(t)	Equivalent Static Load, F_{es}
1. Spring-Mass Damper With Oscillating Force 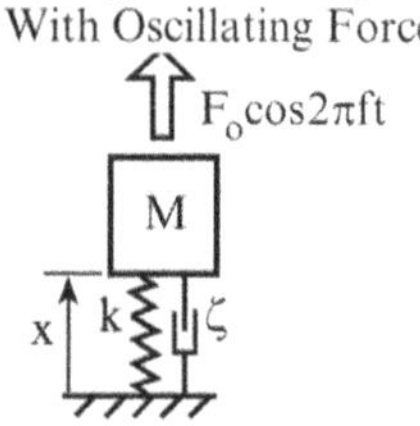 Case 1 of Table 7.1	$M\ddot{x} + 2M\zeta\omega_n\dot{x} + kx = F_o \cos 2\pi f t$ $x(t) = \frac{F_o}{k} \frac{\cos(2\pi f t - \varphi)}{\sqrt{\left(1 - \frac{f^2}{f_n^2}\right)^2 + \left(2\zeta \frac{f}{f_n}\right)^2}}$ $\tan\varphi = \frac{2\zeta f / f_n}{1 - (f/f_n)^2}, \ 0 \le \varphi \le \pi$ $f_n = \frac{\omega_n}{2\pi} = \frac{1}{2\pi}\sqrt{\frac{k}{M}}$ Hz, $k = M\omega_n^2$	$F_{es} = \frac{1}{2\zeta} F_o$
2. Spring Mass Damper Rotating Imbalance m 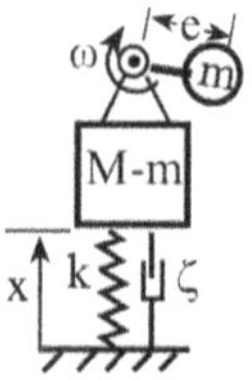 M = total mass non-rotating mass + unbalance.	$x(t) = \frac{m f^2}{M f_n^2} \frac{e \cos(2\pi f t - \varphi)}{\sqrt{\left(1 - \frac{f^2}{f_n^2}\right)^2 + \left(2\zeta \frac{f}{f_n}\right)^2}}$ $\tan\varphi = \frac{2\zeta f / f_n}{1 - (f/f_n)^2}, \ 0 \le \varphi \le \pi$ $f_n = \omega_n/(2\pi) = (1/2\pi)\sqrt{k/M}$ Case 2 of Table 7.1	$F_{es} = \frac{e}{2\zeta} m (2\pi f_n)^2$, for $\zeta << 1$ equivalent static displacement $X_{es} = \frac{1}{2\zeta} e$, for $\zeta << 1$
3. Spring Supported Mass with Base Displacement 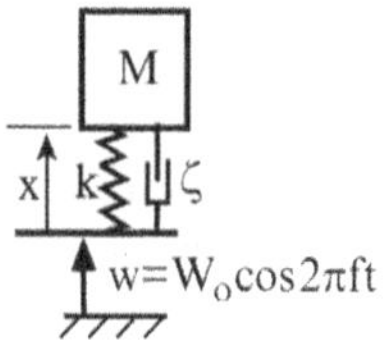 Case 2 of Table 7. $\ddot{W}_o = (2\pi f)^2 W_o$ = amp accl.	$M\ddot{x} + 2M\zeta\omega_n\dot{x} + kx$ Ref. 1 $= -M d^2 w / dt^2 = -M\ddot{W}_o \cos 2\pi f t$ $x(t) = \frac{\ddot{W}_o}{\omega_n^2} \frac{\cos(2\pi f t - \varphi)}{\sqrt{\left(1 - \frac{f^2}{f_n^2}\right)^2 + \left(2\zeta \frac{f}{f_n}\right)^2}}$ $\tan\varphi = \frac{2\zeta f / f_n}{1 - (f/f_n)^2}, \ 0 \le \varphi \le \pi$ $f_n = \omega_n / 2\pi = (1/2\pi)\sqrt{k/M}$, Hz	$F_{es} = \frac{m\ddot{W}_o}{2\zeta}$, for $\zeta << 1$ equivalent static acceleration of mass $\ddot{X}_{es} = \frac{1}{2\zeta} \ddot{W}_o$, for $\zeta << 1$
4. Random External Force On Spring-Mass 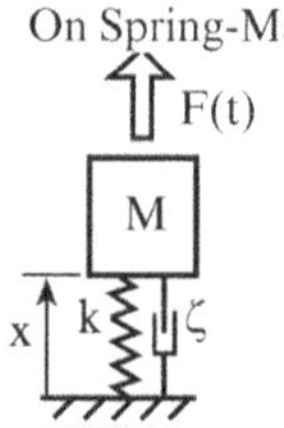	$x_{rms} = \frac{1}{k}\sqrt{\frac{\pi f_n S_F(f_n)}{4\zeta}}$ natural frequency, $f_n = \frac{1}{2\pi}\sqrt{\frac{k}{M}}$, Hz	$F_{es-rms} = \sqrt{\frac{\pi f_n S_F(f_n)}{4\zeta}}$ $S_F(f)$ = power spectral density of force F(t) at natural frequency f_n. Case 1 of Table 7.5

Table 7.6 Equivalent static load and approximate response of beams, continued

Notation: E = modulus of elasticity (Chapter 8); F_o = force amplitude; F = force per unit length; f = frequency of excitation, Hz; f_i = natural frequency, Hz, ith mode; i = integer mode index; I = area moment of inertia about neutral axis, Table 1.5; P, C, F, G = pinned, clamped, free, or guided boundary condition, Tables 4.2, 4.4; L = length of beam; y = transverse displacement; m = mass of beam mass per unit length; t = time; W_o = amplitude of displacement of base; x = axial coordinate; y = transverse displacement; $\tilde{y}_i(x)$ = mode shape, Table 4.2; ζ_i = damping factor of ith mode; φ_i = phase of displacement relative to forcing, $\tan\varphi_i = 2\zeta_i(f/f_i)/(1-(f/f_i)^2)$, 0 to π radian; rms = root mean square. Table 1.2 has consistent units.

Forced Beam	Approx Displace't, $y_i(x,t)$	Equivalent Static Load F_{es}
5. Uniform Beam with Oscillating Force F_o at point x_o 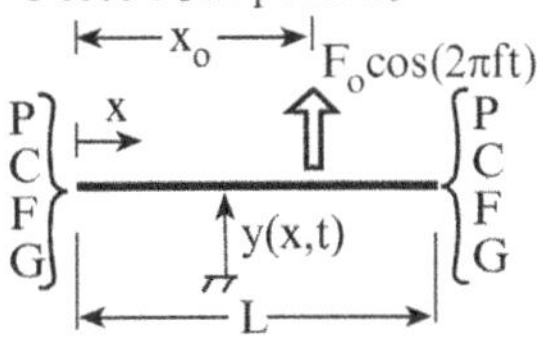	$y_i(x,t) = \frac{F_o}{m(2\pi f_i)^2}\frac{\tilde{y}_i(x_o)}{\int_0^L \tilde{y}_i^2(x)dx} \times \frac{\tilde{y}_i(x)\cos(2\pi ft-\varphi_i)}{\sqrt{\left(1-\frac{f^2}{f_n^2}\right)^2+\left(2\zeta\frac{f}{f_n}\right)^2}}$ nat. freq. $f_i = \frac{\lambda_i^2}{2\pi L^2}\sqrt{\frac{EI}{m}}$, Hz. λ_i, $\tilde{y}_i(x)$ in Tables 4.2 and 4.4	$F_{es} = \frac{1}{2\zeta}F_o$, applied at x_o $\int_0^L \tilde{y}_i^2(x)dx$ from Eq. 4.17 Solution in middle column is exact solution from case 4 of Table 7.2
6. Uniform Beam with Distributed Oscillating Force 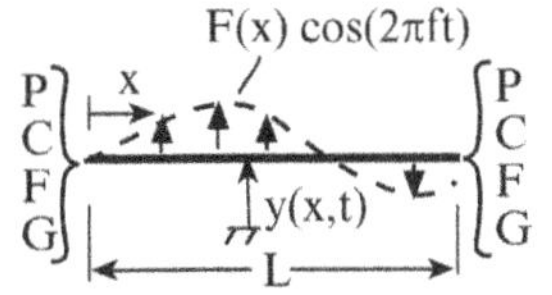 Exact solution in case 9 Table 7.2	$y_i(x,t) = \frac{F_{max}}{m(2\pi f_i)^2\tilde{y}_{imax}} \times \frac{\tilde{y}_i(x)\cos(2\pi ft-\varphi_i)}{\sqrt{\left(1-\frac{f^2}{f_n^2}\right)^2+\left(2\zeta\frac{f}{f_n}\right)^2}}$ nat. freq. $f_i = \frac{\lambda_i^2}{2\pi L^2}\sqrt{\frac{EI}{m}}$, Hz. λ_i, $\tilde{y}_i(x)$ in Tables 4.2 and 4.4	$F_{es} = \frac{1}{2\zeta}F(x)$ = force per unit length Approximate displacement, left, assumes force is proportional to ith mode shape. F_{max} = max F(x) $F(x) = F_{max}(\tilde{y}_i(x)/\tilde{y}_i(x)_{max})$ $\tilde{y}_{imax}$ = max mode shape, Eq. 4.18
7. Uniform Beam with Oscillating Base Acceleration 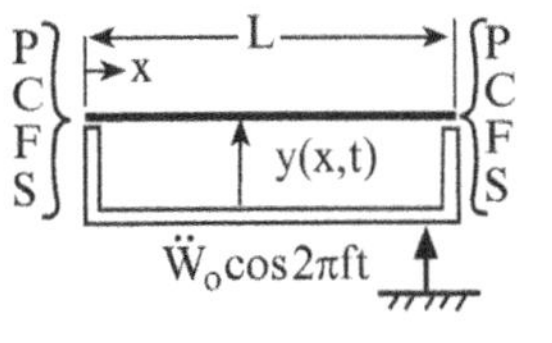 Exact solution case 14 of Table 7.2	$y_i(x,t) = \frac{\ddot{W}_o}{(2\pi f_i)^2\tilde{y}_{imax}(x)} \times \frac{\tilde{y}_i(x)\cos(2\pi ft-\varphi_i)}{\sqrt{\left(1-\frac{f^2}{f_n^2}\right)^2+\left(2\zeta\frac{f}{f_n}\right)^2}}$ nat. freq. $f_i = \frac{\lambda_i^2}{2\pi L^2}\sqrt{\frac{EI}{m}}$, Hz. λ_i, $\tilde{y}_i(x)$ in Tables 4.2 and 4.4	$F_{es} = m\ddot{W}_o\frac{1}{2\zeta}$, force per length, or, $\ddot{W}_{es} = \ddot{W}_o\frac{1}{2\zeta}$ = steady accel. $\tilde{y}_{imax}$ = max value mode shape, Eq. 4.18
8. Uniform Beam with Distributed **Random** Force 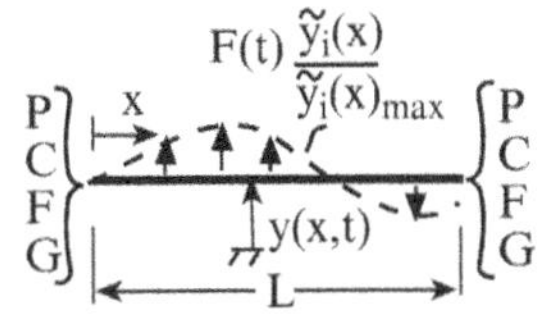 Approximate solution assumes force proportional to ith mode shape.	$y_{i-rms}(x) = \frac{\tilde{y}_i(x)}{m(2\pi f_i)^2\tilde{y}_{imax}} \times \sqrt{\frac{\pi f_n S_F(f_n)}{4\zeta}}$ nat. freq. $f_i = \frac{\lambda_i^2}{2\pi L^2}\sqrt{\frac{EI}{m}}$, Hz. λ_i, $\tilde{y}_i(x)$ in Tables 4.2 and 4.4	$F_{es-rms} = \sqrt{\frac{\pi f_n S_F(f_n)}{4\zeta}}$ = rms force per unit length $S_F(f_n)$ = power spectral density of force per unit length at nat.freq. f_n $\tilde{y}_{imax}$ = max value mode shape, Eq. 4.18

Table 7.6 Equivalent static load and approximate response of plates and shells, continued

Notation: p = pressure; p_{max} = maximum absolute value of pressure over surface of plate or shell; f = frequency of excitation, Hz; f_{ij} = natural frequency, Hz, i-j mode; i, j = integer mode indices; w = out of plane displacement; γ = mass per unit area; t = time; x = axial coordinate; w = transverse displacement; $\tilde{w}_{ij}(x, y)$ = mode shape, Table 5.2; ζ_{ij} = damping factor of ith mode; φ_{ij} = phase of displacement relative to forcing, $\tan \varphi_{ij} = 2\zeta_i (f/f_{ij})/(1-(f/f_{ij})^2)$, 0 to π radian; rms = root mean square random vibration. Table 1.2 has consistent units.

Plate and Shell excited by Oscillating Pressure	Approximate Displacement $w_{ij}(x,y,t)$	Equivalent Static Pressure p_{es}
9. Plate or Shell with Oscillating Pressure 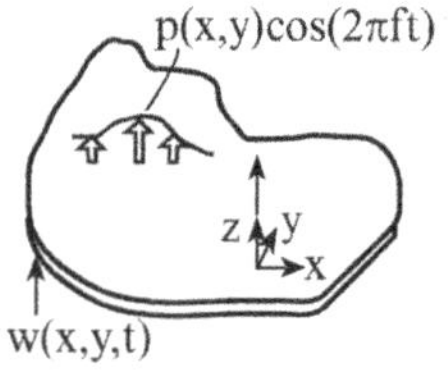	$w_{ij}(x,y,t) = \frac{p_{max}}{(2\pi f_{ij})^2 [\gamma(x,y)\tilde{w}_{ij}(x,y)]_{max}} \times \frac{\tilde{w}_{ij}(x,y)}{\sqrt{\left(1-\frac{f^2}{f_{ij}^2}\right)^2 + \left(2\zeta\frac{f}{f_{ij}}\right)^2}} \cos(2\pi f t - \varphi_{ij})$ $\tilde{w}_{ij}(x,y)_{max}$ = max absolute value of mode shape; mode shapes $\tilde{w}_{ij}(x,y)$ and natural frequencies f_{ij} from Tables in Chapter 5. γ = mass per unit area.	$p_{es} = \frac{1}{2\zeta} p(x,y)$ Approximate displacement assumes pressure over shell is proportional to ij mode shape. $p(r,\theta) = p_{max} \frac{\tilde{w}_{ij}(x,y)}{\tilde{w}_{ij}(x,y)_{max}}$
10. Cylindrical Shell Oscillating Pressure 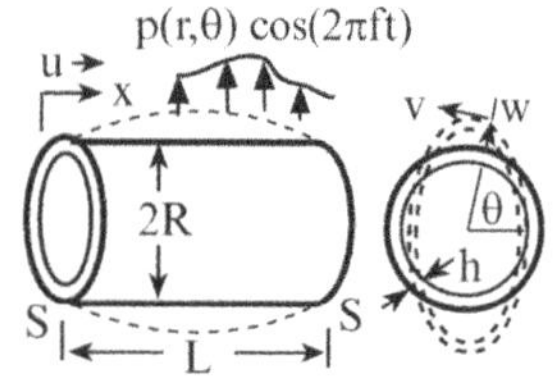	$w_{ij}(r,\theta,t) = \frac{p_{max}}{\gamma \tilde{w}_{ij,max}(2\pi f_{ij})^2} \times \frac{\tilde{w}_{ij}(r,\theta)}{\sqrt{\left(1-\frac{f^2}{f_{ij}^2}\right)^2 + \left(2\zeta\frac{f}{f_{ij}}\right)^2}} \cos(2\pi f t - \varphi_{ij})$ $\tilde{w}_{ij}(r,\theta)_{max}$ = max absolute value of mode shape; mode shapes $\tilde{w}_{ij}(r,\theta)$ and nat. freqs. f_{ij} from Table 5.7. γ = mass per area, f_{ij} =natural freqs. of i,j mode, Hz, Table 5.7. Ref. [12]	$p_{es} = \frac{1}{2\zeta} p(r,\theta)$ Approximate displacement left, assumes pressure over shell is proportional to ij mode shape. $p(r,\theta) = p_{max} \frac{\tilde{w}_{ij}(r,\theta)}{\tilde{w}_{ij}(r,\theta)_{max}}$
11. Random Oscillating Pressure on Plate or shell 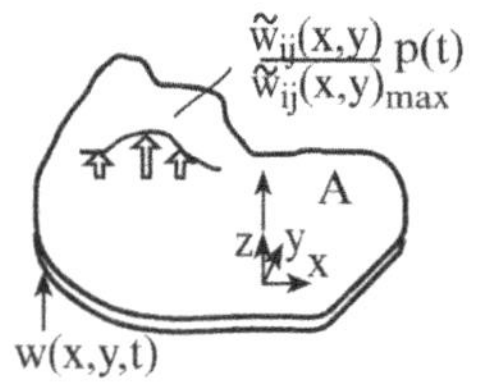	$w_{ij}(x,y)_{ij-rms} = \frac{\tilde{w}_{ij}(x,y)}{\gamma(2\pi f_{ij})^2 \tilde{w}_{ij}(x,y)_{max}} \times \sqrt{\frac{\pi f_{ij} S_p(f_{ij})}{4\zeta}}$ $\tilde{w}_{ij}(x,y)_{max}$ = max absolute value of ij mode shape; mode shapes $\tilde{w}_{ij}(x,y)$ and natural frequencies f_{ij} from Tables in Chapter 5. γ = mass per unit area.	$p_{es-rms} = \sqrt{\frac{\pi f_{ij} S_p(f_{ij})}{4\zeta}}$ $S_p(f_{ij})$ = power spectral density of pressure at natural frequency f_{ij}. Approximate displacement assumes pressure over shell is proportional to ij mode shape. Ref. [12]

7.5.1 Equivalent Static Loads

Equivalent static load is an acceleration (g) or force load that can be applied statically that produces approximately the same deformation, stress, and internal load distribution as a full dynamic analysis. Equivalent static loads envelope resonance dynamic amplification, which is usually the critical case. They are, in concept, similar to the response spectra for transient loads, as described in Section 7.3.

Consider the sinusoidal forced vibration shown in Figure 7.18. The exact dynamic, steady-state, solution is given in case 1 of Table 7.1 and Equation 7.4. The static response to static load F_{es} is simply F_{es}/k, where k is the stiffness (this result can be obtained by setting f to 0 in Eq. 7.4). The equivalent static load is the static load that produces the same maximum dynamic response as the dynamic load.

$$X_{\substack{\max\\ \text{dynamic}}} = \frac{F_{es}}{k} = \frac{F_o}{k}\left.\frac{1}{\sqrt{\left(1-f^2/f_n^2\right)^2+(2\zeta f/f_n)^2}}\right|_{\substack{\max\\ \text{over } f}} = \frac{1}{2\zeta}\frac{F_o}{k},\quad \zeta \ll 1$$

$$F_{\substack{\text{equvalent}\\ \text{static}}} = \left.\frac{F_o}{2\zeta}\right|_{\zeta\ll 1} = \begin{cases} 50F_o, & \zeta = 0.02 \\ 10F_o, & \zeta = 0.0.05 \end{cases} \tag{7.64}$$

In order to include the possibility of resonance, which is the worst case, the dynamic load is multiplied by the maximum dynamic amplification at resonance, $1/(2\zeta)$, which is typically a factor of 10 or more.

7.5.2 Scaling Mode Shapes to Load

Often mode shapes are available, or the static deflection can be used to estimate the first-mode natural frequency and mode shape (Appendix A). Mode shapes can be used to express a load in a convergent series of modes over the domain of the system (Ref. [5, 6], p. 143). If one conservatively assumes that load distribution over the structure in each mode is (1) proportional to the mass weighted mode shape, $\overline{F}(P) \sim M(P)\widetilde{w}_i(P)$ [13], which is useful for distributed pressure and inertia loads, and (2) the maximum distributed force is applied at the point of the maximum mass weighted mode shape then the spatial integrals are resolved.

$$\overline{F}(P) \approx \frac{M(P)F(P)}{|M(P)\widetilde{w}_i(P)|_{\max}}\widetilde{w}_i(P))_{\max}$$

$$W_{\text{i-static}} = \frac{\int_D \widetilde{w}_i(P)\overline{F}(P)dD}{\omega_i^2\int_D M(P)\widetilde{w}_i^2(P)dD} = \frac{F(P)_{\max}}{|M(P)\omega_i^2\widetilde{w}_i(P)|_{\max}} \tag{7.65}$$

Substituting this result into Equation 7.31 gives the amplitude of the dynamic modal coordinate in the ith mode.

$$|W_i(t)| = \frac{F(P)_{max}}{|M(P)\omega_i^2\tilde{w}_i(P)|_{max}}\left[\left(1-\frac{f^2}{f_i^2}\right)^2+\left(\frac{2\zeta f}{f_i}\right)^2\right]^{-1/2} \quad \text{forced}$$

$$= \frac{\ddot{W}_o}{|\omega_i^2\tilde{w}_i(P)|_{max}}\left[\left(1-\frac{f^2}{f_i^2}\right)^2+\left(\frac{2\zeta f}{f_i}\right)^2\right]^{-1/2} \quad \text{base excitation} \qquad (7.66)$$

Thus, if the mode shapes and natural frequencies are known, and distributed loading is proportional to the mode shape, then the displacement of the continuous structure in a mode can be determined immediately by scaling the mode shapes by the ratio of the maximum load to mass times the circular natural frequency square times the maximum values of the mode shape. Equation 7.66 is applied in the second column in Table 7.6. Physical displacement is the modal coordinate displacement times the mode shape.

Example 7.5 Use modal scaling (case 7 of Table 6) to find the amplitude of resonant vibration of the pined–pined tube in Example 7.2 with base excitation from Mil-Std-810C shown in Figure 7.19. Calculate the first-mode resonance.

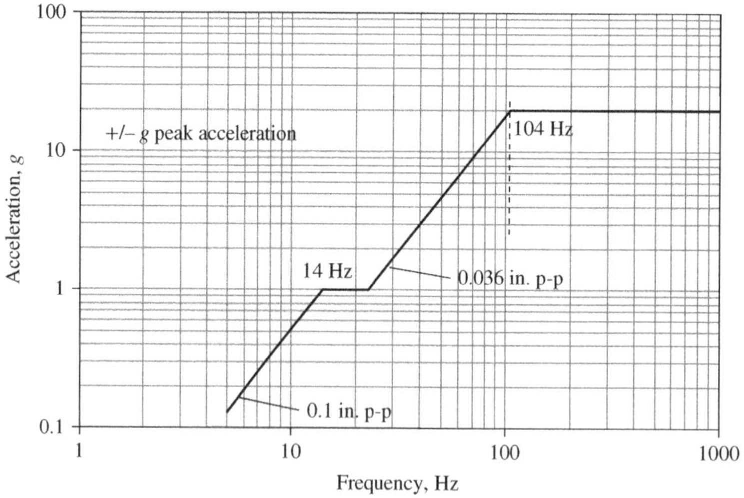

Figure 7.19 Base excitation section in g as a function of frequency from MIL-STD-810C

Solution: The mode shape is sinusoidal, and its maximum value is unity, that is, $\tilde{w}_{1max} = 1$. The acceleration load in Figure 7.19 is 20 g at the first-mode natural frequency of 112 Hz. The amplitude of base acceleration is 196.1 m/s^2, and the response time history is also sinusoidal $g(t)$, which includes the resonant amplification factor. The mode scale is used.

The predicted modal amplitude is given by Eq. 7.66, or case 7 of Table 7.6.

$$w_{1\max} = \frac{\ddot{W}_{bo}}{|\omega_i^2 \widetilde{w}_i(P)|_{\max}} \frac{\widetilde{w}_i(P)}{2\zeta}$$

$$= \frac{196\ \mathrm{m/s^2}}{(2\pi \times 112/s)^2 \times 1} \frac{1}{2(0.02)} = 0.01\ \mathrm{m}$$

This approximate solution is 20% less than the exact solution of 0.0126 m in Example 7.2 because the approximate solution models the g-load as varying over the tube.

The equivalent static load for the 112 Hz fir mode is 20 g (from Figure 7.19) times the magnification factor at resonance, $1/(2\zeta) = 1/[2(0.02)] = 25$ is $25 \times 20\ g = 500\ g$. This is a fatigue load, so both the plus and minus values will occur.

References

[1] Thomson, W. T., Theory of Vibration with Applications, 3rd ed., Prentice Hall, 1988.

[2] Den Hartog, J. P., Mechanical Vibrations, 1928, Dover, New York, 1985, Chapter III.

[3] Biggs, J. M., Introduction to Structural Dynamics, McGraw-Hill, New York, 1964.

[4] Press, W.H.; B.P. Flannery; S. A. Teukolsky, and W. T. Vetterling "Gauss-Jordan elimination" and "Gaussian elimination with backsubstitution." in Numerical Recipes in FORTRAN: The Art of Scientific Computing, 2nd ed. Cambridge University Press, Cambridge, England, pp. 27–32 and 33–34, 1992.

[5] Courant, R., and D. Hilbert, Methods of Mathematical Physics, vol. 1, Interscience Publishers, New York, pp. 359–361, 1953.

[6] Meirovitch, L. Analytical Methods in Vibrations, Macmillan, p. 383,398. 432, 1967.

[7] Caughey, T. K., Classical Normal Modes in Damped Linear Dynamic Systems, Journal of Applied Mechanics, vol. 27, pp. 269–271, 1960.

[8] Clough, R. W., and J. Penzien, Dynamics of Structures, McGraw-Hill, New York, 1993, p. 559.

[9] Spyrakos, C., Finite Element Modeling in Engineering Practice, Algor Publishing, Pittsburgh, 1995.

[10] American Society of Civil Engineers, 2006, Minimum Design Loads for Buildings and Other Structures, ASCE, 7–05, para 12.9.1.

[11] Weaver, W., S. P. Timoshenko, and D. H. Young, Vibration Problems in Engineering, 5th ed., Wiley, New York, 1990.

[12] Soedel, W., 2004, Vibrations of Shells and Plates, 3rd ed., Marcel Dekker, New York.

[13] Blevins, R.D., An Approximate Method in Sonic Analysis of Plates and Shells, Journal of Sound and Vibration, vol. 129, pp 51–71, 1989.

[14] Miles, J. W., On Structural Fatigue under Random Loading, Journal of Aeronautical Sciences, vol 21, pp. 753–762, 1954.

[15] Ayre, R. S., Transient Response to Step and Pulse Functions, Chapter 8 in, Shock and Vibration Handbook, C. M. Harris (ed), 4th ed., McGraw-Hill, 1996.

[16] ASME Boiler and Pressure Vessel Code, 2007, Section 3, Division 1, Appendix N.

[17] Snowdon, J. C., Vibration and Shock in Damped Mechanical Systems, Wiley, New York, 1968, Chapter 4.

[18] Warburton, G. B., Optimum Absorber Parameters for Various Combinations of Response and Excitation Parameters, Earthquake Engineering and Structural Dynamics, vol. 10, pp. 381–401, 1982.

[19] Korenev, B. G. and L. M. Reznikov, Dynamic Vibration Absorbers, John Wiley, New York, pp. 7–14, 1996.

[20] Rivin, E. I., Passive Vibration Isolation, ASME, New York, 2003

[21] Paz, M., Structural Dynamics Theory and Computation, 3rd ed., Van Nostrand Reinhold, New York, 1991, p. 52.

[22] Moreno, C. P. and P. Thomson, Design of an Optimal Tuned Mass Damper for a System with Parametric Uncertainty, Annals of Operation Research, paper 10.1007/s10479-010-0726-x, 2010.

[23] Blevins, R. D., Probability Density of Finite Fourier Series with Random Phases, Journal of Sound and Vibration, vol 208, pp. 617–652, 1997.

[24] Bendat, J. S., and A. G. Piersol, Engineering Applications of Correlation and Spectral Analysis, Wiley-Interscience, New York, 1980.

[25] Crandall, S.H. and W.D. Mark, Random Vibration in Mechanical Systems, Academic Press, New York, 1963.

[26] Blevins, R.D., Flow-Induced Vibration, 2nd ed., Krieger, 1990.

8

Properties of Solids, Liquids, and Gases

This chapter presents material properties of solids, liquids, and gases that support the formulas given in Chapters 1 through 7. The data is given in both SI (metric) and US customary (ft-lb) units.

8.1 Solids

Material data for density, modulus of elasticity, and Poisson's ratio of metals is provided in Tables 8.1–8.3, wood in Table 8.4, and plastic and composites in Table 8.5. The properties are one atmosphere pressure and room temperature, 20 °C (68 °F) unless otherwise noted, and they are based on Refs [1–6]. The density of structural solids is largely independent of temperature and pressure, but the modulus of elasticity (stiffness) declines with increasing temperature. Metals and polymers creep as their temperature approaches their melting or curing temperature.

Poisson's ratio, being more difficult to measure and of less importance than modulus of elasticity, is not known with better than two-digit accuracy for most materials. The theoretical maximum value for Poisson's ratio is $\nu = 0.5$. The theoretical minimum value is -1; however, materials with negative Poisson's ratio are unknown [7]. A reasonable estimate for Poisson's ratio of most engineering metals such as steel or aluminum is $\nu = 0.3$. Rubber and lead have higher values, $\nu \approx 0.45$, and beryllium has a lower value, $\nu \approx 0.01$. Graphite–epoxy composites have $\nu \approx 0.3$.

Densities of nickel, iron and chromium are within 10% of 8 g per cubic centimeter (cc).

Iron	7.86 g/cc (0.284 lb/in.^{3})
Nickel	8.91 g/cc (0.322 lb/in.^{3})
Chromium	7.20 g/cc (0.260 lb/in.^{3})

Thus, densities of nickel, iron, and chrome alloys are within 10% of 8 g/cc (0.29 lb/in.^{3}).

Formulas for Dynamics, Acoustics and Vibration, First Edition. Robert D. Blevins.
© 2016 John Wiley & Sons, Ltd. Published 2016 by John Wiley & Sons, Ltd.

Table 8.1 Modulus of elasticity and Poisson's ratio of ferrous and nickel alloys

Temperature (°C)	304,316,321,347 Stainless steel		Ni–Fe–Cr Alloy 800		Ni–Cr–Fe–Cb Alloy 718		Ni–Cr–Fe Alloy 600		A286	
	E (GPa)[a]	*ν* (Dimensionless)	*E* (GPa)	*ν* (Dimensionless)	*E* (GPa)	*ν* (Dimensionless)	*E* (GPa)	*ν* (Dimensionless)	*E* (GPa)	*ν* (Dimensionless)
−100	204		207	0.336						
25	194	0.265	196	0.339	200	0.293	218	0.293	201	0.206
100	190	0.272	193	0.342	196	0.286	213	0.298	196	0.309
150	187	0.276	190[9]	0.347	193	0.283	210	0.303	192	0.312
200	184	0.28	186	0.353	190	0.279	207	0.398	188	0.314
250	181	0.284	183	0.355	188	0.276	205	0.313	184	0.316
300	177	0.288	179	0.357	185	0.273	202	0.318	181	0.318
350	173	0.292	175	0.360	182	0.272	199	0.323	177	0.321
400	168	0.295	169	0.362	179	0.271	1195	0.329	173	0.323
450	164	0.299	164	0.365	176	0.271	191	0.334	169	0.325
500	159	0.302	159	0.369	173	0.272	187	0.340	165	0.327
550	154	0.306	156	0.368	170	0.274	183	0.346	161	0.339
600	149	0.309	152	0.373	167	0.277	179	0.352	157	0.332
650	144	0.313	146	0.377	163		175	0.358	175	0.358
700	139	0.317	140	0.381	159		170	0.365		
750	133	0.321	133	0.388	155		165	0.372		
800	128	0.324	127	0.396			161	0.379		

Table 8.1 Modulus of elasticity and Poisson's ratio of ferrous and nickel alloys, continued

Temperature	304,316,321,347 Stainless steel		Ni–Fe–Cr Alloy 800		Ni–Cr–Fe–Cb Alloy 718		Ni–Cr–Fe Alloy 600		A286	
°F	E (lb/in.^{2})	ν (Dimensionless)	E (lb/in.^{2})	ν (Dimensionless)	E (lb/in.^{2})	ν (Dimensionless)	E (lb/in.^{2})	ν (Dimensionless)	E (lb/in.^{2})	ν (Dimensionless)
−200	29.7×10^6	—	30.0×10^6	0.335						
50	28.1×10^6	0.265	28.4×10^6	0.339	29.0×10^6	0.293	31.8×10^6	0.290	29.2×10^6	0.306
100	28.0×10^6	0.266	30.4×10^6	0.339	28.8×10^6	0.292	31.4×10^6	0.292	29.0×10^6	0.307
200	27.7×10^6	0.271	27.7×10^6	0.341	28.4×10^6	0.287	30.9×10^6	0.298	28.5×10^6	0.309
300		0.276	27.1×10^6	0.347	28.0×10^6	0.283	30.4×10^6	0.303	27.9×10^6	0.312
400	26.6×10^6	0.281	26.6×10^6	0.353	27.6×10^6	0.279	30.0×10^6	0.308	27.3×10^6	0.314
500	26.1×10^6	0.285	26.1×10^6	0.355	27.1×10^6	0.275	29.6×10^6	0.314	26.6×10^6	0.307
600	25.4×10^6	0.289	25.4×10^6	0.357	26.7×10^6	0.273	29.2×10^6	0.320	26.0×10^6	0.319
700	24.8×10^6	0.293	24.8×10^6	0.361	26.3×10^6	0.271	28.6×10^6	0.325	25.4×10^6	0.321
800	24.1×10^6	0.297	24.1×10^6	0.363	25.8×10^6	0.271	28.0×10^6	0.331	24.8×10^6	0.324
900	23.3×10^6	0.301	23.3×10^6	0.367	25.3×10^6	0.271	27.4×10^6	0.338	24.1×10^6	0.326
1000	22.7×10^6	0.305	22.5×10^6	0.367	24.8×10^6	0.273	26.7×10^6	0.344	23.5×10^6	0.329
1100	22.0×10^6	0.309	21.7×10^6	0.372	24.2×10^6	0.277	26.0×10^6	0.351	22.8×10^6	0.331
1200	21.3×10^6	0.313	20.9×10^6	0.377	23.6×10^6	0.282	25.3×10^6	0.358	22.2×10^6	0.333
1300	20.7×10^6	0.317	20.1×10^6	0.381	23.0×10^6	—	24.6×10^6	0.365		

Table 8.1 Modulus of elasticity and Poisson's ratio of ferrous and nickel alloys, continued

Temperature (°C)	2-1/4 Cr–1Mo		Medium low carbon steel (C ≤ 0.3)	Higher carbon steel (C > 0.3)	C–Mo low Cr (Cr ≤ 3)	Intermediate Cr (5 Cr–9 Cr[b])	Straight chromium[c]	15-5 Ph[d]
	E (GPa)[a]	*ν* (Dimensionless)	*E* (GPa)	*E* (GPa)	*E* (GPa)	*E* (GPa)	*E* (GPa)	*E* (GPa)
−200			207	214	214	213	212	
−100			202	210	210	194	208	
25	218	0.291	191	206	206	189	201	196
100	213	0.298	190	203	203	187	197	195
150	210	0.303	188	200	200	185	195	193
200	207	0.308	185	195	197	182	191	191
250	205	0.313	182	190	194	179	187	189
300	202	0.318	176	185	190	175	181	189
350	199	0.323	172	179	185	173	174	178
400	195	0.329	166	167	166	169	165	175
450	191	0.334	147	158	147	165	154	167
500	187	0.340		142	121	169	141	
550	183	0.346		124	103	157	124	
600	179	0.352		102	89	159	108	
650	175	0.358						

Table 8.1 Modulus of elasticity and Poisson's ratio of ferrous and nickel alloys, continued

Temperature	2-1/4 Cr–1Mo		Medium low carbon steel ($C \leq 0.3$)	Higher carbon steel ($C > 0.3$)	C–Mo low Cr ($Cr \leq 3$)	Intermediate Cr (5 Cr–9 Cr[b])	Straight chromium[c]	15-5 Ph[d]
°F	E (lb/in.^{2})	ν (Dimensionless)	E (lb/in.^{2})	E (lb/in.^{2})	E (lb/in.^{2})	E (lb/in.^{2})	E (lb/in.^{2})	E (lb/in.^{2})
−325			30.0×10^6	31.0×10^6	31.0×10^6	29.4×10^6	30.8×10^6	
−100			29.0×10^6	31.0×10^6	30.4×10^6	28.1×10^6	29.8×10^6	
70	31.8×10^6	0.290	27.9×10^6	30.4×10^6	29.9×10^6	27.4×10^6	29.2×10^6	28.5×10^6
200	30.9×10^6	0.298	27.7×10^6	29.5×10^6	29.5×10^6	27.1×10^6	28.7×10^6	28.2×10^6
300	30.4×10^6	0.303	27.4×10^6	29.0×10^6	29.0×10^6	26.8×10^6	28.3×10^6	27.9×10^6
400	30.0×10^6	0.308	27.0×10^6	28.3×10^6	28.6×10^6	26.6×10^6	27.7×10^6	27.6×10^6
500	29.6×10^6	0.314	26.4×10^6	27.4×10^6	28.0×10^6	26.0×10^6	27.0×10^6	27.2×10^6
600	29.2×10^6	0.320	25.7×10^6	26.7×10^6	27.4×10^6	25.4×10^6	26.0×10^6	26.8×10^6
700	28.6×10^6	0.325	24.8×10^6	25.4×10^6	24.8×10^6	24.9×10^6	24.8×10^6	25.8×10^6
800	28.0×10^6	0.321	23.4×10^6	23.8×10^6	23.4×10^6	24.2×10^6	23.1×10^6	24.9×10^6
900	27.4×10^6	0.338	18.5×10^6	21.5×10^6	18.5×10^6	23.5×10^6	21.1×10^6	23.6×10^6
1000	26.7×10^6	0.344		18.8×10^6	15.4×10^6	22.8×10^6	18.6×10^6	
1100	26.0×10^6	0.351		15.0×10^6	13.0×10^6	21.9×10^6	15.6×10^6	

[a]GPa = 10^9 Pa.
[b]Poisson's ratio $\nu \approx 0.29$ and density $\rho \approx 7.97$ g/cm^3 (0.288 lb/in.^{3}) for steels.
[c]12 Cr, 17 Cr, and 27 Cr.
[d]Density $\rho = 7.83$ g/cm^3 (0.283 lb/in.^{3}) and $\nu = 0.27$ for 15-5 Ph.

Table 8.2 Modulus of elasticity and Poisson's ratio of nonferrous alloys

	Aluminum						
	3003, 3004, 6063	5052, 5154, 5454, 5456	5083 5086	6061	2014, 2024, 2124	2219	7075 7050
Temperature (°C)	*E* (GPa)[a]	*E* (GPa)	*E* (GPa)	*E* (GPa)	*E* (GPa)	*E* (GPa)	*E* (GPa)
−200	76.5	77.9	78.6	76.5	81.4	79.7	79.5
−100	72.7	74.1	74.8	72.7	77.0	73.7	73.8
25	68.9	70.3	71.0	68.9	73.0	72.4	71.0
100	65.8	67.0	68.6	65.8	71.5	70.3	67.5
150	62.7	62.1	65.5	62.7	69.4	68.9	64.3
200	57.7	55.7	60.4	57.7	65.8	69.0	58.2
250					59.2	66.2	49.0
300					51.2	55.2	36.9
350						46.2	
400							
°F	lb/in.^{2}	lb/in.^{2}	lb/in.^{2}	lb/in.^{2}	lb/in.^{2}	lb/in.^{2}	lb/in.^{2}
−325	11.1×10^6	11.4×10^6	11.4×10^6	11.1×10^6	11.8×10^6	11.9×10^6	11.6×10^6
−100	10.4×10^6	10.7×10^6	10.7×10^6	10.4×10^6	11.1×10^6	11.1×10^6	10.8×10^6
70	9.6×10^6	10.2×10^6	10.3×10^6	10.0×10^6	10.6×10^6	10.5×10^6	10.3×10^6
200	6.3×10^6	9.9×10^6	10.0×10^6	9.6×10^6	10.4×10^6	10.3×10^6	9.9×10^6
300	5.5×10^6	9.7×10^6	9.5×10^6	9.1×10^6	10.1×10^6	10.0×10^6	9.4×10^6
400	5.1×10^6	9.1×10^6	8.7×10^6	8.3×10^6	9.54×10^6	9.4×10^6	8.3×10^6
500	5.2×10^6	8.1×10^6			8.48×10^6	8.8×10^6	7.0×10^6
600	3.8×10^6	7.1×10^6			7.42×10^6	7.8×10^6	5.2×10^6
700						6.6×10^6	
800							
Density							
g/cc	1.77	2.66		2.71	2.80	2.85	2.80
lb/in.^{3}	0.0639	0.096		0.098	0.100	0.103	0.101
Poisson ν	0.35	0.33	0.33	0.33	0.33	0.33	0.33

	Magnesium	Copper	Titanium		
	AZ31B	Pure	Unalloyed[b]	Ti6Al4V	Ti6Al2Sn4Zr2Mo
Temperature (°C)	*E* (GPa)	*E* (GPa)	*E* (GPa)	*E* (GPa)	*E* (GPa)
−200		117		119	
−100		115		114	
25	44.8	110	107	110	114
100	43.0	107	103	105	108
150	38.1	106	99.3	101	107
200	35.85	104	95.5	98.2	105
250	30.0	102	91.8	94.3	102
300	24.6	98.9	87.5	91.0	100
350		95.8	83.2	87.2	96.7
400			79.3	83.8	93.3
450		—		79.4	87.6

Table 8.2 Modulus of elasticity and Poisson's ratio of nonferrous alloys, continued

	Magnesium	Copper	Titanium		
	AZ31B	Pure	Unalloyed[b]	Ti6Al4V	Ti6Al2Sn4Zr2Mo
°F	lb/in.^{2}	lb/in.^{2}	lb/in.^{2}	lb/in.^{2}	lb/in.^{2}
−325		17.0×10^6		17.3×10^6	
−100		16.5×10^6		16.5×10^6	
70	6.5×10^6	16.0×10^6	15.5×10^6	16.0×10^6	16.5×10^6
200	$\times 10^6$	15.6×10^6	15.0×10^6	15.7×10^6	15.8×10^6
300	9.1×10^6	15.4×10^6	14.4×10^6	15.3×10^6	15.5×10^6
400	8.3×10^6	15.1×10^6	13.8×10^6	14.7×10^6	15.0×10^6
500		14.7×10^6	13.2×10^6	14.1×10^6	14.7×10^6
600		14.2×10^6	12.5×10^6	13.6×10^6	14.2×10^6
700		13.7×10^6	11.8×10^6	12.6×10^6	13.7×10^6
800			11.2×10^6	11.9×10^6	13.4×10^6
900				10.9×10^6	12.2×10^6
Density					
g/cc	2.74	8.97	4.51	4.42	4.54
lb/in.^{3}	0.099	0.324	0.163	0.160	0.164
Poisson ν	0.33	0.33	0.3–0.34	0.31	0.32

[a]GPa = 10^9 Pa.
[b]Commercially pure.

The natural frequency of continuous structures is proportional to the square root of the stiffness to density (Chapter 3). This implies that structural natural frequencies are proportional to modulus of elasticity divided by density. The ratio of modulus of elasticity to density is nearly constant for steel, titanium, and aluminum and many common engineering structural metals.

Material	Modulus (Pa)	Density (g/cc)	10^9 modulus/density
Steel	200×10^9	7.87	24
Aluminum	72×10^9	2.80	26
Titanium	107×10^9	4.54	23.5
Brass	110×10^9	8.4	13

Thus, the natural frequency of an aluminum beam is nearly the same as that of a steel beam with the same dimensions. The steel beam is three times as stiff as the aluminum beam but it is also three times heavier.

The modulus of elasticity and density of wood vary with moisture content. The values given in Table 8.5 apply for 12% moisture content, which is weight relative to the weight

Table 8.3 Modulus of elasticity and density of various metals

Metal	Symbol	Density, ρ		Modulus of elasticity, E	
		g/cc	lb/in.^{3}	GPa	lb/in.^{2}
Aluminum	Al	2.70	0.0975	69	10.0×10^6
Antimony	Sb	6.62	0.239	78	11.3×10^6
Barium	Ba	3.60	0.13	12	1.8×10^6
Beryllium	Be	1.82	0.066	290	42.0×10^6
Bismuth	Bi	9.80	0.354	32	4.6×10^6
Brass, alum.		8.33	0.301	110	16.0×10^6
Brass, naval		8.41	0.304	100	15.0×10^6
Brass, red		8.44	0.305	120	17.0×10^6
Bronze, alum.		7.39	0.267	120	17.0×10^6
Bronze, leaded		8.83	0.319	120	17.0×10^6
Bronze, phos.		8.86	0.320	120	17.0×10^6
Boron	B	2.30	0.083	410	60.0×10^6
Cadmium	Cd	8.66	0.313	55	8.0×10^6
Calcium	Ca	1.55	0.056	21	3.0×10^6
Carbon	C	2.21	0.080	4.8	0.7×10^6
Chromium	Cr	7.20	0.260	25	36.0×10^6
Cobalt	Co	8.86	0.320	21	30.0×10^6
Copper	Cu	8.97	0.324	110	16.0×10^6
30% cupronickel		8.94	0.323	150	22.0×10^6
20% cupronickel		8.94	0.323	140	20.0×10^6
Copper, beryllium		8.80	0.318	135	20.0×10^6
Gallium	Ga	5.98	0.216	6.9	1.0×10^6
Germanium	Ge	5.31	0.192	79	11.4×10^6
Gold	Au	19.3	0.698	74	12.0×10^6
Hafnium	Hf	13.1	0.473	140	20.0×10^6
Indium	In	7.31	0.264	11	1.6×10^6
Iridium	Ir	22.5	0.831	520	75.0×10^6
Iron, pure	Fe	7.86	0.284	210	29.8×10^6
Iron, gray cast C.E. = 4.8	(a)	7.86	0.284	55	8.0×10^6
Iron, gray cast C.E. = 3.3	(a)	7.86	0.284	140	20.0×10^6
Lead	Pb	11.3	0.409	18	2.6×10^6
Lithium	Li	0.53	0.019	12	1.7×10^6
Nickel	Ni	10.2	0.36	280	40×10^6
Osmium	Os	22.5	0.284	550	$80. \times 10^6$
Palladium	Pd	12	0.434	120	18×10^6
Platinum	Pt	21.5	0.775	140	21×10^6
Potassium	K	0.85	0.031	3.4	0.5×10^6
Rhenium	Re	21.0	0.369	280	40×10^6
Rhodium	Rh	12.5	0.45	370	54×10^6
Silicon	Si	2.33	0.084	110	16×10^6
Silver	Ag	10.5	0.379	76	11×10^6
Sodium	Na	0.97	0.035	9	1.3×10^6
Tantalum	K	0.85	0.031	3.4	0.5×10^6
Thallium	Tl	11.9	0.428	8.3	1.2×10^6

Table 8.3 Modulus of elasticity and density of various metals, continued

Metal	Symbol	Density, ρ		Modulus of elasticity, E	
		g/cc	lb/in.^{3}	GPa	lb/in.^{2}
Thorium	Th	11.7	0.422	79	11×10^6
Tin	Sn	7.31	0.264	41	6×10^6
Titanium	Ti	4.54	0.164	120	16.8×10^6
Tungsten	W	19.3	0.697	340	50×10^6
Vanadium	V	6.00	0.217	130	19×10^6
Zinc	Zn	7.14	0.528	83	12×10^6
Zirconium	Zr	6.37	0.230	76	11×10^6

[a]C.E. = % carbon 0.33 (% silcon + % phosphorus) and GPa = 10^9 Pa.

of oven dry wood after prolonged drying at 100 °C (212 °F). Wood properties vary with the direction of load relative to the grain. Typically, the maximum allowable stress parallel to the fibers is about four times the allowable stress perpendicular to the fibers.

Table 8.6 presents data for resins, fibers, and laminates. Fiberglass, a laminate of glass fibers in a resin matrix, is the most widely used. The strength of resins, plastics, and laminates is dependent on the manufacture methods as well as the resin and reinforcing fibers. Polyester and epoxy resins have relatively low strength and stiffness, but they can be strengthened considerably when used as a binder matrix for high-strength fibers. Randomly oriented chopped graphite or glass fibers are introduced into a resin to strengthen and stiffen injection molded parts. Molding orients the chopped fibers so final properties of a molded part are dependent on the molding process. Tape lay-up machines and skillful hands are used to layup continuous graphite (carbon) fibers in wet epoxy and polyester resins. The resin-impregnated fibers are cured at temperature to form carbon fiber-reinforced polymer (CFRP), which is also called graphite fiber-reinforced plastic (GFRP).

All the graphite fibers are parallel in graphite–epoxy tape. Tape is strongest and stiffest when loaded parallel to the fibers. The modulus for uniaxial tape is 20 million psi (138 GPa) for loads parallel to tape fibers, and it falls to 1 million psi (6.9 GPa) for loads perpendicular to the fibers. Fibers in woven fabric have 0 and 90 degree fiber orientations. Laminates are often laid up with lamina of tape and fabric at orientations of 0, ±45°, and 90°. These quasi-isotropic laminates that have stiffness and strength about 40% those of tape parallel to its fibers.

Resins and their laminates are sensitive to temperature. The modulus of polyethylene pipe decreases with increasing temperature and with increasing time ([2], Section III, NC). For 30 min duration, $E = 526$ MPa (76000 psi) at 23 °C (73 °F) and 328 MPa (48000 psi) at 49 °C (120 °F). For 100 hr duration, these drop to 353 and 220 MPa, respectively. Resins oxidize, degrade, and rubberize as temperature reaches the polymer glass transition temperature or cure temperature. The glass transition temperature and the cure temperature are lower than the melting temperature, if one exists. Typically, nylon, epoxy, and polyester composite laminates are molded and cured at 120–180 °C (250–350 °F), which is an upper limit for their use.

Table 8.4 Modulus and density of wood[a]

Wood	Density, ρ		Modulus of elasticity, E	
	g/cc	lb/in.^{3}	GPa (10^9 Pa)	lb/in.^{2}
Hardwoods				
Ash, white	0.59	0.021	12	1.7×10^6
Ash, black	0.48	0.017	9.0	1.3×10^6
Basswood	0.37	0.014	100	1.5×10^6
Beech	0.63	0.023	12	1.7×10^6
Birch, yellow	0.60	0.022	13	1.9×10^6
Cherry, black	0.53	0.019	9.7	$1.4. \times 10^6$
Chestnut	0.45	0.016	8.3	1.2×10^6
Cottonwood, east.	0.39	0.014	8.3	1.2×10^6
Elm, American	0.5	0.018	9.0	1.3×10^6
Elm, rock	0.63	0.023	10	1.3×10^6
Hickory, shagbark	0.71	0.026	14	2.1×10^6
Locust	0.69	0.025	14	2.1×10^6
Mahogany, c am.	0.51	0.026	10	1.5×10^6
Maple, black	0.57	0.023	12	1.7×10^6
Maple, sugar	0.63	0.023	12	1.7×10^6
Oak, red	0.63	0.023	11	1.6×10^6
Oak, white	0.66	0.024	11	1.6×10^6
Poplar, yellow	0.43	0.016	10	1.5×10^6
Sweet gum	0.50	0.018	9.7	1.4×10^6
Tupelo, black	0.51	0.08	8.3	1.2×10^6
Walnut, black	0.56	0.020	11	1.6×10^6
Softwoods				
Cedar, western	0.35	0.013	7.6	1.4×10^6
Cypress	0.46	0.017	9.7	1.8×10^6
Douglas fir	0.48	0.017	12	1.6×10^6
Fir, balsam	0.36	0.013	7.6	1.3×10^6
Hemlock, western	0.45	0.016	10	1.5×10^6
Larch	0.52	0.019	12	1.8×10^6
Pine, eastern white	0.37	0.013	8.3	1.2×10^6
Pine, ponderosa	0.41	0.015	9.0	1.3×10^6
Pine, southern	0.52	0.019	12	1.8×10^6
Pine, western	0.39	0.014	10	1.5×10^6
Redwood	0.38	0.014	8.3	1.2×10^6
Spruce, eastern	0.40	0.014	10	1.5×10^6
Spruce, Sitka	0.40	0.014	10	1.5×10^6
Tropical				
Balsa	0.14	0.005	3.5	0.5×10^6
Ebony	1.08	0.039	19	2.7×10^6
Lemonwood	0.78	0.028	16	2.3×10^6
Lignum vitae	1.09	0.039		
Mahogany, c am.	0.50	0.018	10	1.5×10^6
Teak	0.65	0.023	12	$1.7. \times 10^6$
Composition board				
Hardwood, fibrous	1.0–1.3	0.04–0.05	5–8	$0.8–1.2 \times 10^6$
Tempered				
Particle board	0.4–1.3	0.01–0.05	1–7	$0.15–1 \times 10^6$

[a]Dry wood, 12% moisture content.

Table 8.5 Density and modulus of elasticity of plastics, glass, fibers, and laminates

Material	Density, ρ		Modulus of elasticity, E	
	g/cc	lb./in.^{3}	GPa	lb./in.^{2}
Thermoplastic plastics				
Cellulose acetate	1.3	0.047	0.7–2.0	$0.1–0.3 \times 10^6$
Cellulose acetate Butyrate	1.2	0.043	0.7–2.0	$0.1–0.3 \times 10^6$
Cellulose nitrate	1.4	0.051	1.0–3.0	$0.2–0.4 \times 10^6$
Ethyl cellulose	1.1	0.040	0.7–3.0	$0.1–0.5 \times 10^6$
Methyl methacrylate	1.2	0.043	2.0–3.0	$0.3–0.5. \times 10^6$
Nylon, 6-6	1.1	0.040	1.0–3.0	$0.02–0.04 \times 10^6$
Vinyl chloride	1.4	0.049	2.0–3.0	$0.3–0.5 \times 10^6$
Polyethylene	0.9–1.0	0.034	0.4–0.7	$0.06–0–0.1 \times 10^6$
Polymethyl methacrylate	1.2	0.043	3.1	0.45×10^6
Polypropylene	0.9	0.033	0.7–2.0	$0.1–0.4 \times 10^6$
Polystyrene	1.0–1.1	0.038	0.3–1.0	$4–5 \times 10^6$
Polytetrafluoro	2.1–2.2	0.076–0.079	0.4	0.06×10^6
Stanyl™ (nylon + fiber)	1.4	0.051	6–10	$0.9–1.4 \times 10^6$
Thermosetting plastics				
Epoxy resin	1.1–1.2	0.04–0.043	1.0–4.0	$0.2–0.6 \times 10^6$
Phenolic resin	1.3	0.047	2.0–3.0	$0.3–0.5 \times 10^6$
Polyester resin	1.1–1.5	0.04–0.054	2.0–4.0^9	$0.3 - 0.6 \times 10^6$
Laminates				1.5×10^6
E-glass fiber/epoxy resin	1.8–2.2	0.065–0.079	20–60	$3–8 \times 10^6$
E-glass fiber/phenolic	1.7–1.9	0.061–0.069	10–30	$2–5 \times 10^6$
E-glass fiber/polyester	1.5–1.9	0.054–069	7–30	$1–4 \times 10^6$
Graphite fiber/epoxy Uniaxial tape, parallel	1.6	0.06	133	20×10^6
Graphite fiber/epoxy Uniaxial tape, transverse	1.6	0.06	7	1×10^6
Graphite fiber/epoxy Woven fabric	1.6	0.06	50	8×10^6
Glass				
Glass sheet	2.3–2.6	0.083–0.094	70	9×10^6
Quartz sheet	2.2	0.080	70	2×10^6
Fibers				
E-glass	2.54	0.092	72	10.5×10^6
S-glass	2.49	0.090	86	12×10^6
D-glass	2.16	0.078	52	7.5×10^6
Boron on tungsten	2.6	0.094	410	60×10^6
Graphite	1.4–1.6	0.05–0.06	170–340	$25–50 \times 10^6$
Quartz	2.2	0.079	70	10×10^6

8.2 Liquids

The density, speed of sound, and kinematic viscosity of liquids are given in Tables 8.6–8.8 that were compiled from Refs [8, 9] in part. The kinematic viscosity (viscosity divided by density) and density of most liquids decrease with increasing temperature. The density,

Table 8.6 Properties of freshwater

Temperature (°C)	Density (kg/m³)	Speed of sound (m/s)	Kinematic viscosity (m²/s)
0.01	999.8	1402	17.91×10^{-7}
5	999.9	1426	15.18×10^{-7}
10	999.7	1447	13.06×10^{-7}
15	999.1	1466	11.38×10^{-7}
20	998.2	1482	10.03×10^{-7}
25	997.0	1496	8.925×10^{-7}
30	995.6	1509	8.005×10^{-7}
35	994.0	1519	7.233×10^{-7}
40	992.2	1528	6.577×10^{-7}
45	990.2	1536	6.016×10^{-7}
50	988.0	1542	5.530×10^{-7}
55	985.7	1547	5.109×10^{-7}
60	983.2	1551	4.739×10^{-7}
65	980.6	1553	4.414×10^{-7}
70	977.8	1555	4.127×10^{-7}
75	974.8	1555	3.871×10^{-7}
80	971.8	1554	3.643×10^{-7}
85	968.6	1553	3.438×10^{-7}
90	965.3	1550.4	3.254×10^{-7}
95	961.9	1547	3.088×10^{-7}
99	959.1	1544	2.967×10^{-7}
°F	lb/ft³	ft/s	ft²/s
32.01	62.42	4601	1.929×10^{-5}
40	62.43	4670	1.663×10^{-5}
50	62.41	4748	1.406×10^{-5}
60	62.37	4815	1.208×10^{-5}
70	62.30	4874	1.051×10^{-5}
80	62.22	4924	0.9259×10^{-5}
90	62.11	4967	0.8232×10^{-5}
100	61.99	5003	0.7381×10^{-5}
110	61.86	5033	0.6667×10^{-5}
120	61.71	5057	0.6064×10^{-5}
130	61.55	5075	0.5547×10^{-5}
140	61.38	5088	0.5102×10^{-5}
150	61.19	5097	0.4716×10^{-5}
160	61.00	5101	0.4379×10^{-5}
170	60.795	5102	0.4082×10^{-5}
180	60.58	5098	0.3821×10^{-5}
190	60.35	5091	0.3589×10^{-5}
200	60.12	5080	0.3382×10^{-5}
211	59.85	5064	0.3180×10^{-5}

Table 8.7 Properties of seawater[a]

Temperature (°C)	Density (kg/m^3)	Speed of sound (m/s)	Kinematic viscosity (m^2/s)
0	1028	1449	1.831×10^{-6}
5	1028	1471	1.56×10^{-6}
10	1027	1490	1.35×10^{-6}
15	1026	1507	1.19×10^{-6}
20	1025	1521	1.05×10^{-6}
25	1023	1534	0.946×10^{-6}
30	1022	1546	0.853×10^{-6}
°F	lb/ft^3	ft/s	ft^2/s
32	64.18	4754	1.97×10^{-5}
40	64.16	4818	1.71×10^{-5}
50	64.11	4889	1.46×10^{-5}
60	64.04	4949	1.26×10^{-5}
70	63.95	5002	1.11×10^{-5}
80	63.85	5047	0.983×10^{-5}
90	63.72	5086	0.877×10^{-5}

[a]At one atmosphere pressure. *Salinity* = 35 parts per thousand. Also see Table 6.1.

Table 8.8 Properties of various liquids[a]

Liquid	Density		Kinematic viscosity		Speed of sound	
	kg/m^3	lb/ft^3	m^2/s	ft^2/s	m/s	ft/s
Alcohol, ethyl	170	49.3	15×10^{-7}	1.6×10^{-5}	1150	3770
Benzene	878	54.8	7.4×10^{-7}	0.8×10^{-5}	1322	4340
Diesel	900	58				
Gasoline	670	42	4.6×10^{-7}	10.5×10^{-5}		
Glycerin	1260	78.7	1.19×10^{-3}	1.28×10^{-2}	1980	6500
Kerosene	804	50.2	23×10^{-7}	12.5×10^{-5}	1450	4770
Mercury	13,600	849	1.2×10^{-7}	0.13×10^{-5}	1450	4760
Oil, caster	950	59	1×10^{-3}	1.09×10^{-2}	1540	5050
Oil, crude	850	53	70×10^{-7}	7.5×10^{-5}		
Oil, SAE 30	920	57	7.3×10^{-5}	17.8×10^{-4}	1290	4240
Turpentine	870	54	17×10^{-7}	1.9×10^{-5}	1250	4100
Liquid O_2, −30°C	1.67	0.102	11×10^{-6}	11×10^{-5}	297	965
Liquid H_2, −30°C	6.46	0.028	5.4×10^{-6}	6.3×10^{-5}	603	2019
Liquid CO_2,−30°C	2.22	0.147	5×10^{-6}	5.9×10^{-5}	239	795

[a]At one atmosphere pressure.

kinematic viscosity, and speed of sound of most liquids are nearly independent of pressure for moderate pressures and temperatures well below the boiling point. Extreme pressure can significantly increase liquid density. For example, the speed of sound and density of freshwater as a function of pressure at 20 °C (68 °F) are as follows.

Pressure (atm)	1	10	100	500	1000
Sound speed (m/s)	1482	1484	1499	1566	1651
Density (kg/m^3)	998.2	998.6	1002	1020	1040

Formulas for the speed of sound are given in Table 6.1. References [9–13] present additional data for properties of liquids.

8.3 Gases

8.3.1 Ideal Gas Law

The molecular kinetic theory of gases provides the ideal gas equation which relates the gas density ρ to pressure p and absolute temperature T [13]. Ideal gas density increases with pressure and decreases with molecular weight and absolute temperature:

$$\rho = \frac{PW}{RT} \tag{8.1}$$

R is the universal gas constant. Values of R and consistent units of ρ, R, P, and T are given in the following table.

Units for ρ	Units for P	Units for T	Gas constant R
kg/m^3	Pa	Degree Kelvin	8315 Pa-m^3/kg-mol-°K
g/L^3	Atmospheres	Degree Kelvin	0.08206 atm-L/g-mol-°K
slug/ft^3	lb/ft^2	Degree Rankine	49,720 lb-ft/slug-mol-°R

The absolute temperature scales are degree Kelvin (°K) and degree Rankine (°R):

$$\begin{aligned} ^\circ\mathrm{K} &= {}^\circ\mathrm{C} + 273.16 \\ ^\circ\mathrm{R} &= {}^\circ\mathrm{F} + 459.69 \end{aligned} \tag{8.2}$$

The molecular weight, W, is expressed per mole. The molecular weights of gases are given in Table 8.8. The equivalent molecular weight of a mixture of N gases is the sum of the mole fraction components times their molecular weight which is the law of mixtures:

$$W_{\text{mixture}} = \sum_{i=1}^{N} x_i W_i \tag{8.3}$$

Each component has molecular weight W_i and mole fraction or volume fraction x_i of the gas mass mixture. A mole of gas is the molecular weight of the gas in grams.

Example 8.1 Calculation of Gas Density

Use the ideal gas law to calculate the density of air at room temperature and pressure.

Solution: The molecular weight of air is computed with Equation 8.3, the mole fractions of the component gases: 78.03 N_2, 20.99 O_2, 0.94 Ar, and 0.03 CO_2. Their molecular weights are given in Table 8.9:

$$W_{\text{air}} = 0.7803(28.02) + 0.2099(32) + 0.01(39.94) = 28.97$$

The ideal gas law (Eq. 8.1) predicts the density of air at room temperature, $T = 20\,°\text{C}$, 68 °F, and absolute pressure, 101,300 Pa, 14.7 psia. First, the temperature is converted to absolute temperature with Equation 8.2:

$$T = 20°\text{C} + 273.16 = 293\ °\text{K}\ \ (527.7\ °\text{R})$$

Then the ideal gas law is applied with a unit-consistent universal gas constant:

$$\rho_{\text{air}} = \frac{PW}{RT} = \frac{101{,}300\ \text{Pa} \times 28.97}{8315 \times 293\ °\text{K}} = 1.20\ \text{kg/m}^3$$

In ft-lb units, the pressure is $2116\ \text{lb/ft}^2$ and the density is $\rho = 2116\ \text{lb/ft}^2 \times 28.97/(49{,}720 \times 527\ °\text{R}) = 0.0023\ \text{slug/ft}^3 = 0.0752\ \text{lb/ft}^3$. These computations agree with the data given in Table 8.10.

Table 8.9 Properties of various gases

Gas	Molecular formula	Molecular weight, W	Ratio of specific heats, γ^b
Air	(a)	28.97	1.400
Argon	Ar	39.94	1.667
Butane	C_4H_{10}	58.12	1.09
Carbon dioxide	CO_2	44.01	1.283
Carbon monoxide	CO	28.01	1.399
Ethane	C_2H_6	28.01	1.399
Ethylene	C_2H_4	30.07	1.183
Helium	He	4.003	1.667
Hydrogen	H_2	2.016	1.404
Methane	CH_4	16.04	1.32
Neon	Ne	20.18	1.400
Nitrogen	N_2	28.02	1.400
Octane	C_8H_{18}	114.2	1.044
Oxygen	O_2	32.00	1.395
Propane	C_3H_8	44.10	1.124
Steam	H_2O	18.02	1.329
Sulfur dioxide	SO_2	64.06	1.613

[a] 78.03 N_2, 20.99 O_2, 0.94 Ar, 0.03 CO_2, and 0.01 other, percent by volume.
[b] At 27 °C (80 °F) [14].

Table 8.10 Properties of air – metric units, 1 atm pressure

Temperature (°C)	Density, ρ (kg/m^3)	Speed of sound, c (m/s)	Kinematic viscosity, ν (10^{-5} m^2/s)	Ratio of specific heat, γ
−50	1.584	299.5	0.9248	1.405
−40	1.516	306.2	1.002	1.404
−30	1.453	312.7	1.082	1.404
−20	1.395	319.1	1.164	1.403
−10	1.342	325.4	1.248	1.403
0	1.293	331.5	1.335	1.403
5	1.269	334.5	1.379	1.403
10	1.247	337.5	1.424	1.402
15	1.225	340.5	1.469	1.402
20	1.204	343.4	1.515	1.402
25	1.184	346.3	1.562	1.402
30	1.164	349.2	1.609	1.402
40	1.127	354.9	1.704	1.401
50	1.092	360.5	1.802	1.401
60	1.059	366.0	1.902	1.400
70	1.028	371.3	2.003	1.399
80	0.9993	376.7	2.107	1.399
90	0.9717	381.9	2.213	1.399
100	0.9456	387.0	2.321	1.398
150	0.8338	411.7	2.889	1.394
200	0.7456	434.7	3.502	1.390
250	0.6743	456.2	4.158	1.385
300	0.6155	476.6	4.856	1.379
350	0.5661	495.9	5.593	1.373

Temperature (°F)	Density, ρ (lb/ft^3)	Speed of sound, c (ft/s)	Kinematic viscosity, ν (10^{-4} ft^2/s)	Ratio of specific heats, γ
−100	0.1105	929.5	0.8121	1.406
−80	0.1046	955.3	0.8976	1.405
−60	0.09938	980.3	0.9864	1.405
−40	0.09462	1004	1.079	1.404
−20	0.09029	1028	1.174	1.404
0	0.08634	1052	1.273	1.403
10	0.08450	1063	1.323	1.403
20	0.08273	1074	1.374	1.403
30	0.08103	1085	1.426	1.403
40	0.07941	1096	1.479	1.403
50	0.07784	1107	1.533	1.402
60	0.07634	1118	1.587	1.402
70	0.07489	1129	1.642	1.402
80	0.07350	1139	1.698	1.402

Table 8.10 Properties of air – metric units, 1 atm pressure, continued

Temperature (°F)	Density, ρ (lb/ft^3)	Speed of sound, c (ft/s)	Kinematic viscosity, ν (10^{-4} ft^2/s)	Ratio of specific heats, γ
90	0.07216	1150	1.754	1.401
100	0.07087	1160	1.811	1.401
120	0.06842	1181	1.928	1.401
140	0.06613	1201	2.047	1.400
160	0.06399	1220	2.169	1.400
180	0.06199	1239	2.293	1.399
200	0.06011	1259	2.421	1.398
250	0.05587	1305	2.750	1.396
300	0.05219	1349	3.095	1.394
350	0.04896	1391	3.455	1.392
400	0.04611	1432	3.831	1.389
450	0.04358	1472	4.212	1.387
500	0.04131	1510	4.623	1.384
550	0.03926	1548	5.039	1.381
600	0.03741	1584	5.469	1.377

Speed of sound for an adiabatic propagation of sound waves is given in Table 6.1:

$$c = \left(\frac{\gamma p}{\rho}\right)^{1/2} = \left(\frac{\gamma RT}{W}\right)^{1/2} \tag{8.4}$$

The ambient pressure is p and ρ is the ambient density. γ is the *ratio of specific heats* at constant pressure to that at constant volume. γ is dimensionless and it varies only within narrow limits. The kinetic theory of gases [14] predicts that $\gamma = 1$ for a monatomic gas such as helium or argon and $\gamma = 1.4$ for a diatomic gas such as nitrogen or oxygen. Larger molecules have larger values. γ varies somewhat with temperature. Values are given in Tables 8.9 and 8.10.

The second formula in Equation 8.4 is the ideal gas law. The ideal gas low implies that sound speed is nearly independent of pressure for low pressure and the speed of sound increases with the square root of absolute temperature. For example, for air,

$$\begin{aligned} c_{\text{air}} &= 20.05T^{1/2}\ \text{m/s} \quad \text{where } T = {}^\circ\text{K} = {}^\circ\text{C} + 273.2 \\ &= 49.02T^{1/2}\ \text{ft/s} \quad \text{where } T = {}^\circ\text{R} = {}^\circ\text{F} + 459.69 \end{aligned} \tag{8.5}$$

Sound waves at very high frequencies are transmitted isothermally rather than adiabatically owing to the large thermal gradients that encourage heat transfer. The isothermal sound speed can be obtained from Equation 8.4 by setting $\gamma = 1$, and so the isothermal sound speed is generally lower than the adiabatic sound speed. Significant deviation from the ideal gas law occurs near saturation temperature and very high pressures as discussed in [14, 15].

Absolute viscosity μ of a fluid is the ratio of shear stress to shearing strain. At low pressure, absolute viscosity is nearly independent of pressure and increases with temperature. *Kinematic viscosity*, denoted by the symbol ν, is absolute viscosity divided by density:

$$\nu = \frac{\mu}{\rho} \tag{8.6}$$

Kinematic viscosity has units of length squared per time. Values of kinematic viscosity are given in Tables 8.9 and 8.10 for various gases. Since gas density decreases with increasing temperature at constant pressure (Eq. 8.1), the kinematic viscosity increases significantly with temperature as can be seen in Table 8.10.

The viscosity of air is nearly independent of pressure at moderate pressures. The kinematic viscosity of air decreases as pressure and density increase. For example, the density, speed of sound, viscosity and kinematic viscosity of air as a function of pressure at 20 °C (68 °F) are as follows:

Pressure (atm)	1	10	100	500	1000
Sound speed (m/s)	343.3	344.5	365.9	588.0	831.5
Density (kg/m^3)	1.204	12.08	121.8	453.6	623.5
Viscosity (10^{-6} Pa-s)	18.25	18.39	20.54	36.8	56.2
Kinematic viscosity (10^{-5} m^2/s)	1.515	0.1522	0.0168	0.0081	0.0090

Pressurization decreases kinematic viscosity of air.

Additional data for viscosity and procedures for measuring viscosity are given in Refs [8, 9, 12, 15].

References

[1] Baumeister, T. (ed.), Standard Handbook for Mechanical Engineers, 7th ed., McGraw-Hill, N.Y. 1967.
[2] The ASME Boiler and Pressure Vessel Code, The American Society of Mechanical Engineers, N.Y.
[3] Aerospace Structural Metals Handbook, Purdue University, 1998.
[4] Metals Handbook, 8th ed., American Society for Metals, Metals Park, Novelty Ohio, 1961.
[5] American Institute of Timber Construction, Timber Construction Manual, John Wiley, N.Y., 1974.
[6] Metallic Materials and Elements for Aerospace Vehicle Structures. Mil-HDBK-5J, US Department of Defense, 2003.
[7] Fung, Y. C., Foundations of Solid Mechanics, Prentice-Hall, Englewood Cliffs, N.J., 1965, p. 353.
[8] Reid, R. C. and T. K. Sherwood, The Properties of Gases and Liquids and their Correlation, 2nd ed., McGraw-Hill, N.Y., 1966.
[9] NIST, REFPROP, Reference Fluid Thermodynamic and Transport Properties, Version 9.1, 2014.
[10] Weast, R. C. (ed.), Handbook of Chemistry and Physics, The Chemical Rubber Company, Cleveland, Ohio,1978.
[11] Perry, J. H., Chemical Engineers Handbook, 4th ed., McGraw-Hill, 1963.
[12] Touloukian, Y. S., and Purdue University, Thermophysical Properties of Matter: The TPRC Data Series: A Comprehensive Compilation of Data. Y. S. Touloukian (ed.), IFI/Plenum, New York, 1970.
[13] Shapiro, A. H., The Dynamics and Thermodynamics of Compressible Fluid Flow, vol. 1, The Ronald Press, N.Y., 1953.
[14] Van Wylen, J. and R. E. Sonntag, Fundamentals of Classical Thermodynamics, John Wiley, N.Y., 1973, p. 683.
[15] Hilsenrath, J., et al., Tables of Thermodynamic and Transport Properties, Pergamon Press, N.Y., 1960.

Appendix A

Approximate Methods for Natural Frequency

Table A.1 presents approximate techniques for determining natural frequencies of systems that do not have exact solutions [1–9]. These simple methods can provide a check for a computer model and gage the effect of a parameter change on natural frequency.

A.1 Relationship between Fundamental Natural Frequency and Static Deflection

One measure of the strain energy is the static deflection of the structure under the acceleration due to gravity. Thus, it is also reasonable to believe that a relationship exists between static deformation of a structure under its own weight and its fundamental natural frequency. The static deflection due to gravity of the mass in the one-degree-of-freedom elastic spring–mass system shown in Figure A.1(a) is

$$\delta_s = \frac{Mg}{k} \tag{A.1}$$

M is the mass, g is the acceleration due to gravity, and k is the spring constant. The natural frequency of this system is (case 1 of Table 3.3)

$$f = \frac{1}{2\pi}\left(\frac{k}{M}\right)^{1/2} \quad \text{Hz} \tag{A.2}$$

Incorporating Equation A.1 into Equation A.2 to eliminate the spring constant k gives

$$f = \frac{1}{2\pi}\left(\frac{g}{\delta_s}\right)^{1/2} \quad \text{Hz} \tag{A.3}$$

Equation A.3 allows the natural frequency of the structure to be expressed solely in terms of the acceleration due to gravity and the maximum static deflection, δ_s, that gravity produces. Of course, it is not necessary to limit the application of Equation A.3 to systems that vibrate

Formulas for Dynamics, Acoustics and Vibration, First Edition. Robert D. Blevins.
© 2016 John Wiley & Sons, Ltd. Published 2016 by John Wiley & Sons, Ltd.

Table A.1 Approximate methods for computing natural frequencies

Notation: j = imaginary constant; L[] = stiffness operator, or equation of motion of free vibration of the system; ω = natural frequency in radians per second = $2\pi f$ where f is natural frequency in Hz, cycles per second; g = acceleration due to gravity, Table 1.2; $\tilde{x}_i(x)$ = assumed (trial) mode shape of *i*th mode that generally satisfies the boundary conditions. Consistent sets of units are given in Table 1.2 [1–8].

Estimated Natural Frequency	Comments
1. **Rayleigh Quotient** $\omega_i^2 = \dfrac{PE}{(KE/\omega^2)} = \dfrac{\{\tilde{x}_i\}^T[K]\{\tilde{x}_i\}}{\{\tilde{x}_i\}^T[M]\{\tilde{x}_i\}}$ $PE = \dfrac{1}{2}\{\tilde{x}_i\}^T[K]\{\tilde{x}_i\}$, $KE = \dfrac{1}{2}\{\omega\tilde{x}_i\}^T[M]\{\omega\tilde{x}_i\}$	Rayleigh quotient is based on an energy balance between maximum kinetic energy KE, which is proportional to ω^2, and maximum potential energy PE during and setting them equal. The assumed mode shape $\tilde{x}$ is usually taken to be static deflection under 1-g load, mass matrix [M] and stiffness matrix [K]. Rayleigh quotient is an upper bound estimate of natural frequency if he assumed mode shapes satisfy the boundary conditions. Refs [1, 2].
2a. **Dunkerley's Method** $\dfrac{1}{\omega_1^2} = \dfrac{1}{\omega_{11}^2} + \dfrac{1}{\omega_{21}^2} + ..\dfrac{1}{\omega_{N1}^2}$ 2b. **Southwell Method** $\omega_1^2 = \omega_{11}^2 + \omega_{21}^2 + ..\omega_{N1}^2$	Dunkerley's and Southwell's equations provide lower and upper bound estimates, respectively, for the fundamental natural frequency, ω_1, of a multi-degree-of freedom system in terms of the natural frequencies of sub components, each with all the mass but only one of the N stiffening (spring) elements. The first subscript is the spring and second subscript is the mass. Refs [3, 4].
3. **Gravity Displacement Method** $\omega_1 = \sqrt{\dfrac{g}{\delta}}$	This method applies to the fundamental mode. The max deflection of the system, δ, due to a 1-g static gravity load is required. Refs [5, 6, 7]
4. **Galerkin Method** $\omega_i^2 = \dfrac{\int_D \tilde{w}_i(P,t)L[\tilde{w}_i(P)]dD}{\int_D M(P)\tilde{w}_i^2(P)dD}$, i = 1,2,3.	L[] is the stiffness operator of the liner equation of motion and $\tilde{x}(P)$ is the assumed mode shape. In this form it is usually applied to continuous systems where integration of point P is made over domain D of system. Exact result if mode shape is exact. Ref. [8].
5. **Rayleigh-Ritz** $\dfrac{\partial}{\partial c_i}\int_S \tilde{x}(x)L[\tilde{x}(x)e^{j\omega t}]dx = 0$ solve for natural frequency ω.	L[]=0 is the equation(s) of motion and $\tilde{x}_i(x)$ are the assumed mode shapes. c_1, c_2, c_3,. are constants in modal sum: Refs [1, 2] $\tilde{x}(x) = \sum_{i=1}^{N} c_i\tilde{x}_i(x)$ Same as Galerkin for multiple mode.
6. **Numerical Matrix Solution** $\left[-\omega^2[M]+[K]\right]\{\tilde{x}\}=0$	[M] is the N by N symmetric mass matrix equation. [K] is the N by N symmetric stiffness matrix equation. Solutions of the eigenvalue problem are sought by numerically for N natural frequencies ω and orthogonal mode shapes $\{\tilde{x}_i\}$, i = 1..N.
7. **Holzer Method** for Torsion of Shafts 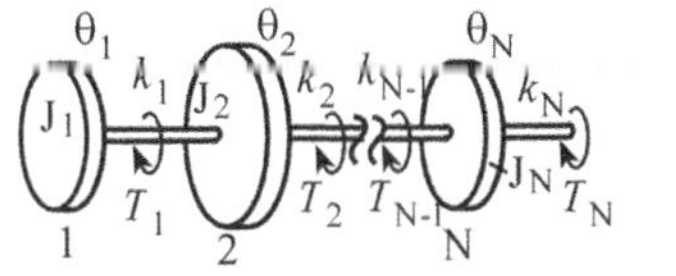 $\theta_1 = 1,\ T_1 = \omega^2 J_1 - k_0,\ n = 1,2,3,...N$ $\theta_{n+1} = \theta_n - T_n/k_n,\ T_{n+1} = T_n + \omega^2 J_{n+1}\theta_n$	A shaft is a series of N massive disks with polar mass moment of inertial J_n (Table 1.6) connected by massless torsional springs with torsion constant k_n (Table 3.2). The modal rotation of the first disk on the left hand side of the shaft is θ_1=1. If there is a fixed-ended shaft to its left then torque is k_0. If first disk is free then k_0=0. Trial ω is chosen. Formulas at left are used to proceed from left to right. θ_N and T_N at right hand end of shaft are computed. Freq ω is such that T_N=0 if Nth disk is free or θ_{N+1}=0 if Nth shaft is fixed. Ref. [1].

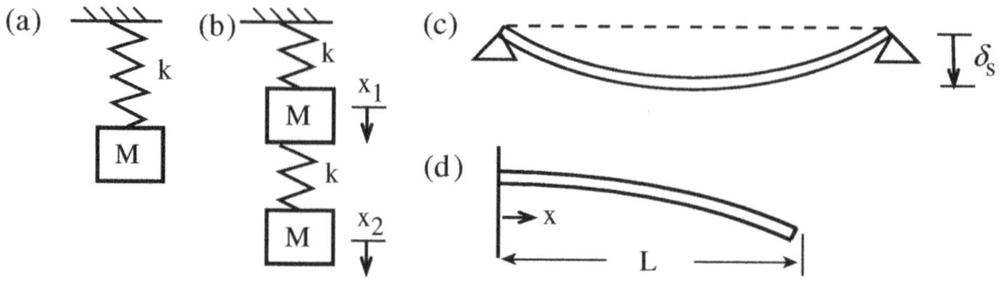

Figure A.1 Elastic systems

in the vertical plane. All that is required is an estimate of the maximum deflection produced by a constant acceleration field in the plane of vibration.

The accuracy of Equation A.3 depends on the degree to which the static deformation under gravity conforms to the mode shape of vibration. Equation A.3 is exact for the simple spring–mass system shown in Figure A.1(a) and generally underestimates the natural frequency of more complex structures [5–7].

Example A.1 Two-Spring, Two-Mass Natural Frequency

Consider the system shown in Figure A.1(b) with two equal masses and two equal springs. The maximum static deflection of the system under gravity is the sum of the deflections of the upper spring, which is loaded by $2Mg$, and the lower spring, loaded by Mg:

$$\delta_s = \frac{2Mg}{k} + \frac{Mg}{k} = \frac{3Mg}{k} \tag{A.4}$$

The fundamental natural frequency and model shape predicted from this deflection are given by Eq. A.3.

$$f = \frac{1}{2\pi}\frac{1}{3^{1/2}}\left(\frac{k}{m}\right)^{1/2} = \frac{0.5774}{2\pi}\left(\frac{k}{m}\right)^{1/2}, \quad \begin{pmatrix}\tilde{x}_1\\ \tilde{x}_1\end{pmatrix} = \begin{pmatrix}1\\ 1.5\end{pmatrix} \tag{A.5}$$

The exact fundamental natural frequency and mode shape are given in case 2 of Table 3.3:

$$f = \frac{0.6180}{2\pi}\left(\frac{k}{m}\right)^{1/2}, \quad \begin{pmatrix}\tilde{x}_1\\ \tilde{x}_1\end{pmatrix} = \begin{pmatrix}1\\ 1.618\end{pmatrix} \tag{A.6}$$

Thus, Equation A.3 underestimates the natural frequency of this system by about 6.6%; it cannot estimate the natural frequency of the second, higher, mode of this system.

Example A.2 Beam Natural Frequency

Consider the slender uniform pinned–pinned beam shown in Figure A.1(c). The static deflection of this beam at midspan due to gravity is [10]

$$\delta_s = \frac{5}{384}\frac{mg\,L^4}{EI} = \frac{1}{(2.960)^4}\frac{mg\,L^4}{EI} \tag{A.7}$$

m is the mass per unit length of the beam, E is the modulus of elasticity, I is the area moment of inertia of the cross section, and L is the length of the beam. Using this result, the fundamental natural frequency of the beam can be estimated from Equation A.3:

$$f = \frac{(2.960)^2}{2\pi L^2}\left(\frac{EI}{m}\right)^{1/2} \quad \text{Hz} \tag{A.8}$$

The exact result presented in case 5 of Table 4.2 has the same form as this equation but with the factor of 2.960 replaced by the factor π; the approximate solution is 11% low.

Example A.3 Plate Natural Frequency

Mazumdar [11] found the maximum deflection of a thin, uniform elliptical plate with a clamped edge due to its own weight:

$$\delta_s = \frac{\rho g h a^4 b^4}{8(3a^4 + 2a^2b^2 + 3b^4)} \frac{12(1-\nu^2)}{Eh^3} \tag{A.9}$$

where h is the plate thickness, ν is Poisson's ratio, ρ is the density of the plate material, E is the modulus of elasticity, and a and b are the major and minor axes of the ellipse. Using this result, Equation A.3 predicts the fundamental natural frequency of the elliptical plate:

$$f = \frac{2.828}{2\pi}\left(\frac{3a^4 + 2a^2b^2 + 3b^4}{\rho g h a^4 b^4}\right)^{1/2}\left[\frac{Eh^3}{12\left(1-\nu^2\right)}\right]^{1/2} \tag{A.10}$$

A more nearly exact analysis gives an expression of the same form but with the 2.828 replaced with 3.612 [5], The approximate solution is 22% low.

A.2 Rayleigh Technique

As noted in Chapter 3, the natural frequency of a structure is the rate of the exchange of kinetic and potential energy within the structure. The kinetic energy is associated with motion of structural mass, and the potential energy is stored by strains during deformation. Rayleigh set these two forms of energy equal and solved for natural frequency with an assumed mode shape as shown in case 1 of Table A.1.

For example, the kinetic energy of the mass in the spring–mass system shown in Figure A.1(a) is

$$\text{Kinetic energy} = \frac{1}{2}MV^2 = \frac{1}{2}M(\omega X)^2 \tag{A.11}$$

M is the mass, ω is the circular natural frequency (radians per second), X is the amplitude of the motion, and $V = \omega X$ is the maximum velocity. Potential energy is stored in the spring by elastic deformation:

$$\text{Potential energy} = \int_0^X kx\,dx = \frac{1}{2}kX^2 \tag{A.12}$$

Equating A.11 and A.12 for conservation of energy during free vibration leads to an expression for the natural frequency:

$$\frac{1}{2}M(\omega X)^2 = \frac{1}{2}kX^2$$

$$\omega^2 = \frac{k}{M}$$

$$f = \frac{\omega}{2\pi} = \frac{1}{2\pi}\sqrt{\frac{k}{M}} \quad \text{Hz} \tag{A.13}$$

This reproduces the exact result (Equation A.2). In general, the Raleigh frequency squared, called the *Rayleigh quotient*, is the ratio of the maximum potential energy in the deformed elastic structure to its maximum kinetic energy:

$$\omega^2_{\text{Rayleigh}} = \frac{\text{PE}}{(\text{KE}/\omega^2)} \tag{A.14}$$

For a general discrete spring–mass elastic system with mass matrix $[M]$, stiffness matrix $[K]$, and deformation matrix $[X]$ [1, p. 293],

$$\omega^2_{\text{Rayleigh}} = \frac{[X]^T[K][X]}{[X]^T[M][X]} \tag{A.15}$$

The result is exact if the mode shape is exact. However, the usual situation is that the deformation is approximate. If the assumed deformations satisfy the boundary conditions, then the Rayleigh quotient will overestimate, that is, provide an upper bound of the exact solution. However, very good results have been obtained even when the assumed mode shape satisfies only some of the boundary conditions [8].

Example A.4 Rayleigh Method

Consider the two-mass system shown in Figure A.1(b). Under gravity, the first mass displaces two units and the second mass displaces three units (Equation A.4), and this will be used for the assumed mode shape. The kinetic energy is the sum of kinetic energy of the two masses. The potential energy is stored in the two springs. The upper spring lengthens two unit and the lower spring lengthens one unit.

$$\text{Kinetic energy} = \frac{1}{2}MV_1^2 + \frac{1}{2}MV_2^2 = \frac{1}{2}M(2\omega X)^2 + \frac{1}{2}M(3\omega X)^2 = \frac{13}{2}M\omega^2X^2$$

$$\text{Potential energy} = \frac{1}{2}kX_1^2 + \frac{1}{2}kX_2^2 = \frac{1}{2}k(2X)^2 + \frac{1}{2}kX^2 = \frac{5}{2}kX^2 \tag{A.16}$$

Equating the potential and kinetic energies gives the estimated fundamental natural frequency:

$$f = \frac{1}{2\pi}\left(\frac{5}{13}\right)^{1/2}\sqrt{\frac{k}{M}} = \frac{0.6201}{2\pi}\sqrt{\frac{k}{M}} \tag{A.17}$$

The exact solution, Equation A.6, has the same form as this equation but with the factor of 0.6201 replaced with 0.6180; the Rayleigh approximation is slightly high.

A.3 Dunkerley and Southwell Methods

The *Dunkerley and Southwell methods* (case 2 of Table A.1) estimate the fundamental natural frequency from the subsystems that have the entire mass but a single stiffening element of the larger system. We identify the first degree-of-freedom subsystem of the two-spring, two-mass system shown in Figure A.1(b) by placing all the mass (2*M*) on the first spring and deleting the second, lower, spring. The second subsystem is created by placing all the mass at the end of second spring. Neglecting the first spring produces zero natural frequency mode for the second subsystem. However, the concept is to include the potential energy of the first spring in the second subsystem. Its stiffness is assigned one half the stiffness of two springs in series (*k*/4):

$$\begin{aligned} f_{11} &= \left(\frac{1}{2\pi}\right)\left(\frac{k}{2M}\right)^{1/2} = \left(\frac{1}{2\pi}\right)0.707\left(\frac{k}{M}\right)^{1/2} \\ f_{12} &= \left(\frac{1}{2\pi}\right)\left[\frac{k/2}{2M}\right]^{1/2} = \left(\frac{1}{2\pi}\right)0.5\left(\frac{k}{M}\right)^{1/2} \end{aligned} \tag{A.18}$$

The Southwell and Dunkerley frequencies are upper and lower bounds, respectively, for the exact solution:

$$\begin{aligned} f_{\text{Southwell}} &= \sqrt{f_{11}^2 + f_{12}^2} = \left(\frac{1}{2\pi}\right)0.866\left(\frac{k}{M}\right)^{1/2} \\ f_{\text{Dunkerley}} &= \frac{1}{\sqrt{(1/f_{11}^2) + (1/f_{12}^2)}} = \left(\frac{1}{2\pi}\right)0.408\left(\frac{k}{M}\right)^{1/2} \end{aligned} \tag{A.19}$$

The exact solution, $(1/2\pi)\,0.6180\ (k/M)^{1/2}$ (Equation A.6), lies between these two estimates.

A.4 Rayleigh–Ritz and Schmidt Approximations

Ritz [1, 2] and Schmidt [12] minimize the Rayleigh quotient frequency with respect to parameters embedded in deformation shape to determine natural frequency. In the Rayleigh–Ritz technique, the deformation is expressed in a series:

$$\tilde{y}(x) = C_1\tilde{y}_1(x) + C_2\tilde{y}_2(x) + C_3\tilde{y}_3(x) + \cdots + C_N\tilde{y}_N(x) \tag{A.20}$$

This series is substituted into the Rayleigh quotient frequency, which is then minimized after integrating by setting the N derivatives $\partial\omega/\partial C_i = 0$; $n = 1, 2, 3 \ldots\ N$ to produce N equations for the N optimal values of the constants $C_1\ \ldots\ C_N$.

Schmidt's technique minimizes the Rayleigh quotient with respect to a power parameter n. For example, consider the first mode of a uniform cantilever beam (Figure A.1(d)). The boundary conditions at the fixed end, $x = 0$, are no displacement and no slope, and on the free end, $x = L$, there are no shear and no moment (Table 4.1):

$$at\ \ x = 0, \quad y(x,t) = y'(x,t) = 0, \quad at\ \ x = L,\ y''(x,t) = y'''(x,t) = 0 \tag{A.21}$$

Consider three approximate shapes for the first mode of the cantilever beam:

$$\tilde{y}(x) = \begin{cases} (x/L)^n & \text{power law} \\ 1 - \cos(\pi x/2L) & \text{quarter wave} \\ (x/L)^2 - (x/L)^3/3 & \text{static deflection to tip load} \end{cases} \tag{A.22}$$

All three of these satisfy the geometric boundary conditions at $y(0) = 0$ at $N = 0$. However, none satisfy the zero shear condition, $y'''(L) = 0$ at $X = L$. The potential energy and kinetic energy of a beam that deform 0 in the mode shape $\tilde{y}(x)$ are [1, p. 294]

$$\text{Potential energy} = \int_0^L EI[\tilde{y}''(x)]^2 dx, \quad \text{kinetic energy} = \omega^2 \int_0^L m[\tilde{y}(x)]^2 dx \tag{A.23}$$

Equating the potential and kinetic energies, inserting the mode shapes of Equation A.22, integrating, and solving for the circulator natural frequency with Rayleigh's technique for the fundamental natural frequency of the cantilever beam gives the following results:

$$\omega^2 = \frac{\int_0^L EI[\tilde{y}''(x)]^2 dx}{m[\tilde{y}(x)]^2 dx} = \left\{ \begin{array}{c} \dfrac{(2n+1)\,n^2(n-1)^2}{(2n-3)} = 19.82, \; (n=2) \\ \dfrac{\pi^5}{[2^6(3\pi/4 - 2)]} = 13.424 \\ \dfrac{840}{31} = 27.09 \end{array} \right\} \frac{EI}{mL^4} \tag{A.24}$$

All three of these estimates exceed the exact result, which has the non dimensional parameter of $1.8751^4 = 12.36$ (Table 4.2). The quarter sine wave is the closest. Minimizing the power law mode frequency, the first result, with respect to n, which is the Schmidt procedure [12], improves its estimate from 19.82 at $n = 2$ to 15.46 at $n = 1.73$.

A.5 Galerkin Procedure for Continuous Structures

The equation of motion of free vibration of a continuous linear structure with displacement $w(P,t)$, where P is a point in the structure, and distributed mass M is [8]

$$L[w(P,t)] + M(P)\frac{\partial^2 w(P,t)}{\partial t^2} = 0 \tag{A.25}$$

As discussed in Section 7.2, this equation can be solved for the mode natural frequencies in terms of integrals of mode shapes over the domain D of the structure:

$$\omega_i^2 = \frac{\int_D \tilde{w}_i(P,t) L[\tilde{w}_i(P)] dD}{\int_D M(P)\tilde{w}_i^2(P) dD}, \quad i = 1, 2, 3 \ldots \tag{A.26}$$

For slender beams (Chapter 4), this equation is

$$\omega_i^2 = (2\pi f_i)^2 = \frac{EI}{m} \frac{\int_0^L \tilde{y}_i{}'''' \tilde{y}_i dx}{\int_0^L \tilde{y}_i^2 dx} \tag{A.27}$$

For plates (Chapter 5), this requires two dimensional integration.

$$\omega_{ij}^2 = (2\pi f_{ij})^2 = \frac{Eh^3}{12(1-\nu^2)} \frac{\int_A (\nabla^4 \tilde{w}_{ij}) \tilde{w}_{ij} dxdy}{\int_A \gamma \tilde{w}_{ij}^2(x,y) dxdy}, \quad i,j\,1,2,3\ \ldots \tag{A.28}$$

For discrete systems, the integrals are summations over the mass $[M]$ and stiffness $[K]$ matrices:

$$\omega_i^2 = \{\tilde{x}\}_i^T [K] \{\tilde{x}\}_i / \{\tilde{x}\}_i^T [M] \{\tilde{x}\}_i, \quad i = 1,2,3\ \ldots \tag{A.29}$$

The result is exact if the assumed mode shape is exact.

References

[1] Thomson, W. T., Vibration Theory and Applications, Prentice Hall, 1988.

[2] Leissa, A. W., Historical Bases of Rayleigh and Ritz Methods, Journal of Sound and Vibration, vol. 287, pp. 961–978, 2005.

[3] Stephen, N. G., On Southwell's and a Novel Dunkerley's Method, Journal of Sound and Vibration, vol. 18, pp. 179–184, 1995.

[4] Jacobsen, L. S. and Ayre, R. S., Engineering Vibrations with Applications to Structures and Machinery, McGraw-Hill, pp. 115–121, 1958.

[5] Jones, R., An Approximate Expression for the Fundamental Frequency of Vibration of Elastic Plates, Journal of Sound and Vibration, vol. 38, 503–504, 1975.

[6] Johns, D. J., Comments on 'An approximate Expression for the Fundamental Frequency of Vibration of Elastic Plates', Journal of Sound and Vibration, vol. 41, 385–387, 1975.

[7] Bert, C. W., Relationship Between Fundamental Natural Frequency and Maximum Static Deflection for Various Linear Vibratory Systems, Journal of Sound and Vibration, vol. 162, 547–557. See Discussion by Maurizi, vol. 177, pp. 140–143; Stephens, vol. 171, pp. 285–287; and Xie, vol. 186, pp. 689–693, 1993.

[8] Meirovitch, L., Analytical Methods in Vibrations, Macmillan, p. 235, 1967.

[9] Stephens, N. G., Extended Dunkerley's Method for Additional Flexibility, Journal of Sound and Vibration, vol. 191, pp. 345–352, 1991.

[10] Roark, R. J., Formulas for Stress and Strain, 4th ed., McGraw-Hill, New York, p. 106, 1965.

[11] Mazumdar, J., Transverse Vibration of Elastic Plates by the Method of Constant Deflection, Journal of Sound and Vibration, vol. 18, 147–155, 1971.

[12] Schmidt, R., Technique for Estimating Natural Frequencies, ASCE Journal of Engineering Mechanics, vol. 109, pp. 654–657, 1984.

Appendix B

Numerical Integration of Newton's Second Law

Newton's second law (Eq. 2.4) can be numerically integrated with time-stepping finite difference methods. Dividing Equation 2.4 through by the mass produces a generic second-order ordinary differential equation in displacement (x) as a function of time (t):

$$\frac{d^2x}{dt^2} = f\left(t, x, \frac{dx}{dt}\right) \tag{B.1}$$

The function $f(t, x, dx/dt)$ is force on the structure divided by mass; it is assumed to be a function of time t and possibly position x and velocity dx/dt as well. For a simple falling object, $f(t, x, dx/dt) = g$, the acceleration of gravity. For a forced spring-mass system, Eq. 7.1, Eq. B.1 becomes $d^2x/dt^2 = f(t, x, dx/dt) = F(t)/M - 2\zeta\omega_n dx/dt - \omega_n^2 x$ where $F(t)$ is the force applied to mass M, which has natural frequency ω_n and damping factor ζ.

Table B.1 presents three explicit numerical solution methods of increasing complexity and accuracy: Euler's method, the central difference method, and the fourth-order Runge–Kutta formulation. These are explicit methods, they provide the full solution to Eq. B.1 at each time step in terms of the previous step(s) [1, 2]. Each can be programmed in nine lines of code.

Euler's method replaces the infinitesimal differential time step dt presented in Chapter 2 (Eq. 2.9) with a small finite time step Δt. The dynamic quantities of acceleration, $a = d^2x/dt^2$, and velocity, $v = dx/dt$, are constant over each time step:

$$\begin{aligned}
&\text{Acceleration:} && a_n = f(t_n, v_n, x_n)\\
&\text{Velocity:} && v_{n+1} = a_n \Delta t + v_n\\
&\text{Displacement:} && x_{n+1} = v_n \Delta t + x_n\\
&\text{Acceleration:} && a_{n+1} = f(t_{n+1}, v_{n+1}, x_{n+1})\\
&\text{Time:} && t = n\Delta t \quad \text{and} \quad n = 0, 1, 2, 3 \ldots
\end{aligned} \tag{B.2}$$

Formulas for Dynamics, Acoustics and Vibration, First Edition. Robert D. Blevins.
© 2016 John Wiley & Sons, Ltd. Published 2016 by John Wiley & Sons, Ltd.

Table B.1 Numerical integration of Newton's second law

Notation: dt = time step; $FM(x, \dot{x}, \ddot{x})$ = force/mass, a user supplied function; t = time; i = integer time step; Steps = integer number of steps; x = displacement; $\dot{x} = dx/dt$ = velocity; $\ddot{x} = d^2x/dt^2$ = acceleration; $*$= multiplication [2, 6].

Basic Subroutines for Time Step Integration of d^2xdt^2 = FM(t, x, dx/dt)

1. Subroutine Euler

```
'Read-in initial time t,  NSteps , time step dt,  initial position x, initial velocity xdot, initial acceleration xdotdot
'time stepping loop
For I = 1 to NSteps
xdot = xdotdot * dt + xdot
x = xdot * dt + x
t = t + dt
xdotdot = FM(t, x, xdot)     ' user supplied acceleration function
'Print t, x, xdot, x
Next I; End Sub
Function FM(t, x,xdot)  ' user supplied acceleration. FM = force/mass
FM= 1  'code acceleration here; End Function
```

2. Subroutine Central_Difference

```
Dimension matrices x(1000), xdot(1000), xdotdot(1000), t(1000)
'Read-in time step dt,NSteps, initial position x(0), initial velocity xdot(0), initial time t(0)
'calculate starting values at i=1 from initial values at i=0
t(1) = t(0) + dt
x(1) = 2 * x(0) - x(0) + xdot0 * dt + dt ^ 2 * FM(t(0), x(0), xdot(0))
xdot(1) = (x(1) - x(0)) / dt
'time stepping loop
For I = 2 to NSteps
x(i) = 2 * x(i - 1) - x(i - 2) + dt ^ 2 * FM(t(i - 1), x(i - 1), xdot(i - 1))
xdot(i) = (x(i) - x(i - 1)) / dt
t(i) = t(i - 1) + dt
xdotdot(i) = FM(t(i), x(i), xdot(i)) ' user supplied acceleration function
'Print t(i), xdotdot(i), xdot(i), x(i)
Next i ; End Sub
Function FM(t, x,xdot)  ' user supplied acceleration. FM = force/mass
FM= 1  'code acceleration here; End Function
```

3. Subroutine Runga_Kutta

```
'Read-in time step dt, NSteps, initial position x, initial velocity xdot, initial time t
'time stepping loop
For I = 1 to NSteps
k1 = (dt ^ 2 / 2) * FM(t, x, xdot)  'user supplied  function FM(t,x,xdot) in case 1
k2 = (dt ^ 2 / 2) * FM(t + dt / 2, x + (1 / 2) * xdot * dt + k1 / 4, xdot + k1 / dt)
k3 = (dt ^ 2 / 2) * FM(t + dt / 2, x + (1 / 2) * xdot * dt + k1 / 4, xdot + k2 / dt)
k4 = (dt ^ 2 / 2) * FM(t + dt, x + xdot * dt + k3, xdot + 2 * k3 / dt)
t = t + dt
x = x + xdot * dt + (1 / 3) * (k1 + k2 + k3)
xdot = xdot + (1 / dt) * (1 / 3) * (k1 + 2 * k2 + 2 * k3 + k4)
xdotdot = FM(t, x,xdot)  ' user supplied acceleration function
'Print t, xdotdot, xdot, x
Next I ; End Sub
Function FM(t, x,xdot)  ' user supplied acceleration. FM = force/mass
FM= 1  'code acceleration here; End Function
```

The integration marches forward from the initial velocity v_0 and displacement x_0 at $t = 0$, by repetitively evaluating this sequence of equations, given the force F_n at each time t_n, $n = 0, 1, 2, \ldots$. This is the simplest and easiest to program time-stepping integration algorithm.

Numerical accuracy increases as the time step size decreases and as the number of digits retained in the calculation increases. For systems that have a natural period T, which is often the cases with Newton's second law systems, Thomson [3] recommends that the time step be no greater than one tenth the period, $\Delta t < T/10$, Bath and Wilson [2] recommend $\Delta t < T/\pi$, and Xie [4] recommends time steps 10 times smaller for nonlinear systems.

The central difference method looks across two time steps to increase accuracy to order Δt^2 [1, 2]. The velocity at $n+1$ step is $v_{n+1} = (x_{n+1} - x_n)/\Delta t$ and at the nth step $v_n = (x_n - x_{n-1})/\Delta t$. The central difference between these two velocities, divided by the time step, is the nth step acceleration, $a_n = (x_{n+1} - 2x_n + x_{n-1})/\Delta t^2$, which equals $f(t_n, v_n, x_n)$ (Eq. B.1). This equation is solved for $n+1$ step displacement as a function of force over mass at the previous time steps [2, 3]:

$$
\begin{aligned}
&\text{Displacement:} && x_{n+1} = 2x_n - x_{n-1} + \Delta t^2 f(t_n, v_n, x_n) \\
&\text{Velocity:} && v_{n+1} = (x_{n+1} - x_n)/\Delta t \\
&\text{Acceleration:} && a_{n+1} = f(t_{n+1}, v_{n+1}, x_{n+1}) \\
&\text{Time:} && t = n\Delta t \quad \text{and} \quad n = 0, 1, 2, 3 \ldots
\end{aligned}
\tag{B.3}
$$

Because it requires the backward step (x_{n-1}), the central difference does not self start from the initial conditions at $n = 0$. This can be patched. x_1 can be estimated from the initial velocity and acceleration at $n = 0$ ([3]; case 1 of Table 2.1), and the central difference then starts from $n + 1 = 2$:

$$
x_1 = x_0 + v_0\Delta t + \frac{a_0 \Delta t^2}{2} \tag{B.4}
$$

The finite difference solutions can include three initial conditions in the first step: position, velocity, and acceleration.

The fourth-order Runge–Kutta formulation, like the central difference, is based on the Taylor series expansion. It makes four evaluations of $f(t, x, dx/dt)$ per time step, two of these at the half step $t_n + \Delta t/2$, to achieve accuracy of order Δt^4 [1, 5]. It is applicable to cases where half-step evaluations are possible. It is a self-starter from an initial condition and it is widely used in engineering. The following sequence of calculations is evaluated at each step [6].

$$
\begin{aligned}
k_{1n} &= \frac{\Delta t^2}{2} f(t_n,\ x_n,\ v_n) \\
k_{2n} &= \frac{\Delta t^2}{2} f\left(t_n + \frac{\Delta t}{2}, x_n + \frac{\Delta t}{2} v_n + \frac{k_{1n}}{4}, v_n + \frac{k_{1n}}{\Delta t}\right) \\
k_{3n} &= \frac{\Delta t^2}{2} f\left(t_n + \frac{\Delta t}{2}, x_n + \frac{\Delta t}{2} v_n + \frac{k_{1n}}{4}, v_n + \frac{k_{2n}}{\Delta t}\right)
\end{aligned}
$$

$$k_{4n} = \frac{\Delta t^2}{2} f\left(t_n + \Delta t, x_n + v_n \Delta t + k_{3n}, v_n + 2\frac{k_{3n}}{\Delta t}\right)$$

$$t_{n+1} = t_n + \Delta t$$

$$x_{n+1} = x_n + v_n \Delta t + \frac{1}{3}(k_{1n} + k_{2n} + k_{3n})$$

$$v_{n+1} = v_n + \frac{1}{3\Delta t}(k_{1n} + 2k_{2n} + 2k_{3n} + k_{4n})$$

$$a_{n+1} = f(t_{n+1}, x_{n+1}, v_{n+1}), \qquad n = 0, 1, 2, \ldots \tag{B.5}$$

The displacement terms k_{1n} ... k_{4n} are of order Δt^2, and they sample acceleration at t_n, $t_n + \Delta t/2$, and $t_n + \Delta t$.

Example B.1 Consider constant acceleration, $f(t, x, dx, dt) = 1$, and 10 steps with a time step $\Delta t = 0.1$ with initial $t = 0$ conditions of zero displacement and zero velocity. The exact solution to Equation B.1 is $x = t^2/2$ (case 1 of Table 2.1).

Solution: Starting from zero initial acceleration, Euler's method gives 0.045, and the error is $-0.5\Delta t$ for both Euler's and central difference methods. With an initial acceleration of unity, the error is $+0.5\Delta t$ for Euler's method and 0 for the central difference method with Equation B.4. The Runge–Kutta formulation samples the acceleration function at the initial time and reproduces the exact solution.

Implicit methods, unlike explicit methods, solve a nondifferential equilibrium equation at $t + \Delta t$ containing both the current and previous steps at each step which increases accuracy and stability but adds to complexity [1, 2]. Implicit numerical integration methods for Equation B.1 include Newmark acceleration method, Wilson's theta method, Nastran method, and Houbolt's method [2, 7, 8]. These implicit methods solve a set of linear simultaneous equations at each time stop.

References

[1] Ferziger, J. H., Numerical Methods for Engineering Application, Wiley, N.Y., pp. 76–79, 1981.
[2] Bath, K-J, and Wilson, E. L., Numerical Methods in Finite Element Analysis, Prentice Hall, New Jersey, pp. 310–313, 1976.
[3] Thomson, W. T., Theory of Vibrations with Applications, 3rd ed., Prentice Hall, p. 102, 1988.
[4] Xie, Y. M., An Assessment of Time Integration Schemes for Non-Linear Dynamic Equations, Journal of Sound and Vibration, vol. 192, pp. 321–331, 1996.
[5] Wood, W L., Practical Time-Stepping Schemes, Clarendon Press, Oxford, p. 197, 1990.
[6] Jennings, W., First Course in Numerical Methods, Macmillan, N.Y., pp. 147–149, 1964.
[7] American Society of Mechanical Engineers, ASME Boiler and Pressure Vessel Code, Section III Appendix N, para N-1222, 2013.
[8] Zienkiewicz, O. C., The Finite Element Method, 3rd Edition, McGraw Hill, N.Y., 1977.

Appendix C

Standard Octaves and Sound Pressure

C.1 Time History and Overall Sound Pressure

The superposition of several independent sound sources produces multifrequency *noise*:

$$p(t) = \sum_{i=1}^{N} p_i(t) = \sum_{i=1}^{N} P_i \cos(2\pi f_i t + \varphi_i) \tag{C.1}$$

Here, t = time, f = frequency in Hertz, P_i = pressure amplitude of the ith frequency and φ_i is phase. Noise is usually modeled as a *stationary random process*, which is valid if time-averaged statistical measures of noise, including *root-mean-square (rms) pressure*, *spectral density*, and *probability distribution*, are independent of the sample length.

The fundamental measure of noise amplitude is *rms pressure*, the square root of the average over time period T of the square of pressure:

$$p_{\text{rms}} = \sqrt{\left(\frac{1}{T}\right) \int_0^T p^2(t)\, dt} \tag{C.2}$$

The rms of a pure tone ($P \cos[2\pi ft + \varphi]$) is its amplitude divided by the square root of 2: $p_{\text{rms}} = P/2^{1/2}$. *Overall rms sound pressure* is the sum of the component mean square pressures [1–3]:

$$p^2_{\substack{\text{overall}\\\text{rms}}} = \frac{1}{T}\int_0^T p^2(t)\, dt = \sum_{i=1}^{N} p^2_{\substack{i\text{th-independent}\\\text{component,rms}}} = \int_0^\infty S_p(f)\, df \tag{C.3}$$

provided 1) the components are *independent* (randomly phased) with respect to each other, so their cross products, $p_i(t)\ p_j(t)$ with $i \neq j$, average to zero over many samples

Formulas for Dynamics, Acoustics and Vibration, First Edition. Robert D. Blevins.
© 2016 John Wiley & Sons, Ltd. Published 2016 by John Wiley & Sons, Ltd.

Table C.1 Decibel scales of sound

Notation: p_{rms} = root mean square pressure; ρ = density; c = speed of sound, T = time period [4–6].

Decibel Scale	Inverse	Reference Value (a)	Relationships
1. Sound Pressure Level in Gases $SPL = 20\log_{10}(p_{rms}/p_{ref})$, dB	$p_{rms} = p_{ref}10^{SPL/20}$	$p_{ref} = 20 \times 10^{-6}$ Pa = 20μPa (2.90×10^{-9} psi)	$p^2_{\substack{overall\\rms}} = \sum_{i=1}^{N} p^2_{i,rms}$
2. Sound Pressure Level in Liquid $SPL = 20\log_{10} p_{rms}/p_{ref}$, dB	$p_{rms} = p_{ref}10^{SPL/20}$	$p_{ref} = 10^{-6}$ Pa (1.45×10^{-10} psi)	$p^2_{\substack{overall\\rms}} = \sum_{i=1}^{N} p^2_{i,rms}$
3. Sound Intensity Level $L_i = 10\log_{10}(I/I_{ref})$, dB	$I_r = I_{ref}10^{L_i/10}$	$I_{ref} = 10^{-12}$ N/(m-s) (5.71×10^{-15} lb/in.-s)	$I_{ref} \approx p^2_{ref}/(\rho c)$ air $\rho c \approx 413$ kg/m^{2}s
4. Sound Power Level $L_w = 10\log_{10}(W/W_{ref})$, dB	$W = W_{ref}10^{L_w/10}$	$W_{ref} = 10^{-12}$ N-m / s (8.85×10^{-12} in.-lb/s)	$W_{ref} \approx (1\,m^2)I_{ref}$
5. Sound Exposure Level $L_E = 10\log_{10}(E_A/E_{ref})$, dB	$E_A = E_{ref}10^{L_E/10}$	$E_{ref} = (20\times10^{-6}\,Pa)^2$ s (2.90×10^{-9} lb^2/in.)^{2}s	$E_A = \int_0^T p_A^2(t)dt$

(a) In air reference pressure is 20 μPa. In water reference pressure is 1 μPa [7].

(*ergodic* average), or 2) their periods are nonequal submultiples of the sample time *T*, as in a Fourier series. The *overall sound pressure level* (OASPL) is generally expressed in decibels (Table C.1), and the summing can be done in decibels (see Beranek [3], Pierce ([1], pp. 69–71) and Example C.1).

C.2 Peaks and Crest

Peaks in time dominate acoustic damage accumulation. The *peak* (maximum) of a pure sinusoidal tone is its amplitude. The maximum possible peak value of the sum of *N* sine waves (Eq. C.1) is the sum of their amplitudes $\sum P_i$. A measure of randomness is its *peak-to-rms ratio* (also called *crest factor*), which is the peak of a sample divided by the rms of the sample (Eq. C.3):

$$\frac{\text{Peak}}{\text{rms}} = \begin{cases} \dfrac{\sum |P_i|}{p_{\substack{overall\\rms}}}, \\ (2N)^{1/2}, \quad \text{equal peaks}, \quad P_1 = P_i,\ i = 1, 2, \ldots, N \end{cases} \tag{C.4}$$

The peak-to-rms ratio of a single sine wave is $2^{1/2}$. It is $(2N)^{1/2}$ for random noise that is the sum of $N = 1, 2, 3, \ldots$ sine waves with random phases and equal amplitudes (Eq. C.1 with $P_i = P_1 = P$, $i = 1, 2, \ldots, N$) [8]. The peak-to-rms ratio approaches infinity for a Gaussian random process. Normally operating machines usually have peak-to-rms ratios between 1.414 and 4.

C.3 Spectra and Spectral Density

A plot of sound level against its frequencies is called a *noise spectrum*. The spectrum of the time history is a plot of the component amplitudes P_i versus their frequencies f_i. The single-sided acoustic *pressure spectral density* is defined as the mean square of oscillating pressures at frequencies between f_1 and f_2 divided by bandwidth $\Delta f = f_2 - f_1$ in Hertz [1, 4, 9]; it has units of pressure^2/Hertz:

$$S_p(f) = \frac{p_{\text{rms}}^2}{\Delta f} \tag{C.5}$$

If discrete frequencies in the time history are spaced at 1 Hz frequency intervals (1 Hz bandwidth), then the spectral density is the mean square pressure at each frequency.

The integral relationship on the right-hand side of Equation C.3 is called *Parseval's equation*. Overall mean square is the integral of the spectral density over its frequency range (Equation C.3). (Other definitions of spectra used in the literature include the *two-sided* spectrum with frequencies from minus infinity to plus infinity with one half the values of single-sided spectrum, spectrum with frequency in radian per second instead of Hertz, and rms and peak spectra rather than mean square spectra. Anyone of these spectra can be converted to another at constant bandwidth.)

C.4 Logarithmic Frequency Scales and Musical Tunings

Based on historical developments in the tuning of stringed instruments [1, 9–12], the audible frequency range is divided into proportional frequency bands called *octaves*. The upper frequency limit, f_a, of an octave is twice the lower frequency limit, f_b, $f_b/f_a = 2$. Octaves are not linear scales; higher-frequency bands are wider than lower-frequency bands. The logarithmic center frequency, $f_c = (f_a f_b)^{1/2}$, is always less than the arithmetic mean frequency, $\frac{1}{2}(f_a + f_b)$. The third-octave, tenth-octave, and twelfth-octave bands are subintervals of one octave. An octave is spanned by three 1/3-octave bands and 12 1/12-octave bands. Center frequencies of successive 1/3-octave bands are approximately in the ratio 5:4 [1].

Frequency band	1 octave	1/3 octave	1/10 octave	1/*n* octave
Upper frequency/ lower frequency, f_b/f_a	2	$2^{1/3}$	$2^{1/10}$	$2^{1/n}$
Center frequency, $f_c = (f_a\ f_b)^{1/2}$	$2^{1/2} f_a$	$2^{1/6} f_a$	$2^{1/20} f_a$	$2^{1/2n} f_a$
Bandwidth, $(f_b - f_a)/f_c$	0.7071	0.2315	0.0693	$(2^{1/n} - 1)2^{-1/2n}$

To convert a measurement from a wider to a narrower frequency band, one usually assumes the spectral densities or the rms pressures are equal in the smaller bands. A one-octave band SPL is converted to three 1/3-octave bands by subtracting $10\ \log_{10}(3) = 4.77$ dB from the one-octave band SPL. Equation C.3 is applied to convert several smaller bands to a single larger band, if the pressures in each of the smaller bands are independent.

Standard 1-octave and 1/3-octave bands in Table C.2 are endorsed by the Acoustical Society of America [9]. The one-third-band limit frequencies are nice integers, approximately equal to $10^{3+n/10}$, where n is a positive or negative integer. One thousand hertz is a band center, whereas classical musical scales are based on the note $A_4 = 440$ Hz and do not include a 1000 Hz band frequency [1, 9].

Table C.2 Standard one and one-third octave bands

The frequency bands can be extended by multiplication or division by powers of 10 [9].

Band	Octave			A-Scale	One-Third Octave			A-Scale	C-Scale
	Center Hz	Lower Limit, Hz	Upper Limit, Hz	Weight dB	Center Hz	Lower Limit, Hz	Upper Limit, Hz	Weight dB	Weight dB
11					12.5	11.2	14.1	-63.4	-11.2
12	16	11	22	-57.6	16	14.1	17.8	-56.7	-8.5
13					20	17.8	22.4	-50.5	-6.2
14					25	22.4	28.2	-44.7	-4.4
15	31.5	22	44	-39.4	31.5	28.2	35.5	-39.4	-3.0
16					40	35.5	44.7	-34.6	-2.0
17					50	44.7	56.2	-30.2	-1.3
18	63	44	88	-26.2	63	56.2	63	-26.2	-0.8
19					80	70.8	89.1	-22.5	-0.5
20					100	89.1	112	-19.1	-0.3
21	125	88	177	-16.1	125	112	141	-16.1	-0.2
22					160	141	178	-13.4	-0.1
23					200	178	224	-10.9	0
24	250	177	355	-8.6	250	224	282	-8.6	0
25					315	282	355	-6.6	0
26					400	355	447	-4.8	0
27	500	355	710	-3.2	500	447	562	-3.2	0
28					630	562	708	-1.9	0
29					800	708	891	-0.8	0
30	1000	710	1420	0	1000	891	1122	0	0
31					1250	1122	1413	+0.6	0
32					1600	1413	1778	+1.0	-0.1
33	2000	1420	2840	1.2	2000	1778	2239	+1.2	-0.2
34					2500	2239	2818	+1.3	-0.3
35					3150	2818	3548	+1.2	-0.5
36	4000	2840	5680	1.0	4000	3548	4467	+1.0	-0.8
37					5000	4467	5623	+0.5	-1.3
38					6300	5623	7079	-0.1	-2.0
39	8000	5680	11360	-1.1	8000	7079	8913	-1.1	-3.0
40					10000	8913	11220	-2.5	-4.4
41					12500	11220	14130	-4.3	-6.2
42	16000	11360	22720	-6.6	16000	14130	17780	-6.6	-8.5
43					20000	17780	22390	-9.3	-11.2

Classical music uses 12 proportional frequencies (notes) per octave in the ratio of small integers, $f \sim m/n$, where m and n are integers. Seven of these notes are the familiar *do*, *re*, *mi*, *fa*, *so*, *la*, and *ti* (*do*) that are seven successive white keys per octave on a piano [2, 4, 9].

C	D	E	F	G	A	B	(C)
do	re	mi	fa	so	la	ti	do
1:1	9:8	5:4	4:3	3:2	5:3	15:8	2:1

The frequency ratio 1:1 is called unison, 3:2 is a perfect fifth, 5:4 is a major third, 6:5 is a minor third, 4:3 is a fourth, 8:5 is a minor sixth, and 2:1 is an octave [2, 7, 8]. The ratio between notes is not quite consistent. For example, D:C is 9:8 but E:D is 10:9. However, it is found by a slight tampering with the exact ratios that the tuning requirements of stringed musical instruments, which sound at discrete frequencies, can be fairly well met by twelve notes per octave approximately 1/12th octave apart in the ratio of $2^{1/12} = 1.0595$, which is called a *half step*. Two half steps $2^{2/12} \sim 9/8$, five $2^{5/12} \sim 4/3$, and so on. *Temperament* is adjustment in the exact frequencies (above) to produce a 12-note musical scale for stringed instruments that is pleasing to the human ear [1, 12].

C.5 Human Perception of Sound (Psychological Acoustics)

Sound pressures impinging on the eardrum membrane are transmitted by vibration of a mechanical linkage of small bones to the fluid-filled canals of the inner ear. The brain processes electrical impulses from vibrating hair cells that line these canals to register sound. Humans are most sensitive to sound at 1000 Hz. The acoustic reference pressure in air of 20 mPa rms (2.9×10^{-9} psi rms, 0 dB) is the onset of hearing of a young healthy human adult at 1000 Hz (Eq. 6.7; [1, 5, 10]). The A-scale decibel weighting factors shown in Table C.2 mimic the frequency sensitivity of the human ear to moderate and loud sounds. A-scale weightings are numerically added to the SPL levels in the frequency bands to produce sound levels in dB (A) [5, 6].

Community and job noise standards are based on sound in A-scale decibels. The US Environmental Protection Agency (EPA) gives maximum permissible daily sound exposure levels and their durations [14].

Exposure, h	8	6	4	3	2	1.5	1	1/2	1/4
SPL, dB (A)	90	92	95	97	100	102	105	110	115

For multiple levels in 1 day, the sum $\sum \mathrm{SPL}_i/\mathrm{exposure}_i$ must be less than unity. These levels accept the possibility of some hearing loss above 4000 Hz ([2], p. 300). Acceptable noise annoyance levels are much lower, typically less than 55 dB(A) outdoors [15].

Example C.1 An unweighted noise spectrum is equal to 100 dB in each of six 1/3-octave bands from 100 to 630 Hz. Convert this spectrum to 1-octave bands with A-scale weighting and calculate the overall sound pressure in dB(A).

Solution: One-third-octave bands and their A-weighting are provided in Table C.2. Equation 6.8b is used to convert one-octave band levels from decibels to pascals. Mean square pressures are summed over three adjacent 1/3-octave bands (Equation C.3). OASPL is the sum of the bands' mean square pressures (Equation C.3).

1/3-octave ctr frequency (Hz)	100	125	160	200	250	315	400	500	630	Overall
SPL decibels in 1/3 octaves	100	100	100	100	100	100	100	100	100	109.54
A-scale weight (dB)	−19.1	−16.1	−13.4	−10.9	−8.6	−6.6	−4.8	−3.2	−1.9	
SPL (A) (dB)	80.9	83.9	86.6	89.1	91.4	93.4	95.2	96.8	98.1	102.96
p^2_{rms} Pa in 1/3 octaves	0.049	0.098	0.183	0.325	0.552	0.875	1.325	1.915	2.583	7.90
p^2_{rms} Pa per octave		0.33			1.75			5.82		7.90
SPL, dB(A) per octave		89.17			96.42			101.63		102.96

References

[1] Pierce, A. D., Acoustics, McGraw-Hill, N.Y., 1981.

[2] Kinsler, L. E., Frey, A., Coppens, A. B., and Sanders, J. V., Fundamentals of Acoustics, 3rd ed., John Wiley, N.Y., 1982.

[3] Beranek, L. Noise and Vibration Control, Institute of Noise Control Engineering, Washington, DC, Revised edition, 1988.

[4] American National Standard ANSI S1.1-1994, Acoustic Terminology, N.Y., Reaffirmed, 1999. Also American National Standard C634-02, Standard Terminology Relating to Environmental Acoustics, 2002.

[5] American National Standard ANSI S1.8-1998, Reference Quantities for Acoustical Levels, N.Y., Reaffirmed 2001. Also, Internationals Standards Organization, ISO 1683.1983, Acoustics –Preferred Reference Quantities for Acoustic Levels.

[6] American National Standard ANSI S1.4-1983, Sound Level Meters, N.Y. Also European Standard EN 60651, Sound Level Meters and IEC 651, January 1979 and British Standard BS EN 60651, Specification for Sound Level Meters, 1994.

[7] Carey, W. M., Standard Definitions for Sound Levels in the Ocean, IEEE Journal of Ocean Engineering, vol. 20, pp. 109–113, 1995.

[8] Blevins, R. D., Probability Density of Finite Fourier Series with Random Phases, Journal of Sound and Vibration, vol. 208, pp.617–652, 1997.

[9] American National Standard ANSI S1.6-1984, Preferred Frequencies, Frequency Levels, and Band Numbers for Acoustical Measurements, N.Y., Reaffirmed 2001.
[10] Helmholtz, H., On the Sensations of Tone, Dover, N.Y., 1954. Revised Edition of 1877.
[11] Lamb, H., The Dynamical Theory of Sound, 2nd ed., Dover, N.Y., 1960. Reprint of 1925 edition.
[12] Isacoff, S., Temperament, The Idea that Solved Music's Greatest Riddle, Alfred Knopf, N.Y., 2001.
[13] American National Standard C634-02, Standard Terminology Relating to Environmental Acoustics, 2002.
[14] USA Code of Federal Regulations, Occupational Safety and Health Standards, 1970, CFR 1910.95 (b).
[15] Crocker, M. J., Noise Control, in Handbook of Acoustics, M. Crocker (ed.), John Wiley, N.Y., 1998.

Appendix D

Integrals Containing Mode Shapes of Single-Span Beams

The formulas in this appendix are from Ref. [1], with the exception of the authors' addition of the pinned–pinned beam. These formulas are very useful in the analysis of dynamic systems whose modes can be described in terms of the mode shapes of single-span beams presented in Table 4.2. The following notation applies:

φ_n, ψ_n = nth beam mode shape presented in Table 4.2
σ_n = nondimensional parameter presented in column 4 of Table 4.2
$\beta_n = \lambda_n/L$, where λ_n = nondimensional frequency parameter presented in column 2 of Table 4.2
L = span of beam
x = coordinate along span of the beam

Reference

Felgar, R. P., Formulas for Integrals Containing Characteristic Functions of a Vibrating Beam, University of Texas Circular No. 14, Bureau of Engineering Research, Austin, Texas, 1950.

1.

$$\int_0^L \varphi_n(x)dx = \frac{1}{\beta_n^4}\left[\frac{d^3\varphi_n}{dx^3}\right]_0^L$$

Clamped–free $\int_0^L \varphi_n(x)dx = 2\sigma_n\beta_n^{-1} = 0.31039\,L,\ n = 1$

Clamped–pinned $\int_0^L \varphi_n(x)dx = \beta_n^{-1}[(-1)^{n+1}\sqrt{\sigma_n^2+1} - \sqrt{\sigma_n^2-1} + 2\sigma_n] = 0.86141\,L,\ n = 1$

Formulas for Dynamics, Acoustics and Vibration, First Edition. Robert D. Blevins.

© 2016 John Wiley & Sons, Ltd. Published 2016 by John Wiley & Sons, Ltd.

Clamped–clamped $\int_0^L \varphi_n(x)dx = 2\sigma_n\beta_n^{-1}[1-(-1)^n] = 0.83086\,L,\ n=1$

Free–pinned $\int_0^L \varphi_n(x)dx = \beta_n^{-1}[(-1)^{n+1}\sqrt{\sigma_n^2+1} - \sqrt{\sigma_n^2-1}] = 0.35167\,L,\ n=1$

Free–free $\int_0^L \varphi_n(x)dx = 0$

Pinned–pinned $\int_0^L \varphi_n(x)dx = \beta_n^{-1}[1-(-1)^n] = (L/n\pi)\,[1+(-1)^n]$

2.

$$\int_0^L \frac{d\varphi_n}{dx}(x)dx = \varphi_n|_0^L$$

Clamped–free $\int_0^L \left(\frac{d\varphi_n}{dx}\right) dx = (-1)^{n+1}2$

Clamped–pinned $\int_0^L \left(\frac{d\varphi_n}{dx}\right) dx = 0$

Clamped–clamped $\int_0^L \left(\frac{d\varphi_n}{dx}\right) dx = 0$

Free–pinned $\int_0^L \left(\frac{d\varphi_n}{dx}\right) dx = -2$

Free–free $\int_0^L \left(\frac{d\varphi_n}{dx}\right) dx = 2[1+(-1)^n]$

Pinned–pinned $\int_0^L \varphi_n(x)dx = 0$

3.

$$\int_0^L \frac{d^2\varphi_n}{dx^2}(x)dx = \left.\frac{d\varphi_n}{dx}\right|_0^L$$

Clamped–free $\int_0^L \left(\frac{d^2\varphi_n}{dx^2}\right) dx = (-1)^{n+1}2\sigma_n\beta_n$

Clamped–pinned $\int_0^L \left(\frac{d^2\varphi_n}{dx^2}\right) dx = \beta_n[(-1)^n\sqrt{\sigma_n^2+1} - \sqrt{\sigma_n^2-1}]$

Clamped–clamped $\int_0^L \left(\frac{d^2\varphi_n}{dx^2}\right) dx = 0$

Free–pinned $\int_0^L \left(\frac{d^2\varphi_n}{dx^2}\right) dx = \beta_n[(-1)^{n+1}\sqrt{\sigma_n^2+1} - \sqrt{\sigma_n^2-1} + 2\sigma_n]$

Free–free $\int_0^L \left(\frac{d^2\varphi_n}{dx^2}\right) dx = 2\sigma_n\beta_n[1-(-1)^n]$

Pinned–pinned $\int_0^L \left(\frac{d^2\varphi_n}{dx^2}\right) dx = \beta_n[1-(-1)^n]$

4.

$$\int_0^L \frac{d^3\varphi_n}{dx^3}(x)dx = \left.\frac{d^2\varphi_n}{dx^2}\right|_0^L$$

Clamped–free $\int_0^L \left(\frac{d^3\varphi_n}{dx^3}\right) dx = -2\beta_n^2$

Clamped–pinned $\int_0^L \left(\frac{d^3\varphi_n}{dx^3}\right) dx = -2\beta_n^2$

Clamped–clamped $\int_0^L \left(\frac{d^3\varphi_n}{dx^3}\right) dx = -2[\beta_n^2(-1)^n + 1]$

Free–pinned $\int_0^L \left(\frac{d^3\varphi_n}{dx^3}\right) dx = 0$

Free–free $\int_0^L \left(\frac{d^3\varphi_n}{dx^3}\right) dx = 0$

Pinned–pinned $\int_0^L \left(\frac{d^3\varphi_n}{dx^3}\right) dx = \beta_n^2[1-(-1)^n]$

5.

$$\int_0^L \varphi_n^2(x)dx = \frac{1}{4}\left[\frac{3}{\beta_n^4}\varphi_n\frac{d^3\varphi_n}{dx^3} + x\varphi_n^2 - \frac{2x}{\beta_n^4}\frac{d\varphi_n}{dx}\frac{d^3\varphi_n}{dx^3} - \frac{1}{\beta_n^4}\frac{d\varphi_n}{dx}\frac{d^2\varphi_n}{dx^2} + \frac{x}{\beta_n^4}\left(\frac{d^2\varphi_n}{dx^2}\right)^2\right]_0^L$$

$$= \begin{cases} L/2, & \text{for P–P} \\ L, & \text{for C–F, C–P, C–C, F–P, F–F} \end{cases}$$

6.

$$\int_0^L \left(\frac{d\varphi_n}{dx}\right)^2 dx = \frac{1}{4}\left[3\varphi_n\frac{d\varphi_n}{dx} + x\left(\frac{d\varphi_n}{dx}\right)^2 - 2x\varphi_n\frac{d^2\varphi_n}{dx^2} - \frac{1}{\beta_n^4}\frac{d^2\varphi_n}{dx^2}\frac{d^3\varphi_n}{dx^3} + \frac{x}{\beta_n^4}\left(\frac{d^3\varphi_n}{dx^3}\right)^2\right]_0^L$$

Clamped–free $\int_0^L \left(\frac{d\varphi_n}{dx}\right)^2 dx = \sigma_n\beta_n(2+\sigma_n\beta_n L)$

Clamped–pinned $\int_0^L \left(\frac{d\varphi_n}{dx}\right)^2 dx = \sigma_n\beta_n(\sigma_n\beta_n L - 1)$

Clamped–clamped $\int_0^L \left(\frac{d\varphi_n}{dx}\right)^2 dx = \sigma_n\beta_n(\sigma_n\beta_n L - 2)$

Free–pinned $\int_0^L \left(\frac{d\varphi_n}{dx}\right)^2 dx = \sigma_n\beta_n(\sigma_n\beta_n L + 3)$

Free–free $\int_0^L \left(\frac{d\varphi_n}{dx}\right)^2 dx = \sigma_n\beta_n(\sigma_n\beta_n L + 6)$

Pinned–pinned $\int_0^L \left(\frac{d\varphi_n}{dx}\right)^2 dx = \frac{\beta_n^2 L}{2}$

7.

$$\int_0^L \left(\frac{d^2\varphi_n}{dx^2}\right)^2 dx = \frac{1}{4}\left[3\frac{d^2\varphi_n}{dx^2}\frac{d\varphi_n}{dx} + x\left(\frac{d^2\varphi_n}{dx^2}\right)^2 - 2x\frac{d\varphi_n}{dx}\frac{d^3\varphi_n}{dx^3} - \varphi_n\frac{d^3\varphi_n}{dx^3} + x\beta_n^4\varphi_n^2\right]_0^L$$

$$\int_0^L \left(\frac{d^2\varphi_n}{dx^2}\right)^2 dx = \begin{cases} \frac{\beta_n^4 L}{2}, & \text{for P–P} \\ \beta_n^4 L, & \text{for C–F, C–P, C–C, F–P, F–F} \end{cases}$$

9.

$$\int_0^L \varphi_n \frac{d\varphi_n}{dx} dx = \frac{1}{2}\varphi_n^2|_0^L$$

Clamped–free $\int_0^L \frac{\varphi_n d\varphi_n}{dx} dx = 2$

Clamped–pinned $\int_0^L \frac{\varphi_n d\varphi_n}{dx} dx = 0$

Clamped–clamped $\int_0^L \frac{\varphi_n d\varphi_n}{dx} dx = 0$

Free–pinned $\int_0^L \frac{\varphi_n d\varphi_n}{dx} dx = -2$

Free–free $\int_0^L \frac{\varphi_n d\varphi_n}{dx} dx = 0$

Pinned–pinned $\int_0^L \varphi_n\left(\frac{d\varphi_n}{dx}\right) dx = 0$

10.

$$\int_0^L \varphi_n \frac{d^2\varphi_n}{d^2x} dx = \frac{1}{4}\left[\varphi_n \frac{d\varphi_n}{dx} - x\left(\frac{d^2\varphi_n}{dx^2}\right)^2 + 2x\varphi_n \frac{d\varphi_n}{dx} + \frac{1}{\beta_n^4}\frac{d^2\varphi_n}{dx^2}\frac{d^3\varphi_n}{dx^3} - \frac{x}{\beta_n^4}\left(\frac{d^3\varphi_n}{dx^3}\right)^2\right]_0^L$$

Clamped–free $\int_0^L \varphi_n \left(\frac{d^2\varphi_n}{dx^2}\right) dx = \sigma_n\beta_n(2 - \sigma_n\beta_n L)$

Clamped–pinned $\int_0^L \varphi_n \left(\frac{d^2\varphi_n}{dx^2}\right) dx = \sigma_n\beta_n(1 - \sigma_n\beta_n L)$

Clamped–clamped $\int_0^L \varphi_n \left(\frac{d^2\varphi_n}{dx^2}\right) dx = \sigma_n\beta_n(1 - \sigma_n\beta_n L)$

Free–pinned $\int_0^L \varphi_n \left(\frac{d^2\varphi_n}{dx^2}\right) dx = \sigma_n\beta_n(1 - \sigma_n\beta_n L)$

Free–free $\int_0^L \varphi_n \left(\frac{d^2\varphi_n}{dx^2}\right) dx = \sigma_n\beta_n(2 - \sigma_n\beta_n L)$

Pinned–pinned $\int_0^L \varphi_n \left(\frac{d^2\varphi_n}{dx^2}\right) dx = \left(\frac{2\beta_n}{3}\right)[(-1)^n - 1]$

11.

$$\int_0^L \varphi_n \frac{d^3\varphi_n}{d^3x} dx = \frac{1}{2\beta_n^4}\left[\left(\frac{d^3\varphi_n}{dx^3}\right)^2\right]_0^L$$

12.

$$\int_0^L \frac{d\varphi_n}{dx}\frac{d^2\varphi_n}{d^2x} dx = \frac{1}{2}\left[\left(\frac{d\varphi_n}{dx}\right)^2\right]_0^L$$

13.

$$\int_0^L \frac{d\varphi_n}{dx}\frac{d^3\varphi_n}{d^3x} dx = \frac{1}{4}\left[\frac{d\varphi_n}{dx}\frac{d^2\varphi_n}{dx^2} - x\left(\frac{d^2\varphi_n}{dx^2}\right)^2 + 2x\frac{d\varphi_n}{dx}\frac{d^3\varphi_n}{dx^3} + \varphi_n\frac{d^3\varphi_n}{dx^3} - \beta_n^4 x\varphi_n^2\right]_0^L$$

14.

$$\int_0^L \frac{d^2\varphi_n}{dx^2}\frac{d^3\varphi_n}{d^3x} dx = \frac{1}{2}\left[\left(\frac{d^2\varphi_n}{dx^2}\right)^2\right]_0^L$$

15.

$$\int_0^L \varphi_n\varphi_m dx = \frac{1}{\beta_n^4 - \beta_m^4}\left[\varphi_m\frac{d^3\varphi_n}{dx^3} - \varphi_n\frac{d^3\varphi_m}{dx^3} - \frac{d\varphi_m}{dx}\frac{d^2\varphi_n}{dx^2} + \frac{d\varphi_n}{dx}\frac{d^3\varphi_m}{dx^3}\right]_0^L = 0, \quad \text{for } m \neq n$$

See case 5 for $m = n$.

16.

$$\int_0^L \frac{d\varphi_n}{dx}\frac{d\varphi_m}{dx} dx = \frac{1}{\beta_n^4 - \beta_m^4}\left[\beta_n^4\varphi_n\frac{d\varphi_m}{dx} - \beta_m^4\varphi_m\frac{d\varphi_n}{dx} - \frac{d^2\varphi_m}{dx^2}\frac{d^2\varphi_n}{dx^2} + \frac{d^2\varphi_n}{dx^2}\frac{d^3\varphi_m}{dx^3}\right]_0^L, \quad \text{for } m \neq n$$

See case 6 for $m = n$.

17.

$$\int_0^L \frac{d^2\varphi_n}{dx^2}\frac{d^2\varphi_m}{dx^2}dx$$

$$= \frac{1}{\beta_n^4-\beta_m^4}\left[\beta_n^4\varphi_n\frac{d\varphi_n}{dx}\frac{d^2\varphi_m}{dx^2} - \beta_m^4\frac{d\varphi_m}{dx}\frac{d^2\varphi_n}{dx^2} - \beta_n^4\varphi_n\frac{d^3\varphi_m}{dx^3} + \beta_m^4\varphi_m\frac{d^3\varphi_n}{dx^3} + \frac{d^2\varphi_n}{dx^2}\frac{d^3\varphi_m}{dx^3}\right]_0^L = 0,\ \text{for } m \neq n$$

See case 7 for $m = n$.

25.

$$\int_0^L x\varphi_n(x)dx = \frac{1}{\beta_n^4}\left[x\frac{d^3\varphi_n}{dx^3} - \frac{d^2\varphi_n}{dx^2}\right]_0^L$$

Clamped–free $\int_0^L x\varphi_n dx = 2/\beta_n^2$

Clamped–pinned $\int_0^L x\varphi_n dx = \beta_n^{-2}(2 - \beta_n L[(-1)^n\sqrt{\sigma_n^2+1} + \sqrt{\sigma_n^2-1}])$

Clamped–clamped $\int_0^L x\varphi_n dx = 2\beta_n^{-2}[1 + (-1)^n - (-1)^n\sigma_n\beta_n L]$

Free–pinned $\int_0^L x\varphi_n dx = L\beta_n^{-1}[(-1)^n\sqrt{\sigma_n^2+1} - \sqrt{\sigma_n^2-1}]$

Free–free $\int_0^L x\varphi_n dx = 0$

Pinned–pinned $\int_0^L x\varphi_n dx = L\beta_n^{-1}(-1)^{n+1}$

29.

$$\int_0^L x^2\varphi_n(x)dx = \frac{1}{\beta_n^4}\left[x^2\frac{d^3\varphi_n}{dx^3} - 2x\frac{d^2\varphi_n}{dx^2} + 2\frac{d\varphi_n}{dx}\right]_0^L$$

Clamped–free $\int_0^L x^2\varphi_n dx = -4(-1)^n\sigma_n\beta_n^{-3}$

Clamped–pinned $\int_0^L x^2\varphi_n dx = \beta_n^{-3}(\beta_n^2L^2[-(-1)^n\sqrt{\sigma_n^2+1} - \sqrt{\sigma_n^2-1}] + 2[(-1)^n\sqrt{\sigma_n^2+1} - \sqrt{\sigma_n^2-1}]$

Clamped–clamped $\int_0^L x^2\varphi_n dx = 2\beta_n^{-2}L(-1)^n[2 - \sigma_n\beta_n L]$

Free–pinned $\int_0^L x^2\varphi_n dx = \beta_n^{-3}(\beta_n^2L^2[(-1)^n\sqrt{\sigma_n^2+1} - \sqrt{\sigma_n^2-1}] - 2[(-1)^n\sqrt{\sigma_n^2+1} + \sqrt{\sigma_n^2-1}] + 4\sigma)$

Free–free $\int_0^L x^2\varphi_n dx = 4\sigma_n\beta_n^{-3}[1 - (-1)^n]$

Pinned–pinned $\int_0^L x^2\varphi_n dx = \beta_n^{-3}(\beta_n^2L^2 - 2)(-1)^{n+1}$

Appendix E

Finite Element Programs

This appendix lists general-purpose finite element programs that are available for dynamic and vibration analyses as of this printing. Finite element programs mesh a structural geometry into linear elastic elements, usually beams, triangles, tetrahedrons, plates, or bricks, whose properties can be modeled by mass and elastic springs. Their mass and stiffness matrices are assembled and then numerically solved for natural frequencies and mode shapes. Programmed software controls the meshing, the matrix solution, and postprocessing. These are large programs that require extensive numerical input.

They are grouped into professional/commercial programs and open source/low-cost programs. The professional/commercial programs offer extensive graphics, meshing, and element libraries. Support is provided as a part of the purchase or lease. The open source program does not offer free support but it can be downloaded at little or no cost.

Software for the formulas in this book with a simple single graphical interface is available from Aviansoft.com. These lists were compiled in part by Hugh McCutchen.

E.1 Professional/Commercial Programs

1. MSC NASTRAN
   NASTRAN is a finite element analysis (FEA) program that was originally developed for NASA in the late 1960s under US government funding for the aerospace industry. Widely used with extensive modal and dynamic analysis capability. Integrated pre- and postprocessor.
   http://www.mscsoftware.com/product/msc-nastran
2. Abacus Simulia
   Abaqus/Simulia solves traditional implicit FEA, including static, dynamic, and thermal analyses, all powered with the widest range of contact and nonlinear material options. Dassault Systèmes acquired ABAQUS, Inc. and announced SIMULIA, the brand encompassing all DS simulation solutions, including Abaqus and CATIA Analysis applications.
   www.3ds.com/products/simulia

Formulas for Dynamics, Acoustics and Vibration, First Edition. Robert D. Blevins.
© 2016 John Wiley & Sons, Ltd. Published 2016 by John Wiley & Sons, Ltd.

3. ANSYS
   Founded by John Swanson, ANSYS has grown to be the largest of the finite-element-based software companies. In 1994, it was sold to TA Associates. ANSYS is extensively used in petroleum, auto, and manufacturing industries. Integrated capabilities include computational fluid dynamics (CFD).
   www.ansys.com
4. LS-DYNA
   LS-DYNA is a transient dynamic explicit solution. It is widely used for transient dynamic automotive crash simulation, impact, and forming. It does not perform modal analysis but step by step integrates time history finite elements with nonlinear effects including large deformation, nonlinear materials, and contact. Supports Unix, Windows, IRIX, and Linux.
   http://www.lstc.com/products/ls-dyna
5. NX NASTRAN
   NX version of original NASTRAN is marketed by Siemens automation. Often run on portable Windows computers coupled with Femap preprocessor/mesh generator. Computer-aided design (CAD) integration.
   http://www.plm.automation.siemens.com/en_us/products/nx/
   http://www.plm.automation.siemens.com/en_us/products/femap/
6. LapFEA
   LapFEA is a mechanical analysis application solving medium-size static and dynamic structural analysis problems. It includes a preprocessor, a solver, and a postprocessor. Often run on portable computers. Supports windows.
   www.lapcad.com

E.2 Open Source /Low-Cost Programs

1. FEAPpv
   FEAPpv is a general-purpose FEA program that is designed for research and educational use. It is available in both source and executable forms. Based in part on *The Finite Element Method: Its Basis and Fundamentals*, 6th ed., by O.C. Zienkiewicz, R.L. Taylor and J.Z. Zhu, Elsevier, Oxford, 2005, (www.elsevier.com). Unix/Linux/WindowsPC.
   http://www.ce.berkeley.edu/projects/feap/feappv/
   http://www.ce.berkeley.edu/projects/feap/feappv/manual.pdf
2. FlexPDE 6
   A solver for partial differential equations. The student version of FlexPDE 6 is limited to 100 nodes in 1D, 800 nodes in 2D, and 1600 nodes in 3D.
   http://www.pdesolutions.com/student6.html
3. LISA 8.0.0
   LISA is a user-friendly FEA package for Windows with an integrated modeler, multithreaded solver, and graphical postprocessor. Free to use, 1300 node limit. Static, thermal, and modal vibration.
   http://lisafea.com

4. CALCULIX
   With CALCULIX, finite element models can be built, calculated, and postprocessed. The pre- and postprocessor is an interactive 3D tool using the openGL API. The solver has linear and nonlinear calculations. Static, dynamic, and thermal solutions are available. Supports Unix, Linux, Irix, and MS-Windows.
   http://www.calculix.de
5. Code_Aster
   Code_Aster is mainly a solver for mechanics, based on the theory of finite elements. This tool covers thermal and mechanical analyses in linear and nonlinear statics and dynamics, modal analysis, for machines, pressure vessels, and civil engineering structures. Supports Linux and FreeBSD.
   http://researchers.edf.com/software/code-aster-44332.html
6. Z88 Aurora
   Z88Aurora is a free finite element software package for static calculation in mechanical engineering. Besides linear static analysis, you can use it for large displacement analysis, steady-state thermal analysis, and natural frequency analysis. Supports Windows, Linux, and Mac.
   http://www.z88.de/z88os/english.html
7. SAP-32-64-2010
   A revised 32-/64-bit port of original SAP and PC SAP software to Fortran. Supports Windows, Unix, Linux, and Mac platforms. FORTRAN compiler suggested.
   http://nisee.berkeley.edu/documents/SWSC/SAP-32-64.zip (6 MB)
8. Cast3M
   Cast3M is a computer code for the analysis of structures by the finite element method (FEM) and the CFD. Cast3M is a flexible analysis and optimization program for mechanical linear elastic problem in statics and dynamics (vibration and extraction of eigenvalue), thermal and heat transfer problem, nonlinear problem (elastic, plastic, and creep materials), step-by-step dynamic problem. Supports Windows.
   http://cast3m2012.software.informer.com/1.0/
   http://www-cast3m.cea.fr
9. Tdyn RamSeries (DYNSOL)
   DYNSOL includes dynamic analyses including step-by-step direct integration, modal/vibration, and seismic analysis. The module supports static and dynamic load definition, either by analytical functions or data tables, response spectra, and several modes of damping definition.
   http://www.compassis.com/compass/en/Productos/RamSeries

Index

Formulas for Dynamics, Acoustics and Vibration, First Edition. Robert D. Blevins.
© 2016 John Wiley & Sons, Ltd. Published 2016 by John Wiley & Sons, Ltd.

Printed and bound by CPI Group (UK) Ltd, Croydon, CR0 4YY
16/04/2025
14658384-0003